INTRODUCTION TO Semiconductor Manufacturing Technology

SECOND EDITION

INTRODUCTION TO
Semiconductor Manufacturing Technology
SECOND EDITION

Hong Xiao

SPIE PRESS
Bellingham, Washington USA

Library of Congress Cataloging-in-Publication Data

Xiao, Hong.
Introduction to semiconductor technology / Hong Xiao. – 2nd ed.
p. cm.
Includes bibliographical references and index.
ISBN 978-0-8194-9092-6
1. Semiconductors–Design and construction. 2. Semiconductor industry. I. Title.
TK7871.85.X53 2012
621.3815'2–dc23

2012013231

Published by

SPIE
P.O. Box 10
Bellingham, Washington 98227-0010 USA
Phone: +1 360.676.3290
Fax: +1 360.647.1445
Email: Books@spie.org
Web: http://spie.org

Copyright © 2012 Society of Photo-Optical Instrumentation Engineers (SPIE)

All rights reserved. No part of this publication may be reproduced or distributed in any form or by any means without written permission of the publisher.

The content of this book reflects the work and thought of the author(s). Every effort has been made to publish reliable and accurate information herein, but the publisher is not responsible for the validity of the information or for any outcomes resulting from reliance thereon.

Printed in the United States of America.
First printing

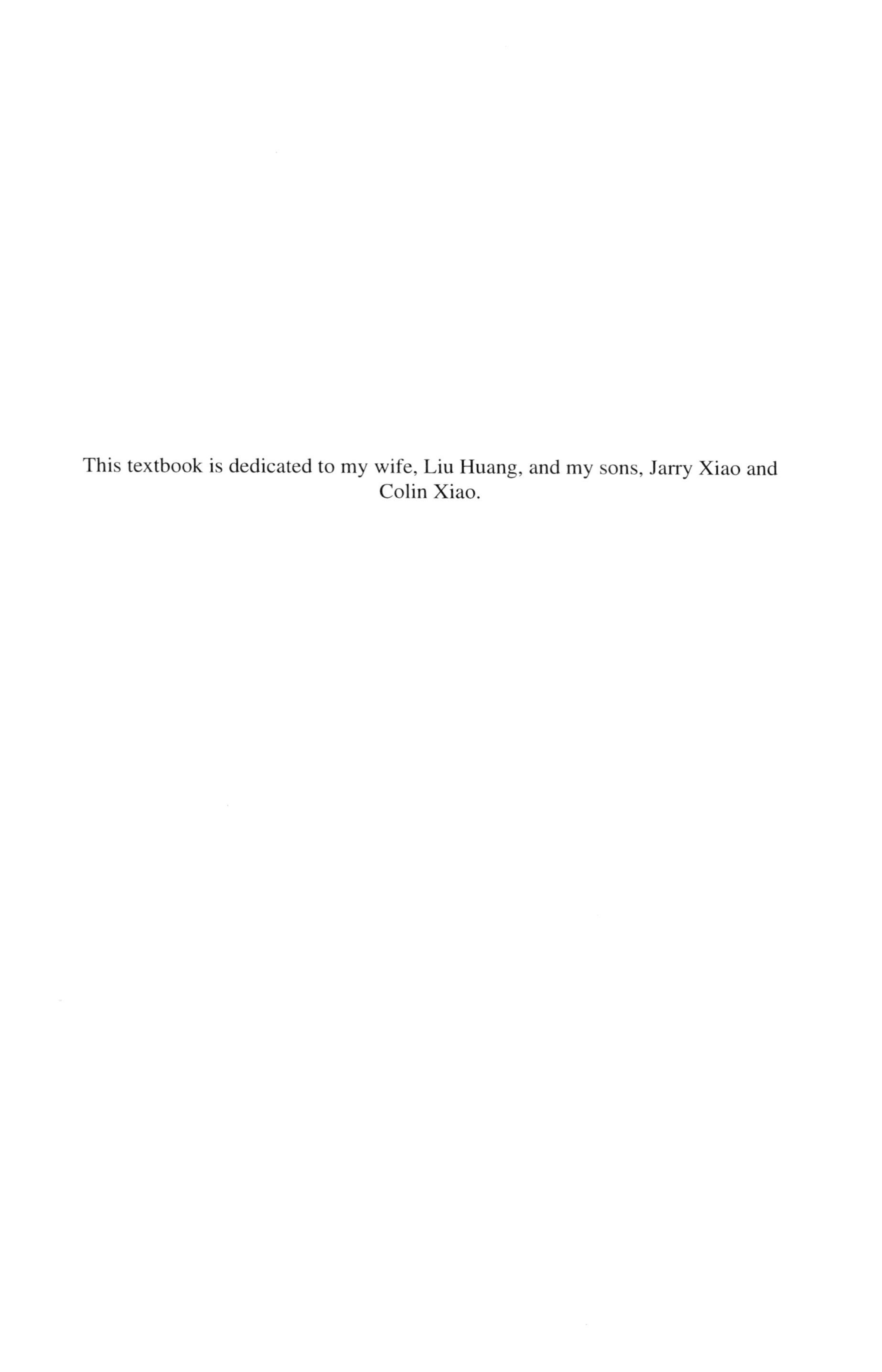

This textbook is dedicated to my wife, Liu Huang, and my sons, Jarry Xiao and Colin Xiao.

Contents

Preface to the First Edition

The semiconductor industry is developing rapidly with new technology introduced almost on a daily basis. The device feature size is shrinking continuously and the number of transistors on an integrated circuit (IC) chip is increasing rapidly, as predicted by Moore's law. Compared with only a decade ago, IC fabrication processing technology has become more complicated.

This book thoroughly describes the complicated IC chip manufacturing processes in a semiconductor fab, using minimum mathematics, chemistry, and physics. It covers the advanced technologies while keeping the contents simple and easy to understand for readers without science and engineering degrees. It focuses on the newest IC fabrication technologies and describes the older technologies to provide better understanding of the historical development. The processes chosen for the book are very close to those used in real fabs, especially process troubleshooting and process and hardware relations.

This book is intended for technical and college students who need an in-depth understanding of the technology as they prepare to find a job in the field. It is also intended as a reference book for engineering students and to provide a more realistic picture of the semiconductor industry. Industry operators, technicians, engineers, and personnel in sales, marketing, administration, and management can also benefit. This book can help them to learn more about their jobs, improve their troubleshooting and problem-solving skills, and raise their career development potential.

Chapter 1 briefly reviews the history of the semiconductor industry and describes semiconductor manufacturing processes. Chapter 2 introduces basic semiconductor fabrication including yield, cleanroom, semiconductor fab, and IC chip test and packaging. Chapter 3 gives a brief review of fabrication semiconductor devices, IC chips, and early technologies in semiconductor processing. Crystal structure, single-crystal silicon wafer manufacturing, and epitaxial silicon growth are described in Chapter 4. Chapter 5 lists and discusses thermal processes, including oxidation, diffusion, annealing, alloying, and reflow processes. Rapid thermal processes (RTPs) and conventional furnace thermal processes are discussed. Chapter 6 details the photolithography process. Fundamentals of plasmas used in semiconductor processing are covered in Chapter 7, which introduces plasma applications, DC bias, and plasma-process relations. Chapter 8 discusses the ion implantation process. Chapter 9 gives a detailed description of etch processes including wet and dry etches; chemical, reactive ion etch (RIE), and physical etches; and patterned and blanket etch processes. Basic chemical vapor deposition (CVD) and dielectric thin-film deposition processes, including dielectric CVD processes, process trends, troubleshooting, and future trends are discussed in detail in Chapter 10. Chapter 11 covers metallization, metal CVD, and physical vapor deposition (PVD) processes. It also describes the copper metallization process. Planarization processes including chemical mechanical polishing (CMP) are discussed in Chapter 12. Chapter 13 discusses process integration. Chapter 14 diagrams CMOS process flows including an advanced CMOS process flow with copper and low-κ interconnection. Chapter 15 predicts the future development of the semiconductor industry.

Many people helped me to write this book. I especially appreciate the useful information provided by my current and former colleagues: Lou Frenzel, Thomas E. Thompson, Ole Krogh, Tony Shi, Alberto Quiñonez, Lance Kinney, Scott Bolton, and Steve Reedy. Many of my students helped me by proofreading and improving the book. I especially express my thanks to Wayne Parent, Jeffrey Carroll, Boyd Woods, and Ronald Tabery.

I would also like to thank the following reviewers for their valuable suggestions: Professor Dave Hata, Portland Community College; Professor Fred Lavender, Albuquerque Technical Institute; Professor Gene Stouder, Southwest Texas State University; Professor Bassam Matar, Glendale Community College; Professor Carlo Sapijaszko, DeVry Institute of Technology; Professor George Shaiffer, Pikes Peak Community College; Professor Val Shires, Gwinnett Technical Institute; and Professor Devinder Sud, DeVry Institute.

Hong Xiao
2001

Preface to the Second Edition

When the first edition was published in 2001, the leading-edge IC technology node was about 130 nm. Shortly after the publication of the first edition, I attended an international IC technology conference where 90 nm was the leading-edge technology node of IC manufacturing. Former Honda CEO Hiroyuki Yoshino gave a keynote speech in which he talked about ASIMO, the robot Honda introduced in 2000. At that time, ASIMO could understand some simple words and follow few verbal instructions to walk slowly and speak simple words. Mr. Yoshino envisioned that in the future, humanoids like ASIMO would be able to run, walk forward and backward, and navigate stairs. Not only would they be able to understand speech, but they could also understand the speaker's mood. They would be able to identify individuals and human emotions through facial-image recognition software. However, 90-nm technology was not sufficient to achieve these technologies, and Mr. Yoshino believed that 22-nm-technology IC chips would be needed.

A decade later, cutting-edge IC technology has reached 22 nm. An all-new ASIMO was introduced in 2011 that can run, dance, and use sign language. Whereas the older version of ASIMO is an "automatic machine" that needs an operator, the new ASIMO is an "autonomous machine," which means it can make its own decisions and actions based on environmental sensors. It has the intelligence to walk among a group of people without collision by adjusting its movement through the observation and prediction of others. It has the capability to recognize the voices of multiple people, and its image sensor is capable of facial recognition. Although the new ASIMO has made a huge improvement, it is still far from having the envisioned emotion and mood recognition capabilities; to achieve that, perhaps sub-10-nm-technology IC chips are needed.

There have been many changes in the semiconductor industry and its manufacturing technologies over the last decade. Although Intel still keeps the bulk of its IC manufacturing technology in North American fabs, the center of IC manufacturing has shifted to East Asia, in nations such as Taiwan, North and South Korea, and China. The IC manufacturing fabs in Europe, Japan, and North America are declining at an alarming rate.

The biggest challenge for IC-technology node scaling has always been patterning technology. As suggested by my predictions in the first edition, unforeseen innovations such as immersion lithography and multiple patterning have extended the application of optical lithography and further delayed the development of next-generation lithography.

Many people helped me write this second edition. The book would not have been possible without the encouragement and support from my wife, Liu (Lucy) Huang. My elder son, Jarry Xiao, proofread several chapters. My current and former colleagues provided many useful suggestions; I would like to express my deepest appreciation to Paul MacDonald, Pierre Lefebvre, and Alan Liang.

Hong Xiao
October 2012

List of Acronyms

3MS	trimethylsilane or $(CH_3)_3SiH$
4MS	tetramethylsilane or $(CH_3)_4Si$
α-Si	amorphous silicon
AA	active area
AAPSM	alternating aperture PSM
ADC	automatic diameter control
ADI	after-development inspection
AFM	atomic force microscope
ALD	atomic layer deposition
ALU	arithmetic-logic unit
AMU	atomic mass unit
APCVD	atmospheric pressure CVD
AR	adsorption rate
ARC	antireflective coating
ASIC	application-specific integrated circuits
AttPSM	attenuated PSM
BARC	bottom antireflective coating
BEoL	back-end of the line
BL	bitline
BLC	bitline contact
BOE	buffered oxide etch
BOX	buried oxide
BPSG	borophosphosilicate glass
BSG	borosilicate glass
BST	$Ba_{0.6}Sr_{0.4}TiO_3$
BTBAS	bis-(tertiary-butylamino)silane] or $SiH_2[NH(C_4H_9)]_2$
bWL	buried wordline
C2C	cell-to-cell (inspection)
CB	contact to bit line
CBH	carborane or $C_2B_{10}H_{12}$
CD	critical dimension
CHO	cyclohexene oxide or $C_6H_{10}O$
CISC	complete instruction set computers
CMOS	complementary metal-oxide semiconductor
CMP	chemical mechanical polishing/planarization
CNT	carbon nanotube
CPU	central processing unit
CR	chemical reaction rate

CS	contact to source line
CVD	chemical vapor deposition
CZ	Czochralski (method)
D2D	die-to-die (inspection)
D2DB	die-to-database (inspection)
DEMS	diethoxymethylsilane or $C_5H_{14}Si$
DFM	design for manufacturing
DHF	diluted HF
DI	deionized
DMAH	dimethylaluminum hydride or $Al(CH_3)_2H$
DOF	depth of focus
DPP	discharge-produced plasma
DPT	double-patterning technology
DR	deposition rate
DRAM	dynamic random access memory
DSC	dichlorosilane (or dichloride silane) or SiH_2Cl_2
DSP	digital signal processing
DUV	deep ultraviolet
EBDW	electron beam direct write
EBI	electron beam inspection
EBR	edge-bead removal
ECD	electrochemical deposition
ECMP	electrochemical mechanical polishing
ECP	electrochemical plating
ECR	electron cyclotron resonance
EDA	electronic design automation
EEPROM	electric erasable programmable read-only memory
EGS	electronic-grade silicon
EMR	electromigration resistance
EOT	effective oxide thickness
EPD	electroplating deposition
EPROM	erasable programmable read-only memory
ESL	etch-stop layer
EUV	extreme ultraviolet
F/C	fluorine-to-carbon (ratio)
FEM	focus exposure matrix
FEoL	front-end of line
FET	field effect transistor
FOUP	front-opening unified pod
FPD	flat-panel display
FPGA	field-programmable gate array
FSG	fluorinated silicate glass
FTES	fluorotriethyloxysilane or $FSi(OC_2H_5)_3$

FTIR	Fourier transform infrared
FZ	floating zone
GPS	global positioning system
HDD	hard disc drive
HDP	high-density plasma
HDP-CVD	high-density plasma CVD
HEPA	high-efficiency particulate air (filter)
HKMG	high-κ and metal gate
HMDS	hexamethyldisilazane or $(CH_3)_3SiNHSi(CH_3)_3$
HOT	hybrid orientation technology
HP	half-pitch
IC	integrated chip
ICP	inductively coupled plasma
IDLH	immediate danger to life and health
IDM	integrated device manufacturer
ILD	interlayer dielectric
IMD	intermetal dielectric
IR	infrared
KGD	known good die
LAT	large angle tilt
LCD	liquid crystal display
LDD	lightly doped drain
LED	light-emitting diode
LELE or LE^2	litho-etch-litho-etch
LER	line edge roughness
LFLE	litho-freeze-litho-etch
LOCOS	local oxidation of silicon
LPC	landing pad contact
LPC	large-particle count
LPCVD	low-pressure chemical vapor deposition
LPP	landing pad poly
LPP	laser-produced plasma
LTE	low-temperature epitaxy
LWR	line width roughness
MBE	molecular beam epitaxy
MEMS	microelectromechanical systems
MEoL	mid-end of line
MERIE	magnetically enhanced RIE
MFC	mass flow controler
MFP	mean free path
MGS	metallurgical-grade silicon
MOCVD	metal organic CVD

MOS	metal-oxide semiconductor
MOSFET	metal-oxide-semiconductor field effect transistor
MRAM	magnetoresistive random access memory
MuGFET	multiple-gate field effect transistor
MW	microwave
NA	numerical aperture
NGL	next-generation lithography
NIL	nanoimprint lithography
nMOS	n-type MOS
NU	nonuniformity
NVM	nonvolatile memory
OMS	optical measurement system
OPC	optical proximity correction
OSG	organosilicate glass
OSHA	Occupational Safety and Health Administration
PBL	poly-buffered LOCOS
PC	personal computer
PCB	printed circuit board
PD	passivation dielectric
PEB	postexposure bake
PECVD	plasma-enhanced CVD
PEL	permissible exposure limit
PFC	perfluorocarbon
PIII	plasma immersion ion implantation
PLAD	plasma doping
PM	preventive maintenance
PMD	premetal dielectric
PMMA	polymethylmethacrylate
pMOS	p-type MOS
POCL	phosphorus oxychloride or $POCl_3$
PR	photoresist
PSG	phosphorous silicate (or phosphosilicate) glass
PSM	phase shift mask
PVA	poly-vinyl-alcohol
PVD	physical vapor deposition
RC	resistance × capacitance
RCA	Radio Corporation of America
RG	recessed gate
RIE	reactive ion etch
RISC	reduced instruction set computer
RP	remote plasma
RPCVD	remote plasma CVD

RTA	rapid thermal annealing
RTCVD	rapid thermal CVD
RTO	rapid thermal oxidation
RTP	rapid thermal process
SAC	self-aligned contact
SA-CVD	subatomic CVD
SAPD	spacer (self-) aligned double patterning
SA-STI	self-aligned shallow trench isolation
SC	standard cleaning
SCE	short channel effect
S/D	source/drain
SDE	S/D extension
SEG	selective epitaxial growth
SEM	scanning electron microscope
SIMOX	separation by implantation of oxygen
SIMS	secondary ion mass spectroscopy
SMIF	standard mechanical interface
SMO	source-mask optimization
SMT	stress memory technique
SN	storage node
SNC	storage node contact
SOC	system on chip
SOD	spin-on dielectric
SOG	spin-on glass
SOI	silicon-on-insulator
SPC	statistic process control
SRAM	static random access memory
SSD	solid-state disk/drive
SSOI	strained silicon-on-insulator
STI	shallow trench isolation
STM	scanning tunneling microscope
TCE	trichloethylene
TCP	transformer-coupled plasma
TCS	trichloride silane (or trichlorosilane) or $SiHCl_3$
TDEAT	tetrakis-diethylamidotitanium-titanium or $Ti[N(C_2H_5)_2]_4$
TDMAT	tetrakis-dimethylamido-titanium or $Ti[N(CH_3)_2]_4$
TEM	transmission electron microscope
TEOS	tetraethoxysilane (or tetraethylorthosilicate) or $Si(OC_2H_5)_4$
TMA	trimethylaluminum or $Al(CH)_3$
TMAH	tetramethyl ammonium hydroxide or $(CH_3)_4NOH$

TMB	trimethylborate or $PO(OCH_3)_3$
TMP	trimethylphosphite or $P(OCH_3)_3$
TSV	through-silica via
TW	thermal wave
UHV	ultrahigh vacuum
UHV-CVD	ultrahigh-vacuum CVD
ULK	ultralow-κ
ULSI	ultralarge-scale integration
USG	undoped silicate glass
USJ	ultrashallow junction
UV	ultraviolet
VLSI	very large-scale integration
WCVD	tungsten CVD
WERR	wet etch rate ratio
WID	within die
WIW	within wafer
WL	wordline
WTW	wafer to wafer
XRR	x-ray reflectometry

INTRODUCTION TO

Semiconductor Manufacturing Technology

SECOND EDITION

Chapter 1
Introduction

Integrated circuit (IC) chip technology has been changing our lives dramatically for more than 50 years. Since their introduction in the 1960s, IC chips have developed in complexity and usefulness to the point where hundreds, if not thousands, can be found in the average households of developed countries. IC chips instigated a technical revolution considered to be more significant than any other period of invention in human history. IC chips are the backbone of the computer industry and have spurred related technologies such as software, mobile electronics, and the Internet. Every product of the information age is an offspring of IC technology.

IC chips can be found in automobiles, televisions, Blu-ray and DVD players, digital cameras, smartphones, appliances, and game consoles. The list of applications and seemingly miraculous behind-the-scene functions of IC chips could easily fill an entire book.

The future of IC technology is full of possibilities, such as humanized robotics. It is not hard to imagine in the near future a robot that assists the elderly or disabled, with the ability to react to voice command, do chores, hold a conversation, and respond to facial expressions.

When dealing with 3D graphics and artificial intelligence, the thirst for computational power and memory space will never end and will further drive demands for more-powerful microprocessors and memory chips with larger storage size.

Objectives

After finishing this chapter, the reader will be able to:

- name the three scientists who invented the first transistor
- identify the two people who shared the patent for the invention of integrated circuits
- explain the difference between a discrete device and an IC device
- describe Moore's law
- explain the effects of feature size and wafer size on IC chip manufacturing
- define the term semiconductor technology node.

1.1 Brief History of Integrated Circuits

1.1.1 First transistor

The semiconductor era started in December 1947. Two AT&T Bell Laboratories scientists, John Bardeen and Walter Brattain, demonstrated a solid-state electrical device made from germanium, a semiconductor material. They observed that when electrical signals were applied to contacts on a crystal of germanium, the output power was larger than the input. These results were published in 1948. This first transistor, a point contact type, is shown in Fig. 1.1. The word transistor comes from the combination of two words: transfer and resistor.

Bardeen and Brattain's supervisor William Shockley, not wanting to be left out of such an important invention, was determined to make his imprint on the discovery. Shockley worked very hard that December to find out how the bipolar transistor functioned. He solved the problem and published his theory in 1949. He also predicted another kind of transistor, the junction bipolar transistor, which would prove easier to mass produce. In 1956 William Shockley, John Bardeen, and Walter Brattain shared the Nobel Prize in physics for the invention of the transistor. Figure 1.2 shows these three coinventors.

Driven by military and civilian demands for electronic devices, the semiconductor industry developed rapidly in the 1950s. Germanium-based transistors quickly

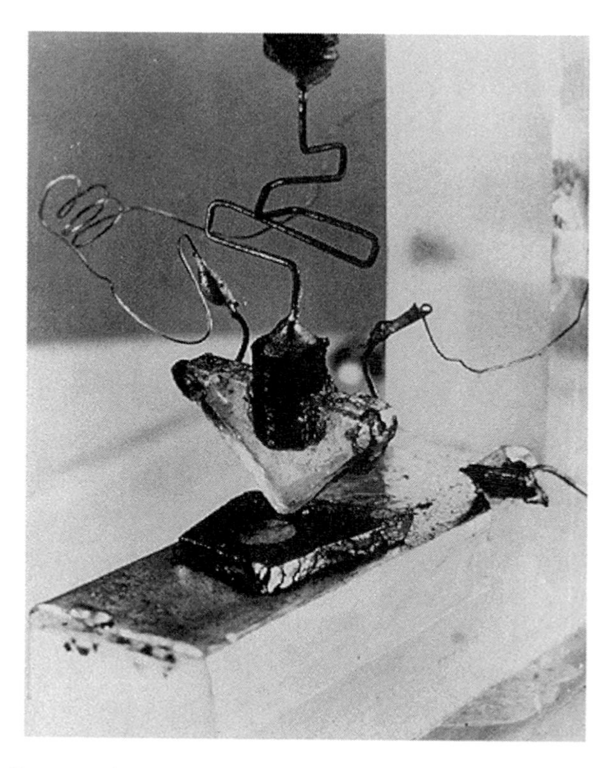

Figure 1.1 The first transistor made in AT&T Bell Laboratories (AT&T Archives, reprinted with permission from AT&T).

Figure 1.2 Three coinventors: William Shockley (front), John Bardeen, (back left), and Walter Brattain (right).

replaced vacuum tubes in most electronics equipment because of their smaller size, lower power consumption, lower operating temperature, and shorter response time. Transistor manufacturing was significantly accelerated by the introduction of technologies to manufacture pure, single-crystal semiconductor materials. The first single-crystal germanium was made in 1950, and the first single-crystal silicon followed in 1952. Throughout the 1950s, the semiconductor industry provided discrete devices that the electronics industry used to make radios, computers, and many other civilian and military products. A *discrete device* is an electronic device, such as a resistor, capacitor, diode, or transistor, that contains only one device per piece. They are still widely used in today's electronic products. One can easily find discrete devices on almost every printed circuit board (PCB) in any electronic system.

1.1.2 First integrated circuit

In 1957, Jack Kilby, one of the attendees at a workshop commemorating the tenth anniversary of the invention of the transistor, noticed that most discrete devices such as resistors, capacitors, diodes, and transistors could be made from a piece of semiconductor material like silicon. Therefore, he thought it would be possible to make them on the same piece of semiconductor substrate and connect them to form a circuit. This would make a much smaller circuit and reduce the cost of the electronic circuit. Kilby joined Texas Instruments (TI) in 1958 to pursue his new idea. As a newly hired employee, he did not have vacation time, thus, when most of his coworkers took the mandatory summer vacation, he worked to consolidate his idea of an IC in the deserted research-and-development laboratory. When his coworkers came back from summer vacation, he presented his idea and started to put it into action. Because no silicon substrate was readily available, he used what material he could find: a strip of germanium with one transistor already built onto it. He added a capacitor and used a germanium bar to form three resistors. By connecting the transistor, capacitor, and three resistors from the 1.5-in. long germanium bar to fine metal wires, Kilby made the first IC device, as shown in Fig. 1.3. At TI, IC devices have been called bars instead of chips or dies for a long time because of the shape of the first IC device made by Kilby.

About the same time, Robert Noyce of Fairchild Semiconductor was working on the same idea: making more for less. Different from Kilby's IC chip (or bar), where real metal wires were used to connect different components, Noyce's chip employed aluminum patterns etched from an evaporated aluminum thin film coated onto the wafer surface to form the metal interconnections between the different devices. By using silicon instead of germanium, Noyce applied planar technology, developed by his colleague Jean Horni, to make junction transistors, which took advantage of the silicon and its natural oxide, silicon dioxide. Highly stable silicon dioxide can be easily grown on a silicon wafer surface in a high-temperature oxidation furnace and can be used as the electric isolation and diffusion mask.

One of the first silicon IC chips Noyce designed in 1960 (shown in Fig. 1.4) was made from a 2/5-in. (~10-mm) silicon wafer. Noyce's chip had the basic processing techniques of modern IC chips and served as a model for all future ICs.

In 1961, Fairchild Semiconductor made the first commercially available IC, which consisted of only four transistors, and sold them for $150 each. Ironically, it was much more expensive than purchasing four transistors and connecting them into the same circuit on a circuit board. NASA was the main customer for the newly available IC chips because rocket scientists and engineers were willing to pay higher prices to reduce even a gram of weight from a space rocket.

Figure 1.3 The first integrated circuit made by Jack Kilby (Texas Instruments).

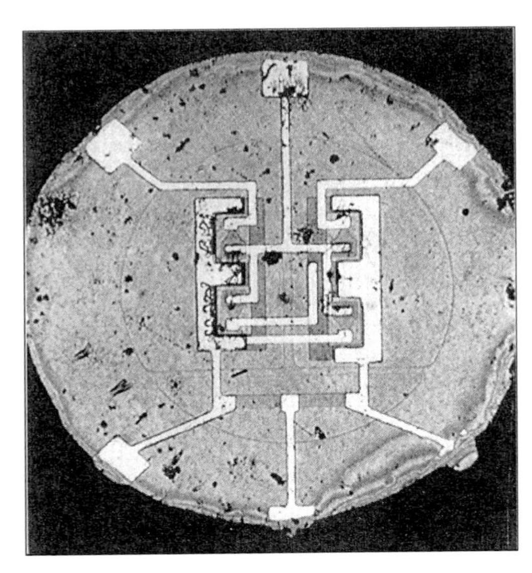

Figure 1.4 The first IC made on a silicon wafer by Fairchild Semiconductor (Fairchild Semiconductor International).

After several years of legal battle, TI and Fairchild Semiconductor settled their case by agreeing to cross-license their technologies. Kilby and Noyce shared the title of coinventor of the IC. In 2000, Kilby (seen in Fig. 1.5) was awarded the Nobel Prize in physics for the invention of the IC. Noyce (seen in Fig. 1.6) left Fairchild Semiconductor and cofounded Intel Corporation with Andrew Grove and Gordon Moore in 1968. He later served as chief executive officer of SEMATECH, an international consortium of semiconductor manufacturers, in Austin, Texas.

1.1.3 Moore's law

In the 1960s, the IC industry developed very rapidly. In 1964, Gordon Moore, one of the cofounders of Intel Corporation, noticed that the number of components on a computer chip doubled every 12 months while the price stayed the same. He predicted that trend would hold in the future. His vision has become well known in the semiconductor industry as Moore's law. Amazingly, Moore's law has been proven accurate for more than 40 years, with only a slight adjustment made in 1975 when he changed 12 months to 18 months. Figure 1.7 shows Moore's prediction in 1965, and Fig. 1.8 is Moore's law for microprocessors. The scales of integration level for IC chips used in the semiconductor industry are listed in Table 1.1.

1.1.4 Feature and wafer size

Before 2000, the semiconductor industry feature size was usually measured in microns, which is equal to one-millionth (10^{-6}) of a meter and is also noted as μm. As a reference, the diameter of a human hair is about 50 to 100 μm. After 2000, semiconductor technology advanced to nanometer (nm) technology nodes.

Courtesy Texas Instruments

Figure 1.5 Jack St. Clair Kilby (November 8, 1923–June 20, 2005). (See http://media.aol.hk/drupal/files/images/200811/25/kilby_jack2.jpg.)

Table 1.1 Scales for different integration levels for IC chips.

Integration level	Abbreviation	Number of devices on a chip
Small-scale integration	SSI	2 to 50
Medium-scale integration	MSI	50 to 5,000
Large-scale integration	LSI	5,000 to 100,000
Very large-scale integration	VLSI	100,000 to 10,000,000
Ultra-large-scale integration	ULSI	> 10,000,000

A nanometer is one-billionth (10^{-9}) of a meter. In less than 50 years, the minimum feature size of IC chips has shrunk dramatically, from about 50 μm in the 1960s to 32 nm in 2010. By reducing the minimum feature size, a smaller device can be made that allows more chips per wafer, or a more powerful chip can be made with the same die size. Both ways help IC fabrication facilities (fabs) gain a greater profit in the manufacturing of IC chips, profit being the most important driving force of IC technology development.

For example, when the technology node shrank from 28 to 20 nm, the size of a chip shrank by a factor of $(20/28)^2 \sim 0.51$. This means that the number of chips

Figure 1.6 Robert Norton Noyce (December 12, 1927–June 3, 1990). (See http://download.intel.com/museum/research/arc_collect/history_docs/pix/noyce1.jpg.)

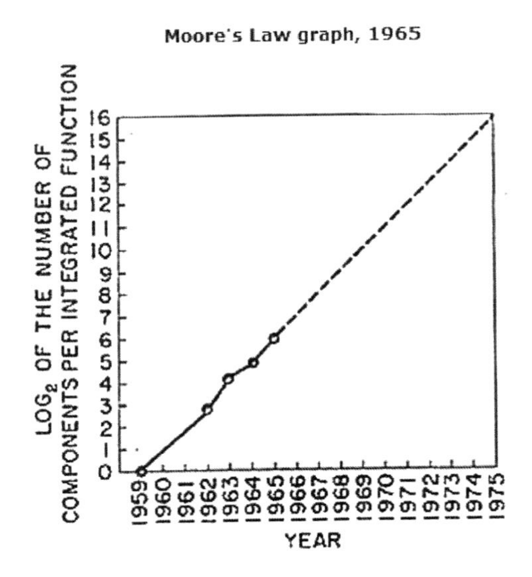

Figure 1.7 Moore's observation and prediction in 1965 (source: http://www.intel.com/technology/mooreslaw/).

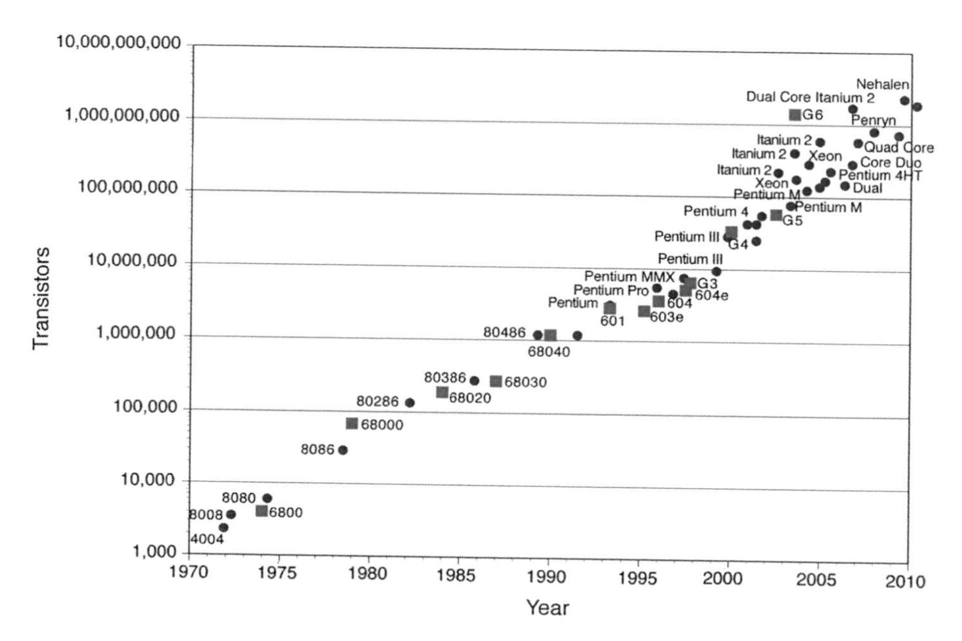

Figure 1.8 Moore's law of microprocessors (source: http://www.is.umk.pl/~duch/Wyklady/komput/w03/Moores_Law.jpg).

almost doubles if both the chip and wafer are squares. Since a silicon wafer is round, the edge effect can cut the increase to about 50%. Similarly, by further shrinking the feature size to 14 nm, the number of chips almost quadruples compared with the 28-nm technology, as shown in Fig. 1.9.

Figure 1.10 shows the smallest known operational metal-oxide-semiconductor (MOS) transistor with an effective gate length of 0.004 μm, or 4 nm, made by NEC and published in the proceedings of the 2003 International Electron Devices Meeting of IEEE).

Question: Is there a limit for minimum feature size?

Answer: Yes. The minimum feature size of a microelectronic device made on a single-crystalline silicon substrate cannot be smaller than the silicon lattice spacing, which is 5.43 Å. Notice that 1 Å = 0.1 nm = 1×10^{-10} m.

Question: What is the achievable minimum feature size of an IC chip?

Answer: History has disproved many predictions on the achievable minimum feature size of an IC device, and the prediction given here may also fall into that category. One silicon atom is not enough to form the feature of an electronic device. Therefore, if one needs ten silicon lattices to form a minimum feature, the achievable minimum feature size of an IC device might be about 50 Å, or 5 nm.

There will be many technological challenges to overcome before minimum feature size can reach its final physical limit. Most notable is the patterning process, which is the fundamental IC manufacturing step used to transfer the

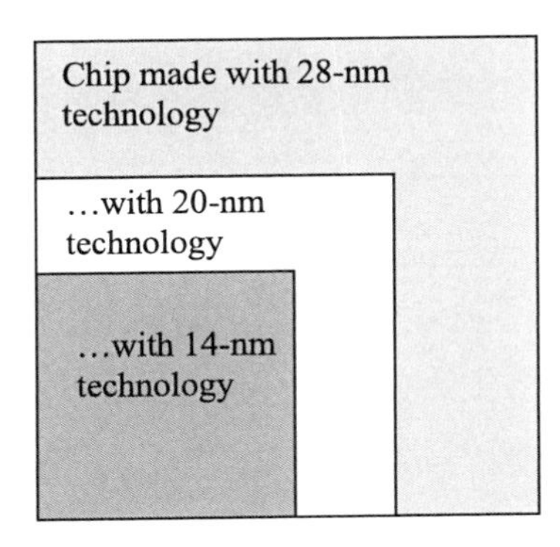

Figure 1.9 Relative chip size with different technology nodes.

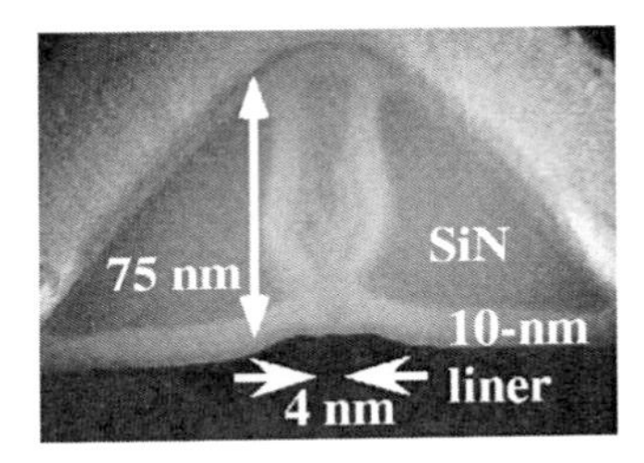

Figure 1.10 The world's smallest known metal-oxide-semiconductor transistor (H. Wakabayashi, et al., *IEEE Proc. IEDM*, 2003).

designed pattern to a wafer surface and form IC devices. Most likely, optical photolithography currently being used will have to switch to alternative lithography technologies—such as extreme ultraviolet (EUV) lithography, nano-imprint lithography (NIL), or electron beam direct write (EBDW) lithography—before minimum feature size can reach its final physical limit. This is discussed in more detail in Chapter 7.

While minimum feature size is shrinking, wafer size is also increasing. Starting from 10 mm (2/5 in.) in the 1960s, wafer size has increased to 300 mm (12 in.). By increasing wafer size, one can put more chips on a wafer. From 200 to 300 mm, the wafer area increases by a factor of $(3/2)^2 = 2.25$, which means the number of chips per wafer can be doubled on a 300-mm wafer. Figure 1.11 illustrates relative wafer sizes of 150, 200, 300, and 450 mm.

Question: What is the maximum wafer size?

Answer: No one knows for sure yet. Because flat-panel display (FPD) manufacturing equipment has already started to handle a tenth-generation glass substrate of size 2850 × 3050 mm, mechanical handling of 1000-mm-diameter (1 m) silicon wafers should not be a big problem. However, the size of a wafer is limited by many factors, such as crystal pulling, wafer slicing technologies, process equipment development, and most

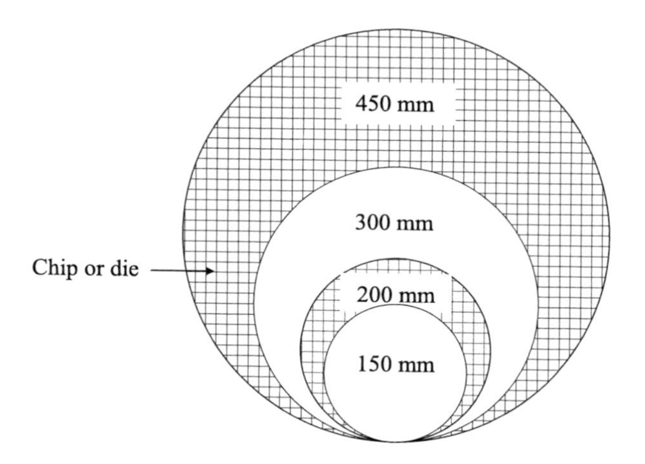

Figure 1.11 Relative wafer sizes.

importantly, the demands of IC fabrication. Because huge amounts of initial capital investment is needed for research and development to increase wafer size, not every IC manufacturer is enthusiastic about doing it, especially doing it first. Currently, the largest wafer used in IC production is the 300-mm wafer, and this is becoming the mainstream for advanced IC fabs. The largest wafer on display is 450 mm in diameter, which was proposed for use in IC fabrication starting in 2012. However, the majority of semiconductor manufacturers and most semiconductor equipment manufacturers are not excited about investing the large amount of capital needed for 450-mm wafer processing, and the targeted year for implementing 450-mm wafers in IC production almost certainly will be delayed. If 450-mm wafers eventually do go into IC production, they will most likely be the largest wafer size.

1.1.5 Definition of the integrated circuit technology node

Technology nodes such as 45, 40, 32, 28, 22, 20 nm, etc., are not defined as the minimum critical dimension (CD) of the devices. The definition of technology node is related to the pitch of the gate pattern, as shown in Fig. 1.12. Although one can easily reduce pattern CD with techniques such as photoresist trimming [shown in Fig. 1.12(b)], this technique will not reduce the pitch of the pattern. To reduce pattern pitch, the pattern technologies (which include photolithography and etch processes) need to be upgraded.

Different IC devices have different relationships of technology node to pattern pitch. For example, in NAND flash devices (which do not have contact between gates), the technology node is defined as a half pitch of the gate pattern. Therefore, a 20-nm NAND flash device has gate patterns of 40-nm pitch (gate CD plus gap) or a 20-nm half pitch. For a logic IC device with contact between gates (called

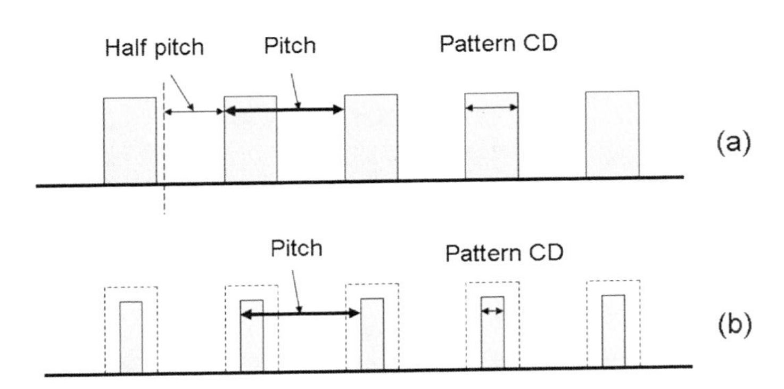

Figure 1.12 Relationship of pattern CD and pattern pitch. (a) Initial photoresist pattern and (b) photoresist pattern after trimming. Although pattern CD is reduced by the trimming process, pattern pitch remains the same.

a contact gate), the technology node usually is one quarter of the gate pitch. For example, a 20-nm logic device usually has an 80-nm gate pitch.

1.1.6 Moore's law or the law of more

Since the invention of the IC, its manufacturing technology has developed rapidly and matched Moore's law very well, as shown in Fig. 1.8. However, the actual driving force of semiconductor technology advancement is not Moore's law, but the law of more (profit). By shrinking the minimum feature size, one can place more chips onto a wafer or place more devices in a chip. In the "good old days," reducing the size of the device improved device speed, reduced power consumption, and enhanced overall device performance. Therefore, scaling down the minimum feature size would reduce manufacturing cost, improve profit margin, and strengthen competitive position. When the cost of research and development was justified by the benefit of feature size reduction, IC manufacturers had strong incentive to heavily invest in new technologies and push device scaling. For the past 50 years, Moore's law overlapped with the law of more, giving the illusion that IC technology was driven by Moore's law. However, when the IC technology node reached the nanometer range, simply shrinking the minimum feature size no longer improved device performance due to the leakage issue, unless very expensive high-κ gate dielectric and metal gates were used in the MOS device. In the nanometer technology era, research and development costs increased almost exponentially with the scaling of the IC technology node. At the 45-nm technology node, some semiconductor companies could not afford the estimated $1.5 billion in research and development costs alone. As IC technology nodes progress to 32/28 nm, 22/20 nm, 14 nm, and beyond, fewer and fewer IC manufacturers will be able to afford research and development costs all by themselves. In the foreseeable future, Moore's law will become history, and IC technology advancement will be determined by the law of more. The semiconductor industry will become a mature industry, like the automobile industry, where technology continues to improve at a more moderate rate.

1.2 Brief Overview of Integrated Circuits

IC chip manufacturing is a very complex system. It involves the manufacture of materials and wafers, circuit design, cleanroom technology, processing equipment, metrology tools, wafer processing, die testing, chip packaging, and final testing.

1.2.1 Manufacturing materials

Semiconductor manufacturing requires a great amount of raw materials for wafer processing. To ensure IC production yield, gaseous, liquid, and solid materials used for wafer processes such as chemical vapor deposition (CVD), etch, physical vapor deposition (PVD), and chemical mechanical polishing (CMP) need to be of ultrahigh purity and have extremely low particle density.

Many semiconductor process materials are poisonous, flammable, explosive, and/or corrosive, and a few are strong oxidizers. Many of the chemicals require that employees receive special training to ensure proper handling.

1.2.2 Processing equipment

Semiconductor processing requires highly specialized tools for a variety of processes. There are epitaxial silicon deposition reactors, CVD and etch equipment, ion implanters, furnaces and rapid thermal process systems, metal deposition reactors, chemical mechanical polishers, and photolithography tools. These process tools are very sophisticated, complicated, and expensive. They require that technicians receive a great deal of special training before being properly operated, maintained, and diagnosed. Because these tools are so expensive and the square footage of a cleanroom is also very costly, semiconductor fabs attempt to keep them running 24 hours a day, seven days a week, nonstop until scheduled preventive maintenance or a breakdown occurs. It is very important to reduce tool downtime to boost productivity and increase throughput to improve profit margin. Well-trained and experienced engineers and technicians play a crucial role in this aspect.

Before the 1970s, most IC manufacturers made their own process equipment. Now semiconductor equipment companies provide the majority of manufacturing equipment for IC manufacturers. They provide not only the sophisticated equipment, but also the production-ready processes. Batch systems, which process many wafers simultaneously, are still widely used, although single-wafer, multichamber cluster process tools are now receiving more attention and application. Cluster tools with multiple process integration capabilities can help improve process throughput and process yield. Another trend is to stack process chambers or process stations in a vertical direction to reduce the footprint of a tool and save cleanroom space. Cleanroom space is very expensive, especially the high-grade cleanrooms of advanced IC fabs. Integration of metrology tools on production equipment for in-situ measurement and real-time process control is another trend of process equipment development.

1.2.3 Metrology tools

Every semiconductor manufacturing step requires specialized tools to measure, monitor, maintain, and control the process. There are tools to measure thin-film characteristics, such as thickness, uniformity, stress, refractivity, reflectivity, and sheet resistance. There are tools to measure device characteristics, such as current-voltage (I–V) curve, capacitance-voltage (C–V) curve, and breakdown voltage. Optical and electron microscopes are widely used to check patterns, profiles, and alignments. Infrared and x-ray radiation are used in some metrology tools to measure and analyze chemical components and their concentrations.

It is very important to keep metrology tools in proper working condition, otherwise false readings can be obtained from time to time, inducing unwanted process troubleshooting and unnecessary tool downtimes, both of which process throughput. Therefore, training employees to understand the basic principles of metrology tools will keep the system well calibrated and prevent unnecessary process interruptions.

The development of semiconductor processes poses the greatest challenge to the expansion of metrology tools. Faster, more accurate measurements, ultrathin-film (<10 Å) measurements, nondestructive pattern and profile measurements, and real-time, in-situ measurements are required for better process monitoring and control.

To enhance and maintain manufacturing yield, defect inspection and monitoring techniques also need to be developed rapidly. Optical inspection systems use photons to detect and capture physical defects on both blank and patterned wafers. Electron beam inspection systems use electrons to capture both tiny physical defects and some electrical defects, such as open or short circuits. To control defect density and maintain production yield, one must have the capability to capture a defect that is half the size of the technology node. For example, in the 14-nm technology node, the capability to capture a 7-nm defect becomes necessary.

1.2.4 Wafer manufacturing

Wafer manufacturing begins with common quartz sand. First, carbon reacts with sand at a high temperature to generate crude silicon or metallurgical-grade silicon (MGS) with purity of at least 98%. Then the MGS is ground into a powder to react with hydrogen chloride to form liquid silicon hydrochloride, such as $SiHCl_3$ (TCS), which can be purified to a level of up to 99.9999999%. Then TCS reacts with hydrogen at high temperatures to deposit high-purity polycrystalline silicon, called electronic-grade silicon (EGS). Placed in a rotating quartz crucible, EGS is melted at 1415 °C, and a rotating seed crystal is slowly pushed in and pulled out to attain an ultrapure single-crystal silicon ingot. Single-crystal silicon wafers are then obtained by sawing the ingot into slices. Because of this process, single-crystal silicon wafers are always round in shape. The wafers are then lapped, cleaned, etched, polished, labeled, and shipped to IC chip manufacturers. Many wafer manufacturers even deposit a thin layer of single-crystal silicon, called epitaxial silicon, on the single-crystal silicon wafer surface for IC chip manufacturers. The processes of wafer fabrication and epitaxy silicon deposition are covered in more detail in Chapter 4.

1.2.5 Circuit design

When Kilby made the first IC with five components, he drew the design circuit by hand, as shown in Fig. 1.13. At the 22-nm technology node, a test chip with 364 million bits (Mb) of static random access memory (SRAM) with more than 2.9 billion transistors has been demonstrated. 64-Gb NAND flash memory chips, which have more than 64 billion components, have also been fabricated with 19-nm technology. It is impossible to design these chips without the help of powerful computer design tools. Even with computer design tools, a complicated IC such as a high-end microprocessor chip takes dozens, if not hundreds, of engineers and designers several months for design, test, and layout.

Main design considerations are performance, die size (cost of chip manufacture), design time (cost of IC designing and scheduling), and testability (cost of testing and scheduling). IC design is always the tradeoff among those factors to achieve adequate performance and profitable results. Figure 1.14(a) illustrates the circuit of a complementary metal-oxide-semiconductor (CMOS) inverter. Figure 1.14(b) is a textbook layout of a CMOS inverter. The advantage of this layout is that it can give the device cross section of both n-type MOS (nMOS) and p-type MOS (pMOS) on the same surface, as shown in Fig. 1.14(c).

In a normal IC design, a CMOS inverter layout is more compact, as shown in Fig. 1.15. It basically rotates the pMOS in Fig. 1.14(b) 180 deg and places it above the nMOS. By doing that, the shared gate of the nMOS and pMOS is shortened and straightened. Compared with the U-shaped gate shown in Fig. 1.14(b), the benefit of this layout is obvious. Of course, this layout will not present both nMOS and pMOS in one cross section. To achieve that, the textbook-style layout is needed.

IC design includes architectural, logic, and transistor level design. Architectural design defines the application operation system and divides modules for the system. Logic design puts logic units such as adders, gates, inverters, and registers into each module. It also runs subroutines in each module. In transistor level design,

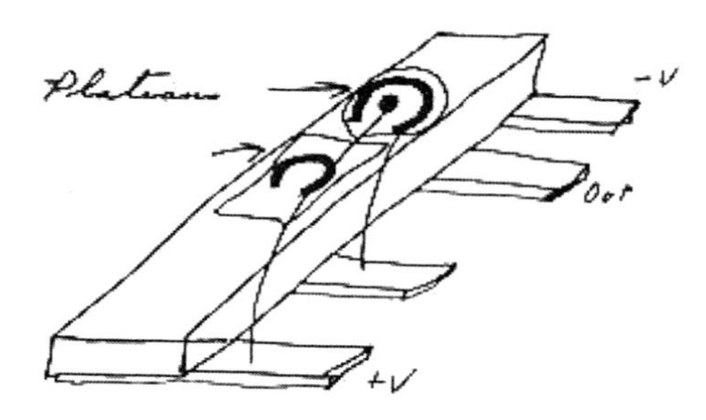

Figure 1.13 The first IC design Kilby sketched in his notebook on September 12, 1958 (Texas Instruments).

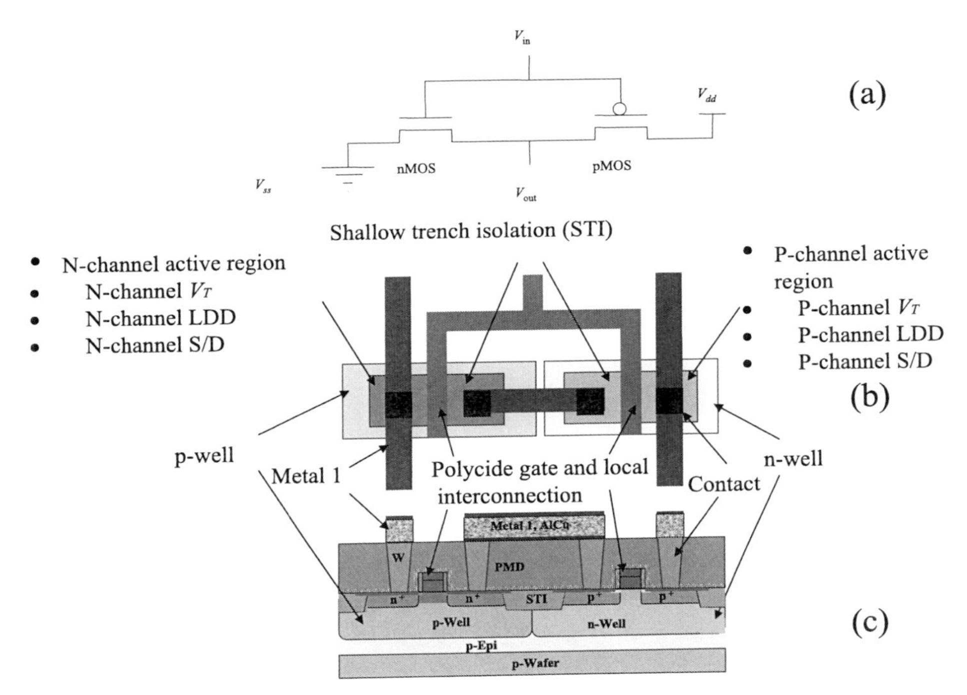

Figure 1.14 (a) The circuit of a CMOS inverter, (b) an example of a textbook-style design layout of a CMOS inverter, and (c) the cross section of the textbook layout.

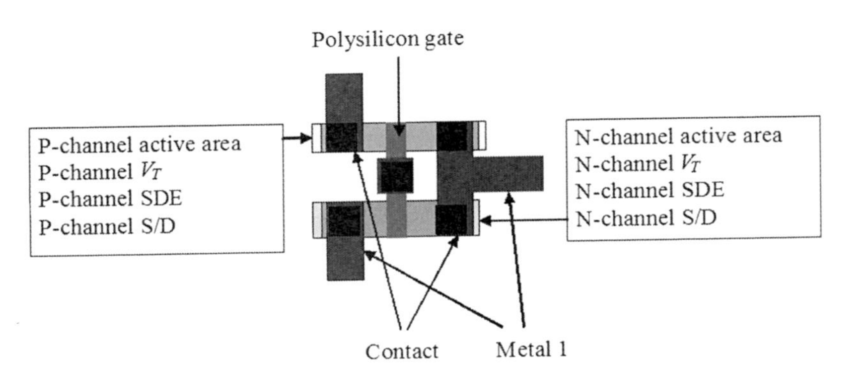

Figure 1.15 CMOS inverter in a real-life IC layout.

individual transistors are laid out in each logic unit. Binary instructions (0 and 1) are given to the logic units to test the circuit design.

After test procedures eliminate design faults or bugs to make certain that there are no design rule violations, and have verified that the design meets yield rule requirements, the layouts of the design can be taped out. In a mask shop, the designed patterns are precisely printed on a piece of chromium glass to make the masks or reticles. Masks or reticles are used in the photolithography process to transfer designed patterns to the photoresist, which is temporarily coated on the semiconductor wafers by means of a photochemical reaction during the exposure process.

Before the 1980s, most semiconductor companies designed, fabricated, and tested IC chips by themselves. These traditional semiconductor companies are called integrated device manufacturers (IDMs). During the 1990s, two kinds of semiconductor companies rapidly developed in the IC industry. One is called a foundry company, which owns wafer processing fabs and does not design its own chips. It receives orders from other companies, makes the masks/reticles or acquires the masks/reticles from mask shops, processes the wafers, and manufactures chips for its customers. Another type is called a fabless semiconductor company, which designs chips for its customers based on customer specifications. These fabless semiconductor companies, or IC design houses, usually place wafer orders with foundry companies to manufacture chips based on their design. Some design houses have their own test center to test the chips after the foundry fabricates and delivers them. Some other fabless companies rely completely on their vendors to do all of the fabrication and testing.

IC design has a direct impact on IC processing. For example, if someone designs metal interconnections on a chip, with one area densely packed with metal lines and another area with very little or no metal lines, this variation could cause what is called a loading effect in the etch process and a dishing effect in the CMP process. Production engineers in the design for manufacturing (DFM) group of a fab usually work closely with both design and process teams to solve these types of problems.

When IC technology arrived in the nanometer era, because printed patterns on the wafer were significantly smaller than the wavelength used to expose these patterns, an optical proximity correction (OPC) had to be used, and design-related process issues became more significant. Design tool companies, which provide electronic design automation (EDA) software for designers to plan IC chips, are now working more closely with foundries to make sure that their products help designers create IC chips with high manufacturability and high production yields in the first silicon run.

1.2.6 Mask formation

After the IC design is finished, the layout image generated by EDA software is printed on a piece of quartz glass coated with a layer of chromium. A computer-controlled laser beam projects the layout image onto the photoresist-coated chrome glass surface. Photons change the chemistry of the exposed photoresist via photochemical reaction, and the photoresist is later dissolved in a base developer solution. A patterned etching process removes the chromium at the location where the photoresist has been dissolved. Therefore, the image of the IC layout is transferred to the chromium on the quartz glass.

To keep the surface of the mask clean, a thin sheet of plastic called a pellicle is mounted a short distance away from the chrome glass surface. The pellicle provides some protection for the tiny chrome patterns. More importantly, particles that fall onto the pellicle of the mask are out of focus during the pattern exposure process and do not print on the wafer surface (causing repeating defects). Figure 1.16(a) illustrates the basic structure of a conventional or binary photomask, and Fig. 1.16(b) illustrates the basic structure of an attenuation phase shift mask. Figure 1.17 shows the relationship of the IC layout and photomasks for a CMOS inverter. It can be seen that to make a functional CMOS inverter, one needs at least ten masks.

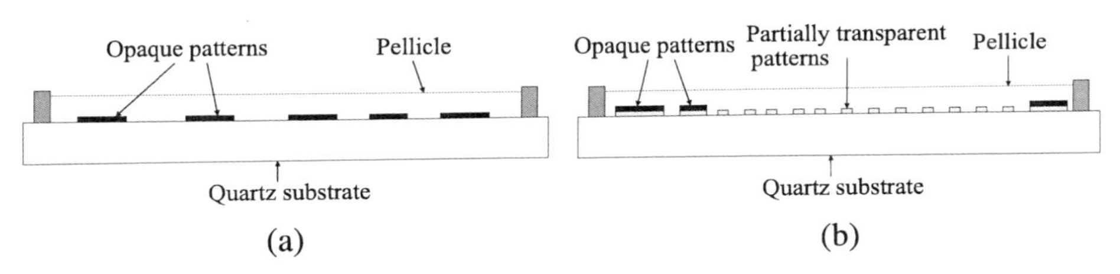

Figure 1.16 (a) Binary mask and (b) an attenuated phase shift mask.

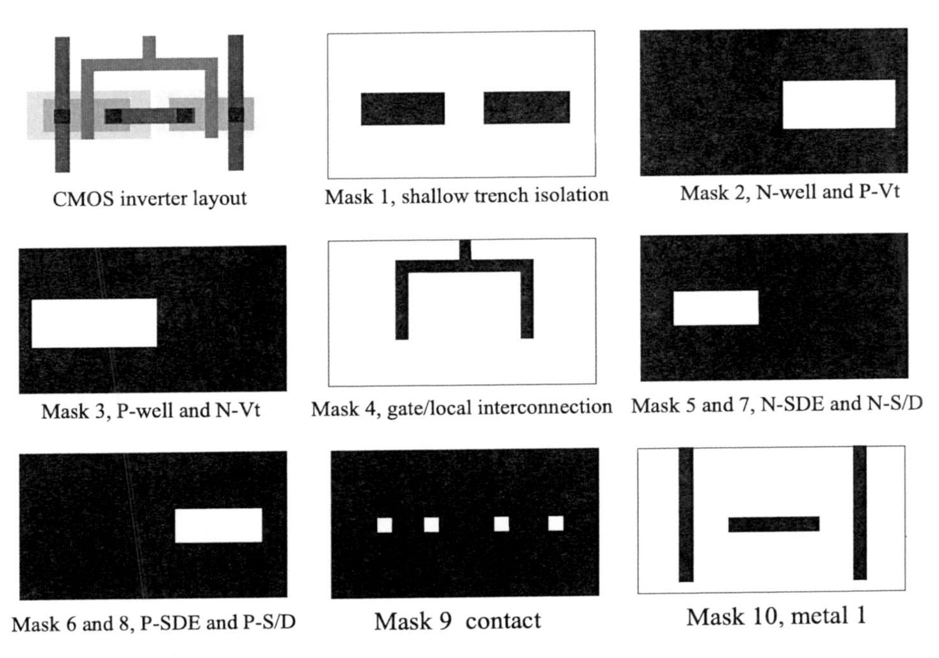

Figure 1.17 Layout and binary masks of a CMOS inverter.

A computer-controlled electron beam (e-beam) is also used to expose photoresist in order to achieve pattern transfer. Since an energetic e-beam has a shorter wavelength than UV light, it has higher resolution and generates a sharper image in the photoresist coated on the chrome glass. As feature size shrinks, more and more masks require an e-beam writer to pattern.

Normally, when the chrome glass has an image covering the entire wafer, it is called a mask. A mask normally transfers an image to the wafer surface at a 1:1 ratio. Exposure systems such as projection, proximity, and contact printers use masks. The highest resolution for a mask is about 1.5 μm.

When the image on the chrome glass covers only part of a wafer, it is called a reticle. A reticle has a larger image and feature size than the image it projects onto the wafer surface, usually with a 4:1 (4×) reduction ratio. Exposure systems that utilize reticles need to expose the reticle multiple times to cover the entire wafer. The process is called step-and-repeat or step-and-scan, and the alignment/exposure systems are called steppers. Some people in fabs also call the system a scanner because of the way it exposes the reticle. Advanced semiconductor fabs usually use steppers for exposure during the photolithography process to achieve the resolution required for patterning. Figure 1.18 illustrates both a mask and reticle. In an IC fab, the reticle is usually referred to as a mask; therefore, the subject of Fig. 1.18(b) can be called both a mask and a reticle, while the subject of Fig. 1.18(a) can only be called a mask.

Partially transparent patterns are formed to allow light to pass through them, with phase shifting that causes destructive interference with the light passing through neighboring clear patterns. This improves photolithography resolution for patterning the tiny features of nanometer technology nodes, as described in Chapter 6.

At least five masks are required to make the simplest MOS transistor. It can take more than 30 masks/reticles to make an advanced IC chip.

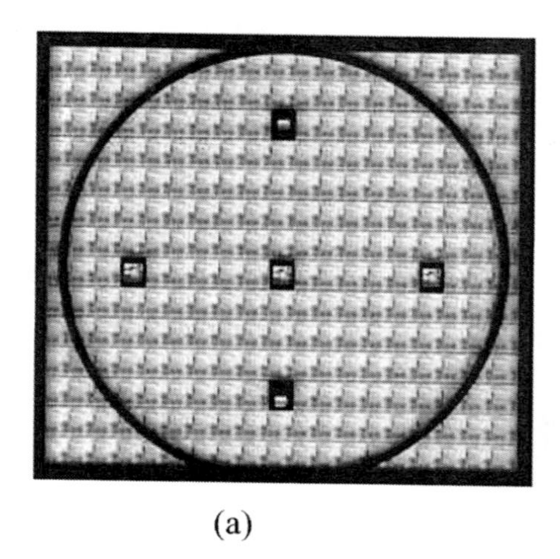

(a)

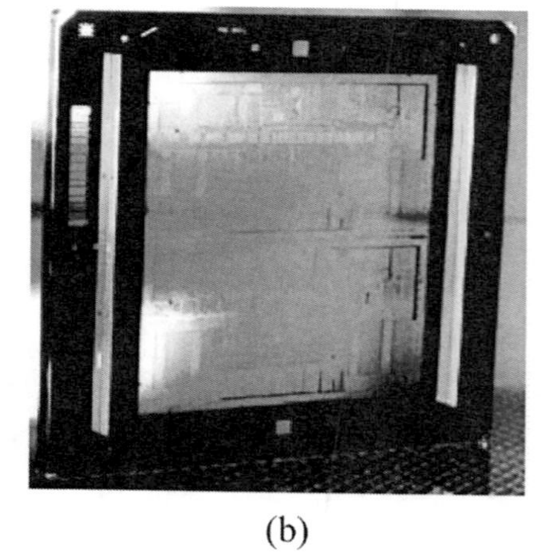

(b)

Figure 1.18 (a) A mask and (b) reticle. (SGS Thompson).

1.2.7 Wafer processing

First, IC designers design the circuits with the help of EDA tools. Then the mask shop uses the layout files provided by the designers to print the designed features on the photoresist coated onto the chrome glass with a laser or e-beam writer, and then etch the chromium to make the masks. The masks are then shipped to the photo bay of the fab. Wafer manufactures produce and provide single-crystal silicon wafers with different crystalline orientations, different dopant types, and different dopant concentrations, with or without epitaxial layers, based on requests from the IC fabs. Material manufactures make a variety of ultrapure materials, which are needed in IC manufacturing.

Once wafers are inside a fab, they usually are laser scribed, cleaned, and thermally grown on a thin layer of silicon dioxide. In the front-end of the line (FEoL) process, where transistors are made, wafers go through photolithography many times, and most of them are followed by different ion implantations to form well junctions, source/drain (S/D) extension junctions, polysilicon gate doping, and S/D junctions. Only two FEoL photolithography processes are followed by patterned etch processes—one forms shallow trench isolation and another forms the gate electrodes.

In the back-end of the line (BEoL) process, all photolithography processes are followed by etch processes. In copper metallization, this dual damascene process is repeated multiple times, depending on how many metal layers there are: dielectric CVD, photolithography, dielectric etch, photoresist strip and clean, photolithography, dielectric etch, photoresist strip and clean, deposition of metal layers, metal anneal, and CMP. After all of the metal layers are formed, CVD silicon oxide and silicon nitride are deposited as the passivation layer, and the last photolithography process defines the pads for wire bonding or bump formation. After that, the chips are tested, die sawed, sorted, packaged, burned-in, and shipped to customers.

It takes an IC fab a few weeks and several hundreds of processing steps to create the microelectronic devices of ICs on silicon wafers. Wafer processing includes wet clean, oxidation, photolithography, ion implantation, rapid thermal process (RTP), etch, photoresist stripping, CVD, PVD, CMP, etc. Figure 1.19 shows the flow chart of IC chip fabrication in an advanced semiconductor fab. These processes in the wafer fabs are covered in detail in later chapters with a minimum of math, chemistry, and physics.

1.3 Summary

- Three scientists who invented the first transistor are William Shockley, John Bardeen, and Walter Brattain.
- Jack Kilby and Robert Noyce are the coinventors of the IC.
- A discrete device is an individual electronic device such as resistor, capacitor, diode, or transistor. An IC device is a functional circuit built on a single piece of substrate; IC devices exist in many electronics products.

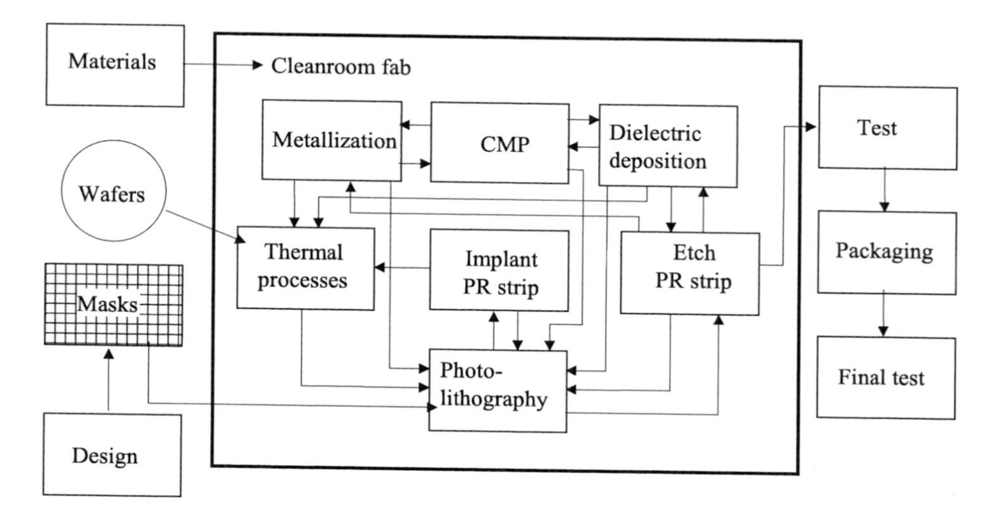

Figure 1.19 Flow chart of a wafer process.

- Moore's law predicts that the number of components on a chip doubles every 12 to 18 months, while the price stays the same.
- By reducing feature size, die size is reduced, which allows more dies on a wafer. By increasing wafer size, more chips can be made on a wafer. Both allow IC chip manufacturers to make a greater profit when research and development costs are manageable.

1.4 Bibliography

J. Bardeen and W. H. Brattain, "The Transistor, A Semiconductor Triode," *Physics Rev.* **74**, 230–231 (1948).

C. Y. Chang and S. M. Sze, *ULSI Technologies*, McGraw-Hill, New York (1996).

J. S. Kilby, "Miniaturized electronic circuit," US Patent No. 3,138,743, filed Feb. 6, 1959, granted June 23, 1964.

G. E. Moore, "Cramming more components onto integrated circuits," *Electron. Mag.* **38**, 114 (1965).

R. N. Noyce, "Semiconductor device-and-lead structure," U.S. Patent No. 2,981,877, filed Jul. 30, 1959, granted Apr. 25, 1961.

M. Riordan and L. Hoddeson, *Crystal Fire*, W. W. Norton, New York (1997).

W. Shockley, "The theory of p-n junctions in semiconductor and p-n junction transistors," *Bell Cyst. Tech. J.* **28**, 435–489 (1949).

H. Wakabayashi, S. Yamagami, N. Ikezawa, A. Ogura, M. Narihiro, K. Arai, Y. Ochiai, K. Takeuchi, T. Yamamoto, and T. Mogami, "Sub-10-nm planar-bulk-CMOS devices using lateral junction control," *Proc. IEEE IEDM Tech. Digest* **20**(7), 1–3 (1993).

N. H. E. Weste and K. Eshraghian, *Principles of CMOS VLSI Design*, second edition, Addison-Wesley, Reading, MA (1993).

1.5 Review Questions

1. When and where was the first transistor made?
2. What is the difference between discrete devices and the IC?
3. Who were the coinventors of the IC?
4. What are the main differences between TI's first IC bar and Fairchild's first IC chip? Which one is closer to a modern IC chip?
5. How many components did the first IC have? Can you identify them in Fig. 1.13?
6. What is the difference between a mask and a reticle? Which one is used with a stepper for higher-resolution photolithography?

Chapter 2
Introduction to Integrated Circuit Fabrication

This chapter introduces the basics of semiconductor fabrication, including the basics of cleanroom, contamination control, yield, IC fab layout, testing, and packaging processes.

Objectives

After finishing this chapter, the reader will be able to:

- define yield
- explain the importance of yield
- describe the basic structure of a minienvironment cleanroom
- explain the importance of cleanroom protocols
- list four basic operations of IC processing
- name at least six processing areas of an IC fab
- list the commonly used facility systems in IC fabrications
- explain the purposes of chip packaging
- compare ceramic and plastic packaging
- describe standard wire bonding and flip-chip bump bonding processes
- list temperature requirements for packaging processes
- describe the purpose of an induced failure test.

2.1 Introduction

IC chip fabrication is a very complicated and time-consuming process. It begins with an IC design, formed with the help of powerful electronic design automation (EDA) software. The layouts of the design are then verified and taped out. In a mask shop, a photoresist (PR) layer is coated onto a chrome glass plate, and the designed pattern is printed by an electron beam writer or a laser writer. After PR development, a chromium etch process is used to transfer the PR pattern to the chrome glass to form the mask or reticle. After cleaning and inspection, the masks are ready to be shipped to IC fabs. A single IC chip requires 20 to 30 masks, depending on the type of device and the technology node.

Initially, wafer providers obtain rough silicon from quartz sand, then purify the silicon, pull it into a single-crystal ingot, and slice the ingot into wafers. After edge rounding, wet etching, surface polishing, and sometimes epitaxial silicon growth, the silicon wafers are ready to be shipped to the semiconductor fabs. A typical 300-mm IC fab can process more than 10,000 wafers per month. For a "mega lab," production can be more than 50,000 300-mm wafers per month.

Many chemical materials with ultrahigh purity are needed as consumables for IC manufacturing. Gases such as oxygen, nitrogen, and hydrogen; liquids such as deionized water, sulfuric acid, nitric acid, and hydrofluoric acid; and solids such as phosphorus, boron, aluminum, and copper are heavily used in IC fabs.

After wafers are shipped to the IC fab, they are cleaned and go through many processing steps, such as thermal processes, photolithography, etch, ion implantation, dielectric thin-film deposition, chemical mechanical polishing (CMP), metallization, etc. After IC fabrication, the processed wafers are shipped to a testing and packaging house, where they are tested, packaged, and given a final test. A generic IC fabrication process flow is illustrated in Fig. 2.1.

Wafers are always kept in a cleanroom, where the size and amount of free particles are carefully controlled and maintained at very low levels. Even in a cleanroom, wafers are usually stored in specially designed containers to minimize possible contamination. For 200-mm or smaller wafers, many fabs use open cassettes with carrier boxes that can hold a cassette with wafers in the slots in a vertical direction. Some advanced 200-mm fabs use containers called standard mechanical interface (SMIF) boxes. Containers for 300-mm wafers are called front-opening unified pods or FOUPs.

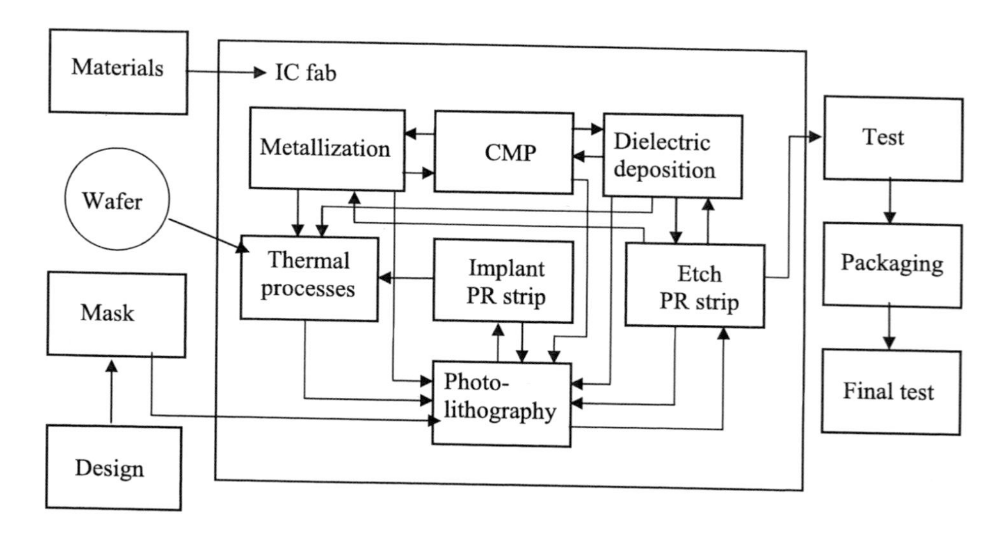

Figure 2.1 IC manufacturing process flow.

2.2 Yield

Yield is one of most important factors of IC fabrication. It can determine whether a fab is making a profit or losing money. Yield relates to many factors, including environment, materials, equipment, processes, and people working in the fab. Improving yield is so critical that semiconductor fabs always hire yield engineers (yield-enhancement engineers) to work in their IC fabs.

2.2.1 Definition of yield

There are three different yields in IC chip fabrication.

Wafer yield: the ratio between the number of good wafers after process completion and total number of wafers used for IC chip fabrication:

$$Y_W = \frac{wafers_{good}}{wafers_{total}}.$$

Die yield: the ratio between the number of good dies per good wafer after process completion and total number of dies on the wafer:

$$Y_D = \frac{dies_{good}}{dies_{total}}.$$

Packaging yield: the ratio between the number of good chips after packaging completion and total number of chips packaged:

$$Y_C = \frac{chips_{good}}{chips_{total}}.$$

Wafer yield mainly relates to processing and wafer handling. Careless human wafer handling, robot malfunction, and robot miscalibration can break the brittle silicon wafers. Incorrect processing, such as misaligned photolithography followed by etch or ion implantation, wrong dopant concentration, nonuniformity of film thickness in thin-film deposition, large amounts of particles on a wafer, etc., can also cause scrapping of the wafers. Die yield relates to many factors such as particle contamination, process maintenance, total processing steps, design-related processing window, etc. Chip yield relates to the wire bonding quality and specification difference between the die test and the chip final test:

$$Y_T = Y_W \times Y_D \times Y_C. \tag{2.1}$$

Overall yield Y_T of the fab is the product of the three, which is a very important factor in semiconductor fabrication because it can determine whether a fab is making or losing money.

2.2.2 Yield and profit margin

The definition of fab yield is the ratio of the number of good chips to the total chips at the start of the process. A simple example can help explain why yield is so important to semiconductor fabs.

The cost of a 300-mm wafer varies from time to time, mainly determined by the condition of supply and demand. Depending on circuit requirements, a wafer requires several-hundred processing steps before it can be sent to a packaging house to be tested and packaged. Each processing step adds some cost, typically about $1 USD per wafer. If we assume a wafer costs $200, and it takes 500 steps to complete the processing, the total cost per processed wafer is $200 (wafer cost) + $500 (processing cost) = $700. We can assume 100% wafer yield, which means no wafer scrap during the 500 processing steps. We can also assume the cost of test and package for each good die is $10, and no dies failed during final testing after package completion (100% packaging yield). In the case of 500 chips per wafer, and each chip selling for $30, 35 good dies per wafer or a 7% die yield is needed to achieve breakeven. This can be expressed as:

$700 (wafer and processing costs) + 35 (good dies) × $10 (test and package cost)
= $1050 = 35 (good dies) × $30 (revenue/die).

If the die yield increases to 50% while wafer yield and package yield both stay at 100%, the total cost per wafer will be $700 (wafer and processing cost) + 250 × $10 (test and packaging cost) = $3200, while the revenue per wafer will be 250 × $30 = $7500. The profit margin per wafer is $7500 – $3200 = $4300. This means that if a wafer fab can process 10,000 wafers a month with 100% wafer yield, 50% die yield, and 100% packaging yield, it can make $43 million in profit a month.

While $43 million a month is indeed a very impressive number, we must remember the approximate $3 billion it costs to build a 300-mm wafer fab, and the thousands of employees necessary to keep it running 24 hours a day, 7 days a week (24/7) nonstop. A total yield of 50% with 10,000 wafers per month might not be high enough to generate sufficient cash flow to pay all the bills! Therefore, improving yield and increasing throughput is vital for an IC fab.

Question: For the same IC fab, if the die yield is 90%, wafer and packaging yield are 100%, and capacity is 20,000 wafers per month, what is the total profit margin per month? (The hardest math in this book.)

Answer: 20, 000 × [500 × 90% × $30 (revenue per wafer) – ($700 + 500 × 90% × $10) (cost per wafer)] = $166 million/month.

We can see that the higher the yield and capacity, the larger the profit margin. Of course, the math becomes much more complicated when the wafer yield and packaging yield are no longer 100%, as they are in the real world of IC manufacturing.

As mentioned earlier, advanced IC chip fabrication involves about 500 processing steps. To achieve a reasonably high overall yield, every step has to have very high yield, close to 100%.

Question: If the die yield for every processing step is 99%, and there are 500 processing steps for IC fabrication, what is the overall die yield?

Answer: The overall die yield is 99% × 99% for 500 times, which equals $0.99^{500} = 0.0066 = 0.66\%$.

For advanced IC chips, the yield of most processing steps needs to be 100%, or very close to a perfect value to ensure overall yield. Yield improvement is a nonstop process in semiconductor fabs. At the beginning of a new processing technology development, the launch of a brand new product, or the introduction of a new set of tools, overall yield is normally not very high. After a few wafer production cycles, systematic factors that affect the yield are mined out and eliminated, and the yield is continuously improved and stabilized when it is limited by random factors such as a few airborne particles. Excursion of yield can happen from time to time due to some unforeseen issues such as a slight mismatch of processing equipment, human error, etc. The cycle starts all over when new processes, new products, or new tools are introduced. Figure 2.2 illustrates a typical yield ramp curve of IC technology.

2.2.3 Defects and yield

Equation (2.2) shows the relationship of overall yield and the killer defect density, chip size, and number of processing steps:

$$Y \propto \frac{1}{(1 + DA)^n}, \tag{2.2}$$

where Y is the overall yield, D is the killer defect density, A is the chip area, and n is the number of processing steps. From Eq. (2.2), we can see that to achieve 100% yield, the killer defect density must be zero for every processing step. At the same defect density and chip size, the more processing steps there are, the lower the yield. This equation also indicates that at the same defect density level, the larger the chip size is, the lower the yield will be, as is illustrated in Fig. 2.3.

Equation (2.2) assumes that the defect density of each processing step is the same, making it an obviously oversimplified model. However, it provides a simple

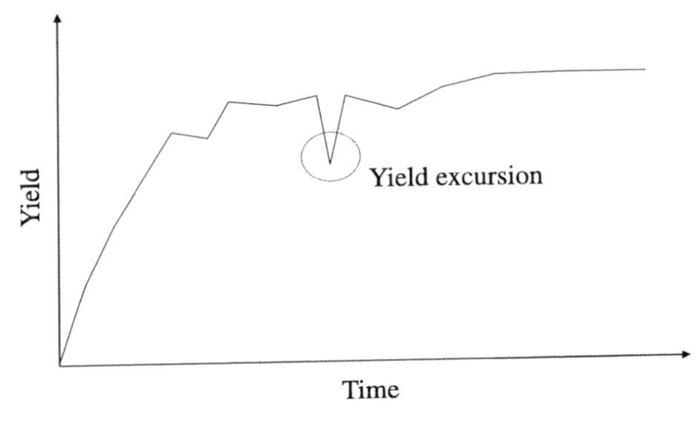

Figure 2.2 Typical yield ramp curve of IC technology.

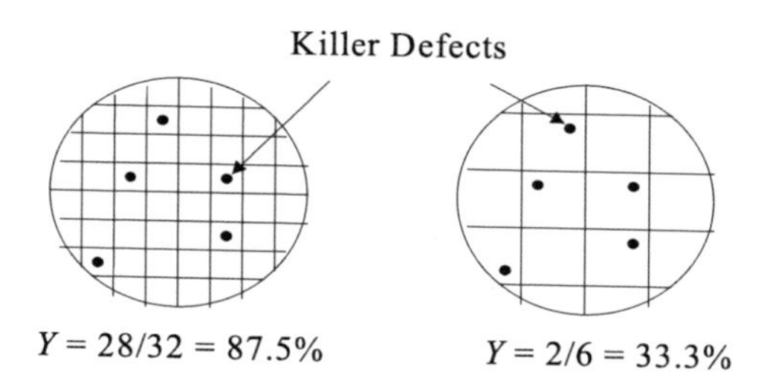

Figure 2.3 Relationship between die size and die yield.

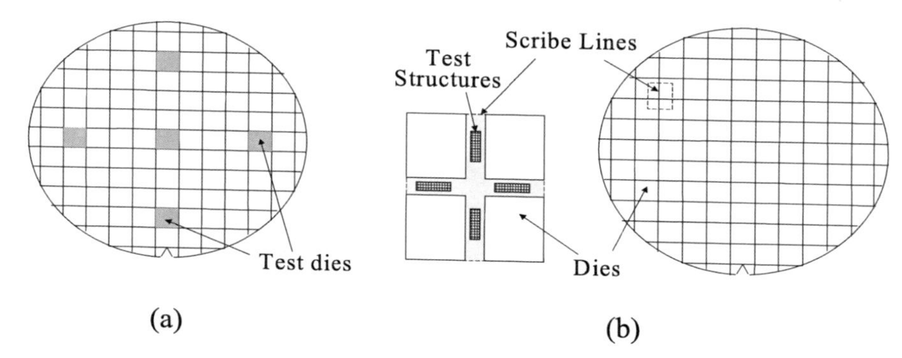

Figure 2.4 Wafer with (a) test dies and (b) test structures in scribe lines.

and straightforward explanation of the relationship between defect and yield. (For more detail on yield modeling, see Chang and Sze in the Bibliography at the end of this chapter.)

Some production wafers are designed with several test dies, in which transistors and test circuits are built during wafer processing, as shown in Fig. 2.4(a). Due to the improvement of technology and shrinking of feature size, test structures—including devices and circuits—are built on the scribe line between the dies to save space on the silicon wafer for making more chips, as shown in Fig. 2.4(b). Sampling tests are performed on these test structures throughout wafer processing to ensure the yield of IC fabrication. If the results of the electrical tests confirm that most transistors and circuits in the test structures do not function as designed, transistors in the real devices will not likely meet design requirements. Wafer processing is stopped, and usually the whole batch of affected wafers is scrapped, which contributes to the loss of wafer yield. A physical failure analysis is performed quickly to pinpoint the failure mechanism and the root cause.

2.3 Cleanroom Basics

The requirement of a cleanroom is one reason IC fabs are so expensive. Since tiny particles can cause defects on a microelectronic device and circuit (affecting chip yield), IC chips must be manufactured inside a cleanroom to achieve acceptable yield. When feature size shrinks, so does the size of killer particles. A smaller feature size requires a higher grade of cleanliness in the cleanroom.

Since particles have a significant impact on yield, considerable efforts have been made to improve cleanroom conditions, which in turn reduces particulate counts and improves IC production yield. Developments of cleanroom furniture, gowns, and gowning procedures have all contributed to this improvement. Tight cleanroom protocol helps keep contaminants out and prevent yield degradation due to particulate and other contamination.

2.3.1 Definition of a cleanroom

A cleanroom is a manmade minienvironment that has much lower particle counts than the normal environment. The first cleanrooms were built for hospital surgery rooms to control airborne bacteria contamination and reduce postsurgery infection. Soon after the semiconductor industry started, people realized the importance of contamination control, and cleanroom technology was adapted in transistor and IC fabrication. The first silicon IC chip made by Noyce of Fairchild Camera (see Fig. 1.4) shows much particle contamination.

The standard definition of cleanroom class is a strange combination of metric and English units: class 10 is defined as less than 10 particles with diameters larger than 0.5 μm per cubic foot. Class 1 is defined as less than 1 particle with diameters larger than 0.5 μm per cubic foot. A fab making IC chips with a minimum feature size of 0.25 μm needs a class-1 cleanroom to achieve an acceptable yield. For comparison, there are more than 500,000 particles with diameters larger than 0.5 μm per cubic foot inside an ordinary "clean" house. Figure 2.5 illustrates the number of particles in a cubic foot of air of different cleanroom classes.

The cleanest class of cleanroom, M-1, only uses metric units. According to the definition of U.S. Federal Standard 209E, the M-1 class cleanroom allows less than 10 particles larger than 0.5 μm in a cubic meter, or less than 0.28 particles larger than 0.5 μm per cubic foot. Table 2.1 gives the definition of cleanroom classes.

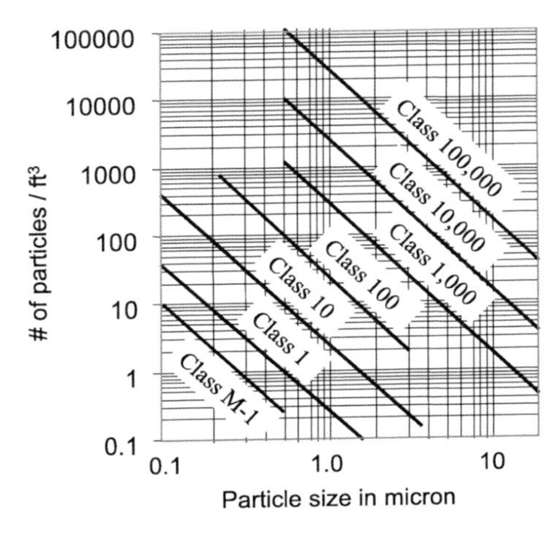

Figure 2.5 Number of airborne particles in the cleanroom.

Table 2.1 Definition of airborne particulate cleanliness class per U.S. Federal Standard 209E.

Class	Particles/ft^3				
	0.1 μm	0.2 μm	0.3 μm	0.5 μm	5 μm
M-1	9.8	2.12	0.865	0.28	
1	35	7.5	3	1	
10	350	75	30	10	
100		750	300	100	
1,000				1,000	7
10,000				10,000	70

Table 2.2 Particle count and profit margin.

Wafer starts/month	10,000	10,000
Fab yield	0.85	0.85
Wafer diameter (mm)	100	100
Edge exclusion (mm)	4	4
Particle count	**20**	**19**
Monthly fixed costs $M	$0.53	$0.53
Various costs/wafer	$76.11	$76.11
Total cost/wafer	$129.00	$129.00
Mask steps	7	7
Defect density (cm^2)	0.30	0.29
Die Size (cm^2)	0.5	.05
Random yield	0.37	0.39
System yield	0.70	0.70
Die yield	0.26	0.27
Die/wafer	113	113
Good die	30	31
Cost/die	$4.35	$4.15
Package cost	$1.00	$1.00
Burn in/test	$0.50	$0.50
Test yield	0.90	0.90
Circuit cost	$6.50	$6.28
Circuit price	$12.00	$12.00
Price/cost ratio	1.85	1.91
Sales/wafer	$320.32	$335.37
Annual sales ($M)	$32.67	$34.21
Annual manufacturing cost ($M)	$17.70	$17.91
Annual gross margin ($M)	$14.98	$16.30
Annual gross margin difference ($M)		**$1,321,943**

2.3.2 Contamination control and yield

Particles on a wafer can cause defects, which can reduce yield and affect the profitability of an IC fab. Table 2.2 gives a good example of the effects of particles. It shows that just one more particle count on each wafer could have cost a 4-in. wafer fab more than $1.3 million a year in the early 1980s. Although these data are old, they still demonstrate the effects of particle contamination on the die yield and the profit margin of an IC fab.

Particles can cause many different defects for different processes. For instance, imaging of particles on the clear area of a mask or reticle can cause pinholes for

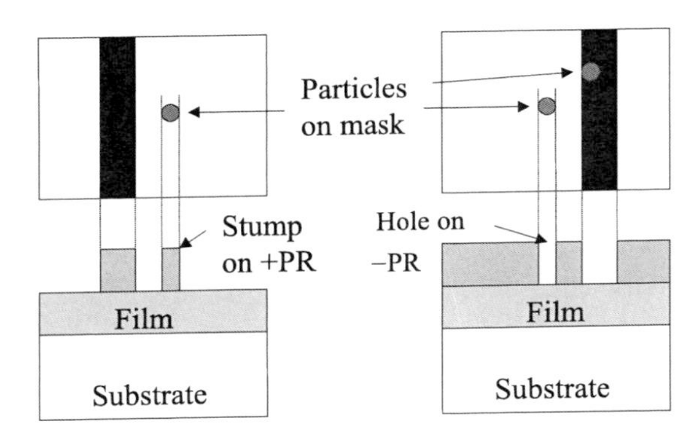

Figure 2.6 Effects of particle contamination on a photomask.

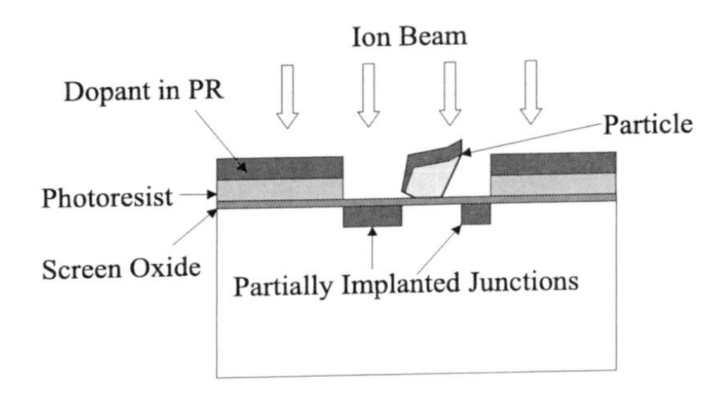

Figure 2.7 Effects of particle contamination during the ion implantation process.

negative photoresist, or stumps for positive photoresist, during photolithography processing. During etch processing, these pinholes and stumps are transferred to the wafer surface and cause defects. Since masks/reticles are used for every wafer for many times during IC fabrication, particle contamination of masks can cause serious problems with chip yield. Masks or reticles must be kept in the cleanest environment to prevent particle contamination. The effects of a photomask particle contaminant on the photolithography process are shown in Fig. 2.6. Note that particles on the dark part of the mask do not affect the photolithography process.

Particle contamination can cause other problems in different processes; for example, it can cause a broken metal line, or a short between neighboring metal lines. Particles can block implanting ions and create an incomplete junction during ion implantation, possibly affecting device performance, as shown in Fig. 2.7.

A particle with half the size of the technology node could be a killer particle. For example, an 11-nm particle can be a killer for a 22-nm IC chip. If a particle falls onto a critical area, it could be a killer even if it is smaller than half the size of the technology node. As feature size reduces, so does the size of the killer particle. Particles of different sizes behave quite differently; therefore, cleanrooms of different classes need different design concepts and require different protocols.

For example, large particles (>1 μm in diameter) can be blown away with high-pressure airflow, while smaller particles cannot. Therefore, high-pressure air or nitrogen blowers are widely used in older, 100-mm (4-in.) fabs, and they are seldom used in the advanced 300-mm fabs. An air blower can remove large particles from a wafer surface; however, it adds more small particles to it. These small particles cause no problem when the feature size is larger than several microns, but they can cause many problems when the feature size shrinks to submicron.

2.3.3 Basic cleanroom structure

The basic structure of a cleanroom for a 300-mm wafer fab is illustrated in Fig. 2.8. A cleanroom is usually located on a raised floor with grid panels, allowing airstream to flow vertically from the ceiling to the facility area underneath the cleanroom. Airflow returns to the cleanroom through a high-efficiency particulate air (HEPA) filter, which removes most particles carried by the airflow. To cut costs, only equipment interfaces are designated as the highest class of cleanroom (where wafers are loaded into processing or metrology equipment), while fabs are built with lower-class cleanroom designs, and most facilities are installed in a subfab beneath the cleanroom floor. Wafers are carried inside a sealed FOUP and are only exposed to airflow in the processing or metrology equipment.

For a less advanced IC fab (illustrated in Fig. 2.9), wafers are carried in an open cassette inside a carrier box. Because wafers will be exposed to airflow before they are loaded into the processing or metrology equipment in an open cassette, the wafer processing areas are designed as a high-class cleanroom. To save costs, equipment areas are made with lower-class cleanroom designs, and facilities are installed in a subfab beneath the cleanroom. For the 0.13-μm technology node, a class-1 or higher cleanroom is required for processing areas. Normally, a class-1000 cleanroom is good enough for the equipment area and costs significantly less to build than a class-1 cleanroom. However, with the shrinking IC feature size,

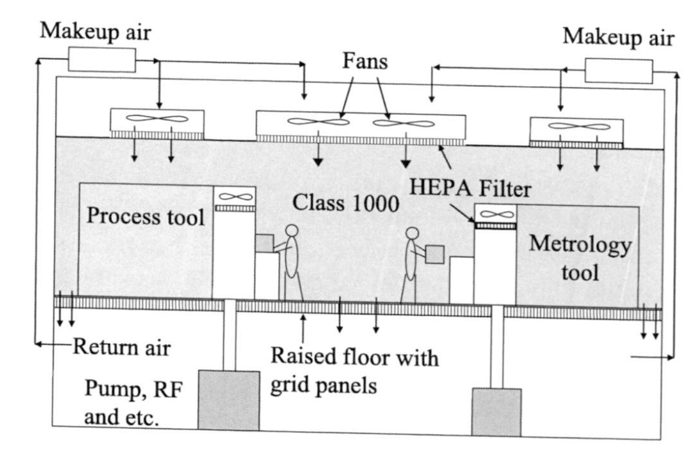

Figure 2.8 Basic structure of an advanced cleanroom.

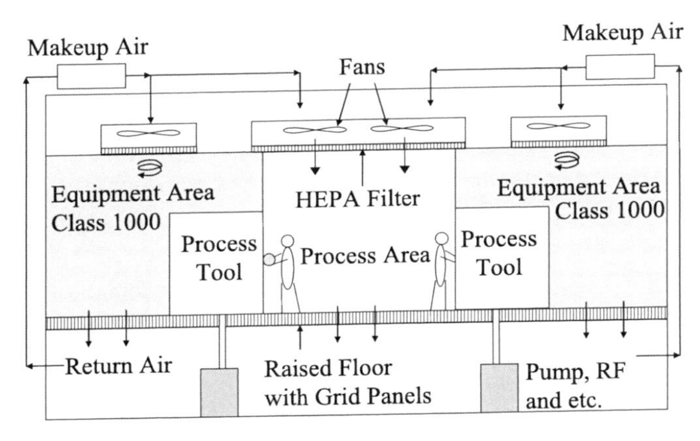

Figure 2.9 A less advanced fab.

it becomes less cost effective to further improve the class level of the processing areas in this kind of cleanroom; thus, the minienvironment cleanroom, illustrated in Fig. 2.8, has become the mainstream for advanced 300-mm fabs.

To achieve better than class-100 cleanliness, it is very important to keep a linear airflow, usually called laminar flow, and avoid air turbulence. Air turbulence can drive out particles sticking to the surfaces of walls, ceilings, tables, and tools, causing them to become airborne, and helps these particles stay airborne for a longer time, increasing the number of particles in the air. With a laminar flow, airborne particles are carried away quickly by the airflow. Class 100 is the dividing line between laminar flow and turbulent flow, a choice that is a very important factor of cleanroom cost. A lower class than class 100 can be achieved with turbulent flow, which costs much less than achieving and maintaining laminar flow.

A cleanroom is always kept at a higher pressure than nonclean areas; this allows continual outward airflow, preventing the entry of airborne particles from the outside when a door is opened to let people, tools, or other materials enter the cleanroom. The same principle is also applied to the different class areas inside the cleanroom. The higher-class areas have a higher pressure than the lower-class areas. Variation of temperature, airflow rate, and humidity can also disturb particles from a surface and cause them to become airborne; therefore, temperature, airflow rate, and humidity are strictly controlled in the cleanroom.

2.3.4 Basic cleanroom gowning procedures

It is very important to maintain rigid cleanroom protocols to minimize the yield loss due to contamination. Human beings emit a great deal of particles, and people are also the major source of sodium, which can cause mobile ion contamination. Therefore, people working inside the cleanroom must wear specially designed garments. Because people are the main contamination sources, some companies have even limited the number of people working in the cleanroom to control contamination. Improvement of cleanroom garments and strict cleanroom gowning procedures have sharply reduced contamination from people working inside the cleanroom. Proper gowning and degowning procedures are important parts of cleanroom protocol.

Although different companies, even different fabs within the same company, have different gowning procedures, the basic idea is the same: to prevent particles and other contaminants from being carried into the cleanroom by people.

Some fabs require people to put on linen gloves before they even enter the gowning room. Linen gloves are usually made of composite fibers that do not tend to attract particles. They can prevent sodium and particles on the hands from contaminating cleanroom garments, and they are also more comfortable when putting on an outer layer of latex gloves.

The bottoms of shoes carry the largest number of particles. Some fabs use sticky pads or shoe brushes to remove dirt from shoes before putting on shoe covers or booties. It is important to keep bare shoes at the entrance side of the bench and covered shoes on the other side of the bench to prevent large numbers of particles from being carried into the gowning room. Figure 2.10 shows the gowning area of a cleanroom.

Hair covers are required before entering a growing room. Human hairs become positively charged via friction, which attracts negatively charged particles. Certain pressure, humidity, and temperature discharge can neutralize these particles so that they are no longer attracted by hairs. As a consequence, discharged particles from hair could easily become airborne and cause contamination. Hair covers block the emission of these particles.

Generally the first cleanroom garment to put on is the head hood with mask, which further covers the hair and face to prevent particles and other contaminants associated with breathing, coughing, and sneezing. Badges, two-way pagers, cell phones, and radios must be removed before putting on the cleanroom suit, so that they will be on the outside of the suit (otherwise they will be sealed inside the suit and cannot be used in the cleanroom). It is forbidden to unzip a cleanroom suit inside the cleanroom, since that activity can release many particles originally sealed

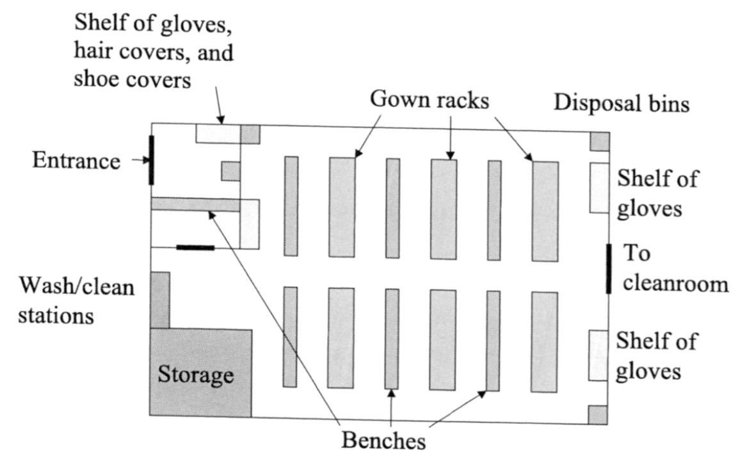

Figure 2.10 Cleanroom gowning area.

in by the suit. After putting on the suit and attaching it to the hood, a pair of boots is put on and attached over the calf of the legs to completely seal particles on the shoes. Safety glasses are required, since IC fabs always use corrosive chemicals in wafer processing, and some processing tools have moving parts, which pose hazards for eye injury. Before entering a cleanroom, a pair of latex gloves is put on top of the linen gloves to prevent particle and mobile-ion contamination. They can also provide minimum protection for the hands from corrosive chemicals. For some processes, such as the CVD chamber and implanter beam line wet clean, two layers of latex gloves are required for better protection from corrosive materials, such as hydrofluoric (HF) acid. For some wet chemical processes, full acid gear is mandatory for protection. Before entering a cleanroom, a person should look in the mirror to make sure hair and street clothes are not exposed.

Before entering a cleanroom, some fabs require people to go through an air shower, in which high-pressure airflow blows away particles on the surface of the garments. To achieve the best result, a person should raise both arms and rotate the body slowly during the air shower. Now the person is ready to enter the cleanroom.

The degowning procedure is almost the reverse of the gowning procedure. First the boots are taken off, then the suit and the head hood are removed. They are usually hung on the garment rack at the next entry. Cleanroom suits are normally washed every week at a specialty laundry, and some fabs are now using disposable cleanroom garments that can be thrown away after being used. Once outside the gowning room, hair covers and gloves are taken off and thrown into trash cans. Shoe covers are the last items to be removed after exiting the cleanroom. Some fabs recycle linen gloves for washing and reusing, while other fabs do not require linen gloves at all.

2.3.5 Basic cleanroom protocols

While cleanroom protocols are different from company to company, the basic concepts are the same: keeping particles where they are and not letting them become airborne, and preventing other contaminants from reaching the wafers.

Once inside a cleanroom, people need to walk steadily. Running or jumping is prohibited because it can disturb particles sticking to the surfaces of the floor, walls, and ceiling. There are very few chairs inside a cleanroom, since particles on the surface of a chair can easily become airborne when a person sits down or stands up. For the same reason, sitting on a table or leaning against a wall is not allowed in a cleanroom, since these actions can dislodge particles from the surfaces.

Ordinary paper carries many small fiber fragments, which can cause particle contamination. Therefore, only specially made cleanroom paper is allowed in a cleanroom. For a class-1 or better cleanroom, even cleanroom paper is not allowed; all all data must be recorded electronically, by using hand-held or notebook computers.

There are other types of contamination, such as mobile ions. For example, a trace amount of sodium contamination can cause the malfunction of MOS transistors and affect the reliability of the integrated circuits. Sodium contamination must be tightly controlled. Therefore, total isolation between people who process the wafer and the wafer itself is required. The human body is a sodium carrier (sodium chloride, NaCl); thus, if one's glove touches the skin of the face, the latex gloves should be replaced immediately with a fresh pair, or a clean pair can be placed over the dirty pair. Similarly, workers should replace or cover latex gloves as soon as possible after covering their mouth during sneezing or coughing, because high-pressure airflow from the mouth can force small amount of saliva through the facemask. Ideally, one should attempt to leave the processing area and go to equipment areas before sneezing or coughing. It goes without saying that eating is strictly forbidden in the cleanroom.

People who work in cleanrooms are not allowed to use cosmetics, perfume, or cologne, because these materials can emit particles and cause contamination. Technicians cannot wear contact lenses, since trace amounts of chlorine in the fab can react with the lenses and cause eye injury. Smoking is forbidden in the cleanroom and adjacent buildings, and smokers working in cleanrooms are strongly encouraged to quit smoking. Even after they have finished smoking, they can emit particles they have inhaled.

Question: In a 200 mm fab, after finishing the scheduled preventive maintenance (PM) of the processing equipment, a technician knocks on the wall of the processing area to warm another technician in the equipment area that the tool is going to be turned on. Is this act against cleanroom protocol?

Answer: Yes. Knocking on the wall can disturb particles adhering to the wall and ceiling surface, and cause them to become airborne. A radio is a better tool for communication between coworkers in the fab. Some processing tools have secondary terminals, so that the system can be turned on in the equipment area after finishing PM while the terminal on the processing side is locked.

2.4 Basic Structure of an Integrated Circuit Fabrication Facility

Semiconductor manufacturing has many different processes, which can be characterized by the following four basic operations: adding, removing, patterning, and heating. Doping, layer growth, and deposition are adding processes; etch, clean, and polish are removing processes; photolithography is the patterning process; and annealing, alloying, and reflow are the heating processes. Figure 2.11 illustrates the IC processing flow module.

An IC fab usually consists of offices, facilities, storage areas, equipment, and wafer and processing areas. Some fabs have their own chip test and packaging facilities. Wafer processing, equipment, chip testing areas, and packaging are located in a cleanroom, while other areas such as offices, facilities, and process-materials storage are located elsewhere on site. Wafers are always in the wafer processing area, which maintains the highest class of all cleanrooms. A large number of processing equipment is installed in a lower class cleanroom, sometimes

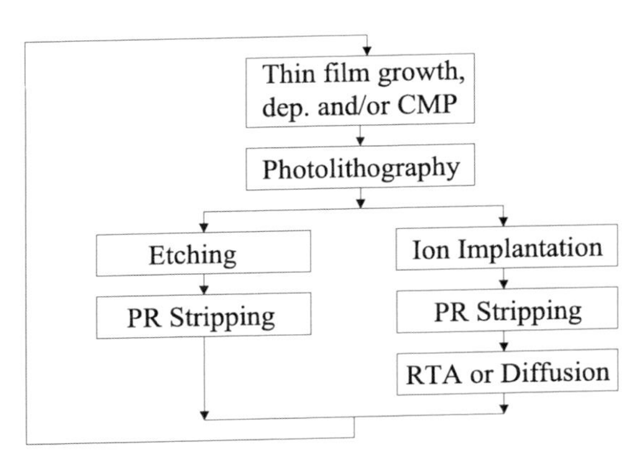

Figure 2.11 IC process flow.

called a gray area, while still interfaced with the high-class cleanroom so that wafers can be transferred into it for processing. Chip testing and packaging are normally performed in an even lower-class cleanroom, since these processes deal with much larger feature sizes that are much less sensitive to particles compared to wafer processing.

2.4.1 Wafer processing area

The cleanroom used for wafer processing always has the highest class in an IC fab, and a wafer does not leave this area unless all of the processing steps are finished. After a wafer is sealed with the final passivation layer, the last masking and nitride/oxide etching processes open the bonding pads (or bump connections) and the photoresist is stripped. The wafers then are sent out to the testing and packaging area to finish the chip fabrication.

Wafer processing areas are usually separated into several processing areas, or bays, as shown in Fig. 2.12. They are: the wet bay, diffusion bay, photo bay, etch bay, diffusion bay, thin-film bay, and chemical mechanical polishing (CMP) bay. In a traditional fab, process engineers, process technicians, and production operators mainly work in the processing areas, while equipment engineers and maintenance technicians mainly work in the equipment areas. Application engineers from the process-equipment manufacturers also work in the processing area during tool start-ups, as they can help to troubleshoot processes when they do not perform as expected.

2.4.1.1 Wet bay

The wet bay is where wet processes take place. Photoresist stripping, wet etching, and wet chemical clean processes are the most common processes in the wet bay. Corrosive chemicals and strong oxidizers such as hydrofluoric acid (HF), hydrochloric acid (HCl), sulfuric acid (H_2SO_4), nitric acid (HNO_3), phosphoric acid (H_3PO_4), and hydrogen peroxide (H_2O_2) are commonly used in this area. Huge amounts of high-purity deionized (DI) water are used for wafer rinsing after the wet processes. Most acids used in the wet bay are corrosive. Nitric acid and

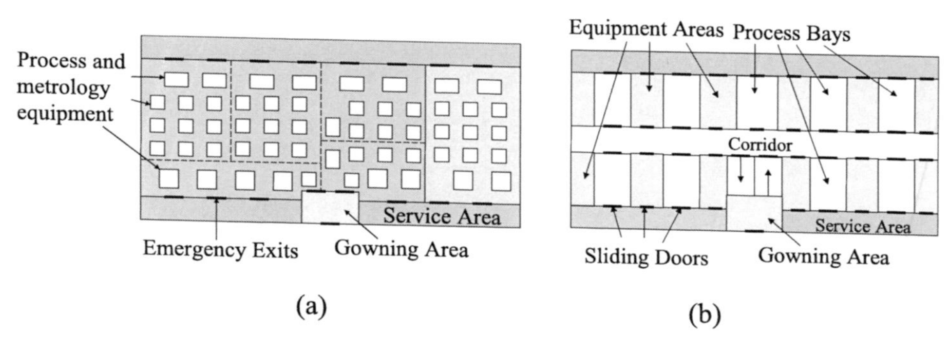

Figure 2.12 Floor plan of a semiconductor fab with (a) a minienvironment and (b) traditional layout.

hydrogen peroxide are strong oxidizers. In an IC fab, it should never be assumed that a clear drop of liquid is water. To be safe, a clear droplet should be treated as HF, because HF has no smell, looks like water, and feels like water. Wet bays always have shower stations and eye washers nearby so that people can use them immediately in case of accidental contact, such as from chemical spilling.

Wet processes are removal processes that usually take three steps—process, rinse, and dry—as illustrated in Fig. 2.13. Wet process tools usually are batch process tools, which can process whole cassettes of 25 wafers at the same time. Robots take the cassette with wafers from the loading station and dip it into a processing solution. After the required process time, robots pick the cassette up and put it into a rinse sink, where DI water washes away the chemicals from the wafer surface. Then the cassette is picked up and sent into a dryer, where wafers are spun dry with the cassette. After that, the cassettes are picked up and sent back to the loading station for wafer unloading. Some wet stations can process several cassettes of wafers at one time. Depending on the chemical used in the process, sometimes wafers need to be picked up from the cassettes and put on a quartz boat for the wet process and rinse.

To reduce chemical usage, single-wafer wet processing tools have also been developed. By processing wafers one at a time, only a small amount of chemical needs to be sprayed onto the wafer surface; thus, the amount of chemical used in the wet process can be significantly reduced in comparison to the batch system, in which the whole cassette is submerged in the wet chemical.

Some fabs do not have a separate wet bay; wet chemical stations are simply placed alongside the oxidation and LPCVD tools, since these processes require a wet clean beforehand.

2.4.1.2 Diffusion bay

The diffusion bay is where thermal processes are performed. These processes can be adding processes, such as oxidation, low-pressure chemical vapor deposition (LPCVD), and diffusion doping, or heating processes, such as postimplantation annealing, dopant drive-in, alloy annealing, and dielectric reflow. High-temperature furnaces are installed for oxidation, LPCVD, diffusion/drive-in

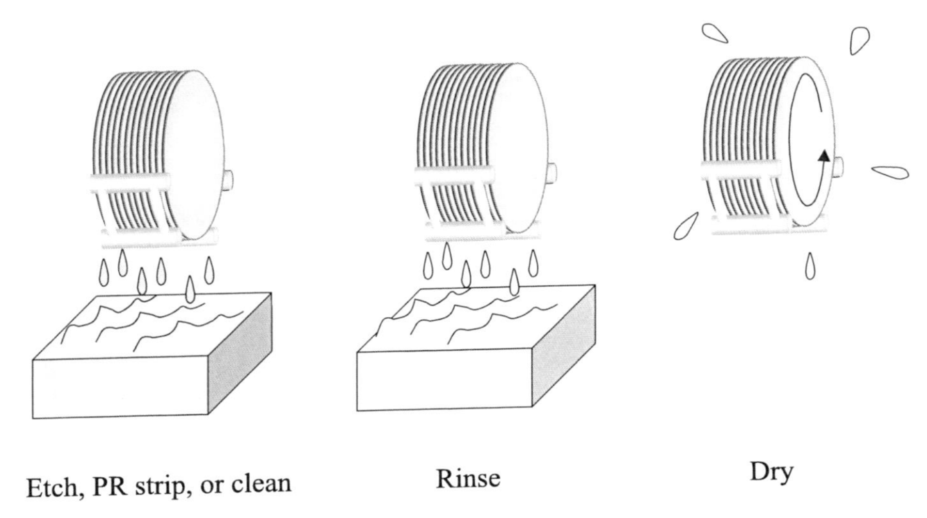

Figure 2.13 Example of wet processes.

processes, and annealing processes. Some fabs also have epitaxy reactors in their diffusion bays. Oxidation and diffusion doping in high-temperature furnaces were the most frequently used processes of IC fabrication before the introduction of ion implantation in the mid-1970s. Although very little diffusion doping processes are performed in advanced IC processing, the name diffusion bay is still used for historical reasons.

Furnaces are batch process tools with the ability to process more than 100 wafers at a time. Figure 2.14(a) gives the schematics of a vertical furnace, and Fig. 2.14(b) shows a horizontal furnace. Due to a smaller footprint and better contamination control, only vertical furnaces are used in advanced 200- and 300-mm IC fabs. Horizontal furnaces are still used in some older fabs with smaller wafer sizes. Most 300-mm fabs also have cluster tools with single-wafer chambers for deposition of polysilicon and silicon nitride films, and silicide annealing processes in their diffusion bays.

Gases commonly used in diffusion bays are oxygen (O_2), nitrogen (N_2), anhydrous hydrogen chloride (HCl), hydrogen (H_2), silane (SiH_4), dichlorosilane (DSC, SiH_2Cl_2), trichlorosilane (TSC, $SiHCl_3$), phosphine (PH_3), diborane (B_2H_6), and ammonia (NH_3). Nitrogen is a safe gas. Oxygen is an oxidizer, which can cause fire or explosions if mixed with other flammable, explosive materials at certain conditions. HCl is corrosive; hydrogen is flammable and explosive; silane is pyrophoric (self-igniting), explosive, and toxic; both DSC and TSC are flammable; and ammonia is corrosive. Both phosphine and diborane are very poisonous, flammable, and explosive. Nitrogen is used in almost every processing tool in an IC fab as the purge gas. O_2 and anhydrous HCl are used for the dry oxidation process, while H_2 and O_2 are used for the wet oxidation process. Silane, DCS, or TCS can be used as a silicon precursor for polysilicon deposition, and they can also be used with NH_3 for silicon nitride deposition. Phosphine and diborane are used as dopant gases for polysilicon deposition.

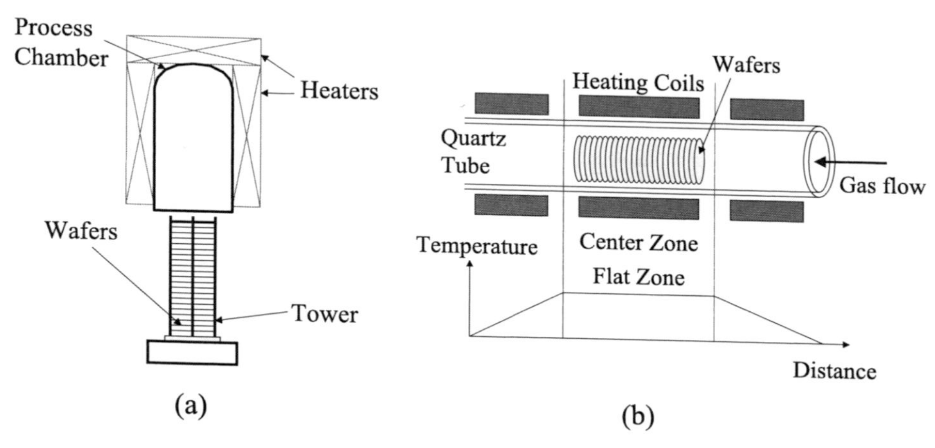

Figure 2.14 Schematics of (a) vertical and (b) horizontal furnaces.

2.4.1.3 Photo bay

The photolithography process is one of the most important processes in IC fabrication. It transfers the designed pattern from a mask or reticle to the photoresist coated onto the wafer surface. In the photo bay are located the integrated track-stepper systems, which perform primer and photoresist coating, baking, alignment and exposure, and photoresist development processing steps. A stepper is the most commonly used tool to pattern the photoresist coated onto the wafer surface by exposing the photoresist to ultraviolet (UV) light or deep UV (DUV) light, thereby inducing photochemical reactions. It is the most expensive tool in advanced IC fabs. For example, an advanced 193-nm immersion scanner could cost more than $40 million. There are also many metrology tools in photo bays, such as ellipsometry and reflectometry systems, for measuring the thickness and thickness uniformity of transparent films such as photoresists. There are also optical microscopes for visual inspection, and scatterometry systems for CD measurements. Scanning electron microscopes (SEMs) are also widely used for CD measurements, and optical overlay systems are used to measure the alignment between different mask layers. Figure 2.15 shows the schematic of an integrated track-stepper system, which is also called a photocell in an IC fab.

In an advanced IC fab, the track systems do not look like the schematic in Fig. 2.15. Many fabs prefer to use a stacked track system, which makes a much smaller footprint by stacking the hot plates and chill plates instead of placing them on the same plain. Some systems even stack the spin coaters and development stations to further reduce the footprint and save precious cleanroom space.

2.4.1.4 Etch bay

After the photoresist is patterned and passes inspection, the wafers are sent either to the implant bay or to the etch bay. In an etch bay, patterns are etched based on the pattern defined by the photoresist, permanently transferring the designed pattern to the layer on the wafer surface. Etching is a removal process in which materials on the wafer surface are selectively removed chemically, physically, or

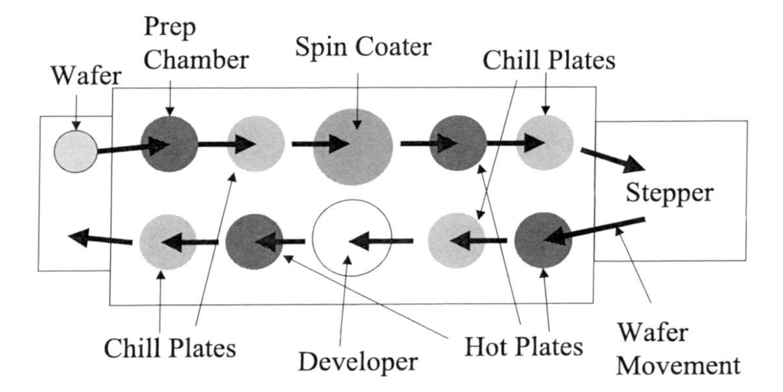

Figure 2.15 Schematic of a track-stepper integrated system in a photo bay.

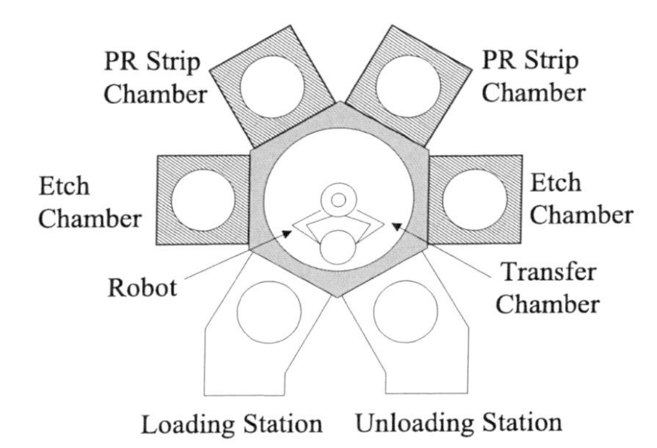

Figure 2.16 Schematic of a cluster tool with etch and photoresist strip chambers.

most commonly by a combination of both, to create patterns on the wafer surface. In the etch bay, only plasma etchers are used for the patterned etch, since wet etching cannot be used to etch features smaller than 3 μm. A plasma etcher usually consists of a vacuum chamber, rf system, wafer transfer mechanism, and gas delivery system. Many etch systems have in-situ photoresist strip chambers, which allow photoresists to be stripped within the system before exposure of the wafer to the atmosphere. Figure 2.16 illustrates a cluster tool with both etch and photoresist strip chambers.

There are four different kinds of etch processes commonly used in semiconductor fabrication. The dielectric process etches silicon oxide and nitride film to form contact via a bonding pad or silicon etch hard mask. The silicon etch process etches single-crystal silicon to form shallow trench isolation (STI). It also has been used to form deep trenches for capacitors in deep trench dynamic random access memory (DRAM). The poly process etches polysilicon or silicide and polysilicon stacked films to form gate and local interconnections. And the metal process etches metal stacks to form metal lines for global interconnections.

Because more fabs have started using copper metallization, less metal etch is required.

Each etch process has different requirements and chamber designs and uses different chemical gases. Carbon fluorides such as CF_4, C_2F_6, C_3F_8, and CHF_3 are commonly used along with argon for dielectric etch processes. Hydrogen bromide (HBr) is widely used to etch single-crystal silicon, and chlorine (Cl_2) is commonly used for both polysilicon and metal etch processes. None of these chemicals are user friendly. Although carbon fluoride gases are very stable, they are considered global warming gases; hydrogen bromide is corrosive; and chlorine is an oxidizer and poisonous.

2.4.1.5 Implant bay

Besides the etch bay, the implant bay is the only other place wafers are sent after the photolithography process. Ion implantation and rapid thermal annealing (RTA) systems can be found in this bay. Ion implantation is an adding process in which dopant is added to the semiconductor substrate to change its conductivity. RTA is a heating process that repairs lattice damage of the substrate at high temperatures while neither removing nor adding materials to the wafer surface. The ion implanter is usually the largest and heaviest processing tool in a semiconductor fab. It also involves many safety hazards, such as high voltage (to 100 kV), a strong magnetic field (which can affect pacemakers), and generation of strong x-ray radiation. It also uses poisonous, flammable, and explosive gases such as arsine (AsH_3) and phosphine (PH_3), poisonous solid materials such as boron (B), phosphorus (P), and antimony (Sb), and the corrosive gas boron trifluoride (BF_3).

In the implant bay, there are several different kinds of implanters used for different processes. For example, well formation in the CMOS process needs a high-energy, low-current implanter. To form the source and drain of the metal-oxide-semiconductor field effect transistor (MOSFET) requires a low-energy, high-current implanter. Other applications require a mid-current, mid-energy implanter. One of the properties of the ion implanter is that it can run different processes with different chemicals in the same implanter, without too much concern regarding cross-contamination due to high-purity ion beam selection by the magnetic analyzer. Metrology tools commonly used in the implant bay are the four-point probe, thermal wave system, and optical measurement system (OMS).

2.4.1.6 Thin-film bay

The main processes performed in the thin-film bay are dielectric and metal thin-film deposition for interconnection applications. Some fabs separate these two processes; therefore they have dielectric and metal bays instead of a single thin-film bay. Obviously, thin-film depositions are additive processes. CVD processes are always used for dielectric thin-film deposition. Plasma-enhanced CVD (PECVD) is commonly used due to the lower temperature requirement for dielectric layers in multilevel interconnection applications. O_3-TEOS [ozone tetra ethyl oxy silane $Si(OC_2H_5)_4$]-based CVD processes are also widely used to deposit silicate glasses

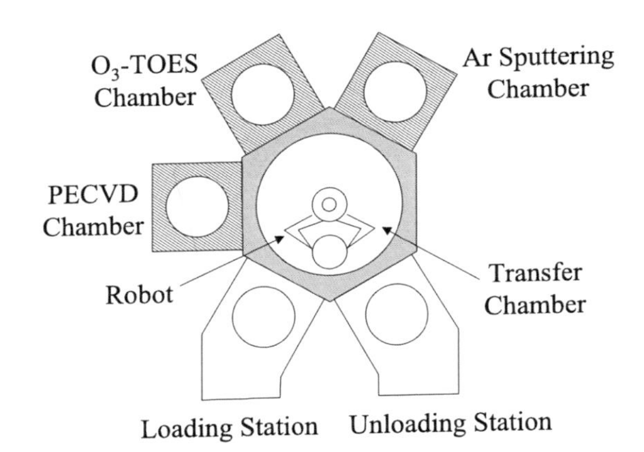

Figure 2.17 Schematic of cluster tool with dielectric CVD and etchback chambers.

due to its excellent gap-fill abilities. An argon (Ar) sputtering etch process is also performed in the thin-film bay to help gap-fill in dielectric deposition. Figure 2.17 illustrates a cluster tool with PECVD, O_3-TEOS, and Ar sputtering chambers.

Before 2000, metals used for connecting the millions of transistors in most IC chips were aluminum-copper alloy (Al·Cu), tungsten (W), titanium (Ti), and titanium nitride (TiN). Still underway is a transition from traditional interconnections to copper interconnections, in which copper (Cu) with a tantalum (Ta) or tantalum nitride (TaN) barrier layer is used in IC chips for interconnection. Copper replaced the aluminum-copper alloy in logic IC and microprocessors after the 0.18-μm technology node because it helps improve the speed and reliability of IC chips due to its lower resistivity and higher electromigration resistance. Copper is replacing aluminum-copper alloys in DRAM and flash memory for metal interconnects.

In metal bays, PVD tools (or mainly sputtering deposition tools) are used to deposit AlCu, Ti, and TiN, and CVD tools are widely used for W and TiN deposition. PVD tools are also used to deposit the Ta or TaN barrier layer and copper seed layer, while electrochemical plating deposition (EPD) tools are used to deposit bulk copper. PVD processes usually operate in a vacuum chamber, with a very high base vacuum to get rid of moisture in the chamber and minimize metal oxidation. Figure 2.18 illustrates a cluster tool with a PVD chamber for a Ta or TaN barrier layer and Cu seed layer deposition.

The most commonly used metrology tools in the thin-film bay are the reflectospectrometer, ellipsometer, stress gauge, laser scattering particle mapping tool for dielectric thin-film measurement, and four-point probe and profilometer for metal thin-film monitoring.

TEOS is the most commonly used silicon precursor for silicon oxide deposition, while silane (SiH_4) is commonly used for silicon nitride PECVD processes. Oxygen, ozone, and nitrous oxide (N_2O) are commonly used oxygen source gases, and nitrogen and ammonia are commonly used nitrogen sources. Nitrogen trifluoride (NF_3), or one of the carbon fluoride gases such as CF_4, C_2F_6, and C_3F_8,

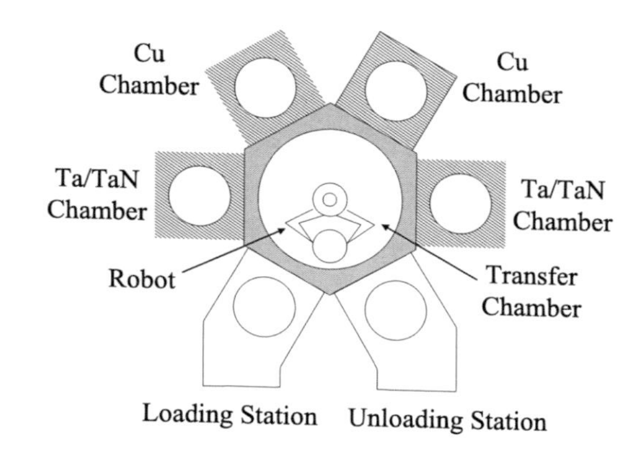

Figure 2.18 Schematic of a cluster tool with AlCu, Ti, and TiN PVD chambers.

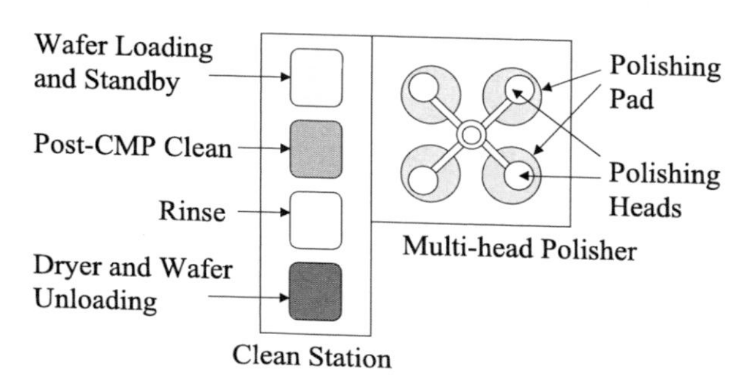

Figure 2.19 Schematic of a dry-in/dry-out multihead CMP system.

is widely used along with oxygen or nitrous oxide for dielectric CVD chamber cleaning. Silane is pyrophoric, explosive, and toxic; TEOS is combustible; O_3 and NF_3 are strong oxidizers; N_2O (laughing gas) can cause numbness; and carbon fluoride gases are considered global warming gases.

Metal PVD only uses argon and nitrogen, and neither of them is considered a safety hazard. Tungsten CVD widely uses tungsten hexafluoride (WF_6), silane (SiH_4), and hydrogen (H_2). WF_6 is corrosive, as it forms HF when it reacts with water (H_2O).

2.4.1.7 Chemical mechanical polishing bay

CMP is a removal process in which materials are moved from a wafer surface by the combination of mechanical grinding and wet chemical reaction. Commonly used CMP processes are silicate glass CMP, tungsten CMP, and most recently copper CMP. Timely post-CMP cleaning is a critical step to ensure the yield of CMP processing; therefore, some CMP tools are integrated with wet clean stations to form the so-called dry-in/dry-out CMP systems. Figure 2.19 illustrates a dry-in/dry-out multipolishing head CMP system.

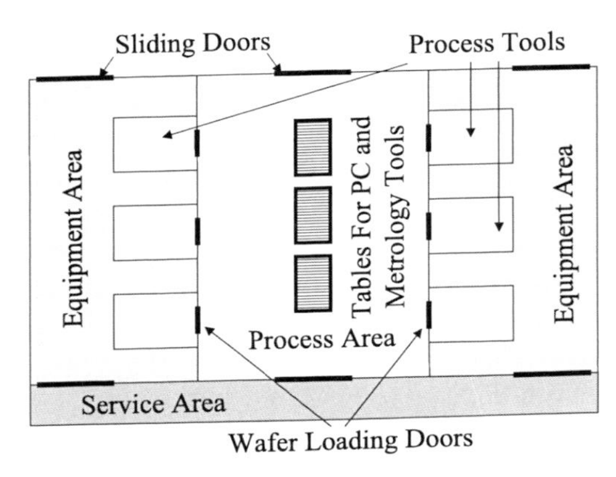

Figure 2.20 Process and equipment areas.

Fine particles, such as silica (SiO_2) or celica (CeO_2) in silicate glass CMP slurries and alumina in metal CMP slurries, are used as abrasives and play very important roles in CMP processes.

2.4.2 Equipment area

Many processing tools such as ion implanters, plasma etchers, CVD reactors, and PVD and CMP tools are located in gray areas of a fab, which are cleanrooms with higher airborne particle counts, or lower-class cleanrooms, typically at class 1000. By putting processing tools in a lower-class cleanroom instead of a high-class cleanroom such as the processing area, both fab building and maintenance costs can be significantly reduced.

Figure 2.20 illustrates the location of the processing and equipment areas in a semiconductor fab. Equipment engineers and technicians mainly work in the equipment area. Equipment support personnel from tool manufacturers also work in this area to help to install, start up, and maintain processing tools.

For the minienvironment fab illustrated in Fig. 2.12(a), there is no such separation of process and equipment areas. The entire cleanroom is laid out as one big ballroom.

2.4.3 Facility area

The facility area hosts facilities that support the processing tools and metrology instruments. It is not located in the cleanroom, and some facilities are not even in the same building as the cleanroom. Facilities needed for the wafer processes are gases, water, electric power, and subsystems that support the processing equipment. Facility engineers, technicians, and support personnel from facility providers mainly work in this area. They also need to work inside the cleanroom during installation and maintenance of the processing equipment.

Gases include process gases with ultra high purity, purge gas such as nitrogen with a lower purity than process gases, and dry air used for driving the pneumatic systems of the processing tools. A large amount of clean air is needed to constantly feed into the cleanroom to make up for the loss of air while maintaining the positive pressure of the cleanroom. Large amounts of gases are consumed every day in IC fabrication, especially nitrogen, which is the most heavily used purge gas for purging process reactors and gas delivery lines. High-purity nitrogen is also used as a process gas in IC processing. Many fabs must have their own nitrogen plants to manufacture and purify the nitrogen needed from the atmosphere by freezing and vaporization of compressed air. Most process gases are held in high-pressure gas bottles, which are stored in specially designed gas cabinets located in small rooms built with thick concrete walls and doors that open only toward the outside. This design can prevent major damage to the fab in case of a gas explosion.

A large quantity of water is used in IC manufacturing. High-purity DI water is used for rinsing during wafer wet clean and wet etch processes. DI water is also used for the cooling of processing systems. IC fabs must have their own plants to generate the DI water needed in wafer processing. A large amount of city water is also used to cool the subsystems and other facility equipment, such as air conditioners. An IC fab also requires strong air-conditioning systems to maintain a constant temperature. Temperature variation can cause a drift of measurement results in metrology systems and induce yield loss due to process drift.

Electric power breaks and distribution systems are located in the facility area. Electric power consumption of an IC fab is enormous. For example, a diffusion furnace can consume about 28.8 kW (480 V by 60 A). Keep in mind that a furnace is always on unless it requires PM, and an IC fab usually has more than 50 of these furnaces.

Question: Assuming the average power of a furnace is 20 kW and a fab has 50 such furnaces, calculate the electricity bill for a year to keep these 50 furnaces operating if electric power costs 10 cents per kilowatt-hour.

Answer: $20 \times 50 \times 24 \times 365 \times 0.1 = \$876{,}000$.

Many subsystems of the processing equipment, such as vacuum pumps, rf power generators, gas and liquid delivery systems, and heat exchangers, are installed in the facility area, usually directly beneath the processing and equipment area, as illustrated in Fig. 2.8. Vacuum pumps are needed to evacuate air from the reactors and keep them under vacuum environments required by the processes. Radio frequency generators are needed to strike and maintain stable plasmas in many PECVD and plasma etch processing reactors. Gas and liquid delivery systems flow processing gases or liquids to the processing tools as required by the process. Heat exchangers provide chilled or heated water to the processing systems to maintain constant temperatures required by the different processes.

2.5 Testing and Packaging

After wafer processing is finished, wafers are sent out for testing and packaging. Some fabs do testing and packaging on site, and some fabs send the finished wafers out to a packaging plant. Die testing is a very labor-intensive process, since it requires operators to carefully contact the fine pins of the test tool to tiny bonding pads by hand (about 100 × 100 μm^2) under an optical microscope for every die on the wafer. Some IC manufacturers build testing and packaging plants in developing countries to take advantage of lower labor costs. This could change due to rapid improvement of automation technology in IC testing equipment and the continuous shrinking of the percentage of labor costs in the total cost of IC manufacturing.

2.5.1 Die testing

It is very important to detect a faulty chip as early as possible in the IC manufacturing process. The earlier a problem is caught, the lower the costs for the manufacturer. First, a design team must verify the design and make sure there are no design defects before the layouts are sent out to the mask shop to make masks or reticles. Detecting a design fault costs a semiconductor fab nothing. Mask/reticles also need to be carefully inspected, and their condition is constantly monitored in IC fabs during wafer processing.

Throughout wafer processing, specially designed test devices and test circuits on the testing structures are randomly sampled and tested, to ensure yield for certain key processing steps, and to catch any process specification shift that affects the yield. Detecting a bad chip at the wafer level costs the semiconductor chip manufacturer about $0.10 per chip. The cost for detecting a malfunctioning chip at the packaging level is about $1. This cost increases rapidly from $1 to $10 per chip if a faulty chip is used at the electric board level. The cost further increases at the electronics system level ($10 to $100 per chip) and customer level (>$100 per chip). In the early 1990s, some malfunctioning microprocessors caused errors in certain division calculations running on their personal computers (PC), and it cost millions of dollars for the manufacturer to replace these faulty microprocessors.

After finishing all of the processes in a high-class cleanroom, wafers are transferred to a lower-class cleanroom (typically about a class 1000) for die testing, and even a lower-class cleanroom (typically about a class 10,000) for chip packaging. Specially designed test tools are used to test each die on the wafer where tiny pins are connected to the bonding pads or bumps on the chip, and the test programs are carried out to test and verify whether the IC chip meets the design requirements. Faulty die are inked with a dot, as shown in Fig. 2.21, and are not picked up for packaging after die separation. Advanced test-sorting tools store the information of die condition in an on-board computer and do not use ink to mark the bad dies because ink can cause contamination and must be constantly refilled. Mistakenly ink-marking good dies or missing the mark on bad dies due to loss of ink can cause unwanted costs for the IC manufacturer.

2.5.2 Chip packaging

There are four main reasons for chip packaging: to provide physical protection for the IC chip, to provide a barrier layer against chemical impurities and moisture,

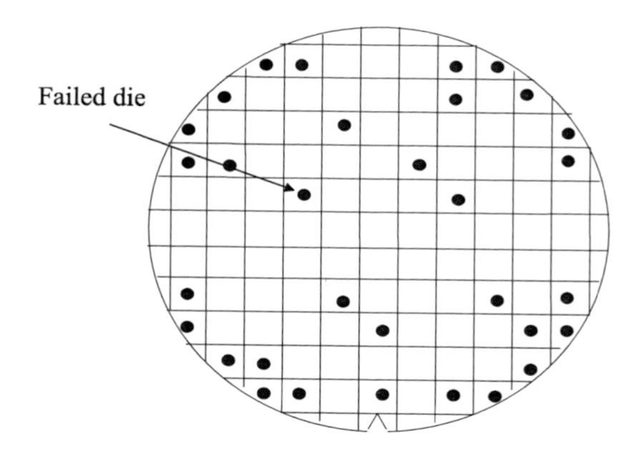

Figure 2.21 Tested wafer.

to allow the IC chip to connect to an electric circuit with a sturdy set of leads, and to dissipate heat generated during chip operation. The main concerns with chip packaging are process cost, process temperature, and packaging materials, including their thermal conductivity, coefficient of thermal expansion, and Young's elastic modulus.

After chip testing, wafers are normally coated with a protective layer on the surface and mechanically polished on the backside to reduce wafer thickness. The backside thinning process can help remove the backside coating from the wafer process and improve the contact of metal coating to the silicon substrate. It also reduces the thickness of the chip, generally from 600 to 775 μm to 250 to 350 μm, so that chips can easily fit into the shallow pocket on the lead frame. Some chips need to be thinned to about 100 μm to fit into a smart card. Other techniques, such as wet etch and plasma etch, are also developed for the wafer thinning process. A thin metal layer usually is coated onto the backside of the wafer after wafer thinning. Gold is commonly used for backside metallization, and both sputtering and evaporation processes can be used for gold metallization.

After removal of the protective layer from the wafer surface, the wafer is taped onto a solid frame from the backside with a sticky, flexible material such as Mylar. The tape helps prevent wafer movement and holds the wafer in one piece during die separation. Diamond-impregnated saws operating at very high speeds (up to 20,000 rpm) with constant coolant flow are used to separate the individual die from the wafer along the scribe lines. Other technologies such as scribing and mechanical breaking have been used; however, they are no longer in use because they can cause chipping and cracking.

In the die sorting process, good chips (without dots) are picked up, while bad ones (with dots) are left behind. Good dies are attached to a packaging socket, which is a good conductor for both heat and electricity. Metal or metal-coated ceramic are commonly used as die-attaching materials. In a thermal process, solder between the metal-coated chip backside and metal-coated substrate surface melts and attaches the chip to the substrate when it cools down. Temperature for the die-attaching process is limited by the fact that tiny aluminum alloy wires are

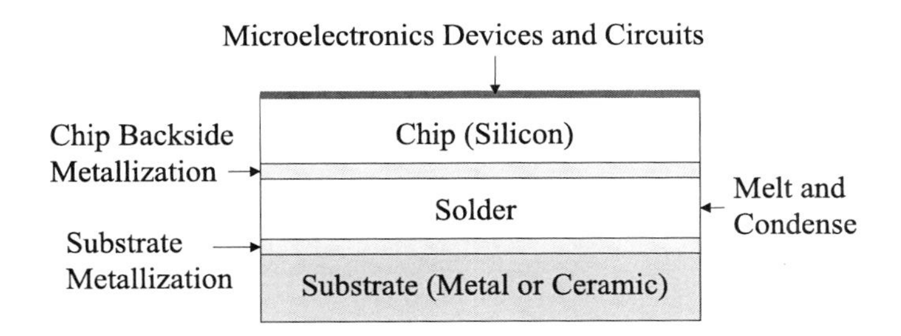

Figure 2.22 Chip bond structure.

used in many IC chips, and aluminum alloy cannot sustain a temperature higher than 550 °C due to the eutectic of aluminum and silicon. Rule of thumb is that no thermal process with a temperature higher than 450 °C is allowed after aluminum metallization. A gold-silicon eutectic from the heat is commonly used in the IC industry, because a mixture of gold and silicon creates a low melting temperature of the alloy. A mixture of 97.15% Au and 2.85% Si by weight melts at 363 °C. Figure 2.22 illustrates the basic structure of die attachment.

After the chip is attached to the lead frame, a soldering machine is used to make the connection between bonding pads on the chip and lead pins on the lead frame with thin metal wires. A traditional wire bonding process is illustrated in Fig. 2.23.

First, a molten ball is formed by a high-temperature heater at the tip of the metal wire in the soldering head. Then the soldering head presses the molten metal ball to the bonding pad surface to solder the metal wire with the bonding pad. The head then retreats and bends the wire to the pin lead, heats, and then presses the wire to solder it with the pin lead. The clamp closes as the soldering head moves away, breaking the metal wire due to excessive extension stress. After a short time, the tip of the broken metal wire forms a molten ball, again, due to surface tension, and the soldering head is ready for the next soldering step. Gold wire is commonly used in the wire bonding process. It is important that the wire bonding temperature is not higher than the chip bonding solder melting point. Figure 2.24 illustrates a chip with bonding pads, and the traditional packaging technology with wire bonding of the IC chip to the lead pins of the chip socket.

An alternative packaging technology has been developed and is becoming widely used in IC manufacturing. Different from the traditional wire bonding process, flip-chip technology forms metal bumps instead of bonding pads on the IC chip surface after passivation dielectric etch. After die separation, the chip is placed into a socket face down, and metal bumps on the chip surface and metal leads on the socket surface are precisely aligned and contacted. In a heated environment, the metal bumps and pin leads melt and join together after the chip is cooled, forming connections between the IC chip and the pins on the socket. This technology is called flip-chip packaging because in traditional wire bonding packaging, chips are placed face up in the lead frame, while in flip-chip packaging, the chip must be mounted face down in the sockets. One advantage of flip-chip packaging is that it

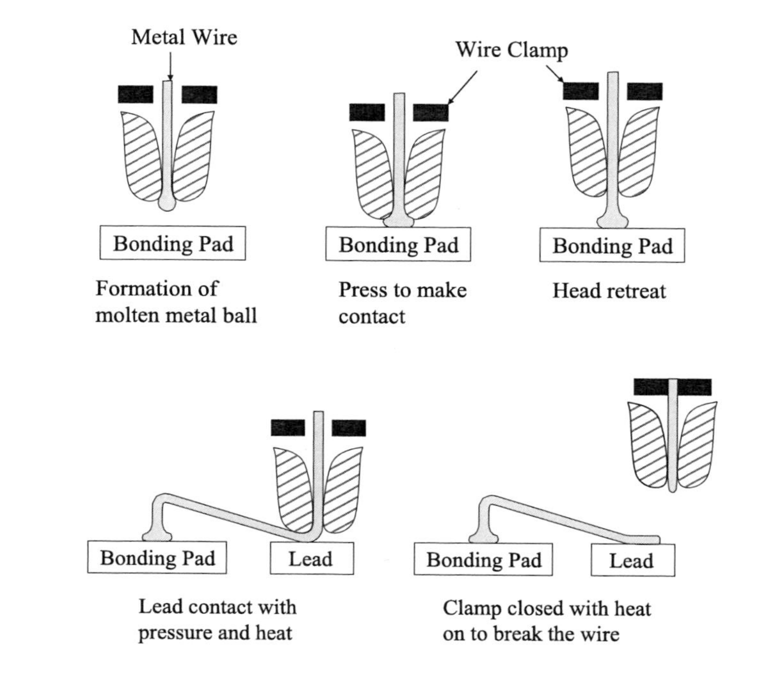

Figure 2.23 Wire bonding process steps.

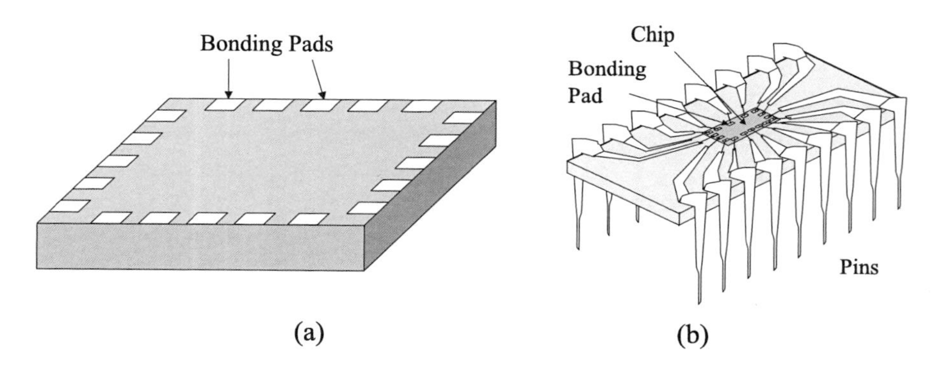

Figure 2.24 IC chip with (a) bonding pads and (b) wire bonding.

can significantly reduce package size. Figure 2.25 illustrates the chip with bumps, and Fig. 2.26 illustrates flip-chip packaging process steps.

After the IC chip is connected to the lead pins of the chip socket through wire bonding or bump connection, the chip and lead frame are ready to be sealed. A preseal inspection is conducted to eliminate chips broken during the mechanical handling and thermal processes of the die attaching and wire bonding (or bumps) connection. There are two kinds of chip sealing processes: ceramic and plastic. Ceramic is a much better barrier material for impurities such as mobile ions and moisture, and it also has much higher thermal stability, higher thermal conductivity,

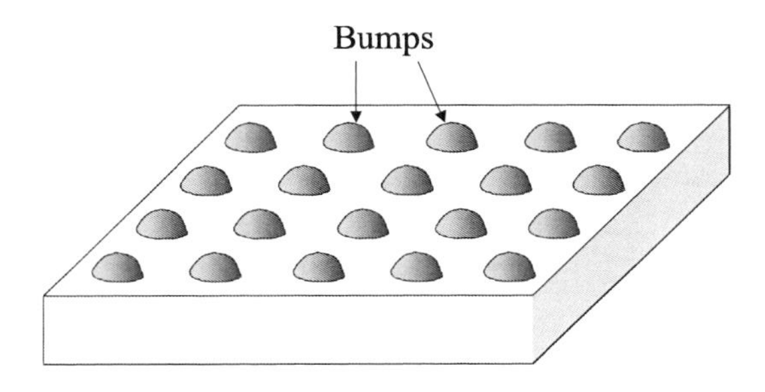

Figure 2.25 IC chip with metal bumps.

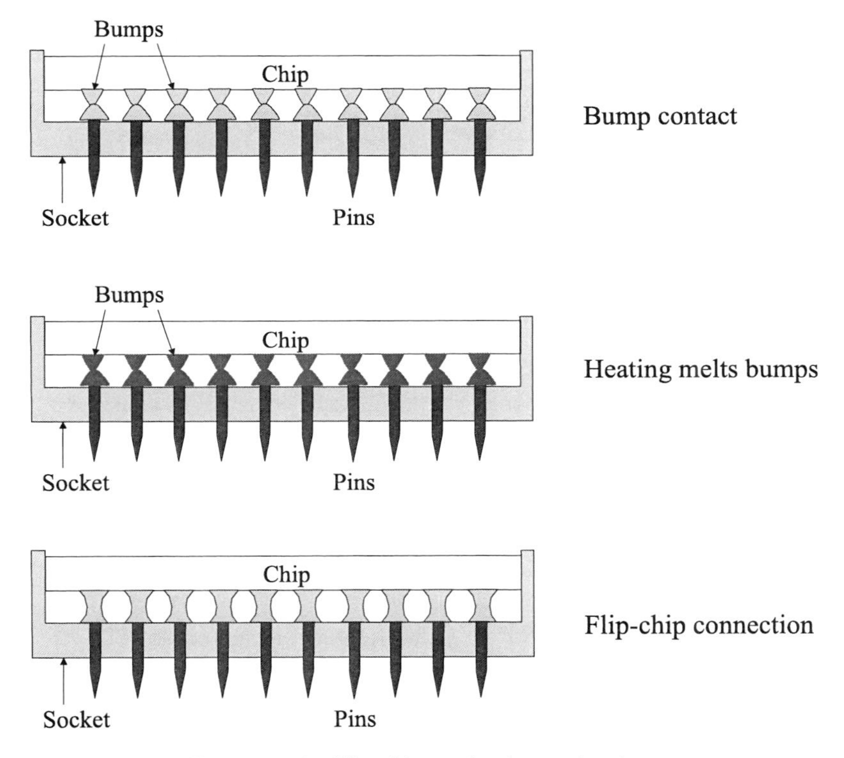

Figure 2.26 Flip-chip packaging technology.

and lower coefficients of thermal expansion. However, it is much more expensive than the plastic packaging, by about a factor of 5 per chip packaging, and it is also much heavier. Therefore, to reduce packaging costs, people like to use plastic packaging whenever it is applicable. For example, most memory chips and logic chips use plastic packaging. Some IC chips, especially microprocessors for computers, or CPUs, require ceramic packaging because of the large amount of heat generated during their operation. Ceramic packaging was previously the standard in the IC industry; now, more and more chips use plastic packaging.

Ceramic packaging usually uses metal solder to seal the cap and the lead frame. During the sealing process, the ceramic cap is heated and pressed to the lead frame.

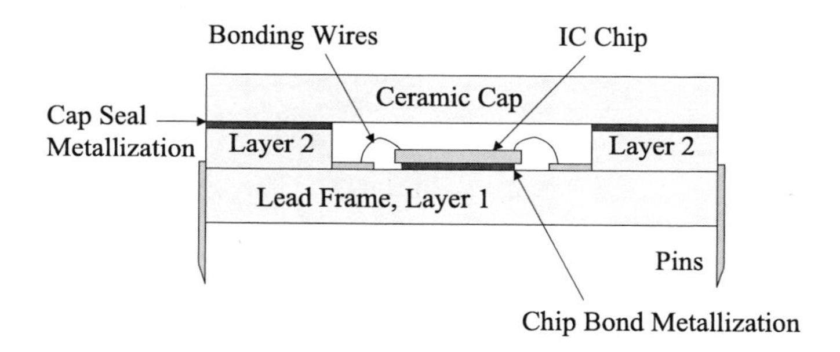

Figure 2.27 Cross-section of IC chip ceramic packaging.

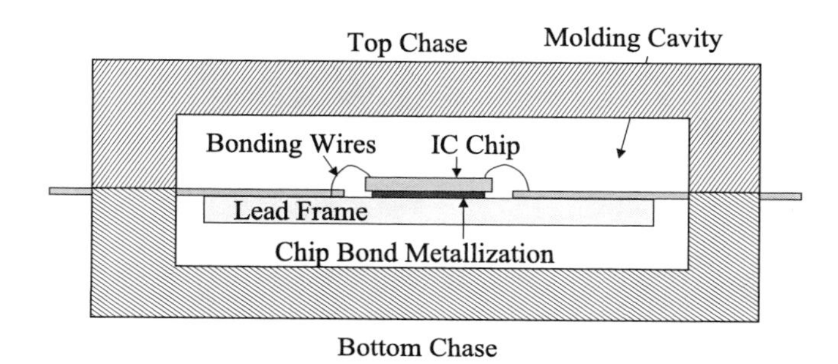

Figure 2.28 Cross section of the molding cavity for plastic packaging.

The heat melts the metal layers coated on surfaces and solders them together. Figure 2.27 illustrates the cross section of the ceramic packaging of the IC chip.

Plastic packaging uses a molding technique to seal the IC chip and lead frame with plastic. After wire bonding, the lead frame is placed between the top and bottom chases of the packaging tool. The heated chases close to form a cavity surrounding the chip and the lead frame. Molten plastic flows into the cavity and seals the chip after cooling and plastic condensation. Figure 2.28 shows the molding system for chip plastic packaging.

Plastic is not a very good barrier for moisture and other impurities such as mobile sodium ions. These materials can penetrate the plastic seal and affect IC performance, which can affect IC chip reliability. However, the improvement of packaging technology helps extend the lifetime of a plastic packaged IC chip to more than 5000 h in the failure-inducing test, a time period that is equivalent to more than ten years of life in normal conditions, enough for most electronic applications.

2.5.3 Final test

After packaging is finished, chips are sent to a final test, which places them in an unfavorable environment to force any unreliable chips to fail. High acceleration in a spinner can mechanically cause any loosely bonded wires to break off from the bonding pad, and high-temperature thermal stress can accelerate the electrical failure of a chip. Chips then are put on racks to run their normal operation continuously for several days, in what is called a burn-in process. Some microprocessor chips that fail to operate at their designed frequency can work fine at lower frequencies and hence are often sold as lower-speed processor chips at lower prices. Chips that pass the burn-in process can be shipped to customers (mainly the electronic industry) for placement on electronic boards to build electronic systems.

2.5.4 3D packaging

In the future, 3D packaging technology, which allows IC manufacturers to stack multiple chips in a package, will be developed and implemented in production. Stacking two chips is almost the equivalent of reducing feature size by a technology node and doubling device density. By stacking four or eight chips, the effect is more significant. When the cost of scaling feature size skyrockets, 3D packaging will become more attractive and could be more cost effective.

3D packaging has already been used in CMOS image sensors and other devices. Through-silicon via (TSV) is one of these enabling technologies that could be used in the implementation of 3D packaging. There are several options of TSV 3D packaging processes in development. Figure 2.29 illustrates one of them, in which the TSV is formed in an IC fab after transistor formation and before the forming of the interconnection, as shown in Figs. 2.29(a) to 2.29(e). After wafers are sent to a packaging house, the backside of the wafer can be thinned to expose the buried TSV plugs used to form the backside bumps. The thickness of the wafer after thinning depends on the depth of the TSV holes. Figure 2.29(f) shows the wafer with backside bumps, which is ready for 3D packaging.

If the die yield of the wafers is high, fabs have the option to perform wafer-to-wafer stacking, which aligns wafers so that the backside bumps of the top wafer accurately connect with the front-side bond pad of the wafer underneath. A heating process melts the solder bumps to form an electrical connection, bonding the two wafers together, as shown in Fig. 2.29(g). In case die yield of the wafers is not very high, die to known-good-die (KGD) stacking is more cost effective. In this case, a good die is picked up and aligned with a KGD on the handling wafer so that the backside bump of the die can accurately contact with the front-side bond pad of the KGD. A heating process melts the solder bumps to form an electrical connection and bond the two dies.

2.6 Future Trends

In the near future, IC processing materials will have some significant changes. High-dielectric-constant (κ) dielectrics, or high-*k* dielectrics, will replace silicon dioxide and silicon oxynitride as the gate dielectrics in advanced logic IC chips. Low-κ dielectrics have replaced commonly used silicon oxide as insulators for

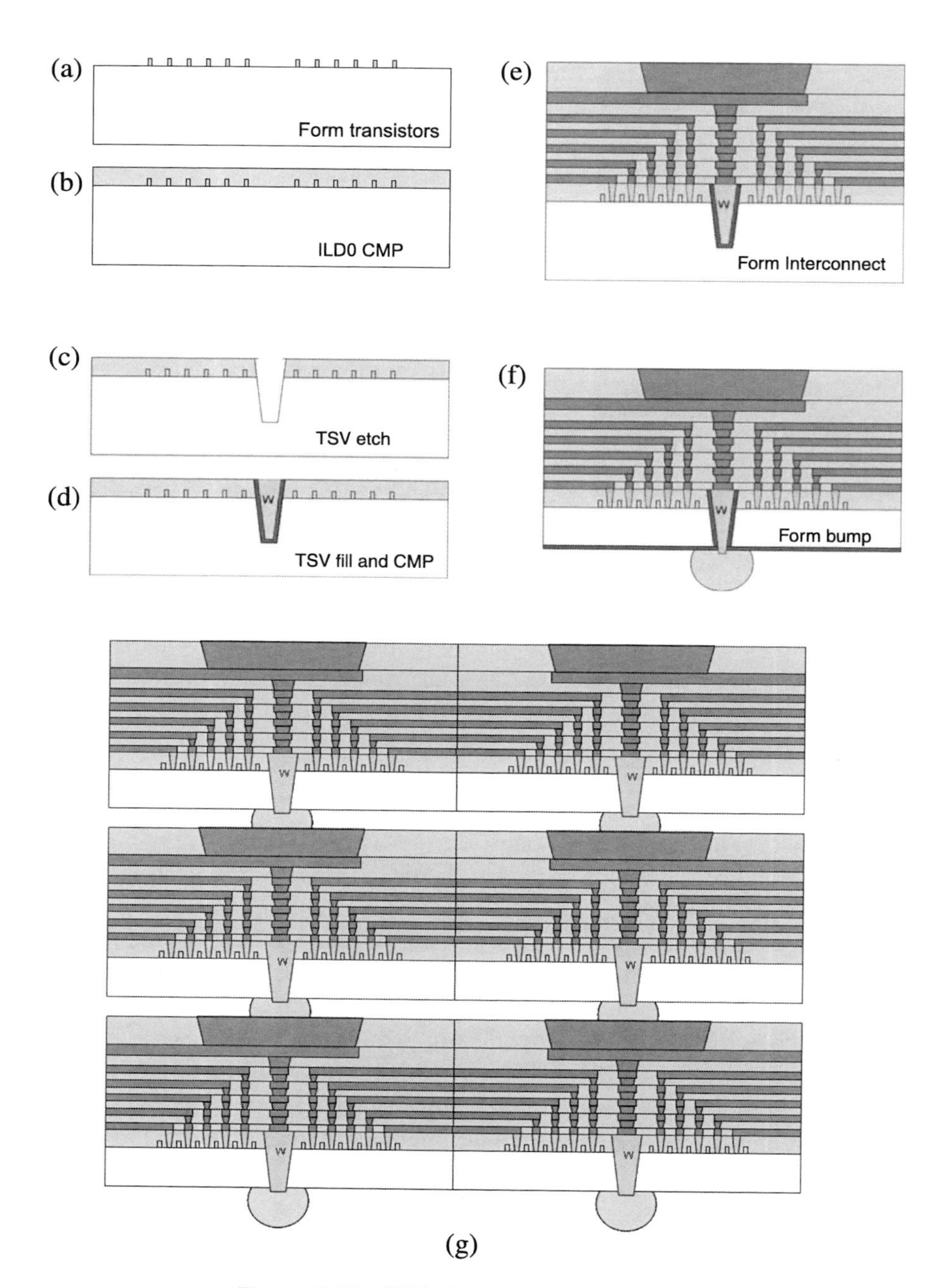

Figure 2.29 TSV 3D packaging processes.

the interconnection of IC chips since the 90-nm technology node, and they are being used to replace silicon oxide in memory devices. The standard κ-value to determine high-κ or low-κ is 3.9, which is the dielectric constant of silicon dioxide. Copper interconnects have become mainstream for the interconnection of logic devices since the 0.18-μm technology node and are gradually replacing aluminum alloy interconnects in memory devices in advanced nanometer technology nodes. Immersion lithography, which can improve patterning resolution, became widely used after the 45-nm technology node. A double-patterning technique that can further improve pattern density has been adopted in the 32-nm technology node and beyond. CMP processes will be more widely used in IC manufacturing, especially when memory devices start adopting copper metallization. The combination of copper and low-κ dielectrics can help increase the speed of IC chips. Facilities that support IC processes will change accordingly.

Faster, more precise, more reliable, and fully automatic IC test tools will be developed for more-complicated IC testing in the future. More IC manufacturers will adopt the flip-chip packaging technique for chip packaging processes. Multilayer lead frames for multichip packaging will be more commonly used, especially stacked multichip packaging. 3D packaging will see more development, and very likely 3D packaging using TSV technology will go to production in the near future.

2.7 Summary

- Overall yield is the ratio between the total number of good chips after the final test and total number of dies on all of the wafers used to manufacture this chip.
- Yield determines whether IC chip fabrication is losing money or making a profit.
- The cleanroom is a controlled environment in which particle concentration in air is kept very low by using filters and controlling airflow, pressure, temperature, and humidity.
- It is very important to strictly follow cleanroom protocols to reduce airborne particulates and other contaminants; otherwise, these contaminants can affect chip yield and fab profit.
- An IC fab normally has a photo bay, diffusion bay, implantation bay, etch bay, thin-film bay, CMP bay, and wet bay.
- The processing area consists of several processing bays, and is a high-grade cleanroom. For manufacturing IC chips with 0.25-μm feature size, the processing area requires a class-1 cleanroom. For equipment areas, class 1000 is good enough and costs much less to build and maintain. To achieve better than a class-100 cleanroom, laminar flow is required. Turbulent airflow can be used for a class-1000 cleanroom.
- Gas delivery systems, DI water systems, electric power distribution systems, pump and exhaust systems, and rf power systems are the most commonly used facility systems in IC fabrications.
- Chip packaging provides physical and chemical protections for IC chips, provides the fine wiring that connects the chip to the lead pins on a sturdy base, and helps to dissipate the heat generated during chip operation.

- Compared with plastic packaging, ceramic packaging provides better protection for chemical and moisture contamination and better physical protection. It also has much better thermal stability and higher thermal conductivity. The main advantage for plastic packaging is its significantly lower cost. Most IC chips now use plastic packaging.
- Standard wire bonding uses fine metal (gold) wire to connect chip bonding pads to lead pin bonding pads. The flip-chip uses bump bonding to connect chips to lead pins.
- For standard wire bonding packaging processes, the temperature for chip sealing should be low enough to not affect wire bonding and die attachment. Wire bonding temperature should not affect die attachment, and die attachment temperature is limited by aluminum's melting point.
- In final testing, a packaged chip is tested in some abnormal conditions such as higher temperatures, high acceleration, high humidity, etc., to intentionally cause unreliable IC chips to fail before they are sent to customers.
- 3D packaging has already been used in CMOS image sensors. Development of technologies such as TSV will help more IC products adopt 3D packaging in the near future.

2.8 Bibliography

R. Carranza, *Silicon Run II* (video tape), Ruth Carranza Productions, Mountain View, CA (1993).

C. Y. Chang and S. M. Sze, *ULSI Technologies*, McGraw-Hill, New York (1996).

S. M. Sze, *VLSI Technology, Second Edition*, McGraw-Hill, New York (1988).

S. Wolf and R. N. Tauber, *Silicon Processing for the VLSI Era, Vol. 1, Process Technology, Second Edition*, Lattice Press, Sunset Beach, CA (2000).

See http://www.set3.com/papers/, paper 209e (last accessed Sept. 2012).

2.9 Review Questions

1. How many particles larger than 0.5 μm in a cubic foot are allowed in a class-1000 cleanroom?
2. Sodium is a mobile ion, and trace amounts of sodium ions can damage microelectronic devices. What is the main source of sodium?
3. Explain the relationship between die yield and defect density, die size, and number of process steps by using Eq. (2.2).
4. What is the yield of the wafer shown in Fig. 2.21? (Total dies should exclude those on the wafer edge; these should be treated as incomplete dies.)
5. Why does a class-1 cleanroom need laminar flow?
6. List at least two pieces of processing equipment that can be found in a photo bay.

7. List at least two pieces of metrology equipment that can be found in photo bay.
8. Explain why a class-1000 cleanroom is usually used in a minienvironment IC fab instead of a higher class-1 cleanroom.
9. List at least two processing systems in a diffusion bay.
10. CVD and PVD are _____ processes.

    (a) adding (b) removing (c) patterning (d) heating
11. Photolithography is a _____ process.

    (a) adding (b) removing (c) patterning (d) heating
12. Anneal is a _____ process.

    (a) adding (b) removing (c) patterning (d) heating
13. CMP is a _____ process.

    (a) adding (b) removing (c) patterning (d) heating
14. Nitrogen can be used both as a purge gas and processing gas in IC fabrication. Does nitrogen for these two applications require the same purity?
15. Explain why people no longer use ink to mark the failed dies.
16. List the main purposes of chip packaging.
17. Describe the chip bonding and wire bonding processes.
18. What are the advantages and disadvantages of ceramic packaging compared with plastic packaging?
19. What is the purpose for the failure-inducing final test?
20. What is the benefit of using multichip packaging?
21. Describe flip-chip packaging.
22. Describe the TSV process.

Chapter 3
Semiconductor Basics

This chapter provides the basics of a semiconductor, semiconductor devices, and semiconductor processes.

Objectives

After finishing this chapter, the reader will be able to:

- identify at least two semiconductor materials from the periodic table of elements
- list n-type and p-type dopants
- describe a diode and MOS transistor
- list three kinds of chips made in the semiconductor industry
- list at least four basic processes required for chip manufacturing.

3.1 What Is a Semiconductor?

Semiconductors are materials with electrical conductivity between good conductors such as metal (including copper, aluminum, tungsten, etc.) and good insulators such as rubber, plastic, and dry wood. The most commonly used semiconductor materials are silicon (Si) and germanium (Ge), both of which are located in column IVA of the periodic table of elements, shown in Fig. 3.1. Some compounds such as gallium arsenate (GaAs), silicon carbide (SiC), and silicon germanium (SiGe) are also semiconductor materials. One of the most important properties of a semiconductor is that its conductivity can be controlled by intentionally adding certain impurities, a process called doping, and by applying an electric field.

3.1.1 Bandgap

The fundamental difference between a semiconductor and an insulator or a conductor is the bandgap. Atoms form all materials, and every atom has its own orbital structure [see Fig. 3.2(a)]. Electron orbits of the atom are called shells because electrons orbit around the nucleus in a 3D shell. Thus, the orbits in Fig. 3.2(a) can be seen as the cross section of these shells. The outermost shell is called a *valence shell*. Electrons in a valence shell cannot conduct an electric

IA	IIA		IIIB	IVB	VB	VIB	VIIB	VIIIB			IB	IIB	IIIA	IVA	VA	VIA	VIIA	VIIIA
1 H																		2 He
3 Li	4 Be												5 B	6 C	7 N	8 O	9 F	10 Ne
11 Na	12 Mg												13 Al	14 Si	15 P	16 S	17 Cl	18 Ar
19 K	20 Ca		21 Sc	22 Ti	23 V	24 Cr	25 Mn	26 Fe	27 Co	28 Ni	29 Cu	30 Zn	31 Ga	32 Ge	33 As	34 Se	35 Br	36 Kr
37 Rb	38 Sr		39 Y	40 Zr	41 Nb	42 Mo	43 Tc	44 Ru	45 Rh	46 Pd	47 Ag	48 Cd	49 In	50 Sn	51 Sb	52 Te	53 I	54 Xe
55 Cs	56 Ba	*	71 Lu	72 Hf	73 Ta	74 W	75 Re	76 Os	77 Ir	78 Pt	79 Au	80 Hg	81 Tl	82 Pb	83 Bi	84 Po	85 At	86 Rn
87 Fr	88 Ra	**	103 Lr	104 Rf	105 Db	106 Sg	107 Bh	108 Hs	109 Mt	110 Ds	111 Uuu	112 Uub	113 Uut	114 Uuq	115 Uup	116 Uuh	117 Uus	118 Uuo

*Lanthanoids	*	57 La	58 Ce	59 Pr	60 Nd	61 Pm	62 Sm	63 Eu	64 Gd	65 Tb	66 Dy	67 Ho	68 Er	69 Tm	70 Yb
**Actinoids	**	89 Ac	90 Th	91 Pa	92 U	93 Np	94 Pu	95 Am	96 Cm	97 Bk	98 Cf	99 Es	100 Fm	101 Md	102 No

Figure 3.1 Periodic table of elements.

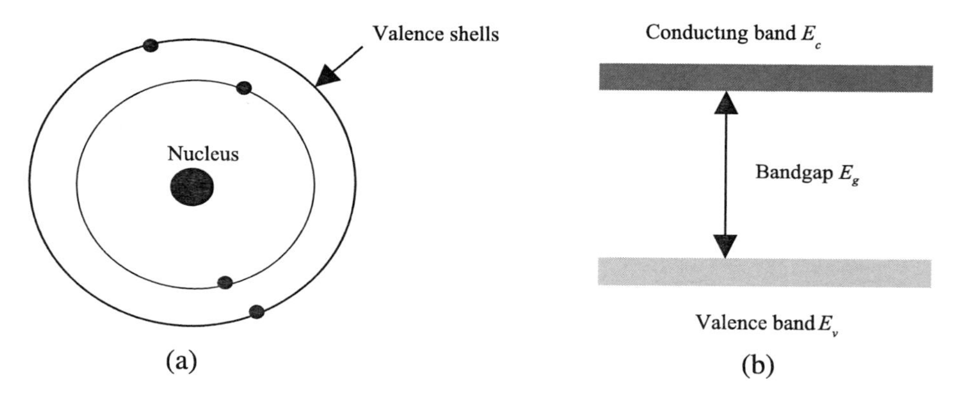

Figure 3.2 (a) Orbital structure of a single atom and (b) an energy band.

current. When an electron escapes the constriction of the nucleus and leaves the valence shell, it becomes a free electron and can conduct an electric current.

When many atoms bond together to make solid materials, their orbits overlap and form bands, as shown in Fig. 3.2(b). Electrons in conducting bands can move relatively freely inside solid materials and can conduct electric currents when an electric field is applied to the solid material. Electrons in the valence band are bound with the nuclei and cannot move freely; therefore, they cannot conduct electric currents. Since the valence band has lower electric potential, electrons always tend to stay in the valence band.

Resistivity is the ability of a material to resist an electric current. A good conductor has a very low resistivity, and a good insulator (or dielectric) has a very high resistivity. The most commonly used unit of resistivity is micro-ohm

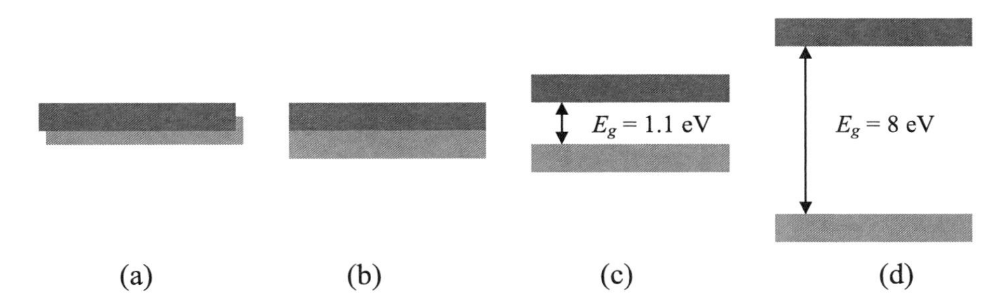

Figure 3.3 (a) Energy bands and bandgap of Al, (b) Na, (c) Si, and (d) SiO_2.

times centimeter ($\mu\Omega \cdot$ cm). At room temperature, the resistivity of aluminum is 2.7 $\mu\Omega \cdot$ cm, sodium is 4.7 $\mu\Omega \cdot$ cm, intrinsic silicon is about 10^{11} $\mu\Omega \cdot$ cm, and silicon dioxide is $>10^{18}$ $\mu\Omega \cdot$ cm.

For most metals, the conducting band and valence band overlap or have a very small bandgap, so small that electrons with thermal energy at room temperature (300 K $\approx$ 0.0259 eV) can jump across it. One electron volt (1 eV) is the energy gained by an electron when it passes through two points with one volt (1 V) of voltage difference. Therefore, a conducting band has many electrons, explaining why metals are always good electrical conductors. For dielectrics such as glass and plastic, the bandgap is so large that electrons cannot jump across it from the valance band, so the conducting band has very few electrons to conduct electric currents.

For semiconductors, the bandgap is somewhere between that of conductors and insulators; for instance, silicon is 1.1 eV, germanium is 0.67 eV, and gallium arsenate is 1.40 eV. While most electrons stay in the valance band, there are always some thermal electrons (from the tail of the Boltzmann distribution, discussed in Chapter 7) that can jump into the conducting band and conduct electric current. For intrinsic silicon at room temperature (300 K), there are about 1.5×10^{10} per cubic centimeter (cm^{-3}) electrons in the conducting band. This means at room temperature, only about one in ten trillion electrons is in the conducting band, while the majority of electrons remain in the valence band. Therefore, intrinsic semiconductors can conduct electric currents at room temperature better than dielectrics, but not as well as conductors.

3.1.2 Crystal structure

From their position in column IVA of the periodic table of elements, we can see that the most commonly used semiconductor materials—silicon and germanium—both have four electrons in the outermost shell. In a single-crystal structure, every atom is bonded with four atoms and shares a pair of electrons which each of them, (see Fig. 3.4). Since most IC chips are made from a silicon substrate, this text mainly covers the silicon process and uses silicon to explain doping.

3.1.3 Doping semiconductor

By intentionally adding dopants into pure single-crystal semiconductor materials such as silicon, conductivity can be improved. There are two kinds of dopants:

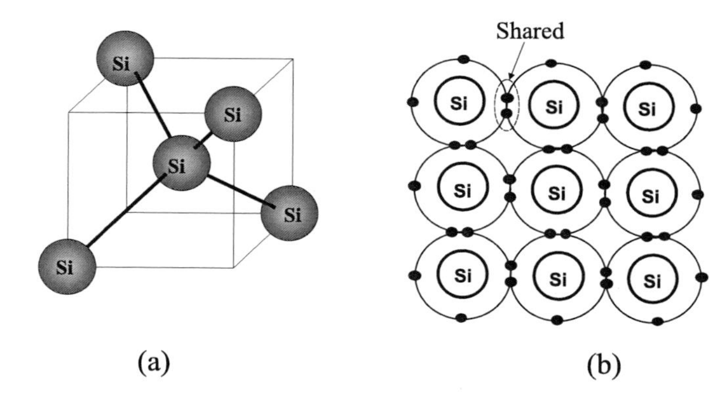

Figure 3.4 Actual tetrahedron silicon single-crystal structure and (b) its 2D simplification.

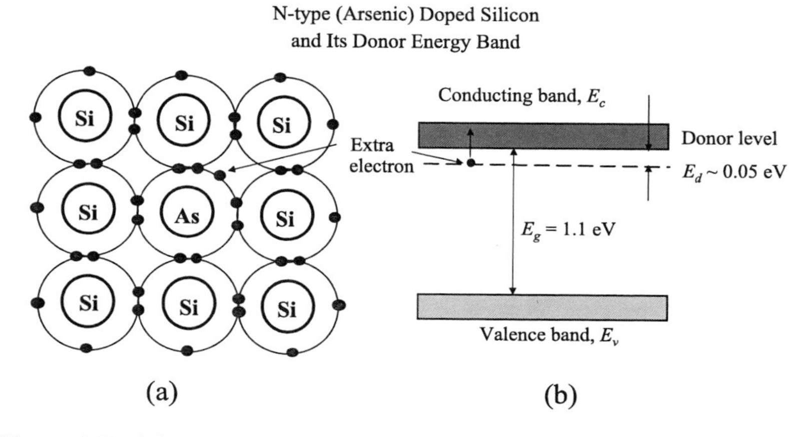

Figure 3.5 (a) n-type (arsenic) doped silicon and (b) its donor energy band.

p-type such as boron (B) from column IIIA of the periodic table, and n-type such as phosphorus (P), arsenic (As), and antimony (Sb) from column *VA*. Since P, As, and Sb provide an electron in semiconductor materials, they are called donors. Boron provides a hole, which allows other electrons to jump in and create a hole elsewhere. It is called an acceptor.

Phosphorus and arsenic have five electrons in their outermost shell. When P or As is doped into pure single-crystal Si or Ge, there is an extra electron in the outermost shell [see Fig. 3.5(a)]. This extra electron can easily jump into the conducting band and become a free electron to conduct an electric current. In this case, the majority carriers of electric currents are electrons, and the semiconductor is an n-type dopant (n stands for negative) because electrons have negative charges. P, As, and Sb are n-type dopants. The more n-type dopant atoms enter the semiconductor substrate, the more free electrons they will provide to conduct electric current, and therefore the better the conductivity of the semiconductor.

Figure 3.6(a) shows the empty spot (or hole) in the outermost shell left when boron is doped into pure single-crystal silicon or germanium. Figures 3.6(c),

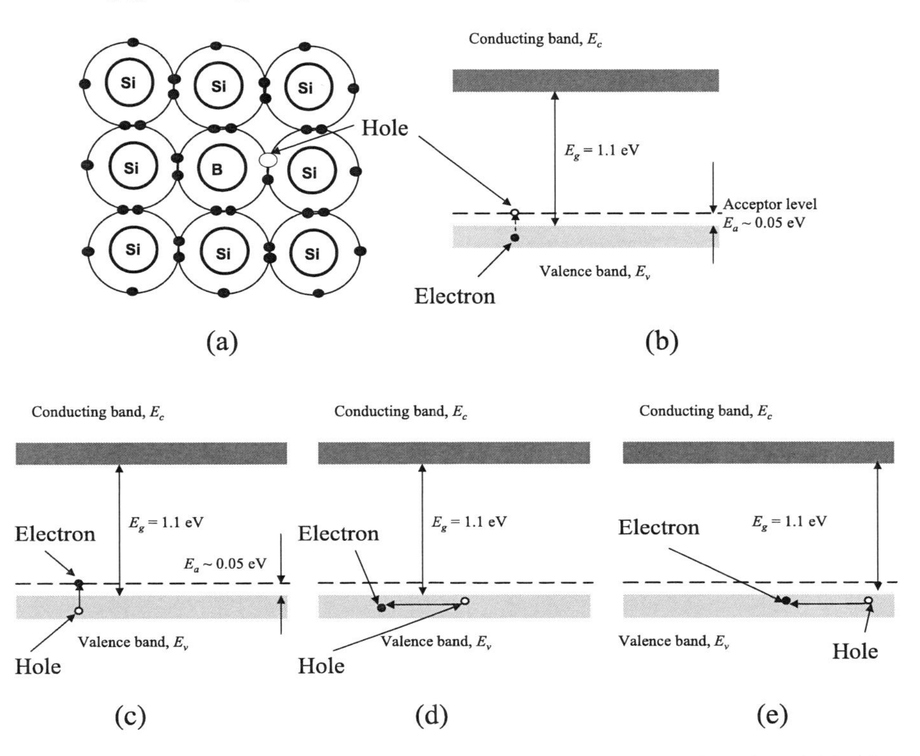

Figure 3.6 (a) p-type (boron) doped silicon, (b) its acceptor energy band, and (c), (d), and (e) illustrations of hole movement.

3.6(d), and 3.6(e) show that electrons in the valence band can easily jump to the acceptor energy band, creating holes in the valance band. In the electric field, other electrons in the valance band move and jump into these holes, while creating new holes where these electrons originated, allowing other electrons to jump in. The subsequent hole movement carries an electric current just like the movement of positive charges. A semiconductor with holes as majority carriers is a p-type dopant (p stands for positive). Boron is the main p-type dopant used in the IC industry.

3.1.4 Dopant concentration and conductivity

Figure 3.7 shows the relation between silicon resistivity and dopant concentration. From it we can see that the higher the dopant concentration, the lower the resistivity. This is because the more dopant atoms there are in silicon, the more carriers (electrons or holes, determined by type of dopant) they will provide to conduct electric current.

Figure 3.7 also shows that at the same dopant concentration, phosphorus-doped silicon (n-type) has lower resistivity than boron-doped silicon (p-type). This is because electrons move significantly faster in the conducting band than holes move in the valance band.

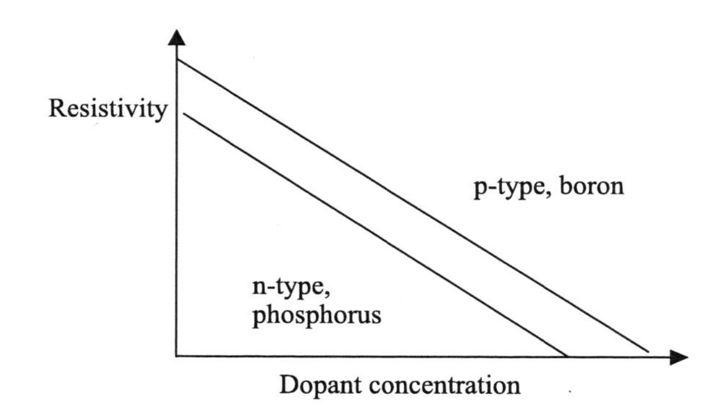

Figure 3.7 Dopant concentration and resistivity for silicon.

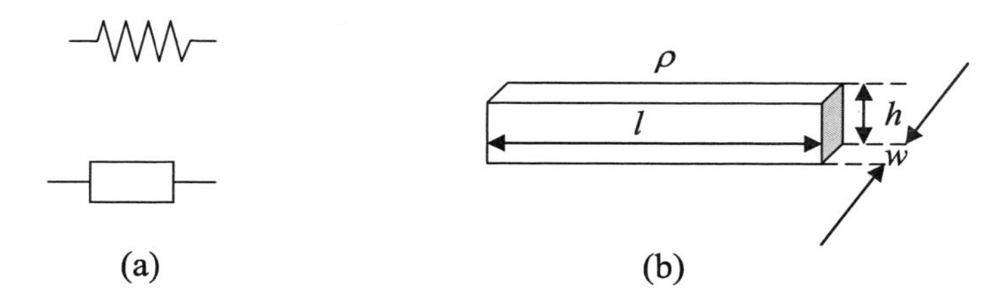

Figure 3.8 (a) Symbols and (b) structure of a resistor.

3.1.5 Summary of semiconductors

- Semiconductors have conductivity between conductors and insulators.
- Semiconductor conductivity can be controlled by dopant concentration: the higher the dopant concentration, the lower the semiconductor's resistivity.
- Holes form the majority carriers in p-type semiconductors. Boron is a p-type dopant.
- The majority carriers of n-type semiconductors are electrons. Phosphorus, arsenic, and antimony are p-type dopants.
- At the same dopant concentration and temperature, n-type semiconductors have lower resistivity than p-type semiconductors because their electrons have higher mobility than their holes.

3.2 Basic Devices

The basic devices in an IC chip are a resistor, capacitor, diode, transistor, and MOSFET.

3.2.1 Resistor

A resistor is the simplest electronic device. Its symbol in an electronic circuit and its structure are shown in Fig. 3.8(a).

Resistance of a resistor is:

$$R = \rho \frac{l}{wh}, \tag{3.1}$$

where R is the resistance, ρ is the resistivity, and w, h, and l are the conductor line width, height, and length, respectively [see Fig. 3.8(b)]. This equation is used again in Chapter 11.

In the past, patterned doped silicon was used to make resistors, with resistance determined by the length, line width, junction depth, and dopant concentration. Currently, doped polysilicon lines are used to make resistors on an IC chip, with resistance determined by the height, width length, and dopant concentration of the poly line.

Question: Many fabs use polysilicon to form gates and local interconnections. Resistivity of polysilicon is determined by dopant concentration, which is very high, about 10^{22} cm^{-3}, and $\rho \sim 200\ \mu\Omega \cdot \text{cm}$. Assuming that the polysilicon gate and local interconnection line width, height, and length are 1, 1, and 100 μm, respectively (note: 1 μm = 10^{-6} m = 10^{-4} cm), what is the resistance?

Answer: $R = \rho \frac{l}{wh} = 200 \times \frac{100\times10^{-4}}{10^{-4}\times10^{-4}} = 2 \times 10^{8}\ \mu\Omega = 200\ \Omega.$

Question: From the late 1980s to the late 1990s, the minimum feature size (gate width) shrank from 1 to 0.25 μm. What is the resistance of a polysilicon line with width, height, and length equal to 0.25, 0.25, and 25 μm, respectively?

Answer: $R = \rho \frac{l}{wh} = 200 \times \frac{25\times10^{-4}}{0.25\times10^{-4}\times0.25\times10^{-4}} = 8 \times 10^{8}\ \mu\Omega = 800\ \Omega.$

It is evident that while device dimension shrinks, resistance increases! It is very important to keep resistance low, otherwise, it will slow down the device speed and increase power consumption and heat generation. Therefore, it is necessary to use a better conducting material such as silicide, which has a much lower resistivity, from 13 to 50 $\mu\Omega \cdot \text{cm}$. In advanced CMOS IC chips, fabs normally use a polysilicon-silicide stack, also known as polycide, to form the gate and local interconnections.

Question: At one time, aluminum-copper alloy was the most commonly used material for metal interconnection in the IC industry. Assuming the resistance of aluminum-copper alloy $\rho \sim 3.2\ \mu\Omega \cdot \text{cm}$, and metal line width, height, and length are 1, 1, and 100 μm, respectively, what is the resistance of the metal interconnection?

Answer: $R = \rho \frac{l}{wh} = 3.2 \times \frac{100\times10^{-4}}{10^{-4}\times10^{-4}} = 3.2 \times 10^{6}\ \mu\Omega = 3.2\ \Omega.$

Question: Copper has been used for metal interconnections in semiconductor manufacturing since the late 1990s. Copper has a significantly lower resistivity than aluminum copper alloy, $\rho = 1.7\ \mu\Omega \cdot \text{cm}$. What is the resistance of a copper line with the same geometry as the first question?

Answer: $R = 1.7\ \Omega.$

By using copper for metal interconnection, resistance can be reduced by almost half, which can be translated into higher device speed, lower power consumption, and less heat generation.

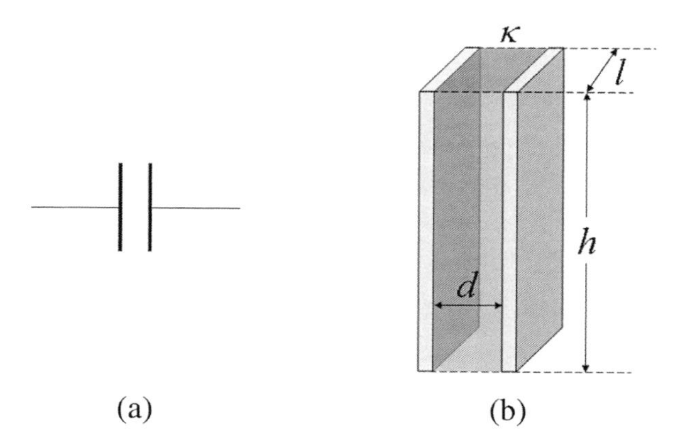

Figure 3.9 (a) Symbol and (b) structure of a capacitor.

3.2.2 Capacitor

The capacitor is one of the most important components in an IC chip, especially for a dynamic random access memory (DRAM) chip. Its symbol in an electronic circuit and its structure are shown in Fig. 3.9.

Capacitors are very important for DRAM chips that store electrical charges and maintain memory. When two conducting surfaces are separated by a dielectric, a capacitor is formed, and its capacitance can be expressed as

$$C = \kappa\varepsilon_0 \frac{hl}{d}, \tag{3.2}$$

where C is the capacitance of the capacitor, h and l are the conducting plate height and length, respectively, and d is the distance between the two parallel conducting plates. $\varepsilon_0 = 8.85 \times 1010^{-12}$ F/m is the absolute permittivity of a vacuum, and κ is the dielectric constant of the dielectric material between the two parallel plates. (In most academic disciplines, a dielectric constant is denoted as ε, but since this text is orientated toward industrial work, we use κ instead of ε.)

In an IC chip, the conductor part of the capacitor is mainly made with polysilicon. Dielectric materials vary from silicon dioxide and silicon nitride to more recent high-k dielectrics such as titanium dioxide (TiO_2), hafnium dioxide (HfO_2), and aluminum oxide (Al_2O_3). The purpose of using a high-κ dielectric is to shrink capacitor size while keeping the same capacitance. Capacitors can be made in planar, stack, and deep trench styles, as shown in Fig. 3.10. Stack and deep trench styles are favored by DRAM manufacturers.

Question: What is the capacitance for a capacitor with the same structure as Fig. 3.9(b) with $h = l = 10$ μm? (The dielectric between two conducting plates is silicon dioxide, with $\kappa = 3.9$, and $d = 1000$ Å.)

Answer:

$$C = \kappa\varepsilon_0 \frac{hl}{d} = 3.9 \times 8.85 \times 10^{-12} \times \frac{10 \times 10^{-6} \times 10 \times 10^{-6}}{1000 \times 10^{-10}}$$

$$= 3.45 \times 10^{-14}\text{F} = 34.5 \text{ fF}.$$

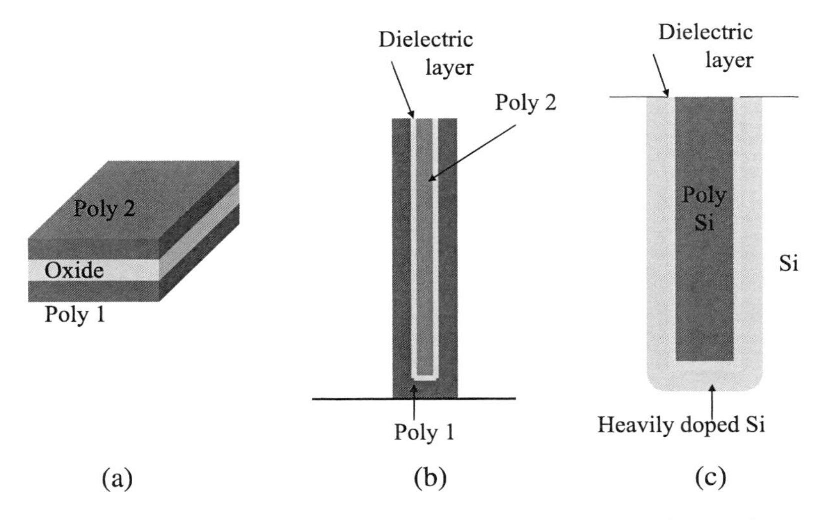

Figure 3.10 (a) Parallel plate, (b) stack, and (c) deep trench capacitors.

Question: By reducing the distance d between the two plates, capacitor dimensions h and l can be reduced while keeping the same capacitance. For $h = l = 1\ \mu\text{m}$, what is d using the same capacitance as the first question?

Answer:

$$d = \kappa\varepsilon_0 \frac{hl}{C} = 3.9 \times 8.85 \times 10^{-12} \times \frac{10^{-6} \times 10^{-6}}{3.45 \times 10^{-14}} = 10^{-9} m = 10\ \text{Å}.$$

Unfortunately, a 10 Å silicon dioxide layer is too thin to work reliably, even at 1 V, because the dielectric strength of SiO_2 is about 10^7 V/cm, equivalent to 0.1 V/Å. Therefore, a high-κ dielectric such as HfO_2 ($\kappa \sim 25$) or ZrO_2 ($\kappa \sim 25$) is needed to reduce capacitor dimension while keeping a certain value of capacitance and preventing dielectric breakdown.

Question: What is the required dielectric constant κ for the capacitor in the previous question with $d = 100$ Å?

Answer: $\kappa = \dfrac{Cd}{\varepsilon_0 hl} = \dfrac{3.45 \times 10^{-14} \times 100 \times 10^{-10}}{8.85 \times 10^{-12} \times 10^{-6} \times 10^{-6}} = 39.$

On an IC chip, metal interconnection lines (shown in Fig. 3.11) form unwanted capacitors, called parasite capacitors. Currently, IC device speed is mainly limited by an RC (resistance multiplied by capacitance) time delay, which is the time electrons need to charge up the parasite capacitors and metal line resistance. Using low-κ dielectric materials and a better conduction metal (such as copper) can reduce RC time delay and increase device speed.

The time it takes for charges to reach the capacitor is about $t = Q/I$ for the first order of approximation. Here, charge $Q = CV$ is the amount of charge needed to

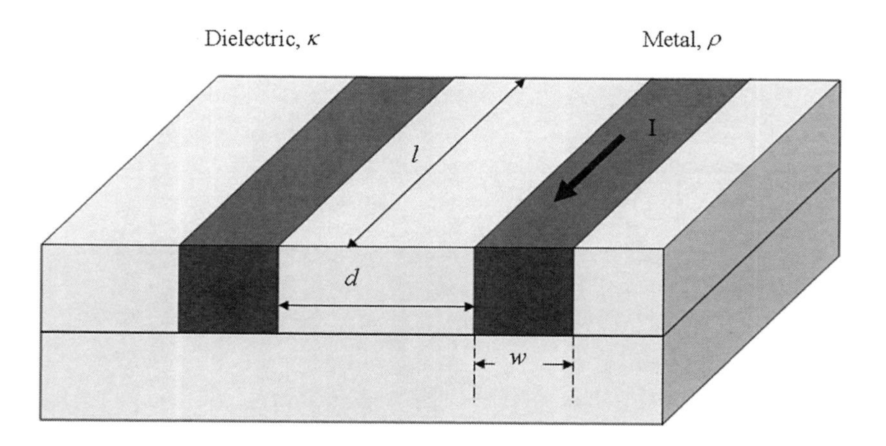

Figure 3.11 Metal interconnection.

fill the capacitor, V is the voltage drop on the metal line, and $I = V/R$ is the current in the metal line. Therefore, time delay $t = CV/(V/R) = \text{RC}$. An electrical signal with a frequency higher than 1/RC cannot pass through the metal interconnection lines. To achieve high device speed, it is necessary to reduce RC time.

Question: Most IC chips use aluminum-copper alloy metal interconnections. Assume resistivity $\rho = 3.2\ \mu\Omega \cdot \text{cm}$, and metal line geometry width w, height h, length l, and line spacing d are 1 μm, 1 μm, 1 cm, and 1 μm, respectively. CVD silicon oxide is between the metal line with dielectric constant $\kappa = 4.0$. What is the time delay RC?

Answer:

$$\text{RC} = \rho\frac{l}{wh}\kappa\varepsilon_0\frac{hl}{d} = \rho\kappa\varepsilon_0\frac{l^2}{wd} = 3.2\times10^{-8}\times4.0\times8.85\times10^{-12}$$
$$\times\frac{0.01^2}{10^{-6}\times10^{-6}} = 1.133\times10^{-10}\,\text{sec}.$$

Therefore, an IC chip with this kind of interconnection cannot operate for frequencies higher than 1/RC = 8.83 GHz. To increase circuit speed, the RC time delay needs to be reduced by changing the interconnection line geometry and/or changing the conducting and dielectric materials. One way is to reduce l, which requires shrinking the device dimension. Another way is to increase w and d, which means more layers of metal interconnection. State-of-the-art technology in the late twentieth century employed more than nine layers of metal interconnection. Another option is to reduce both ρ and ε, by using a better conducting metal (such as copper) to replace the commonly used Al-Cu alloy, and employing low-κ dielectric materials to replace commonly used silicate glass.

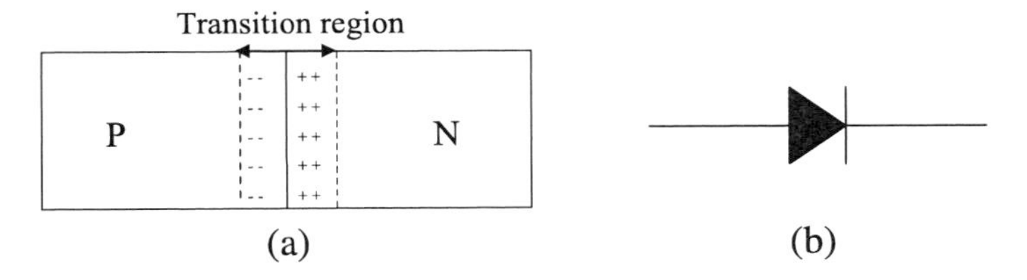

Figure 3.12 (a) p-n junction diode and (b) its symbol.

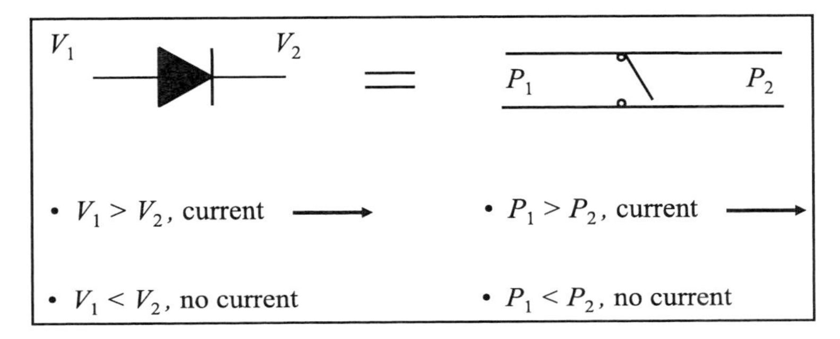

Figure 3.13 Mechanism of a diode and a one-way flow value.

3.2.3 Diode

Unlike the resistor and capacitor, a diode is a nonlinear device, so the response of current to voltage is no longer a linear relation. Figure 3.12 shows a p-n junction diode and its symbol.

A diode allows electric currents to pass through in only one direction when the voltage is forward biased. It does not allow currents to pass through if the applied potential is reverse biased. When applied voltage V_1 is higher than V_2, it is called forward bias, as the electric current goes through the diode (in the direction marked by the symbol in Fig. 3.13) with little resistance. When V_1 is lower than V_2, it is called reverse bias, as the resistance is very high and almost no electric current can pass through the diode. The mechanism is similar to a tire valve. When you pump air into the tire, and pump pressure P_1 is higher than tire pressure P_2; the pressure opens the valve and air flows into the tire. Without the pump, tire pressure P_2 is higher than atmospheric pressure P_1; therefore, the pressure difference closes the valve, and air stays inside the tire (see Fig. 3.13).

When p- and n-type semiconductors join together, they form a p-n junction diode, as shown in Fig. 3.14. Holes in the p-type region will diffuse to the n-type region, and electrons in the n-type region will diffuse to the p-type region. The charge separation causes electrostatic force, which eventually stops the minority carrier diffusion. The area dominated by minority carriers is called the transition region.

Voltage crossing the transition region can be expressed as

$$V_0 = \frac{\kappa T}{q} \ln \frac{N_a N_d}{n_i^2}, \tag{3.3}$$

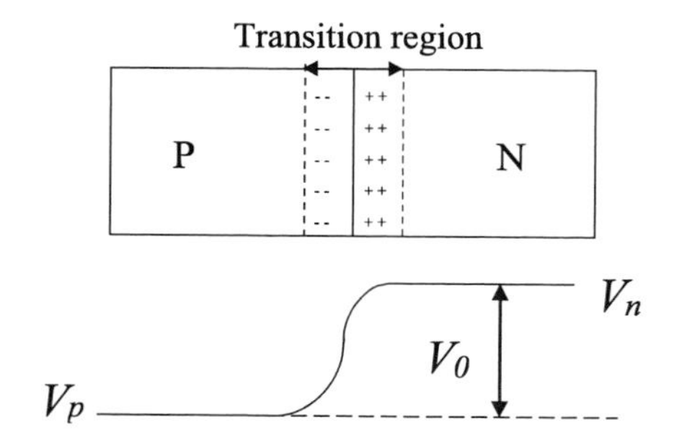

Figure 3.14 A p-n junction diode and its intrinsic voltage.

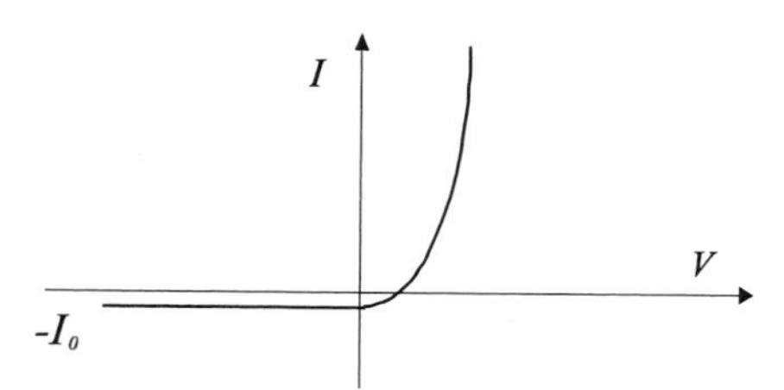

Figure 3.15 I-V curve of a diode.

where κ is the Boltzmann constant, T is the temperature, q is the charge, N_a is the acceptor (p-type dopant) concentration, N_d is the donor (n-type dopant) concentration, and n_i is the intrinsic carrier concentration.

For silicon at room temperature, $\kappa T/q = 0.0259$ V, $n_i = 1.5 \times 10^{10}$ cm^{-3}, and if $N_a = N_d = 10^{16}$ cm^{-3}, we can calculate that $V_0 \sim 0.7$ V. From Eq. (3.1), we can see that V_0 is not very sensitive to the dopant concentrations. Therefore, it always requires about 0.7 V of forward bias to flow current through a p-n junction. For the same reason, it requires about 0.7 V to turn on a bipolar transistor.

The I–V curve of a diode is illustrated in Fig. 3.15. When the diode is forward biased and V is larger than 0 (especially for $V > 0.7$ V), the current I going through the diode increases exponentially with V with very little resistance. When the diode is reverse biased, the current going through it is very low. It is almost a constant: I_0 when the voltage increases. Resistance in this case is very high. If the reverse bias voltage is too high, it could irreversibly break down the diode and cause it to no longer have its characteristic I–V curve.

3.2.4 Bipolar transistor

The symbols and bar structure of npn and pnp bipolar transistors are shown in Fig. 3.16.

The first transistor made in the Bell Laboratories was a point-contact bipolar transistor. Most bipolar transistors made in IC chips today are planar junction transistors, as shown in Fig. 3.17.

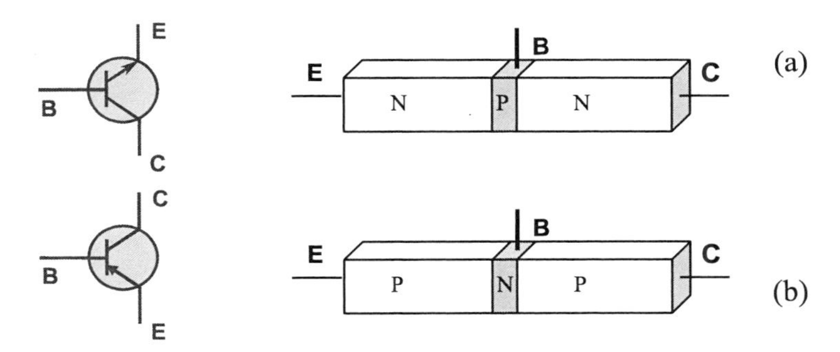

Figure 3.16 Symbols and bar structure of (a) npn and (b) and pnp bipolar transistors. E stands for emitter, B for base, and C for collector.

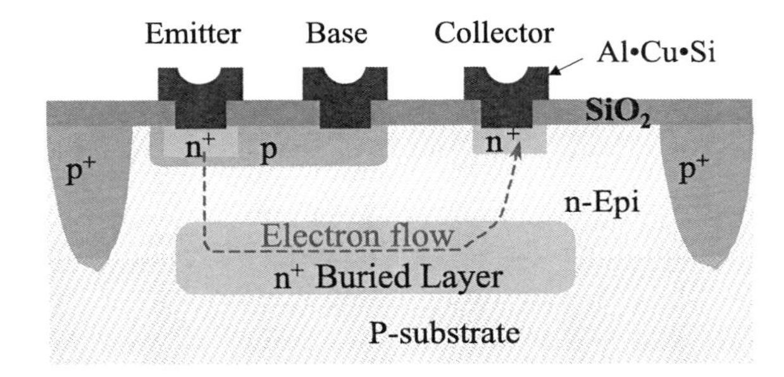

Figure 3.17 Cross section of a conventional npn planar bipolar transistor.

A bipolar transistor can be used as a switch because its emitter-to-collector current can be controlled by a base-to-emitter bias. For an npn transistor, when the base emitter is forward biased and $V_{BE} > 0.7$ V, electrons in the emitter can overcome the emitter-base n-p junction barrier potential, inject into the base, sweep across the thin base layer, and reach the collector.

When $V_{BE} = 0$ V, there is no electron injection from the emitter. Thus, there is no current between the emitter and collector, no matter how biased it is between the collector and emitter.

Most fabs use bipolar transistors for their ability to amplify an electric signal. Under the right conditions, a current between the emitter and collector equals β times the current flow into the base, or $I_{EC} = \beta I_B$, where β is the amplifying coefficient, normally between 30 to 100.

Figure 3.18 shows a cross section of an advanced sidewall base contact npn bipolar transistor. Bipolar transistors are mainly used for high-speed devices, analog circuits, and high-power devices.

From 1950 to 1980, bipolar transistors and bipolar-transistor-based IC chips were dominant in the semiconductor industry. After 1980, driven by the demands of logic circuits (especially demands for DRAM), MOS-transistor-based IC chips developed rapidly and surpassed bipolar-based IC chips. Today, CMOS-based IC chips dominate the semiconductor industry.

3.2.5 Metal-oxide-semiconductor field effect transistor

The symbols for n-channel MOS (nMOS) and p-channel MOS (pMOS) are shown in Figs. 3.19(a) and 3.19(b), respectively.

An nMOS structure is shown in Fig. 3.20. It has a conducting gate, silicon substrate, and a very thin layer of gate dielectric, usually silicon dioxide, sandwiched in between. For nMOS, the substrate is p-type, and the source and drain are heavily doped with an n-type dopant (which is why they are noted as n^+). Source and drain are symmetric; normally, the grounded side is called the source and the biased side is called the drain.

When no bias voltage is applied to the gate, no matter how the source/drain is biased, no current can pass through from source to drain, or vice versa. When the gate is biased positively, it generates positive charges at the metal-oxide surface. The gate dioxide is a very thin dielectric layer between the metal and semiconductor. Like a capacitor, the positive charge on the metal-oxide surface expels positive charges (holes, majority carriers) from the silicon-oxide surface and attracts negative charges (electrons, minority carriers) to that surface. When the gate voltage is higher than the threshold voltage, $V_G > V_T > 0$, the silicon-oxide surface accumulates enough electrons to form a channel and allow electrons from the source and drain to flow across it. This is why nMOS is also called an n-channel MOSFET or nMOSFET. This process is illustrated in Fig. 3.21. By controlling gate voltage, the electric field will affect the conductivity of the semiconductor device and switch the MOS transistor on and off. This is the reason it is called a field effect transistor (FET).

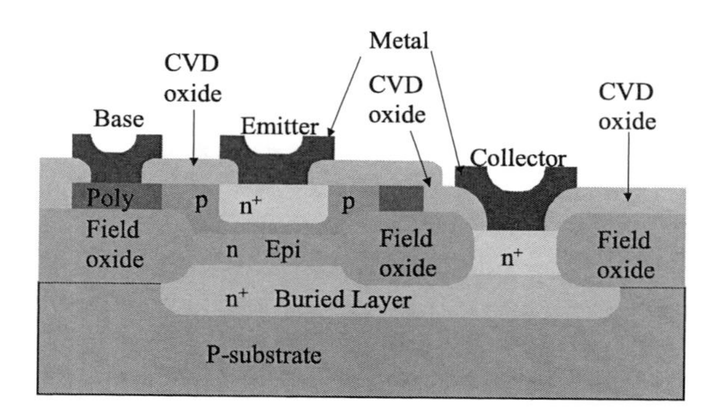

Figure 3.18 Sidewall base contact npn bipolar transistor.

(a) (b)

Figure 3.19 Symbols for (a) nMOS and (b) pMOS.

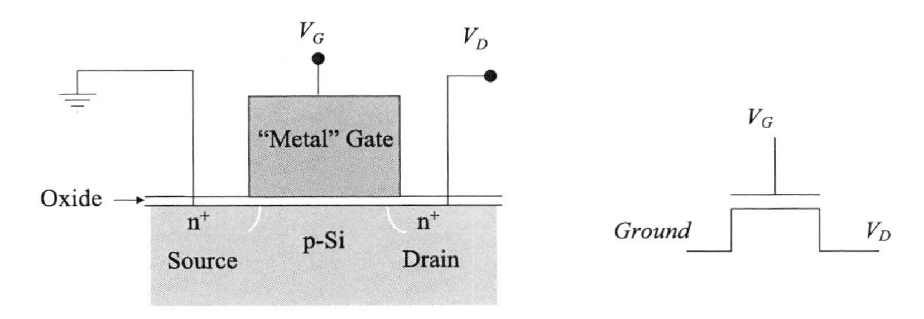

Figure 3.20 An nMOS transistor.

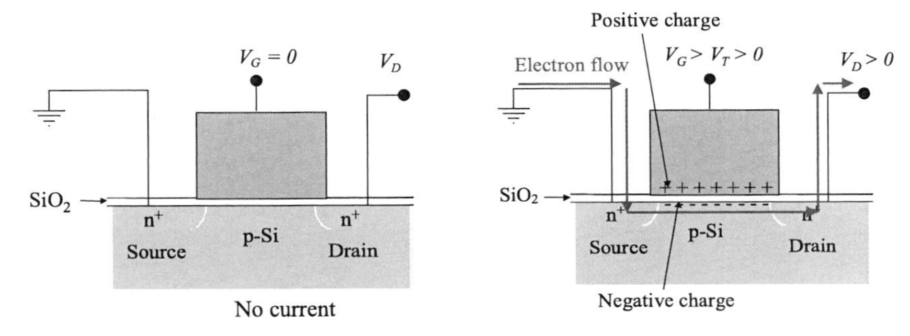

Figure 3.21 Switching process of the nMOS.

For a pMOS, the substrate is an n-type semiconductor, and the source and drain are heavily p-type doped. The pMOS uses a negative gate bias to generate negative charges (electrons) on the metal-oxide surface. This in turn drives away electrons (majority carriers of the substrate) and attracts holes (minority carriers) to the silicon-oxide surface to form a hole-channel underneath the gate. Holes from the source and drain then flow through the channel and conduct electric current between the source and drain (see Fig. 3.22). With positive bias, the pMOS transistor turns off.

By adjusting the dopant concentration of the substrate with ion implantation underneath the gate, the threshold voltage V_T can be adjusted. Therefore, the MOSFET can be set as normal-on and normal-off. In a CMOS circuit, both nMOS and pMOS are used; usually nMOS is set as normal-off, and pMOS is set as normal-on.

MOSFETs are mainly used for logic circuits such as microprocessors and memory chips. Figure 3.23 shows a tunneling electron microscope (TEM) image of a MOSFET manufactured with 32-nm technology by Intel Corporation.

The idea of the FET was proposed in 1925. In their effort to create it, Bell Laboratories scientists unintentionally made the first point-contact bipolar transistor on a piece of polycrystalline germanium. Lack of single-crystal semiconductor materials hampered early efforts to develop the FET. The inventions of single-crystal germanium in 1950 and especially single-crystal silicon in 1952 changed that situation. Eventually a group led by M. M. Atalla at Bell Laboratories made the first practical MOSFET in 1960. MOSFET technology developed rapidly after that.

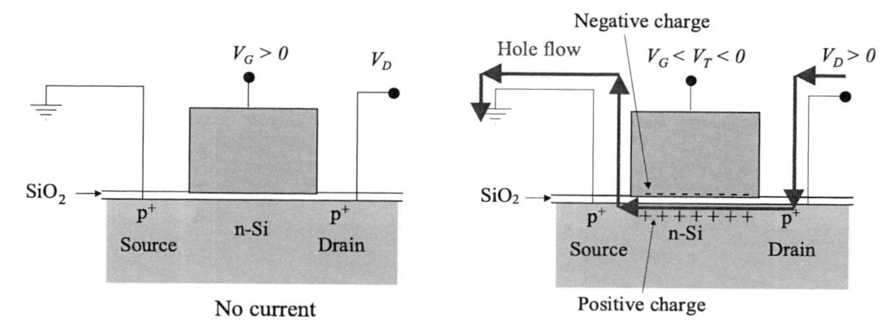

Figure 3.22 Structure and switching process of a pMOS.

Figure 3.23 TEM image of a MOSFET manufactured with 32-nm technology (Chipworks).

Beginning in the 1980s, IC chips based on MOS transistors dominated the semiconductor industry. Since most IC chips are MOS-based, this text focuses mainly on the MOS IC processes.

In addition to silicon and germanium, there are many compound semiconductors. Gallium arsenate (GaAs) is commonly used to make high-frequency, high-speed IC devices for electronics instruments of defense and scientific applications. Other important applications of compound semiconductors are light-emitting diodes (LEDs), which are used in almost every electronic product as indicators and are widely used in traffic lights. Because of their highly efficient light emission and long lifetimes, it is possible that white LED lights will replace incandescent light bulbs and fluorescent lighting in homes and businesses in the future, when mass production brings down the cost of LEDs.

3.3 Integrated Circuit Chips

There are many different kinds of IC chips made in the world. They can be categorized into three groups: memory, microprocessor, and application-specific integrated circuits (ASICs).

3.3.1 Memory

Memory chips can memorize digital data by storing and discharging electrical charges. They are widely used in computers and other electronic devices for data storage. The largest portions of chips manufactured are memory chips, notably DRAM and NAND flash.

3.3.1.1 Dynamic random access memory

DRAM stands for dynamic random access memory. Random access indicates that each memory cell in the chip can be accessed to read or write in any order, in contrast to sequential memory devices where data have to be read or written in a certain order. For instance, hard disks and compact disks employ random access, while a cassette tape employs sequential access.

DRAM is the most commonly used memory chip, particularly inside computers for data storage. When people purchase a personal computer, one important factor is how many bytes of memory it has—that is, the data storage capability the DRAM chips have. The basic memory cell of a DRAM is formed by an nMOS and a capacitor, as shown in Fig. 3.24.

The n-type MOSFET serves as a switch. It allows electrons to flow and become stored in the capacitor, which holds the memory. The capacitor needs to be recharged periodically by the power supply V_{dd} to compensate for electron losses. That is why it is called dynamic RAM or DRAM. When power is removed from a DRAM, data are lost. While editing a file on a computer, all of the input materials, including words, graphs, and symbols, are stored in the DRAM main memory of the computer before the "save" command writes them permanently onto a hard disk or a universal serial bus (USB) flash drive.

The demand for more memory and faster memory chips in computer-related applications is one of the most important driving forces of technology development within the IC industry.

3.3.1.2 Static random access memory

SRAM stands for static random access memory and uses latching transistors to keep the instruction or data memory. It requires at least six components, either four transistors and two resistors or six transistors. Figure 3.25 shows the circuit of a six-transistor SRAM. We can see that it has four nMOS and two pMOS. SRAM

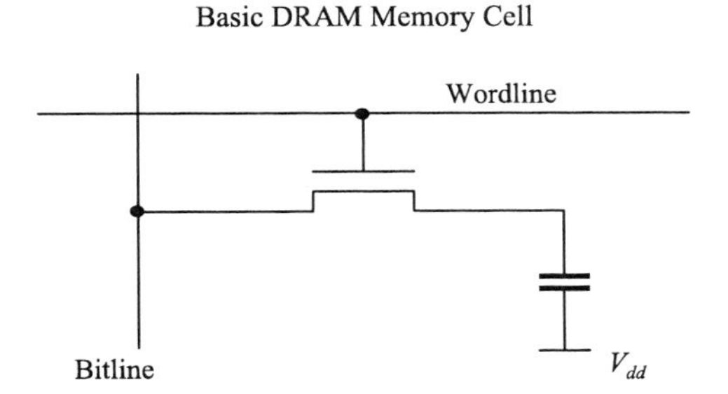

Figure 3.24 Circuit of a DRAM memory cell.

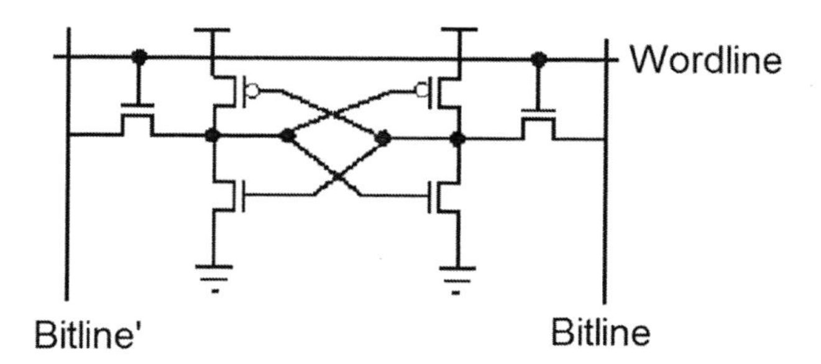

Figure 3.25 Circuit of a six-transistor SRAM.

is much faster than DRAM, since it does not need to recharge the capacitor to store data. However, for the same memory with the same processing technology, a SRAM chip is much larger and more expensive than a DRAM chip. SRAM is mainly used in computers as cache memory to store the most frequently used instructions. DRAM is used as the main memory to store less frequently used instructions and data. Flash drives and hard disk drives (HDDs) are nonvolatile memories, which are commonly used to permanently store the data files.

Most logic IC chips such as microprocessors, ASIC chips, and field programmable gate array (FPGA) chips have embedded cache memory, which integrates SRAM into their chip design. In these chips, SRAM arrays always have the highest device density. Almost all logic IC fabs use SRAM in their test vehicles for technology development and qualification.

3.3.1.3 Erasable programmable read-only memory, electric erasable programmable read-only memory, and flash

DRAM and SRAM need a power supply to keep the data. Without a power supply, the memory will be lost, thus it is called volatile memory. EPROM stands for erasable programmable read-only memory, and EEPROM stands for electric erasable programmable read-only memory. Both EPROM and EEPROM have nonvolatile memory. Their main applications are for permanent storage of data and instructions without power supply.

The structure of an EPROM cell is similar to an nMOS, as shown in Fig. 3.26. The fundamental difference is that it has a floating gate, which connects to nowhere and can permanently store electrons as digital data. By biasing the control gate with $V_G > V_T > 0$, electrons are attracted to the silicon–gate oxide interface and form an n-type channel underneath the floating gate. Electrons from the source and drain pass through the channel and conduct electric currents. While most electrons pass through the channel, some electrons are intentionally injected into the floating gate through the thin gate oxide by the electron tunneling effect, known as the hot electron effect. When electrons are injected into the floating gate, they can stay there for years because they have no way to escape. Thus, the gate can keep that memory, regardless of whether it is powered on or off. The memory writing process is shown in Fig. 3.27(a).

One can erase the memory from EPROM by shining UV light through the passivation dielectric. The UV light excites the electrons in the floating gate,

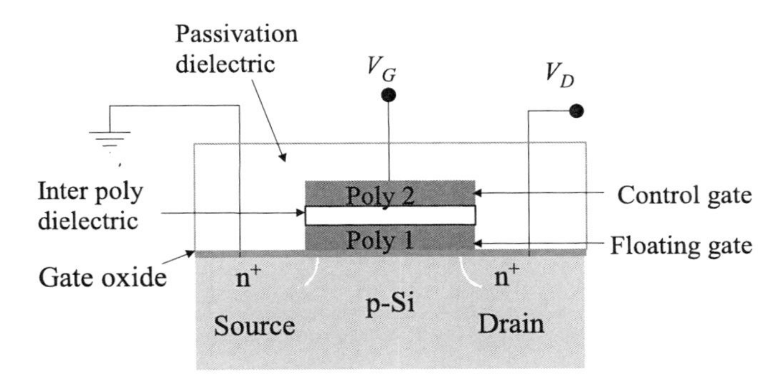

Figure 3.26 Cross section of an EPROM memory cell.

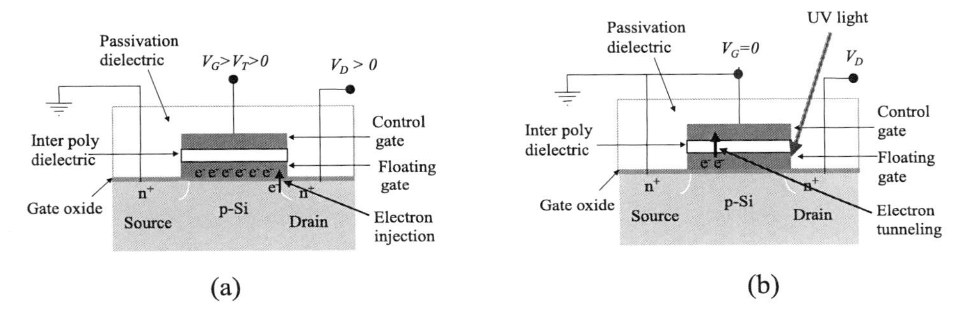

Figure 3.27 EPROM (a) programming and (b) erasing processes.

causing them to tunnel through the interpoly dielectric and drain to the ground at the control gate [see Fig. 3.27(b)].

Different from EPROM, which uses UV light to erase memory from the entire chip, EEPROM erases memory from individual cells electrically by biasing the control gate and causing electron tunneling from the floating gate to the control gate.

Flash memory is a specific EEPROM that costs significantly less to manufacture. Figure 3.28(a) shows a 64-bit NAND flash circuit. WL stands for word line, of which there are 64, numbered from 0 to 63. SG stands for select gate, which is an nMOS transistor. SG0 is the nMOS to select the string to the ground (source line), and SG1 is the nMOS to allow the string to select to the bit line. Figure 3.28(b) illustrates the cross section of a 64-bit NAND flash along the bit line direction. 64-bit NAND has been used in mass production in advanced NAND flash fabs with 19-nm technology.

NAND flash has been widely used in USB flash drives, solid state disks (SSDs), memory cards for mobile phones, digital cameras, digital camcorders, and other mobile devices. SSDs are faster and more reliable and also consume less power than HDDs. For example, a 256-Gb, 2.5-in. notebook SSD consumes about 20% of the power of a 250-Gb HDD of the same size. SSDs have been used in high-end mobile electronics such as notebook computers to replace HDDs for high volume data storage. The main advantage of HDDs is the cost, which is one of the most important factors for consumer electronics. For the same price as an SSD,

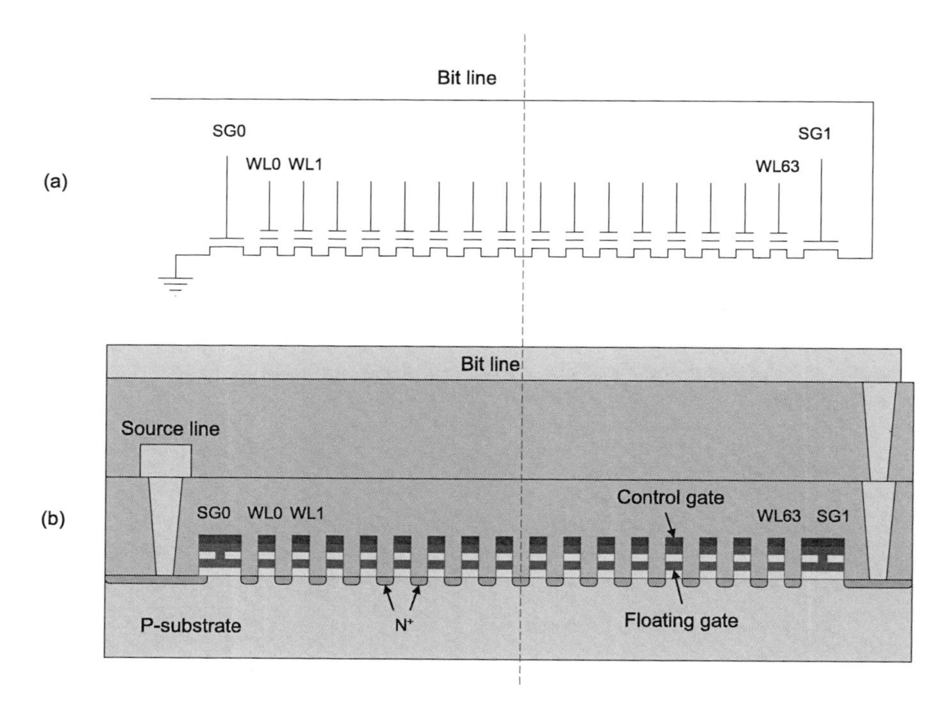

Figure 3.28 (a) Circuit of a 64-bit NAND flash and (b) the illustration of its cross section in the bit line direction.

a person can buy an HDD with more than ten times the storage. With the shrinking of feature size, it is possible that the cost per byte of flash memory could be reduced to a level comparable to a HDD. However, with the feature size shrinking, device standby leakage will increase. Also, as more devices are packed into one SSD with smaller feature size, power consumption of SSDs could increase rapidly. If flash memory can solve the leakage-induced power consumption issue, in the future, SSDs could replace HDDs for data storage in most computers and most other digital electronics, especially mobile devices. Otherwise, flash memory would only play a supporting role in a hybrid disk drive that integrates both SSD and HDD.

3.3.2 Microprocessor

The microprocessor is also called a central processing unit (CPU). It has two components: a control system and an arithmetic-logic unit (ALU). Most microprocessors also have a built-in cache memory. It is the brain of the computer and other control systems. Two kinds of microprocessor architectures are widely used: complete instruction set computers (CISCs) and reduced instruction set computer (RISCs).

A graphics processing unit (GPU) is a processor specially designed for graphics processing with high-speed floating point computation. It can free up the main CPU from computation-intensive 3D graphics processing and is widely used in embedded systems, personal computers (PCs), mobile phones, and game consoles.

3.3.3 Application-specific integrated circuits

Many chips belong in the ASIC category, including digital signal processing (DSP) chips, power devices, and chips for TV, radio, Internet, automobiles, wireless communications, telecommunications, etc. As the feature size shrinks, the cost of photomasks required for IC chip manufacturing increases dramatically. For high-volume products such as microprocessors and memory chips, the cost of the masks can be shared by thousands of wafers. However, ASIC chips with limited production volume have to face the tough challenge of cost effectiveness. FPGA chips, which allow users to configure system applications after chip manufacturing, become attractive alternatives.

3.4 Basis Integrated Circuit Processes

IC processing technology began with Kilby's germanium bar and then quickly switched to a single-crystal silicon chip with planar technology, first developed by Noyce. After 1960, the IC industry developed exponentially at a rate that matches well with Moore's law.

Bipolar-transistor-based IC chips dominated the semiconductor industry from the 1960s to the 1980s. Demands for electronic watches, calculators, computers, and other digital electronic products rapidly drove MOSFET-based IC process development and chip fabrication. The requirements for low-power-consumption circuits pushed the rapid development of CMOS IC chips. CMOS-based IC chips have dominated the IC industry since the 1980s. Many people predict that the semiconductor industry will keep the development pace described by Moore's law for the next ten years; then photolithography will become obsolete.

Basic IC processes can be divided into the following four categories:

- adding
- removing
- heating
- patterning.

Adding indicates that atoms are added into a (doped) wafer or a layer is added onto the wafer surface (thin-film growth or deposition). Ion implantation and diffusion processes are used to add dopants into the semiconductor substrate. In oxidation and nitridation processes, oxygen and nitrogen are added to chemically react with silicon to form silicon dioxide and silicon nitride. Thin-film depositions add thin layers of semiconductors, dielectric or metal via CVD, PVD, spin-on, or electrical plating processes.

Removing involves removing material from the wafer, either chemically, physically, or by a combination of both. Wafer cleaning, which uses a chemical solution to remove contaminants and particles from the wafer surface, is a removal process; patterned and blanket etching are as well. CMP, which removes materials from the wafer surface by both chemical reaction and mechanical grinding to achieve surface planarization, also belongs in this category.

In heating or thermal processes, wafers are heated to specific temperatures in a certain amount of time to achieve desirable physical or chemical results. Very little or no materials are added or removed from the wafer during heating processes, which include postimplantation annealing (high temperature > 1000 °C) and metal annealing (low temperature < 450 °C), alloying, and thermal reflow processes. Two kinds of thermal processes are commonly used in semiconductor manufacturing: conventional (furnace) process and rapid thermal process (RTP). The heating process is sometimes performed in situ, with oxidation and high-temperature deposition processes in a furnace or RTP chamber for oxide or thin-film annealing.

All adding, removing, and heating processes are employed in the patterning process, which transfers circuit-design layouts from the mask or reticle to photoresist on the wafer surface. The pattering process is the most frequently used process in IC fabrication, so it is very important. Most fabs use the photolithography process for pattern transferal in present-day IC chip manufacturing. Extreme ultraviolet (EUV) lithography, nanoimprint lithography (NIL), and electron beam direct write (EBDW) lithography could be used in the future when minimum feature size becomes too small to use photolithography.

The rest of this section briefly summarizes the basic processing steps involved in IC manufacturing. Since the majority of IC chips manufactured are MOSFET-based IC chips, this text focuses on MOS, especially CMOS processing steps.

3.4.1 Conventional bipolar transistor process

The main processing steps for bipolar-transistor-based IC chip manufacturing are: buried layer doping, epitaxial silicon growth, isolation and transistor doping, interconnection, and passivation.

Silicon-based bipolar-transistor ICs normally use a p-type wafer to make npn transistors. The basic processing steps originated with technology from the mid-1970s (ion implantation for the doping process). The process consists of seven mask steps, starting with a p-type doped wafer. The buried layer is formed to reduce the series collector resistance and to improve device speed. Its processing steps include wafer cleaning, oxidation, photolithography, ion implantation, photoresist stripping, and silicon dioxide stripping. Epitaxial silicon growth is a high-temperature CVD process. During epitaxial growth, the implanted buried layer anneals and diffuses slightly, as shown in Figs. 3.29(a) and 3.29(b). After a thin layer of oxide grows, repeated photolithography and ion implantation processes define isolations, as well as the emitter, base, and collector of the transistor [Figs. 3.28(c) and 3.29(d)]. Next, the screen oxide is stripped, and a thick layer of SiO_2 is grown. Photolithography and oxide etch define the contact holes. After photoresist stripping, an aluminum alloy metal layer is deposited to fill the contact holes and cover the entire wafer. Photolithography and metal etch form the metal line interconnection and make contacts with transistors [Fig. 3.29(e)]. After stripping the photoresist, a layer of CVD silicon oxide is deposited to protect the transistors and metal wires [Fig. 3.29(f)]. The final processes are photolithography, bonding pad etching, and photoresist stripping.

More-advanced process technologies such as dielectric trench fill isolation, deep collector, selective epitaxy growth, and self-aligned emitter base have been developed and applied in bipolar-transistor-based IC chip fabrications. Since the main focus of this text is MOSFET-based IC, these topics are not covered in detail.

The main applications of bipolar-transistor ICs are in analog systems such as chips for TVs, sensors, and power devices. They are also used with CMOS to form BiCMOS ICs for a variety of applications, including some microprocessors.

3.4.2 p-Channel metal-oxide-semiconductor process (1960s technology)

In the early stages of the IC industry, most fabs made bipolar-transistor-based IC chips. When they started to make MOSFET-based IC chips, they used p-channel MOSFET or pMOS because of the limitations of technology. Although it has exactly the same design (same gate material, geometry, substrate dopant concentration, and source/drain dopant concentration), nMOS is significantly faster than pMOS because electron mobility is two to three times higher than holes. However, difficulties associated with the nMOS process could not be solved

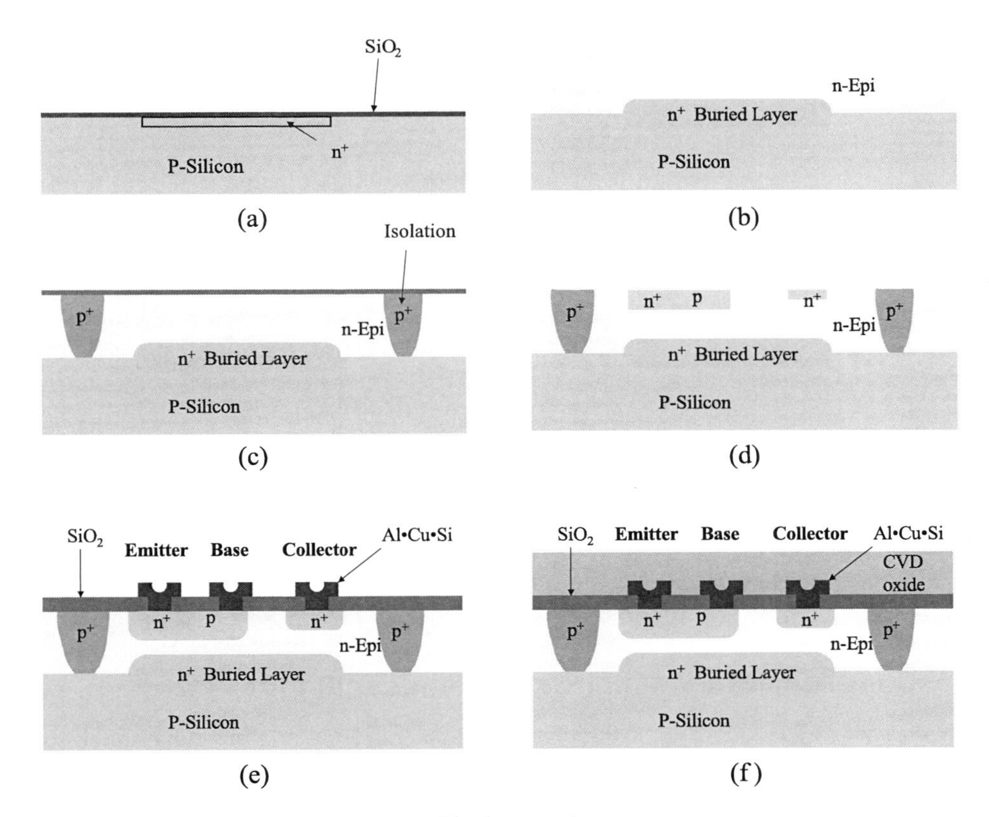

Figure 3.29 Bipolar-transistor process.

without ion implantation technology. It was much easier and more cost effective to make pMOS than nMOS using diffusion techniques for silicon doping.

A pMOS process flow with technology from the 1960s is given in Figs. 3.30 and 3.31 and Table 3.1. It uses blanket-field oxide for isolation, boron diffusion for source/drain doping, and aluminum-silicon alloy for gate and interconnection. About 1% of silicon was added into the aluminum to saturate the dissolving of the silicon, preventing junction spiking of the aluminum caused by the silicon dissolution. Minimum feature size was about 20 μm.

The entire pMOS process has five masking steps. Every masking step is a photolithography step, including wafer cleaning, prebake, primer application, photoresist coating, soft bake, alignment and exposure, development, pattern inspection, and hard bake. From the process flow it is evident that the IC process always repeats the removing, adding, pattering, and heating processes.

3.4.3 n-Channel metal-oxide-semiconductor process (1970s technology)

IC processing technology changed significantly after the introduction of ion implantation into mass production in the mid-1970s. Ion implantation processing replaced diffusion processing for silicon doping because it offered the advantages of independently controlled dopant concentration and junction depth, as well as an anisotropic doping profile. Polysilicon replaced aluminum as the gate material and local interconnection. It was used to form self-aligned sources and drains by taking advantage of the anisotropic profile of ion implantation and the high-temperature stability of polysilicon. Minimum feature size was about 7.5 μm. Figure 3.31 shows the self-aligned source/drain implantation process; this technique is still used in advanced MOSFET IC chip manufacturing.

Phosphorus silicate glass (PSG) was used as a (premetal dielectric) PMD material. PSG can trap mobile ions such as sodium and prevent them from diffusing to the gates and damaging the MOSFETs. Its use represents one of the major IC technology breakthroughs of the late 1960s. PSG can flow thermally at 1100 °C, at which temperature it can smooth and planarize the dielectric surface. A planarized surface is favorable for subsequent metallization and photolithography processes. A thin layer of undoped silicate glass (USG) was used as a barrier layer between the source/drain (S/D) and PSG, aluminum-copper-silicon alloy was used as a global interconnection, and CVD nitride was used for passivation dielectrics.

After the mid-1970s, IC manufacturers made nMOS-based IC chips because nMOS is faster than pMOS for the same geometry and dopant concentrations. An nMOS process flow is given in Table 3.2 and shown in Fig. 3.32.

3.5 Complementary Metal-Oxide Semiconductor

Starting in 1970, electronic watches and hand-held calculators developed very quickly, and LEDs were used for the displays. Because LEDs consume a great deal of power and consequently limit battery lifetime, great efforts were made to find their replacement for electronic watch and calculator applications. The liquid

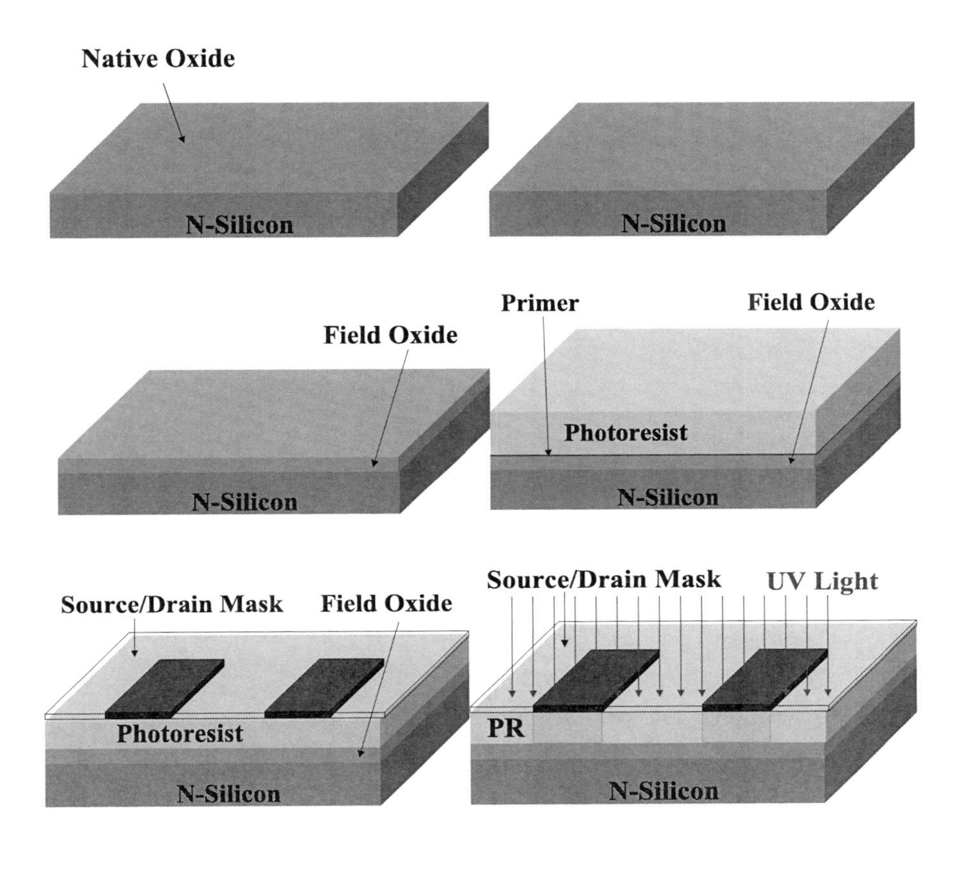

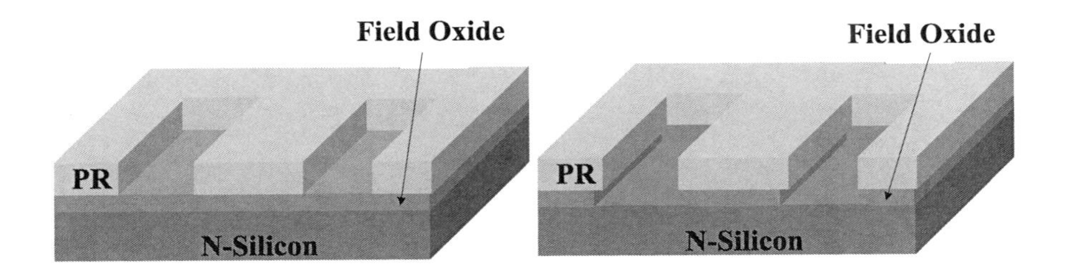

Figure 3.30 Processing steps of pMOS IC fabrication (*continued on next page*).

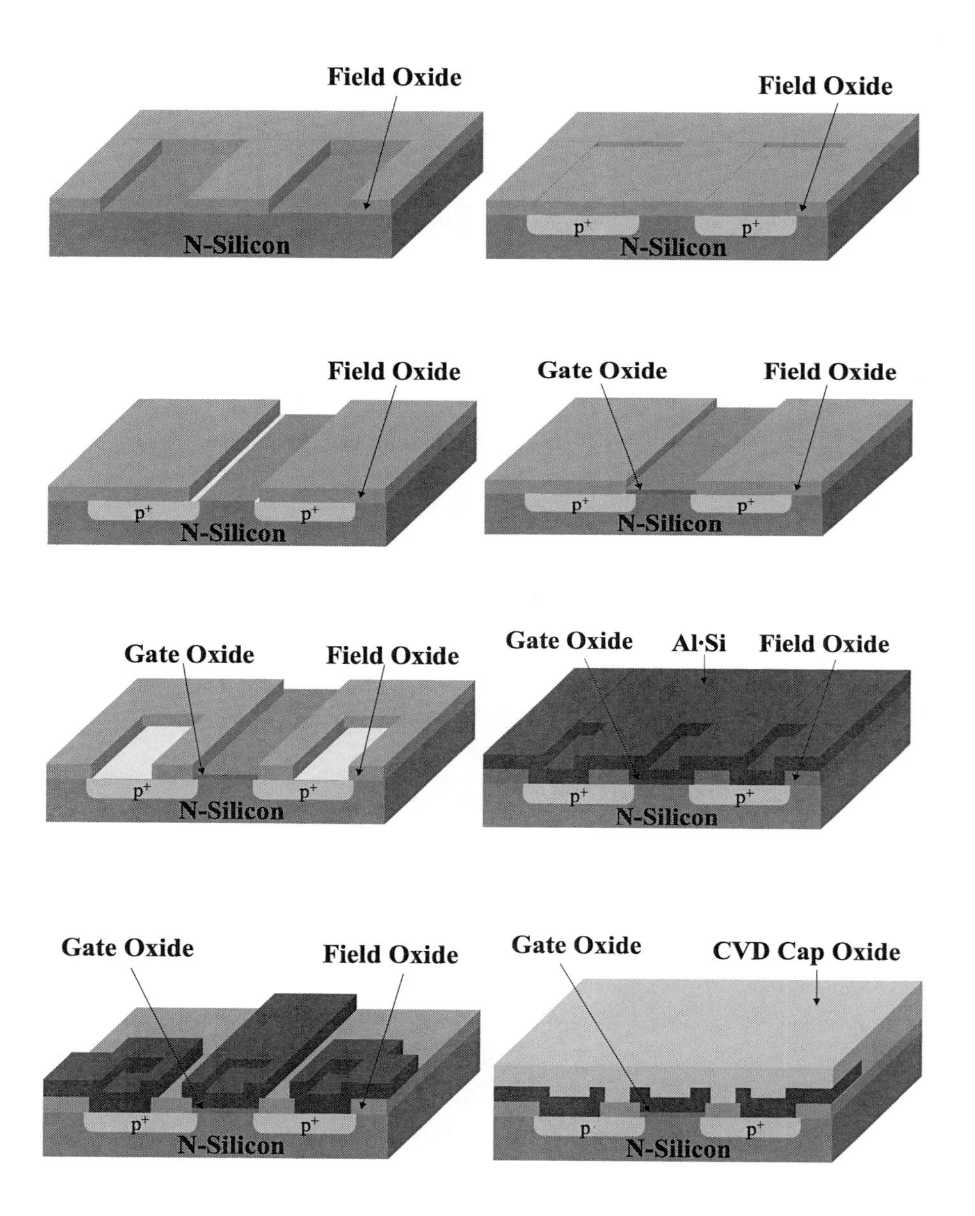

Figure 3.30 (*continued*)

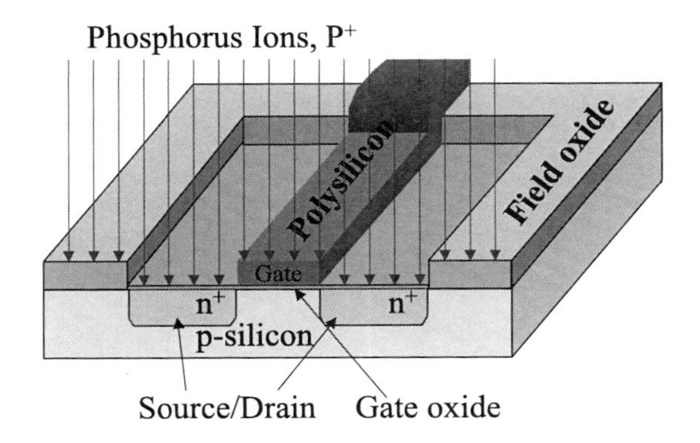

Figure 3.31 Self-aligned source/drain implantation.

Table 3.1 pMOS processing sequence (1960s); A stands for adding, H for heating, P for patterning, and R for removing.

Wafer cleaning	(R)	Etch oxide	(R)
Field oxidation	(A)	Strip photoresist	(R)
Mask 1: source/drain	(P)	Al deposition	(A)
Etch oxide	(R)	**Mask 4:** metal	(P)
Strip photoresist/clean	(R)	Etch aluminum	(R)
S/D diffusion (B)/oxidation	(A)	Strip photoresist	(R)
Mask 2: gate	(P)	Metal anneal	(H)
Etch oxide	(R)	CVD oxide	(A)
Strip photoresist/clean	(R)	**Mask 5:** bonding pad	(P)
Gate oxidation	(A)	Etch oxide	(R)
Mask 3: contact	(P)	Testing and packaging	

crystal display (LCD), which consumes much less power, quickly replaced LEDs for those applications after its introduction in early 1980.

The need to reduce power consumption of the circuits for calculators and electronic watches was one of the major driving forces for CMOS-based IC chip development. CMOSs are used in logic and memory chips, and have dominated the IC market since the 1980s.

3.5.1 Complementary metal-oxide-semiconductor circuit

Figure 3.33(a) shows a CMOS inverter circuit. Two MOSFETs can be seen: one is nMOS and the other is pMOS. If the input is high voltage, or logic 1, the nMOS is turned on and the pMOS is turned off. Therefore, the output voltage is ground voltage V_{ss}, and thus V_{out} is low voltage, or logic 0. Conversely, if the input is low voltage, or logic 0, the nMOS is switched off, and the pMOS is switched on. Output voltage is high voltage V_{dd}, thus V_{out} is high voltage, or logic 1. Because it inverts the input signal, it is called an inverter. This design is one of the basic logic gates used in logic circuits. Table 3.3 is the digital logic table of an inverter.

From Fig. 3.33(a), it is evident that when the nMOS is on, the pMOS is off, and vice versa. This is why the circuit is called the complementary MOS, or CMOS.

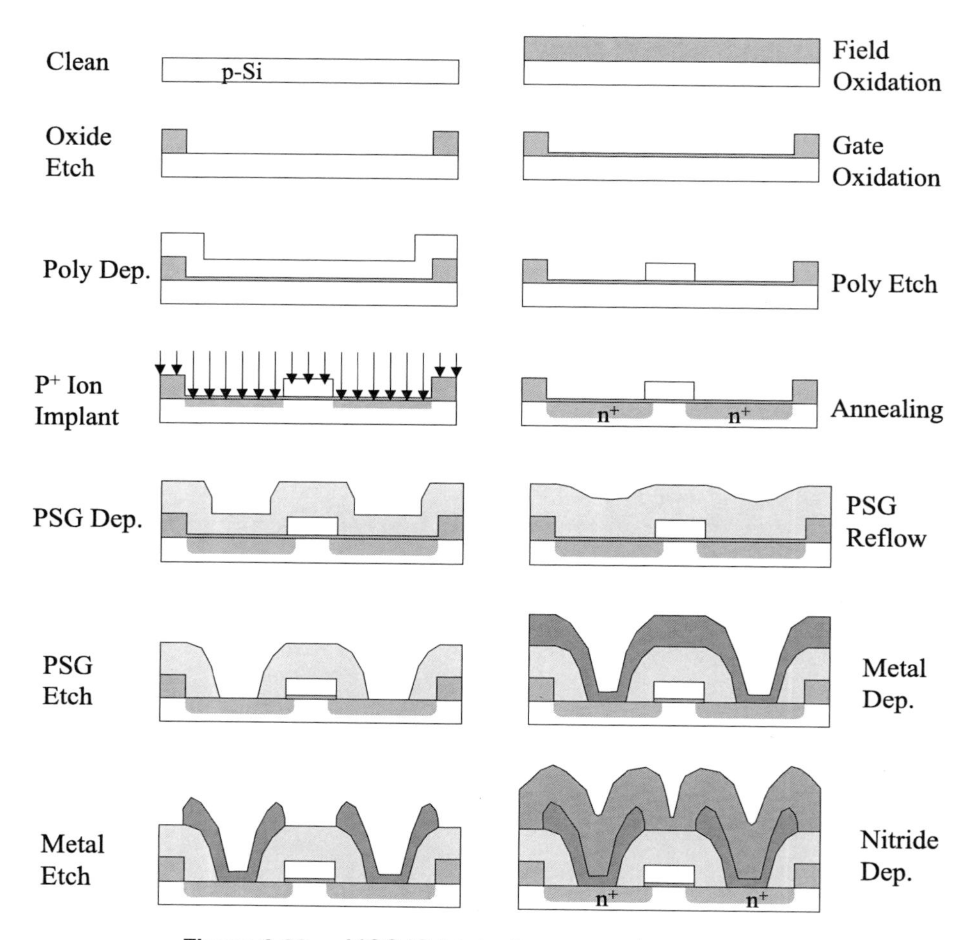

Figure 3.32 nMOS IC fabrication processing sequence.

Table 3.2 nMOS processing sequence (mid-1970s).

Wafer cleaning	PSG reflow
Grow field oxide	**Mask 3:** contact
Mask 1: active area	Etch PSG/USG
Etch oxide	Strip photoresist/clean
Strip photoresist/clean	Al deposition
Grow gate oxide	**Mask 4:** metal
Deposit polysilicon	Etch aluminum
Mask 2: gate	Strip photoresist
Etch polysilicon	Metal anneal
Strip photoresist/clean	CVD oxide
S/D and poly dope implant	**Mask 5:** bonding pad
Anneal and poly reoxidation	Etch oxide
CVD USG/PSG	Testing and packaging

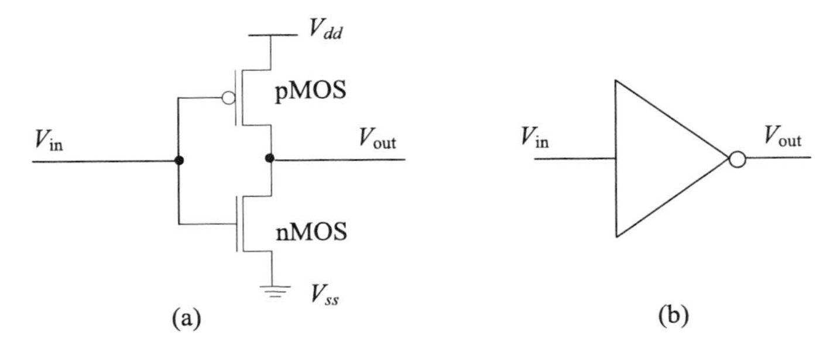

Figure 3.33 Circuit of (a) a CMOS inverter and (b) its logic symbol.

Table 3.3 Logic table of an inverter.

Input	Output
1	0
0	1

For the CMOS inverter, the circuit is always open between high-voltage-biased V_{dd} and grounded V_{ss}. Ideally, there is no current flow between V_{dd} and V_{ss}, so CMOS has very low inherent power consumption. The main power consumption of a CMOS inverter comes from leakage during switching, which has a very high frequency. Other advantages of CMOS over nMOS are that it has higher noise immunity, a lower chip temperature, wider operation temperature range, and less clocking complexity.

By combining both CMOS and bipolar technology, the BiCMOS IC developed rapidly in the 1990s. The CMOS circuit is used for the logic part, and the bipolar is used for input/output to increase device speed. Since the BiCMOS is not a mainstream product and is likely to lose its applications when IC power supply voltage drops below 1 V, this process is not covered in detail.

3.5.2 Complementary metal-oxide-semiconductor circuit process (1980s technology)

CMOS processing was developed from nMOS processing. It takes at least three more mask steps to form a CMOS than it does for an nMOS. One is for n-well formation, one is for pMOS source/drain implantation, and the third is for nMOS source/drain implantation.

Figure 3.34 and Table 3.4 show CMOS manufacture processing steps with technology from the 1980s. Local oxidation of silicon (LOCOS) replaced blanket-field oxidation for isolation between transistors. Borophosphosilicate glass (BPSG) was used for the PMD or interlayer dielectric (ILD0) to reduce the required reflow temperature. Because of the shrinkage of the critical dimension (CD), plasma etch (dry etch) replaced wet etch for most patterned etching. One layer of metal line was no longer enough to route all of the components on the IC chip with required conductance; therefore, a second metal layer was used. Deposition and planarization of dielectrics between metal lines [intermetal dielectric (IMD)],

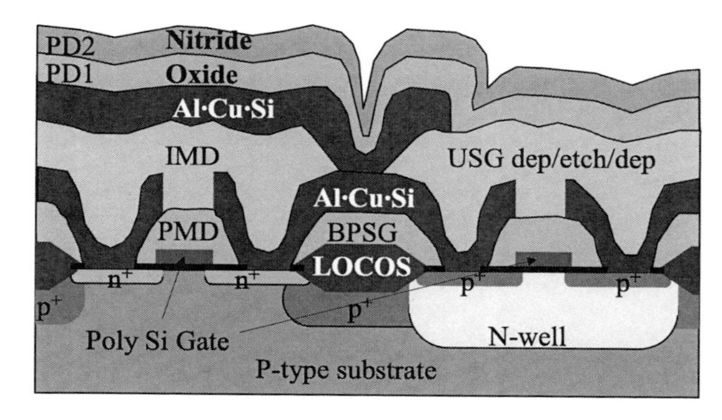

Figure 3.34 Cross section of a CMOS IC chip with two layers of metal.

Table 3.4 CMOS processing sequence (1980s).

Wafer cleaning	Strip photoresist/clean	USG deposition
Pad oxidation	Polysilicon annealing/oxidation	Etch etchback
Nitride deposition	**Mask 4:** p-channel S/D	USG deposition
Mask 1: LOCOS	Boron implant	**Mask 8:** via
Etch nitride	Strip photoresist	Etch IMD
Field V_T implant (boron)	**Mask 5:** n-channel S/D	Strip photoresist
Strip photoresist/clean	Phosphorous implant	Premetal clean
Field oxidation	Strip photoresist/clean	Sputter Al alloy
Strip nitride	Deposit USG (barrier layer)	**Mask 9:** metal 2
Mask 2: n-well	Deposit BPSG (ILD0)	Metal etch
n-well implant	BPSG reflow	Strip photoresist
Strip photoresist/clean	**Mask 6:** contact	Metal annealing
n-well drive-in	Etch oxide	CVD oxide/nitride
Strip pad oxide	Strip photoresist	**Mask 10:** bonding pad
Wafer cleaning	Premetal clean	Etch nitride/oxide
Gate oxidation	Sputter Al alloy	Strip photoresist
Polysilicon deposition	**Mask 7:** metal 1	
Poly dope implantation	Metal etch	
Mask 3: Gate	Strip photoresist/clean	*Testing and packaging*
Polysilicon etch	Metal anneal	*Final test*

posed a great challenge from the 1980s to the 1990s. The minimum feature size shrank from 3 to 0.8 μm in the 1980s.

Basic CMOS processing steps are wafer preparation, well formation, isolation formation, transistor making, interconnection, and passivation. Wafer preparation includes epitaxial silicon deposition, wafer cleaning, and alignment mark etch. Well formation defines the substrate for nMOS and pMOS transistors. Depending on the technology development, there are different techniques for well formations: single well, self-aligned twin well (also called a single-photo twin well), and double-photo twin well. The isolation process builds insulation blocks to electrically isolate the neighboring transistors. In the 1980s, LOCOS replaced blanket-field oxide as the main technique for device isolation. Transistor production involves gate oxide growth, polysilicon deposition, photolithography, polysilicon etch, ion implantation, and thermal annealing. These are the most crucial steps in

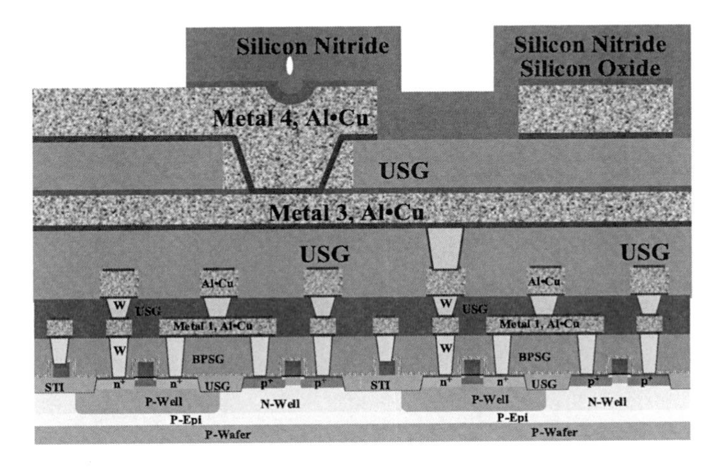

Figure 3.35 Cross section of a CMOS IC chip with four Al-Cu metallization layers.

IC processing. Interconnection processes use the combination of deposition, photolithography, and etch processes to define contact holes and metal wires to connect millions of transistors built on the silicon surface. Finally, passivation dielectric deposition, photolithography, and etch processing seal the IC chip from the outside world, leaving only the bonding pad open for testing and soldering.

3.5.3 Complementary metal-oxide-semiconductor process (1990s technology)

During the 1990s, the feature size of IC chips continued to shrink, from 0.8 μm to smaller than 0.18 μm, and several new technologies were adapted in IC manufacturing. Shallow trench isolation (STI) replaced LOCOS for isolation formation when feature size was smaller than 0.35 μm. Silicides were widely used to form gates and local interconnections, and tungsten (W) was widely used for metal interconnection as the plug between different metal layers. More fabs started using CMP to form STI and tungsten plugs, and they began to planarize the ILDs. High-density plasma etch and CVD processes became more popular, and copper metalization started ramping up in the production line. Figure 3.35 illustrates the cross section of a CMOS IC with four Al-Cu alloy metal interconnection layers. Figure 3.36 illustrates the cross section of a CMOS IC with four Cu metal interconnection layers and an Al-Cu alloy bond pad layer.

3.6 Technology Trends after 2000

CMOS IC technology entered nanometer technology nodes at the start of the new millennium. Technology nodes shrank from 130 to 32 nm. 193-nm became the dominant wavelength of optical photolithography. Immersion photolithography, which uses water as the medium between the objective lens of a scanner and photoresist on the wafer to further improve patterning resolution, has become widely used in IC manufacturing at the 45-nm node and beyond. The double-patterning technique has been used in IC manufacturing since the 45-nm

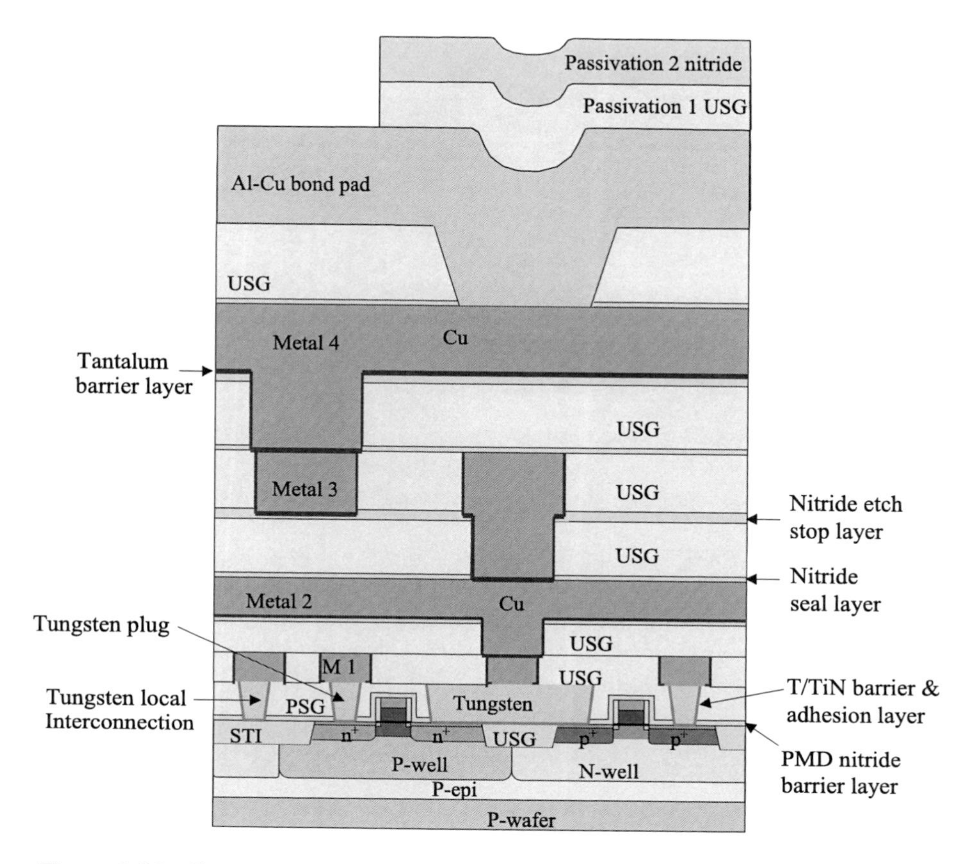

Figure 3.36 Cross section of a CMOS IC chip with four copper metallization layers.

technology node. The combination of immersion photolithography and double patterning has helped IC manufacturers to further shrink feature size without next-generation lithography. Starting from the 65-nm node, nickel silicide has replaced cobalt silicide as the material of choice for self-aligned silicide. High-κ and metal gates started to replace silicon dioxide and polysilicon as gate dielectric and gate electrode materials. Substrate engineering such as strain silicon is widely used to enhance device performance by improving carrier mobility. Techniques such as dual stress liners and selective epitaxial silicon germanium (eSiGe) growth are commonly used methods to strain the channel of the MOSFET to enhance carrier mobility and device speed. Figure 3.37 illustrates the cross section of a 32-nm CMOS with selective epitaxial SiGe and SiC, gate-last integration of a high-κ metal gate, nine layers of copper interconnection, and a lead-free bump.

3.7 Summary

- Semiconductors are materials with conductivity between the conductor and dielectric. Their conductivity can be controlled by dopant concentration and applied voltage.

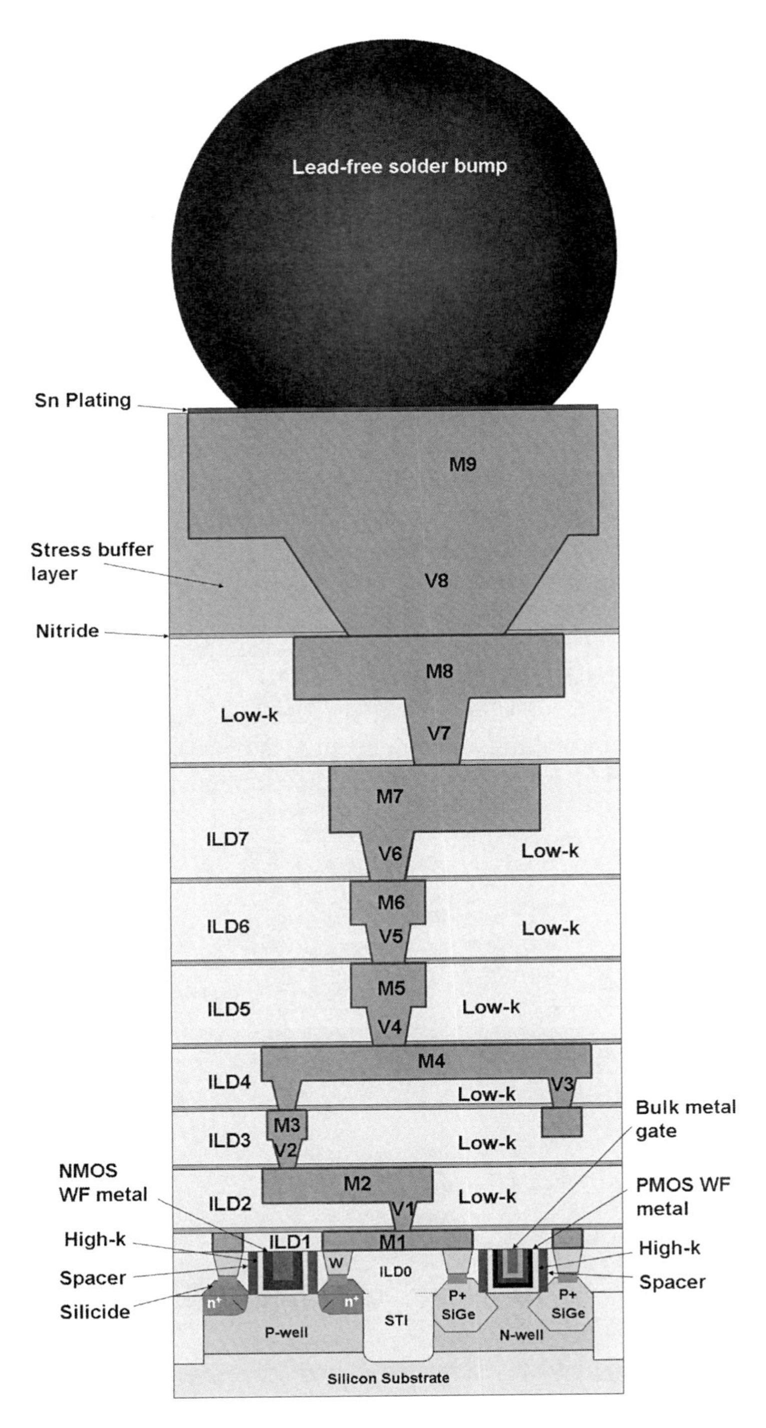

Figure 3.37 A 32-nm CMOS with selective epitaxial SiGe and SiC, gate-last high-κ metal gate, nine copper/low-κ layers, and a lead-free bump.

- Silicon, germanium, and gallium arsenide are the most commonly used semiconductor materials.
- p-type semiconductors are doped with atoms from column IIIA of the table of elements, mainly boron. The majority carriers are holes.
- n-type semiconductors are doped with atoms from column VA of the table of elements, principally P, As, and Sb. The majority carriers are electrons.
- The higher the dopant concentration, the lower the semiconductor resistivity.
- Electrons move faster than holes; therefore, n-type silicon has lower resistivity than p-type silicon for the same dopant concentration.
- Resistors are mainly made with polysilicon. Their resistance is determined by polysilicon line geometry and dopant concentration.
- Capacitors are used in DRAM to store charges and keep data memory.
- Bipolar transistors can amplify electric current and can also be used as switches.
- MOSFET can be turned on and off by biasing the gate at different voltages.
- MOSFET-based IC chips have dominated the semiconductor industry since 1980, and their market share is still increasing.
- Memory, microprocessor, and ASIC chips are the three most commonly made IC chips in the semiconductor industry.
- The advantages of CMOSs are low power consumption, low heat generation, noise immunity, and clocking simplicity.
- The basic CMOS process steps are wafer preparation, isolation formation, well formation, transistor making, interconnection, and passivation.
- The basic semiconductor processes are adding, removing, heating, and patterning (see Fig. 3.38).

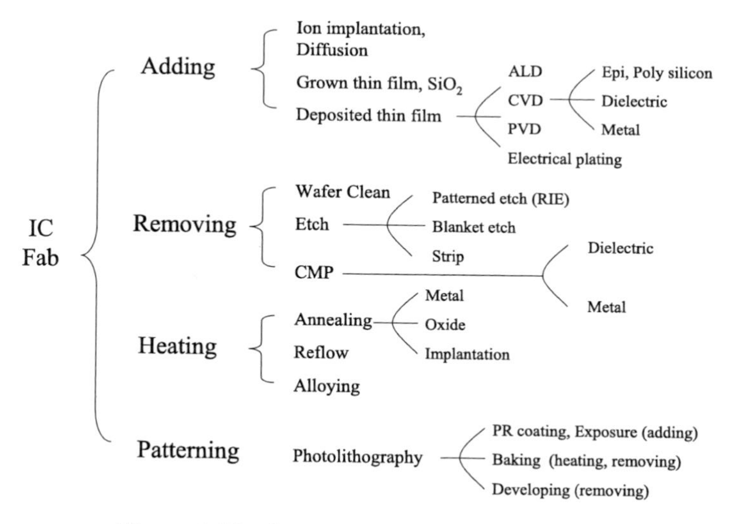

Figure 3.38 Summary of basic IC processes.

3.8 Bibliography

D. Manners, "50 Not Out," *Electronics Weekly*, Dec. 17, 1997.

P. Packan et al., "High Performance 32-nm Logic Technology Featuring 2nd Generation High-k + Metal Gate Transistors," *IEDM Tech. Dig.*, 659 (2009).

G. Stix, "Trends in Semiconductor Manufacturing: Toward 'Point One'," *Sci. Am.* **727**(2), 90–95 (1995).

S. M. Sze, *Physics of Semiconductor Devices, Second Edition*, John Wiley & Sons, New York (1981).

P. J. Zdebel, "Current Status of High Performance Silicon Bipolar Technology," *14th Ann. IEEE GaAs IC Symp. Tech. Dig.*, 15–18 (1992).

3.9 Review Questions

1. What is a semiconductor? List the three most commonly used semiconductor materials.
2. What is the majority carrier of a p-type semiconductor? What material is a p-type dopant?
3. List three n-type dopants. What does "n" stand for?
4. How does semiconductor resistivity change with dopant concentration?
5. At the same dopant concentration, which has higher electrical conductivity, phosphorus-doped silicon or boron-doped silicon?
6. What materials are commonly used to make resistors in IC chips? What factors determine the resistance of the resistor?
7. What kind of IC chip needs many capacitors? Why does it need them?
8. List the two kinds of memory chips that dominate memory IC manufacturing.
9. Describe how to turn an nMOS on and off.
10. List the four basic IC processes. Ion implantation and postimplant RTA (rapid thermal annealing) represent which ones? How about photolithography and etch?
11. What are the minimum mask steps required to make a functional pMOS? How about a bipolar transistor?
12. What was the major technological breakthrough in the mid-1970s in the IC industry?
13. Describe a self-aligned source/drain process. Why it is still used today?
14. Why are CMOS circuits widely used in semiconductor chips?
15. List the basic processes for CMOS IC chip manufacturing.
16. Why do advanced IC chips use multilayer metal interconnections?

Chapter 4
Wafer Manufacturing, Epitaxy, and Substrate Engineering

Objectives

After finishing this chapter, the reader will be able to:

- give two reasons why silicon is more commonly used than any other semiconductor material
- list at least two preferred orientations of single-crystal silicon
- list the basic steps for making silicon wafers from sand
- describe the Czochralski (CZ) and the floating zone methods
- explain the purpose of epitaxial silicon layer deposition
- describe the epitaxial silicon deposition process
- list two ways to make a silicon-on-insulator (SOI) wafer
- identify the advantage of strained silicon
- describe the selective epitaxial process and its application in generating strained silicon.

4.1 Introduction

Single-crystal silicon wafers are the most commonly used semiconductor material in IC manufacturing. This chapter explains why silicon wafers are used the most, and how to make them.

All materials are made from atoms. According to the arrangement of atoms inside solid materials, there are three different material structures: amorphous, polycrystalline, and single crystal. In an amorphous structure, there are no repeated patterns of atoms. A polycrystalline structure has some repeated patterns of atoms, which form grains. All atoms are arranged in the same repeated pattern in a single-crystal structure. Figure 4.1 shows cross sections of the three different structures.

In nature, most solid materials are either amorphous or polycrystalline in structure. Only a few solids have a single-crystal structure, and they are usually gem stones, such as quartz (single-crystal silicon dioxide), ruby (single-crystal aluminum oxide with the presence of the element chromium), sapphire (single-crystal aluminum oxide with different impurities other than chromium), and diamond (single-crystal carbon).

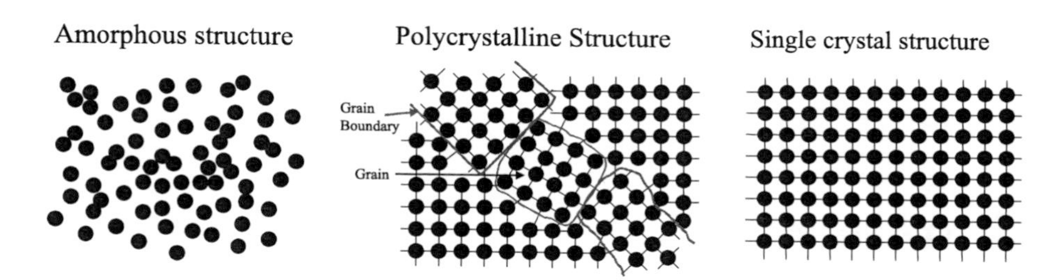

Figure 4.1 The three different structures of solid materials.

The first transistor was made from polycrystalline germanium. However, to make a miniature transistor, a single-crystal semiconductor substrate is required. Electron scattering from the grain boundary in a material can seriously affect p-n junction characteristics. Although they are not good for IC chip manufacturing, both polycrystalline silicon and amorphous silicon can be used to make solar panels, which directly convert photon energy from sunlight into electrical power.

4.2 Why Silicon?

In the early years of the semiconductor industry, germanium was the major semiconductor material used for making electronic devices, such as transistors and diodes. When Kilby made the first IC, it was built on a piece of single-crystal germanium substrate. However, starting in the 1960s, silicon quickly replaced germanium and dominated the IC industry.

About 26% of earth's crust is silicon, making it one of the most abundant elements in the earth's crust, second only to oxygen. It is not necessary to mine silicon. Quartz sand is mainly silicon dioxide, and it can be found in large quantities in many places. As silicon crystal technology developed, the cost of a single-crystal silicon wafer dropped, becoming much lower than single-crystal germanium wafers or any other single-crystal semiconductor material.

Another major advantage of a silicon substrate is that a layer of silicon dioxide can easily grow on its surface in a thermal oxidation process. Silicon dioxide is a strong and stable dielectric, whereas germanium dioxide is difficult to form, unstable at high temperatures (>800 °C), and worst of all, it is water soluble. In fact, in 1947 at Bell Laboratories, a technician mistakenly washed away the anodized germanium oxide layer from the germanium sample surface in the postanodization clean process. This paved the way for Bardeen and Brattain to make the first point-contact bipolar transistor. Otherwise, they might have made the first MOSFET, as they intended to do.

Compared with germanium, silicon has a larger bandgap. Therefore, it can tolerate higher operating temperatures and wider doping ranges. It also has a higher breakdown voltage than germanium. For dopants of interest such as phosphorus and boron, silicon dioxide can serve as a doping mask, since most dopants diffuse much more slowly in silicon dioxide than in silicon. For metal-insulator-semiconductor (MIS) transistors, the SiO_2–Si interface has superior electrical characteristics for MOS. Information on silicon is summarized in Table 4.1.

Table 4.1 Facts about silicon.

Symbol	Si
Atomic number	14
Atomic weight	28.0855
Discoverer	Jöns Jacob Berzelius
Discovery location	Sweden
Discovery date	1824
Origin of name	From the Latin word silicis, meaning flint
Bond length in single-crystal Si	2.352 Å
Density of solid	2.33 g/cm^3
Molar volume	12.06 cm^3
Velocity of sound in substance	2200 m/sec
Hardness	6.5
Electrical resistivity	100,000 $\mu\Omega \cdot$ cm
Reflectivity	28%
Melting point	1414 °C
Boiling point	2900 °C
Thermal conductivity	150 W m^{-1} K^{-1}
Coefficient of linear thermal expansion	2.6×10^{-6} K^{-1}
Etchants (wet)	HNO_4 and HF, KOH, etc.
Etchants (dry)	HBr, Cl_2, NF_3, etc.
CVD precursor	SiH_4, SiH_2Cl_2, $SiHCl_3$, and $SiCl_4$

4.3 Crystal Structures and Defects

4.3.1 Crystal orientation

Figure 4.2 shows the atomic structure of a basic lattice cell of single-crystal silicon. It is a unit cell of the zincblende lattice structure, in which every silicon atom is bonded with another four silicon atoms. Diamond, which is single-crystal carbon, also has this type of crystal structure.

Crystal orientations are defined with Miller indexes, which show a cross section of the orientation plane to the x, y, and z axes. Figure 4.3 illustrates the $\langle 100 \rangle$ orientation plane and the $\langle 111 \rangle$ orientation plane of a cubic crystal. Note that the $\langle 100 \rangle$ plane has a square shape [see Fig. 4.3(a)], and the $\langle 111 \rangle$ plane is a triangle [see Fig. 4.3(b)].

The $\langle 100 \rangle$ and $\langle 111 \rangle$ planes are the most popular orientations used in single-crystal wafers for IC processing. The $\langle 100 \rangle$ wafer is commonly used for MOSFET IC chip manufacturing. The $\langle 111 \rangle$ orientation wafers are employed to make bipolar transistors and bipolar IC chips, because the $\langle 111 \rangle$ surface is stronger due to its higher atom surface density, which makes it ideal for higher-power devices. $\langle 100 \rangle$ and $\langle 111 \rangle$ orientation planes are illustrated in Fig. 4.4. When a $\langle 100 \rangle$ wafer breaks, the fragments normally have right angles (90 deg). If a $\langle 111 \rangle$ wafer breaks, the fragments are 60-deg triangles.

Crystal orientation can be determined by numerous methods. Visual identification depends on distinguishing features such as etch pits and growth facets, or x-ray diffraction. Single-crystal silicon can be wet etched. If there are some defects on

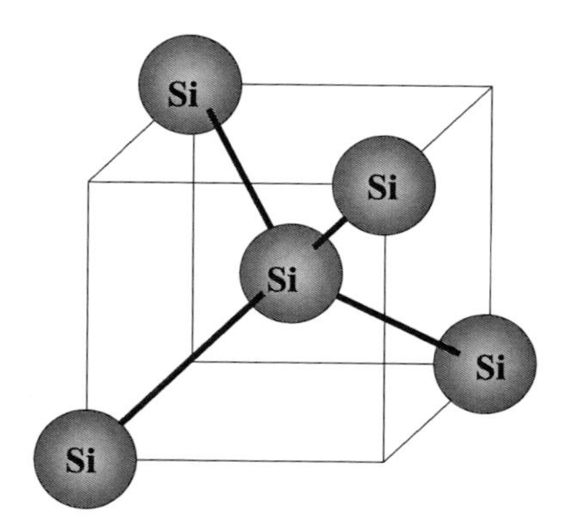

Figure 4.2 A unit cell of the single-crystal silicon lattice structure.

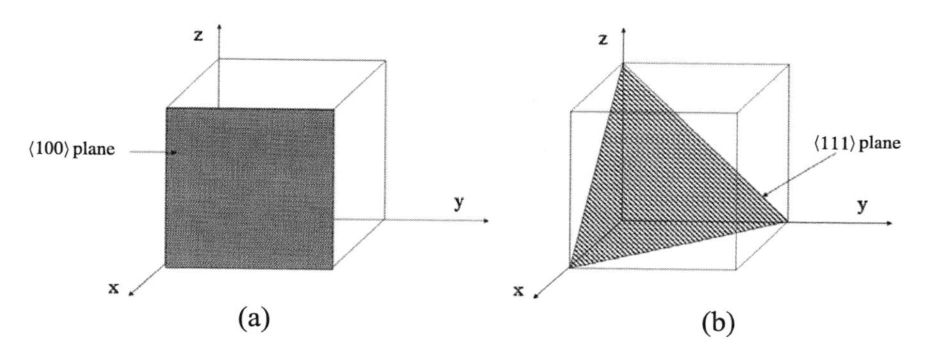

Figure 4.3 Definition of (a) ⟨100⟩ and (b) ⟨111⟩ orientation planes.

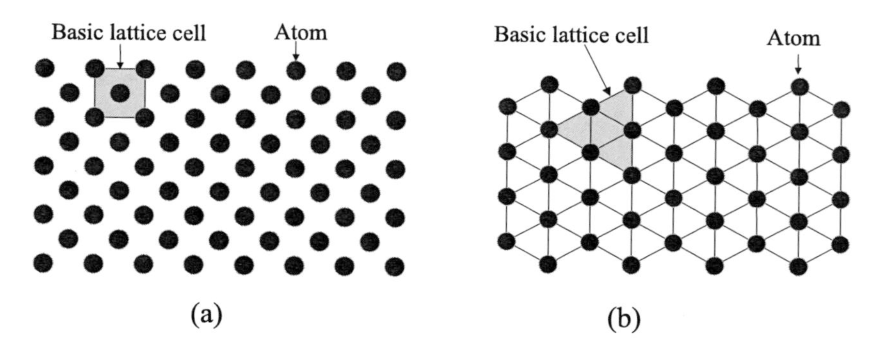

Figure 4.4 Lattice structure of ⟨100⟩ and ⟨111⟩ orientation planes.

the surface where the etch rate is higher, this can create etch pits. For the ⟨100⟩ wafer, an etch pit looks like an inverted four-sided pyramid when it is selectively etched with solutions such as potassium hydroxide (KOH), which etches much faster on the ⟨100⟩ plane than on the ⟨111⟩ plane. For the ⟨111⟩ wafer, an etch pit is a tetrahedron, or three-side inverse pyramid (see Fig. 4.5).

4.3.2 Crystal defects

There are numerous crystal defects that occur during the growth and subsequent processing of silicon crystals and wafers. The simplest point defect is vacancy, which is also called the Schottky defect, in which an atom is missing from the lattice (Fig. 4.6). Vacancy can affect doping processes, since the dopant diffusion rate in single-crystal silicon is a function of the vacancy number.

When an extra atom occupies the space between normal lattice sites, it is called an interstitial defect. If an interstitial defect and a vacancy occur in proximity, the pair is called a Frenkel defect (see Fig. 4.6).

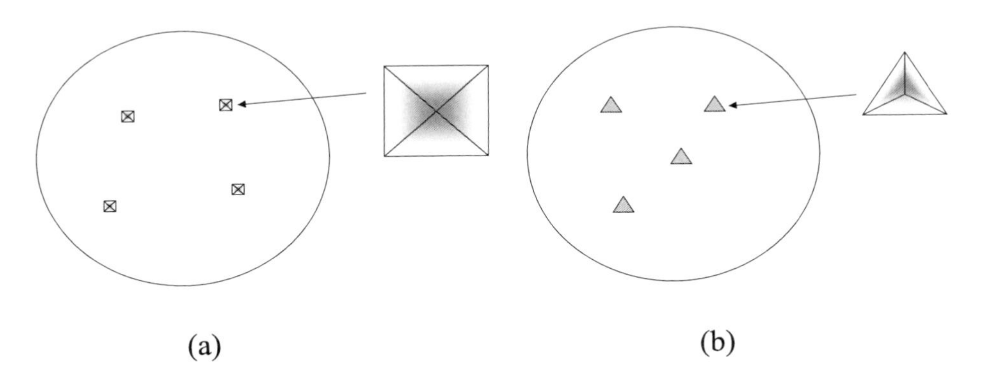

Figure 4.5 Etch pits on (a) ⟨100⟩ wafer and (b) ⟨111⟩ wafer.

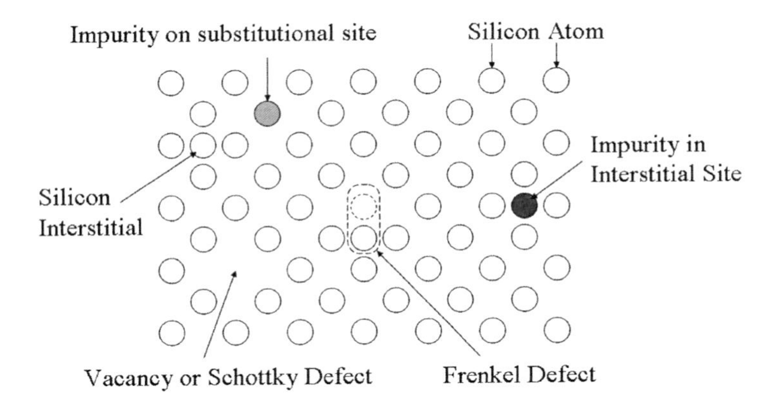

Figure 4.6 Potential silicon crystal defects.

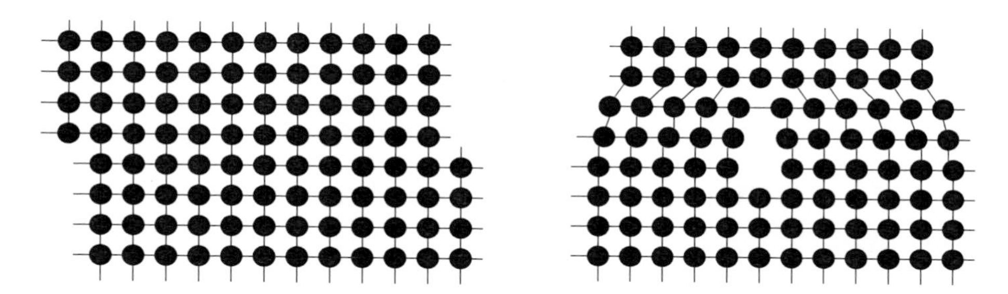

Figure 4.7 Dislocation defects.

Dislocations are geometric faults in the lattice, which can be caused by the crystal pulling process. During the wafer process, dislocation is mainly associated with excessive mechanical stress, often from uneven heating or cooling, dopant diffusion into the lattice, or film deposition, or from external forces such as tweezers. Figure 4.7 shows two examples of dislocation defects.

Dislocation and other defect densities must be minimized on the polished side of wafers, where transistors and other microelectronic devices are made. Defects on the silicon surface can scatter electrons, increasing resistivity and affecting device performance. Defects on a wafer surface can reduce the production yield of IC chips.

Each defect has some dangling silicon bonds, with which impurity atoms tend to bind and become immobilized. On the backside of the wafer, defects are intentionally created to trap unwanted mobile impurities inside the wafer. This prevents them from affecting the normal function of microelectronic devices on the front side.

4.4 Sand to Wafer

4.4.1 Crude silicon

Common quartz sand is mainly silicon dioxide, which can react with carbon at high temperatures. Carbon then replaces silicon to form silicon and carbon monoxide, or carbon dioxide. Because the silicon-oxygen bond is strong, silicon dioxide is very stable. Therefore, this carbon-reducing reaction requires very high temperatures. Manufacturers put quartz sand with carbon into a high-temperature furnace. Carbon used in this reaction does not need to be of very high purity; in fact, it could be in the form of coal, coke, or even pieces of wood. At high temperatures, carbon starts to react with silicon dioxide to form carbon monoxide. This process generates polycrystalline silicon with about 98 to 99% purity, called crude silicon or metallurgical-grade silicon (MGS). The chemical reaction to form MGS can be expressed as

$$\underset{\text{quartzite}}{SiO_2} + \underset{\text{coal}}{2C} \xrightarrow{\text{heat}} \underset{\text{MGS}}{Si} + \underset{\text{carbon monoxide}}{2\,CO}$$

Crude silicon has a high impurity concentration and must be purified for use in semiconductor device manufacturing.

4.4.2 Silicon purification

The purification of silicon has several steps. First, crude silicon is ground into a fine powder. Then the silicon powder is introduced into a reactor to react with hydrogen chloride (HCl) vapor to form a trichlorosilane (TCS, $SiHCl_3$) vapor at about 300 °C. The chemical reaction can be expressed as:

$$\underset{\text{MGS}}{\text{Si}} + \underset{\text{hydrochloride}}{3\,\text{HCl}} \xrightarrow{\text{heat (300 °C)}} \underset{\text{TCS}}{\text{SiHCl}_3} + \underset{\text{hydrogen}}{\text{H}_2}$$

TCS vapor then passes through a series of filters, condensers, and purifiers to reach ultrahigh-purity liquid TCS, with purity as high as 99.9999999% (nine 9s!), or less than one impurity per billion. Figure 4.8 gives the schematic of the high-purity TCS formation process.

High-purity TCS is one of the most commonly used silicon precursors for silicon deposition. It is widely applied to amorphous silicon, polycrystalline silicon (polysilicon for short), and epitaxial silicon depositions. At high temperatures, TCS can react with hydrogen and deposit high-purity polysilicon. The deposition reaction is

$$\underset{\text{TCS}}{\text{SiHCl}_3} + \underset{\text{hydrogen}}{\text{H}_2} \xrightarrow{\text{heat (1100 °C)}} \underset{\text{EGS}}{\text{Si}} + \underset{\text{hydrochloride}}{3\text{HCl}}$$

The high-purity polycrystalline silicon is called electronic-grade silicon (EGS). Figure 4.9 illustrates the deposition process of high-purity EGS, and Fig. 4.10 shows actual photographs of it. The EGS is now ready to be pulled into a single-crystal silicon ingot for creating wafers for IC processing.

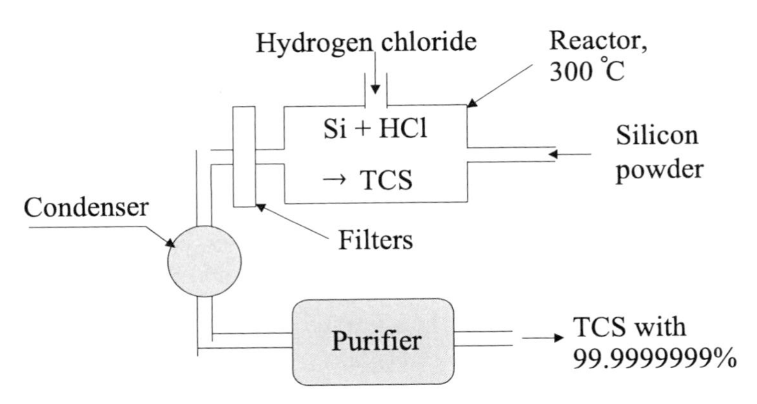

Figure 4.8 Schematic of the process for obtaining high-purity TCS from crude silicon.

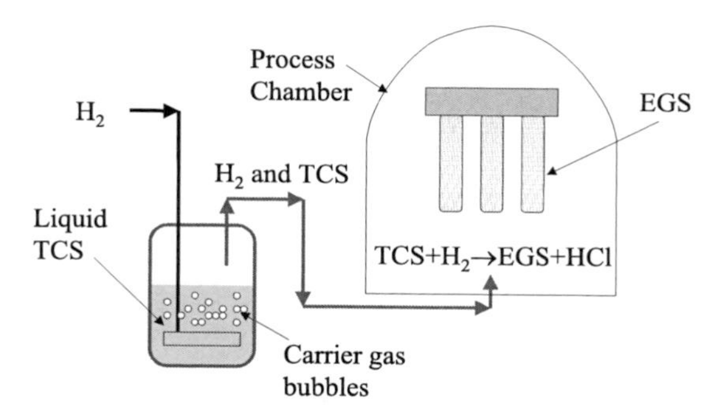

Figure 4.9 EGS deposition.

Figure 4.10 EGS photographs (courtesy of MEMC Electronic Materials).

4.4.3 Crystal pulling

To make a single-crystal silicon ingot, high temperatures are required to melt the EGS and a single-crystal silicon seed. The molten silicon then can recondense with the same crystal structure as the seed crystal. There are two methods commonly used in the semiconductor industry to generate single-crystal silicon: the Czochralski (CZ) method and the floating zone (FZ) method.

4.4.3.1 Czochralski method

The CZ method, which was invented by Polish chemist Jan Czochralski, is the more popular method in wafer manufacturing because it has many advantages over the only alternative, the floating zone method. Only the CZ method can make wafers with a diameter larger than 200 mm, and it has a relatively low cost, since pieces of single-crystal silicon and polycrystalline silicon can be used. Finally, it can create heavily doped single-crystal silicon by melting and recondensing the dopant with the silicon. A schematic of the CZ process is given in Fig. 4.11; the entire system exists in a sealed chamber with an argon ambient to control contamination.

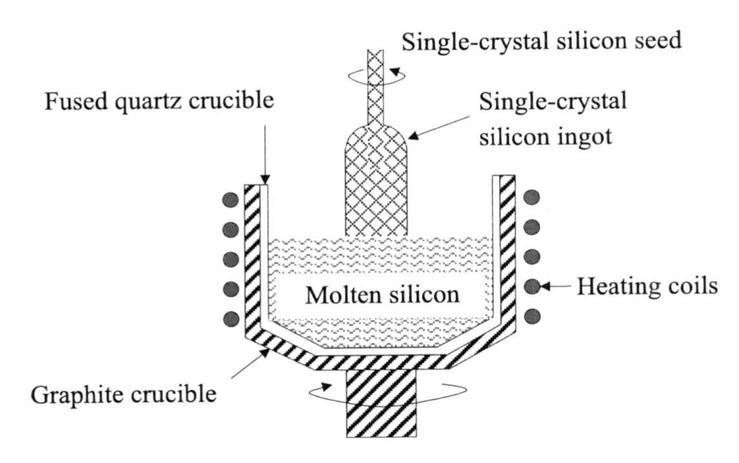

Figure 4.11 Schematic of the CZ crystal-pulling method.

In the CZ method, high-purity EGS is melted in a slowly rotating quartz crucible at 1415 °C, just above the silicon melting point of 1414 °C. It is heated by rf or resistive heating coils. A single-crystal silicon seed rod mounted on a slowly rotating chuck is gradually lowered into the molten silicon, and the surface of the seed crystal submerges into the molten silicon and starts to melt. Seed crystal temperature is precisely controlled at just below the silicon melting point. When the system reaches thermal stability, the seed crystal is withdrawn very slowly, dragging along some molten silicon to recondense around it with the same crystal orientation. The ingot, a whole piece of single-crystal silicon, is formed after up to 48 h of pulling. Rotation and melting of the silicon helps improve dopant uniformity across the crystal ingot, which is sawed into wafers in a later process. In some cases (40–100-$\Omega \cdot$ cm n-type wafers and 100–200-$\Omega \cdot$ cm p-type wafers), a magnetic field is also used to further improve radial uniformity of the dopant. Figures 4.12(a) through 4.12(e) illustrate the crystal pulling process, and Fig. 4.12(f) shows a photograph of a single-crystal silicon ingot pulled with the CZ method.

The diameter of crystals created using the CZ method can be controlled by the temperature and pulling rate. An automatic diameter control (ADC) system is commonly used to control crystal diameter. The ADC system uses infrared sensors to monitor the bright radiation ring at the crystal and molten silicon interface [see Fig. 4.12(c)], and the feedback signals control the pulling rate. If the ring is in the sensor range, it increases the pulling rate, reducing the crystal diameter; if the ring is out of the sensor range, it reduces the pulling rate, increasing the crystal diameter. The grooves on the side of the single-crystal ingot in Fig. 4.12(f) are caused by this feedback diameter control process.

Single-crystal silicon ingots pulled by the CZ method always have trace amounts of impurities of oxygen and carbon, which come from the crucible materials. Typically, the oxygen concentration in silicon crystal grown by the CZ method is approximately from $1.0 \times 10^{16}/\text{cm}^3$ to $1.5 \times 10^{18}/\text{cm}^3$. The carbon content varies from about $2.0 \times 10^{16}/\text{cm}^3$ to about $1.0 \times 10^{17}/\text{cm}^3$. Oxygen and

Figure 4.12 Single-crystal silicon ingot and its CZ pulling process (courtesy of MEMC Electronic Materials).

carbon concentrations in the silicon are the result of ambient pressure during crystal growth, pulling and rotation rates, and the ratio of the crystal diameter to its length.

4.4.3.2 Floating zone method

The floating zone method is the only other practical way to make single-crystal ingots. Figure 4.13 illustrates the floating zone crystal pulling method. As with the CZ method, the process takes place in a sealed furnace chamber with an Ar ambient.

The process starts with a polycrystalline silicon rod, about 50 to 100 cm in length, placed vertically in a furnace chamber. The heating coil heats and melts the lower end of the polysilicon bar, and the seed crystal fuses into the molten zone. The melt is suspended between the seed and polysilicon rod by the surface tension of the molten silicon. Then the heating coils move slowly upward, melting the polysilicon rod just above the molten silicon. The molten silicon at the seed end starts to freeze with the same crystal orientation as the seed crystal. After the heating coils move over the entire polysilicon rod, the rod converts to a single-crystal silicon ingot.

The diameter of the ingot is controlled by the relative rotation rates of the top and bottom sections. The largest diameter of a silicon wafer is 450 mm (18 in.) with the CZ method. With the floating zone method, the largest wafer diameter is 150 mm (6 in.). The CZ method has been used for mass producing the 300-mm (12-in.) wafer since the 1990s.

Since no crucible is used in this process, the primary advantage of the floating zone method is that it can achieve higher silicon purity due to the lower melt contamination, especially from oxygen and carbon. This method is mainly used to manufacture wafers needed for discrete power devices. These wafers require high-resistance starting materials.

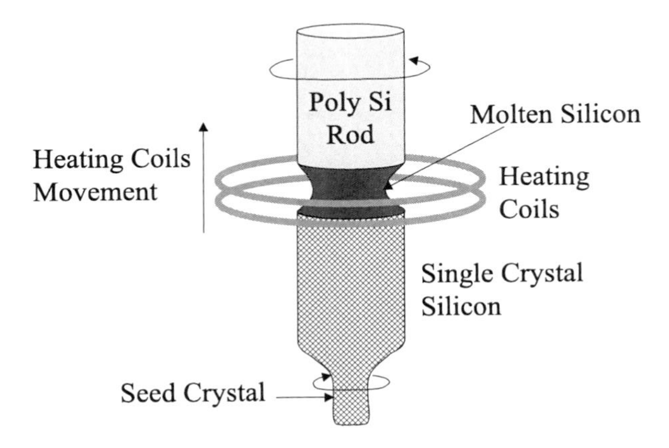

Figure 4.13 Schematic of the floating zone method.

There are two major disadvantages of the floating zone method. One is that the melt–crystal interface is very complex, thus it is difficult to achieve dislocation-free crystals. Another is that it is more expensive, since it requires high-purity polysilicon rods as the starting material, while for the CZ method, any kind of high-purity silicon, such as wafer saw powder, ingot end cuts, and the like, can be used as starting materials.

Because they are formed by freezing the rotating molten silicon, single-crystal silicon ingots—and the wafers sliced from them—are always round. It is possible to make square wafers by grinding the ingot into square columns before slicing wafers from them. However, a square wafer is much harder to handle mechanically. It can be very easy to chip the corners of a square wafer and break it down.

4.4.4 Wafering

After the crystal ingot cools down, machines cut both ends and polish the side to remove the grooves created by the automatic diameter control process. Then a flat (150 mm or smaller) or notch (200 mm or larger) is ground into the ingot to mark the crystal orientation (Fig. 4.14).

Now the ingot is ready to have wafers sliced from it. In a rapid-rotating inward-diameter diamond-coated saw, the ingot moves outward, and wafers are cut, slice by slice, from the ingot. Coolant pours constantly over both ingot and saw to control the considerable heat generated by the sawing process. Wafers are sawed as thin as possible; however, they must be thick enough to sustain the mechanical handling in wafer processing. Larger-diameter wafers require larger thickness. Thicknesses of the different wafer sizes are listed in Table 4.2.

During the sawing process, about one-third of a single-crystal ingot is reduced to sawdust. Fortunately, it can be reused in the CZ crucible as the starting silicon material. The sawing process is illustrated in Fig. 4.15. While sawing, the saw blade must be free of vibration, since any blade vibration will scratch the wafer surface and increase difficulties and costs of the subsequent tapping and polishing processes. Blade withdrawal must be well controlled to prevent retrace damage.

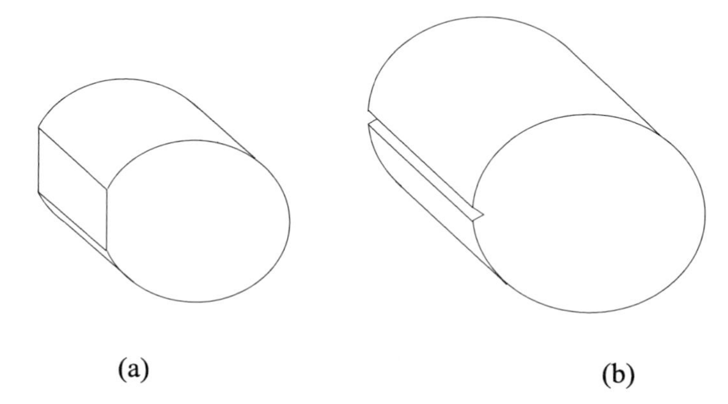

Figure 4.14 Single-crystal ingot after polishing, with (a) a flat or (b) notch.

Table 4.2 Wafer thickness for different wafer sizes. The asterisk indicates the preliminary standard set in October 2008.

Wafer size (mm)	Thickness (μm)	Area (cm^2)	Weight (g)
50.8 (2 in.)	279	20.26	1.32
76.2 (3 in.)	381	45.61	4.05
100	525	78.65	9.67
125	625	112.72	17.87
150	675	176.72	27.82
200	725	314.16	52.98
300	775	706.21	127.62
450	925 ± 25*	1590.43	342.77

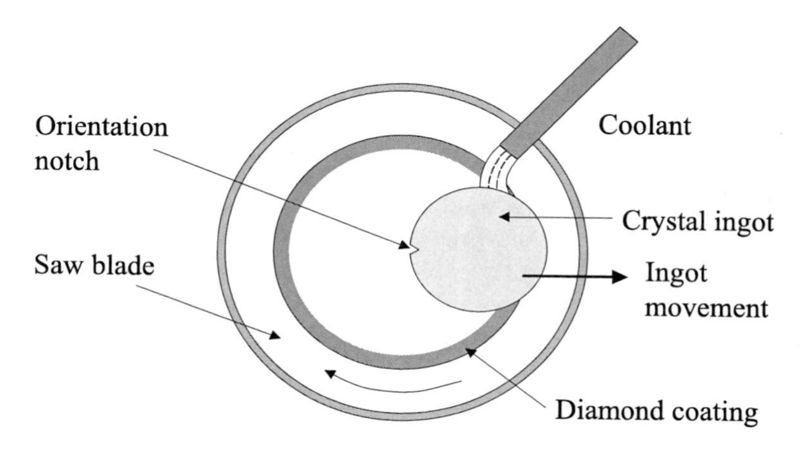

Figure 4.15 Wafer sawing process.

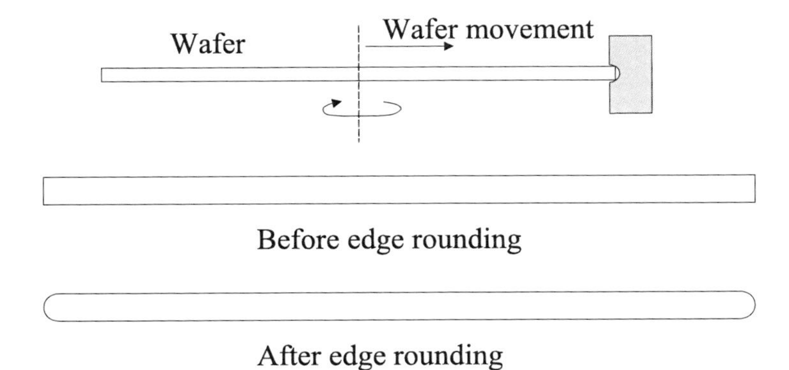

Figure 4.16 Edge-rounding process.

4.4.5 Wafer finishing

After wafer sawing, the wafer edge is ground in a mechanical process to round the sharp edges created in the slicing process. Round edges prevent edge chipping during mechanical handling of wafer processing. Figure 4.16 shows a schematic of the edge-rounding process.

The wafer is then rough polished by the conventional, abrasive, slurry-lapping process to remove the majority of surface damage caused by the wafer-slicing process, and to create the flat surface required for photolithography. The mechanical, double-sided lapping process is performed under pressure using glycerin slurry with fine alumina (Al_2O_3) particulate suspended in it. The lapping process can produce a wafer with a surface flatness within 2 μm and removes about 50 μm of silicon from both sides of a 200-mm wafer.

A wet etch process is used to remove particles and damage created during the sawing, edge-rounding, and lapping processes. Since some sawing damage can be 10 μm deep into the silicon, a wet etch needs to remove about 10 μm of silicon from both sides of a wafer. A commonly used etchant is a mixture of nitric acid (HNO_3), hydrofluoric acid (HF), and acetic acid (CH_3COOH). Nitric acid oxidizes the silicon to form silicon dioxide on the wafer surface. HF dissolves and removes the silicon dioxide. CH_3COOH helps to control the reaction rate. Due to the isotropic etch characteristic, the etch process further smoothes the wafer surface due to the isotropic etch characteristic with an HNO_3-rich solution. The usual formulation is a 4:1:3 mixture of HNO_3 (79 wt% in H_2O), HF (49 wt% in H_2O) and pure CH_3COOH. The combined chemical reaction can be expressed as

$$3\,Si + 4\,HNO_3 + 6\,HF \rightarrow 3\,H_2SiF_6 + 4\,NO + 8\,H_2O.$$

In the CMP process shown in Fig. 4.17, the wafer is held on a rotating wafer holder and pressed onto a rotating polishing pad, with slurry and water in between. The slurry is a colloidal suspension of fine silica (SiO_2) particles with diameters of about 100 Å in an aqueous solution of sodium hydroxide (NaOH). NaOH oxidizes the silicon surface (chemical process) with the help of heat generated by the friction

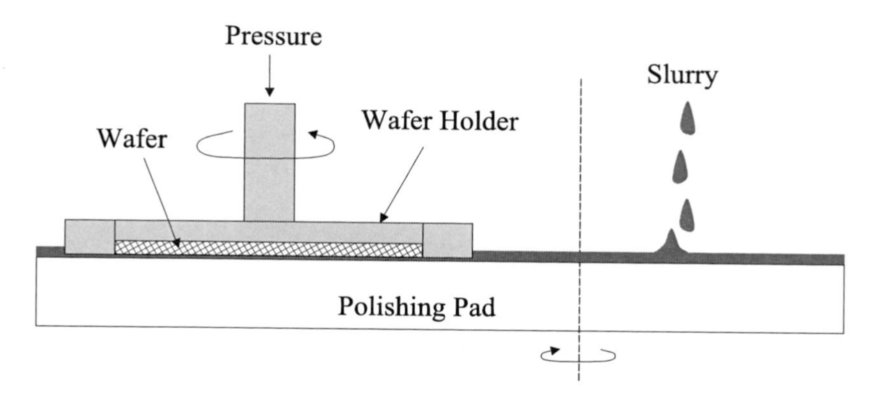

Figure 4.17 Schematic of CMP process.

between the wafer and polishing pad. Then the silica particles abrade the silicon oxide away from the surface (mechanical process). This combined chemical and mechanical process removes silicon from the wafer surface.

Post-CMP cleaning processes use mixtures of acid-oxidizer solutions to remove organic and inorganic contaminants and particles. This finishes the wafer-surfacing process and creates a defect-free surface ready for IC processing. Commonly used clean solutions are a mixture of ammonium hydroxide (NH_4OH) and hydrogen peroxide (H_2O_2), a mixture of hydrochloric acid (HCl) and hydrogen peroxide (H_2O_2), or a mixture of sulfuric acid (H_2SO_4) and H_2O_2.

Figure 4.18 illustrates 200-mm wafer surface roughness changes and wafer thickness changes during the wafer finishing processes. After post-CMP cleaning, inspection, and labeling, the wafer is ready to ship to the customer for semiconductor chip processing.

Manufacturers intentionally create defects and dislocations on the backside of wafers to trap heavy metals, mobile ions, oxygen, carbon, and other contaminants. Backside defects can also be created by argon ion implantation, polysilicon deposition, and heavily doped phosphorus. During IC processing, the backside of a wafer is always deposited with a CVD silicon oxide or silicon nitride layer to prevent any outward diffusion during thermal processes.

Since all transistors and circuits are made on one side of the wafer, most wafers need to be polished on only one side. The unpolished side is used for handling during wafer processing. Some processes, such as Fourier transform infrared (FTIR) measurement of thin-film moisture absorption, requires a double-side polished wafer. Otherwise, scattering of infrared light from the rough surface of the wafer backside will disrupt measurement signals and cause invalid results.

4.5 Epitaxial Silicon Deposition

The word epitaxy comes from two Greek words, *epi* meaning upon, and *taxis* meaning arranged or ordered. The epitaxial deposition process is a process used to grow a thin crystalline layer on a crystalline substrate. Unlike the CZ or floating

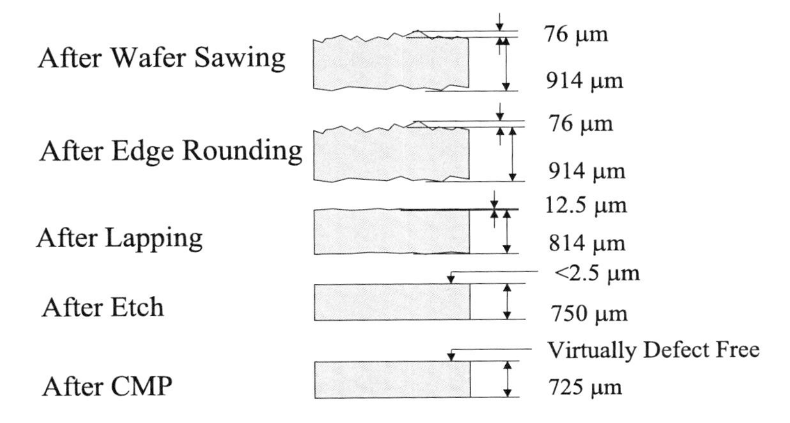

Figure 4.18 Changes of thickness and surface roughness of a 200-mm wafer during wafer manufacturing processes.

zone crystal growing processes, an epitaxy layer grows at a temperature much lower than the melting temperature of silicon.

Earlier development of epitaxial silicon deposition was driven by the requirement of bipolar transistors for high breakdown voltage of the collector. An epitaxial layer can provide a high-resistance layer over the low-resistance substrate, enhancing performance of the bipolar transistor. The epitaxial layer can also enhance the performance of DRAM and CMOS ICs. There are two fundamental advantages of using an epitaxial layer. One is that bipolar transistors require an epitaxial layer to form the heavily doped burial layer deep inside the silicon. This layer cannot be made by existing doping processes such as ion implantation or diffusion. Another advantage is that an epitaxial layer can allow for different physical properties in a substrate wafer. For example, a p-type epitaxial layer grown on an n-type wafer can provide a designer more freedom in the structure of a microelectronic device and circuits. The epitaxial layer is generally free of oxygen and carbon. This cannot be achieved in silicon wafers made with the CZ method because small amounts of oxygen in the quartz (SiO_2) crucible and carbon in the graphite liner can diffuse into the molten silicon during the CZ crystal-pulling process and end up in the silicon crystal.

Bipolar transistors always require an epitaxial layer in order to form a buried layer. For low-speed CMOS and DRAM chips, fabs try to avoid using an epitaxial layer. The epitaxial growth process is the most expensive processing step in IC fabrication, about \$20 to \$100 per wafer, compared with about \$1 per wafer per step for other processes. For a higher-performance IC chip, an epitaxial layer is required because the oxygen impurity in the silicon wafer formed by the CZ method can reduce the carrier lifetime and slow down the device. Figure 4.19 illustrates the application of an epitaxial silicon layer in bipolar and CMOS IC chips.

Currently, epitaxial silicon layers are deposited onto wafers by wafer manufacturers, not IC manufacturers. This is why the epitaxy process is covered in this chapter.

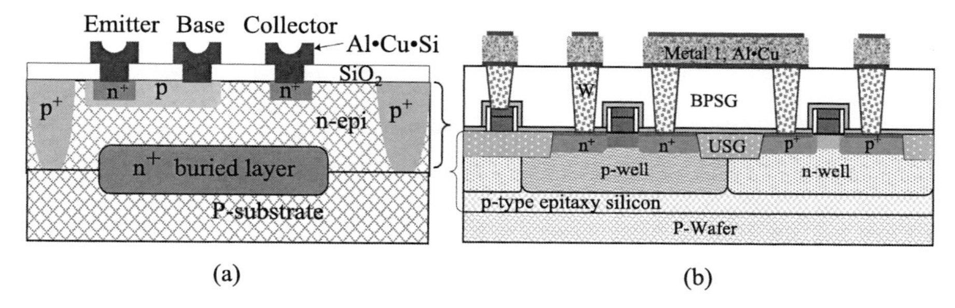

Figure 4.19 Epitaxial silicon layer applications for (a) bipolar and (b) CMOS IC chips.

There are two ways to grow epitaxy silicon layers on silicon wafers: a CVD epitaxy process and a molecular beam epitaxy process.

4.5.1 Gas phase epitaxy

High-temperature (~1000 °C) CVD epitaxial silicon layer growth is the most popular method in the semiconductor industry for growing a single-crystal silicon layer. Silicon source gases used are silane (SiH_4), dichlorosilane (DCS, SiH_2Cl_2), and trichlorosilane (TCS, $SiHCl_3$). The chemical reactions of epitaxial silicon growth are

$$\underset{\text{silane}}{SiH_4} \overset{\text{heat (1000 °C)}}{\longrightarrow} \underset{\text{epi-Si}}{Si} + \underset{\text{hydrogen}}{2\,H_2},$$

$$\underset{\text{DCS}}{SiH_2Cl_2} \overset{\text{heat (1100 °C)}}{\longrightarrow} \underset{\text{epi-Si}}{Si} + \underset{\text{hydrogen chloride}}{2HCl},$$

$$\underset{\text{TCS}}{SiHCl_3} + \underset{\text{hydrogen}}{H_2} \overset{\text{heat (1100 °C)}}{\longrightarrow} \underset{\text{epi-Si}}{Si} + \underset{\text{hydrogen chloride}}{3HCl}.$$

As the film grows, epitaxial silicon can be doped by flow dopant gases, such as arsine (AsH_3), phosphine (PH_3), and diborane (B_2H_6), with silicon source gas in the reactor. At high temperatures, these dopant hydrides dissociate from the heat and release arsenic, phosphorus, and boron into the epitaxial silicon film. In this way, in-situ doping of epitaxial film can be achieved. Chemical reactions of the in-situ doping are

$$AsH_3 \overset{\text{heat (~1000 °C)}}{\longrightarrow} As + 3/2\,H_2,$$

$$PH_3 \overset{\text{heat (~1000 °C)}}{\longrightarrow} P + 3/2\,H_2,$$

$$B_2H_6 \overset{\text{heat (~1000 °C)}}{\longrightarrow} 2\,B + 3\,H_2.$$

All three hydride dopant source gases mentioned are highly poisonous, flammable, and explosive. During epitaxial-layer growth, the dopant inside the substrate, driven by the high temperature, diffuses into the epitaxial layer. If the film growth is slower than the dopant diffusion, the entire epitaxial layer becomes doped by the substrate dopant. This effect, called autodoping, is undesirable, since it can affect dopant concentration in the epitaxial layer. To avoid autodoping, the deposition rate of the epitaxial layer must outpace the dopant diffusion rate in the epitaxial layer so that the dopant concentration of the epitaxial film is determined by the dopant induced in the gas phase during deposition. High deposition temperatures can help meet this requirement.

4.5.2 Epitaxial growth process

An example of the epitaxial silicon growth and doping process is illustrated in Fig. 4.20. First, precursors such as DCS and AsH_3 are introduced into the reactor. Precursor molecules diffuse to the wafer surface, adsorb on the surface, dissociate, and react on the surface. Solid byproduct atoms called adatoms migrate onto the surface and bond with other surface atoms in the same crystal structure as the substrate crystal, while volatile byproducts desorb from the hot surface and diffuse out.

Figure 4.21 shows the relationship between temperature and epitaxial silicon growth rate for different silicon precursors. We can see that there are two deposition regimes: one at lower temperature, with growth rates highly sensitive to temperature, and another at higher temperatures with growth rates less sensitive to temperature. The first regime is called the surface-reaction-limited regime, and the second is called the mass-transport-limited regime. We discuss these in detail in Chapter 10.

The silane process is in the surface-reaction-limited regime when the temperature is below 900 °C. It changes to the mass-transport-limited regime when the temperature is higher than 900 °C.

Question: For the silane process, if the temperature continues to increase to 1300 °C, how will the growth rate change?

Answer: Silane is a very reactive gas. When the temperature is higher than 1200 °C, it starts to react in the gas phase (gas phase nucleation). This reduces the epitaxy growth rate and generates huge amounts of particles (a regime to be avoided at all costs). Dichlorosilane and trichlorosilane are less reactive than silane; they have higher reaction temperatures and can deposit at higher temperatures.

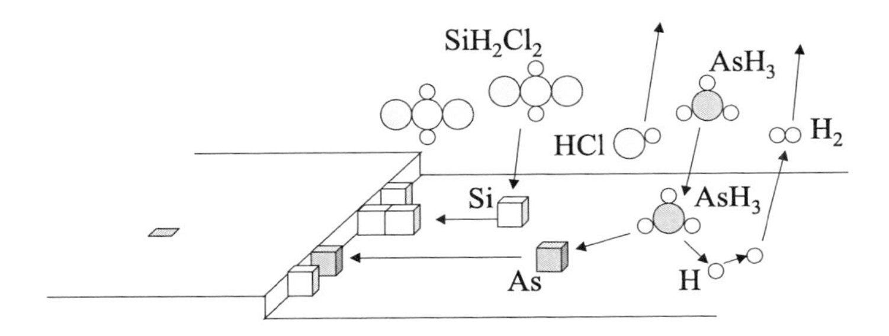

Figure 4.20 Schematic of epitaxial silicon layer growth and doping process.

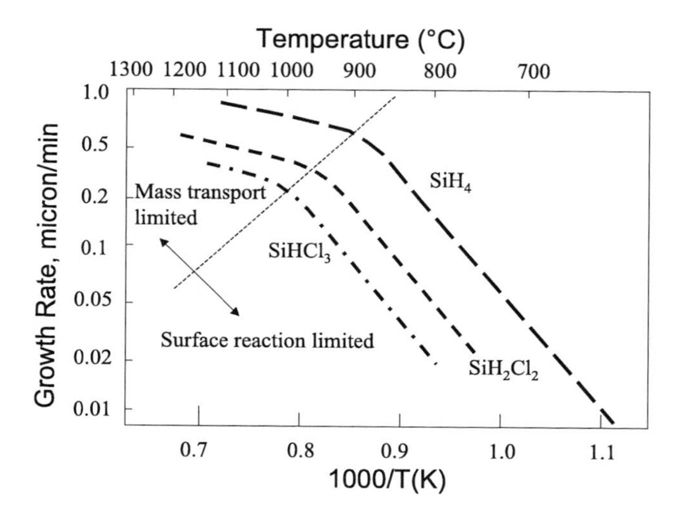

Figure 4.21 Epitaxial silicon film growth rate versus temperature [redrawn from F. C. Everstyn, *Phillips Research Reports* **29** (1974)].

At lower temperatures (550 to 650 °C) in a low-pressure reactor, a silane-based reaction can be used to deposit polycrystalline silicon on a single-crystal silicon wafer surface. This is because at lower temperatures, surface mobility of the adatoms is lower. Multiple nucleation sites form on the surface, causing different crystal grains to grow and form a polycrystalline silicon layer. At even lower temperatures (<550 °C), the silane-based process deposits amorphous silicon because of the extremely low surface mobility of the thermally dissociated radicals from silane molecules, SiH_3, SiH_2, and SiH.

4.5.3 Epitaxy hardware

There are two kinds of epitaxy systems, the batch system and the single-wafer system. The batch system can process multiple wafers at a time. As a result, it has a higher throughput. Three different batch reactors are widely used in the semiconductor industry; they are the barrel reactor, vertical reactor, and horizontal reactor, as illustrated in Fig. 4.22.

All three reactors have their own advantages and disadvantages. Barrel systems have very good uniformity but need extensive preventive maintenance when operating temperatures are greater than 1200 °C. This makes them unsuitable for the >1200 °C high-temperature processes. The horizontal system is simpler and less expensive. However, using this system, it is difficult to control wafer-to-wafer uniformity because it is hard to control the process over the entire susceptor. The pancake vertical reactor normally has good uniformity but has formidable mechanical complexity. All of the batch systems have problems with large-size wafers, especially within wafer uniformity and wafer-to-wafer uniformity on 300-mm wafers. Therefore, the single-wafer epitaxy reactors introduced in the 1990s are becoming more popular. Figure 4.23 gives a schematic of a single wafer epitaxy reactor.

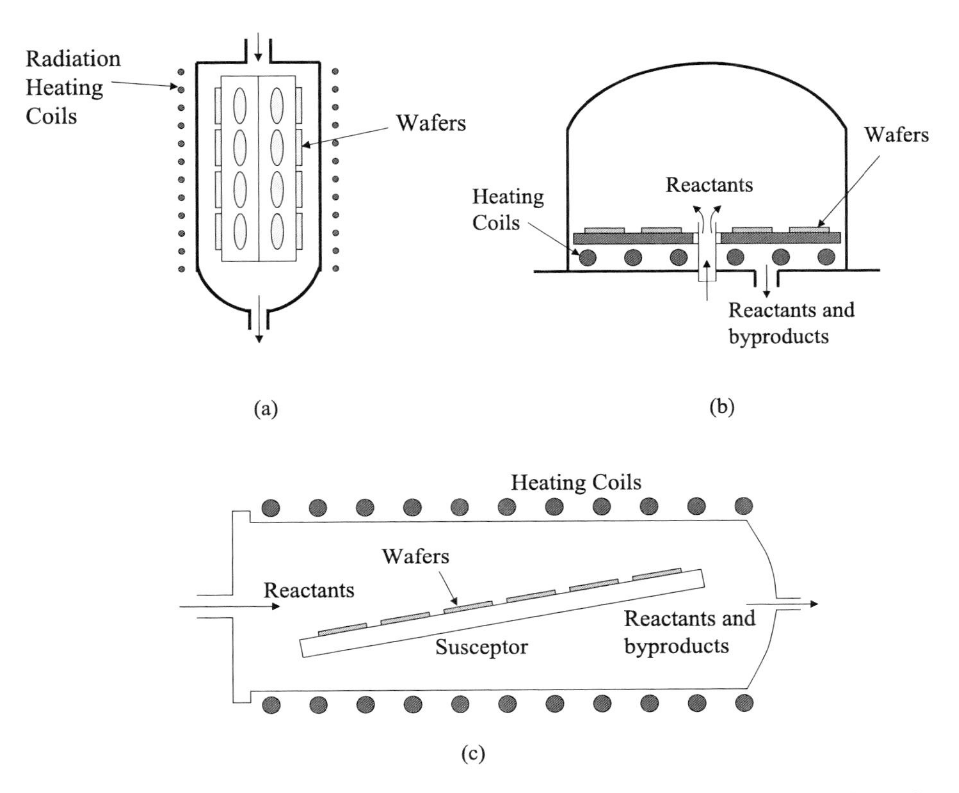

Figure 4.22 Three batch epitaxy reactors: (a) barrel, (b) vertical, and (c) horizontal.

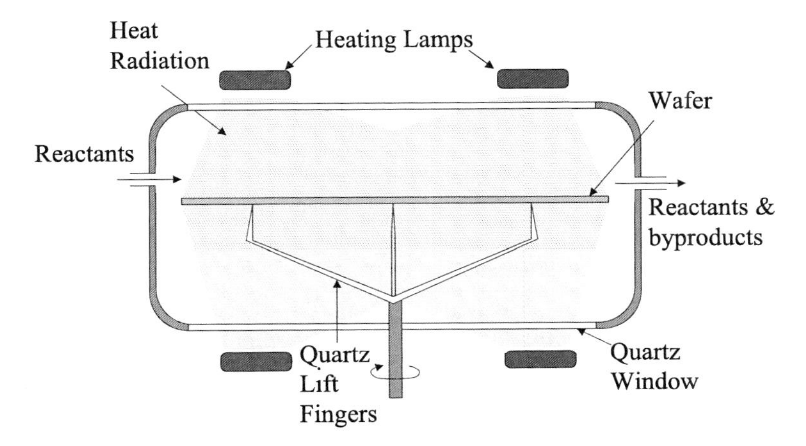

Figure 4.23 Schematic of single-wafer epitaxy system.

The single-wafer epitaxy reactor has a high epitaxial-layer growth rate and better reliability and repeatability than batch systems. It can deposit high-quality, low-cost film at both atmospheric and reduced pressures.

4.5.4 Epitaxy process

Epitaxy processes for these batch reactors are similar and have several processing steps. First, wafers are loaded into the chamber on the susceptor, then the chamber is closed and purged with hydrogen. After the hydrogen purge, the temperature is raised to 1150 to 1200 °C, and HCl then flows into the chamber for about three minutes to clean the chamber surface and etch the wafer surface to remove native oxide, particles, and surface defects. This helps minimize the mobile ions, especially those due to sodium contamination. Next, the chamber temperature is adjusted to the required process temperature. After the temperature is stabilized, silicon and dopant precursor gases flow into the chamber to grow the epitaxial silicon layer at a growth rate of 0.2 to 4.0 μm/min, depending on the process conditions (mainly pressure, gas flow rate, and temperature). After the epitaxial film is grown, process gases are eliminated, and the heater power is turned off. Hydrogen again flows into the chamber to purge the processing gases. When the temperature drops low enough, nitrogen purges the chamber to an ambient temperature, and the chamber is ready to open, unload, and reload. The whole process takes about an hour and can process 10 to 28 wafers at one time, depending on wafer size and reactor type.

For a single-wafer system, the process is similar. One difference is that the chamber does not need to cool down to an ambient temperature for loading and unloading. Because there is only one wafer, thermal capacity is small. When heated by an array of heating lamps, the wafer temperature can ramp-up very quickly in the single-wafer epitaxial system. After epitaxial-layer deposition, the wafer is removed from the deposition chamber by a transfer robot and sent to a cool-down chamber before it is placed into a plastic wafer cassette or a front-opening unified pod (FOUP).

Question: Nitrogen is the most commonly used purge gas in every type of processing equipment in semiconductor fabs. Why does epitaxy silicon processing use hydrogen instead of nitrogen as the main purge gas?

Answer: Nitrogen is very stable and abundant; 78% of the atmosphere is nitrogen. This makes it the most cost-effective gas to use for purging chambers and gas lines. However, at >1000 °C, nitrogen is no longer inert and can react with silicon to form silicon nitride, which can affect the epitaxial silicon deposition process. Therefore, hydrogen is used for the epitaxy chamber purge because it helps clean the wafer by forming gaseous hydride with contaminants on the wafer surface. Some facts of hydrogen are listed in Table 4.3.

Several possible defects are shown in Fig. 4.24. Dislocation of the epitaxial layer can be caused by dislocation of the substrate. This shows up as a slip or particulate

Table 4.3 Facts about hydrogen.

Symbol	H
Atomic number	1
Atomic weight	1.00794
Discoverer	Henry Cavendish
Discovery location	England
Discovery date	1766
Origin of name	From the Greek words hydro and genes, meaning water and generator.
Molar volume	11.42 cm^3
Velocity of sound	1270 m/sec
Refractive index	1.000132
Melting point	−258.99 °C
Boiling point	−252.72 °C
Thermal conductivity	0.1805 W m^{-1} K^{-1}
Main applications in IC processing	Epitaxial deposition, wet oxidation, premetal deposition reactive cleaning, and tungsten CVD
Main source	H_2

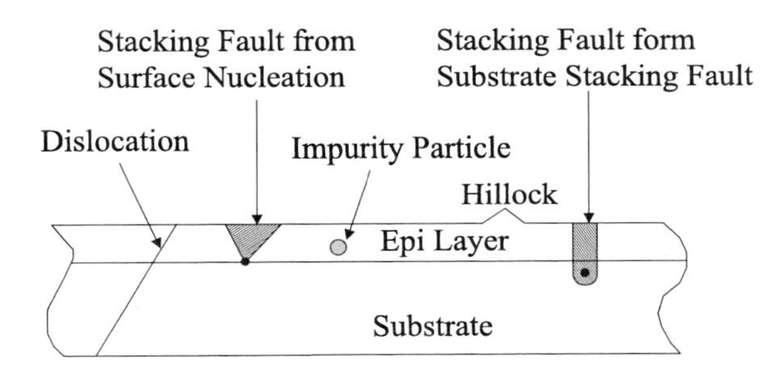

Figure 4.24 Different defects.

contamination on the wafer surface and can cause nucleation and stacking faults in the epitaxial layer. In film, particles can be generated during the epitaxy growth process due to spreading of stacking defects from the substrate surface to the epitaxial film.

4.5.5 Future trends of epitaxy

While device geometry continues to shrink and performance improves, higher quality, defect-free epitaxial layers with minimal thickness are required. Conventional epitaxial-layer growth requires a very high processing temperature, which can cause an autodoping effect and limit the minimum epitaxial-layer thickness. By reducing the epitaxy growth temperature, an abrupt transition between the epitaxial layer and substrate can be achieved. Therefore, low-temperature epitaxy (LTE) growth is of great importance. One way to reduce the epitaxy growth temperature is by reducing the processing pressure. Currently, reduced-pressure epitaxial growth processes operate at 40 to 100 torr and require about a 1000 °C processing temperature. Operating temperatures can be reduced to 750 to 800 °C when the processing pressure is further reduced to 0.01 to 0.02 torr.

Ultrahigh-vacuum CVD (UHV-CVD) epitaxial growth processes at 10^{-6} to 10^{-9} torr with low temperatures of 550 to 650 °C have been in the research and development phase for quite some time. It is likely to be the most promising LTE technology for the future of IC fabrication. However, atmospheric-pressure CVD (APCVD) epitaxy and molecule beam epitaxy (MBE) technologies could also be used for LTE applications in the future.

Another possible application of the UHV-CVD epitaxy process is growing a silicon-germanium (SiGe) epitaxial layer on a silicon substrate with SiH_4 and GeH_4. SiGe has higher hole mobility and therefore can be used to make faster CMOS IC devices. Silicon-carbine (SiC) and SiGeC epitaxy processes are also being researched and could be applicable for future IC devices.

4.5.6 Selective epitaxy

Selective epitaxy is another promising development. It uses silicon dioxide or silicon nitride as the epitaxy mask; thus, the epitaxial layer only grows on the area where silicon is exposed, a characteristic that can help to increase device packing density and reduce parasitic capacitance. Figure 4.25 illustrates the selective epitaxy process.

The selective epitaxial growth (SEG) process illustrated in Fig. 4.25 has been widely used to deposit SiGe in the pMOS S/D area to help create uniaxial compressive strain in the pMOS channel, increasing hole mobility and improving pMOS drive current and speed. It can also be used to deposit SiC in the nMOS S/D region to help create tensile strain in the nMOS channel, increasing electron mobility and improving nMOS drive current and speed. The SEG process can also be used in hybrid orientation technology, which allows two different silicon orientations to be on the same wafer surface, so that nMOS and pMOS can be built on different crystal orientations, maximizing carrier mobility and device performance. These are described in detail in the following section on substrate engineering.

4.6 Substrate Engineering

With the development of silicon IC technology, many methods have been employed to enhance MOSFET device performance. Global substrate engineering such as silicon-on-insulator (SOI), strained silicon, and strained silicon-on-insulator (SSOI) are some examples. In IC manufacturing, more fabs apply localized substrate engineering such as hybrid orientation technology (HOT), stress liners, selective epitaxial SiGe, and selective epitaxial SiC.

4.6.1 Silicon-on-insulator wafer

An SOI wafer allows semiconductor manufacturers to completely isolate IC devices from their neighbors, thus reducing interference and leakage, and in turn improving device speed and performance. There are two ways to make an SOI wafer: one method uses heavy oxygen implantation and high-temperature annealing, and the other uses hydrogen implantation and wafer bonding.

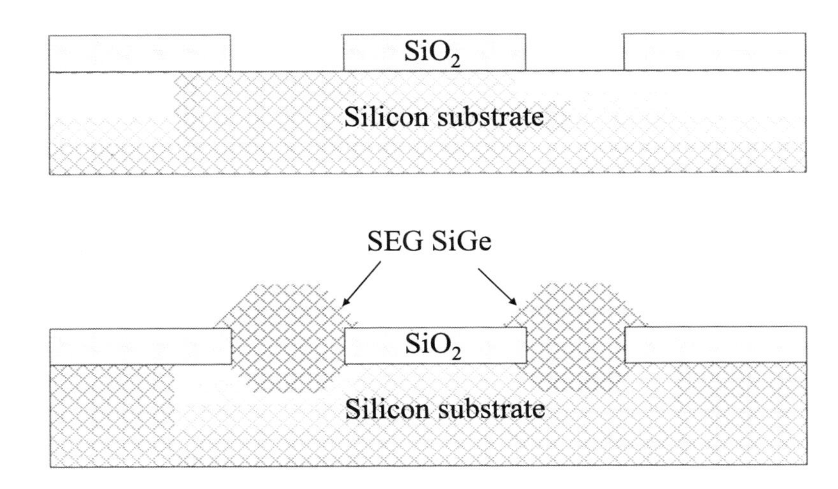

Figure 4.25 The selective epitaxy silicon process.

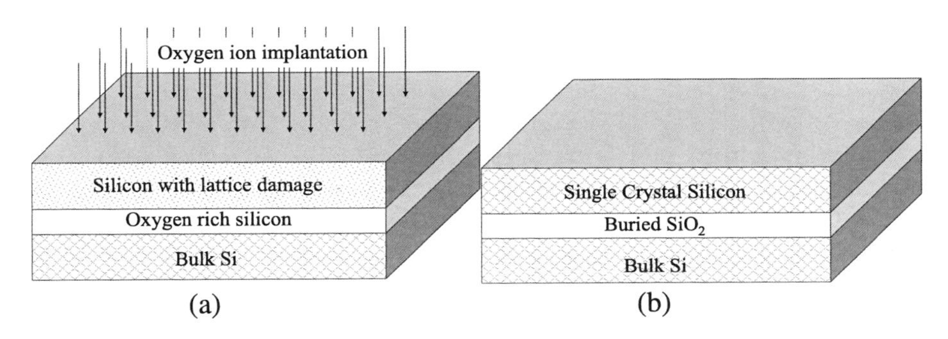

Figure 4.26 SOI wafer formation with (a) oxygen ion implantation and (b) high-temperature annealing.

Figure 4.26 illustrates the SOI formation of the first approach—separation by implantation of oxygen (SIMOX). First, a high-energy, high-current implanter (up to 10^{18} ions/cm^2) is used to implant oxygen ions deep into the silicon substrate to form an oxygen-rich silicon layer. A thermal annealing process at a high-temperature (~1400 °C) causes a chemical reaction between silicon and oxygen atoms and forms the buried silicon dioxide layer, while silicon near the surface relaxes and recovers the single-crystal structure. The thickness of the silicon on top of the oxide is determined by the oxygen implantation energy, and the thickness of the buried oxide is determined by the number of oxygen atoms.

The other method is called bonded SOI, which uses hydrogen implantation and wafer bonding (Fig. 4.27). First, wafer A is cleaned, and a silicon dioxide layer is grown with the thickness required for the buried oxide. Wafer A is then is implanted with hydrogen to form a hydrogen-rich layer [see Figs. 4.27(a) and 4.27(b)]. The depth of the hydrogen-rich layer is determined by the thickness of the silicon on top of the buried oxide. Wafer A is then flipped and bonded with wafer B, the handler wafer [Fig. 4.27(c)]. In a thermal process, silicon dioxide on the surface of wafer

A forms chemical bonds with the silicon surface of wafer B, while the hydrogen in wafer A bubbles up inside the silicon, causing wafer A to split [Figs. 4.27(d) and 4.27(e)]. After CMP and cleaning, the SOI wafer is ready, as shown in Fig. 4.27(f).

After processing, wafer A can be polished, cleaned, and recycled to become a new wafer A or B. The major advantage of bonded SOI is the cost. Because the hydrogen dosage needed to split a silicon wafer is significantly lower than the oxygen dosage needed to form buried oxide, bonded SOI has higher throughput than SIMOX SOI. Most SOI wafers used in IC manufacturing are made with bonded SOI technology.

4.6.2 Hybrid orientation technology

By using a ⟨110⟩ orientation wafer as wafer A in Fig. 4.27 and a ⟨100⟩ orientation wafer as wafer B, a hybrid orientation SOI wafer can be made, as shown in Fig. 4.28(a). Using the selective epitaxial growth technique illustrated in Fig. 4.25, hybrid ⟨110⟩ and ⟨100⟩ orientations can be achieved on the wafer surface [see Fig. 4.28(c)]. Figure 4.28(d) shows CMOS devices in which pMOS is made on a ⟨110⟩ substrate, while nMOS is made on a ⟨100⟩ substrate.

Because hole mobility is higher in the ⟨110⟩ orientation than in the ⟨100⟩ orientation of single-crystal silicon, a CMOS IC made with HOT could have a higher driving current and faster speed than an IC made with a regular ⟨100⟩ substrate. The tradeoff is cost: the CMOS IC would need an SOI wafer and an additional photolithography mask layer. If other techniques such as strain silicon

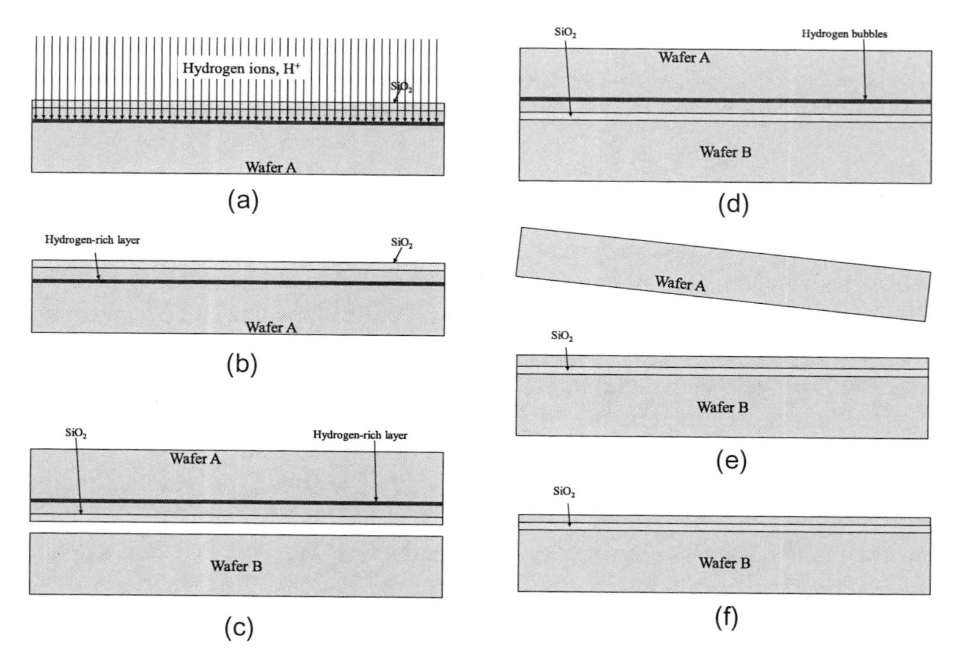

Figure 4.27 Bonded SOI formation process.

can achieve the same performance gain at a lower cost, it would be difficult for HOT to become mainstream in CMOS technology.

4.6.3 Strained silicon wafer

By heavily stressing single-crystal silicon, silicon lattices are stretched or compressed beyond their normally interatomic distance, forming what is called strained silicon. Carrier mobility in strained silicon can be significantly higher than that in strain-free silicon. Strained silicon can be achieved by depositing silicon onto a SiGe surface. Gradient SiGe with gradually increasing Ge concentration allows growth on a silicon surface without much lattice mismatch. After relaxed-SiGe layer deposition, an epitaxial silicon layer is deposited onto the SiGe surface, with a lattice structure that matches the SiGe underneath and the strained silicon layer above (see Fig. 4.29).

4.6.4 Strained silicon-on-insulator wafer

Using the strained silicon wafer described in Fig. 4.29(a) as wafer A of the bonded SOI process (shown in Fig. 4.27), an SSOI wafer can be made. These types of wafers have the high carrier mobility of strained silicon and the high device packing density of SOI (see Fig. 4.30).

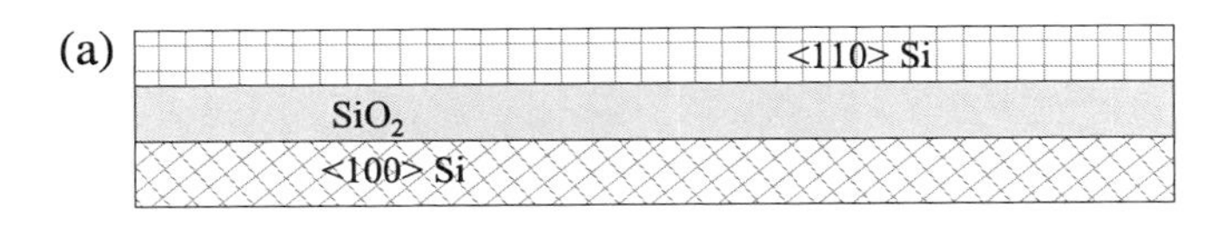

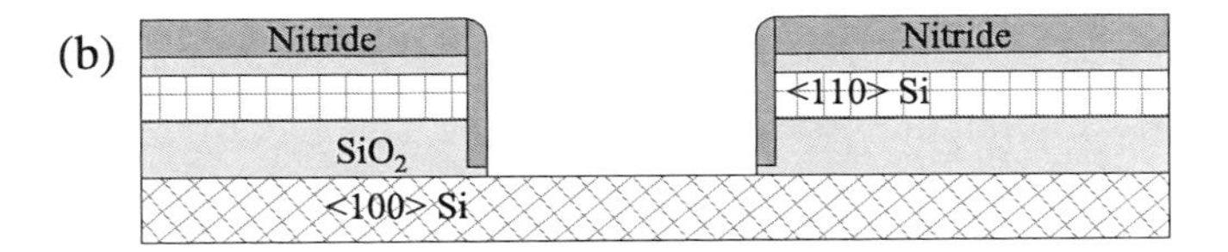

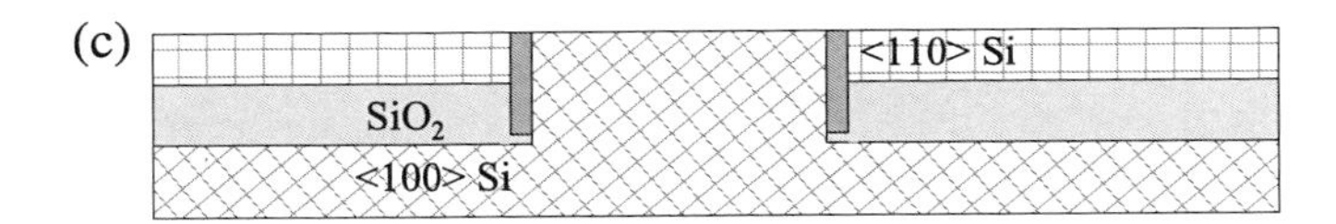

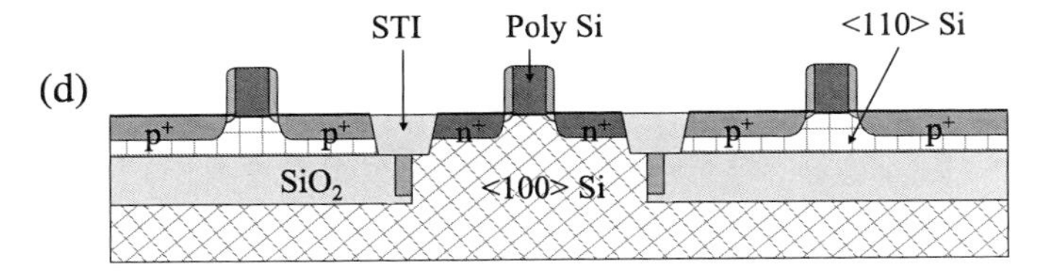

Figure 4.28 Hybrid orientation technology.

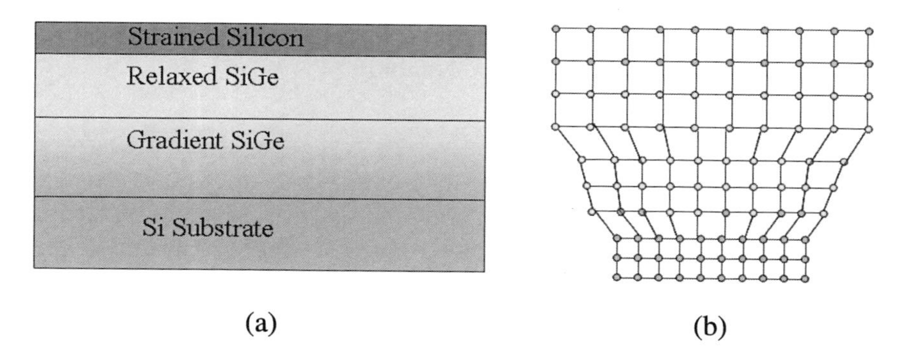

Figure 4.29 (a) Formation and (b) lattice of strained silicon.

The strained silicon and SSOI wafers described in this section are wafers with strained silicon on their top surfaces that is used for making microelectronic and nanoelectronic devices. However, scientists and engineers have found that by using existing process technologies, they can create strained silicon locally on a normal silicon wafer. Because only the channel underneath the gate oxide of a MOSFET needs to be strained, it is not necessary to strain the entire surface of a silicon wafer. Also, pMOS and nMOS channels require different types of strain; pMOS prefers compressive strain to enhance hole mobility, while nMOS needs tensile strain to enhance electron mobility. A strained silicon wafer cannot satisfy both needs.

4.6.5 Strained silicon in integrated circuit manufacturing

IC manufacturers started using strained silicon in MOSFET devices when the feature size shrunk into nanometer technology nodes. Dual stress liners applied compressive stress liners on pMOS and tensile stress liners on nMOS to boost the mobility of holes in the p-channel and electrons in the n-channel, respectively.

Because stress liners place stress mainly on the gate electrode and not directly on the channel, their effectiveness on channel strain is limited. As devices continue to shrink, fabs have found that by using selective epitaxial SiGe, they can generate higher compressive strain on the pMOS channel. Figure 4.31 shows a CMOS cross section with hybrid strain techniques, in which a pMOS channel compressive strain is achieved by SEG SiGe, and an nMOS channel tensile strain is achieved by a tensile stress liner. Recessed nMOS S/G allows for more effective channel strain from the stress liner.

Figure 4.32 illustrates the cross section of a state-of-the-art 22-nm high-κ metal gate CMOS with an SEG SiGe pMOS S/G that strains the p-channel compressively, and an SEG SiC nMOS S/G that strains the n-channel with tensile stress.

4.7 Summary

- Silicon is an inexpensive semiconductor material, and silicon dioxide is a strong and stable dielectric that can be easily grown on silicon surfaces in a thermal process.

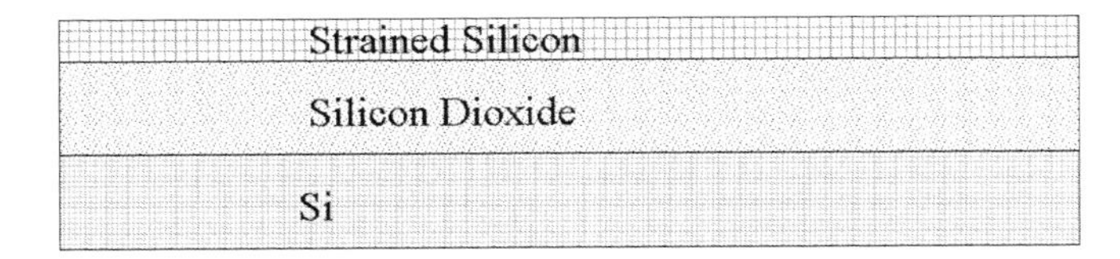

Figure 4.30 Strained SSOI.

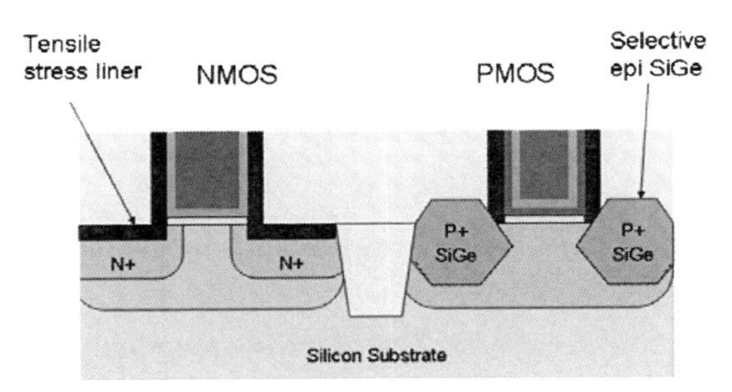

Figure 4.31 Cross section of a CMOS device with selective epitaxial SiGe for pMOS and tensile stress liner for nMOS.

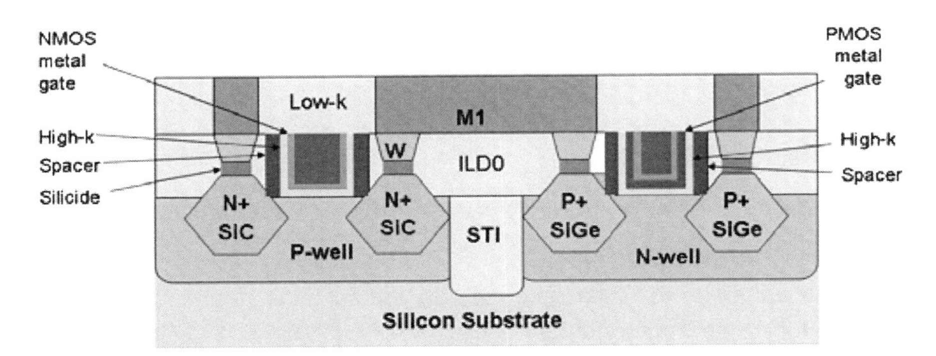

Figure 4.32 Cross section of a CMOS device with SEG SiGe for pMOS and SEG SiC for nMOS.

- The most commonly used silicon crystal orientations are ⟨100⟩ and ⟨111⟩.
- The wafer fabrication process converts sand (silicon dioxide) to MGS, MGS to TCS, TCS to EGS, EGS to single-crystal silicon ingots, and ingots to wafers.
- Both the CZ and floating zone methods are used for wafer manufacturing, with the CZ method being more common.
- The CZ method is cheaper and can produce larger-sized wafers.
- The floating zone method can produce wafers with higher purity.
- The epitaxial silicon layer is needed for bipolar devices and improves performance for CMOS and DRAM.
- The IC industry uses high-temperature CVD processes to grow epitaxial silicon layers.

- Most SOI wafers are made by bonded SOI technology.
- HOT uses SOI wafers and selective epitaxial growth. It allows IC manufacturers to make pMOS on ⟨110⟩ substrate and nMOS on ⟨100⟩ substrate to improve CMOS IC performance.
- Strained silicon techniques are widely applied in IC devices in the nanometer technology node.

4.8 Bibliography

M. S. Bawa, E. F. Petro, and H. M. Grimes, "Fracture strength of large-diameter silicon wafers," *SI Mag.* **18**(11), 115–118 (1995).

C. Y. Chang and S. M. Sze, *ULSI Technologies*, McGraw-Hill, New York (1996).

F. C. Eversteyn, "Chemical-reaction engineering in the semiconductor industry," *Philips Res. Rep.* **29**, 45–66 (1974).

L. Shon-Roy, A. Wiesnoski, and R. Zorich, *Advanced Semiconductor Fabrication Handbook*, Integrated Circuit Engineering Corp., Scottsdale, AZ (1998).

S. M. Sze, *VLSI Technology, Second Edition*, McGraw-Hill, New York (1988).

M. Yang et al., "High-performance CMOS fabricated on hybrid substrate with different crystal orientations," *IEDM Tech. Dig.*, 453–456 (2003).

See http://www.webelements.com for further information on the elements in this chapter.

4.9 Review Questions

1. Why are single-crystal materials needed for IC chip manufacturing?
2. Draw the ⟨100⟩ and ⟨111⟩ planes from a cube.
3. What are the main reasons silicon wafers are far more popular than any other semiconductor wafers in the IC industry?
4. What is the chemical used to purify MSG to ESG? List its safety hazards.
5. How does the CZ puller work? Why do wafers made from the CZ puller have higher oxygen concentration than wafers manufactured with the floating zone method?
6. Explain the purpose of the epitaxy process.
7. What is the autodoping effect? How can it be avoided?
8. List three epitaxial silicon precursors.
9. List the three dopants commonly used for epitaxy doping. What are the safety hazards of these dopant gases?
10. What are the advantages of the single-wafer epitaxy reactor over batch epitaxy systems?
11. What kind of ion implantation is required to make a bonded SOI wafer? What kind of ion implantation is required to make a SIMOX SOI wafer?

12. Explain why most IC manufacturers use local strain instead of strained silicon wafers to make MOSFETs.
13. Explain why more IC manufacturers use bulk silicon wafers and local strain techniques to make advanced IC chips instead of using HOT.

Chapter 5
Thermal Processes

One of the advantages of the silicon substrate over other semiconductor substrates is the high-temperature processing capability of silicon. Silicon wafer processing involves many high-temperature (700 to 1200 °C) procedures. This chapter covers these high-temperature thermal processes, including diffusion, oxidation, deposition, and annealing, in both standard furnace and rapid thermal processing.

Objectives

After finishing this chapter, the reader will be able to:

- list at least three important thermal processes
- describe the basic systems of vertical and horizontal furnaces, and list the advantages of a vertical furnace
- describe the oxidation process
- explain the importance of preoxidation cleaning
- identify the difference between dry and wet oxidation processes and applications
- describe the diffusion process
- explain why ion implantation replaced diffusion for silicon doping
- name at least three high-temperature deposition processes
- explain the reasons for postimplantation annealing
- describe the advantages of rapid thermal processing.

5.1 Introduction

The natural oxide of silicon, silicon dioxide, is a very stable and strong dielectric material that can be easily formed in a high-temperature process. This is one of the important reasons why silicon dominates the IC industry as a semiconductor substrate material. IC fabrication processing starts with an oxidation process, which grows a layer of silicon dioxide to protect the silicon surface. Silicon wafers then go through thermal processing in high-temperature furnaces and using rapid thermal processing (RTP) tools many times during the IC processing flow. Figure 5.1 shows the IC fabrication process flow.

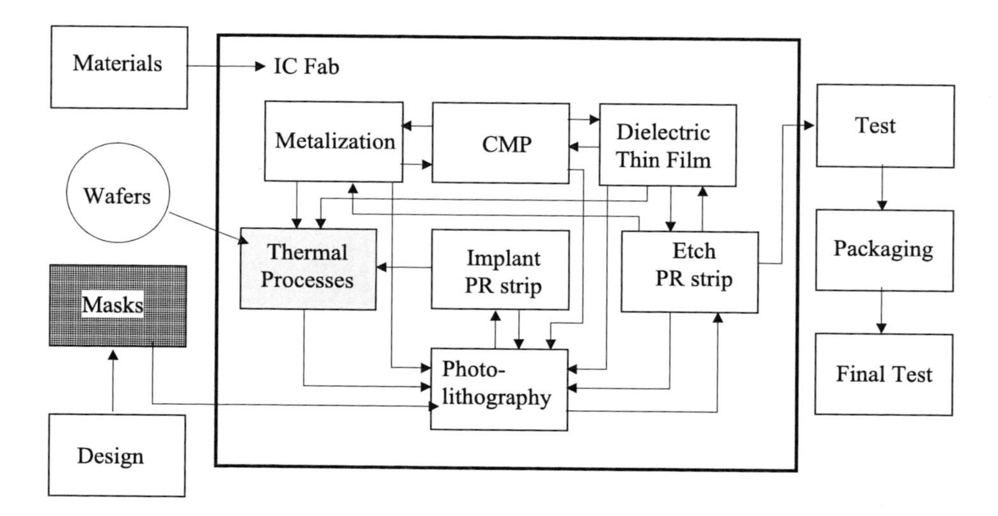

Figure 5.1 IC fabrication process flow.

5.2 Thermal Process Hardware

5.2.1 Introduction

Thermal processes occur in a high-temperature furnace called a diffusion furnace, because it was extensively used in the diffusion doping process in the early years of the semiconductor industry. There are two kinds of furnaces, horizontal and vertical, determined by the way the quartz tubes and heating elements are placed in the system. Typical requirements for a furnace include uniformity, accurate temperature control, low particulate contamination, high productivity, high reliability, and low cost.

A furnace consists of five basic components: a control system, process tube, gas delivery system, exhaust system, and loading system. The furnace used for low-pressure CVD (LPCVD) processing also requires a vacuum system. Figure 5.2 illustrates a schematic for a horizontal furnace.

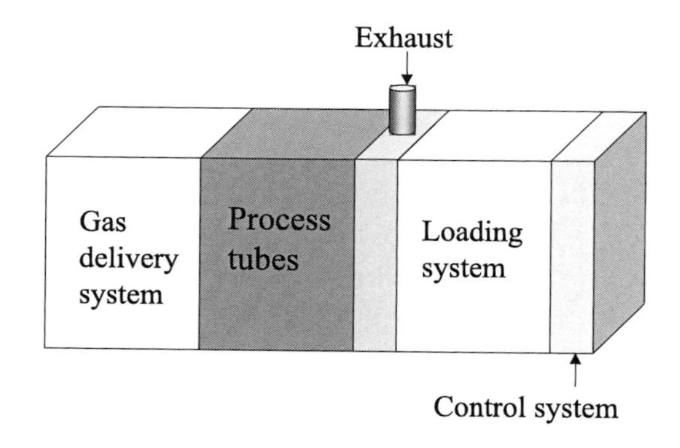

Figure 5.2 Layout of a horizontal furnace.

A vertical furnace has many advantages over a horizontal furnace, such as lower particle contamination, the ability to handle a large number of heavy wafers, better uniformity, lower maintenance cost, and a smaller footprint. It is very important for a processing tool to have a small footprint, since high-grade cleanroom space is very expensive in advanced fabs. Also, because wafers are placed in vertical stacks, large particles fall only onto the top wafer and do not reach the wafers underneath. Vertical furnaces are more commonly used in advanced semiconductor fabs.

Most of the parts in a thermal process furnace, such as the wafer boat, paddle, and wafer tower, are made of fused quartz. Quartz is a single-crystal silicon dioxide, which is a very stable material even at high temperatures. Its drawbacks are fragility and metallic impurities. Because quartz is not a sodium barrier, a trace amount of sodium can always penetrate the tube and cause damage to devices on the wafer. When the temperature is higher than 1200 °C, small flakes can occur and cause particle contamination.

SiC is another material used for high-temperature furnaces. Compared with quartz, SiC has a higher thermal stability and is a better mobile-ion barrier. The disadvantages of SiC are that it is heavier and more expensive. As device dimensions are further reduced, more SiC parts will be used in furnaces to meet processing requirements.

5.2.2 Control system

The controller consists of a computer connected to several microcontrollers. Each microcontroller connects to another system control interface board, which controls processing sequences such as wafer loading and unloading, processing times for each step, processing temperatures and temperature ramp rates, processing gas flow rates, exhaust, etc. It also collects and analyzes processing data, programs the processing sequence, and tracks the lot numbers. Figure 5.3 shows a block diagram of a furnace control system.

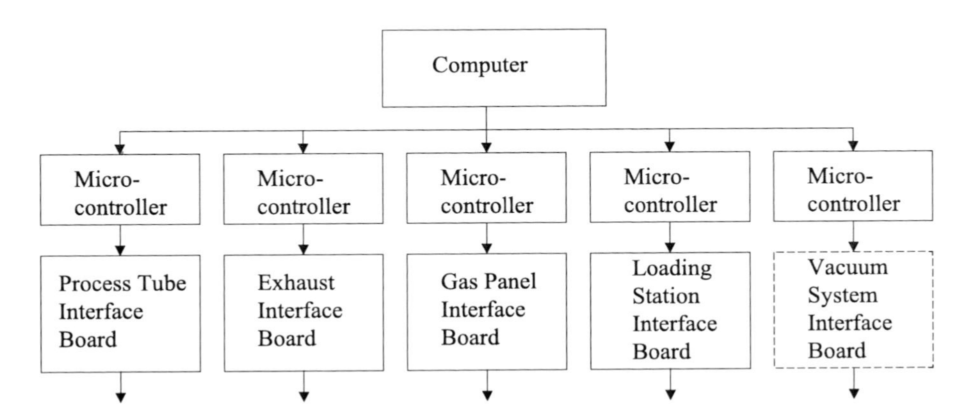

Figure 5.3 Block diagram of the control system of a furnace.

5.2.3 Gas delivery system

The gas delivery system handles processing gases and delivers them to the processing tube as required. A gas panel consists of regulators, valves, mass flow controllers (MFCs), and filters; it distributes processing and purge gases to the processing tube as needed. Processing gases are usually stored in high-pressure (more than 100 psi) bottles in remote gas cabinets. Processing gases are throttled to the gas panel by dozens of pounds per square inch (too high to be directly used in the process). Regulators and valves monitor the processing gas pressure and control pressure. Gas flow rate is precisely controlled by the MFC, which controls the rate by adjusting an internal control valve so that the measured flow rate equals the set value. Filters in the gas panel help minimize particulate contamination by preventing particles from flowing into the processing tube with the gases. Figure 5.4 shows a schematic of a gas delivery system.

5.2.4 Loading system

The loading station is used for wafer loading, unloading, and temporary storage. It loads wafers from cassettes into quartz boats, which sit on a quartz paddle; then the paddle is gently pushed into the processing tube by a loading mechanism, which is controlled by the microcontroller. After thermal processing, the paddle is gradually pulled out of the processing tube.

Several paddle types have been used in the loading systems of horizontal furnaces. Paddles with wheels are no longer used due to the particle generation caused by direct contact between the paddle and quartz tube. Soft-landing paddles, which gradually land on the quartz tube after the paddle is pushed into the desired position inside the processing tube, were developed next. Due to direct contact between the paddle and quartz tube, soft-landing paddles caused particulate contamination from friction between the paddle and tube wall during paddle landing and raising. To avoid this direct contact, suspended or cantilevered paddles are currently used in most systems. They allow the paddle to move in and out of the processing tube without any direct contact with the quartz surface, minimizing particle contamination. However, wafer loading can affect the paddle's suspension.

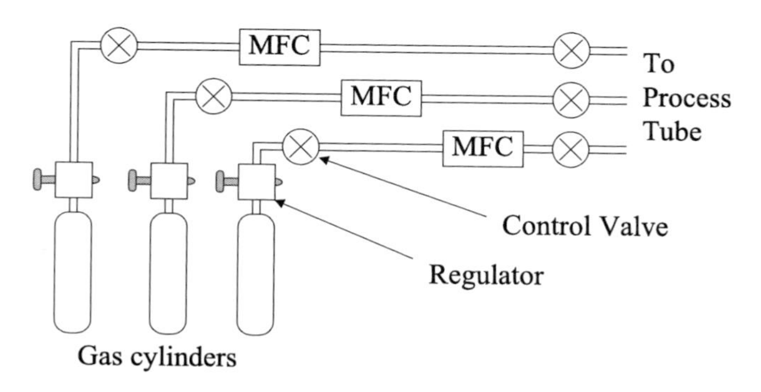

Figure 5.4 Schematic of the gas delivery system.

For a vertical furnace, wafers are loaded from the cassette into the tower by a robot. Then a lift mechanism (controlled by the microcontroller) raises the tower into the processing chamber, or the chamber elevator lowers the chamber until the entire tower is inside the chamber. This eliminates contact between the wafer holder and the quartz chamber wall, and there is no suspension problem for the wafer holder.

5.2.5 Exhaust system

Process byproducts and unused source gases are removed from the processing tube or chamber by an exhaust system. Exhaust gas is evacuated from the processing tube symmetrically, and purge gas is drawn in from the exhaust manifold to prevent backstreaming of the exhaust gas. For furnace processes involving pyrophoric or flammable gases, such as silane (SiH_4) and hydrogen (H_2), an additional chamber called a burn box is required. In the burn box, exhaust gases pass through a controlled combusting process in the presence of oxygen and are reduced to less harmful, less reactive oxide compounds. Filters are used to remove particles generated in the burning process, such as silicon dioxide, after burning silane. A scrubber with water or an aqueous solution absorbs most of the toxic and corrosive gases before the exhaust gas is vented to the atmosphere.

5.2.6 Processing tube

Wafers undergo a high-temperature procedure in the processing tube, which consists of a quartz tube chamber body and several heaters. Thermocouples touch the chamber wall and monitor chamber temperature. Each heater is independently supplied by a high-current power source. The power of each heater is controlled by feedback from the thermocouple data through the interface board and microcontroller and becomes stable when the setting temperature is reached. Temperature in the flat zone at the center of the tube is precisely controlled within 0.5 °C at 1000 °C. Figure 5.5 shows schematics of horizontal and vertical furnaces.

In a horizontal furnace, wafers are placed on quartz boats, which sit on a paddle made of SiC. The paddle (with wafer boats) is then slowly pushed into the quartz tube where it places the wafers in the flat zone of the furnace for thermal processing. After processing, the wafers must be pulled out very slowly to avoid warping caused by the overwhelming thermal stress of sudden temperature change.

In a vertical furnace, wafers are loaded into a wafer tower made of quartz or SiC. The wafers are placed in the tower face up, then the tower is slowly raised into the quartz tube to heat. Afterward, the tower is lowered slowly to avoid wafer warping.

5.3 Oxidation

The oxidation process is one of the most important thermal processes. It is an adding process that adds oxygen to a silicon wafer to form silicon dioxide on the wafer surface. Silicon is very reactive to oxygen; thus in nature, most silicon

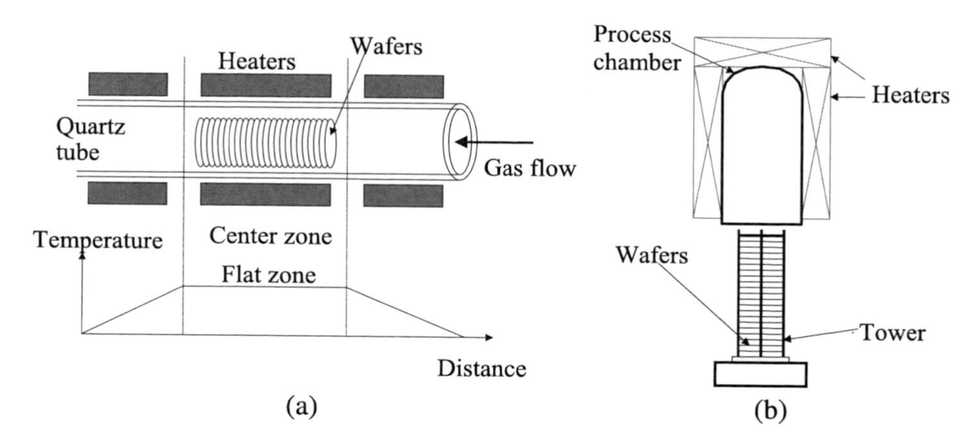

Figure 5.5 Schematics of (a) horizontal and (b) vertical furnaces.

exists in the form of silicon dioxide, such as quartz sand. It reacts with oxygen very quickly, and forms silicon dioxide on the silicon surface. The reaction can be expressed as

$$Si + O_2 \rightarrow SiO_2.$$

Silicon dioxide is a dense material that fully covers the silicon surface. To continue the oxidization of silicon, oxygen molecules have to diffuse across the oxide layer to reach the silicon atoms underneath and react with them. The growing silicon dioxide layer increasingly blocks and slows the oxygen. When bare silicon is exposed to the atmosphere, it reacts almost immediately with oxygen or moisture in the air and forms a thin layer (about 10 to 20 Å) of silicon dioxide, called native oxide. The layer of the native oxide is thick enough to stop further oxidation of the silicon at room temperature. Figure 5.6 illustrates the oxidation process.

In the oxidation process, oxygen comes from the gas phase, and silicon comes from the solid substrate. Therefore, while silicon dioxide is growing, it consumes the substrate silicon, and the film grows into the remaining silicon substrate, as shown in Fig. 5.6. Oxygen is used as an oxidizer in processes such as thermal oxidation, CVD, and reactive sputtering deposition. It is also commonly used in etch and photoresist stripping processes. Oxygen is the most abundant element in the earth's crust and the second most abundant element in the earth's atmosphere, after nitrogen. Some facts about oxygen are listed in Table 5.1.

At high temperatures, thermal energy causes oxygen molecules to move much faster, which can drive them to diffuse across an existing oxide layer and react with silicon to form more silicon dioxide. The higher the temperature, the faster the oxygen molecules, and the quicker the oxide film growth. The oxide film quality is also better than that grown at lower temperatures. Therefore, to get high-quality oxide film and fast growth rate, oxidation processes are always performed in a high-temperature environment, normally in a quartz furnace. Oxidation is a slow

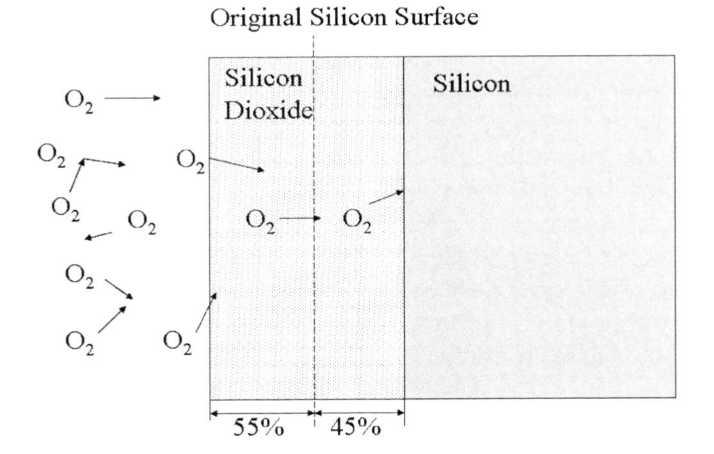

Figure 5.6 Silicon oxidation process.

Table 5.1 Facts about oxygen.

Symbol	O
Atomic number	8
Atomic weight	15.9994
Discoverer	Joseph Priestley, Carl Scheele
Discovery location	England, Sweden
Discovery date	1774
Origin of name	From the Greek words oxy genes, meaning acid (sharp) and forming (acid former)
Molar volume	17.36 cm^3
Velocity of sound	317.5 m/sec
Refractivity	1.000271
Melting point	54.8 K = −218.35°C
Boiling point	90.2 K = −182.95°C
Thermal conductivity	0.02658 W m^{-1} K^{-1}
Applications	Thermal oxidation, oxide CVD, reactive sputtering, and photoresist stripping
Main sources	O_2, N_2O, O_3

process; even in furnaces hotter than 1000 °C, a thick oxide (>5000 Å) still takes several hours to grow. Therefore, oxidation processes usually are batch processes, with a large number (100 to 200) of wafers processing at the same time to achieve reasonable throughput.

5.3.1 Applications

Oxidation of silicon is one of the basic processes throughout the IC process. There are many applications for silicon dioxide; one of them is as diffusion mask. Most dopant atoms used in the semiconductor industry such as boron and phosphorus have much lower diffusion rates in silicon dioxide than they do in single-crystal silicon. Therefore, by etching windows on the masking oxide layer, a silicon substrate can be doped at a designated area by a dopant diffusion process, as shown in Fig. 5.7. The thickness of the masking oxide is about 5000 Å.

Screen oxides are also commonly used for ion implantation processes. They can help prevent silicon contamination by blocking the sputtered photoresist. They can also minimize the channeling effect by scattering incident ions before they enter a single-crystal silicon substrate. The thickness of a screen oxide is about 100 to 200 Å. Figure 5.8 illustrates a screen oxide for application of ion implantation.

Thermally grown silicon dioxide is used as a pad layer for silicon nitride in both local oxidation of silicon (LOCOS) and shallow trench isolation (STI) formation. Without the stress buffer from this pad oxide, LPCVD silicon nitride film could crack and in some cases even break a silicon wafer due to the high tensile stress (up to 10^{10} dynes/cm^2). The thickness of the pad oxide is about 150 Å.

Silicon dioxide is also used as a barrier layer to prevent contamination of the silicon substrate before trench fill in the STI process. Trench fill is a dielectric CVD process in which undoped silicate glass (USG) is deposited to fill the trench for electrical isolation of neighboring transistors. Since the CVD process always brings small amounts of impurities, a dense, thermally grown silicon dioxide barrier layer is necessary to block possible contamination. Figure 5.9 shows the pad and barrier oxides in the STI process.

One of most important applications of thermally grown silicon dioxide in the past was forming isolation blocks to electrically isolate neighboring transistors in an IC chip. Blanket field oxide and LOCOS were two kinds of blocks used

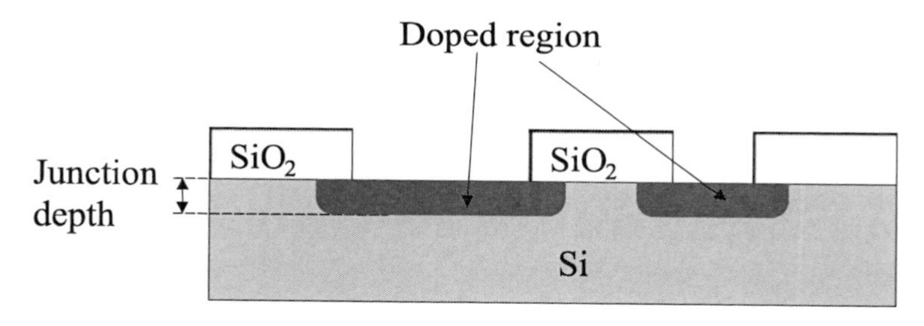

Figure 5.7 Diffusion masking oxide.

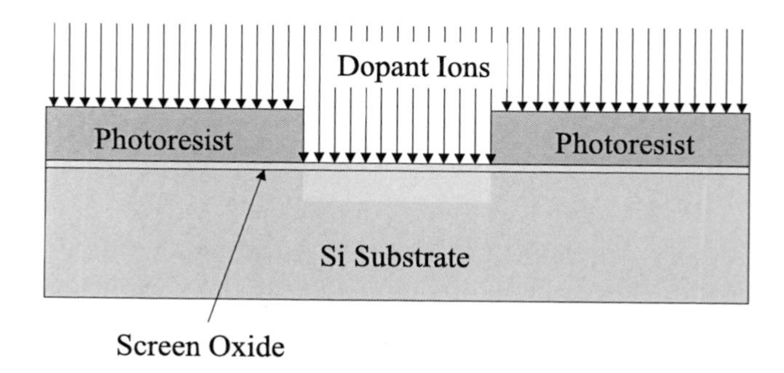

Figure 5.8 Screen oxide for application of ion implantation.

to isolate neighboring devices and prevent them from crosstalking. Blanket field oxide was the simplest isolation process and widely used in the early years of silicon IC manufacturing. By thermally growing a thick layer of silicon dioxide (5,000 to 10,000 Å), patterning it with photolithography, and etching the oxide with hydrofluoric acid (HF), the activation areas could be opened for transistor making, as shown in Fig. 5.10.

LOCOS has better isolation properties than blanket-field oxide. The LOCOS process uses a thin layer of oxide (200 to 500 Å) as the pad layer to buffer the strong tensile stress of LPCVD nitride. After the nitride etch, photoresist strip, and wafer cleaning, a thick layer of oxide (3000 to 5000 Å) is grown on the area not covered by silicon nitride. Silicon nitride is a much better barrier layer than silicon dioxide. Oxygen molecules cannot diffuse across the nitride layer; therefore, the silicon underneath the nitride layer does not oxidize. In the area not covered by the nitride, oxygen molecules continuously diffuse across the silicon dioxide layer,

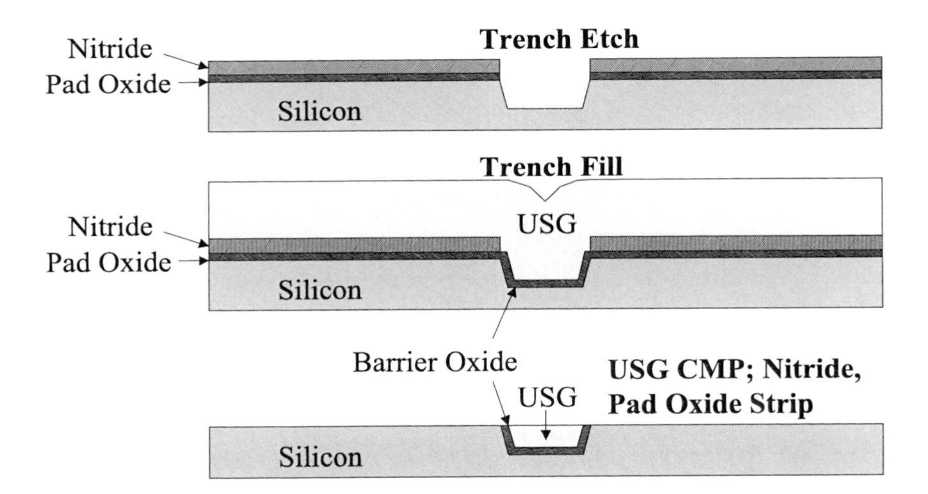

Figure 5.9 Pad oxide and barrier oxide in the STI process.

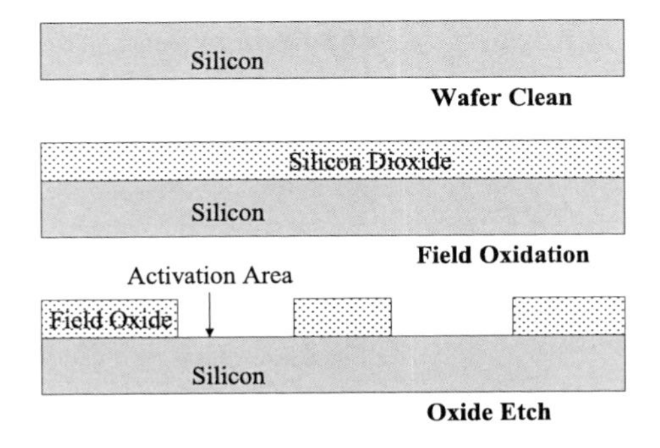

Figure 5.10 Blanket field oxide process.

where they react with silicon underneath to form more silicon dioxide. The LOCOS formation process is illustrated in Fig. 5.11.

Because oxygen diffusion inside silicon dioxide is an isotropic process, oxygen also reaches silicon at the side. This causes the oxide to grow underneath the nitride layer near the etched oxidation window, forming the so-called bird's beak (see Fig. 5.11). An undesirable consequence is that this bird's beak uses up a good deal of space on the wafer surface. Another disadvantage of LOCOS is the surface planarization problem due to the oxide-to-silicon surface step, caused by the oxide's growing characteristics, illustrated in Fig. 5.6.

Many different methods have been used to suppress the bird's beak. The most commonly used method is poly-buffered LOCOS (PBL). A thicker pad oxide allows a longer bird's beak to grow because of the wide oxide path for oxygen diffusion. By using a polysilicon layer (~ 500 Å) to buffer the high tensile stress of the LPCVD silicon nitride, the thickness of the pad oxide layer can be reduced from 50 to 100 Å, significantly reducing oxide encroachment. But, no matter how many different methods were tried, there was still an approximately 0.1- to 0.2-μm bird's beak at both sides of the LOCOS, a thickness that is intolerable when the minimum feature size is smaller than 0.35 μm. The STI process has been developed to avoid the bird's beak problem and has a more planarized surface topology as well. STI gradually replaced LOCOS isolation starting in the mid-1990s when device feature size was smaller than 0.35 μm.

Sacrificial oxide is a thin layer (< 1000 Å) of silicon dioxide grown on activation areas of a silicon surface and stripped in HF solution right after its growth. It is applied before the gate oxidation process to remove damages and defects on the silicon surface. This oxide growth and removal processing sequence helps to create a defect-free silicon substrate surface for growing a high-quality gate oxide layer.

The thinnest and most important silicon dioxide layer in a MOSFET-based IC chip is the gate oxide. While device dimensions have been shrinking, the thickness

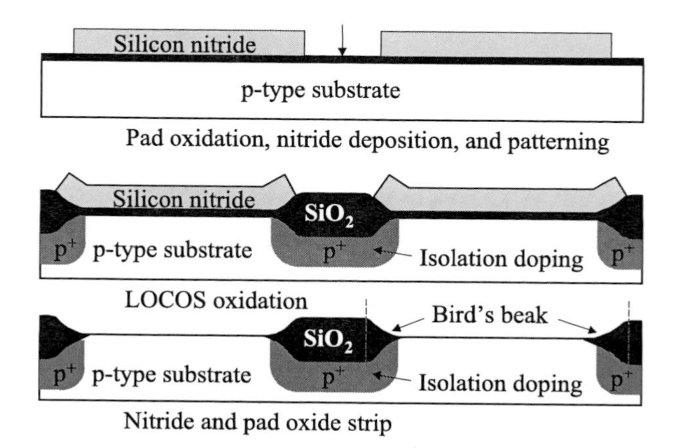

Figure 5.11 LOCOS process.

of gate oxides has been reduced from more than 1000 Å in the 1960s to about 15 Å in high-end chips in the mid-2000s; operation voltage of IC chips has been reduced from 12 to 1.0 V. The quality of gate oxide is vital for proper device function. Any defect, impurity, or particle contamination in a gate oxide can affect device performance and significantly reduce chip yield. Figure 5.12 illustrates the sacrificial oxidation and gate oxidation processes. Applications of thermally grown silicon dioxide in IC fabrication processes are summarized in Table 5.2. For nanometer technology node IC chips, the gate oxide is nitrided to increase its dielectric constant, thus it can increase in thickness with a higher breakdown voltage, while still possessing larger gate capacitance to maintain proper MOSFET switching properties. Also, a nitrogen-rich layer can block the diffusion of boron in heavily p-type doped polysilicon gate electrons into the n-type doped channel of pMOS.

5.3.2 Preoxidation cleaning

Thermally grown silicon dioxide is an amorphous material. It is unstable, and its molecules tend to cross-link to form crystalline structures. This is the main reason silicon dioxide exists in the form of quartz and quartz sand in nature. Since the crystallization of amorphous silicon dioxide takes millions of years at room temperature, amorphous silicon dioxide in IC chips is very stable in their lifetimes. However, the crystallization process is dramatically accelerated at the high temperatures (>1000 °C) required during silicon dioxide growth. If the silicon surface is not free of contaminants, defects and particles can serve as nucleation sites for crystallization during the oxidation process, and silicon dioxide will grow into a polycrystalline structure similar to ice crystals that forms on glass in the winter. Crystallization of silicon dioxide is very undesirable, since it is not uniform and the crystal boundaries provide easy paths for impurities and moisture.

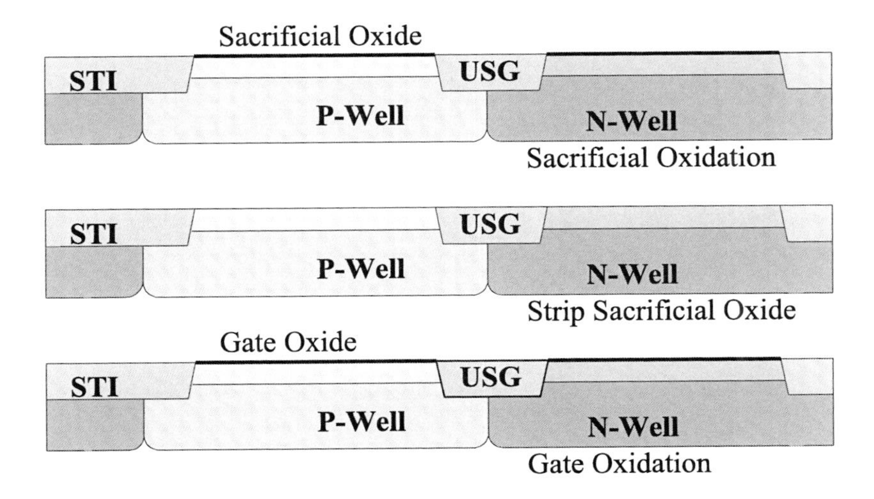

Figure 5.12 Sacrificial and gate oxidation processes.

Table 5.2 Summary of oxide applications.

Oxide name	Thickness	Application	Time of application
Native	15–20 Å	Undesirable	-
Screen	~200 Å	Implantation	Mid-1970s to present
Masking	~5000 Å	Diffusion	1960s to mid-1970s
Field and LOCOS	3000–5000 Å	Isolation	1960s to 1990s
Pad	100–200 Å	Nitride stress buffer	1960s to present
Sacrificial	<1000 Å	Defect removal	1970s to present
Gate	15–120 Å	Gate dielectric	1960s to present
Barrier	100–200 Å	STI	1980s to present

Therefore, proper preoxidation wafer cleaning—removing the particles, organic and inorganic contaminants, native oxide, and surface defects—becomes very important to eliminate crystallization. Figure 5.13 shows a crystalline structure of silicon dioxide grown on a poorly prepared silicon surface.

Wet clean processing is the most commonly used cleaning process in advanced semiconductor fabs. Strong oxidizers, such as the solutions $H_2SO_4 : H_2O_2 : H_2O$ or $NH_4OH : H_2O_2 : H_2O$, can remove particles and organic contaminants. When a wafer submerges into these solutions, particles and organic contaminants oxidize,

Figure 5.13 Oxide structure on a poorly prepared surface (Integrated Circuit Engineering Corporation).

and oxidation byproducts are either gaseous (CO, etc.) or soluble in the solution (such as H_2O). In most IC fabs, the composition of $NH_4OH : H_2O_2 : H_2O$ with a 1:1:5 to 1:2:7 ratio at 70 to 80 °C is widely used. This cleaning process is known as standard cleaning 1 (SC-1) of the Radio Corporation of America (RCA) clean, which was first developed by Kern and Puotinen (from RCA) in 1960. After SC-1, wafers are rinsed by DI water in a dunk tank and dried in a spin dryer.

After SC-1 and DI water rinse and dry, wafers are submerged in a solution with a 1:1:6 to 1:2:8 composition ratio of $HCl: H_2O_2 : H_2O$ at 70 to 80 °C. This is called SC-2 of the RCA clean, which is commonly used to remove inorganic contaminants by forming soluble byproducts in a low-pH solution. In the SC-2 process, H_2O_2 oxidizes the inorganic contaminants, and HCl reacts with the oxides to form soluble chlorides, allowing desorption of contaminants from the wafer surface. SC-2 is also followed by a DI water rinse and spin dry processes.

Native oxides occurring in silicon have poor quality and need to be stripped, especially for gate oxides, which require the highest quality. HF is a chemical that is commonly used to dissolve native silicon dioxide. A native oxide strip can be performed either on a wet bench with $HF : H_2O$ solution, or in a HF vapor etcher, which uses HF vapor to react with silicon dioxide and vaporize the byproduct. After native oxide stripping, some fluorine atoms bind with silicon atoms and form Si-F bonds on the silicon surface.

5.3.3 Oxidation rate

When oxygen starts to react with silicon, it forms a silicon dioxide layer, which prevents silicon atoms from reacting with oxygen molecules. When the oxide first begins to grow and the oxide layer is very thin (< 500 Å), oxygen molecules can penetrate the oxide with very few collisions in the oxide layer and reach the silicon, where they react and continue to grow the silicon dioxide film. This is called a linear growth regime because the oxide thickness linearly increases with the growth time. When the oxide film thickness becomes thicker, oxygen molecules cannot pass through the oxide layer without numerous collisions with the atoms inside the oxide film. They must diffuse across the growing oxide to reach the silicon and react with it to form silicon dioxide. This is called the diffusion-limited regime because the oxide growth rate is slower than the linear growth regime. Figure 5.14 illustrates the two oxide growth regimes.

A and B in the equations in Fig. 5.14 are the two coefficients related to oxide growth rates; they are determined by many factors, such as oxidation temperature, oxygen source (O_2 or H_2O), silicon crystal orientation, dopant type, concentration, and pressure.

Oxide growth rates are very sensitive to temperature because the oxygen diffusion rate in silicon dioxide is exponentially related to temperature: $D \propto \exp(-E_a/k_BT)$. Here D is the diffusion coefficient, E_a is the activation energy, $k_B = 2.38 \times 10^{-23}$ J/K is the Boltzmann constant, and T is the temperature. Increasing the temperature can significantly increase both B and B/A, as well as the oxide growth rate.

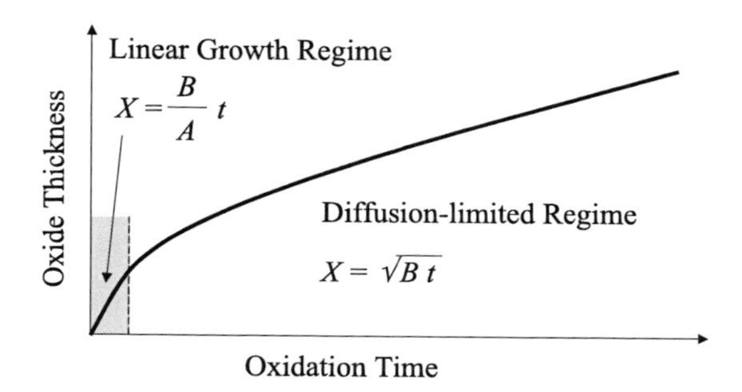

Figure 5.14 The two oxidation rate regimes.

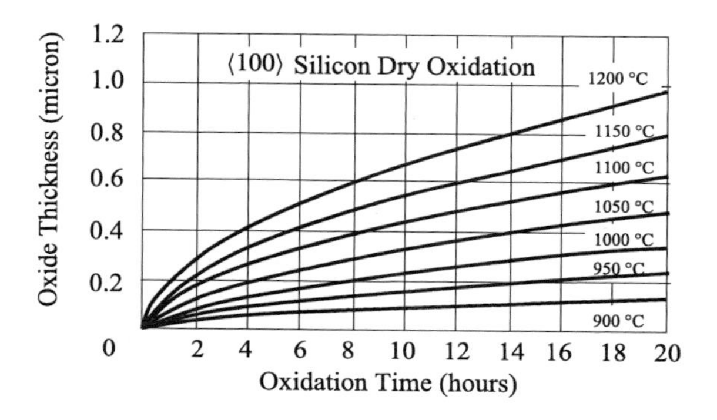

Figure 5.15 Dry oxide growth on a ⟨100⟩ surface.

Oxide growth rate is also related to the oxygen source: dry oxidation with O_2 has a lower oxide growth rate than wet oxidation with H_2O. This is because the diffusion rate of the oxygen molecule O_2 in silicon dioxide is lower than that of the hydroxide HO generated from the dissociation of H_2O molecules at high temperatures. Oxide growth rates of dry and wet oxidation processes are illustrated in Figs. 5.15 and 5.16.

The figures show that wet oxidation is significantly faster than dry oxidation. For example, for ⟨100⟩ silicon at 1000 °C, the wet oxide layer grows to ~ 2.2 μm after 20 h, while the dry oxide layer only grows to 0.34 μm. This is why the wet oxidation process is preferred for growing thick oxide layers such as masking and field oxides.

Oxide growth rates are also related to single-crystal silicon orientation. Normally, ⟨111⟩ orientation silicon has a higher oxide growth rate than ⟨100⟩ orientation silicon. This is because a ⟨111⟩ silicon surface has higher silicon atom density than that of a ⟨100⟩ silicon surface and can provide more silicon atoms to react with oxygen and form thicker silicon dioxide layers. By comparing Fig. 5.17, which shows the wet oxidation rate of ⟨111⟩ silicon, with Fig. 5.16, which shows the wet oxidation rate of ⟨100⟩ silicon, it is evident that ⟨111⟩ silicon has a higher growth rate than ⟨100⟩ silicon.

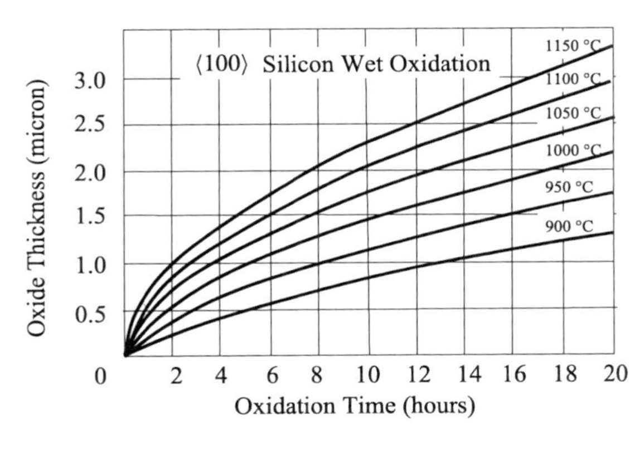

Figure 5.16 Wet oxide growth on a ⟨100⟩ surface.

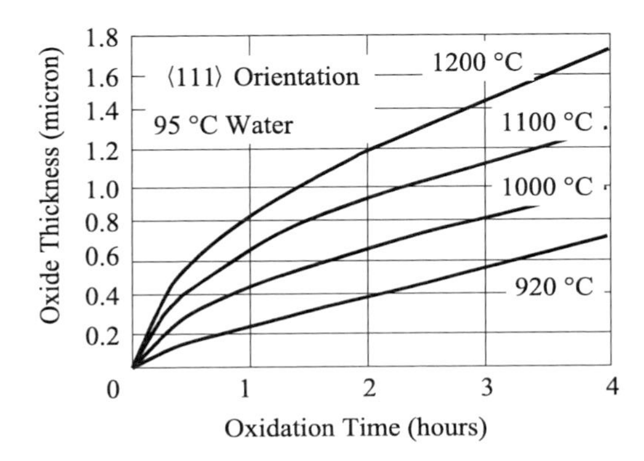

Figure 5.17 Wet oxide growth on a ⟨111⟩ surface.

Oxidation rate is related to dopant and dopant concentrations. Heavily doped silicon generally oxidizes faster than lightly doped silicon. During oxidation, boron in the silicon tends to be drawn up into the silicon dioxide, which causes depletion of the boron concentration at the Si–SiO_2 interface. n-type dopants such as phosphorus, arsenic, and antimony have the opposite effect. When oxidation occurs, these dopants are driven deeper into the silicon. Similar to snow piling up from a snowplow, the n-type dopant concentration in the Si–SiO_2 interface can become significantly higher than its original value. Figure 5.18 illustrates the pile-up effect of an n-type dopant and the depletion effect of a p-type dopant.

Oxidation rate is also related to the addition of gas, such as HCl, that is commonly used in the gate oxidation process to suppress mobile ions. With the appearance of HCl, the oxidation rate can increase by about 10%.

5.3.4 Dry oxidation

Dry oxidation has a lower growth rate than wet oxidation, yet it has better oxide film quality. Therefore, thin oxides such as screen, pad, and especially gate oxides, use a dry oxidation process. Figure 5.19 shows a dry oxidation system.

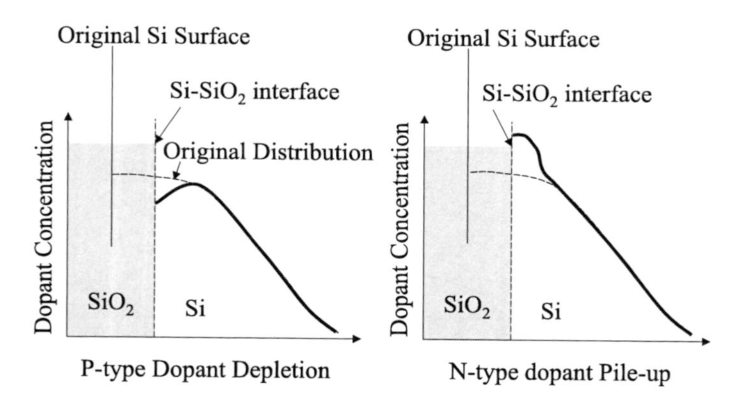

Figure 5.18 Dopant depletion and pile-up effects caused by oxidation.

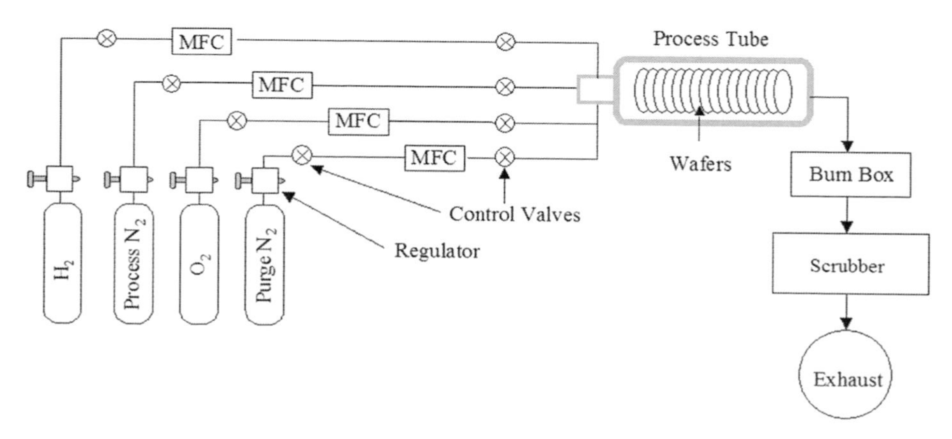

Figure 5.19 Schematic of a dry oxidation system.

There are two nitrogen (N_2) sources in most oxidation systems: one for the process application (with higher purity), and the other for the chamber purge (with lower purity and also lower cost). Because nitrogen is a very stable gas, it is used as an inert gas in the oxidation process during system idle time, wafer loading, temperature ramp, temperature stabilization, and wafer unloading steps. For dry oxidation, high-purity oxygen gas (O_2) is used as the oxygen source during the oxidization step to oxidize silicon. HCl is also used during the oxidation step to reduce mobile ions in the oxide and minimize interface state charge.

The quartz tube starts to sag with temperatures above 1150 °C, so oxidation processes cannot operate at those temperatures for very long. Dry oxidation processes operate at about 1000 °C. In dry oxidation, HCl is commonly used to remove mobile metallic ions, especially sodium, by forming immobile chloride compounds. This is very important, since a trace amount of sodium can cause MOS transistor malfunction and affect IC chip performance and reliability.

When silicon dioxide is forming on a single-crystal silicon surface, there is an abrupt change at the silicon–silicon-dioxide interface. There are always some dangling bonds at the interface due to crystal structure mismatch, as shown in

Fig. 5.20. The dangling bonds induce an interface state charge, which is a positive charge that strongly affects IC chip performance and reliability. This is because hydrogen or other atoms can diffuse to the silicon–silicon-dioxide interface during IC chip application and attach to the dangling bonds. This changes the interface state charge, the threshold voltage V_T of the MOS transistor, and the performance of the IC device. Although dangling bonds are always present at the silicon–silicon-dioxide interface, it is very important to minimize them as much as possible to achieve highly stable and reliable device performance.

Introducing HCl into the oxidation reaction means that some chlorine atoms can integrate into the silicon dioxide film and bind with silicon at the silicon-silicon-dioxide interface. This helps minimize the number of dangling bonds and improve IC device reliability. However, if chlorine concentration in the oxide is too high, device stability can be affected because of the response of the extra chlorine ions to the bias voltage.

Question: Fluorine has one unpaired electron and can attach to the dangling bonds at the Si-SiO_2 interface. Why don't fabs use HF in the oxidation process and integrate fluorine into the oxide to minimize the interface state charge?

Answer: HF can aggressively etch the silicon dioxide layer and the quartz tube, so it is undesirable to use HF in the thermal oxidation process. However, after BF_2 ion implantation, the integration of small amounts of fluorine atoms into the silicon substrate can help reduce the interface state charge.

A typical dry oxidation process sequence for gate oxide growth is as follows:

- idle with purge N_2 flow
- idle with process N_2 flow
- wafer boats pushed in with process N_2 flow
- temperature ramp-up with process N_2 flow

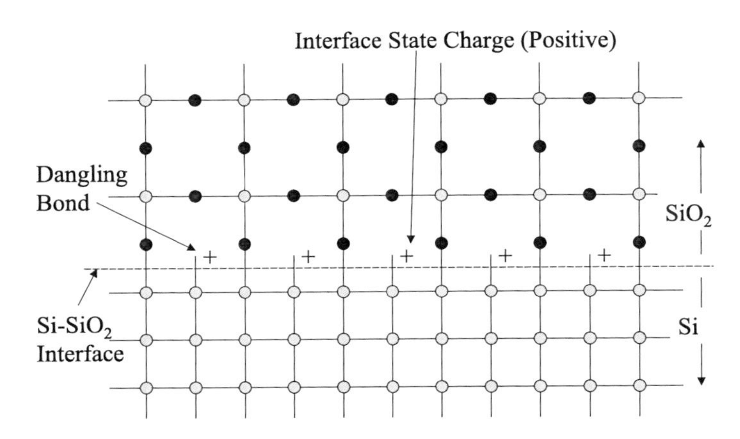

Figure 5.20 Example of dangling bonds inducing interface state charge.

- temperature stabilization with process N_2 flow
- oxidation with O_2, HCl; stop N_2 flow
- oxide annealing, stop O_2, start process flow N_2
- temperature cool down with process N_2 flow
- wafer boats pulled out with process N_2 flow
- idle with process N_2 flow
- bring in next boats and repeat process
- idle with purge N_2 flow.

During system idle time, the furnace is always kept at a high temperature, such as 850 °C, so that temperature ramp-up for processing does not require much time. N_2 purge gas is always used when the system is idle for long periods of time. Before wafer loading, process N_2 starts to flow into the processing tube in order to fill it with higher-purity nitrogen. It needs to slowly push the wafer boats into the processing tube to avoid wafer warping due to the high thermal stress induced by sudden temperature change. When the wafer boats are placed into the flat zone of the processing tube, temperature ramps up at a rate of 10 °C/min. Due to its high thermal capacity, the temperature of a furnace system cannot ramp up too quickly, as oscillations can occur due to over- or undershooting the temperature around the setting point. After the furnace reaches the set temperature required by the process (typically 1000 °C), a temperature-stabilization step with N_2 flow is required to allow for temperature oscillation to settle down and for the furnace to finally reach a steady state with the set temperature.

Now the system is ready for oxidation. By turning on O_2 and HCl flows and turning off the N_2 flow, oxygen begins to react with silicon and forms a thin layer of silicon dioxide on the silicon wafer surface. After the required oxide thickness is reached, the O_2 and HCl flows are terminated, and N_2 flow is resumed. The wafers stay at the high temperature for a while to anneal the oxide. This step improves the quality of the silicon dioxide, makes it denser, reduces the interface state, and increases breakdown voltage. A thin gate oxide (thickness of about 50 Å) can grow in a furnace at a lower temperature such as 700 °C, so that oxidation time will be long enough to control the process. After the oxide film grows, the film is annealed at >1000 °C in a nitrogen ambient to improve oxide quality. After oxide annealing, the furnace temperature is gradually cooled down to its idle temperature, and wafer boats are slowly pulled out from the furnace with a constant nitrogen flow.

5.3.5 Wet oxidation

When using H_2O instead of O_2 as the oxygen source, the silicon oxidation process is called wet oxidation, and the oxide is called steam oxide. The chemical reaction can be expressed as

$$2\,H_2O + Si \rightarrow SiO_2 + 2\,H_2.$$

At high temperatures, H_2O can dissociate and form hydroxide (HO), which can diffuse in silicon dioxide faster than O_2 can. Therefore, a wet oxidation process has significantly higher oxidation rates than a dry oxidation process. Wet oxidation is used to grow thick oxides such as masking, blanket field, and LOCOS oxides. Table 5.3 shows an example of the oxidation time difference to grow a 1000-Å oxide film at 1000 °C.

Several systems have been used to deliver water vapor into the processing tube of a furnace. The boiler system vaporizes water at temperatures higher than 100 °C, and the water vapor flows to the processing tube by means of a heated gas line. In a bubbler system, nitrogen is bubbled through pure DI water, which carries the water vapor into the processing tube. Figure 5.21 gives the schematics of the boiler and bubbler systems.

In a flush system, small droplets of pure DI water are dropped onto a quartz hot plate to vaporize the water. Then an oxygen gas flow carries the steam into the processing tube. Figure 5.22 illustrates a flush system.

A major problem with boiler, bubbler, and flush systems is that they cannot precisely control the H_2O flow rate. The most commonly used wet oxidation system is the dryox system (or pyrogenic steam system), which burns hydrogen at the entrance of the processing tube, so that water vapor can be formed pyrogenically with the chemical reaction between H_2 and O_2:

$$2\,H_2 + O_2 \rightarrow 2\,H_2O.$$

The dryox system eliminates the liquid and vapor handling requirements, and can precisely control the flow rate. The tradeoff is the induction of flammable and explosive hydrogen gas. Figure 5.23 illustrates the schematic of the pyrogenic steam system.

Table 5.3 Time comparison of wet and dry oxidation processes.

Process	Temperature	Film thickness	Oxidation time
Dry oxidation	1000 °C	1000 Å	~ 2 h
Wet oxidation	1000 °C	1000 Å	~ 12 min

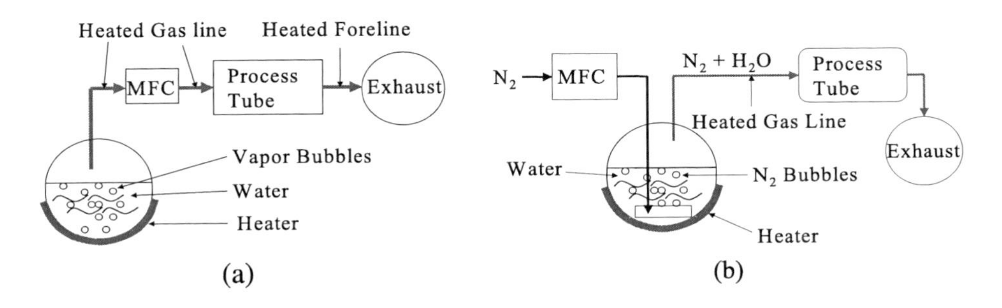

Figure 5.21 Schematics of (a) boiler and (b) bubbler system.

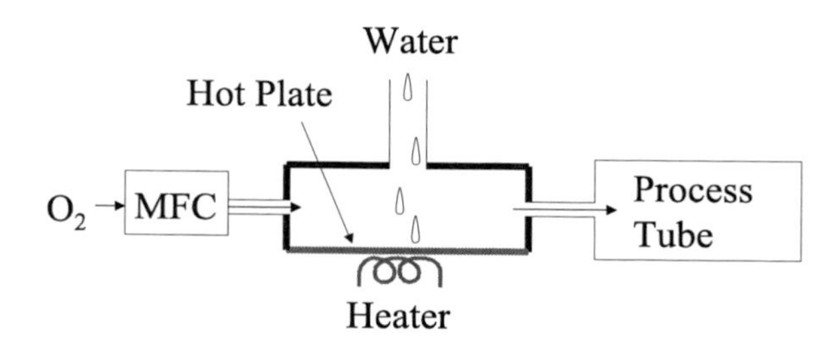

Figure 5.22 Schematics of a flush system.

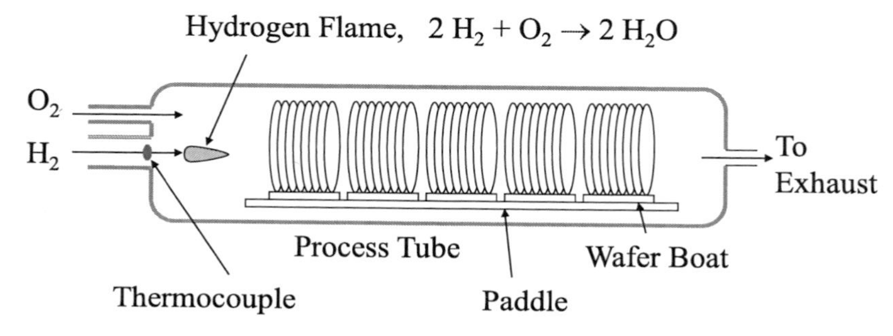

Figure 5.23 Schematics of a pyrogenic steam system.

A wet oxidation system with pyrogenic steam is illustrated in Fig. 5.24. A burn box is required in the exhaust system to burn out any residual hydrogen before releasing the exhaust gases into the atmosphere.

The autoignition temperature of hydrogen is about 400 °C. At a processing temperature with oxygen present, a flow of hydrogen gas into a processing tube will automatically react with the oxygen and form steam inside the tube. In a pyrogenic steam oxidation process, the ratio of the H_2 to O_2 flow rates is very important. The normal flow rate ratio of H_2:O_2 should be slightly lower than 2:1 to ensure the presence of excessive oxygen, which can fully oxidize hydrogen in the pyrogenic reaction. Otherwise, hydrogen can accumulate inside the tube, causing an explosion. The typical H_2:O_2 ratio is 1.8:1 to 1.9:1.

A normal pyrogenic wet oxidation process sequence is as follows:

- idle with purge N_2 flow
- idle with process N_2 flow
- ramp O_2 with process N_2 flow
- wafer boat pushed in with process N_2 and O_2 flows
- temperature ramp-up with process N_2 and O_2 flows
- temperature stabilization with process N_2 and O_2 flows
- ramp O_2, turn off N_2 flow
- stabilize O_2 flow
- turn on H_2 flow, ignition and H_2 flow stabilization

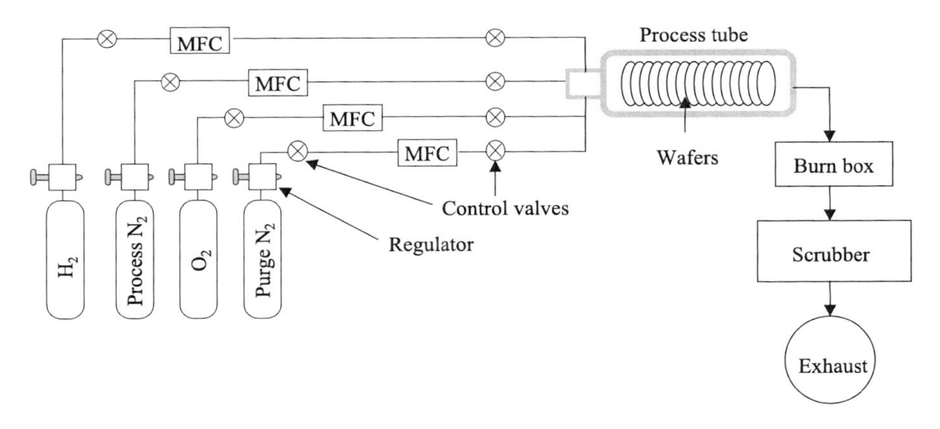

Figure 5.24 Schematics of a pyrogenic steam oxidation system.

- steam oxidation with O_2 and H_2 flow
- hydrogen termination, turn off H_2 while keeping flow
- oxygen termination, turn off O_2, start process N_2 flow
- temperature ramp-down with process N_2 flow
- wafer boat pulled out with process N_2 flow
- idle with process N_2 flow
- bring in next boats and repeat process
- idle with purge N_2 flow.

The appearance of oxygen during wafer push-in, temperature ramp-up, and temperature stabilization steps helps grow a thin layer (a few hundred angstroms) of better-quality dry oxide on the wafer surface. This serves as a barrier layer for the lesser-quality steam oxide and helps reduce defects at the silicon–silicon-dioxide surface. After steam oxidation, hydrogen flow is terminated, while oxygen continuously flows into the processing tube to scavenge any residual hydrogen. This oxygen flow also can help reduce hydrogen integration in the steam oxide.

5.3.6 High-pressure oxidation

Increasing pressure increases the oxygen or steam concentration inside the processing chamber as well as the diffusion rate inside the silicon dioxide, helping to increase the oxidation rate. High-pressure oxidation can reduce oxidation time at the same oxidation temperature, or reduce oxidation temperature at the same oxidation time. Generally, increasing pressure by one atmosphere can reduce oxidation temperature by 30 °C. Tables 5.4 and 5.5 show time and temperature reductions caused by using a high-pressure oxidation process to grow a 10,000 Å (1 μm) oxide film in wet oxidation processes.

The hardware required for high-pressure oxidation is different from that needed for other types of oxidation; Fig. 5.25 illustrates a high-pressure oxidation system. Due to hardware complexity and safety concerns, high-pressure oxidation processes are not very popular in advanced semiconductor fabs.

Table 5.4 Oxidation time to grow 10,000-Å wet oxide.

Temperature	Pressure	Time
1000°C	1 atm	5 h
	5 atm	1 h
	25 atm	12 min

Table 5.5 Oxidation temperature to grow 10,000-Å wet oxide in 5 h.

Time	Pressure	Temperature
5 h	1 atm	1000 °C
	10 atm	700 °C

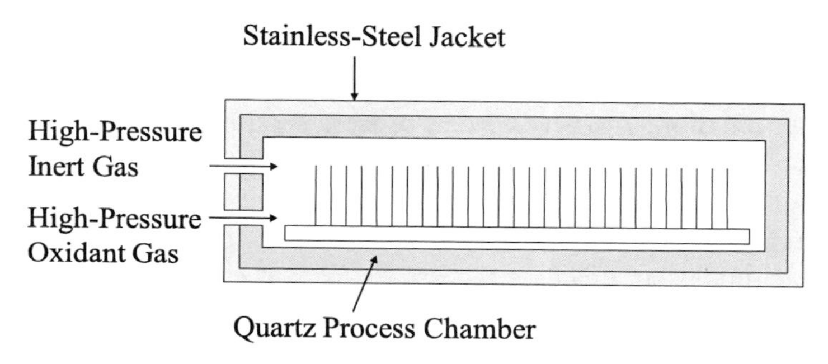

Figure 5.25 Schematics of a high-pressure oxidation system.

5.3.7 Oxide measurement

Monitoring the oxidation process means measuring oxide thickness and its uniformity. Ellipsometry is commonly used to measure dielectric thin-film refractivity and thickness. When a beam of light is reflected from a film's surface, the polarization state changes (see Fig. 5.26). By measuring this change, information about the film's refractive index and thickness can be acquired. An approximate value of film thickness is needed in advance, since the measured ellipsometric quantities are the periodic functions of the thickness. Because the refractive index of silicon dioxide is well known to be 1.46 for a light wavelength of 633 nm (red He-Ne laser), ellipsometry can also be used to measure the oxide film thickness.

After the oxide is grown, the color of the wafer surface changes. Color depends on film thickness, refractive index, and the angle of light. Reflected light from the oxide surface (light 1) and reflected light from the Si-SiO_2 interface (light 2) have the same frequencies, but with different phases, since light 2 travels a longer distance inside the oxide film, as shown in Fig. 5.27. The two reflected lights interfere with each other and cause both constructive and destructive interference at different wavelengths, since the refractive index is a function of wavelength. The color seen on the wafer surface is determined by constructive interference

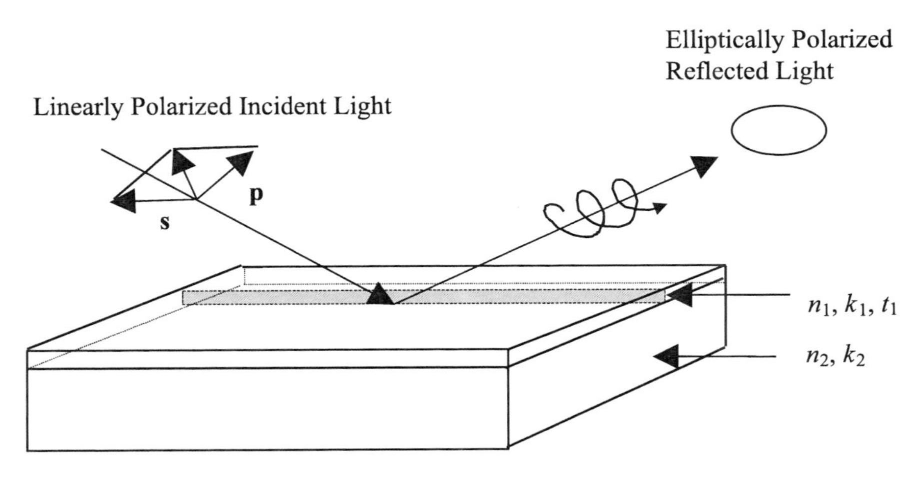

Figure 5.26 Ellipsometry system.

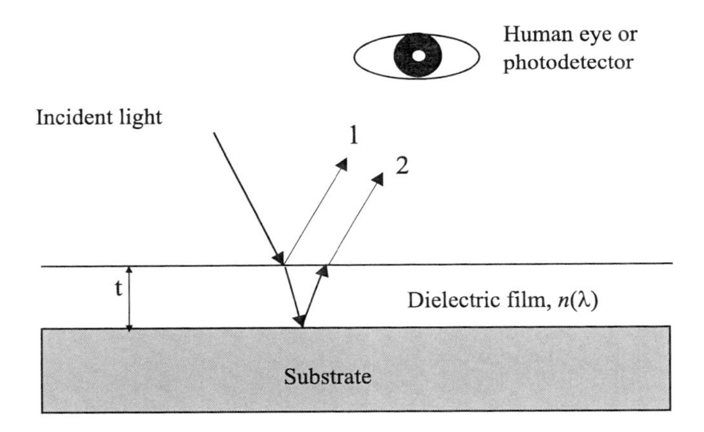

Figure 5.27 Reflected light and phase difference.

frequency,

$$\Delta\phi = 2tn(\lambda)/\cos\theta = 2N\pi,$$

where t is the thin-film thickness, $n(\lambda)$ is the thin-film refractive index, θ is the incident angle, and N is an integral number. When phase shift $\Delta\phi$ is larger than 2π, the color pattern repeats itself.

Color charts are a convenient tool to measure film thickness (see Table 5.6). Although these charts are no longer used for thickness measurements in advanced IC fabs, they are still a useful tool to estimate oxide thickness quickly and to detect obvious nonuniformity problems.

If a wafer with a thick oxide layer is placed into a HF solution, HF will etch silicon dioxide. By pulling back the wafer gradually, the oxide thickness changes gradually due to the different etch times, as does the color of the wafer. In this way,

Table 5.6 Color chart of silicon oxide.

Thickness (Å)	Color	Thickness (μm)	Color
500	Tan	1.0	Carnation pink
700	Brown	1.02	Violet-red
1000	Dark violet to red-violet	1.05	Red-violet
1200	Royal blue	1.06	Violet
1500	Light blue to metallic blue	1.07	Blue violet
1700	Metallic to very light yellow-green	1.10	Green
2000	Light gold or yellow, slightly metallic	1.11	Yellow-green
2200	Gold with slight yellow-orange	1.12	Green
2500	Orange to melon	1.18	Violet
2700	Red-violet	1.19	Red-violet
3000	Blue to violet-blue	1.21	Violet-red
3100	Blue	1.24	Carnation pink to salmon
3200	Blue to blue-green	1.25	Orange
3400	Light green	1.28	Yellowish
3500	Green to yellow-green	1.32	Sky blue to green-blue
3600	Yellow-green	1.40	Orange
3700	Green-yellow	1.45	Violet
3900	Yellow	1.46	Blue-violet
4100	Light orange	1.50	Blue
4200	Carnation pink	1.54	Dull yellow-green
4400	Violet-red		
4600	Red-violet		
4700	Violet		
4800	Blue-violet		
4900	Blue		
5000	Blue-green		
5200	Green		
5400	Yellow-green		
5600	Green-yellow		
5700	Yellow to yellowish (at times appears light gray or metallic)		
5800	Light orange or yellow to pink		
6000	Carnation pink		
6300	Violet-red		
6800	Bluish (appears between violet-red and blue-green, overall looks grayish)		
7200	Blue-green to green		
7700	Yellowish		
8000	Orange		
8200	Salmon		
8500	Dull, light red-violet		
8600	Violet		
8700	Blue-violet		
8900	Blue		
9200	Blue-green		
9500	Dull yellow-green		
9700	Yellow to yellowish		
9900	Orange		

a rainbow wafer can be created with color pattern changes occurring gradually and periodically.

For a precise measurement of silicon dioxide thickness, spectroreflectometry is used. It measures the reflected light intensity at different wavelengths, and the thin-film thickness can be calculated from the relation of reflected light intensity and wavelength of the light.

For gate oxides it is very important to measure breakdown voltage and fixed charge, measurements that can be achieved by depositing a patterned conductor layer onto the oxide layer to form a metal-oxide-semiconductor capacitor. By applying bias voltage, the response of the capacitance to the applied voltage, or the C–V curve, gives the information of the fixed change at the Si–SiO_2 interface. By increasing bias voltage until the silicon dioxide breaks down, the breakdown voltage can also be measured. By using a higher test temperature, such as 250 °C, the accelerated device failure time caused by thermal stress can help predict the device's lifetime. Figure 5.28 illustrates the C–V test system.

5.3.8 Recent oxidation trends

While feature size has continued to reduce, STI gradually has replaced LOCOS for isolation of neighboring transistors. There is no longer a need to grow a thick oxide layer in IC chip manufacturing. Most oxidation processes in fabs grow thin oxide layers, such as pad, screen, barrier, and gate oxides. All of the layers are grown in dry oxidation processes. Wet oxidation is mainly used to produce test wafers needed for process control and development.

As gate oxide thicknesses continue to shrink, rapid thermal oxidation (RTO) will replace furnace oxidation processes. RTPs have better temperature control at high temperatures, and better wafer-to-wafer uniformity. An RTO system can be integrated with a HF vapor etcher in one cluster tool to perform in-situ native oxide stripping, gate oxidation, and gate oxide annealing. Nitridation of a gate oxide can increase the dielectric constant of a film and help reduce effective oxide thickness (EOT), while keeping the physical thickness of the gate oxide thick enough to prevent electrical breakdown. Nitridation can be performed in a thermal annealing process using nitric oxide (NO) as an ambient gas. It can also be accomplished

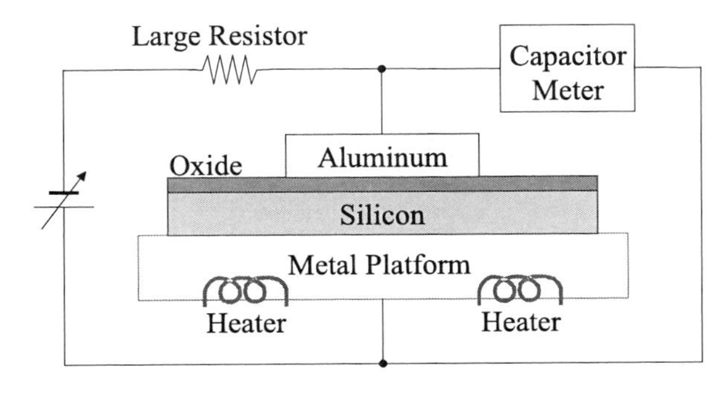

Figure 5.28 C–V test configuration.

with nitrogen plasma followed by annealing in ambient nitrogen. One advantage of this integrated process is that the silicon surface does not need to be exposed to the atmosphere before the oxidation process.

5.4 Diffusion

In the early years of the IC industry, diffusion was widely used for doping semiconductors. Because the most commonly used tool for silicon doping was the high-temperature quartz tube furnace, it was (and still is) called a diffusion furnace. For the same reason, in an IC fab, the area in which high-temperature furnaces are located is called a diffusion bay, although very few diffusion processes are actually performed in an advanced IC fab.

Diffusion is a basic physical phenomenon in which material moves from high-concentration regions to low-concentration regions, driven by the thermal motion of the molecules. The diffusion process happens anywhere and anytime. Perfume is a good example of diffusion in air; sugar, salt, and ink are examples of diffusion in liquid; and wood soaking in water or oil is an example of diffusion in a solid.

Diffusion doping processes dominated the early IC industry. By introducing a high concentration of dopant onto a silicon surface with high temperature, the dopant can diffuse into the silicon substrate, changing the conductivity of the semiconductor. Figure 5.29 illustrates a silicon diffusion doping process. Junction depth is reached when the diffused dopant concentration equals the substrate dopant concentration. Figure 5.30 shows the definition of junction depth.

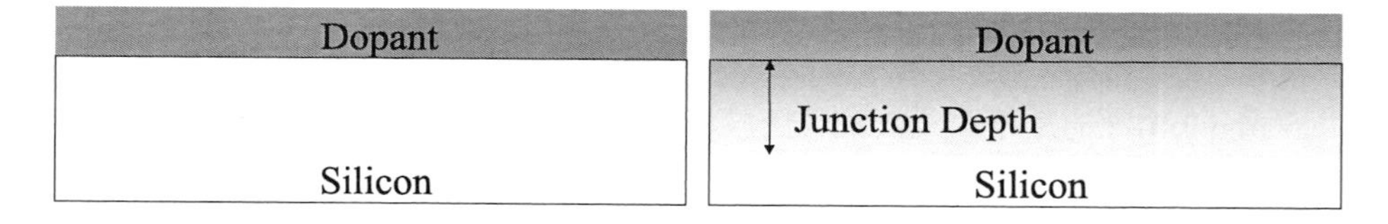

Figure 5.29 Diffusion doping process.

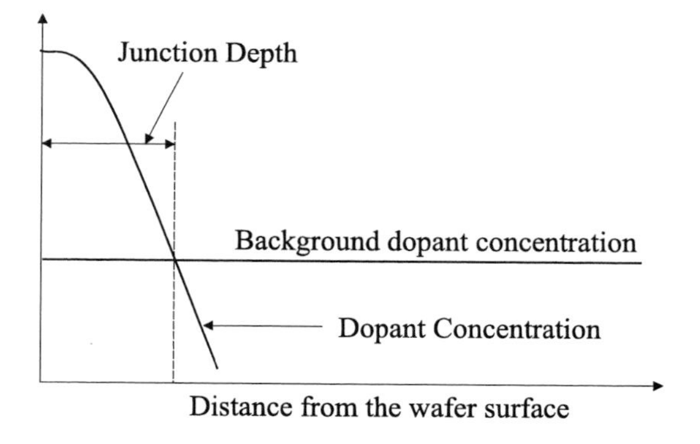

Figure 5.30 Definition of junction depth.

Because the diffusion rate in solid materials is exponentially related to temperature [$D \propto \exp(-E_a/k_B T)$], the diffusion process can be accelerated significantly at a high temperature (E_a is the activation energy, k_B is the Boltzmann constant, and T is the temperature).

For most dopants of interest in semiconductor processes, such as boron and phosphorus, their activation energy in silicon dioxide is higher than in single-crystal silicon, and their diffusion rate in silicon dioxide is much lower than in silicon. Therefore, silicon dioxide can be used as a diffusion mask to dope designated areas of the silicon surface, as shown in Fig. 5.31.

Compared with the ion implantation doping process, the diffusion doping process has several disadvantages. For instance, it cannot independently control dopant concentration and junction depth. Diffusion is an isotropic process; thus, dopants always diffuse underneath the masking oxide, as shown in Fig. 5.31. When used for small feature sizes, diffusion can cause neighboring junctions to short. Therefore, when ion implantation was introduced in semiconductor fabrication in the mid-1970s, it quickly replaced diffusion as a silicon doping process.

One important concept in the IC industry related to diffusion processes is thermal budget. After self-aligned S/D implantation, heavily doped S/Ds are formed separately by the dimension of the gate, as shown in Fig. 5.32. Any high-temperature process after S/D implantation can cause diffusion of the dopants in the S/D and expand the S/D junctions. If they expand too much, device performance can be dramatically affected. The combination of time and temperature that the wafer can spend in thermal processing after S/D implantation is called the thermal budget.

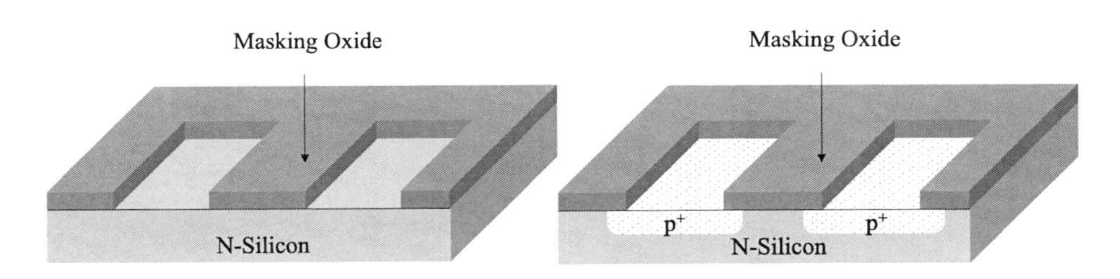

Figure 5.31 Patterned diffusion doping process.

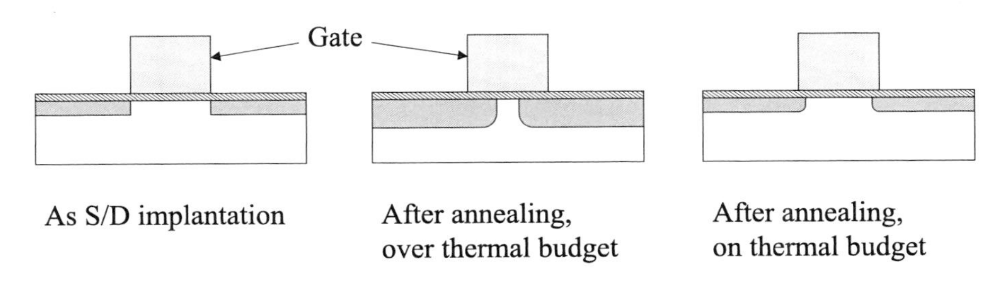

Figure 5.32 S/D diffusion in postimplantation thermal processes.

The thermal budget is determined by gate dimension, which usually is the minimum feature size of the IC chip. A device with smaller gate width has less room for the S/D junction to diffuse and therefore has a smaller thermal budget. As minimum feature size shrinks, a wafer requires less time in high-temperature (> 1000 °C) processes and needs a tighter thermal budget control. Figure 5.33 shows thermal budgets (time at temperature) for devices with different feature sizes at different temperatures. The figure assumes that surface dopant concentration is 10^{20} atoms/cm^3. A smaller device has a smaller thermal budget. For example, 0.25-μm devices only require 24 sec at 1000 °C after S/D implantation, while 2-μm devices require 1000 sec. By reducing temperature, a thermal budget can be significantly increased. For instance, 0.25-μm devices need about 200 sec at 900 °C after S/D extension (SDE) implantation. In comparison, 16-nm devices need only 1 sec at 900 °C after SDE, explaining why 16-nm devices need millisecond annealing, such as spike, laser, and flash annealing.

5.4.1 Deposition and drive-in

The commonly used diffusion doping process sequence is called predeposition and drive-in. First, a layer of dopant oxide, such as B_2O_3 or P_2O_5, is deposited onto the wafer surface at 1050 °C, followed by a thermal oxidation process that consumes the residue dopant gas and grows a layer of silicon dioxide, which caps the dopant and prevents out-diffusion. Chemical reaction of the predeposition and cap oxidation for the most commonly used boron and phosphorus sources [diborane (B_2H_6) and phosphorus oxychloride ($POCl_3$, commonly called POCL)]

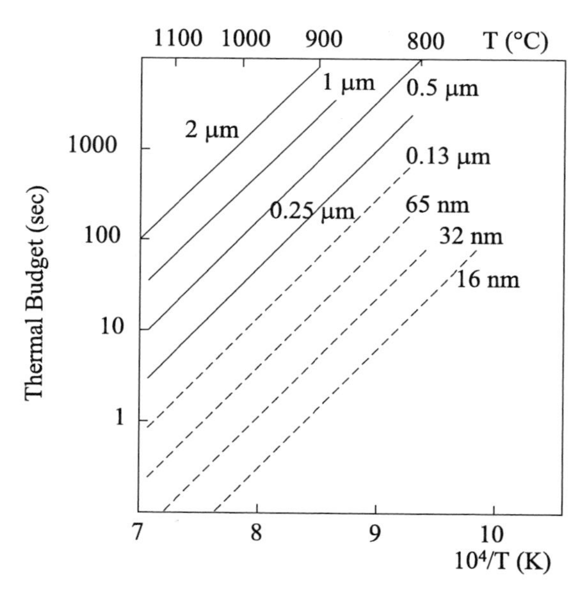

Figure 5.33 Available time-at-temperature (thermal budget) for devices with different feature sizes (after R. B. Fair).

can be expressed as

boron: predeposition $B_2H_6 + 2\ O_2 \rightarrow B_2O_3 + 3\ H_2O$
cap oxidation $2\ B_2O_3 + 3\ Si \rightarrow 3\ SiO_2 + 4\ B$
$2\ H_2O + Si \rightarrow SiO_2 + 2\ H_2$,

phosphorus: predeposition $4\ POCl_3 + 3\ O_2 \rightarrow 2\ P_2O_5 + 6\ Cl_2$
cap oxidation $2\ P_2O_5 + 5\ Si \rightarrow 5\ SiO_2 + 4\ P$.

Diborane (B_2H_6) is a poisonous gas with a burnt chocolate, sickly sweet smell. It can be fatal if absorbed through the skin or inhaled. Diborane is flammable, with an autoignition temperature of 56 °C, and becomes explosive when the concentration in air is higher than 0.8%. POCL is a corrosive liquid, which can cause burns to the skin or eyes. POCL vapor is irritating to the skin, eyes, and lungs, and can cause dizziness, headache, loss of appetite, nausea, and lung damage. Other commonly used n-type dopant chemicals are arsine (AsH_3) and phosphine (PH_3), both being poisonous, flammable, and explosive. Their predeposition and oxidation reactions are very similar to those of diborane (B_2H_6).

Figure 5.34 illustrates a furnace system used for boron predeposition and cap oxidation processes. A dedicated furnace processing tube for each dopant prevents cross-contamination.

After predeposition and cap oxidation, the temperature of the furnace is ramped up to 1200 °C in ambient oxygen, providing enough thermal energy for the dopant to diffuse into the silicon substrate quickly. Drive-in time is determined by the required junction depth and can be easily calculated from the existing theory for each dopant. Figure 5.35 illustrates the predeposition, cap oxidation, and drive-in steps for a diffusion doping process.

The diffusion process cannot independently control dopant concentration and dopant junction depth, since both are strongly related to the processing

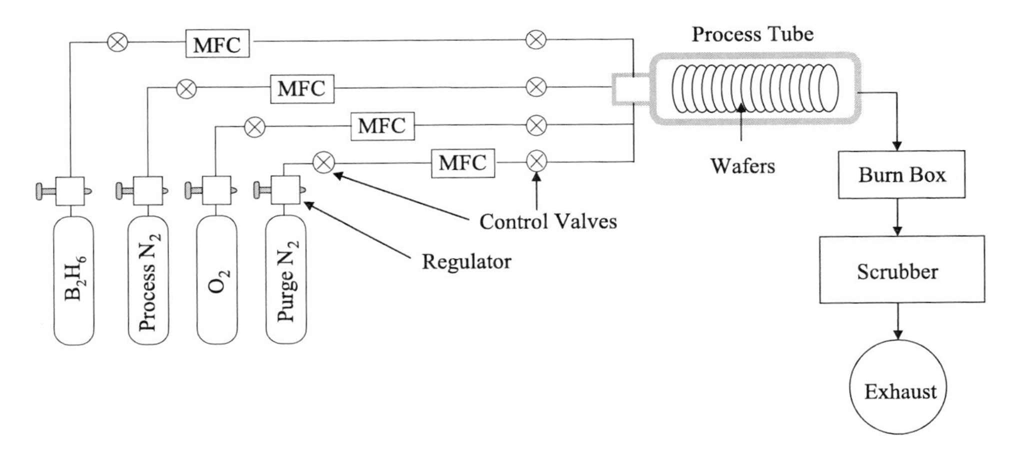

Figure 5.34 Schematic of a furnace system for the boron doping process.

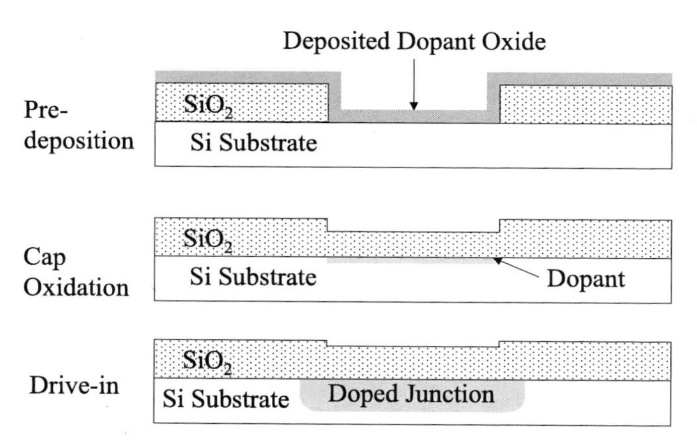

Figure 5.35 Diffusion doping process sequence.

temperature. The diffusion process is an isotropic process, and dopant atoms always diffuse underneath the masking silicon dioxide. In advanced IC processing, almost all semiconductor doping is done with an ion implantation process instead of a diffusion process, because ion implantation has much better control of dopant concentration and profile. The main application of the diffusion process in advanced IC processing is dopant drive-in during the well implantation annealing process.

The diffusion process was revisited by fab research and development departments for ultrashallow-junction (USJ) formation. First, a layer of borosilicate glass (BSG) with very high boron concentration is deposited onto the wafer surface by a CVD process. Then, by using an RTP, boron is driven out of the BSG and diffuses into silicon to form a shallow junction. Figure 5.36 illustrates the USJ formation by deposition, diffusion, and stripping processes.

5.4.2 Doping measurement

The four-point probe is one of the most commonly used measurement tools for monitoring doping processes. Resistivity of silicon is related to dopant concentration; therefore, measuring sheet resistance on the silicon surface can provide some useful information of the dopant concentration. Sheet resistance R_s is a defined parameter. The resistance of a conducting line can be expressed as

$$R = \rho \frac{L}{A},$$

where R is the resistance, ρ is the resistivity of the conductor, L is the length of the conducting line, and A is the area of the line cross section. If the wire is a band with a rectangular cross section, the area of the cross section simply changes to the product of the width W and thickness t. Line resistance can be expressed as

$$R = \rho \frac{L}{Wt}.$$

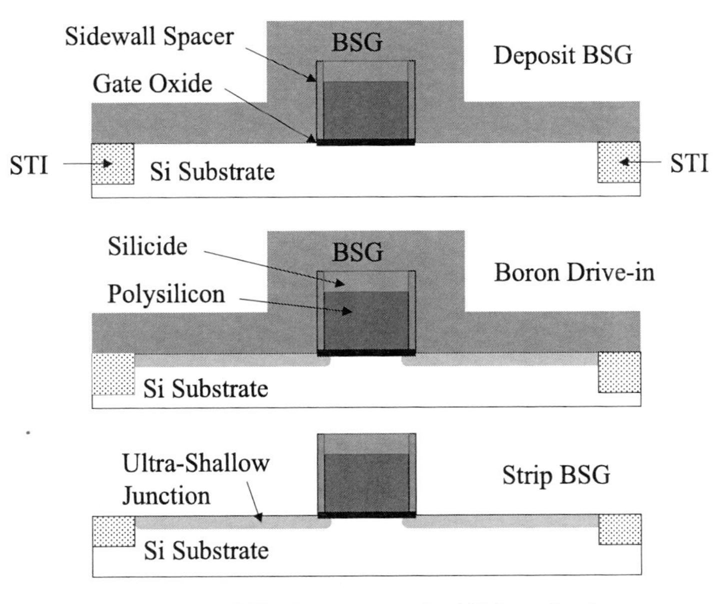

Figure 5.36 Diffusion process for USJ applications.

For a square sheet, the length is equal to the width, $L = W$, thus they cancel each other. Therefore, the resistance of a square conducting sheet, defined as sheet resistance, can be expressed as

$$R_s = \rho/t.$$

Resistivity ρ of doped silicon is mainly determined by dopant concentration, and thickness t is mainly determined by dopant junction depth. Therefore, a sheet resistance measurement can provide information about dopant concentration, since junction depth can be estimated by knowing the ion energy, ion species, and substrate materials.

By applying a certain amount of current between two of the pins and measuring voltage difference between the other two, sheet resistance can be calculated. Typically, the distances between the probes are: $S_1 = S_2 = S_3 = 1$ mm. If a current I is applied between P_1 and P_4, as shown in Fig. 5.37, sheet resistance $R_s = 4.53\ V/I$, where V is the measured voltage between P_2 and P_3. If the current is applied between P_1 and P_3, $R_s = 5.75\ V/I$, where V is measured between R_2 and R_4. These two equations are derived with the assumption of an infinite film area, which is not the case for thin-film measurements on wafers. During the measurement, an advanced tool normally runs four sets of measurements, programmed in a sequence with both configurations and reverse current for each configuration, to minimize the edge effect and achieve an accurate measurement.

Since the four-point probe makes direct contact and creates defects on the wafer surface, it is only used on test wafers for process development, qualification, and control. During measurement, it is important that probes touch the silicon surface

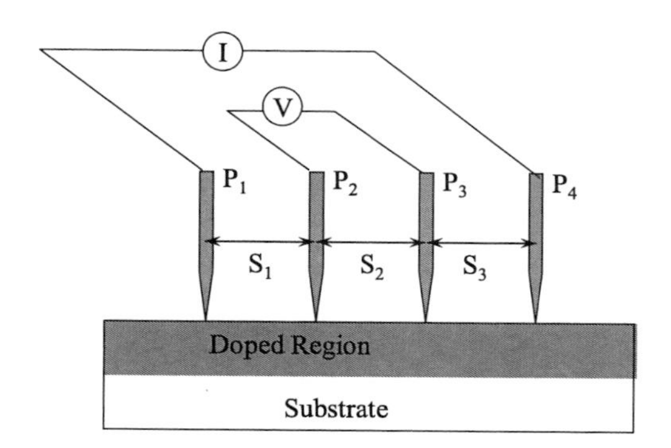

Figure 5.37 Four-point probe.

with enough force that the pins break through the thin (~10 to 20 Å) native oxide to make good contact with the silicon substrate.

5.5 Annealing

Annealing is a heating process in which a wafer is heated to achieve a desired physical or chemical change, with minimal material being added to or removed from the wafer surface.

5.5.1 Postimplantation annealing

In an ion implantation process, energetic dopant ions cause extensive damage to the silicon crystal structure near the wafer surface. To achieve device requirements, the lattice damage must be repaired in a thermal annealing process to restore the single-crystal structure and activate the dopant. Only when dopant atoms are in the form of a single-crystal lattice can they effectively provide electrons or holes as majority carriers for conducting electric current. In a high-temperature process, atoms move quickly, driven by thermal energy, and they find and rest at sites that have the lowest free energy, located at single-crystal lattices; this process restores the single-crystal structure. Figure 5.38(a) illustrates crystal damage after ion implantation, and Fig. 5.38(b) shows the crystal recovery and dopant activation after thermal annealing.

Before the 1990s, furnaces were widely used for postimplantation annealing. A furnace annealing process is a batch process operated from 850 to 1000 °C for about 30 min in ambient nitrogen and oxygen. A small amount of oxygen is used to prevent the formation of silicon nitride on the wafer surface where silicon is exposed. During idle time, a furnace is always kept at a high temperature, between 650 and 850 °C, and wafers must be pushed into and pulled out of the furnace very slowly to avoid wafer warping. Therefore, wafers at opposite ends of the paddle or tower will have different annealing times due to the slow entry and exit; this can cause wafer-to-wafer (WTW) nonuniformity.

Another concern for furnace annealing is the thermal budget, or the dopant diffusion problem during the annealing process. The furnace annealing process

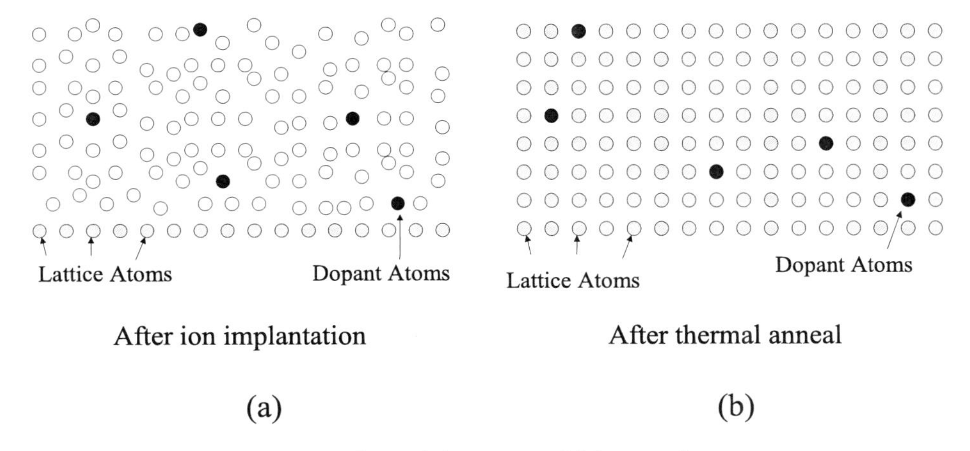

Figure 5.38 (a) Crystal damage and (b) annealing process.

takes a long time and can cause too much dopant diffusion, which is intolerable for a transistor with a small geometry. Therefore, the RTA process is preferred for most postimplantation annealing in an advanced IC fab. An RTA system can ramp up wafer temperature from room temperature to about 1100 °C in a very short time, normally within 10 sec. RTPs can precisely control wafer temperature and within-wafer (WIW) temperature uniformity. At approximately 1100 °C, a single-crystal structure can be recovered in about 10 sec, with minimum dopant diffusion. The RTA process also has much better WTW temperature uniformity control. Thus, only some noncritical postimplantation annealing processes, such as well implantation annealing and drive-in, still use furnaces for advanced IC processing.

5.5.2 Alloy annealing

Alloy annealing is a heating process in which thermal energy helps different atoms chemically bond with each other to form a metal alloy. In semiconductor processing, several alloy processes have been used. Formation of cobalt silicide ($CoSi_2$) in a self-aligned silicide (salicide) process is one of the most common, as shown in Fig. 5.39.

The first anneal at 450 °C forms CoSi, and the second anneal at 700 °C forms $CoSi_2$. With the RTP, annealing occurs at temperatures of 700 to 750 °C to directly form $CoSi_2$ in one step. $CoSi_2$ was widely used in logic IC with technology nodes from 0.25 μm to 90 nm. It is still used for some memory chips.

After the 65-nm node, nickel silicide (NiSi) was introduced as the silicide material for high-speed logic ICs. NiSi can be formed at lower temperatures, around 450°C, with resistance comparable to $CoSi_2$ and $TiSi_2$. This is a very important advantage because it can significantly reduce the thermal budget to meet requirements of multiple logic IC technology nodes, all the way to 10 nm and beyond.

Both furnace and RTP systems are used for titanium silicide and cobalt silicide alloy processes, but the RTP system clearly offers better control of thermal budget and WTW uniformity. For nickel silicide, only RTP systems are used.

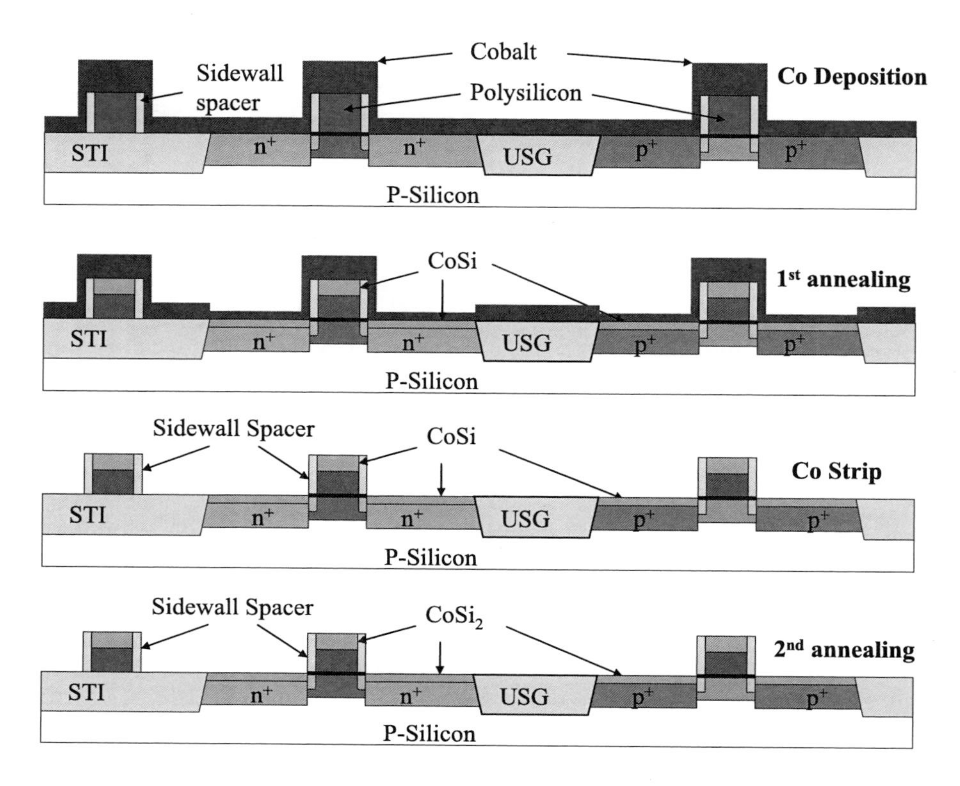

Figure 5.39 Self-aligned cobalt silicide process.

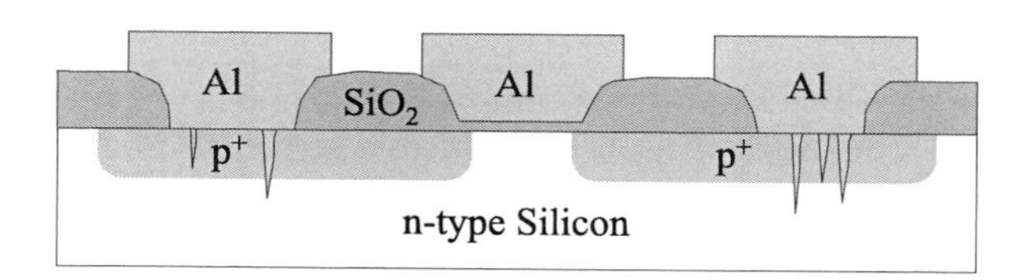

Figure 5.40 Junction spiking.

Furnaces have been used to form an aluminum-silicon alloy at 400 °C in nitrogen and ambient hydrogen to prevent silicon-aluminum interdiffusion, which can cause a junction spike, as shown in Fig. 5.40.

5.5.3 Reflow

When the temperature exceeds the glass transition temperature T_g of silicate glass, the glass becomes soft and starts to flow. This property is widely used in the glass industry to shape glass into different kinds of glassware. It is also applied in wafer processing to smooth silicate glass surfaces in a thermal process commonly called reflow. At 1100 °C, phosphosilicate glass (PSG) becomes soft and starts to flow.

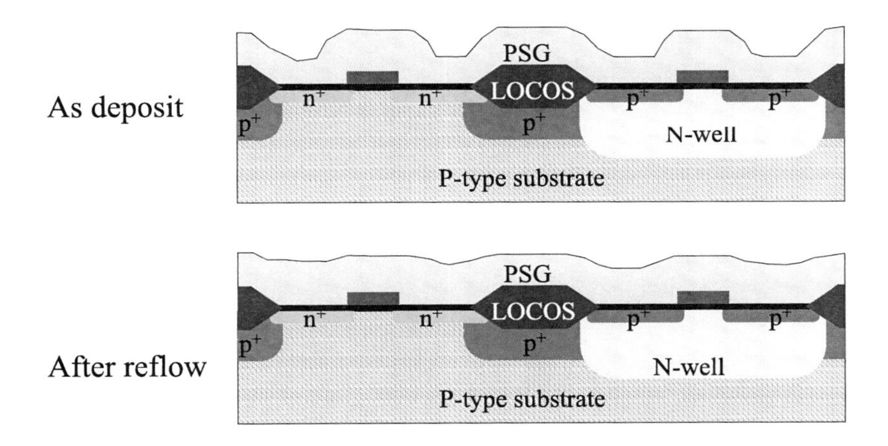

Figure 5.41 PSG reflow process.

The softened PSG follows surface tension, which smoothes and planarizes the dielectric surface. This improves photolithography resolution and makes it easier for metallization processes to succeed. PSG deposition and reflow is shown in Fig. 5.41.

As minimum feature size shrinks, the thermal budget becomes tighter. Borophosphosilicate glass (BPSG) reduces reflow temperature to about 900 °C, which significantly increases thermal budget. The reflow process takes approximately 30 min in a furnace with ambient nitrogen (after the wafer boats are pushed in, and the temperature ramps up and stabilizes).

When the minimum feature size is smaller than 0.25 μm, reflow alone can no longer meet the surface planarization requirement for high photolithography resolution, and the tight thermal budget limits the application of the reflow process. The CMP process then replaces reflow for dielectric surface planarization.

5.6 High-Temperature Chemical Vapor Deposition

CVD is an adding process in which a layer is deposited onto a wafer surface. High-temperature CVD processes include epitaxial silicon deposition, polysilicon deposition, and LPCVD nitride deposition.

5.6.1 Epitaxial silicon deposition

Epitaxial silicon is a single-crystal silicon layer deposited on a single-crystal silicon wafer surface in a high-temperature process. An epitaxial silicon layer is needed for bipolar transistors, BiCMOS IC chips, and high-speed, advanced CMOS IC chips.

Silane (SiH_4), dichloride silane (DCS, SiH_2Cl_2), and trichloride silane (TCS, $SiHCl_3$) are the three most commonly used source gases for silicon-epitaxial growth processes. Chemical reactions of the epitaxial silicon growth process are:

$$\underset{\text{silane}}{SiH_4} \xrightarrow{\text{heat } (1000^{\circ}C)} \underset{\text{epi-Si}}{Si} + \underset{\text{hydrogen}}{H_2},$$

$$\underset{\text{DCS}}{SiH_2Cl_2} \xrightarrow{\text{heat } (1100^{\circ}C)} \underset{\text{epi-Si}}{Si} + \underset{\text{hydrochloride}}{2HCl},$$

$$\underset{\text{TCS}}{SiHCl_3} + \underset{\text{hydrogen}}{H_2} \xrightarrow{\text{heat } (1100^{\circ}C)} \underset{\text{epi-Si}}{Si} + \underset{\text{hydrochloride}}{3HCl}.$$

Epitaxial silicon can be doped as the film grows by flowing dopant gas such as arsine (AsH_3), phosphine (PH_3), and diborane (B_2H_6) with the silicon source gas into the reactor. All three dopant gases are very poisonous, flammable, and explosive. Blanket epitaxial silicon deposition (growth) is normally done outside the IC fab by wafer manufacturers. The epitaxial process is discussed in detail in Chapter 4.

5.6.2 Selective epitaxial growth processes

By patterning silicon oxide or silicon nitride masking film, epitaxial layers can be grown at locations where the masking film is removed and silicon is exposed. This process is called selective epitaxial growth (SEG). Figure 5.42 illustrates the application of SEG SiC to form tensile-strained channels of nMOS and SEG SiGe to form compressive-strained channels of pMOS.

5.6.3 Polycrystalline silicon deposition

Polysilicon has been widely used as a gate material since the introduction of ion implantation as a silicon doping process in the IC industry in the mid-1970s. It is also widely used as contact plugs and capacitor electrodes in DRAM chips. Figure 5.43 illustrates the applications of polysilicon in an advanced DRAM chip.

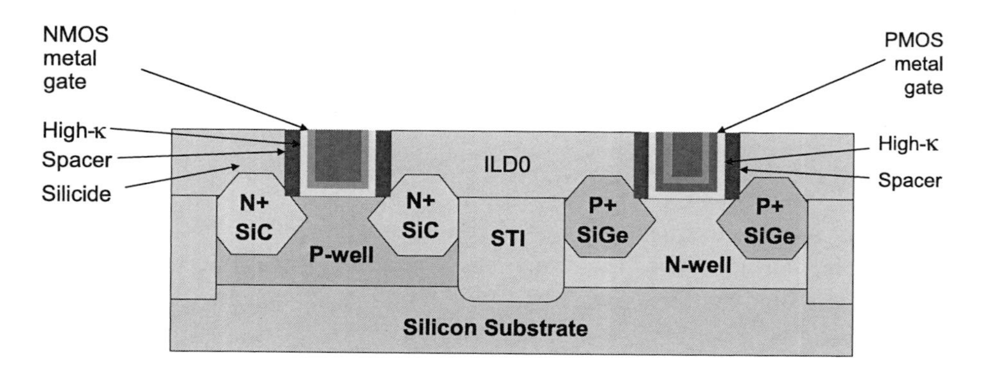

Figure 5.42 SEG applications in advanced CMOS ICs.

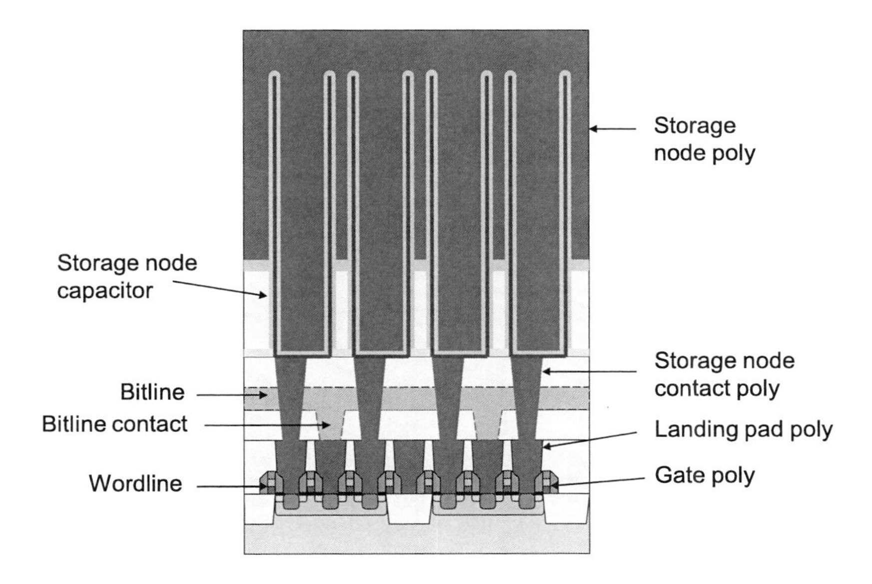

Figure 5.43 Applications of polysilicon in a DRAM chip.

The first poly forms the gate electrodes and wordline and has tungsten on top to reduce resistance. The second poly forms the plugs, which provide landing pads for contact between the S/D and the bitline, and contact between the S/D and the storage capacitor. The third poly forms the contact plugs between the storage capacitors and the landing pad poly. The fourth poly forms the grounding electrode of the storage capacitor of the DRAM, and titanium nitride forms another electrode with high-κ dielectric sandwiches between the two electrodes. By using a high-κ dielectric, the size of the capacitor can be reduced while keeping the required capacitance.

Polysilicon deposition is a LPCVD process that takes place in a furnace with a vacuum system, as illustrated in Fig. 5.44. Polysilicon deposition uses silane (SiH_4) chemistry. At high temperatures, silane will dissociate, and silicon can be deposited on the heated surface. The chemical reaction can be expressed as

$$SiH_4 \rightarrow Si + 2\ H_2.$$

Polysilicon can also use dichlorosilane (DCS, SiH_2Cl_2) chemistry for deposit. At high temperatures, DCS will react with hydrogen and deposit silicon on a heated surface. The DCS process requires a higher deposition temperature than the silane process. The DCS chemical reaction is:

$$SiH_2Cl_2 + H_2 \rightarrow Si + 2\ HCl.$$

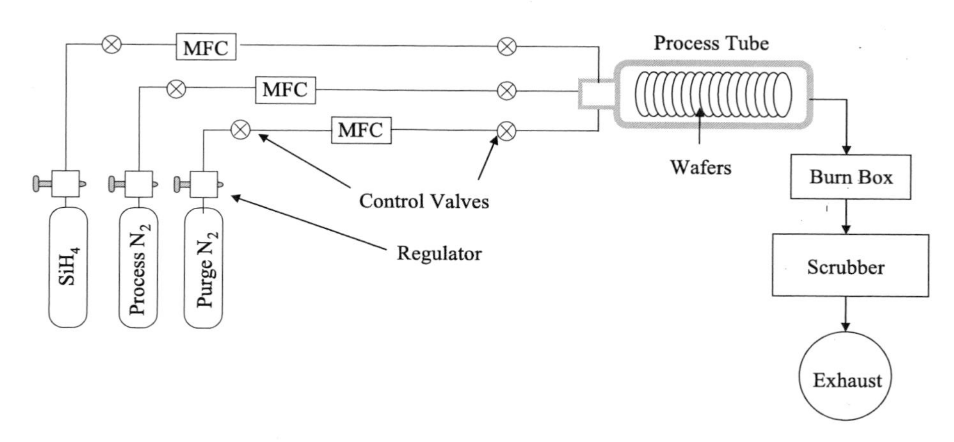

Figure 5.44 Schematic of a LPCVD system for polysilicon deposition.

LPCVD polysilicon can be doped in situ by flowing dopant gas such as arsine (AH_3), phosphine (PH_3), or diborane (B_2H_6) with silane or DCS into the processing chamber (tube).

Polysilicon is deposited at low pressures of 0.2 to 1.0 torr, with deposition temperatures between 600 and 650 °C, using either pure silane or 20 to 30% silane diluted in nitrogen. Deposition rates for both processes are from 100 to 200 Å/min, mainly determined by deposition temperature. Nonuniformity of film thickness within the wafer is usually less than 4%.

The polysilicon deposition processing sequence is as following:

- idle with purge N_2 flow
- idle with process N_2 flow
- wafer loading with process N_2 flow
- process tube (bell jar) lowers with process N_2 flow
- pump down chamber to base pressure (< 2 mtorr) by turning off N_2 flow
- stabilize wafer temperature with N_2 flow and leak check
- pump-to-base pressure (< 2 mtorr) by turning off N_2 flow
- set up process pressure (~250 mtorr) with N_2 flow
- turn on SiH_4 flow and turn off N_2, start deposition
- turn off silane and open gate valve, pump to base pressure
- close gate valve, fill N_2, and ramp-up pressure to atmospheric pressure
- raise bell jar and cool down wafer temperature with process N_2 flow
- unload wafer with process N_2 flow
- idle with purge N_2 flow.

The LPCVD polysilicon deposition process is mainly controlled by process temperature, total process pressure, silane partial pressure for the diluted process, and dopant concentration. Wafer spacing and load size have minimum effect on deposition rate, but they are important for uniformity.

The resistivity of polysilicon film is highly dependent on deposition temperature, dopant concentration, and annealing temperature. Increasing deposition temperature causes resistivity to decrease. Increasing dopant concentration also reduces resistivity, and higher annealing temperatures form large grain sizes, which reduce resistivity as well. Larger polysilicon grain size can cause difficulties in the etch process because it can induce rough sidewalls; therefore, polysilicon is deposited at lower temperatures with smaller grain size and higher resistivity. After polysilicon etch and photoresist stripping, the film is annealed at a high temperature to form large grain size and lower resistivity. In some cases, amorphous silicon (α-Si) is deposited at about 450 °C, then patterned, etched, and annealed to form polysilicon with larger and more uniformly sized grains.

Polysilicon can also be deposited in a single-wafer system. One of the advantages of such deposition is that it can deposit tungsten silicide in situ with the polysilicon. In some DRAM chips, polysilicon-tungsten silicide-staked films are applied to form the gate/wordline and the bitline. The in-situ polysilicon/silicide deposition process eliminates requirements of deglazing and surface cleaning of the polysilicon layer before tungsten silicide deposition, processes that are required in traditional furnace polysilicon deposition and the CVD tungsten silicide processing sequence. By using a polysilicon-tungsten silicide integrated system, throughput can be improved significantly. The single-wafer polysilicon deposition chamber is similar to the single-wafer epitaxial silicon deposition chamber illustrated in Fig. 4.23. A schematic of an integrated polysilicon plus tungsten silicide (WSi_x) deposition system, or polycide system, is shown in Fig. 5.45.

In an integrated polycide system, a wafer is first loaded from the loading station and then sent by a robot to the polysilicon chamber via the transfer chamber. After

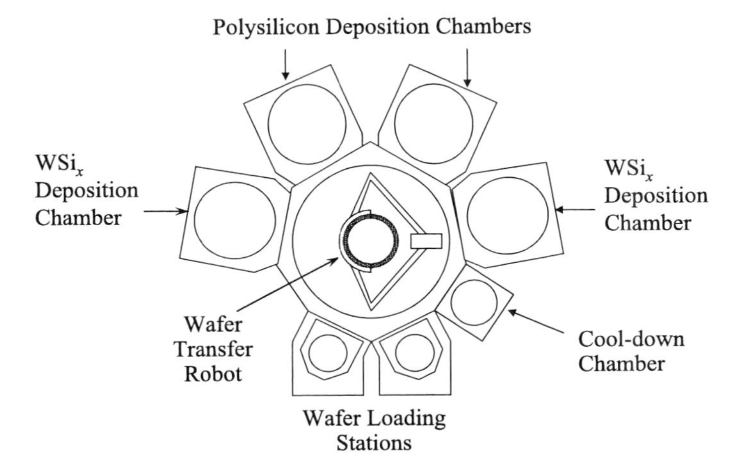

Figure 5.45 Multichamber polycide system.

polysilicon deposition, the wafer is taken out of the polysilicon chamber and sent to the WSi_x chamber to deposit tungsten silicide (Si_x $2 < x < 3$) via the transfer chamber, which is under vacuum conditions. When polycide deposition finishes, the robot picks up the wafer and sends it to the cool-down chamber. Nitrogen gas flow then removes heat from the wafer before the robot places it into a plastic wafer cassette in the loading station for unloading.

For advanced DRAM chips, polysilicon, tungsten silicide, tungsten nitride, and stacked tungsten (poly/WSi_x/WN/W) are commonly used for the gate/wordline, and tungsten nitride and stacked tungsten (WN/W) are used for the bitline. One of the most advanced DRAM chips uses buried wordline (bWL) technology, which uses stacked TiN/W for array transistor gates and wordline, and poly/WSi_x/WN/W for bitline and periphery transistor gate electrodes. A cluster system (illustrated in Fig. 5.45) can be used to deposit stacked poly/WSi_x/WN/W, with four deposition chambers to deposit each film in one processing sequence.

Single-wafer polysilicon deposition uses silane chemistry at a reduced pressure from 10 to 200 torr. Deposition temperature is from 550 to 750 °C, with a deposition rate of up to 2000 Å/min. Hydrogen chloride is used in the chamber dry-clean process to remove polysilicon film deposited on the chamber wall, helping reduce particle generation.

5.6.4 Silicon nitride deposition

Silicon nitride is a dense material widely used as a diffusion barrier layer in IC chips. It is used as an oxidation mask to block oxygen diffusion in the LOCOS formation process, as shown in Fig. 5.11. It is also used in the STI formation process as a CMP stop layer because it has a significantly lower polish rate than USG (see Fig. 5.9).

LPCVD silicon nitride is also used for sidewall spacer formation, either as an etch-stop layer for oxide sidewall spacers or as the spacer itself. Silicon nitride is deposited as a dopant diffusion barrier layer before interlay dielectric (ILD0) PSG or BPSG deposition, to prevent boron and phosphorus from diffusing across the ultrathin gate oxide and penetrating the silicon substrate, possibly causing device malfunction. The nitride barrier layer can also serve as an etch-stop layer for self-aligned contact application, as shown in Fig. 5.46.

These front-end-of-line (FEoL) nitrides can be deposited with LPCVD processes. For the nitride diffusion barrier, some advanced IC chips use plasma-enhanced CVD (PECVD) processes due to thermal budget concerns because PECVD requires significantly lower temperatures than LPCVD. Some advanced CMOS IC chips use nitride stress liners to create different channel strains for pMOS and nMOS. The dual stress liner technique employs PECVD nitride for compressive stress liners to form compressive strain channels of pMOS and uses LPCVD nitride for tensile stress liners to create tensile strain channels of nMOS.

A thin layer of silicon nitride is also commonly used for copper metallization as an ILD etch-stop layer (ESL). A thick layer of nitride is always used as the

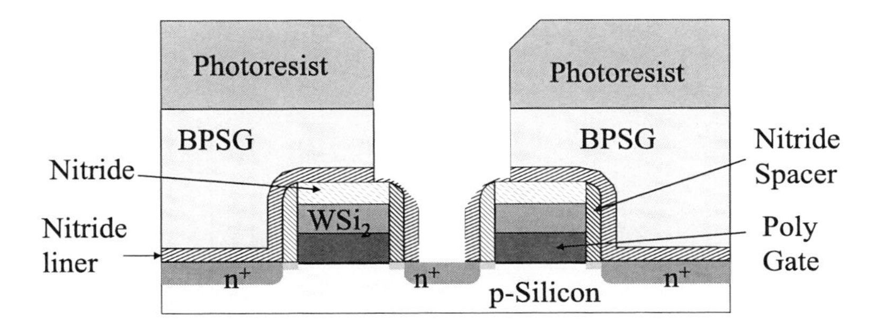

Figure 5.46 Silicon nitride sidewall spacer and self-aligned contact etch stop.

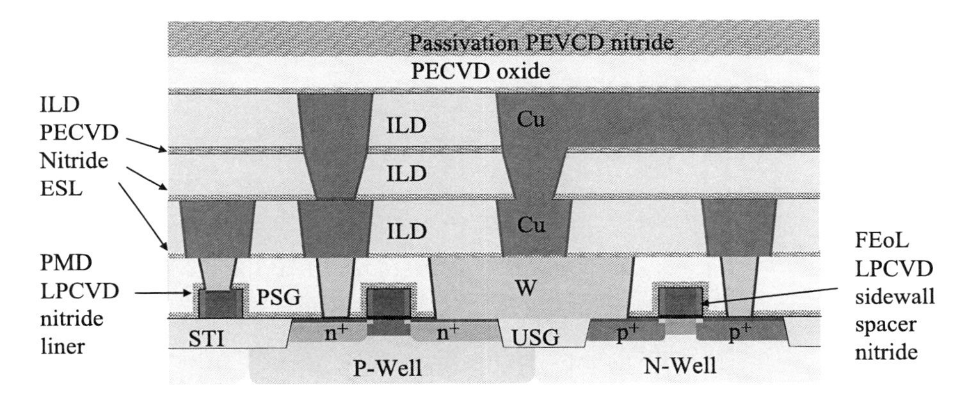

Figure 5.47 Application of silicon nitride in a CMOS IC chip.

passivation dielectric (PD) for IC chips. The ILD0 and passivation applications of silicon nitride in a copper chip are illustrated in Fig. 5.47.

Since wafers cannot tolerate any processing with temperatures higher than 450 °C after the first aluminum-alloy metal layer deposition, most ILD and PD nitride deposition processes are PECVD processes operated around 400 °C. PECVD can achieve high deposition rates at relatively low temperatures because the plasma-induced free radicals dramatically enhance the chemical reaction rate. PECVD processing is discussed in detail in Chapter 10.

LPCVD processes are widely used for LOCOS nitride, STI nitride, spacer nitride, and ILD0 barrier nitride deposition because LPCVD nitride has better film quality and less hydrogen integration than PECVD nitride. In addition, LPCVD nitride is not prone to plasma-induced device damage, which is unavoidable with the PECVD process. LPCVD processing uses a furnace with a vacuum system, as shown in Fig. 5.48. Silicon nitride is deposited with dichlorosilane (SiH_2Cl_2) and ammonia (NH_3) chemistry at 700 to 800 °C. The chemical reaction can be expressed as

$$3\ SiH_2Cl_2 + 4\ NH_3 \rightarrow Si_3N_4 + 6\ HCl + 6\ H_2.$$

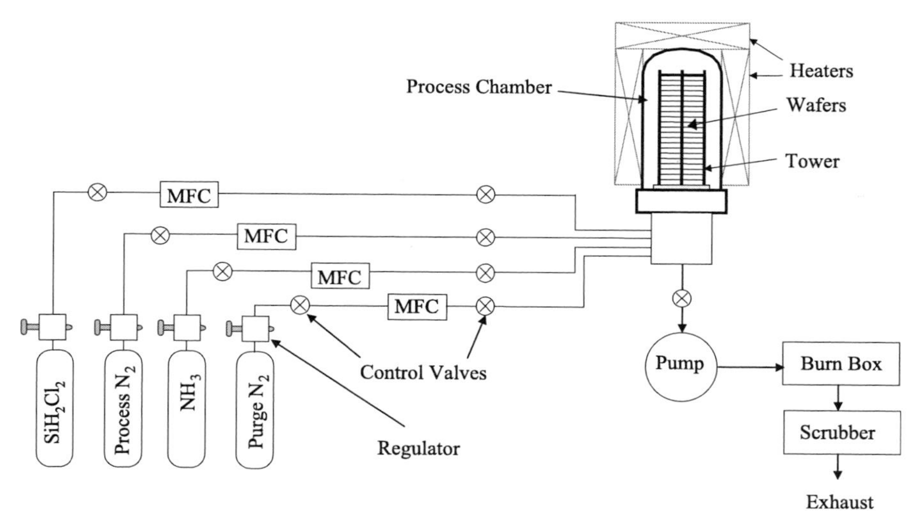

Figure 5.48 Schematic of silicon nitride LPCVD system.

Silicon nitride can also be deposited with silane (SiH_4) and ammonia (NH_3) chemistry at about 900 °C. The chemical reaction is expressed as

$$3\ SiH_4 + 4\ NH_3 \rightarrow Si_3N_4 + 12\ H_2.$$

The process sequence is:

- load wafers
- raise wafer tower with constant nitrogen flow and processing chamber temperature
- pump the chamber down to base pressure without nitrogen flow
- backflow nitrogen to stabilize wafer temperature
- turn off nitrogen and pump down to base pressure
- flow nitrogen and ammonia to set up and stabilize processing pressure
- turn off nitrogen and turn on dichlorosilane to deposit silicon nitride
- pump chamber down to base pressure with all gas flows turned off
- start nitrogen flow into the chamber to raise pressure to atmospheric pressure
- lower wafer tower and unload wafers with nitrogen flow.

The process flow diagram is illustrated in Fig. 5.49.

One of the possible byproducts of a DCS-based nitride LPCVD process is ammonium chloride (NH_4Cl), which is a solid and can cause particle contamination and pump damage. Therefore, researchers are seeking other precursors for the LPCVD nitride process. A possibility is bis(tertiary-butylamino)silane—$SiH_2[NH(C_4H_9)]_2$ or BTBAS, which is a liquid with a boiling point of 164 °C. At 550 to 600 °C, it can react with ammonia to deposit uniform silicon nitride film with high film quality and good step coverage, and free of ammonium chloride.

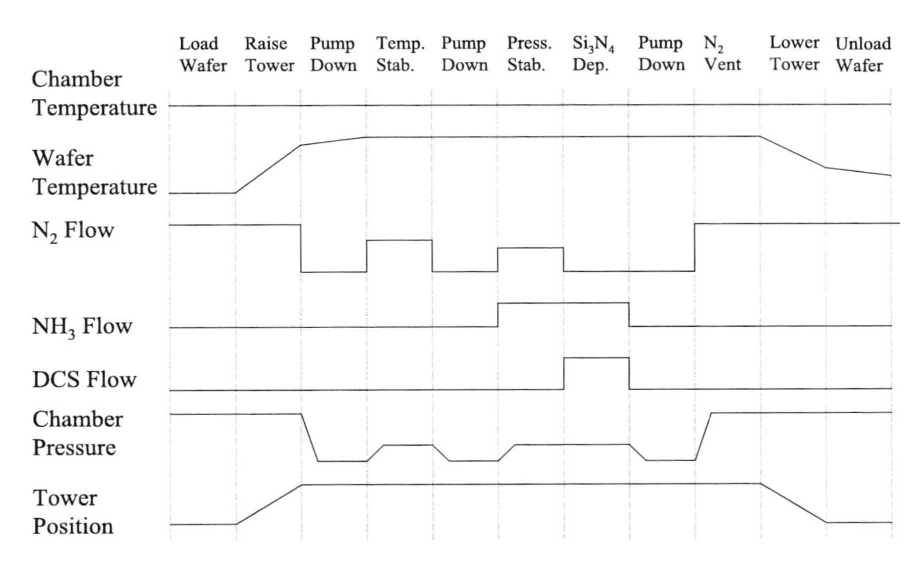

Figure 5.49 Silicon nitride LPCVD process flow diagram.

5.7 Rapid Thermal Processing

Rapid thermal processing is a batch processing tool that processes hundreds of wafers at the same time. Due to the large thermal capacity, the temperature of processing tubes or chambers must ramp up and down slowly. AN RTP system is used for single-wafer processing and can ramp up temperature at rates between 75 to 200 °C/sec. AN RTP system is the favored system for postimplantation annealing, silicide annealing, and ultrathin silicon dioxide layer growth.

RTP systems include rapid thermal annealing (RTA), rapid thermal oxidation (RTO), and rapid thermal CVD (RTCVD) processes. There are several different RTA processes, including postimplantation annealing (>1000 °C), alloy annealing (~700 °C), dielectric annealing (700 °C to 1000 °C), and metal annealing (<500 °C).

RTP uses a quartz chamber and many quartz parts. The heating element is a tungsten-halogen lamp, which can generate intense heat in the form of infrared (IR) radiation. The wafer temperature is precisely measured by an infrared pyrometer. Figure 5.50 illustrates one type of RTP system.

Lamps on the top and bottom are positioned in a vertical array to allow wafers to be heated uniformly by the IR radiation. Wafer temperature is monitored by a pyrometer and feedback is controlled by the lamp power. Figure 5.51 shows the lamp arrangement in an RTP system.

Another type of RTP system places heating lamps in gold-plated lamp houses with a honeycomb structure, as shown in Fig. 5.52. Gold-plated lamp housing improves the power transfer efficiency. During the process, a wafer rotates to improve heating uniformity. Several pyrometers monitor the temperature across wafer and feedback signals to control the heating power of the lamps in different heating zones and achieve very precise, very uniform wafer heating.

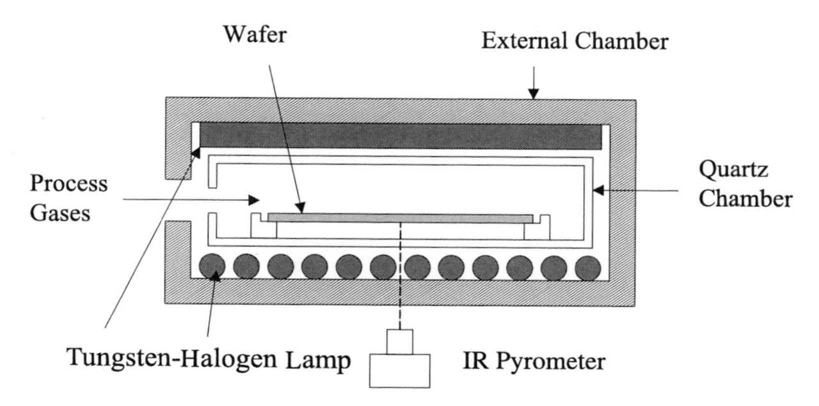

Figure 5.50 Schematic of an RTP chamber.

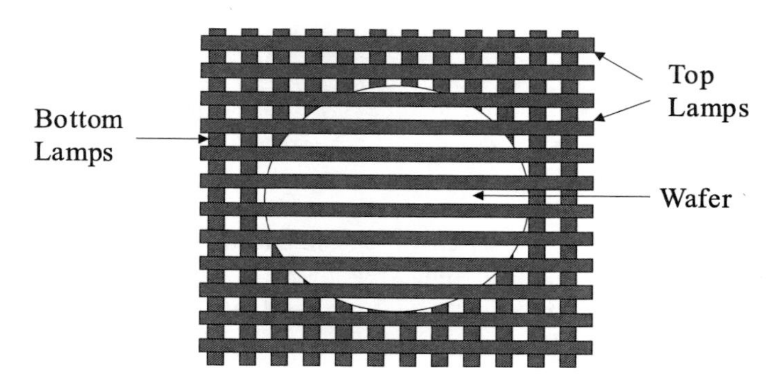

Figure 5.51 Heating lamp array in an RTP chamber.

5.7.1 Rapid thermal annealing

The most common application of RTP is the postimplantation RTA process. After ion implantation, the crystal structure of silicon near the surface is heavily damaged by energetic ion bombardment. A high-temperature process is required to anneal the damage, restore the single-crystal structure, and activate the dopant. During the high-temperature annealing process, dopant atoms diffuse fast, driven by the thermal energy. It is very important to minimize the dopant diffusion during thermal annealing. As device dimensions shrink into nanometer ranges, there is much less room for dopant atoms to diffuse. Therefore, precise control of the thermal budget is critical.

In amorphous silicon, thermal motion of the dopant atoms is unrestricted, whereas the dopant atoms in single-crystal lattices are severely restricted by binding energy. Therefore, dopant atoms can diffuse much faster than those bonded in a single-crystal lattice. At lower temperatures, the diffusion process outpaces the annealing process; at high temperatures (>1000 °C), the annealing process is faster. This is because the annealing activation energy (~5 eV) is higher than the diffusion activation energy (3 to 4 eV). Because furnace processing requires more time and the annealing temperature is relatively low, dopant diffusion cannot be minimized.

Figure 5.52 RTP system with honeycomb heating source (courtesy of Applied Materials, Incorporated).

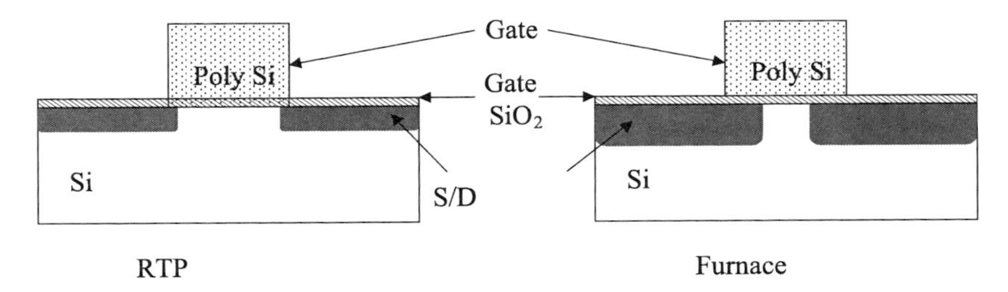

Figure 5.53 Dopant diffusion in RTP and furnace annealing processes.

Diffusion of dopant atoms can be quite significant and becomes intolerable for small devices. For some noncritical implantation processes, such as well implantation, furnace annealing could still be used; but for most postimplantation annealing, the RTA process is required. Both processes are shown in Fig. 5.53.

A RTP system can ramp up and cool down wafer temperature very quickly. Normally, it takes less than 10 sec to reach the required annealing temperature, from 1000 to 1150 °C. Annealing takes about 10 sec, then the wafer is cooled down very quickly by turning the lamps off and flowing nitrogen gas. The faster the temperature ramp, the less the diffusion of the dopant atoms. For sub-0.1-μm device applications, a ramp-up rate of 250 °C/sec can be needed to achieve junction annealing with minimized dopant diffusion.

A postimplantation RTA process sequence is as follows:

- wafer in
- temperature ramp-up, temperature stabilization
- annealing
- wafer cool-down
- wafer out.

Temperature ramp-up rate is normally 75 to 150 °C/sec, and the annealing temperature is approximately 1100 °C. The entire processing sequence takes less than two minutes in ambient nitrogen with a constant nitrogen flow. Figure 5.54 shows the temperature change of the RTP system in a postimplantation annealing process.

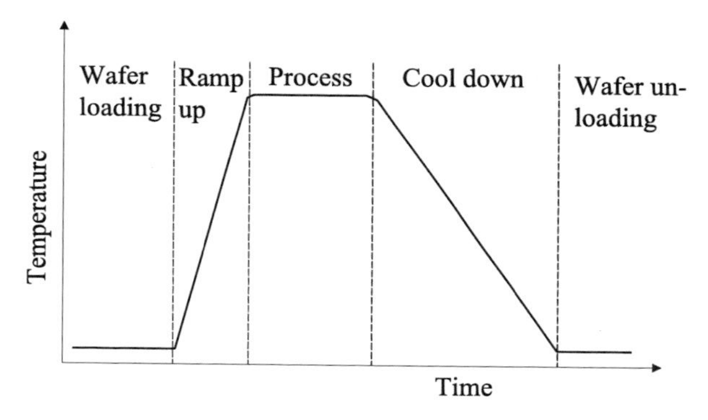

Figure 5.54 Temperature change in an RTA process.

Question: If the pyrometer is not well calibrated, what could be the consequence in the RTA process?

Answer: A poorly calibrated pyrometer can fail to measure wafer temperature accurately; thus, the stabilized processing temperature will not be the set value. If it is too low, sheet resistance will be high due to incomplete dopant activation and damage repair. If it is too high, in extreme cases, it can melt a silicon wafer. Therefore, routine calibration of the pyrometer is an important part of PM procedures.

Other RTP annealing processes include alloy annealing, especially silicide processes such as those for titanium silicide, cobalt silicide, and nickel silicide. $TiSi_2$ and $CoSi_2$ undergo RTA around 700 °C in ambient nitrogen, and NiSi undergoes RTA between 400 to 450 °C. It takes about one minute to anneal and form silicide.

RTP is also used for thermal nitridization processes, in which ammonia reacts with titanium to form titanium nitride on the surface as a barrier and adhesion layer for aluminum metallization. The chemical reaction can be expressed as

$$NH_3 + Ti \rightarrow TiN + 3/2\ H_2.$$

After titanium deposition, the wafer goes into the RTA chamber. After temperature ramp up and stabilization in the ambient nitrogen, the nitrogen is turned off and ammonia is turned on. Soon after the nitridization finishes, ammonia is turned off and nitrogen is turned on. The transfer robot removes the wafer from the processing chamber and sends it to a cool-down chamber until it is ready to be placed into a plastic wafer cassette. The processing temperature is about 650 °C. Figure 5.55 illustrates the titanium nitridization process.

5.7.2 Rapid thermal oxidation

When transistor size decreases, the thickness of the gate oxide also decreases. The thickness of gate oxides has been thinned to about 15 Å, a thickness that can no longer be reduced because of the leakage issue. With such a fine thickness, it is difficult to precisely control oxide thickness and WTW thickness uniformity within a multiwafer batch system, such as an oxidation furnace. There are several advantages to using single-wafer RTP systems to grow this ultrathin, high-quality oxide. Because RTP systems can precisely control wafer temperature and temperature uniformity across a wafer, the rapid thermal oxidation (RTO) process can grow a very thin and uniform oxide layer. Because it is a single-wafer system, the RTO process has better WTW uniformity control than furnace processing, especially for an ultrathin oxide layer. Another advantage is that the RTO chamber can be integrated with a HF vapor etch chamber in one mainframe. After the HF vapor etch process removes native oxide on the silicon wafer surface, the wafer is sent to the RTO/RTA chamber through a transfer chamber, which is always kept under high vacuum. The possibility of reoxidation of the silicon surface is eliminated, since the wafer does not become exposed to the atmosphere

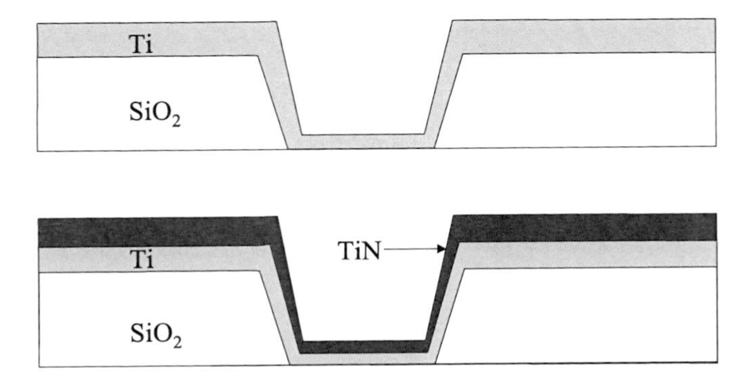

Figure 5.55 The titanium nitridization process.

or moisture before it is sent to the RTO chamber for HCl cleaning, oxidation, and annealing processes.

An RTO process diagram is illustrated in Fig. 5.56. After loading the wafer into the processing chamber, the heating lamp is turned on and the temperature ramps up in two steps, first to about 800 °C, at a higher ramp rate, then to the required oxidation temperature, such as 1150 °C, at a lower ramp rate. This two-step temperature ramp up minimizes ramping time by reducing the required stabilization time at a lower ramp rate to reach the processing temperature. After temperature stabilization, oxygen starts to flow into the processing chamber to react with the silicon and grow silicon dioxide on the wafer surface. HCl is used in the oxidation process to minimize mobile ion contamination and reduce interface state charge. After the oxide is grown, O_2 and HCl flows are turned off and N_2 flows into the chamber. Wafer temperature is ramped up to about 1100 °C to anneal the oxide layer, helping improve the oxide film quality and further reducing the interface state charge. Thermal nitridation with nitric oxide (NO) can be performed in this annealing step. If plasma nitridation is required, the wafer is sent to another chamber designed for that process, before going through the annealing step. When the annealing step is finished, the heat lamp is turned off and the wafer starts to cool down. A robot in the transfer chamber picks up the still-hot wafer and sends it to the cool-down chamber before sending it to the cassette.

The gate dielectric has changed from commonly used silicon dioxide ($\kappa = 3.9$) to nitrated silicon oxide (SiON), and finally to a dielectric with high dielectric constant (high-κ), which allows a thicker gate dielectric layer for preventing gate leakage and gate dielectric breakdown. Atomic layer deposition (ALD) is the most commonly used method for depositing a high-κ dielectric, and RTA processing is used to improve film quality and reduce interface state charge.

RTO processing has been widely used for gate oxidation processes due to better process control, especially WTW uniformity control. Except for wet oxidation,

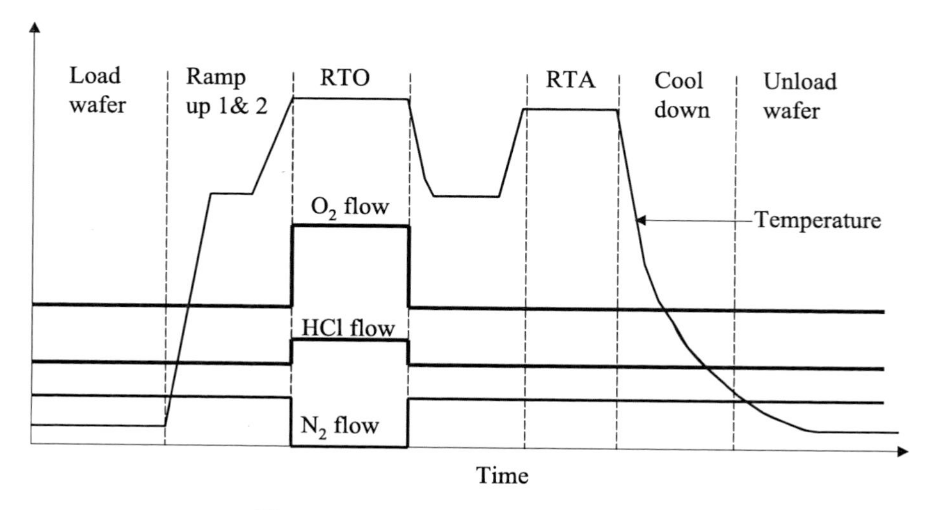

Figure 5.56 RTO processing diagram.

most oxidation processes for advanced IC chips are most likely to be performed in the RTO processing chamber due to better thermal budget control.

5.7.3 Rapid thermal chemical vapor deposition

Rapid thermal CVD (RTCVD) processing is a thermal CVD process performed in a single-wafer, cold-wall processing chamber with the capability of rapid temperature change and precise temperature control (see Fig. 5.57). Because it is a single-wafer system, the deposition rate has to be high enough that thin-film deposition can be finished within 1 to 2 min to achieve a reasonable throughput of 30 to 60 wafers per hour.

Figure 5.58 shows temperature changes of RTCVD and LPCVD processes. Compared with furnace LPCVD processing, RTCVD processing has better control of thermal budget and WTW uniformity. With device dimensions shrinking, the thickness of deposited thin films in the FEoL processes also decreases, typically from 100 to 2000 Å. With deposition rates from 100 to 1000 Å/min, single-wafer RTCVD processing becomes more attractive for FEoL thin-film deposition.

RTCVD processing can be used to deposit polysilicon, silicon nitride, and silicon dioxide, such as the CVD oxide used for trench-fill in shallow trench isolation processes. At high temperatures, tetraethoxysilane (TEOS) or $Si(OC_2H_5)_4$ can be used to deposit USG with high film quality and gap-fill capabilities.

In-situ process monitoring will become more important for better process control and higher throughput. The precise measurement of temperature, ±2 °C at 1000 °C, is the key to success for RTP. In-situ, real-time thickness measurement is required for oxidation and RTCVD processing endpoint detection.

A cluster tool consists of a mainframe and several processing chambers, and can integrate different processes into one system. It can improve throughput due to less interval time between different processes and achieve higher yield, because wafer transfer between different processes occurs in a vacuum environment, which reduces the chance of contamination. Figure 5.59 illustrates a cluster system for complete gate oxidation, nitridation, annealing, polysilicon deposition, and polysilicon annealing processes.

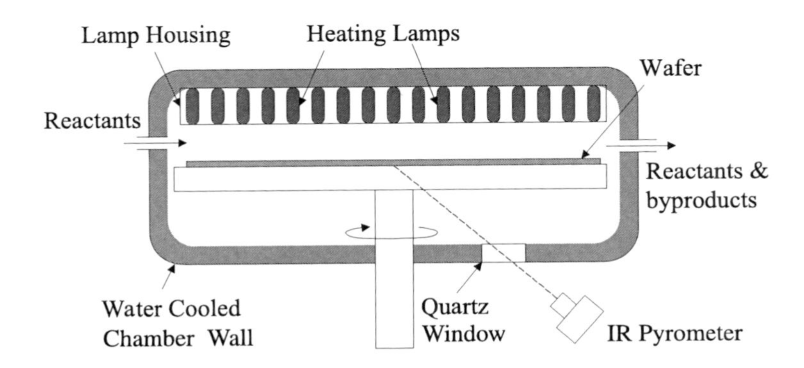

Figure 5.57 Schematic of RTCVD chamber.

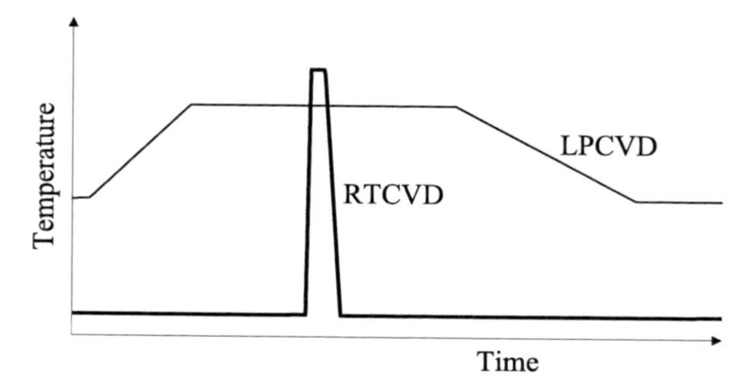

Figure 5.58 Temperature changes in RTCVD and LPCVD.

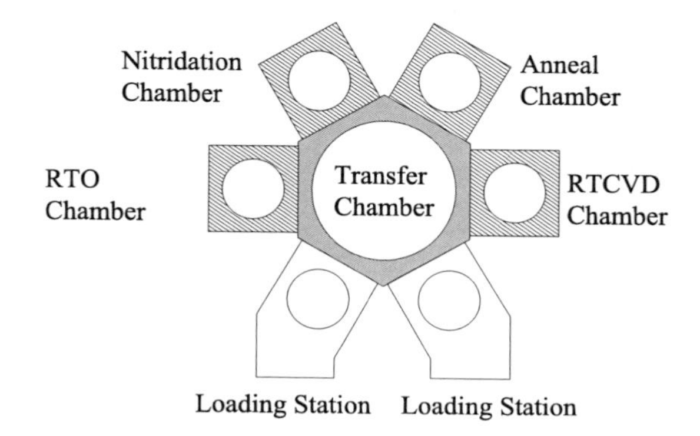

Figure 5.59 A cluster tool for gate oxide/polysilicon processes.

5.8 Recent Developments

Recently, there has been more focus on the development and application of rapid thermal processing (RTP), in-situ process monitoring, and cluster tools, while furnace processes are still used in noncritical thermal processes.

Junction depth becomes extremely shallow as devices grow smaller (less than 200 Å), making postimplantation annealing very challenging. Annealing ion implantation damage requires high temperatures; however, thermal budget requirements limit annealing time to the millisecond range. Spike, laser, and flush annealing techniques have been developed to meet the requirements of ultra-shallow junctions.

Spike annealing is an RTA process with a very short peak time, usually far less than 1 sec. It uses high peak temperatures to maximize dopant activation, a high temperature ramp rate to minimize dopant diffusion, and an aggressive temperature ramp-down rate to minimize dopant deactivation. Figure 5.60 illustrates the temperature variations of a spike annealing process. From the temperature curve, it is easy to see why this process is called spike annealing. Figure 5.61(a) is the simulation result of a dopant profile of a 0.13-μm pMOS with standard 1025 °C and 15-sec RTA, and Fig. 5.61(b) is the same device with 1113 °C spike annealing. It is evident that spike annealing significantly reduces dopant diffusion and becomes

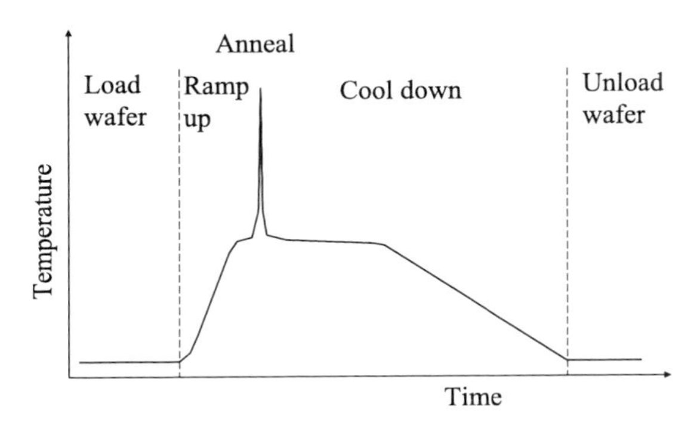

Figure 5.60 Temperature variations in a spike annealing process.

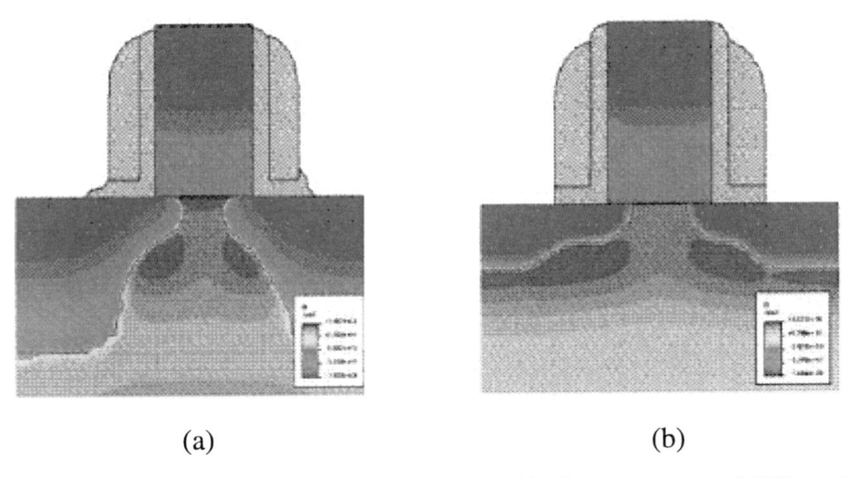

Figure 5.61 Simulated dopant profile of a 0.13-μm pMOS with a standard RTA at 1025 °C and 15 sec, and (b) spike annealing at 1113 °C. (E. Josse, et al., 2002)

more important for devices in nanometer technology nodes because the required junction depths become much thinner, and dopant diffusion during the annealing process must be minimized.

A laser annealing system uses energy from laser light to very quickly heat the wafer surface to the submelting point. Because of the high thermal conductivity of silicon, the wafer surface can be cooled down very quickly, in about a tenth of millisecond. Laser annealing systems can activate a dopant after ion implantation with minimum dopant diffusion and have been used in sub-45-nm process development. Laser annealing systems can be used with spike annealing systems to achieve optimized processing results.

Other postimplantation annealing technologies, such as flush annealing, have also been developed to achieve dopant activation with minimum diffusion for application of ultrathin junctions. Another technique is low-temperature microwave annealing, which uses microwaves to heat only the implantation-damaged area while the rest of the wafer stays in a low-temperature environment, thus achieving dopant activation without dopant diffusion.

5.9 Summary

- Thermal processing is a high-temperature process. In thermal processing, either a layer is being added onto the wafer surface (oxidation, deposition, and doping), or the chemical condition (alloying) or physical condition (annealing, diffusion, and reflow) of wafer materials is changed.
- Oxidation, annealing, and deposition are the three most important thermal processes.
- In the oxidation process, oxygen or water vapor reacts with silicon and forms silicon dioxide.
- It is very important to clean silicon wafer surfaces prior to the oxidation process. Contaminated silicon surfaces can provide nucleation sites that form a layer of silicon dioxide polycrystalline, which has very poor quality.
- Dry oxidation has a lower growth rate than wet oxidation. Dry oxide film quality is better than that of wet oxide. Thick oxide films, such as field oxide, usually use the wet oxidation process, while most thin oxide films use dry oxidation processes to grow.
- Diffusion processing was commonly used in doping processes of the IC industry. It uses silicon dioxide as a diffusion mask because most dopant atoms diffuse much slower in silicon dioxide than in single-crystal silicon.
- Diffusion processing consists of three steps: dopant oxide deposition, oxidation, and drive-in.
- Diffusion doping processes cannot independently control dopant concentration and dopant junction depth. Diffusion processing is an isotropic process; therefore, it always diffuses underneath the masking oxide. It had been replaced by ion implantation, which was introduced in the mid-1970s.
- Polysilicon and front-end silicon nitride depositions are LPCVD processes, which use a furnace with a vacuum system.
- After ion implantation, energetic ions cause extensive crystal structure damage; therefore wafers need postimplantation annealing to recover the single-crystal structure and activate dopants.
- RTP can ramp up temperatures at 50 to 250 °C/sec, compared to 5 to 10 °C/min in a furnace process. RTP has more control over thermal budget than does furnace processing.
- Postimplantation RTA is the most commonly used RTP. It is fast, can minimize dopant diffusion during the annealing process, and has excellent thermal budget control.
- Other RTP applications are RTA for dielectric annealing and silicide alloying, RTO, and RTCVD processes.
- RTP chambers possessing multiple controllable heating zones, in-situ process monitoring, and cluster tools are important trends for future IC fabrication thermal processes.
- Furnaces will continue to be used in future fabs for noncritical thermal processes due to their high throughput and low cost.

- To meet the requirements of diminishing device geometry, millisecond annealing techniques such as spike, laser, and flush annealing are being developed.

5.10 Bibliography

D. G. Baldwin, M. E. Williams, and P. L. Murphy, *Chemical Safety Handbook for the Semiconductor/Electronics Industry*, 2nd edition, OME Press, Beverly, MA (1996).

A. E. Braun, "Thermal Processing Options, Focus and Specialize," *SI* **22**(5), 56 (1999).

C. Y. Chang and S. M. Sze, *ULSI Technologies*, McGraw-Hill, New York (1996).

R. B. Fair, "Challenges in manufacturing submicron, ultra-large scale integrated circuits," *Proc. IEEE* **78**, 1687 (1990).

E. Josse, F. Arnaud, F. Wacquant, D. Lenoble, O. Menut, and E. Robilliart, "Spike anneal optimization for digital and analog high performance 0.13 μm CMOS platform," *Proc. Eur. Solid-State Dev. Res. Conf.*, 207–210 (2002).

J. M. Kowalski, J. E. Kowalski, and B. Lojek, "Microwave annealing for low-temperature activation of As in Si," *Proc. IEEE*, 51–56 (2007).

R. K. Laxman, A. K. Hochberg, D. A. Roberts, and F. D. W. Kaminsky, "Low temperature LPCVD silicon nitride using a chlorine-free organosilicon precursor," VLSI Multilevel Interconnection Conf., 1998, pp. 568.

SEMATECH, *Furnace Processes and Related Topics, Participant Guide* (1994).

R. Sharangpani, R. P. S. Thakur, N. Shah, and S. P. Tay, "Steam-based RTP for advanced processes," *Solid State Technol.* **41**(10), 91 (1998).

L. Shon-Roy, A. Wiesnoski, and R. Zorich, *Advanced Semiconductor Fabrication Handbook*, Integrated Circuit Engineering Corp., Scottsdale, AZ (1998).

See http://www.webelements.com for further information on the elements in this chapter.

5.11 Review Questions

1. List at least three thermal processes.
2. Describe a thermal oxidation process. Why does the oxide film grow into the silicon substrate during LOCOS formation?
3. For field oxide formation, which oxidation process is preferred, wet or dry? Explain your answer.
4. What are the advantages and disadvantages of a pyrogenic wet oxidation system compared with other wet oxidation systems?
5. In a pyrogenic wet oxidation process, why is the flow ratio H_2:O_2 slightly lower than 2:1?

6. List all gases used in a gate oxidation process, and explain the roles of each.
7. When temperature increases, how does the oxide growth rate change? What is the effect on the oxide growth rate when pressure increases?
8. Pad, barrier, gate, screen, and field oxides have been used in IC chip fabrication. Which is the thinnest and which is the thickest?
9. Why are furnaces commonly called diffusion furnaces even though they are not used for diffusion processes?
10. What are the advantages of a vertical furnace versus a horizontal furnace?
11. List three steps of the diffusion doping process.
12. Why can silicon dioxide be used as a diffusion mask?
13. What is junction depth?
14. Describe a titanium silicide process.
15. Why does a wafer need to be annealed at a high temperature after implantation? What are the advantages of using RTA processing in this application?
16. Describe a PSG reflow process. Can USG reflow be used? Explain the answer.
17. List processing gases that can be used for p-type-doped polysilicon LPCVD processing.
18. In LPCVD nitride deposition processes, why is ammonia used as a nitrogen source instead of nitrogen gas?
19. What is the temperature ramp rate of an RTP system? What is the rate of a furnace? Why can an RTP system ramp up temperature faster than a furnace?
20. Compared with the RTP system, what are the advantages of furnace systems?
21. What is the difference between spike annealing and RTA?
22. What is the driving force to develop millisecond annealing techniques?

Chapter 6
Photolithography

Objectives

After finishing this chapter, the reader will be able to:

- list the four components of photoresist
- describe the difference between positive and negative photoresists
- describe a photolithography processing sequence
- list four alignment and exposure systems
- identify the most commonly used alignment and exposure system in IC production
- describe the wafer movement in a track-stepper integrated system
- explain the relationship between resolution and depth of focus with wavelength and numerical aperture
- list at least three candidates for next-generation lithography.

6.1 Introduction

Photolithography is a patterning process that transfers a designed pattern from a mask or reticle to the photoresist on a wafer's surface. It was first used in the printing industry and has long been used to make printed circuit boards (PCBs). Photolithography has been a part of the semiconductor industry for transistor and integrated circuit manufacture since the 1950s. It is the most crucial processing step in IC fabrication because device and circuit designs are transferred to wafers by either etch or ion implantation by means of a pattern; that pattern is defined on a photoresist on a wafer's surface by the photolithography process.

Photolithography is the core of the IC manufacturing process flow, as shown in Fig. 6.1. From a bare wafer to bonding pad etch and photoresist strip, the simplest MOS-based IC chip needs five photolithography processing or masking steps; an advanced IC chip can take more than 30 masking steps. IC processing is very time consuming: even with a 24-7 nonstop working schedule, it takes six to eight weeks to go from bare wafer to finished wafer. The photolithography process takes about 40 to 50% of total wafer processing time.

The requirements of photolithography are high resolution, high sensitivity, precise alignment, and low defect density. The development of IC processing

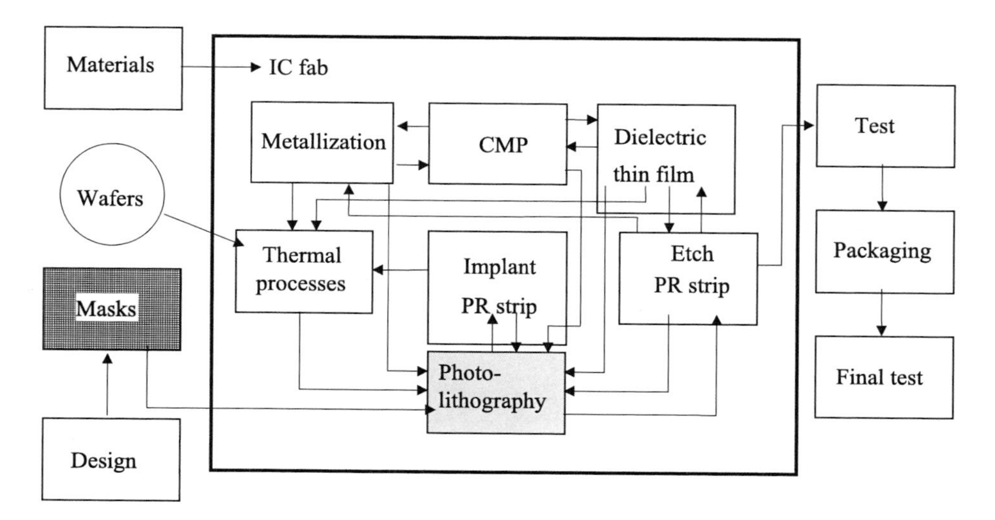

Figure 6.1 IC fabrication process flow.

technology is measured by the half-pitch (HP) of the minimum feature size on the production wafer, which was 25 nm (in 2010), as shown in Fig. 6.2. The smaller the HP, the more devices one can make on a wafer. 25-nm technology in 2010 was limited by photolithography resolution and patterning technology. By improving photolithography resolution and patterning technology, the HP can be further reduced. A photoresist must be very sensitive to exposure light to achieve reasonable throughput because the higher the photoresist sensitivity, the shorter the exposure time, and the higher the throughput. However, if the sensitivity is too high, other photoresist characteristics can be affected, such as line edge roughness (LER) and line width roughness (LWR), and those factors in turn affect the resolution. Thus, a careful balance between resolution and sensitivity is required. An advanced IC chip needs more than 30 patterning process steps; each one must be precisely aligned with the previous one to achieve a successful pattern transfer for the entire chip design. There is very little room for alignment error, since maximum tolerance is 10 to 20% of the CD. For 25-nm technology, misalignment needs to be controlled to within 2.5 to 5 nm, which is why automatic alignment systems are required. Advanced photolithography systems are very challenging systems, since everything must be precisely controlled. For example, a 1 °C difference on a 300-mm wafer can cause a 0.75-μm difference in wafer diameter due to silicon thermal expansion (or contraction when the temperature is lower) at a 2.5×10^{-6}/°C expansion (or contraction) rate. The photolithography process has to control defect density, since defects introduced in this process are transferred to devices and circuits via succeeding etch or ion implantation processes, affecting the yield and reliability of the products.

The photolithography process can be subdivided into three main operations: photoresist coating, alignment and exposure, and photoresist development. First, the wafer is coated with a thin layer of photosensitive material, called photoresist, which is exposed by ultraviolet (UV) light through a mask or reticle, with the

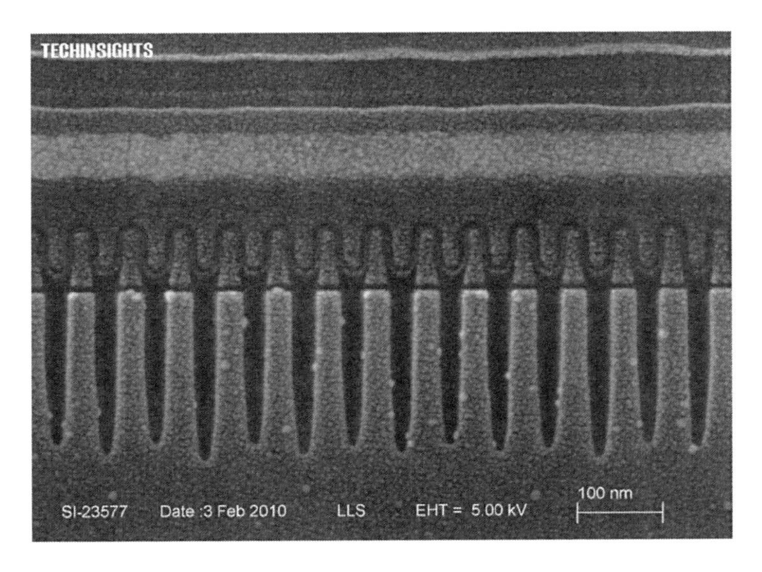

Figure 6.2 Cross-section SEM image of the IM Flash Technologies 25-nm flash memory array along the wordline direction. (Image from www.semiconductorblog.com)

pattern of transparent and opaque areas generated by a mask writer. The chemistry of the exposed photoresist is changed by photochemical reactions under the transparent areas where UV light passes through. For a positive photoresist, which is most commonly used in advanced semiconductor fabs, the exposed area is dissolved in developer solution, leaving the unexposed photoresist on the wafer surface, which reproduces a dark pattern on the mask or reticle.

6.2 Photoresist

Photoresist is a photosensitive material used to temporarily coat a wafer and then transfer an optical image of a chip design onto a mask or reticle to the wafer surface. It is very similar to the photosensitive material that is coated onto the plastic of photographic films, which can transfer optical images from a camera lens onto the film's surface. But unlike these photosensitive coatings on films, photoresists are not very sensitive to visible light, nor are they required to be sensitive to changes in color. Since it is mainly sensitive to UV light and is insensitive to visible light, photolithography processing does not need a darkroom. Because photoresists are insensitive to yellow light, all semiconductor fabs use yellow lights to illuminate photolithography areas, commonly called photo bays.

There are two kinds of photoresists, positive and negative. For the negative photoresist, the exposed parts become cross-linked and polymerized due to a photochemical reaction, then harden and remain on the wafer surface after development; the unexposed parts are dissolved by the developer solution. Positive photoresists normally are cross-linked and polymerized before exposure. After exposure, the exposed photoresist changes from an aqueous-base-insoluble material to an aqueous-base-soluble material in a photochemical reaction called photosolubilization and is dissolved by the developer; the unexposed parts remain on the wafer surface.

Figure 6.3 illustrates the two different kinds of photoresists and their pattern transfer processes. The image of the positive photoresist is the same as the image on the mask or reticle, and the image of the negative photoresist is the reversed image on the mask or reticle. Before digital cameras, a roll of photographic film had to be purchased and loaded into a camera; that film was a negative film. The images left on negative film after development are the reversed images of those images the camera has taken; they require exposure and development again on negative photographic papers to print normal images. For a higher price, a roll of positive film can be purchased that gives the same images taken by the camera after development. Positive films are normally used to make the slides used for slide shows.

Presently, most advanced semiconductor fabs use positive photoresist because it can achieve the high resolution required for nanometer feature sizes. Photoresist has four basic components: polymer, sensitizer, solvent, and additives.

Polymer is the solid organic material that adheres to the wafer surface and withstands etch and ion implantation processes for the masking of the pattern transfer process. It is formed by organic compounds, which are carbon-hydrogen molecules (C_xH_y) with complicated chain and ring structures. A commonly used positive photoresist polymer for 248- and 193-nm photoresists is polymethyl-methacrylate (PMMA). Phenol-formaldehyde or novolac resin are commonly used polymers for photoresist of i-line (365-nm) and g-line (436-nm) lithography. The most commonly used negative photoresist polymer is polyisoprene rubber.

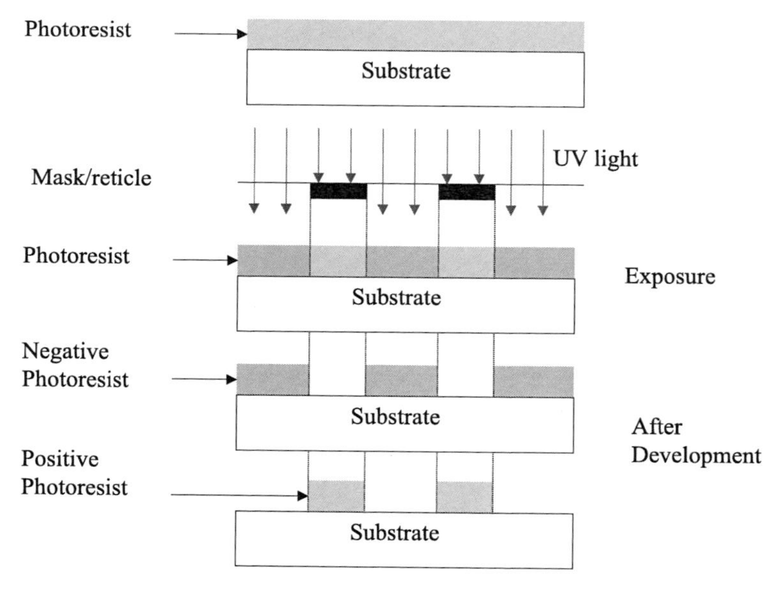

Figure 6.3 Patterning process with negative and positive photoresists.

A sensitizer is an organic compound with very high photoactivity that controls and modifies the photochemical reaction of a photoresist during exposure. The sensitizer for a positive photoresist is a dissolution inhibitor, which is cross-linked within the resin. During the exposure process, energy from the light dissociates the sensitizer and breaks down the cross-links causing the exposed resin to become soluble in an aqueous developer solution. The sensitizer for a negative photoresist is an organic molecule containing the N_3 group. Exposure to UV light liberates N_2 gas, forming free radicals that help cross-link the rubber molecules. The cross-linking chain reaction polymerizes the exposed areas, which have greater bonding strength and higher chemical resistance.

Solvent is the liquid that dissolves the polymer and sensitizer. It makes photoresist easy to apply on a wafer surface as a thin film of 0.5 to 3 μm thickness. The solvent thins the photoresist to allow application of thin layers by spinning, analogous to painting. Before the spin-coating process, about 75% of the photoresist is solvent. Positive photoresists use acetate-type solvents, and negative photoresists use xylene (C_8H_{10}).

Additives control and modify the photochemical reaction of photoresists during exposure to achieve optimized lithography resolution. Dye is a commonly used additive for both positive and negative photoresists.

For a negative photoresist, the developer solution is mainly xylene. The developer solution dissolves the unexposed photoresist, but often some of the developer solvent is absorbed in the exposed, cross-linked photoresist. This causes the photoresist swelling effect, which distorts pattern features and limits resolution to about two to three times that of the thickness of the photoresist. Negative photoresists were widely used in the semiconductor industry before the 1980s, when minimum feature size was larger than 3 μm. They are no long commonly used in advanced semiconductor fabs due to their poor resolution. Positive photoresists do not absorb developer solvents. They can achieve much higher resolution and are widely used for photolithography processes. Figure 6.4 illustrates the resolution of negative and positive photoresists.

Question: Positive photoresists can achieve much higher resolution than negative photoresists. Why were they not used before the 1980s?

Answer: Positive photoresist is more expensive than negative photoresist; therefore, manufacturers used negative photoresist until it had to be

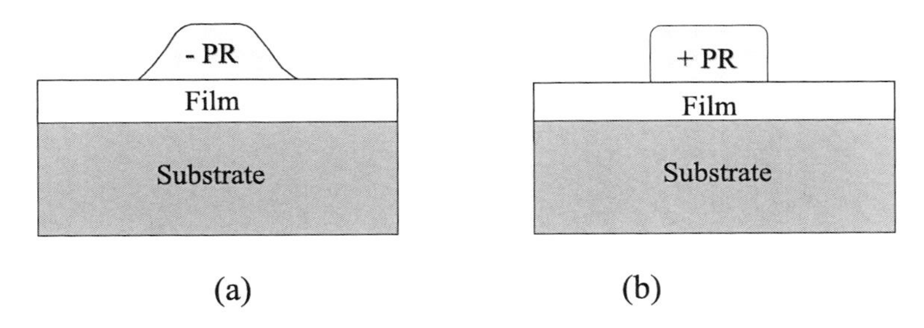

Figure 6.4 (a) Negative and (b) positive photoresists (PRs).

> replaced when the minimum feature size shrank to smaller than 3 μm. Photoresist is one of the most important materials used in the photolithography process. Unless it is absolutely necessary, engineers are extremely reluctant to switch from one photoresist to another.

To pattern a small feature, shorter-wavelength exposure light is required. Photolithography processes using DUV (248 nm) or ArF (193 nm) require photoresist that is different from photolithography processes that use mercury g-line (436 nm) and i-line (365 nm). This is because the intensity of DUV light sources, usually excimer lasers, is much lower than that of mercury lamps. Chemically amplified photoresists are developed for DUV photolithography processes in 0.25-μm or smaller feature patterning applications. A catalysis effect is used to increase the sensitivity of this kind of photoresist. A photoacid is created in the photoresist when it is exposed to DUV light. In the postexposure bake (PEB) process a wafer is heated, and this heat drives acid diffusion and amplification in a catalytic reaction, as shown in Fig. 6.5.

To achieve completed pattern transfer, the photoresist needs to have good resolution, high etch resistance, and good adhesion. High resolution is the key to achieving successful pattern transfer. Without high etch resistance and good adhesion of the photoresist, the next etch or ion implantation process will most likely fail to meet the processing requirements and cause intolerable error. The thinner the photoresist film, the higher the resolution will be. However, it is also true that the thinner the photoresist film is, the lower the etching and ion implantation resistance. There is always a tradeoff between these two conflicting requirements.

The latitude of the photoresist process is tolerance of the photoresist to varying spin rates baking temperature, and exposure flux. The wider the process latitude, the more stable the process. This is an important factor when selecting a photoresist for a given process technology.

6.3 Photolithography Process

The photolithography process includes three major steps: photoresist coating, exposure, and development. To achieve high resolution, photolithography also has many baking and chilling steps. For older, all-manual processing technologies,

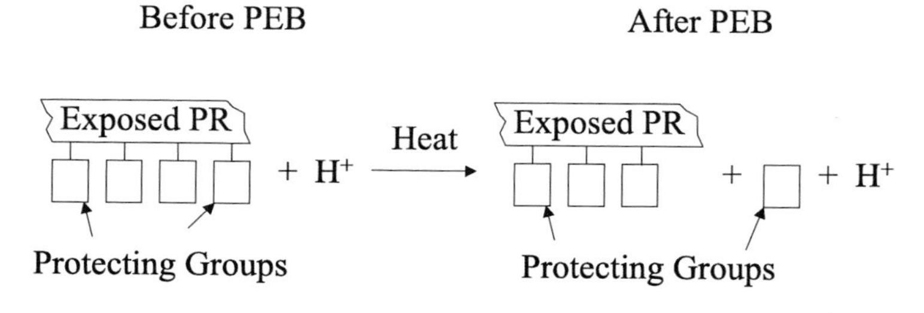

Figure 6.5 Chemically amplified photoresist.

the entire photolithography processing flow had eight steps: wafer cleaning, prebake, spin coating primer and photoresist, soft bake, alignment and exposure, development, pattern inspection, and hard bake. If the wafer failed to pass inspection, it would bypass the hard-bake step, the photoresist would be stripped, and the entire process would be repeated until the wafer passed inspection.

For advanced photolithography processes, three basic steps are the same. Detailed steps are added to achieve higher photolithography resolution. Integrated track-scanner systems are widely used to improve process yield and throughput. Because all of the coating, baking/cooling, exposure, and development processing steps are performed in a track-scanner system, and pattern inspection is performed after the hard bake. Figure 6.6 shows a flow chart of the photolithography process, and Fig. 6.7 gives an advanced photolithography processing sequence on the wafer surface.

6.3.1 Wafer cleaning

Before wafers reach the photolithography process, they have undergone some previous processes such as etch, ion implantation and annealing, oxidation, CVD, PVD, and CMP. The wafer might have some organic contaminants (from the photoresist, etch byproducts, bacteria, and debris of human skin) and inorganic contaminants (such as particles and residues from storage containers, improper wafer handling, and nonorganic materials in the environment, e.g., dust and mobile ions). The wafers need to be cleared of these contaminants before they are ready for the photolithography process. Even if the wafers are contamination free, it has been found that further cleaning is desirable for better adhesion of the photoresist to the wafer surface.

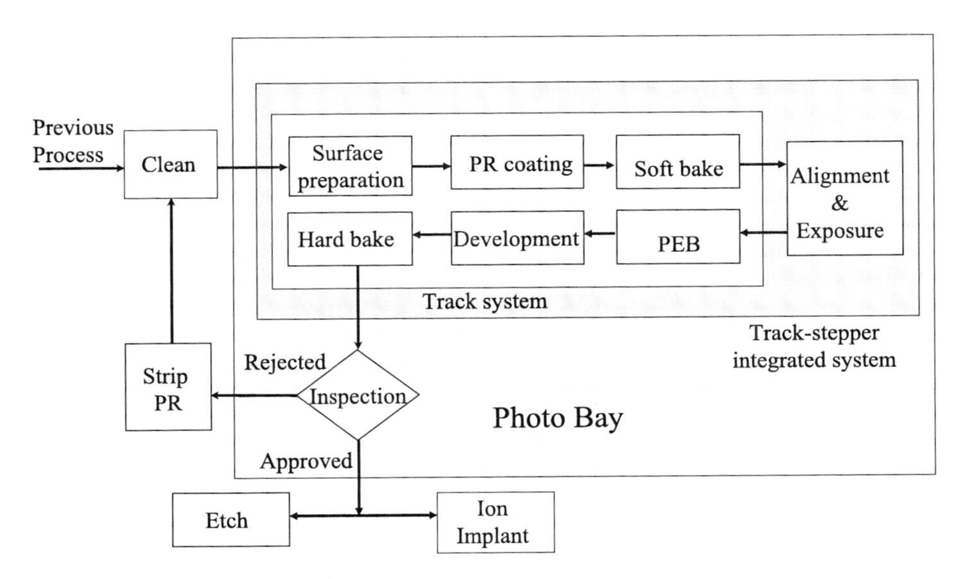

Figure 6.6 Flow chart of photolithography process.

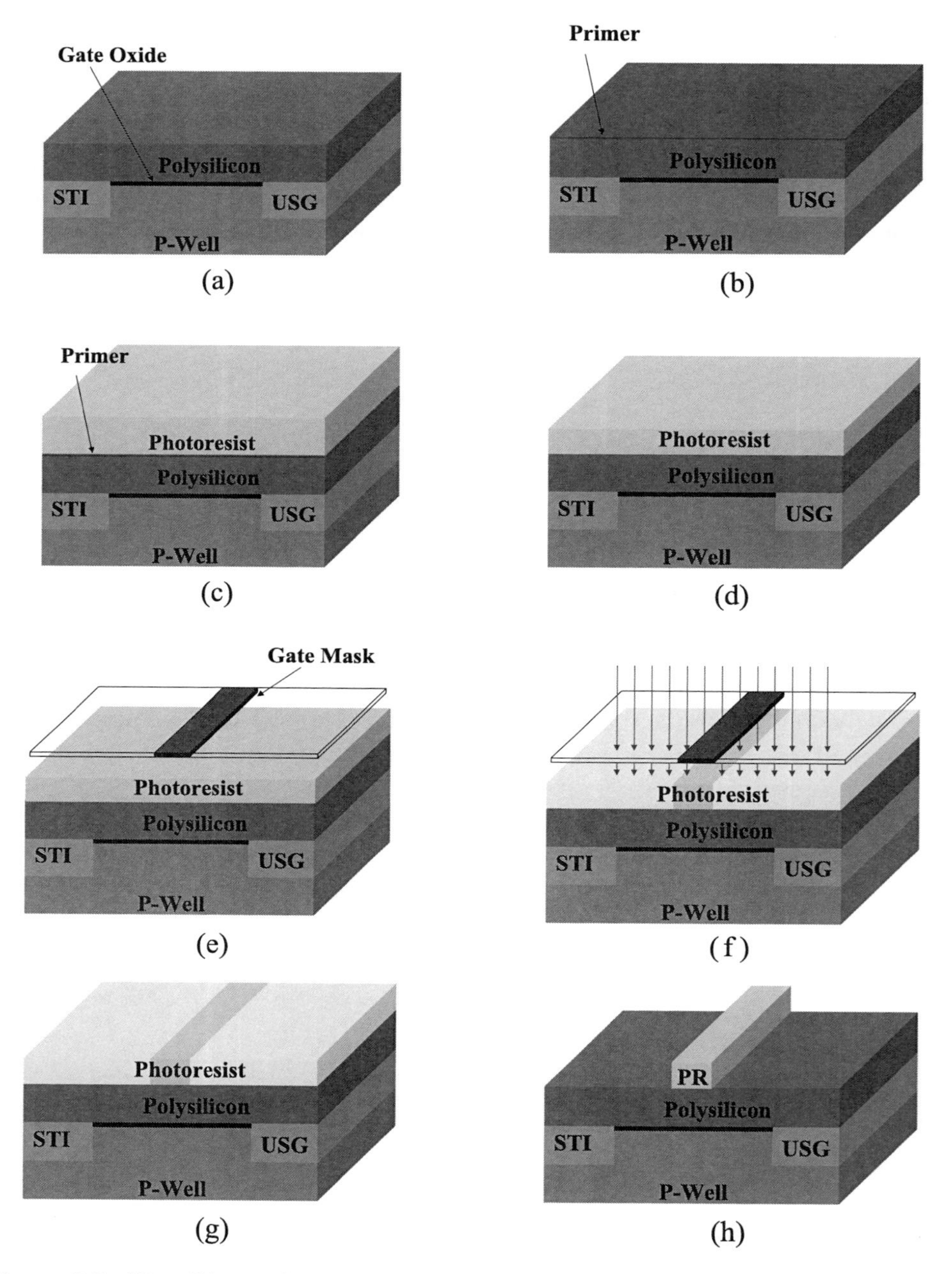

Figure 6.7 Photolithography process: (a) wafer cleaning, (b) prebake and primer vapor coating, (c) photoresist spin coating, (d) soft bake, (e) alignment, (f) exposure, (g) postexposure bake, and (h) develop, hard bake, and pattern inspection.

Chemical cleaning is most commonly used for wafer cleansing, which uses solvents and acids to remove organic and inorganic contaminant residues, respectively. It is usually followed by DI water rinse and spin dry processes, as illustrated in Fig. 6.8.

Other methods such as dry air or nitrogen blow, high-pressure steam blow, oxygen plasma ashing, and mechanical brushing have been used before and could still be used in some fabs. When feature size shrinks, killer particle size also shrinks. These methods worked well for the removal of larger particles, but they failed to remove smaller particles on smaller wafers, even adding more killer particles than they were able to remove.

Particles on a wafer's surface can cause pinholes in the photoresist; organic and inorganic contamination can cause photoresist adhesion problems and device and circuit defects. It is very important to minimize or eliminate these contaminants before the photolithography process begins to ensure process yield.

6.3.2 Preparation

Preparation is a two-step process performed in a sealed chamber commonly called a prep chamber. The first step is a heating process for the removal of moisture adsorbed on the wafer surface, called dehydration bake or prebake. A clean, dehydrated wafer surface is essential for adhesion of the photoresist on the wafer surface. Poor adhesion can lead to failure of photoresist patterning and can cause undercuts during subsequent etch processes. In most cases, the wafer is baked for about one or two minutes on a hot plate, with temperatures from 150 to 200 °C. Baking temperature and time are critical to achieving optimized processing results. If the baking temperature is too low or the baking time too short, insufficient surface dehydration can cause photoresist adhesion problems. If the baking temperature is too high, it can cause primer dissociation, which can cause contamination and affect photoresist adhesion.

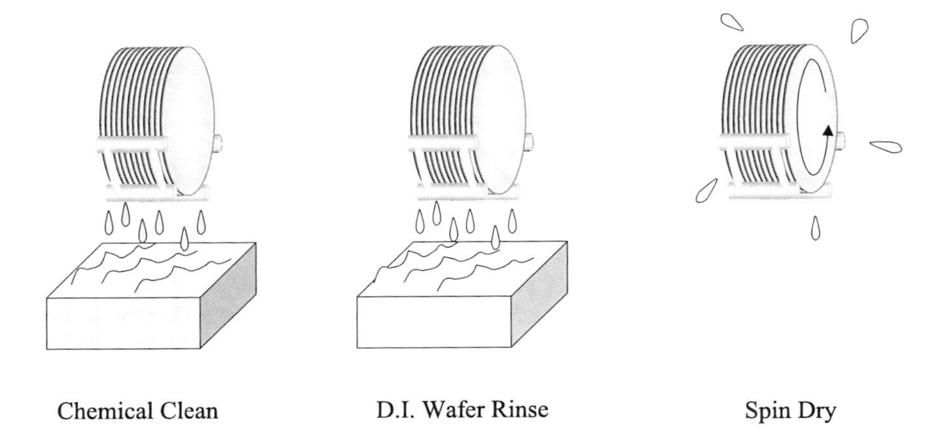

Figure 6.8 Wafer cleaning process.

The second step is a deposition process called priming. In this process, a thin layer of primer, which can wet the wafer surface and promote adhesion between the organic photoresist and the inorganic silicon or silicon compound wafer surface, is coated onto the wafer surface before the photoresist coating. Hexamethyldisilazane [HMDS, $(CH_3)_3SiNHSi(CH_3)_3$] is the most commonly used primer in IC photolithography processes. In advanced photolithography processes, HMDS is vaporized, introduced into the prep chamber, and deposited onto the wafer surface in situ with the prebake process. It is very important to coat the surface with photoresist immediately after the primer coating to prevent surface rehydration. Thus, the prep chamber is always placed in-line with the photoresist coater on a track system. Primer can also be coated by a spin-on process in situ with the photoresist coating, but this is not as popular as vapor priming in advanced IC fabs. Primer vapor coating is preferred over primer spin coating because it uses less HMDS and minimizes the potential for surface contamination by particulate carried in liquid chemicals. Figure 6.9 illustrates the in-situ prebake and vapor priming processes.

If the photoresist is still hot during photoresist spin coating, the solvents in the photoresist evaporate rapidly and cool down the wafer at the same time. This makes a very undesirable processing condition, since both solvent loss and temperature change can affect photoresist viscosity, which affect the thickness and thickness uniformity during the photoresist spin coating. Therefore, after the preparation process, the wafer needs to be cooled to the ambient temperature before applying the photoresist. Then the wafer is cooled on a chill plate, which is water cooled by a heat exchanger, on the same track system.

6.3.3 Photoresist coating

Photoresist coating is a deposition process in which a thin layer of photoresist is applied to the wafer surface. The wafer is placed on a spindle with a vacuum chuck that can hold the wafer during the high-speed rotation. Liquid photoresist

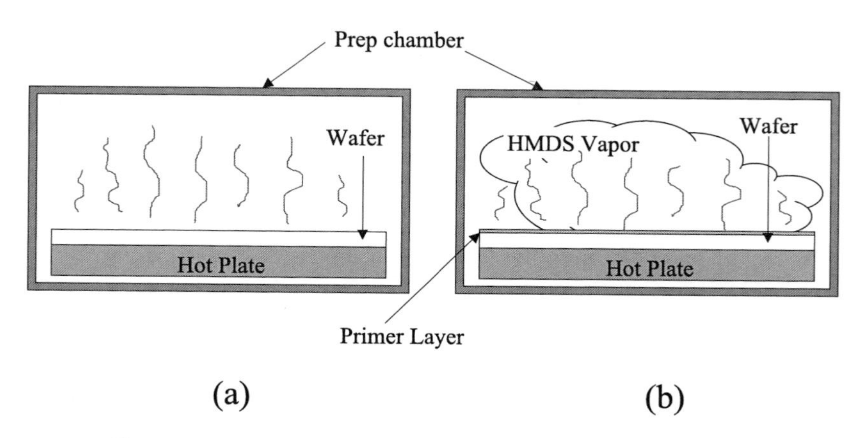

Figure 6.9 (a) In-situ prebake and (b) primer vapor coating.

is applied to the wafer surface, and the centrifugal force from the wafer rotation spreads the liquid across the entire wafer. After the solvents in the photoresist have evaporated, the wafer is left coated with a thin layer of photoresist. Photoresist thickness is related to both photoresist viscosity and wafer spin rate, as illustrated in Fig. 6.10. The higher the spin rate, the thinner the photoresist layer, and the better the thickness uniformity. The photoresist thickness is inversely proportional to the square root of the spin rate. Since photoresists have high viscosity and very high surface tension, a very high spin rate is required for uniform photoresist spin coating. The higher the viscosity, the thicker the photoresist films at the fixed spin rate. Viscosity of the photoresist can be controlled by the solid content of the photoresist solution. A typical photoresist thickness in photolithography processing is between 3,000 to 30,000 Å.

Photoresist can be dispensed with either a static or dynamic dispensing method. For static dispense, the photoresist is distributed onto a stationary wafer surface and allowed to spread across part of the wafer surface. When the photoresist puddle spreads to a certain diameter, the wafer is spun rapidly at a spin rate of up to 7000 rpm to distribute the photoresist evenly across the entire wafer surface. The thickness of the photoresist is related to photoresist viscosity, surface tension, photoresist drying characteristics, the spin rate, the acceleration rate, and the spin time. Photoresist thickness and thickness uniformity is especially sensitive to the acceleration rate.

For dynamic dispense, the photoresist is applied to the center of the wafer while the wafer is rotating at a low spin rate (around 500 rpm). After the photoresist is dispensed, the wafer is accelerated to a high spin rate of up to 7000 rpm to spread the photoresist uniformly across the wafer surface. The dynamic dispense method uses less photoresist; however, static dispense can achieve better

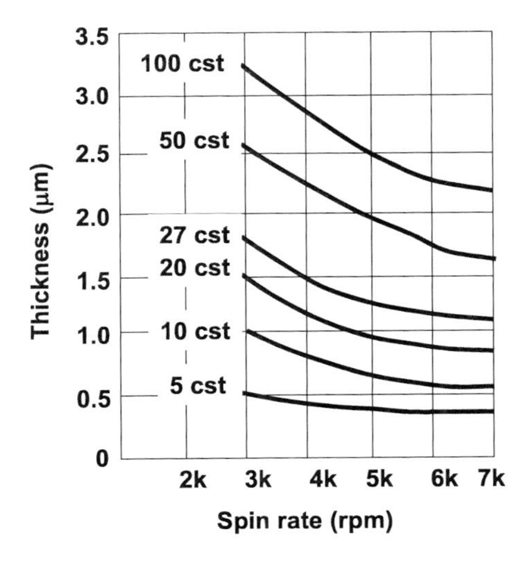

Figure 6.10 Relationship of photoresist thickness and spin rate at different viscosities (Integrated Circuit Engineering Corporation).

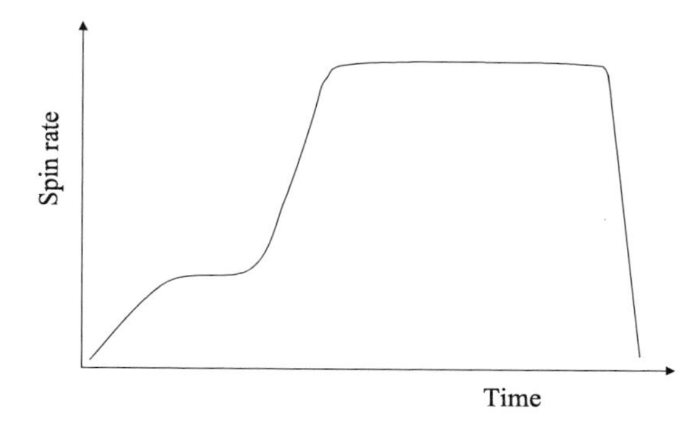

Figure 6.11 Spin rate change for spin coating processes.

photoresist coating uniformity. Figure 6.11 shows the change of spin rate in a dynamic dispense spin coating process.

Solvents in the photoresist evaporate rapidly during spin spreading processes, changing the photoresist viscosity. Therefore, it is very important to increase the spin rate as quickly as possible after the photoresist is applied to the wafer surface to reduce the effects of photoresist viscosity change due to solvent evaporation. In some processes, a thin layer of solvent is spin coated onto the wafer surface before the photoresist coating to improve photoresist adhesion and uniformity.

The photoresist spin coating process is illustrated in Fig. 6.12. The photoresist "suck back" feature is designed to prevent unwanted photoresist droplets from depositing onto the wafer after photoresist dispenses. If not removed, dry photoresist droplets could form at the tip of the dispense nozzle and fall onto photoresist-coated wafers, causing defects in subsequent processing steps.

Spin coating is also used for primer (HMDS) deposition just prior to the photoresist coating. First, liquid HMDS is applied to the wafer surface at a low spin rate to coat the wafer, then the spin rate is ramped up to 3000 to 6000 rpm to dry the HMDS in 20 to 30 sec. The advantage of the primer spin coating is that the primer spin coating occurs *in situ* as part of the photoresist coating process. This can effectively prevent rehydration of the wafer surface before photoresist coating. Currently, most fabs prefer primer vapor coating due to less primer usage (HMDS is very expensive), better coating uniformity, less risk of particulate contamination, and photoresist dissolving by the wet HMDS associated with liquid applications.

A schematic of a photoresist spin coater is illustrated in Fig. 6.13. The photoresist is drawn into the dispenser nozzle by means of a tube with a water sleeve in which water from a heat exchanger is used to maintain the photoresist at a constant temperature, since viscosity is related to temperature. Spin rate and spin rate ramp are precisely controlled; air flow temperature and air flow rate are also controlled in the coater, since they both affect the drying characteristics of the photoresist. The spindle is either nitrogen or water cooled to avoid wafer temperature nonuniformity, since the spindle at the center can become very hot during high-speed spins if it is not properly cooled. Excess photoresist and edge-bead removal (EBR) solution are collected and drained out of the station

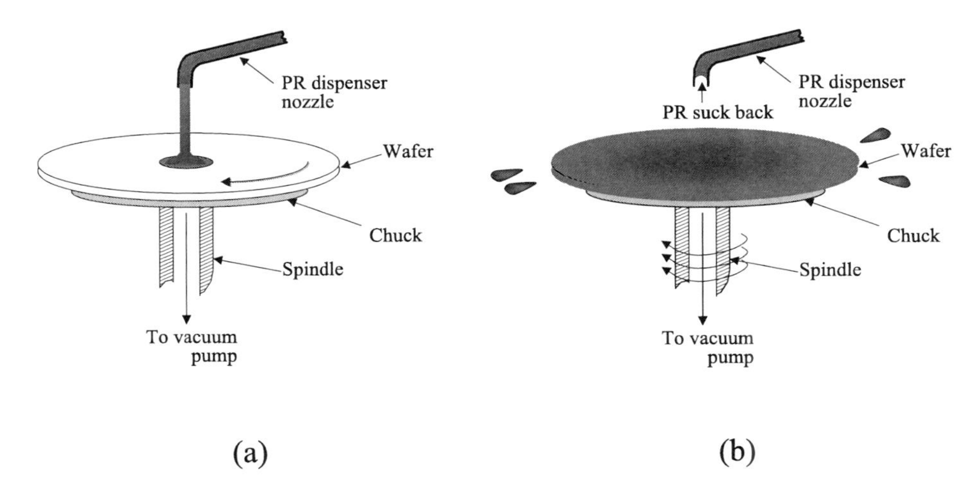

Figure 6.12 (a) Application of photoresist and (b) spin coating.

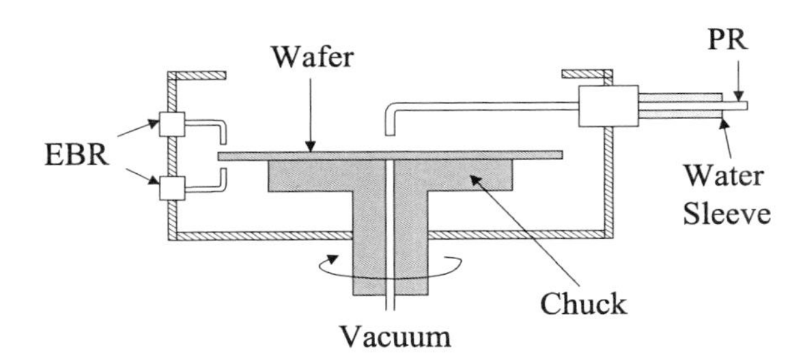

Figure 6.13 Schematic of a photoresist spin coater.

from the bottom, and vaporized solvent is evacuated from the exhaust. Photoresist thickness and uniformity are also related to exhaust gas temperature and gas flow rate, since they can affect a photoresist's drying characteristics. In fact, the ideal thickness uniformity of a photoresist is achieved without exhaust flow. However, without exhaust, the accumulation of solvent vapor fumes can be dangerous. Increasing the exhaust flow rate causes edge-thick photoresist profiles—faster drying near the edge of the wafer increases the viscosity of the photoresist, increasing thickness.

After photoresist spin coating, both sides of the wafer near the edge are covered by photoresist. It is absolutely necessary to perform EBR to eliminate this photoresist buildup. As the wafers move into the next process, e.g., etch or ion implantation, mechanical handlers such as robot fingers or wafer clamps can crack the photoresist buildup left on the wafer's edge and cause particulate contamination. A thick edge bead can also cause focusing problems during the exposure process. Both chemical and optical methods are used in EBR.

Chemical EBR is performed in the spin coater, after photoresist spin coating. In this process, solvents are injected onto the edge of both sides of the wafer while the wafer is rotating; they dissolve the photoresist only at the edge and wash it away, as shown in Fig. 6.14.

Optical EBR is performed after the exposure process, in a specially designed station located on a track system, before the wafer is sent into developer. It uses a light source such as an LED to expose the top side of the wafer edge while the wafer is rotating. The exposed photoresist is removed during the developing process to ensure that there is no particle contamination during wafer clamping (this is especially important for some etch processes). Optical EBR processes are illustrated in Figs. 6.15(a) and 6.15(b).

Other photoresist coating methods, such as moving arm dispensers and roller coaters, could still be used in some less-advanced IC fabs; however, they are not as popular as spin coating.

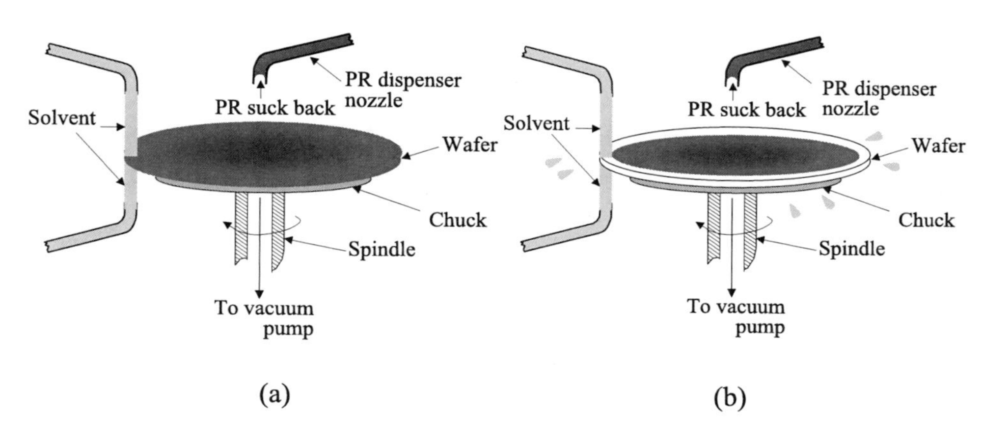

Figure 6.14 Chemical removal of an edge bead.

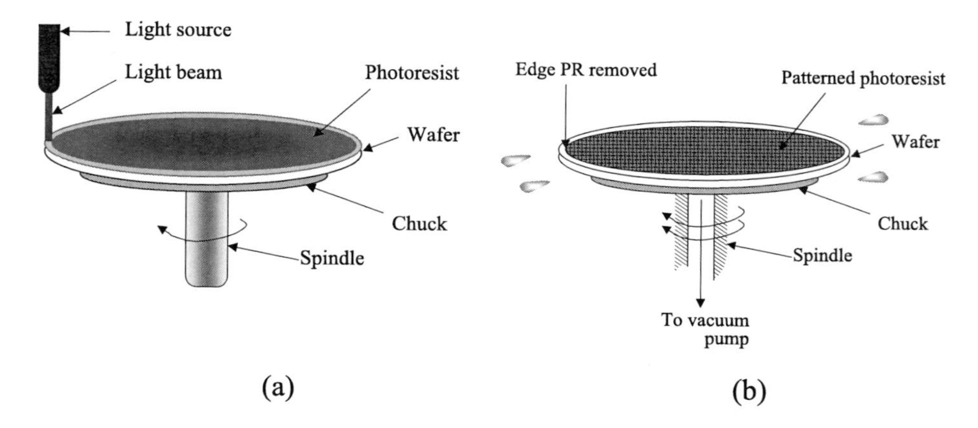

Figure 6.15 Optical EBR (a) after edge exposure and (b) after developer.

6.3.4 Soft bake

After photoresist coating, the wafer is again put through a thermal process to drive out the majority of solvent inside the photoresist and transform the photoresist from a liquid to a solid state. The soft bake also improves photoresist adhesion on the wafer surface; in some fabs it is called a pre-exposure bake. After soft bake, photoresist thickness shrinks about 10 to 20%, and the photoresist consists of about 5 to 20% solvent residue.

Soft-bake temperature and time depend on the type of photoresist (negative or positive) and vary with specific processes. Each photoresist has its own optimized baking time and baking temperature. If the photoresist is underbaked, either because the baking temperature was too low or because baking time was too short, the photoresist could peel off from the wafer surface during subsequent processes due to adhesion failure. Underbaked photoresist also can affect pattern resolution, first because excessive solvents in the photoresist cause insensitivity to light exposure, and second because of the microscopic shaking of the photoresist. Underbaking causes insufficient hardening, and a jelly-like photoresist can shake in a microscopic scale during wafer stepping, enough to create a blurred image on the photoresist (think of shooting a picture with a shaky camera).

Overbaking in the soft-bake step can cause premature polymerization of the photoresist, making it insensitive to light exposure. For chemically amplified photoresists used in DUV lithography processes, some residue solvent in the photoresist is needed for acid diffusion and amplification during the PEB step. Overbake can cause an insufficient catalysis chemical reaction, which could cause image underdevelopment.

There are several methods used for soft-baking processes: convection oven, infrared oven, microwave (MW) oven, and hot plate (see Fig. 6.16). A convection oven uses convection flow of heated nitrogen gas in an oven to heat a wafer at the required temperature, from 90 to 120 °C for about 30 min. An infrared oven can bake a wafer in a shorter time period than convection or MW ovens but can also burn the wafer bottom, since infrared radiation can pass through the photoresist layer and heat the wafer first, before the photoresist. A microwave oven is another method used for the baking process.

A hot plate can heat and dry photoresist on a wafer from the bottom up, since it heats the wafer first through heat transfer between the heated plate and the wafer. Thus, using a hot plate can avoid the photoresist crust problem due to surface heating and drying associated with convection oven baking processes. Unlike convection oven batch systems, a hot plate is a single-wafer system, which can achieve uniform heating within the wafer and, more importantly, stabilize processing results from wafer to wafer. Hot plates can be easily integrated into a track system for in-line coating, baking, and developing. Although it is possible that some less advanced IC fabs still use other methods, the hot plate is the most common method for all baking processes. It can be found in almost all photo-track systems in advanced IC fabs.

After the soft bake, the wafer is put on a chill plate to cool down to ambient temperature. It is very important to keep the wafer temperature constant during

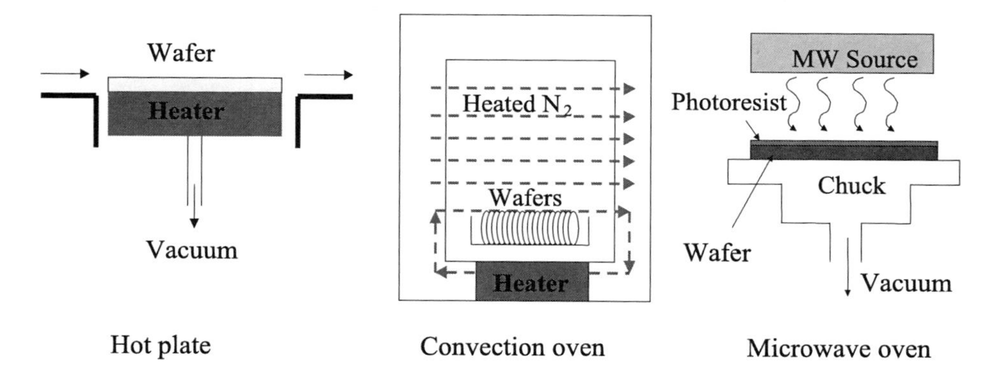

Figure 6.16 Different baking methods.

alignment and exposure, because 1 °C difference in temperature can cause 0.75-μm difference for a 300-mm silicon wafer due to thermal expansion effects.

6.3.5 Alignment and exposure

Alignment and exposure are the most critical steps of the photolithography process and for IC processing as a whole. Alignment and exposure determine the transfer success of the IC design pattern on the mask or reticle to the photoresist on the wafer surface.

The exposure process is very similar to taking a photograph with a camera: the patterned image on the mask or reticle is exposed to the photoresist on the wafer, just as an image is exposed on the film or image sensor inside a camera. Resolution of an IC photolithographic exposure system is much higher than a camera's, thus the IC photo exposure tool, called a stepper or scanner, is much more expensive than the most expensive photographic camera. In addition to the resolution requirement, precise alignment is also vital. Advanced IC chips have more than 30 mask steps, and each mask or reticle needs to be precisely aligned with the original alignment mark. Otherwise, the mask or reticle cannot successfully transfer the designed pattern to the wafer surface, causing device and circuit failures. Other requirements are high repeatability and reliability, high throughput, low defectivity, and low cost of ownership.

6.3.5.1 Contact and proximity printers

In the early years of the semiconductor industry, contact and proximity printers were widely used for the alignment and exposure process. A contact printer was the earliest and simplest tool. In the contact printing process, the mask makes direct contact with the photoresist on the wafer, and UV light passes through the clear pattern from the mask, exposing the photoresist underneath. A contact printer can achieve very good resolution, close to what can be achieved on the mask. However, due to the different curvatures of the mask and wafer, only a few points on the wafer actually make direct contact with the mask, and most of the area on the wafer's surface has about a 1- to 2-μm air gap between the mask and

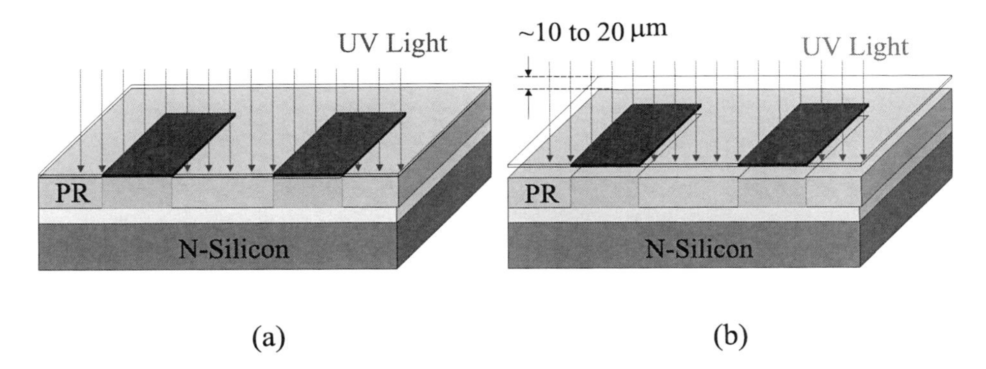

Figure 6.17 (a) Contact printing and (b) proximity printing.

photoresist. Despite this, the highest resolution of a contact printer can still be in the submicron range.

For every contact printing alignment and exposure, the touch and detach between the mask and photoresist can generate particles on both wafer and mask surfaces. Particles quickly accumulate on the mask surface and cause defects on the wafer due to both the particulate contamination and particle image transfer. A mask lifetime is seriously limited by particle contamination. To solve the particle generation problem, engineers adapted another approach that places the mask about 10 to 20 μm away from the photoresist. They called it a proximity printer. Since there is no direct contact, it has much less particle contamination and a much longer mask lifetime than that of a contact printer. The tradeoff is worse resolution, due to more light diffraction in a larger gap. The highest resolution a proximity printer can achieve is about 2 μm. Neither contact nor proximity printers are used in very large-scale integration (VLSI) or ultralarge-scale integration (ULSI) chip fabrication. Figure 6.17 illustrates exposure processes with contact and proximity printing systems.

6.3.5.2 Projection printer

To further improve exposure resolution while maintaining a low particle contamination level, the projection exposure system was developed and became widely used in VLSI semiconductor fabs, as shown in Fig. 6.18.

A projection system works in a manner very similar to an overhead projector. The mask is like a transparency foil, and the image is refocused on the wafer surface in a 1:1 ratio. In comparison, an overhead projector refocuses the image on a transparency foil to the screen in about a 1:10 ratio. In a projection exposure system, mask and wafer are far apart, which eliminates the possibility of particle generation from mask–wafer contact. By taking advantage of the optical characteristics of lenses and mirrors, projection exposure systems can achieve 1-μm minimum feature sizes and were widely used for VLSI device fabrication.

The most commonly used projection system is the scanning projection system (see Fig. 6.19), which uses a slit to block partial light from the light source to reduce light scattering and improve exposure resolution. The light is focused on

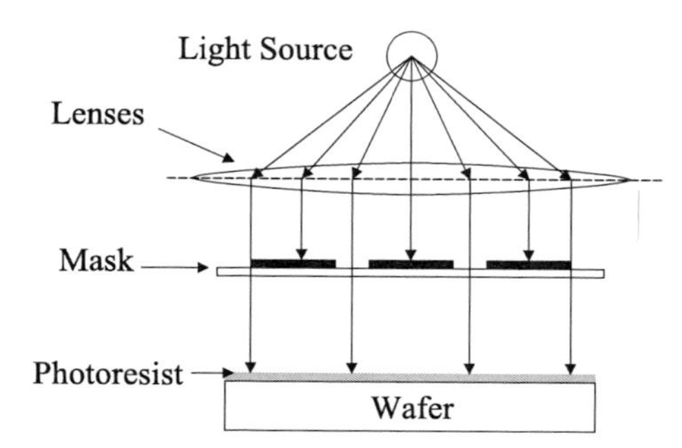

Figure 6.18 Schematic of a projection exposure system.

the mask by a lens and is refocused on the wafer surface by a projection lens as a slit. The mask and wafer move synchronously, allowing UV light to scan across the mask, refocus on the wafer surface, and expose the photoresist across the wafer.

6.3.5.3 Stepper/scanner

When feature size approaches the submicron level, a projection system can no longer meet the resolution requirements. Thus, step-and-repeat systems have been developed for VLSI and ULSI chip manufacturing.

In a projection system, image transfer from mask to wafer is 1:1, and a wafer requires only one exposure for the pattern transfer process. By shrinking and refocusing the image of the mask/reticle to the wafer surface at a ratio of 4:1 or 10:1, patterning transfer resolution can be improved. However, the exposure system must be completely redesigned because it would be impractical to make a mask and high-precision optical system with dimensions more than four to ten times the wafer diameter (which is 800 to 2000 mm for a 200-mm wafer) in order to expose the entire wafer in one exposure. It is also very difficult to find a UV light source

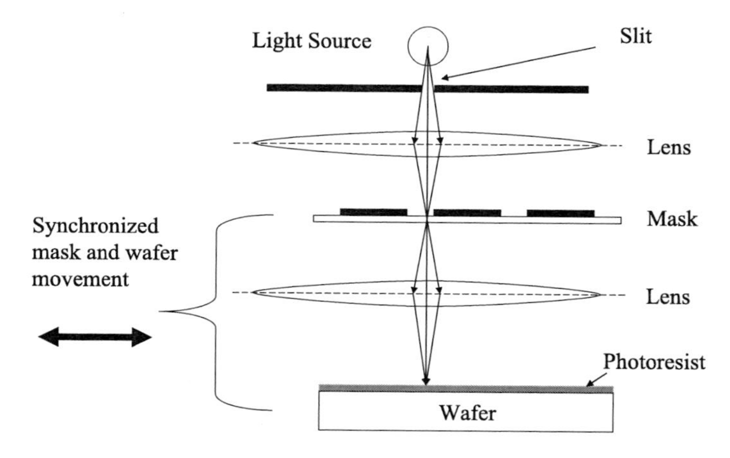

Figure 6.19 A scanning projection exposure system.

intense enough to do such an exposure. Therefore, the design pattern is made from a piece of chromium glass, called a reticle, which is used to expose only part of a wafer. In comparison, masks used in projection, contact, and proximity printers expose an entire wafer. Since the stepper shrinks the image from the reticle, the feature size on the reticle is much larger than that on the wafer surface. For instance, if the minimum feature size is 25 nm on the wafer surface, the minimum feature size on the reticle is 100 nm for a 4:1 shrink ratio. This is much easier to create than a 1:1 ratio photomask.

Question: Why is a 4:1 shrink ratio more popular in the semiconductor industry than a 10:1 shrink ratio?

Answer: Choosing between a 4:1 shrink ratio and a 10:1 shrink ratio involves a tradeoff between resolution and throughput. Obviously, 10:1 image shrinking will have better photolithography resolution than 4:1 image shrinking; however, it only exposes 16% of the area that is exposed by a 4:1 shrink ratio for a reticle. This means that total exposure time will be 6.25 times longer. Keep in mind that area changes as the power of the square of the dimension change, $A \propto d^2$.

Because a stepper only exposes a small portion of a wafer at a time, the exposure step must be repeated multiple times to expose the entire wafer. Figure 6.20 shows the basic structure of a stepper system and two steps of the exposure process. The stepper system is much more complicated than other optical exposure systems. For instance, the stepper needs to be aligned for every step, and each wafer needs 20 to 100 steps (determined by wafer size and product specification) to be entirely covered. In comparison, projection and contact/proximity printers only need one alignment per wafer. In submicron photolithography processes, there is very little room for alignment error. To meet throughput requirements, less than one second for each alignment and exposure step is allowed. Therefore, an automatic alignment system is required for stepper systems. A schematic of a stepper system is illustrated in Fig. 6.21.

After the wafer finishes photoresist coating and soft baking, it is sent into a stepper and put onto the wafer stage. It is first aligned with a previous alignment mark by a computer-controlled reticle and lens mechanism. For the very first photolithography step of the process, alignment is accomplished by using a notch or flat on the wafer, which is designed to indicate wafer crystal orientation and also serves as an alignment mark. The stepper usually adjusts the alignment optically with an automatic laser interferometer positioning system.

To further improve resolution of the image transfer, engineers combined scanning projection printer and stepper technology and developed step-and-scan systems, commonly called scanners, which are widely used in nanometer technology nodes of IC fabrication.

Because of high-precision, high-resolution, and high-throughput requirements for optical, mechanical, and electric systems, a scanner normally is the most expensive processing tool in a semiconductor fab. For example, an advanced 300-mm high-NA 193-nm immersion scanner can reach $30 to $40 million per system.

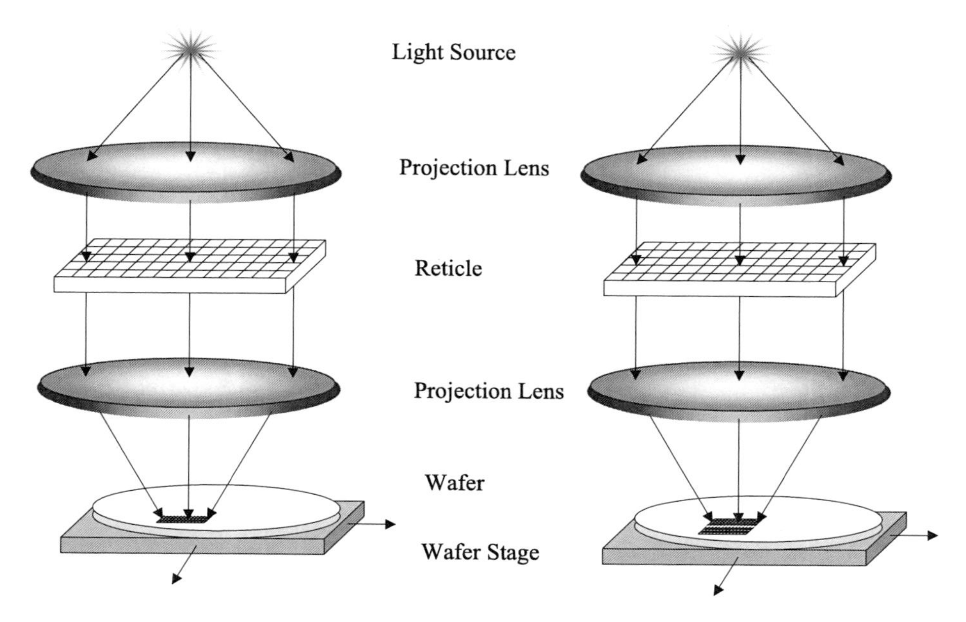

Figure 6.20 Step-and-repeat exposure system.

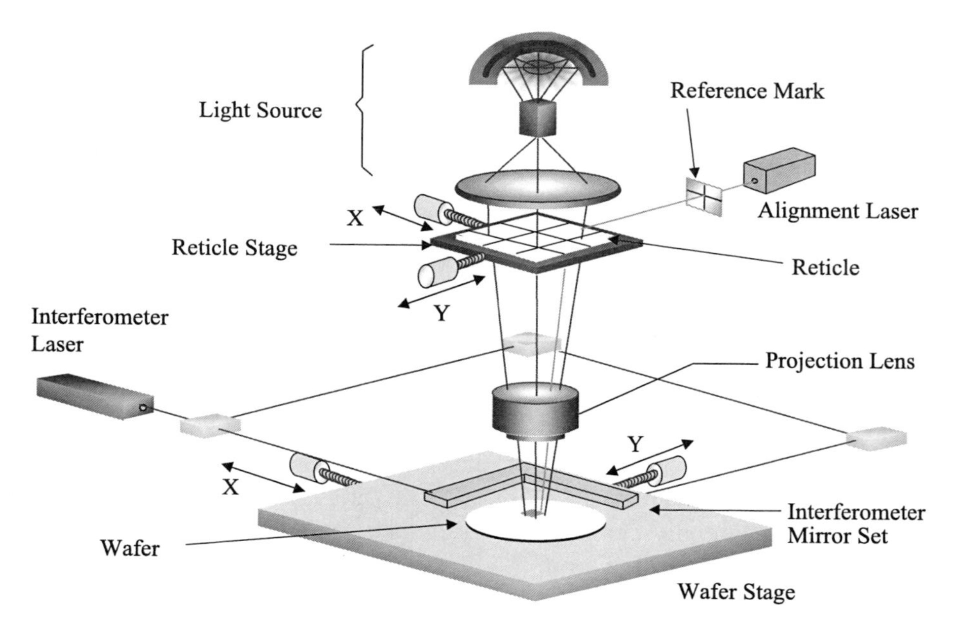

Figure 6.21 Step-and-repeat alignment and exposure system.

Many photoresists used for submicron semiconductor fabrication need to be exposed as soon as they have finished the soft bake; otherwise, the exposure resolution could be affected due to sensitizer decay in the photoresists. Therefore, in most fabs, scanners are integrated with coater-developer track systems.

6.3.5.4 Exposure light sources

The exposure process of photolithography is similar to the exposure of camera film. A picture taken in bright sunlight requires less exposure time and can achieve higher resolution than a picture taken in candlelight. Therefore, high-intensity light sources are important for achieving high resolution and high throughput. The wavelength of the UV light source used to expose the photoresist is a key factor in photolithography. Since certain photoresists are only sensitive to a narrow range of UV wavelengths, fabs choose the exposure light wavelength according to photoresist sensitivity and feature size of the circuit. The shorter the wavelength, the better the patterning resolution. When feature size shrinks, the wavelength of the exposure light also needs to shorten to match the patterning resolution requirement. There are two kinds of light sources widely used for photolithography processes: mercury lamps and excimer lasers. The exposure light source must be stable, reliable, of high intensity, and long in life.

For feature sizes larger than 2 μm, a broadband (multiwavelength) mercury lamp is used as the light source for contact/proximity and projection printers. When the feature size shrinks, single-wavelength light sources are required to achieve the desired resolution. High-pressure mercury lamps were the most commonly used UV light sources for submicron photolithography processes in projection systems and steppers in the 1980s and 1990s. The mercury UV lamp wavelength spectrum is given in Fig. 6.22. For photolithography exposure processes of IC chips with 0.50- and 0.35-μm feature sizes, g- and i-lines are most commonly used, respectively. These systems are still used in back-end processes of advanced IC fabs for layers so that their resolution can meet the requirements.

For photolithography processes with 0.25- and 0.18-μm minimum feature sizes, light sources with even shorter wavelengths are needed. A KrF excimer laser with a DUV wavelength of 248 nm is most commonly employed as a light source for steppers used in 0.25-μm processes, and it is capable of patterning feature sizes as small as 0.13 μm. Steppers using ArF excimer lasers with wavelengths of 193 nm have been used in IC production to pattern features since the 0.18-μm technology node and have extended to the 22-nm technology node, thanks to techniques such as immersion and double patterning. Research and development efforts on 157-nm scanners with F_2 excimer lasers were discontinued after the successful development of immersion technology, which extended applications of ArF 193-nm scanners for multiple technology nodes. Light sources commonly used for semiconductor photolithography are summarized in Table 6.1.

6.3.5.5 Exposure control

Exposure is controlled by exposure light intensity and exposure time. The total exposure light flux is the product of the intensity and exposure time and is very similar to the exposure of a camera.

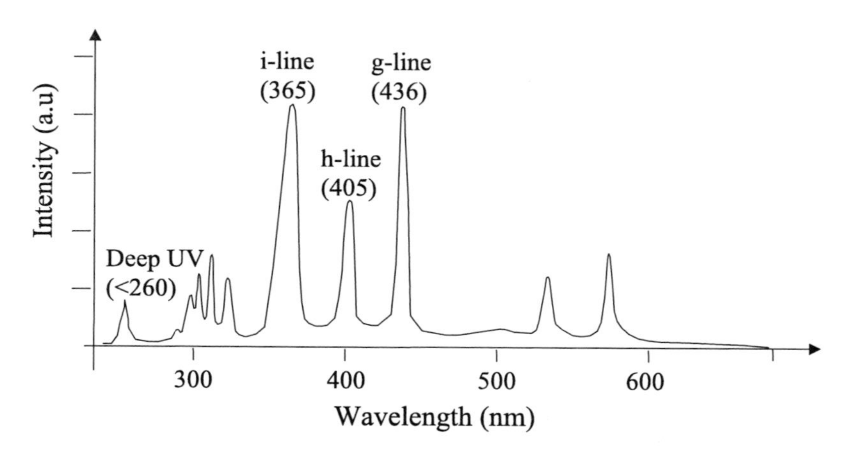

Figure 6.22 Light spectrum of the mercury lamp.

Table 6.1 Light sources for photolithography in fabrication, and research and development.

	Name	Wavelength (nm)	Application technology node (nm)
Mercury lamp	g-line	436	500
	h-line	405	
	i-line	365	350 to 250
Excimer laser	XeF	351	
	XeCl	308	
	KrF (DUV)	248	250 to 130
	ArF	193	180 to 14
Fluorine laser	F_2	157	No longer required
Laser-produced plasma (LPP) or discharge-produced plasma (DPP)	Extreme ultraviolet (EUV)	13.5	14 or less

Light intensity is mainly controlled by the electric power of a lamp or laser. By increasing applied power, the output light intensity can be increased; however, this increase can affect the reliability and lifetime of the lamp or laser.

A scanner must expose wafers with different photoresists and different reticles, so fab technicians need to be able to adjust a scanner's light intensity with precision. Even after a scanner runs the same process for a certain time period, light intensity can shift and cause patterning problems. Frequent light-intensity calibration is required to maintain a stable photolithography process. Illuminator intensity I is normally measured in mW/cm^2 by a photodetector. The amount of exposure is light intensity multiplied by exposure time, which is measured in mJ/cm^2.

Question: A routine illuminator intensity calibration was performed with a reticle still on the reticle stage. What kind of problem will it induce?

Answer: Because the reticle blocks some light from the illuminator, the photodetector on the wafer stage will receive fewer photons. Therefore, it will have a lower intensity than it should. To calibrate the light source to the required level, the applied power will be increased, and the light intensity will be too high. During the proceeding exposure process, the wrongly calibrated illuminator will cause overexposure of the photoresist. Similar to overexposed film, this will cause bad resolution. For a positive photoresist, overexposure can induce critical dimension (CD) loss. Often it is difficult to find the source of this kind of problem; however, a correct recalibration procedure will solve the problem.

6.3.6 Postexposure bake

When exposure light reflects off of the photoresist–substrate interface, it can interfere with incoming exposure light and cause a standing wave effect, due to constructive interference and destructive interference at the different depths. The standing wave pattern is illustrated in Fig. 6.23.

The standing wave effect causes striations of overexposed and underexposed areas throughout the photoresist, as shown in Fig. 6.24. The distance between two peaks is the wavelength of the exposure light λ divided by two times the refractive index of the photoresist $2n_{PR}$.

When feature size is large, the standing wave effect is not a big concern. As minimum feature size has continued to shrink, several methods have been used to reduce the standing wave effect. Dye has been added in the photoresist to reduce reflection intensity. Metallic and dielectric layers have been deposited on the wafer surface as antireflective coatings (ARCs) to reduce and minimize reflection. Another approach is using a spin coater to apply an organic antireflective layer prior to photoresist spin-on. Because the layer is at the bottom of the spin-on photoresist

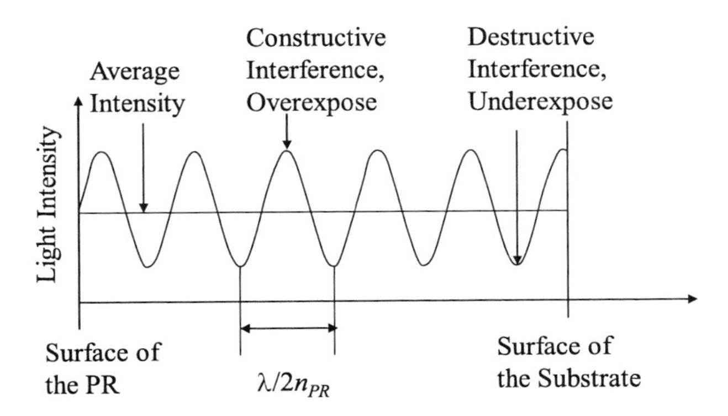

Figure 6.23 The standing wave effect induces light intensity changes.

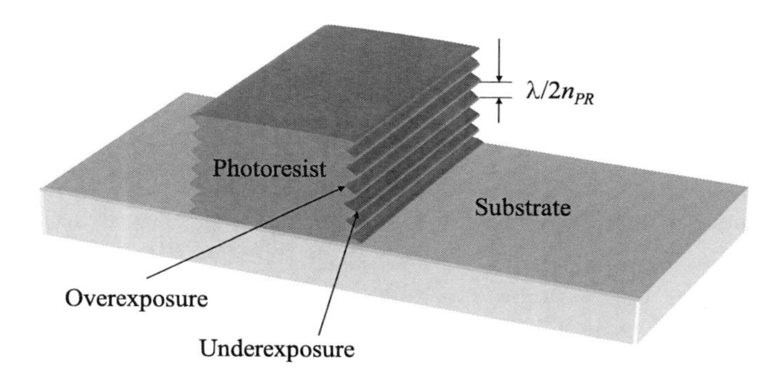

Figure 6.24 The standing wave effect on a photoresist.

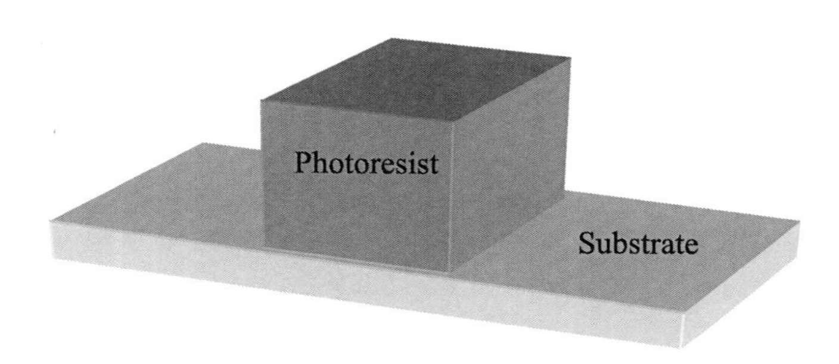

Figure 6.25 PEB minimizes the standing wave effect.

stack, it is commonly called bottom antireflective coating (BARC). PEB between exposure and develop processes also helps minimize the standing wave effect (see Fig. 6.25).

Photoresists have a special property called glass transition. When the temperature is above glass transition temperature T_g, photoresist molecules become more mobile. A baking process with temperatures higher than T_g provides the heat necessary for sufficient thermal movement of molecules in the photoresist. Thermal motion rearranges the over- and underexposed molecules, averaging and smoothing the standing wave effect, and improving photolithography resolution.

For chemically amplified photoresists used for DUV processes, PEB provides the heat necessary for acid diffusion and amplification. After the PEB process, images of the exposed areas appear on the photoresist due to the significant chemical change after acid amplification.

PEB processes use a hot plate at 110 to 130 °C for about 1 min. For the same type of photoresist, a PEB requires a higher baking temperature than a soft bake. Insufficient PEB will not completely eliminate a standing wave pattern, which affects resolution. On the other hand, overbaking causes polymerization of the photoresist and affects the developing process, which can result in failure of the pattern transfer. After PEB, the wafer is placed on a chill plate to cool down to the ambient temperature before continuing to the development process.

6.3.7 Development

After the photoresist-coated wafer proceeds through exposure, PEB, and for some processes, optical EBR, it is sent to a developer station for development. The development process removes unwanted photoresist and forms the desired pattern defined by the mask or reticle. For commonly used positive photoresist, the exposed portions are dissolved in the developer solution.

There are three steps in the development process: develop, rinse, and dry, as illustrated in Fig. 6.26. In the develop step, unwanted photoresist is dissolved by developer solution. A rinse dilutes the developer solution and prevents overdevelopment, and the drying process prepares the wafer for the next process.

Positive photoresists use a weak base solution as the developer solution. Alkaline water-based solutions such as sodium hydroxide (NaOH) and potassium hydroxide (KOH) can be used. However, these solutions can introduce mobile ions such as sodium and potassium, which are very undesirable since they can cause device damage. Therefore, most semiconductor fabs use nonionic base solution for positive photoresist developments. The most commonly used is tetramethyl ammonium hydroxide [TMAH, $(CH_3)_4NOH$].

The most commonly used developer solution for negative photoresists is xylene. Usually, *n*-butylacetate is used for the rinse. Mixtures of alcohol and trichloroethylene (TCE) and milder-acting Stoddart solvent can also be used for a negative photoresist rinse.

Development was previously performed in batch immersion processes in wet sinks, as illustrated in Fig. 6.26. Currently, most fabs use a spinner developer for this process. The main advantage of the spinner is that it can run the developing, rinsing, and drying processes in situ and can be easily integrated into a track system with a coater and baker. A spin developer system is very similar to a spin coater.

Figure 6.27 shows a schematic of a spin developer system. The development process is a chemical process that is sensitive to temperature. Therefore, both

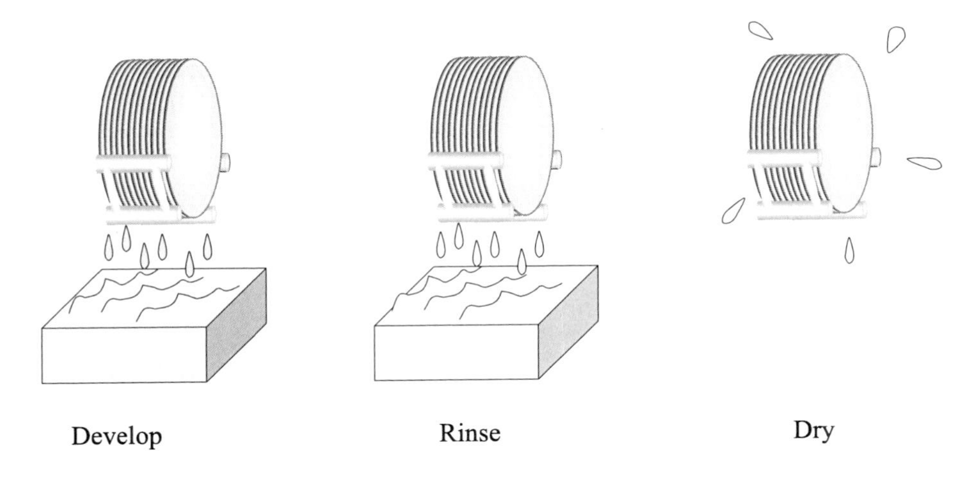

Figure 6.26 Three steps of the development process.

photoresist and wafer temperatures need to remain constant during the process. During the development process, higher temperatures can cause higher chemical reaction rates, which can induce photoresist overdevelopment and cause CD loss. Lower temperatures can cause lower chemical reaction rates, which can induce photoresist underdevelopment and cause CD gain. Both cases affect optimal photolithography resolution (see Fig. 6.28).

First, the developer solution is sprayed onto the wafer surface and spread across it by centrifugal force from the spinning. After development has finished, DI water is sprayed onto the wafer surface for rinsing, then the DI water is turned off and the spin rate increases to dry the wafer. The process is illustrated in Fig. 6.29.

Another method is called puddle development, which is very similar to spray development. Although it uses the same type of spin system, it starts with a certain amount of developer solution being sprayed onto a stationary wafer surface rather than a spinning one. The developer forms a puddle and covers the entire wafer due to surface tension. After required puddle time and the majority of the development process is finished, more developer solution is sprayed on the wafer as it starts to spin, washing away the dissolved photoresist; then the wafer is rinsed and dried at a high spin rate.

During the development process, both exposed and unexposed photoresist are dissolved in the developer solution. Selectivity between the two needs to be high enough to achieve good resolution. For the developing process, temperature

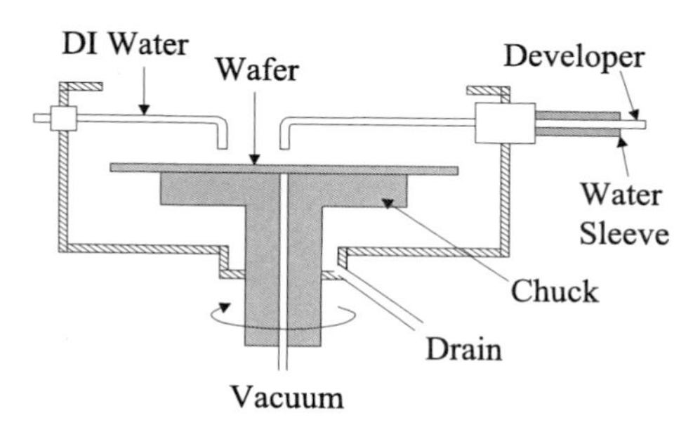

Figure 6.27 Schematic of a spin developer system.

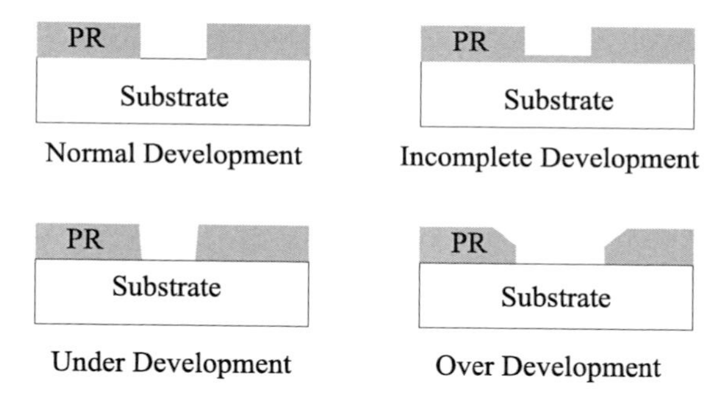

Figure 6.28 Photoresist profile for different developments.

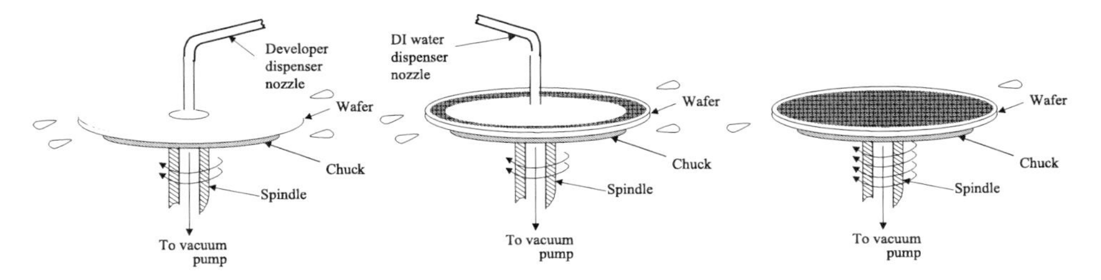

Figure 6.29 Developer spin-on, DI water rinse, and spin-dry processing sequence.

controls for the developer solution and wafer are very important. Different photoresists use different developer solution and require different development temperatures.

6.3.8 Hard bake

After development, the wafer is put through a bake process called hard bake. Hard bake drives out any remaining solvents and strengthens the photoresist, improves photoresist etch and implantation resistance due to further polymerization, and also improves adhesion of the photoresist to the wafer because of thermal dehydration. Several methods can be used for the hard bake process, as described earlier in the soft bake section. The most commonly used method is the hot plate, in which the temperature is usually from 100 to 130 °C and baking time is about 1 to 2 min, depending on the photoresist requirement. The hard bake temperature is higher than the soft bake temperature for the same kind of photoresist. For some applications, UV light is also used with a high-temperature (>100 °C) bake to harden photoresists.

Hard bake time and temperature need to be carefully controlled, since underbaking can cause high photoresist etch rates and affect wafer adhesion. Overbaking can cause bad resolution. If the baking temperature is too low, the photoresist will not achieve the required strength due to insufficient thermal polymerization and less photoresist thermal flow to fill the pinholes. It is important that the baking temperature is slightly higher than the photoresist transition temperature, so the photoresist flows a little bit to fill the pinholes and smooth the edges, as shown in Fig. 6.30. High temperature also can help to further dehydrate the photoresist and improve adhesion.

If the photoresist is overbaked (either the temperature is too high or the baking time is too long), it could flow too much, which will affect the resolution of the photolithography process, as shown in Fig. 6.31.

Question: When a photoresist has been used up or has been in use for too long, it needs to be replaced. If someone uses the wrong bottle, what could be the consequence?

Answer: Different photoresists have different sensitivities to certain wavelengths. They require different spin rates, spin ramp rates, and spin times. They also need different baking times and temperatures, and demand different

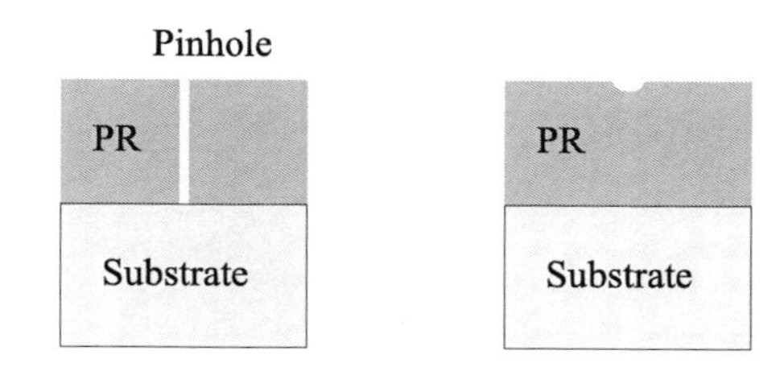

Figure 6.30 Pinhole filling by photoresist thermal flow.

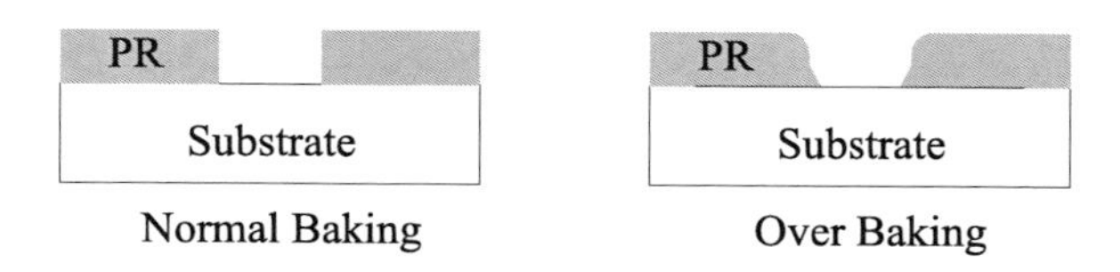

Figure 6.31 Overbake causes photoresist overflow.

exposure intensities and exposure times. They need different developer solutions and development conditions. Therefore, the pattern transfer will fail if the wrong type of photoresist is used. Using the wrong photoresist would significantly affect the throughput due to the required wafer rework and photoresist line clean and purge.

6.3.9 Metrology and defect inspection

In a production line of an IC fab, large numbers of wafers are processed with the highest throughput possible. Therefore, it is critical to capture processing shifts or defects on a wafer as soon as possible to quickly identify the problematic processing equipment causing the shift or defects. Then other wafers can be diverted from that equipment to avoid further loss.

After photoresist patterning, a wafer is processed through metrological and inspection steps to make sure that the controlled parameters of photoresist patterns, such as overlay, CD, and defect density, are within the tolerance limit. If one of them is out of control, the photoresist needs to be stripped and the wafers could be sent back for rework in the photolithography process. It is very important to measure and inspect a wafer before etch and ion implantation processes, since the pattern on the photoresist is only temporary, whereas after etch or ion implantation, it becomes permanent. It is impossible to rework a wafer after etch or ion implantation; if errors are found at that point, the wafer must be scrapped.

More than 30 masks are used to fabricate advanced CMOS IC chips. Overlay measurement is required for every photolithography step to ensure that each mask is accurately aligned. Figure 6.32 illustrates several kinds of misalignment cases, including run-out, run-in, reticle rotation, wafer rotation, misplacement in the x direction, and misplacement in the y direction. If misalignment is over the tolerance limit, the device will fail. For example, if the contact is misaligned with the metal

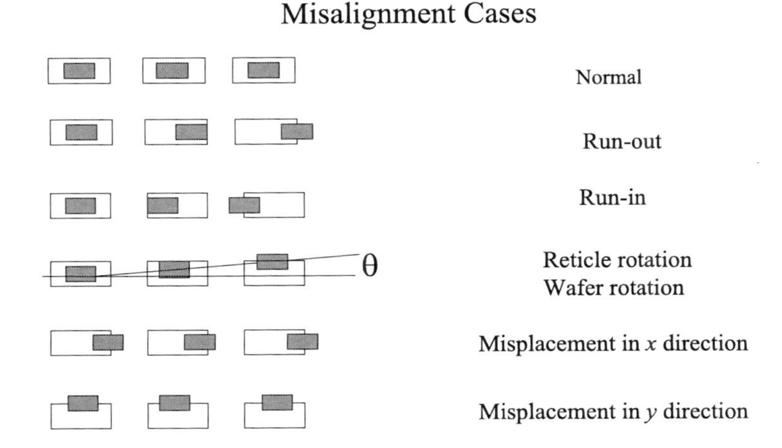

Figure 6.32 Examples of misalignment cases.

pad underneath, contact resistance could be too high and cause failure of the IC chip.

Overlay metrology uses optical measurements on designed alignment marks. The box-in-box alignment mark illustrated in Fig. 6.33 is one of the overlay patterns used in IC fabrication. The overlay shift in the x and y directions can be measured by:

$$\Delta X = X_1 - X_2 \text{ and } \Delta Y = Y_1 - Y_2.$$

Recently, grating-over-grating marks (see Fig. 6.34) have been developed for scatterometry-based overlay measurements in nanometer technology node IC manufacturing. Scatterometry is a technique that uses the spectral change of polarized light to measure pattern properties, which are described in detail later.

With the advancement of technology nodes, overlay metrology on test patterns within scribe lines might not be enough for proper yield management of the photolithography process. Overlay measurements of real devices after pattern etch using an SEM-based metrology system could be necessary in the future. For CD measurement, two systems are commonly used: an SEM-based metrology system called CD-SEM, and an optical system called scatterometry.

A wafer is inspected to determine whether the photolithography produced a desired and usable pattern on the photoresist. This is usually accomplished with microscopes (optical or electron) or in automatic inspection systems. An optical microscope can be used for visual inspection of larger-dimension features, while half-micron or submicron feature inspection requires an SEM.

Question: Can an optical microscope resolve a 0.25-μm feature?

Answer: No, because the feature size (0.25 μm = 2500 Å) is smaller than the wavelength of the visible light, which is from 3900 Å (violet) to 7500 Å (red).

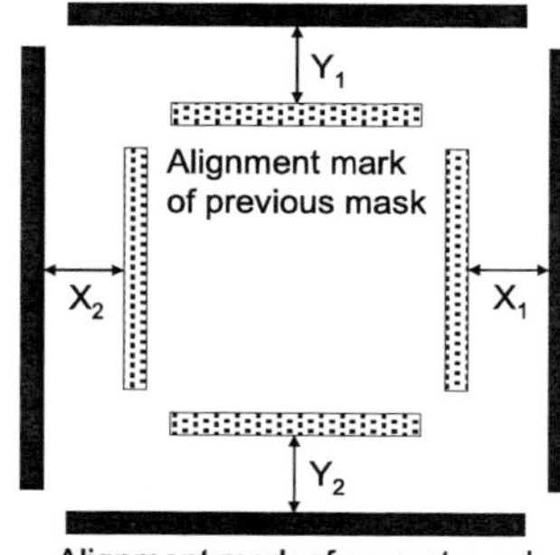

Figure 6.33 Box-in-box alignment mark.

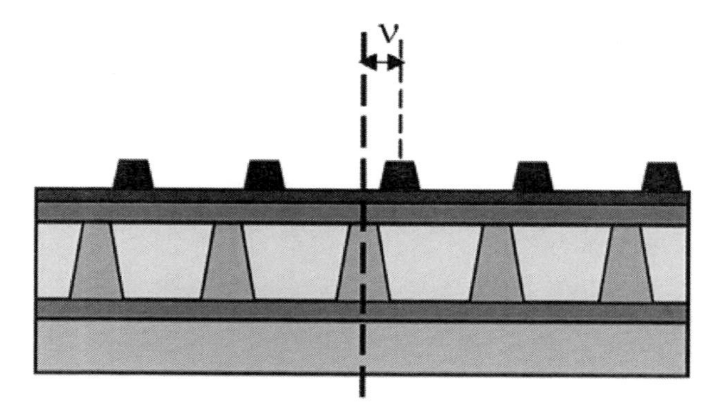

Figure 6.34 Grating-over-grating alignment mark for scatterometry overlay measurement. (B. Dinu et al., *Proc. SPIE*).

In his 1924 doctoral thesis, “Recherches sur la théorie des quanta (research on the quantum theory),” French physicist Louis de Broglie introduced a novel idea of the wave–particle duality of an electron; an electron is a tiny particle, and it is also a wave. His idea was immediately endorsed by Albert Einstein, who discovered the wave–particle duality of light; light is an electromagnetic wave and a particle called a photon. In 1927, experiments of electron diffraction of crystal lattices confirmed the particle–wave duality of an electron. In 1929, de Broglie became the youngest solo winner of the Noble Prize in physics. A material particle wave is called de Broglie’s wave. The wavelength of an electron in de Broglie’s wave is determined by the electron’s momentum, which is related to its energy; the higher the electron’s energy, the shorter its wavelength. Therefore, energetic electron beams can be used to detect very small features. When an electron beam hits material, it can excite secondary electrons. By mapping the detected electrical signal from these secondary electrons, an image of the tiny feature can be observed. Sharp edges on the photoresist, such as feature corners, have higher secondary electron yield rates and therefore higher intensity of the detected signal and a brighter image (see Fig. 6.35).

Scatterometry measures the optical signals of reflectometry or ellipsometry on measurement targets that consist of line/space arrays patterned in transparent or nonopaque films. Incident light strikes the target and causes an interaction of light reflection, diffraction, and refraction (see Fig. 6.36). Reflected light from the target contains information about phase and intensity that can be used (along with computational modeling) to reconstruct the film stack, pattern profile, and pattern CD of a feature on the measurement target. Because it measures the average CD value from an array, scatterometry CD measurements have very high repeatability.

Scatterometry CD measurements are performed on test structures on the scribe line, which are arrays of dense and isolated patterns, as shown in Figs. 6.37(a) and 6.37(b), respectively. CD-SEM measurements are performed on test structures called CD bars, as illustrated in Fig. 6.37(c). The CD value of the test structures usually are close to the CD value of the real device in that layer. There are

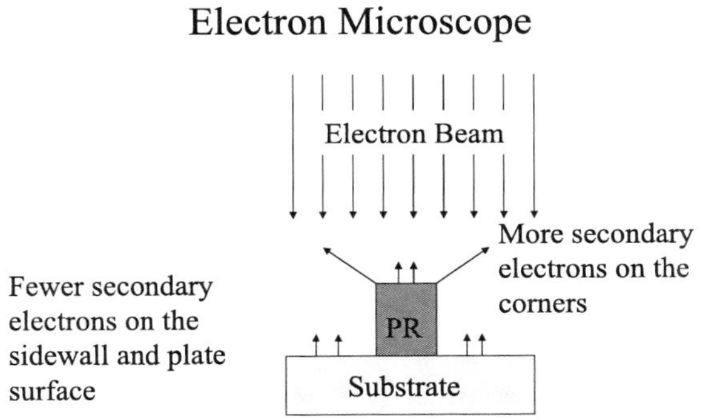

Electron Microscope

Electron Beam

More secondary electrons on the corners

Fewer secondary electrons on the sidewall and plate surface

PR

Substrate

Figure 6.35 Electron beam and secondary electron distribution.

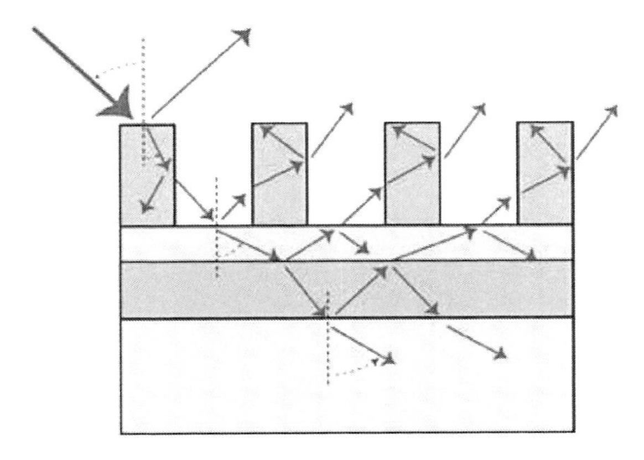

Figure 6.36 Light interaction of reflections, diffractions, and refractions. Scatterometry measurements based on this interaction can provide information of film stack, pattern CD, and profile information. (A. H. Shih)

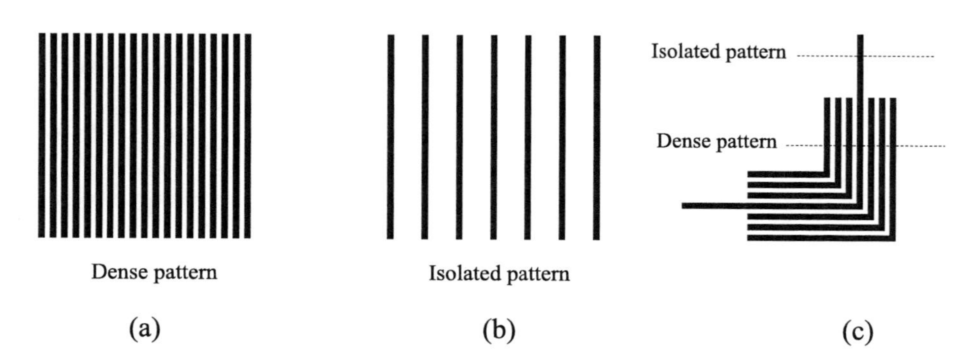

Figure 6.37 Test structures for (a) and (b) scatterometry, and (c) CD-SEM.

currently more requirements of design for manufacturing (DFM) due to smaller feature sizes. CD-SEM is widely used for design-based metrology applications, which directly measure the CDs on specific device patterns. Compared with CD-SEM, scatterometry is faster, more repeatable, and does less damage to the photoresist. On the other hand, CD-SEM shows pattern images and allows an engineer to see problems when they occur. CD-SEM can easily measure isolated devices with 2D structures that scatterometry cannot. Therefore, both scatterometry and CD-SEM coexist in advanced wafer fabs and will continue to coexist in the near future.

In advanced IC fabs, CD loss (or gain) causes many photolithography reworks. For critical layers such as the polysilicon gate of CMOS, less than 10% CD variation is allowed. Figure 6.38 shows some patterning issues in both cross-sectional views and top views that are similar to CD-SEM images. If after-development inspections (ADIs) find any density defects, overlays, or CD values outside of specifications, the photoresist can be stripped from the wafer surface for rework in the photolithography process. However, rework lowers throughput and affects yield, both of which always negatively affect a fab's profitability.

Bright-field optical inspection systems are used for defect inspection after photoresist development. A bright field inspection system projects high-intensity short-wavelength incident light onto the wafer surface and collects the reflected light on the image sensor to form an image of the patterns on the wafer. By comparing images of different dies at the same location, it can detect tiny defects by sensing the differences between each pixel in the images. This is called die-to-die (D2D) inspection. For chips with repeating patterns such as memory cell arrays, defects can be captured by comparing images of repeating cells, which is called cell-to-cell (C2C) inspection, or array mode inspection. Another method for detecting defects is comparing an inspection image to designed features extracted from design database files, which is commonly called die-to-database (D2DB) inspection. Figure 6.39 illustrates a bright-field defect inspection system.

If the wafers pass inspection, they move out of the photo bay and proceed to the next processing step, which is either etch or ion implantation.

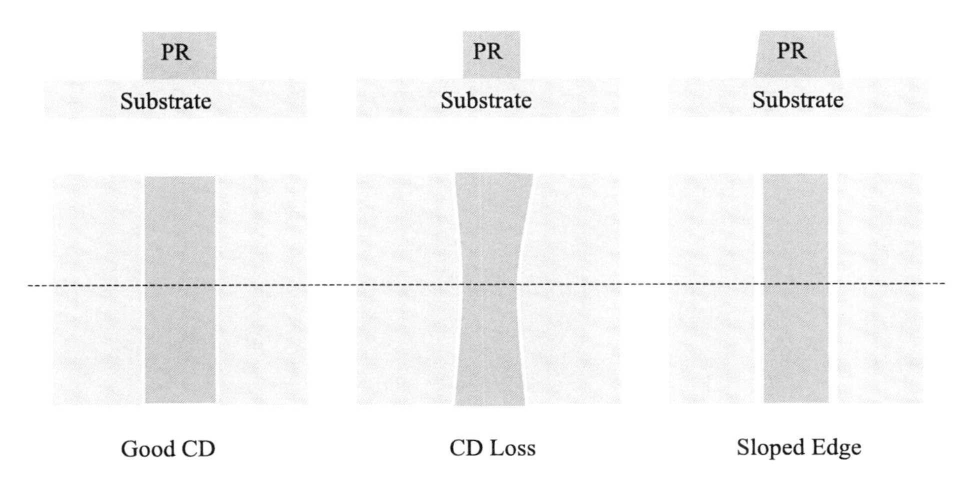

Figure 6.38 CD problems.

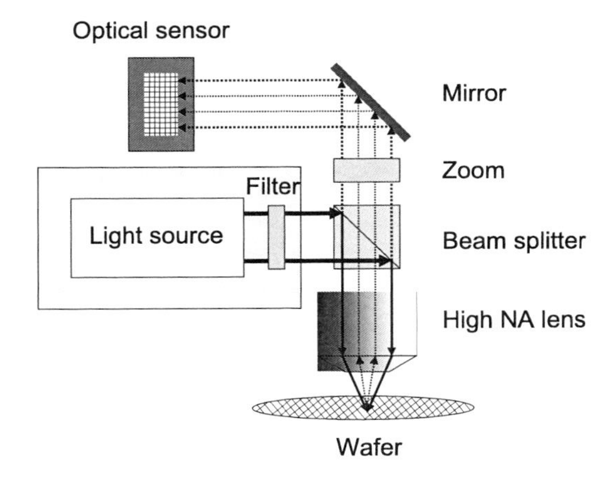

Figure 6.39 Bright-field defect inspection system (H. S. Kim, et al. *Proc. SPIE*).

6.3.10 Track-stepper integration system

All advanced semiconductor fabs use track-stepper integrated systems for photolithography processes. A track system has a wafer loading/unloading deck, prep chambers for dehydration bake and HMDS vapor primer coating, hot plates for the different baking processes, chill plates for cooling the wafer after baking, spin coaters for photoresist coating, and developers for photoresist development. Some systems also have an edge exposure system for optical EBR. A computer-controlled central robot transfers the wafers from one station to another. The in-line process improves throughput and process yield. Figure 6.40 gives a schematic of a track-stepper integrated system and wafer movement.

The track-stepper system improves throughput significantly by reducing wafer handling and increases yield by reducing the time interval between primer and spin coating, and soft bake and exposure. Improved throughput and increased yield are critical to photolithography resolution and photoresist adhesion.

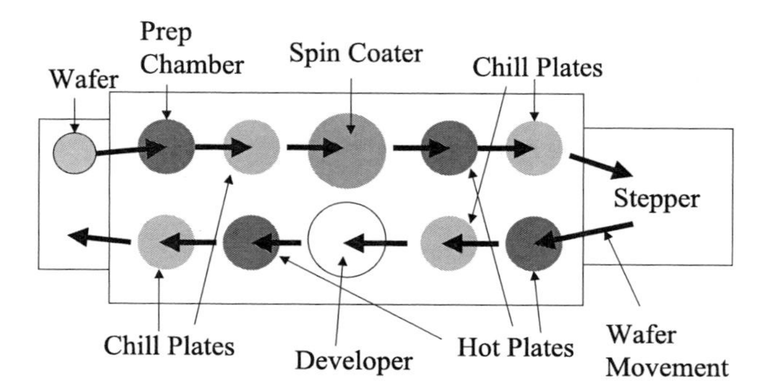

Figure 6.40 Track-stepper integrated system.

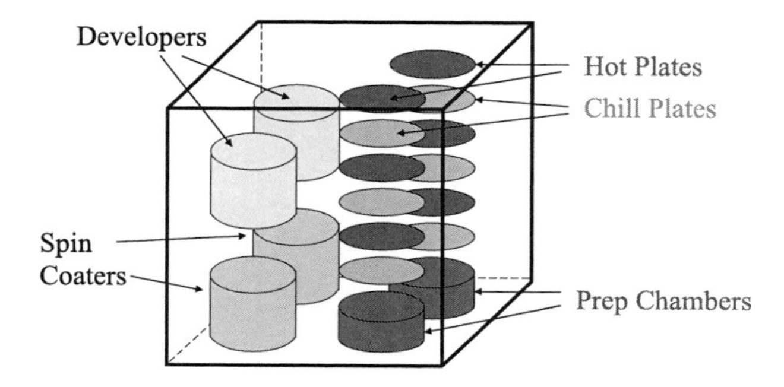

Figure 6.41 Schematic of a stacked track system.

Track systems in advanced semiconductor fabs no longer look like the 2D model in Fig. 6.40. Cleanroom grades have become higher and higher (as has the cost per square foot of the cleanroom) to keep up with shrinking feature sizes. Increasingly expensive cleanroom footage is the major driving force for stacked systems, e.g., hot plates and chill plates that are stacked to make a system more compact. In some advanced systems, coaters and developers are also stacked, saving even more space (see Fig. 6.41).

Some metrology systems, such as overlay and scatterometry CD metrology tools, can be integrated into track-scanner systems, commonly called litho cells, to further improve process throughput.

6.4 Lithographic Technology Trends

Optical photolithography systems have been used for pattern transfer in the semiconductor industry since its inception. In the near future, lithographic technology will change dramatically. This section speculates on possible changes.

6.4.1 Resolution and depth of focus

When a light wave passes through a gap or hole on a mask, it has diffraction, and the projected image will never be as sharp as that on the mask. Using a lens to focus

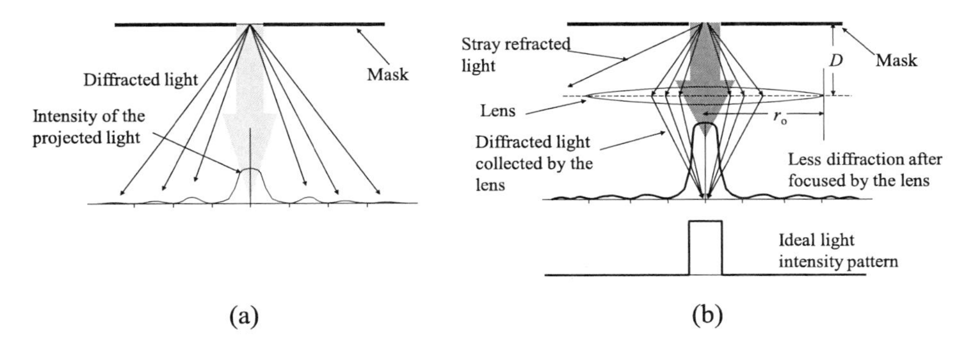

Figure 6.42 Light diffraction (a) without and (b) with a lens.

some of the diffracted light can improve resolution by reducing the light diffraction (see Fig. 6.42).

The minimum resolution an optical system can achieve is determined by the wavelength of the light and the numerical aperture of the system. The resolution can be expressed as

$$R = \frac{k_1 \lambda}{NA}, \tag{6.1}$$

where k_1 is the system constant, λ is the wavelength of the light, and NA $= 2r_o/D$ is the numerical aperture, which indicates the capability of the lens to collect the diffracted light. D is the distance between the object, mask (or reticle), and the lens, and $2r_o$ is the diameter of the lens. Equation (6.1) shows that finer resolution can be achieved with a larger lens diameter, similar to a camera with a larger lens achieving a sharper image. However, an optical system with a large lens is much more expensive, the way a camera with a large lens is more expensive than a compact camera with a small lens. There are also technological limitations when creating a high-precision lens with a very large diameter.

Equation (6.1) also illustrates that resolution improves when a shorter wavelength is used for exposure. This is why the wavelength of an exposure light source becomes shorter and shorter in photolithography processes (Fig. 6.43). However, there is a limit; when a wavelength reduces to a certain value, light leaves the UV range and becomes an x ray, as illustrated in Fig. 6.44. For x rays, most optical equations are no longer valid, including Eq. (6.1).

Another important property of an optical system is depth of focus (DOF), which is the range in which light is in focus and can achieve good resolution of the projected image. DOF can be expressed as

$$DOF = \frac{k_2 \lambda}{2(NA)^2}. \tag{6.2}$$

Equation (6.2) confirms that an optical system with a smaller NA has a larger DOF, which is why point-and-shoot cameras on cell phones have very small lenses.

The focus needs no adjustment when a picture is taken with this kind of camera, since its DOF is so large that almost everything is in focus. However, this kind of camera cannot produce a sharp image, since resolution will be poor with such a small lens. Figure 6.45 gives an illustration of the DOF of an optical system.

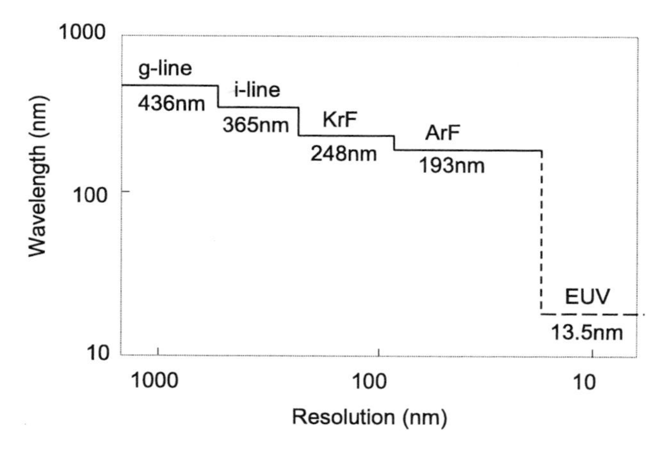

Figure 6.43 Relationship of lithography wavelength and resolution.

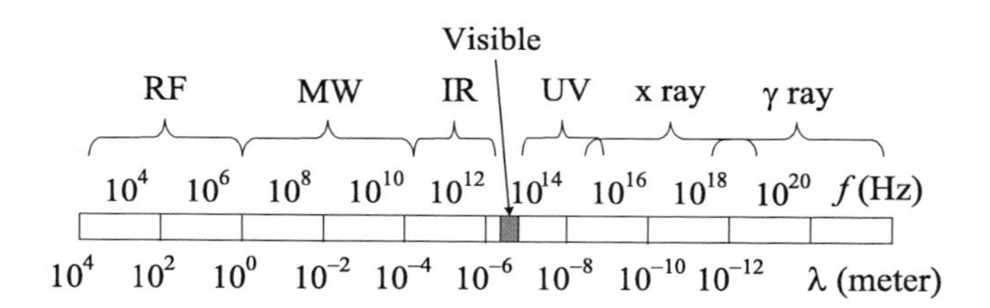

Figure 6.44 Wavelength and frequency of an electromagnetic wave.

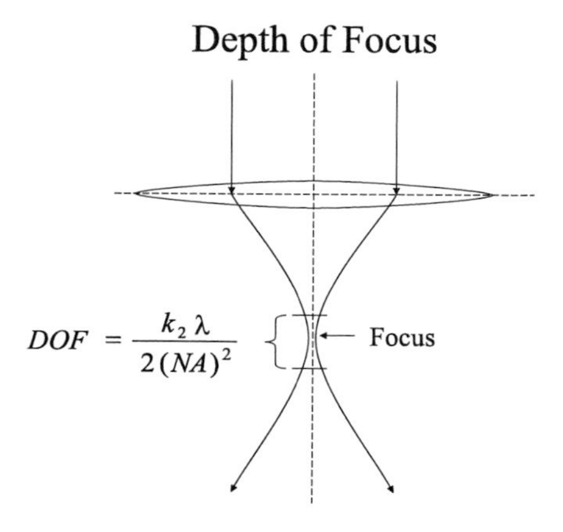

Figure 6.45 DOF of an optical system.

The larger the DOF is on a scanner system, the easier it will be to focus the exposure light in the photoresist on a wafer surface. Unfortunately, the DOF and resolution always work against each other—to improve resolution, a shorter wavelength and larger NA are needed, both of which reduce the DOF.

In advanced photolithography processes, resolution is so high that the DOF becomes very small; thus, the center of the focus must be placed exactly at the mid-plane of the photoresist to achieve the best lithography results, as shown in Fig. 6.46. Photolithography engineers in IC fabs routinely use a focus-exposure matrix (FEM) to check on focusing issues when exposing a photoresist wafer.

Because of DOF requirements in photolithography processes, wafer surfaces need to be highly planarized for the patterning of feature sizes smaller than 0.25 μm. This is one reason that CMP processes are widely used in advanced semiconductor fabs. Only a CMP process can achieve the surface planarization results required by photolithography resolution for quarter-micron (and smaller) pattern geometry.

6.4.2 Mercury lamps and excimer lasers

Since shorter wavelengths achieve higher resolution, stable, high-intensity, and short-wavelength light sources have been developed and applied in exposure systems. Ultrahigh-pressure mercury lamps and excimer lasers are widely used as light sources for steppers.

A mercury vapor lamp has many lines of radiation, of which an i-line at 365 nm is the most commonly used in stepper exposure systems for 0.35-μm feature size IC processes.

KrF excimer lasers at 248 nm have been developed as light sources of DUV scanners for 0.25-μm minimum feature size IC applications, and they have also been used for 0.18- and 0.13-μm IC fabrication. KrF scanners are still heavily used in noncritical layers such as implantations and upper metal layers of advanced nanometer technology IC fabs. Scanners using ArF excimer lasers at 193 nm are widely used for 130- to 65-nm IC chip manufacturing. Photolithography processing tools using F_2 lasers with 157-nm wavelengths have been studied. However, the development of a 157-nm scanner has been discontinued due to

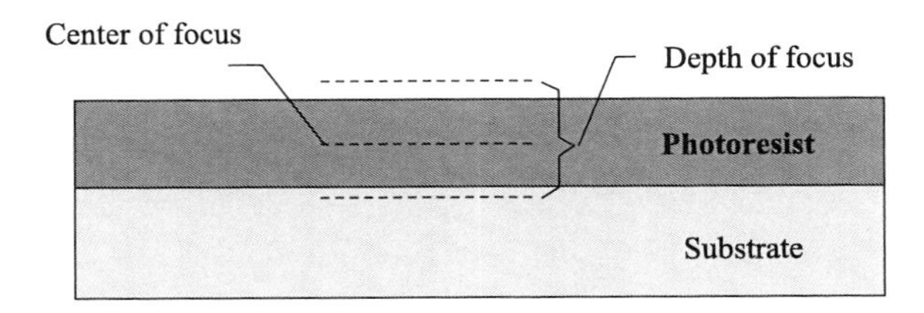

Figure 6.46 Focus light in the mid-plane of a photoresist to optimize resolution.

the development of 193-nm immersion lithography technology, which extends the resolution of ArF systems to the 45-nm technology node. Combined with double-patterning or multiple-patterning techniques, ArF systems could possibly push IC manufacturing beyond the 22-nm node. Next-generation lithography (NGL) technologies, such as EUV lithography, nanoimprint lithography (NIL), and e-beam direct write (EBDW), could start seeing use in IC manufacturing in 16-nm or smaller technology nodes by the mid-2010s. These technologies will not replace optical lithography in the near future.

Different lithography exposure wavelengths use different photoresists because a photoresist is designed to be sensitive only to certain wavelengths. ARC is also related to exposure wavelength. Different wavelengths require different dielectric deposition processes for dielectric ARCs or different spin coating processes for BARCs.

Since the 1980s, there have always been predictions that photolithography processing must be replaced by an alternative lithography process within ten years, due to the resolution limit of the optical exposure system. However, scientists and engineers working in photolithography processing constantly improve resolution, push the limit, and extend the lifetime of photolithography (see Fig. 6.47).

Because of the development of photolithography technologies such as phase shift mask (PSM), optical proximity correction (OPC), off-axis illumination, immersion lithography, and double patterning, engineers can use 193-nm optical lithography systems to pattern 22-nm technology node devices. With the development of multiple-patterning techniques, it is possible to push the limit farther and delay deployment of NGL technologies.

6.4.3 Resolution enhancement techniques

To improve photolithography resolution, several resolution enhancement techniques have been developed and applied in IC chip manufacturing that extend

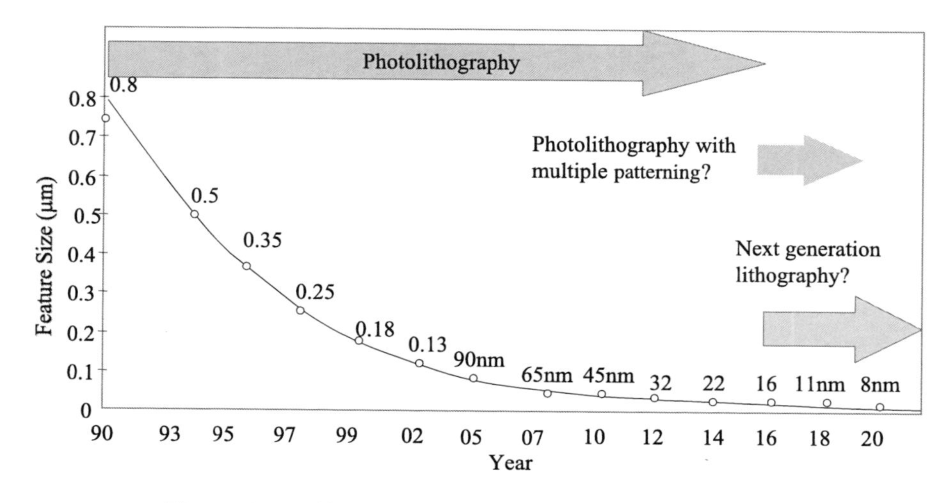

Figure 6.47 IC technology nodes and lithography trends.

optical lithography to the 22-nm technology node and possibly beyond. This section briefly describes some important resolution enhance techniques, such as PSM, OPC, and off-axis illumination.

6.4.3.1 Phase shift mask

Equation (6.1) shows that photolithography resolution can be improved by reducing k_1. One of the methods used to reduce k_1 is the PSM.

It is not difficult to transfer an isolated, small feature pattern to a photoresist; the challenge comes when the small features are densely packed—light diffraction and interference can distort the feature. To solve this problem, a PSM is introduced. The PSM has a phase shifter at every other opening (clear, transparent area) on the mask, as shown in Fig. 6.48. Although this kind of PSM is not used in IC manufacturing, it is still useful for explaining the principle of a PSM.

The thickness and dielectric constant of the dielectric layer are carefully controlled, so that $d(n_f - 1) = \lambda/2$, where d is the thickness of the dielectric, n_f is the dielectric constant, and λ is the exposure light wavelength. Due to opposite phase shift, light passing through an open area without phase shift coating will destructively interfere with light passing through an opening with phase shift coating, resulting in a sharper image in the densely packed area, as shown in Fig. 6.49.

By etching a quartz substrate instead of adding a shifter, an alternating aperture PSM (AAPSM) can be created, as shown in Fig. 6.50. The etch depth needs to be precisely controlled at $d = \lambda/[2(n - 1)]$, where d is the etch depth, λ is the wavelength of exposure light, and n is the refractive index of the quartz substrate. An etch depth of 171.4 nm is needed for a 193-nm exposure wavelength and a quartz refractive index of 1.563 at 193 nm.

Another type of PSM used in IC manufacturing in nanometer node technology is an attenuated PSM (AttPSM). An AttPSM is formed by patterning the partially transparent film deposited on the quartz substrate (see Fig. 6.51).

Molybdenum silicide (MoSi) with transmittance of 6 to 20% is commonly used for this application. Thickness of the MoSi film is carefully controlled so that the small amount of light passing through it has a phase difference 180 deg from the light passing through the quartz at the exposure wavelength (193 nm for an ArF excimer laser). Light intensity in the area covered by MoSi is lower than the photoresist exposure threshold because of the destructive interference of the light, while light intensity in the area without MoSi is higher than the photoresist

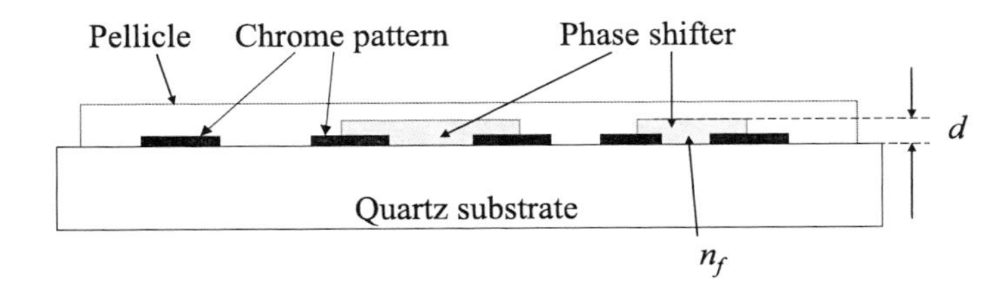

Figure 6.48 A PSM.

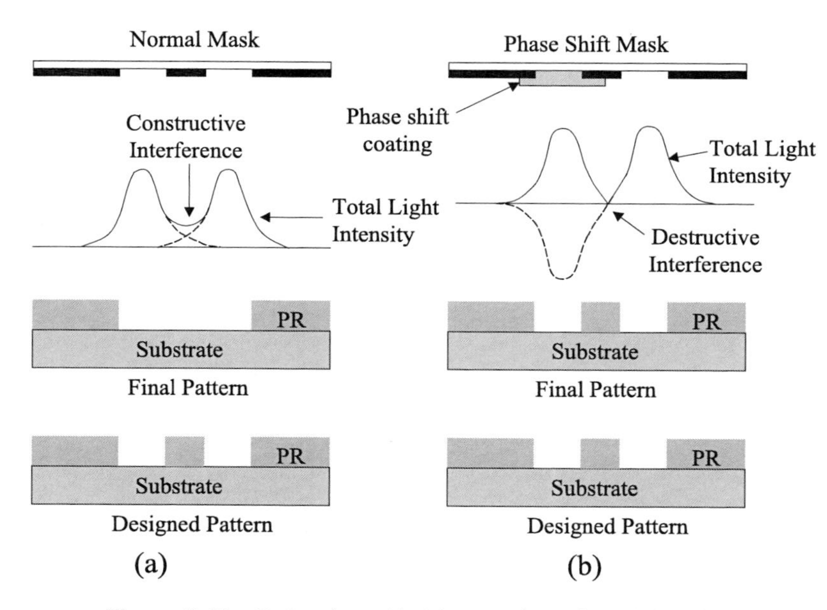

Figure 6.49 Patterning with (a) normal mask and (b) PSM.

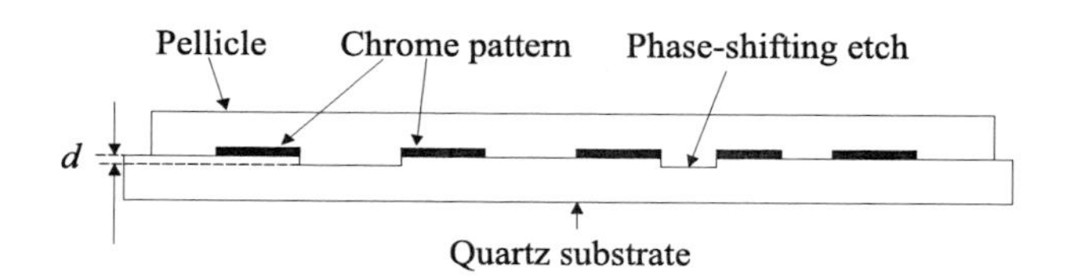

Figure 6.50 AAPSM formed by quartz etch.

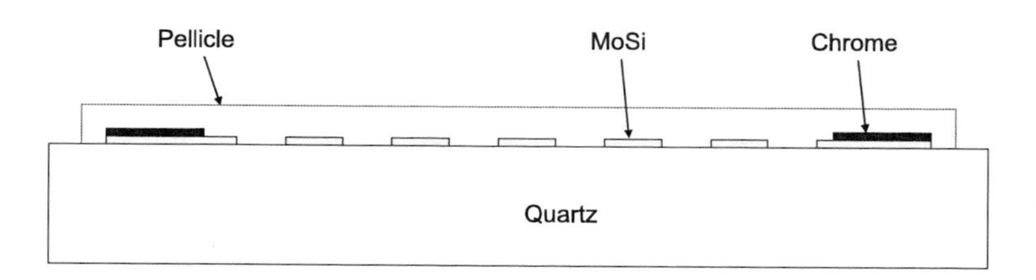

Figure 6.51 The AttPSM.

exposure threshold. This allows high-resolution subwavelength patterning (see Fig. 6.52).

6.4.3.2 Optical proximity correction

In the good old days when feature size was larger than the exposure wavelength, patterns printed on a wafer were almost the same as patterns on the mask, except for some corner-rounding effects caused by light diffraction. When feature size becomes smaller than the wavelength, the light diffraction effect becomes more profound, and patterns printed on a wafer are no longer the same as patterns on the

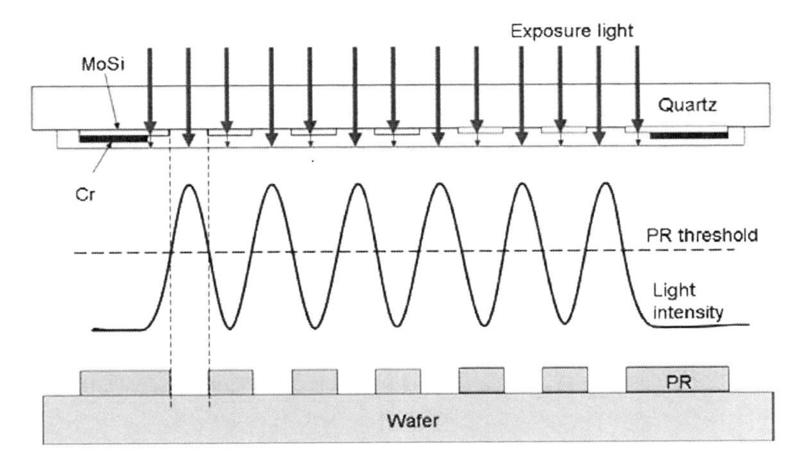

Figure 6.52 Photoresist patterning with AttSPM.

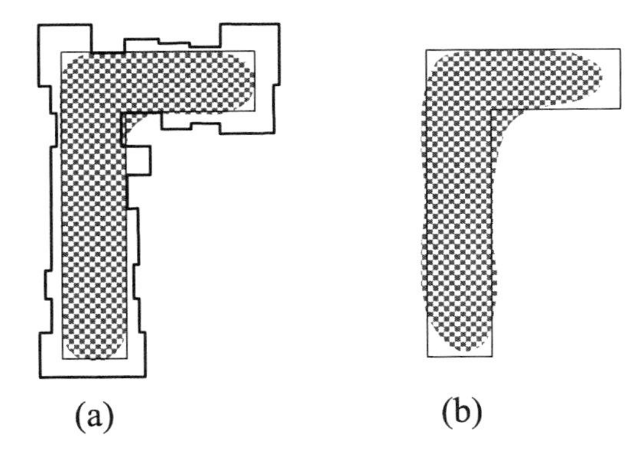

Figure 6.53 Optical lithography pattern (a) with and (b) without OPC.

mask. Therefore, to print the desired patterns on a wafer, miniature features have to be added on the mask pattern to compensate for diffraction effects. These add-on features are called OPC features (see Fig. 6.53).

6.4.3.3 Off-axis illumination

By using an aperture that forces incident light to enter at an angle (off axis) through the lens of an optical system, it is possible to collect the first diffraction order of light from the mask to effectively reduce the k_1 factor in Eq. (6.1) and improve lithographic resolution. Figure 6.54(a) illustrates an on-axis illumination system, and Fig. 6.54(b) shows an off-axis illumination system, which is also called monopole illumination, since incident light comes from a single pupil of the aperture.

A dipole illumination system uses two different illuminations with symmetric angles; this allows zero- and first-order diffraction from both poles to be collected and imaged, resulting in more balance in intensity than with monopole illumination. An illustration of diffracted orders of dipole illumination and a dipole

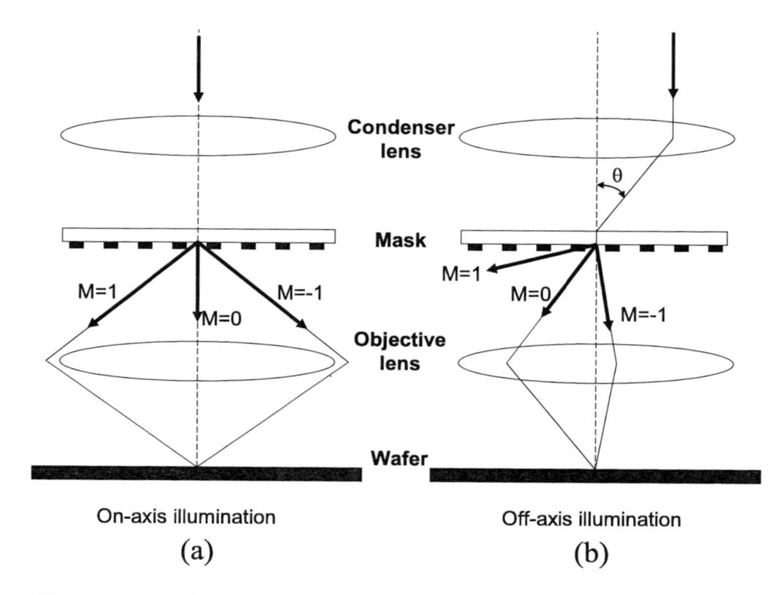

Figure 6.54 Schematics of (a) on-axis and (b) off-axis illumination.

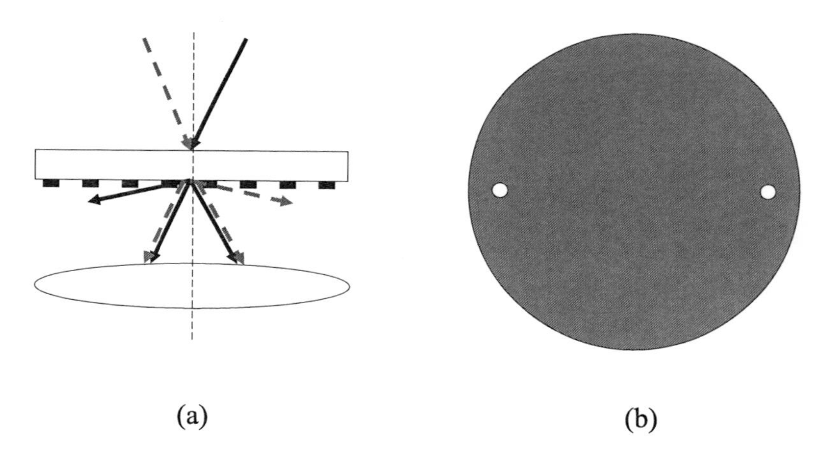

Figure 6.55 (a) Dipole illumination and (b) an aperture of dipole illumination.

illumination pupil can be seen in Figs. 6.55(a) and 6.55(b), respectively. Similarly, a quadruple illumination source can be used for optical lithography.

By optimizing both aperture features and mask patterns, called source-mask optimization (SMO), scientists and engineers can stretch the applications of optical lithography. Figure 6.56 shows an example of SMO. The pattern on the mask [Fig. 6.56(b)] is completely different from the pattern on the wafer [Fig. 6.56(d)], and the source design [Fig. 6.56(c)] is much more complicated than dipole or quadruple designs. Because it would take an excessive amount of computation to optimize both source and mask patterns, this technique is also called computational lithography. Combined with immersion technology, optical lithography can be used

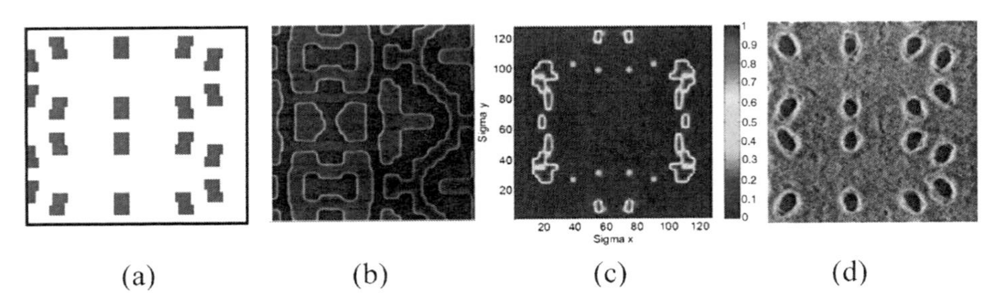

Figure 6.56 Example of SMO of contact photolithography: (a) design layout, (b) mask, (c) designed source, and (d) photoresist pattern on the wafer (K. Lai, et al., *Proc. of SPIE*).

for 22-nm technology node patterning. One of the advantages of this approach is that the mask is still binary, which means that it is easier to create than a PSM.

6.4.4 Immersion lithography

By immersing the gap between the objective lens of a microscope and a sample viewed with water or oil, resolution of the microscope image can be improved. The same idea can be applied to photolithography. By filling the gap between the objective lens and wafer surface with DI water, lithographic resolution can be improved significantly. Resolution of immersion lithography can be expressed by

$$R = \frac{k_1 \lambda}{n_{fluid} NA}. \tag{6.3}$$

The only difference between Eqs. (6.1) and (6.3) is n_{fluid}, the refractive index of the fluid between the objective lens and wafer. When the fluid is air, the refractive index is very close to 1, and Eq. (6.3) becomes Eq. (6.1). Equation (6.3) demonstrates that by adding water (n_{fluid} = 1.44 at 193 nm) between the objective lens and wafer of a 193-nm scanner, resolution can be improved by one technology node without a major change in the optical system.

$$DOF_{immersion} = \frac{1 - \sqrt{1 - (\lambda/p)^2}}{n_{fluid} - \sqrt{n_{fluid}^2 - (\lambda/p)^2}} \frac{k_2 \lambda}{2(NA)^2}, \tag{6.4}$$

where λ is the wavelength of the exposure light, and p is the pitch of the mask pattern. When $n_{fluid} = 1$, Eq. (6.4) becomes Eq. (6.2).

The DOF improvement factor of an immersion system, $\eta = DOF_{immersion}/DOF$, is larger than n_{fluid}, meaning that the DOF increases when applying immersion fluid between the objective lens and wafer. When applying a DI wafer, the DOF increases by at least a factor of 1.46 with 193-nm immersion lithography for large-pitch mask patterns. For smaller-pitch mask patterns, the DOF improvement factor

η increases more (see Fig. 6.57). By increasing the DOF using immersion fluid, the NA can be increased to further enhance resolution while keeping the DOF in a comfortable range. High-NA 193-nm immersion lithography is widely used in IC manufacturing for 45- to 22-nm technology nodes. Figure 6.58 shows that water is only in the gap between the objective lens and wafer in an immersion lithography system.

6.4.5 Double, triple, and multiple patterning

By patterning a single mask layer more than once, patterning resolution will improve because the k_1 factor of Eq. (6.1) is effectively reduced. If the same lithography system is used, the effective k_1 will be k_1/N, where N is the number of the patterning. For commonly used double patterning, $k_{1,\text{eff}} = k_{1/2}$. Double-patterning technology (DPT) has been used in IC chip manufacturing since the 45-nm technology node and is widely used in 32- and 22-nm technology nodes. It is also possible to pattern more than twice, such as three times (triple patterning) or four times (quadruple patterning), to further reduce effective k_1 and to pattern even

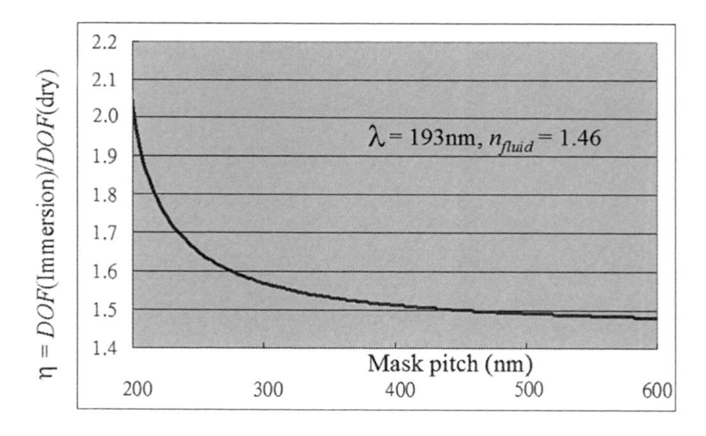

Figure 6.57 Immersion-induced DOF improvement versus mask line/space pitch.

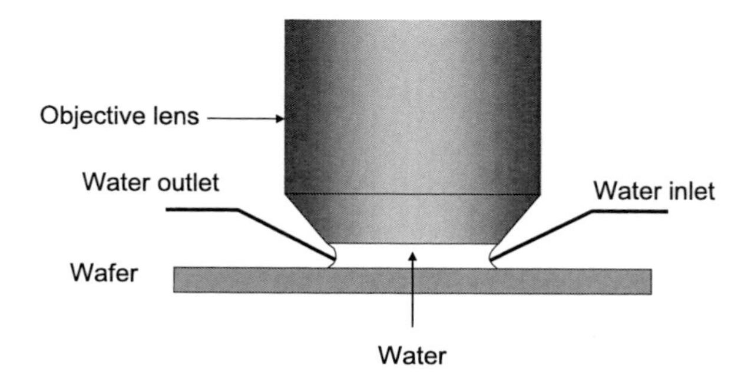

Figure 6.58 Schematic of an immersion lithograph system.

finer features. If NGL (such as EUV lithography) is not ready in time for 16-nm technology nodes and beyond, multiple-patterning technologies could be used.

There are many approaches of DPT, such as litho-freeze-litho-etch (LFLE), litho-etch-litho-etch (LELE), and self-aligned double patterning (SADP). LFLE is very attractive to IC chip manufacturers because it is regarded as the DPT with the lowest cost. By freezing the developed photoresist of the first mask and applying the second photoresist layer then exposing it with the second mask, the pitch density can be doubled using an optical lithography system with resolution that cannot pattern the final pitch density. Figure 6.59 illustrates the LFLE process. The pattern freeze of the first photoresist can be performed with a chemical process that reduces the solubility of the photoresist while it is in the developer solution so that the photoresist will not dissolve when developing the second photoresist. Ion implantation that hardens the photoresist has also been studied for this application.

The LELE (or LE^2) process, illustrated in Fig. 6.60, uses two layers of hard mask. One of the most challenging portions of this process flow is the etch selectivity of the first hard mask to the second hard mask, a process that is discussed in Chapter 9. If polysilicon is the material to be etched on the wafer surface, silicon oxide can be chosen as the material for the second hard mask, and amorphous carbon can be chosen for the material of the first hard mask. Using oxygen plasma ashing, amorphous carbon can be removed when stripping the second photoresist.

LELE double patterning can also be achieved with one hard mask. The drawback of this approach is that the CD of the final pattern is directly related to the overlay between the two masks. Any overlay error will transfer into a CD variation (see Fig. 6.61).

In a critical mask layer, such as a MOSFET gate, the CD control gate limit is plus or minus 10% of the target CD, while the overlay control is a little more relaxed at about 20% of the target CD. Therefore, this LELE double-patterning technique

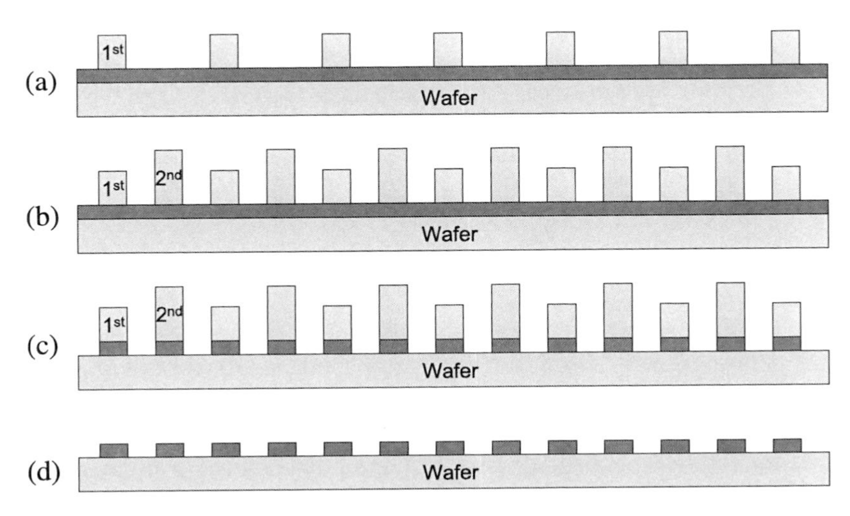

Figure 6.59 Illustrations of the LFLE process: (a) the first photoresist pattern and pattern freeze, (b) the second photoresist pattern, (c) the etch pattern, and (d) the strip photoresist.

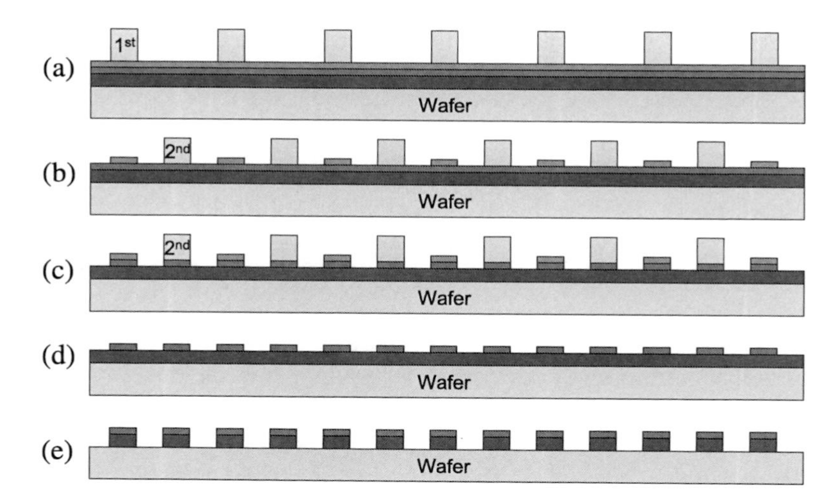

Figure 6.60 Illustrations of the LELE process: (a) first litho, (b) second litho after etch of the first hard mask layer, (c) etch of the second hard mask layer, (d) stripping of the second photoresist and the first hard mask, and (e) etch of the pattern on the wafer.

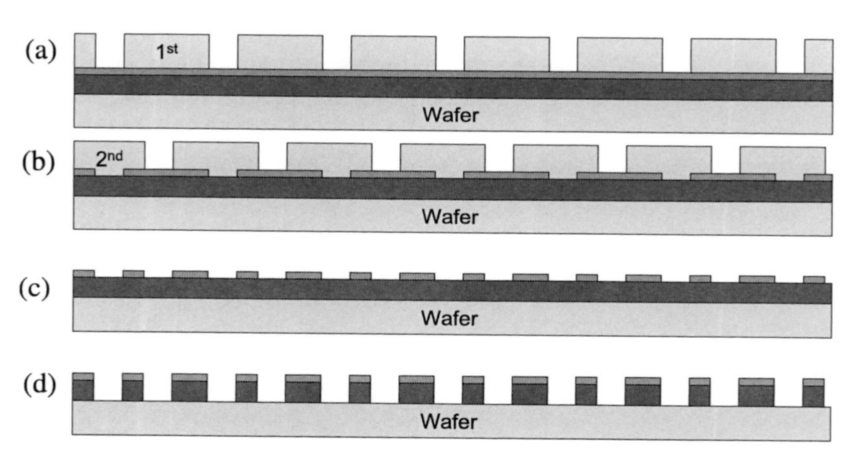

Figure 6.61 Illustrations of the LELE process using only one hard mask. Note the CD variation of the final pattern induced by an overlay error of the first and second masks.

can only be used for noncritical layers, where CD control can also be relaxed to ~20%.

Spacer (self-) aligned double patterning (SADP) is the most promising technique for patterning sub-22-nm device features, especially for NAND flash manufacturing. Figures 6.62 and 6.63 illustrate a cross-sectional view and top view of the SADP process, respectively. The first mask defines the photoresist pattern, as well as the low-temperature oxide CVD and etchback to form the spacer. After removal of the photoresist dummy pattern, the hard mask is etched for the first time. After second photomask patterning, hard mask etch, and photoresist removal, the third etch forms the designed device pattern. The pattern CD is controlled by the

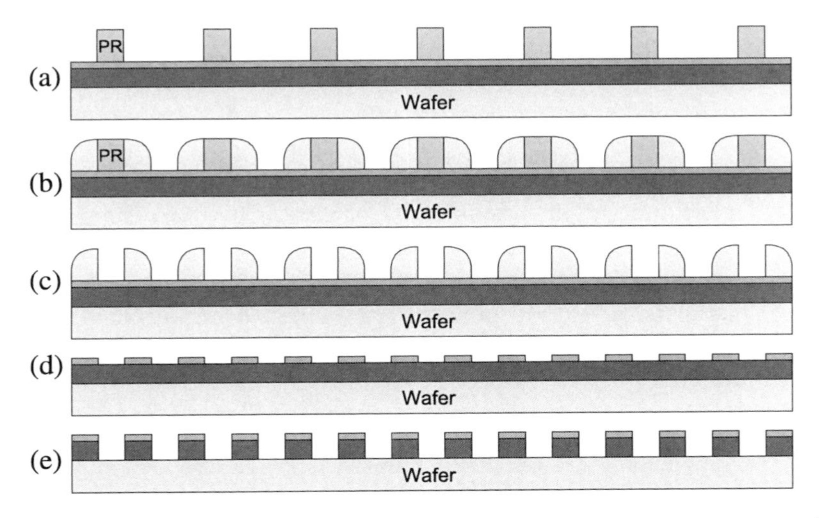

Figure 6.62 Illustrations of the SADP process: (a) first mask photoresist patterning; (b) low-temperature oxide CVD and spacer etch; (c) photoresist strip; (d) hard mask etch, spacer oxide strip, second mask pattern, and hard mask etch; and (e) final pattern etch.

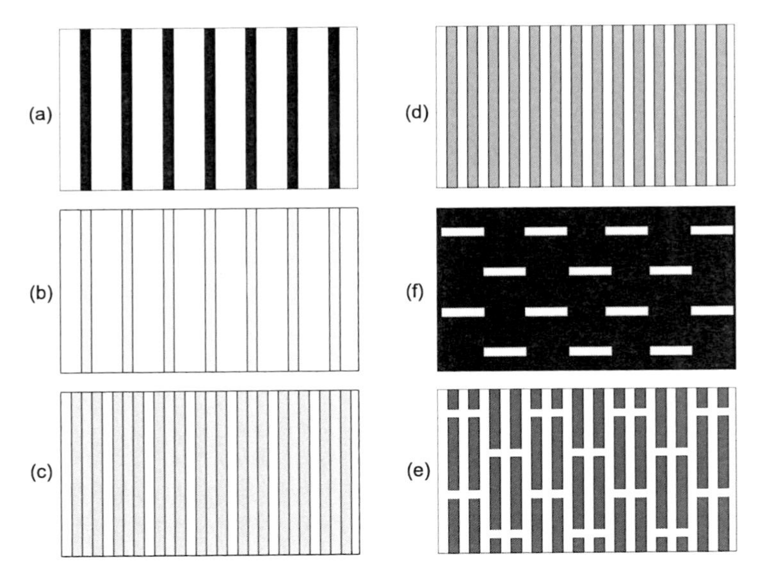

Figure 6.63 Top view of the SADP process: (a) the first mask, (b) the photoresist pattern of the first mask, (c) low-temperature oxide spacer formation, (d) hard mask etch and spacer removal, (e) the second mask, and (f) hard mask and final pattern etch.

thickness of a low-temperature CVD oxide film that forms the spacer, a thickness that can be precisely controlled with a transparent thin-film metrology system.

Because SADP requires more processing steps than other double-pattern processes (such as oxide CVD and spacer etch), it is the most expensive approach of the three DPTs discussed. However, it has many advantages, such as accurate pattern CD and space CD control, less strict overlay requirements for the second mask, low line edge roughness (LER), etc., making it very attractive to some IC manufacturers, especially NAND flash makers. This is because NAND flash has many layers of dense line–space patterns with very tight CD requirements, which is perfect for SADP application. For advanced microprocessor manufacturers, SADP is also very attractive due to its good CD control and low LER.

Although photolithography has its limits, patterning has not reached its limit yet. It is possible to use multiple pattering such as quadruple patterning to achieve even smaller feature size. Directed self-assembly could also be used to multiply pattern frequency and increase pattern density. Fabs could even use double-patterning techniques on the tiny features formed by EUV lithography or NIL to further reduce feature size.

6.4.6 Extreme-ultraviolet lithography

One of the most promising NGL technologies for patterning sub-22-nm technology nodes is EUV lithography with a 13.5-nm wavelength. An electromagnetic radiation wave with wavelength between 1 and 50 nm is in the region between UV and x ray. It can be called EUV, vacuum UV, or soft x ray. At one time it was called soft x-ray projection lithography, but was renamed EUV lithography to avoid confusion with x-ray lithography in the late 1990s.

The basic idea of EUV lithography is that by sharply reducing wavelength λ and moderately reducing NA, lithography resolution can be improved. The process can also operate in the comfort zone of DOF > 100 nm for mass manufacturing. For example, when $k_1 = 0.25, k_2 = 1.0, \lambda = 13.5$ nm, and NA = 0.25, from Eqs. (6.1) and (6.2), resolution R = 13.5 nm, and DOF = 108 nm.

Due to strong absorption at short wavelengths, no material can be used to make a lens for the EUV lithography process; thus, an EUV system must use a mirror-based system. High-intensity EUV light sources have been developed for preproduction systems (~60 wafers/h), and sources with even higher intensity for full production systems (120 wafers/h) are still in development. Figure 6.64 illustrates an EUV exposure system.

To effectively reflect EUV light, multilayer coatings of thin Mo/Si pairs are needed. Figure 6.65 illustrates an EUV mask that uses a quartz substrate. Backside metal coating is needed for electrostatic chucking, which is the preferred chucking mechanism in semiconductor manufacturing because it eliminates mechanical chucking, which is prone to generating particles. About 40 pairs of Mo and Si thin films are deposited on top of the quartz to form a reflective layer of EUV light at a 13.5-nm wavelength. About 70% reflective efficiency can be achieved at a 6-deg incident angle. The absorber usually is boron-doped tantalum nitride (TaBN), and the buffer layer can be chromium nitride (CrN) with a silicon capping layer or ruthenium (Ru). The buffer layer is used to protect the multilayer from damage when pattering the absorber. The ARC usually is boron-doped tantalum oxynitride (TaBON) deposited on top of a TaBN absorbing layer. It can reduce reflection of the absorber area and enhance contrast of the multilayer area at DUV inspection wavelengths to improve the sensitivity of defect inspection using optical

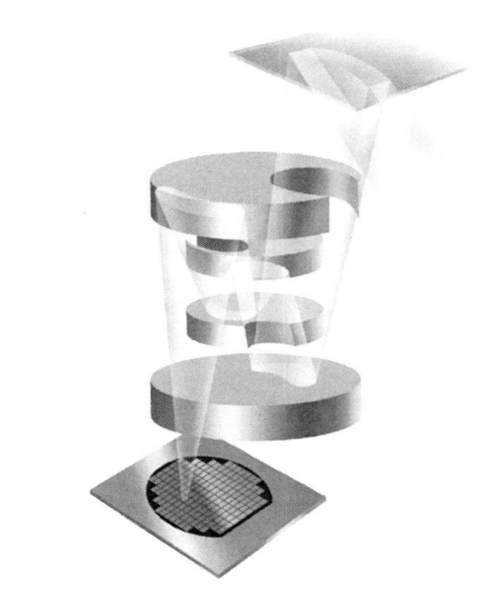

Figure 6.64 EUV lithography system (Carl Zeiss).

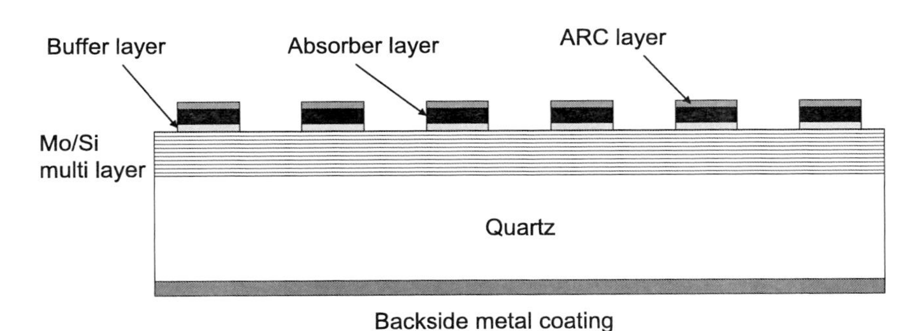

Figure 6.65 An EUV mask.

mask inspection systems. E-beam inspection and actinic inspection do not require ARC.

Ideally, an EUV mask can be inspected at the EUV wavelength of 13.5 nm, which is called actinic inspection. Theoretically, actinic inspection can capture all printable defects caused by both surface and phase defects, both of which are induced by defects embedded or below the multilayer. Since actinic EUV mask inspection systems are still in development, optical mask and e-beam mask inspections can be used for the development and pilot line production of EUV lithography. Due to an inability to capture embedded phase defects in the EUV mask, it is very likely that EUV lithography will be deployed in mask layers insensitive to phase defects, such as bitline contact layers of NAND flash and some via layers of CMOS. Because more than 90% or even 99% of the areas are covered by an absorber, embedded phase defects have very little chance of being printed to affect device performance. For memory chips such as NAND flash, the redundancy design would allow them to withstand some defects without suffering major yield loss.

6.4.7 Nanoimprint lithography

Imprint technology is widely used to make coins, music, videos, and software compact disks. Using an imprint lithography process to generate nanometer IC features was proposed in 1996, and this alternative lithography technology was called NIL. Figure 6.66 illustrates the NIL patterning process. First, a resist is sprayed onto a wafer surface [see Fig. 6.66(a)]; then the template is pressed onto the resist [Fig. 6.66(b)]; and the resist is hardened by applying either thermal or UV energy [Fig. 6.66(c)]. After resist hardening, the quartz template or mold is released from the wafer surface [Fig. 6.66(d)], and an etch process follows to transfer the resist pattern formed by NIL to the thin film on the wafer surface [Fig. 6.66(e)].

Figure 6.67 shows a line–space pattern formed by the NIL process. Notice that the CD of the line is only 11 nm with very low LER.

The advantage of NIL is high resolution, low LER, and a low cost of ownership, especially for the 22-nm technology node. However, to apply NIL in high-volume IC production, several issues must be solved: template defects, overlay, throughput, and management of thousands of working templates. A template might be able to last for about 10,000 prints, which is only about 100 wafers. To reduce the cost of wafer printing, thousands of working templates need to be duplicated from a master template with the nanoimprint technique.

One promising application of NIL is to form pattern media of hard disk drives (HDDs). When HDD storage increases, the size of storage cells reduces. HDD

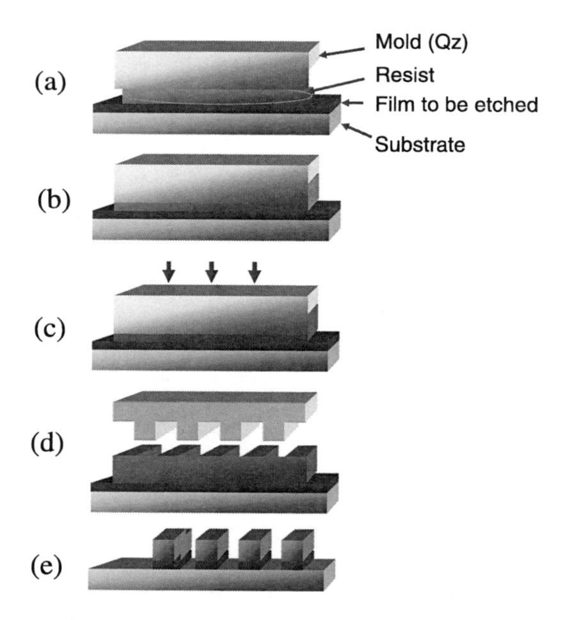

Figure 6.66 NIL patterning process: (a) resist coating, (b) imprint, (c) resist hardening, (d) template release, and (e) pattern etch.

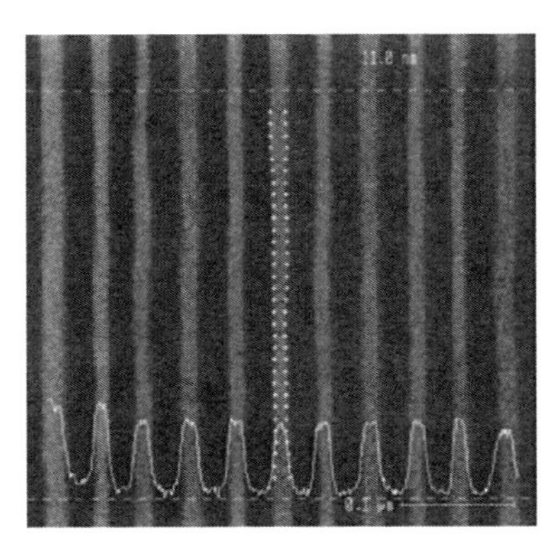

Figure 6.67 This SEM image shows the capability of NIL to form an 11-nm line pattern. (M. Malloy and L. C. Litt, *Proceedings of SPIE*).

storage almost reaches the limit of a storage cell of a nonpatterned magnetic disk, which is called media in the HDD industry. To further reduce the size of storage cells and increase storage density, cell patterning with sub-50-nm pitch is required. Because there is only one layer of HDD media, there is no overlay issue, and working-template management is also significantly simpler than multiple-layer patterning of IC processing.

6.4.8 X-ray lithography

When a wavelength is shorter than 5 nm, an electromagnetic wave becomes an x ray. An x ray has a much shorter wavelength than UV light; therefore, it can achieve higher resolution than photolithography. X-ray lithography has been in research and development since it was proposed in 1972.

Because no material can refract an x ray, and the ray can reflect at only a very small angle (~1 deg), x-ray lithography must be accomplished by a direct printing process, which is very similar to proximity printing, as shown in Fig. 6.68. The x ray passes through the transparent part of the mask and exposes the photoresist on the wafer surface. Because the wavelength is so small that the diffraction effect is almost negligible, it can achieve resolution close to that on the mask.

There are many disadvantages of x-ray lithography. For instance, UV light can be reflected by the mirror and focused by the lens, but an x ray cannot; thus, the exposure system needs to be redesigned. For a commonly used 4:1 scanner system, feature size on the reticle is four times larger than that on the wafer surface. Therefore, to pattern a 25-nm feature size on the wafer surface, feature size on the reticle must be 100 nm. Since it takes less than 100-nm chromium to block UV radiation, the pattern aspect ratio is smaller than 1:1 on the optical reticle. Because an x ray cannot be focused, the mask and wafer feature sizes must be in a 1:1 ratio. To pattern a 25-nm feature size on the wafer surface, feature size on the mask must also be 25 nm. An x-ray mask requires a layer of gold greater than 100-nm thick to block more than 90% of the x ray. Also, the pattern aspect ratio is larger than 4:1,

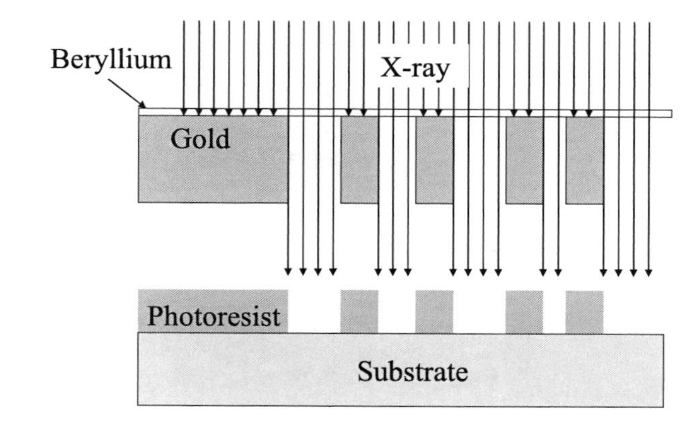

Figure 6.68 X-ray lithography process.

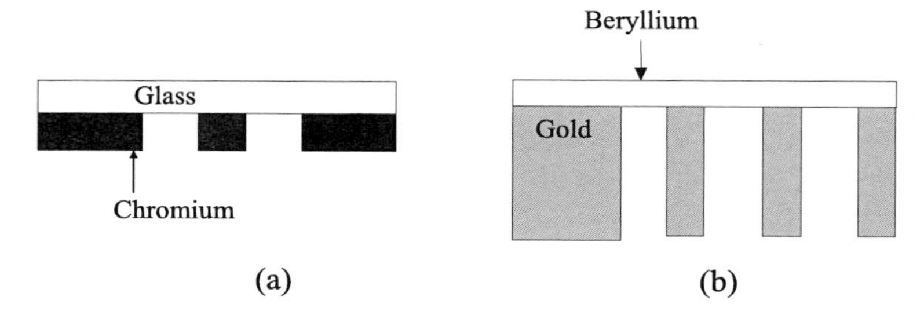

Figure 6.69 Comparison of (a) a photomask and (b) an x-ray mask.

making the x-ray mask much more difficult to produce than an optical photomask. Figure 6.69 illustrates a photomask and an x-ray mask.

Question: What is the pattern aspect ratio of an x-ray mask if that mask is intended to print 10-nm half-pitch line–space patterns on the wafers?
Answer: The aspect ratio is 100 nm:10 nm, or 10:1.

When pattern dimensions go down to the sub-10-nm region, the aspect ratio of the pattern on the x-ray mask further increases, causing it to be too high and very difficult to making the mask produce. Also, x-ray lithography requires a single-wavelength x-ray source, such as a synchrotron, which has a large footprint and is very expensive to build. X-ray lithography is no longer regarded as one of the promising candidates for NGL solutions.

6.4.9 Electron beam lithography

An electron is a very small particle and is also a wave, with its wavelength determined by its momentum, which is related to electron energy. Electron wavelength is inversely proportional to the square root of electron energy; thus, the higher the electron energy, the shorter the electron wavelength. An energetic electron beam (10 to 100 keV) has a much shorter wavelength than UV light, thus electron beam (e-beam) lithography inherently has higher spatial resolution and a wider processing window than optical lithography. E-beam lithography is widely used in mask shops for writing design layout patterns to the resist on a mask. It is one of the most time-consuming and expensive processes in the mask manufacturing process.

An EBDW lithography system uses a fine scanning e-beam to directly write the design patterns stored in a database computer onto the electron-sensitive resist on a substrate surface, similar to a laser printer printing words or images from a computer onto a sheet of paper. Energy from the electrons changes the solubility of the resist where the electron beam hits. A positive resist becomes soluble in the development solution. Because of the nature of serial writing, throughput of EBDW is low. For a single–e-beam direct write system, as illustrated in Fig. 6.70(a), throughput will be too low to be cost effective for large-scale semiconductor wafer production. Therefore, the only option for EBDW applications in IC manufacturing is a multiple beam solution, as shown in Fig. 6.70(b).

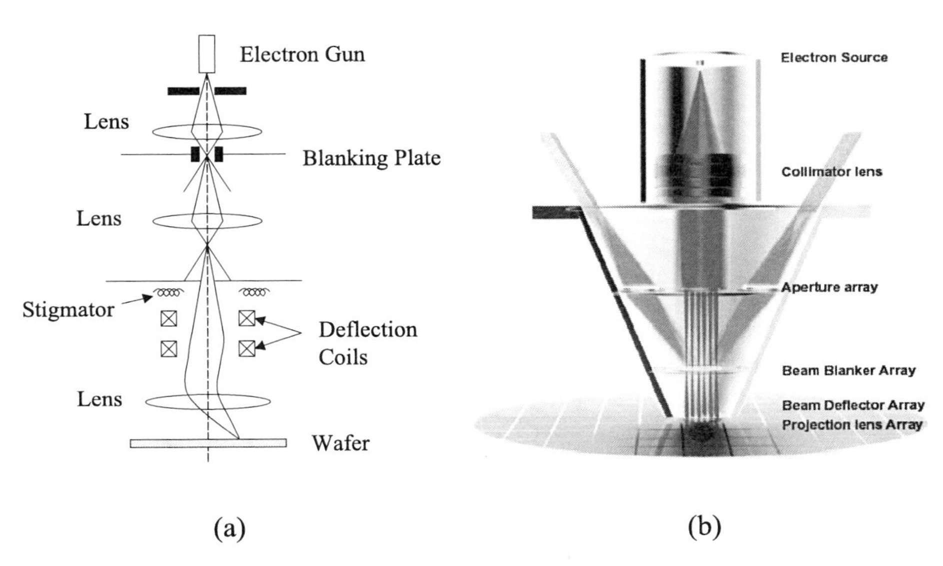

Figure 6.70 EBDW systems: (a) single beam and (b) multibeam (Mapper Lithography, B.V.).

6.4.10 Ion beam lithography

Similar to e-beam lithography, ion beam lithography also can achieve higher resolution than photolithography. An ion beam can also be used in both direct writing and projection resist exposing. One of the advantages of ion beam lithography is direct ion implantation and ion beam sputtering pattern etch, which saves some processing steps. However, throughput of ion beam lithography is very low due to the serial writing for ion beam direct writing; thus, this technology is very unlikely to be used in mass production. One application of ion beam processing is mask/reticle repairing. It can also be used for IC chip device defect detection and repair.

6.5 Safety

Safety is always one of the most important issues in semiconductor fabs. For photolithography processes, the safety concerns are mainly chemical, mechanical, and electrical.

Photolithography uses many chemicals—some of them are flammable and explosive, and some of them are corrosive and toxic. For chemicals commonly used for wet cleaning, sulfuric acid (H_2SO_4) is corrosive and causes skin burns if directly contacted; even a diluted solution can cause a skin rash. Hydrogen peroxide (H_2O_2) is a strong oxidizer, and direct contact with it can cause irritation and burns to the skin and eyes. Xylene, a commonly used solvent and developer for negative photoresists, is flammable with a flash point at 27.3 °C (about room temperature) and is explosive when its concentration reaches 1 to 7% in air. Repeated contact with xylene can cause skin irritation. Xylene vapor has a sweet, airplane-glue-like odor; exposure to it can cause eye, nose, and throat irritation, and inhaled fumes can cause headache, dizziness, loss of appetite, and fatigue. HMDS is the primer most commonly used to promote photoresist adhesion to wafer surfaces. It is flammable with a flash point at 6.7 °C and is explosive when its vapor concentration is 0.8 to 16% in air. It will vigorously react with water, alcohol, and mineral acid and release ammonia. TMAH is widely used as positive photoresist developer solution. It is poisonous and corrosive; swallowing and direct skin contact can be fatal. Contact with its dust or mist can be irritating to the eyes, skin, nose, and throat; high concentration can cause death from fluid in the lungs.

Mercury lamps supply UV light sources for i-line (365 nm) systems, which are used for large feature size patterning. Mercury is a liquid that begins to evaporate at room temperature. Mercury (Hg) vapor is highly toxic; exposure to it can cause coughing, chest pain, headache, difficulty sleeping, loss of appetite, and lung and kidney disorders. Both chlorine (Cl_2) and fluorine (F_2) are used in excimer lasers as DUV light sources. Both gases are toxic, with a greenish color and strong irritating odor. Inhaling at high concentrations can cause death.

UV light can break chemical bonds between atoms and molecules by providing energy to the binding electrons to break free from the bonds. Organic molecules are more vulnerable to damage from strong UV light because of their relatively weak bonds in comparison to the bonds of inorganic compounds. This is why UV light can be used to kill bacteria in food processing. Looking directly at a UV light source in photolithography tools can cause damage to visual cells in the eyes and result in eye injury. UV protection goggles are required when working on optical systems with UV light sources.

High-pressure mercury lamps must be carefully handled. Gloves are required during lamp replacement, since any fingerprint left on the lamp surface can cause uneven heating of the glass, possibly cracking the glass and causing an explosion.

Both mercury lamps and lasers use high-voltage electrical power, which is also a safety hazard. The power supply should always be turned off and static charges grounded before working on these parts. Tag-out and lockout is also very important to avoid a high-voltage electric power supply being turned on while people are still working on these parts.

All moving parts are potential mechanical hazards, especially robotics and slit valves.

6.6 Summary

- Photolithography is a patterning process that uses UV light to transfer a designed pattern on the mask/reticle to the photoresist temporarily coated onto the wafer surface.
- Positive photoresist becomes soluble after exposure to UV light; negative photoresist becomes insoluble due to crosslinking of a photoresist polymer. Positive photoresist is commonly used due to its higher resolution capability.
- A photoresist consists of polymer, sensitizer, solvent, and additive.
- The basic photolithography steps are: wafer cleaning, prebake and HMDS primer vapor coating, photoresist spin coating, soft bake, alignment and exposure, PEB, optical EBR (optional), development, hard bake, and pattern inspection.
- Wafer cleaning reduces contamination and improves photoresist adhesion.
- Prebake dehydrates the wafer surface, and the HMDS primer layer helps photoresist stick to the wafer surface.
- Convection ovens, infrared ovens, microwave ovens, and hot plates can be used for the baking processes. The hot plate is the most commonly used method in advanced semiconductor fabs.
- Spin coating is most commonly used for coating the photoresist.
- Photoresist thickness and uniformity are related to spin rate, spin rate ramp, photoresist temperature, wafer temperature, gas flow rate, and gas temperature.
- Soft bake drives out the majority of solvent from the photoresist and causes it to become solid.
- Overbaking during the soft bake process can polymerize the photoresist and affect its exposure sensitivity.
- Underbaking during the soft bake process can cause a blurred image due to excessive solvents and photoresist lift during etch or ion implantation processes.
- Contact printers, proximity printers, projection printers, and steppers have been used as alignment and exposure systems. A stepper has the highest resolution and is the most commonly used in advanced semiconductor fabs.
- PEB suppresses the standing wave effect due to the thermal movement of photoresist molecules.
- Developer solution dissolves the exposed positive photoresist during the development process. The development process is sensitive to temperature.
- Hard bake drives out the residue solvent from the photoresist and improves etch, implantation resist, and photoresist adhesion. Underbaking can cause

photoresist loss during the etch process. Overbaking can cause photoresist flow and affect resolution.

- Baking, coating, and developing processes are performed in-line with a track system, which is commonly integrated with a stepper.
- Shorter wavelengths have better lithography resolution than longer wavelengths. 193-nm ArF is the most commonly used wavelength of photolithography systems used in advanced IC fabs.
- The combination of immersion lithography and double- or multiple-patterning techniques extended optical lithography to the 22-nm technology node and likely will extend it to 16 nm and beyond.
- EUV lithography, EBDW lithography, and NIL are the three candidates for NGL technologies.
- EUV lithography could replace photolithography in some critical patterning processes in the near future.

6.7 Bibliography

D. G. Baldwin, M. E. Williams, and P. L. Murphy, *Chemical Safety Handbook for the Semiconductor/Electronics Industry*, 2nd ed., OME Press, Beverly, MA, 1996.

R. Bowman, G. Fry, J. Griffin, D. Potter, and R. Skinner, *Practical VLSI Fabrication for the 90s,* Integrated Circuit Engineering Corp., Scottsdale, AZ (1990).

T. H. P. Chang, D. P. Kern, and L. P. Muray, "Arrayed miniature electron beam columns for high throughput sub-100 nm lithography," *J. Vac. Sci. Technol.* **10**(6), 2743–2748 (1992).

L. de Broglie, "Recherches sur la théorie des quanta (Research on the Quantum Theory)," PhD Thesis, Paris Univ. (1924).

B. Dinu, S. Fuch, U. Kramer, M. Kubis, A. Marchelli, A. Navarra, C. Sparka, and A. Widmann, "Overlay control using scatterometry based metrology (SCOM^{TM}) in production environment," *Proc. SPIE* **6922**, 69222S-1 (2008) [doi:10.1117/12.771581].

H. S. Kim, B. H. Lee, E. Ma, F. Wang, Y. Zhao, K. Kenai, H. Xiao, and J. Jau, "After development inspection defectivity studies of an advanced memory device," *Proc. SPIE* **7638**, 76380L (2010) [doi:10.1117/12.848066].

K. Lai, A. E. Rosenbluth, S. Bagheri, J. Hoffnagle, K. Tian, D. Melville, J. Tirapu-Azpiroz, M. Fakhry, Y. Kim, S. Halle, G. McIntyre, A. Wagner, G. Burr, M. Burkhardt, D. Corliss, E. Gallagher, T. Faure, M. Hibbs, D. Flagello, J. Zimmermann, B. Kneer, F. Rohmund, F. Hartung, C. Hennerkes, M. Maul, R. Kazinczi, A. Engelen, R. Carpaij, R. Groenendijk, and J. Hageman, "Experimental result and simulation analysis for the use of pixelated illumination from source mask optimization for 22-nm logic lithography process," *Proc. SPIE* **7274**, 72740A-1 (2009) [doi:10.1117/12.814680].

M. D. Levenson, N. S. Viswanathan, and R. A. Simpson, "Improving resolution in photolithography with a phase-shifting mask," *IEEE Trans. Electron Dev.* **ED-29**(12), 1812–1846 (1982).

J. A. Liddlea, Y. Cui and P. Alivisatos, "Lithographically directed self-assembly of nanostructures," *J. Vac. Sci. Technol.* **22**(6), 3409–3414 (2004).

B. J. Lin, "The k_3 coefficient in nonparaxial 1/NA sealing equations for resolution, depth of focus, and immersion lithography," *J. Micro/Nanolith. MEMS MOEMS* **1**(1), 7–12 (2002) [doi:10.1117/1.1445798].

M. Malloy and L. C. Litt, "Step and flash imprint lithography for semiconductor high volume manufacturing?" *Proc. SPIE* **7637**, 763706-1 (2010) [doi:10.1117/12.846617].

J. A. McClay and A. S. L. McIntyre, "157-nm optical lithography: the accomplishments and the challenges," *Solid-State Tech.* **42**(6), 57–68 (1999).

M. Rothschild, M. W. Horn, C. L. Keast, R. R. Kunz, V. Liberman, S. C. Palmateer, S. P. Doran, A. R. Forte, R. B. Goodman, J. H. C. Sedlacek, R. S. Uttaro, D. Corliss, and A. Grenville, "Photolithography at 193 nm," *Lincoln Lab. J.* **10**(1), 19–34 (1997).

A. H. Shih, *Scatterometry-based critical dimension and profile metrology*, http://www.eetasia.com/ART_8800271012_480200_TA_7589d612.HTM (last accessed on 05/25/2010).

S. M. Sze, *VLSI Technology*, 2nd ed., McGraw-Hill, New York (1988).

P. van Zant, *Microchip Fabrication, a Practical Guide to Semiconductor Processes*, 3rd ed., McGraw-Hill, New York, 1997.

See http://www.semiconductorblog.com/wp-content/uploads/2010/03/IMFT25nmWLSEM.jpg (last accessed April 2010).

6.8 Review questions

1. What is photolithography?
2. What is the difference between positive and negative photoresists?
3. List the four components of photoresist and explain their functions.
4. List the steps of the photolithography process sequence.
5. Why do wafers need to be cleaned before photoresist coating?
6. What is the purpose of prebake and primer coating?
7. List two methods for primer coating. Which one is used the most in advanced IC fabs? Why?
8. What factors can affect photoresist spin coating thickness and uniformity?
9. What is the purpose of soft bake? List the consequences of over- and underbaking.

10. List the four exposure methods. Which one has the highest resolution?
11. What are two main factors that control the exposure process?
12. Explain the purpose of PEB. What can go wrong with overbaking and underbaking in PEB?
13. Name the three steps of development.
14. Explain the purpose of hard bake. What can go wrong if the photoresist is underbaked or overbaked?
15. What are the two processes that follow the photolithography process?
16. Why does a wafer need parametric measurements and defect inspection before it is sent to the next processing step?
17. Explain why a high-intensity, short-wavelength light source is sought out.
18. Why does subquarter-micron IC manufacturing need a CMP process?
19. Explain how the immersion technique improves photolithography resolution.
20. An engineer working on double-patterning process development stated: "Lithography has a limit, patterning does not!" Do you agree with this statement? Explain your answer.
21. List at least two alternative lithography technologies that might replace photolithography in the near future.
22. In your opinion, which method is the most promising candidate for NGL?

Chapter 7
Plasma Basics

Objectives

After finishing this chapter, the reader will be able to:

- define plasma
- list the three basic components of plasma
- list the three key collisions in plasma and identify their importance
- list the benefits of using plasmas in CVD and etch processes
- identify major differences between plasma-enhanced CVD and plasma etch processes
- name at least two high-density plasma systems
- define and explain the mean free path and its relationship to pressure
- define and explain the effect of magnetic fields on plasma
- describe ion bombardment and its relationship to plasma processes.

7.1 Introduction

Plasma processes are widely used in semiconductor processing. For example, all patterned etching in IC fabrication is either a plasma etch or dry etch process. Plasma-enhanced CVD (PECVD) and high-density plasma CVD (HDP-CVD) processes are widely used for dielectric depositions. Ion implanters use plasma sources to generate ions for wafer doping and to provide electrons to neutralize ions on wafer surfaces. Plasmas are also used in physical vapor deposition (PVD) processes, in which metal is sputtered from a target surface by ion bombardment and deposited on a wafer surface. Remote plasma source systems are commonly used in process chamber cleaning, film stripping, and film deposition processes. This chapter covers the basic properties of plasma and its applications in semiconductor processes, focusing mainly on etch and CVD.

7.2 Definition of Plasma

Plasma is widely defined in the semiconductor industry as an ionized gas with equal numbers of positive and negative charges. More precisely, plasma is a quasi-neutral gas of charged and neutral particles that exhibit a collective behavior. Readers can find more information about plasma in the References section at the end of this chapter.

7.2.1 Components of plasma

Plasma consists of neutral atoms or molecules, negative charges (electrons), and positive charges (ions). In a plasma, the electron concentration is about equal to the ion concentration, $n_e = n_i$, which keeps the plasma quasi-neutral.

The ratio of electron concentration to the total concentration is defined as the ionization rate:

$$\text{ionization rate} = n_e/(n_e + n_n),$$

where n_e is electron concentration, n_i is ion concentration, and n_n is neutral concentration. Ionization rate is mainly determined by electron energy; it is also related to the gas species, since different gases require different ionization energies. For example, the sun is a big ball of plasma. At the edge of the sun, ionization rate is very low, or $n_e \ll n_n$, due to the relatively low temperature (about 6000 °C). At the center of the sun, the temperature is so high (~10,000,000 °C) that almost all gas molecules are ionized. Thus, at $n_n \ll n_e$, the ionization rate is almost 100%.

Plasmas used in semiconductor processing have very low ionization rates. For example, the ionization rate of plasma generated in a PECVD reactor with two parallel plate electrodes is about one in one million to one in ten million, or less than 0.0001%. In a plasma etch chamber with two parallel plate electrodes, the ionization rate is a little higher. About one in ten thousand atoms or molecules are ionized, meaning that the ionization rate is less than 0.01%. Even in the two most commonly used high-density plasma sources, inductively coupled plasma (ICP) or electron cyclotron resonance (ECR), ionization rates are still very low, usually from 1 to 5%. High-density plasma sources with almost 100% ionization rates are still in research and development and not used in IC production yet.

The ionization rate of a plasma reactor is mainly determined by electron energy, which is controlled by the applied power. It is also related to pressure, electrode spacing, and the processing gas species.

7.2.2 Generation of plasma

An external energy source must be applied in order to generate plasma. There are several ways to generate plasma in semiconductor processes. A hot filament with potential dc bias is commonly used to generate plasma in ion source and plasma flooding systems in ion implanters. Most PVD systems use a dc power supply to generate plasma. The most commonly used plasma sources in semiconductor processing are rf plasma sources, which are the main topic of this chapter.

In most PECVD and plasma etch chambers, plasmas are generated by rf power applied between two parallel plate electrodes in a vacuum chamber, as shown in Fig. 7.1. This is called a capacitively coupled plasma source, since the two parallel electrodes are exactly like the electrodes of a capacitor.

When rf power is applied to the electrodes, a varying electric field is established between them. If the rf power is high enough, a free electron can be accelerated by the varying electric field until it gains enough energy to collide with an atom or

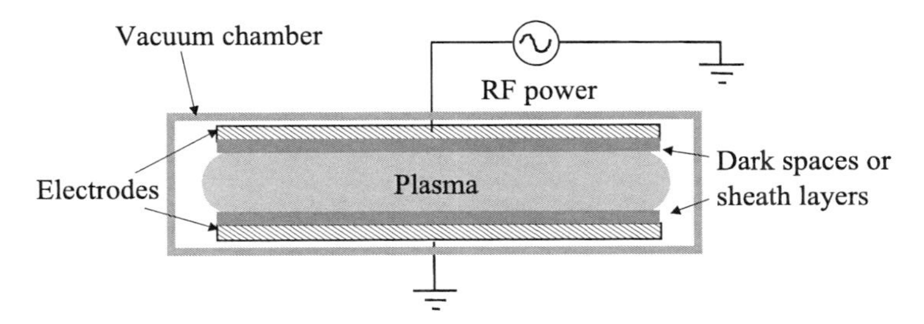

Figure 7.1 Schematic of a capacitively coupled plasma source.

molecule inside the chamber, generating an ion and another free electron. Due to the cascading of ionization collisions, the entire chamber fills quickly with equal numbers of electrons and ions—this is plasma.

In plasma, some electrons and ions are continually lost and consumed by collisions with electrodes and chamber walls, and also by recombination collisions between electrons and ions. When the generation rate of electrons through ionization collisions is equal to the loss rate of electrons, the plasma is said to be stabilized.

Other plasma sources such as dc plasma, ICP, ECR, and microwave (MW) remote plasma sources all have very similar plasma generating processes.

Question: Without the first free electron inside a chamber, there is no way to start plasma. Where does the first free electron come from?

Answer: It is usually generated by a cosmic ray. It also can be generated by heat (thermal electrons) or by native radioactive decay.

7.3 Collisions in Plasma

There are two kinds of collisions inside plasma: elastic and inelastic. Elastic collisions are the most frequent collisions in plasma. Since there is no energy exchange between colliding particles in an elastic collision, it is not important. There are many kinds of inelastic collisions happening simultaneously in plasma: collisions between electrons and neutrals, collisions between ions and neutrals, collisions between ions and ions, and collisions between electrons and ions. In plasma, any possible collision imaginable can happen. Since different collisions have varying probabilities of occurrence, the importance of each type of collision is not the same. For plasmas used in semiconductor processing, three collisions are very important: ionization, excitation-relaxation, and dissociation collisions.

7.3.1 Ionization

When an electron collides with an atom or molecule, it can transfer part of its energy to the orbital electron confined by the nucleus of that atom or molecule. If the orbital electron gains enough energy to break free from the constraint of the nucleus, it becomes a free electron (see Fig. 7.2). This process is called electron

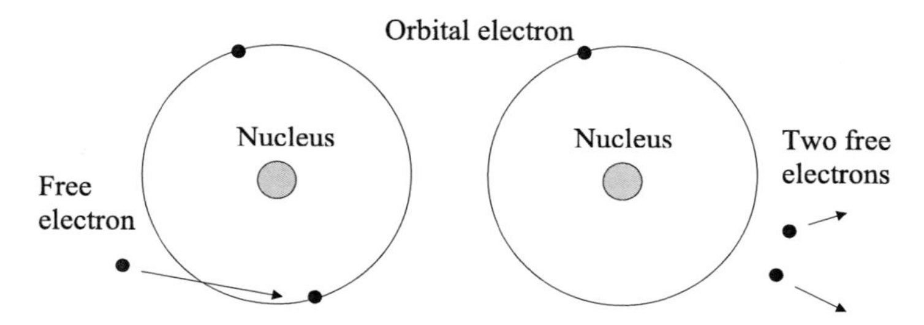

Figure 7.2 Ionization collision before and after electron impact.

impact ionization. Ionization collision can be expressed as

$$e^- + A \rightarrow A^+ + 2e^-,$$

where e^- represents an electron with a negative charge, A represents a neutral atom or molecule, and A^+ represents a positive ion. Ionization is very important because it generates and sustains the plasma.

7.3.2 Excitation-relaxation

Sometimes an orbital electron does not get enough energy from the impact electron to break free from the constraint of the nucleus. The transferred energy to the orbital electron from the collision will cause it to jump to a higher energy level of the orbit. This process is called excitation. An excitation collision can be expressed as

$$e^- + A \rightarrow A^* + e^-,$$

where A^* is the excited state of atom or molecule A, indicating that it has an electron in the higher energy level orbit.

The excited state is not stable and has a short lifetime. An electron in an excited orbit does not stay at that higher energy level for very long and falls back to the orbit with the lowest possible energy level, or ground state. This process is called relaxation. The excited atom or molecule quickly relaxes back to its ground state and releases the extra energy it gained from the electron impact in the form of a photon, which is the light emission. The relaxation process can be expressed as

$$A^* \rightarrow A + h\nu \text{ (photons)},$$

where $h\nu$ is the energy of the photon, h is Planck's constant, and ν is the frequency of the light emission, which determines the color of light emitted from the plasma. Different atoms or molecules have different orbital structures and energy levels; therefore, the light emission frequencies are different. This is why different gases glow in various colors within the plasma. The glow of oxygen is grayish-blue, nitrogen is pink, neon light is red, a fluorine glow is orange-red, etc.

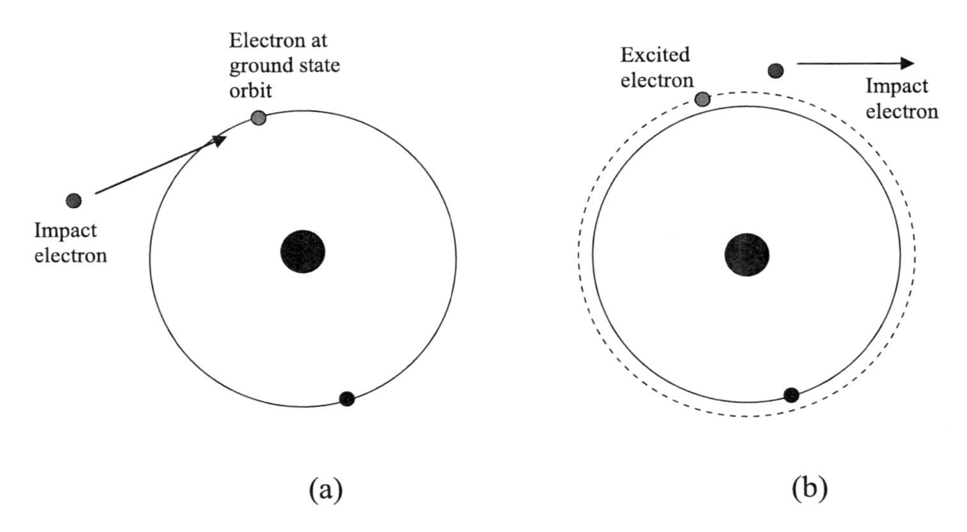

Figure 7.3 (a) Before and (b) after excitation collision.

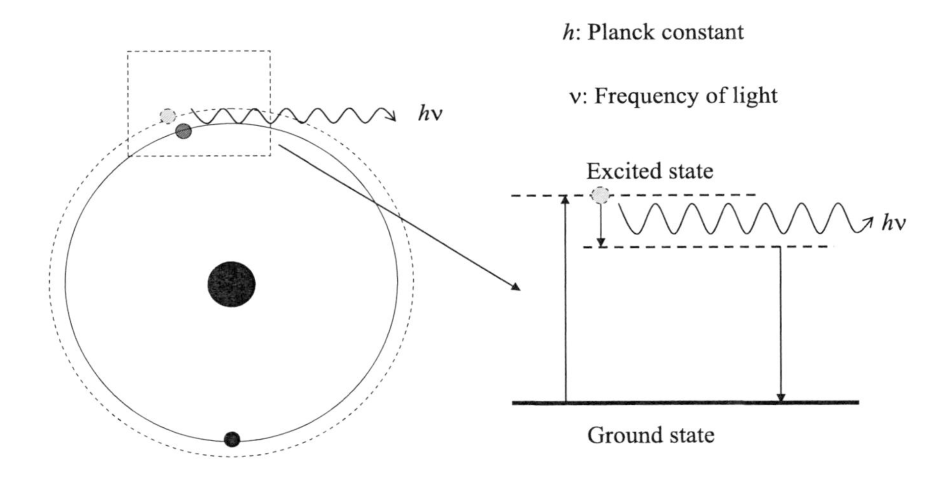

Figure 7.4 Relaxation process.

The excitation-relaxation process is illustrated in Figs. 7.3 and 7.4. Detection of a change in light emission is widely used in semiconductor processes to determine the endpoints of both etch and CVD chamber cleaning processes. This is covered in greater detail in Chapters 9 and 10.

7.3.3 Dissociation

When an electron collides with a molecule, it can break the chemical bond and generate free radicals if the energy transferred by the impact to the molecule is higher than the molecule's bonding energy. This is called a dissociation collision and can be expressed as

$$e^- + AB \rightarrow A + B + e^-,$$

where AB is a molecule, and A and B are free radicals generated by a dissociation collision. Free radicals are molecular fragments with at least one unpaired electron. This makes them chemically very reactive, since they have a strong tendency to grab an electron from other atoms or molecules to form a stable molecule. Free radicals can enhance chemical reaction in both etch and CVD processes. Figure 7.5 illustrates the dissociation collision process.

For example, in both oxide etch and CVD chamber cleaning processes, fluorocarbon gas such as CF_4 is used in plasma to generate free fluorine radical F:

$$e^- + CF_4 \rightarrow CF_3 + F + e^-.$$

In the PECVD oxide process, silicon precursor silane (SiH_4) and oxygen precursor nitrous oxide (N_2O, also known as laughing gas) are used to generate free radicals:

$$e^- + SiH_4 \rightarrow SiH_2 + 2H + e^-,$$
$$e^- + N_2O \rightarrow N_2 + O + e^-.$$

Free radicals such as F, SiH_2, and O are chemically very reactive. This is why plasma can enhance chemical reactions in both CVD and etch processes.

Question: Why are dissociation collisions unimportant in aluminum and copper sputtering processes?

Answer: Aluminum and copper sputtering processes only use argon, which is a noble gas. Unlike other gases, noble gases exist in the form of atoms instead of molecules. Thus, there is no dissociation process in argon plasma.

Question: Is there any dissociation collision in plasma sputtering PVD processes?

Answer: Yes. Both argon (Ar) and nitrogen (N_2) are used in titanium nitride (TiN) sputtering deposition processes. In plasma, N_2 is dissociated to generate a free radical N, which reacts with titanium or tantalum to from TiN

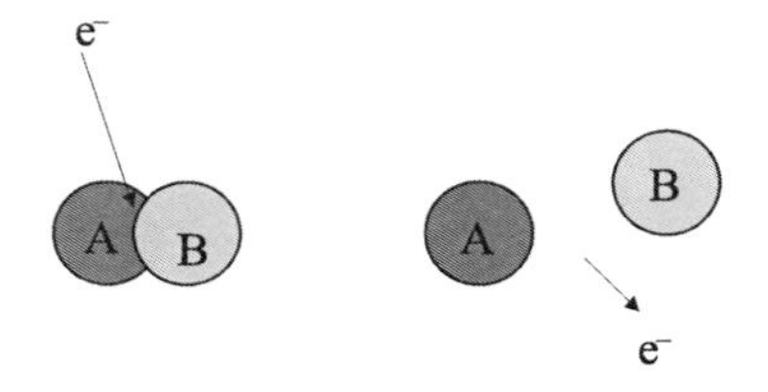

Figure 7.5 Dissociation collision.

on the target surface. Ar^+ ions sputter TiN molecules from the target surface and deposit them on a wafer surface. Sputtering deposition of tantalum nitride (TaN) is very similar to the deposition of TiN.

7.3.4 Other collisions

Other collisions inside plasma, such as recombination, charge-exchange, pitch-angle scattering, and neutral-neutral collisions, etc., are not important in PECVD and plasma etch processes. Some collisions are the combination of two or more different kinds of collisions.

Table 7.1 shows possible collision processes in PECVD silane plasma. It can be seen that some are dissociation, some are a combination of dissociation and excitation, and some are a combination of dissociation and ionization collisions.

Question: Which one of the collisions in Table 7.1 is most likely to happen? Why?

Answer: The collision that requires the least energy is the most likely to happen. It is much easier for an electron to fall to a lower energy level than climb to a higher one. Under conditions of field strength, pressure, and temperature, an electron is more likely to be accelerated for a short distance without a collision to attain enough energy required for that first collision to occur (2.2 eV). It is least likely for an electron to travel a much longer distance without a collision to gain enough energy (15.3 eV) needed for the last reaction in Table 7.1.

7.4 Plasma Parameters

The main plasma parameters are mean free path (MFP), thermal velocity, magnetic field, and the Boltzmann distribution.

Table 7.1 Possible plasma collision processes.

Collisions	Byproducts	Energy of formation
$e^- + SiH_4 \rightarrow$	$SiH_2 + H_2 + e^-$	2.2 eV
	$SiH_3 + H + e^-$	4.0 eV
	$Si + 2H_2 + e^-$	4.2 eV
	$SiH + H_2 + H + e^-$	5.7 eV
	$SiH_2^* + 2H + e^-$	8.9 eV
	$Si^* + 2H_2 + e^-$	9.5 eV
	$SiH_2^+ + H_2 + 2e^-$	11.9 eV
	$SiH_3^+ + H + 2e^-$	12.32 eV
	$Si^+ + 2H_2 + 2e^-$	13.6 eV
	$SiH^+ + H_2 + H + 2e^-$	15.3 eV

7.4.1 Mean free path

The MFP is defined as the average distance a particle can travel before it collides with another particle. MFP (or λ) can be expressed by the following equation:

$$\lambda = \frac{1}{\sqrt{2}n\sigma}, \tag{7.1}$$

where n is particle density, and σ is the collision cross section. Higher particle density causes more collisions, which shorten the MFP. Large particles collide more easily with other particles and therefore have shorter MFPs. As shown in Eq. (7.1), MFP is determined by chamber pressure, which determines particle density. Gas (or gases) in the chamber can also affect MFP because different gas molecules have different sizes or cross-sectional areas.

Question: What is the MFP for a molecule with diameter 3 Å and density 3.5×10^{16} cm^{-3} (the density of an ideal gas at 1 torr or 1 mm Hg)?
Answer: Based on Eq. (7.1), $\lambda = 1/[\sqrt{2} \times 3.5 \times 10^{16} \times \pi \times (3 \times 10^{-8}/2)^2] =$ 0.029 cm.

Figure 7.6 shows that the MFP is shorter when the gas density is higher [Fig. 7.6(a)] and longer when gas density is lower [Fig. 7.6(b)]. Large particles have a larger cross section and sweep through more space. Therefore, they are more likely to collide with other particles and have shorter MFPs than average sized or smaller particles. Changing pressure changes particle density, therefore affecting MFP:

$$\lambda \propto 1/p.$$

As pressure reduces, the MFP increases; as pressure decreases, particle density reduces, and therefore the frequency of collision decreases. An approximation of the MFP of a gas molecule in air is

$$\text{MFP (cm)} \approx 50/p \text{ (mtorr)}.$$

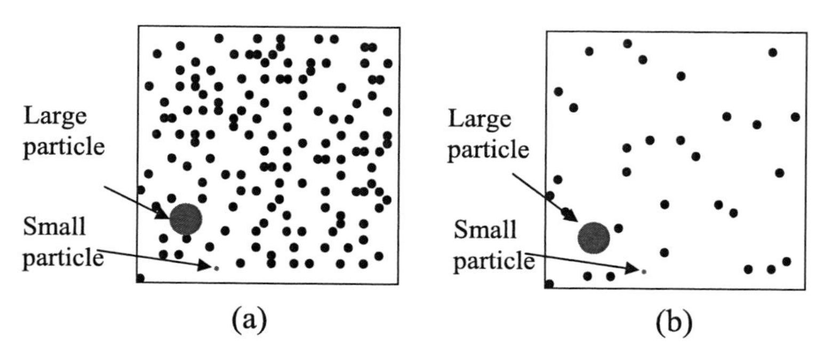

Figure 7.6 (a) High-pressure short MFP and (b) low-pressure long MFP.

For an electron, the MFP can be twice as long, since it has a much smaller size:

$$\lambda_e \text{ (cm)} \approx 100/p \text{ (mtorr)}.$$

PECVD processes normally operate at 1 to 10 torr; therefore, in PECVD chambers the electron MFP λ_e = 0.01 to 0.1 cm. Etch process pressures are much lower, from 3 mtorr to 300 mtorr. Therefore, in an etch chamber, electron MFP λ_e varies from 0.33 to 33 cm.

The MFP is one of the most important concepts of plasma. It can be controlled by chamber pressure, and it can affect processing results, sometimes significantly, especially for plasma etch processes. When pressure in a plasma processing chamber changes, the MFP changes. Likewise, ion bombardment energy and ion direction change, which can affect rate and profile in etch processes, and film stress in PECVD processes. Plasma shape will also change due to electron MFP changes. At higher pressure, plasma is more concentrated near the electrodes, while at lower pressure, plasma can spread throughout the chamber. Pressure can affect plasma uniformity and change the etch rate or deposition rate across a wafer.

Question: Why is a vacuum chamber necessary to generate stable plasma?

Answer: At atmospheric pressure (760 torr or 760 mm Hg), the MFP of an electron is very short. It is very difficult for an electron to acquire enough energy to ionize gas unless it is in an extremely strong electric field. This creates plasma in the form of arcing (lightning) instead of a steady-state glow discharge.

7.4.2 Thermal velocity

In plasma, electrons, ions, and neutrals constantly move due to external electric power and thermal movement. Electrons have the lightest weight and smallest size; therefore, they are more likely to pick up energy from an external power supply compared to ions and neutrals. In plasma, electrons always move faster than ions and neutrals.

The mass ratio of an electron compared to the mass of the lightest hydrogen ion is 1 to 1836. In PECVD, etch, and PVD processes, the most commonly used ions are oxygen, argon, chlorine, and fluorine, all of which are all significantly heavier than a hydrogen ion. Therefore, these ions are much heavier than an electron by at least a factor of 10,000. However, the electric force that provides energy to the plasma is the same for both electrons and ions, since it is only related to electrical charge and electric field:

$$F = qE,$$

where F is the force on a charged particle, q is the electrical charge (electron is negative and ion is positive), and E is the electric field generated by an external power supply such as rf, dc, or microwave. The acceleration of the charged particle can be expressed as

$$a = F/m = qE/m,$$

where m is the mass of the charged particle. Since the mass of electrons is less than a factor of 1/10,000 compared with ions, electrons can be accelerated more than 10,000 times faster than ions! This is similar to a motorcycle, which can accelerate much faster than an 18-wheel truck. It is not very hard to imagine how fast a motorcycle would start up using a powerful engine from an 18 wheeler, or to visualize how slow an 18 wheeler powered by a motorcycle engine would creep away from a traffic stop.

Most etch and CVD plasma sources use rf power. Radio frequency power generates a varying electric field, which changes directions rapidly. Electrons can be accelerated rapidly and start collision processes such as ionization, excitation, and dissociation in the positive cycle of an rf field, and repeat those processes in the negative cycle. Because ions are too heavy to respond quickly enough to the varying electric field, electrons absorb most of the rf energy due to their light weight and quick response. This process is very similar to two vehicles traveling on a road with stop signs at every intersection. The motorcycle can start up and slow down very quickly. The big truck starts very slowly and has to slow down gradually. It is easy to see that the average speed of the motorcycle would be much higher than the big truck on this kind of road.

When subjected to a lower rf power, ions pick up a little more energy than they would with a higher rf power. Lower rf power allows more time for ions to respond; therefore, it can accelerate ions to a higher energy and provide more energy for ion bombardment.

In any case, electrons always have a much higher temperature than ions or neutrals in plasma. Thermal velocity can be expressed as

$$v = (kT/m)^{1/2}, \tag{7.2}$$

where $k = 1.38 \times 10^{-23}$ J/K is the Boltzmann constant, T is temperature, and m is the mass of the particle. For plasma generated by rf power within two parallel electrodes (or capacitively coupled plasma), electron temperature T_e is about 2 eV, where one electron volt (1 eV) is equivalent to 11,594 K or 11,321 °C. Based on Eq. (7.2), electron thermal velocity can be calculated as

$$v_e \approx 4.19 \times 10^7 T_e^{1/2}\ (T \text{ in } eV) \approx 5.93 \times 10^7 \text{ cm/sec} = 1.33 \times 10^7 \text{ mph}.$$

An electron in plasma moves faster than an orbiting space shuttle.

In contrast, the thermal velocity of an argon ion (Ar^+) with $T_{Ar} \approx 0.05$ eV is $v_i = 3.46 \times 10^4$ cm/sec $= 774$ mph. The ions move at the speed of an airplane, much slower than the electrons.

7.4.3 Magnetic field

In a magnetic field, a charged particle experiences a magnetic force that can be expressed as

$$F = qv \times B, \tag{7.3}$$

where q is the charge of the particle, v is the velocity of the particle, and B is the magnetic field strength. Since a magnetic force is always perpendicular to particle velocity, a charged particle will spiral around the magnetic field line. This movement is called gyromotion.

The gyromotion of a charged particle in a magnetic field (see Fig. 7.7) is a very important property of plasma and has many applications in semiconductor processing. A number of capacitively coupled plasma etch chambers have magnetic coils to generate a magnetic field and cause electron gyromotion. This helps to generate and maintain a higher-density plasma at low pressure. One of the most commonly used high-density plasma sources is the ECR plasma source, which uses a magnetic field and microwave power source. When the microwave frequency is equal to the electron gyrating frequency, the microwave power resonantly couples to the electrons to generate high-density plasma at very low pressure.

The ion implanter is another processing tool that uses magnets. The strong magnetic field generated by a dc current in the analyzer magnetic coils of an ion implanter can bend the trajectories of the high-energy ions. Because ions with different charge/mass (q/m) ratios have different trajectories in a magnetic field, they exit the magnetic field at different places. This allows fabs to precisely select the desired ions and terminate the undesired ones.

The frequency of a charged particle cycling around a magnetic field line is called the gyrofrequency Ω and can be expressed as

$$\Omega = \frac{qB}{m}. \tag{7.4}$$

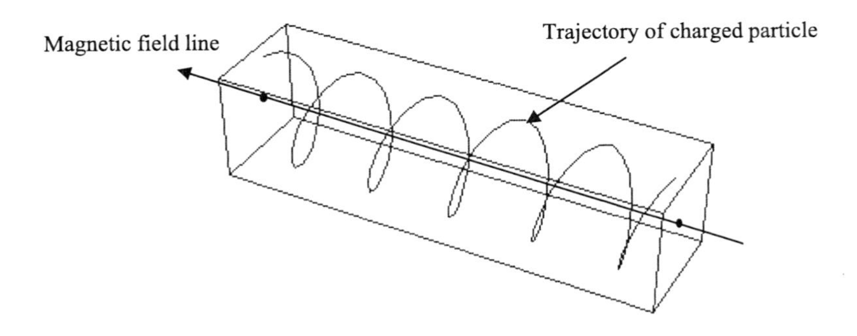

Figure 7.7 Gyromotion of a charged particle in a magnetic field.

For a charged particle with a fixed charge and specific mass, the gyrofrequency is determined only by magnetic field strength B. The gyrofrequency for an electron is Ω_e (MHz) = 2.8B (gauss or G).

The radius of the gyration is called the gyroradius ρ and can be expressed as

$$\rho = v_\perp/\Omega,$$

where $v_\perp$ is the particle speed perpendicular to the magnetic field line. For an electron, the gyroradius is ρ_e (cm) $= 2.38T_e^{1/2}/B$, where T_e is the electron temperature expressed in electron volts eV, and B is in gauss. Similarly, the ion gyroradius can be expressed by ρ_i (cm) $= 102\ (AT_i)^{1/2}/ZB$. Here, A is the atomic weight of the ion, and Z is the number of the ionization charge of the ion; both values are integer numbers. Ion mass $m_{ion} = Am_p$, where m_p is the mass of the proton, which is equal to 1.67×10^{-27} g. Ion charge q = Ze, where $e = 1.6 \times 10^{-19}$ C is the charge of an electron.

Question: In an argon sputter etch chamber, the electron temperature is $T_e \approx 2$ eV, argon ion temperature is $T_i \approx 0.05$ eV, and a magnetic field $B = 100$ G. For an argon ion, $A = 40$ and $Z = 1$. What are the electron and ion gyroradii?

Answer: Electron gyroradius $\rho_e = 2.38 \times 2^{1/2}/100 = 0.034$ cm; and argon ion (Ar^+) gyroradius $\rho_i = 102 \times (40 \times 0.05)^{1/2}/100 = 1.44$ cm.

Question: In an ion implanter with an analyzer magnetic field $B = 2000$ G and argon ion (Ar^+) energy $E_{Ar} = 200$ keV, what is its gyroradius?

Answer: $\rho_i = 102 \times (40 \times 200,000)^{1/2}/2000 = 144$ cm.

7.4.4 Boltzmann distribution

In thermal equilibrium plasma, both electron and ion energy follow the Boltzmann distribution, as shown in Fig. 7.8. The average electron energy in a capacitively coupled plasma source is about 2 to 3 eV. The ion energy in bulk plasma is mainly determined by the chamber temperature, ~200 to 400 °C, or 0.04 to 0.06 eV.

Figure 7.8 demonstrates that most electrons have energy around the average value, 2 to 3 eV. Very few electrons have energy high enough for ionization, which is approximately 15 eV. This explains why the ionization rate is so low in parallel plate plasma sources.

Question: Assume that the electron temperature in a plasma source averages 1 keV (~11,600,000 °C, the temperature in the core of the sun). What is the ionization rate of this plasma?

Answer: Almost 100%.

7.5 Ion Bombardment

Since electrons move much faster than ions, anything close to the plasma, including the chamber wall and electrodes, will be charged negatively as soon as the plasma

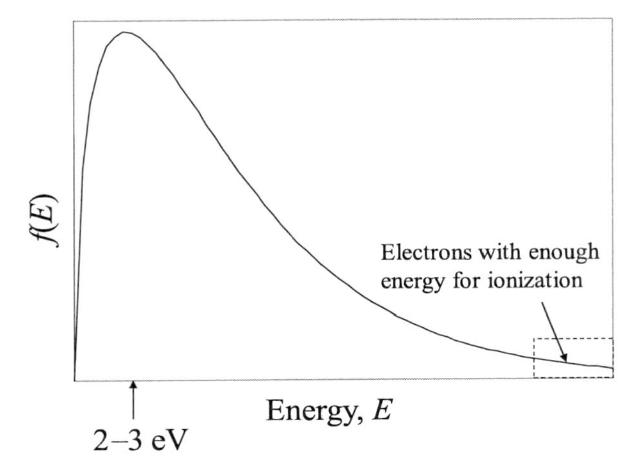

Figure 7.8 Electron energy distribution.

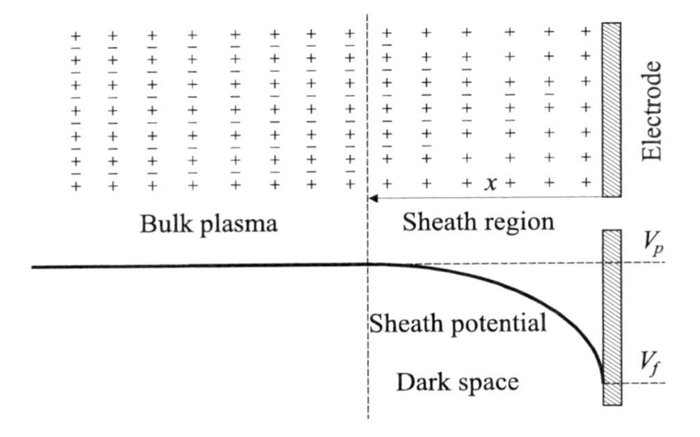

Figure 7.9 Sheath potential nears the plasma surface.

is initiated. The negatively charged electrodes will repel negative electrons and attract positive ions; therefore, in the vicinity of the electrodes, there are fewer electrons than ions.

The difference between the negative and positive charges creates an electric field in that region, called the sheath potential, as illustrated in Fig. 7.9. Light emission from this region is less intense than from the bulk plasma, since fewer electrons are present and fewer excitation-relaxation collisions occur; thus, a dark space can be observed near the electrodes.

The sheath potential accelerates ions toward the electrodes, causing ion bombardment. If a wafer is placed over an electrode, ions accelerated by the sheath potential will bombard its surface.

Ion bombardment is a very important property of plasma. Any conducting materials close to plasma will be subject to ion bombardment. Ion bombardment can affect etch rate, selectivity, and profiles in plasma etch processes. It also can affect deposition rate and film stress in PECVD and HDP-CVD processes.

Ion bombardment has two parameters. One is ion energy, the other is ion flux. Ion energy is related to external power supply, chamber pressure, electrode spacing, and processing gases. Ion flux is related to plasma density; it is also determined by external power supply, chamber pressure, electrode spacing, and processing gases.

In rf plasma systems, rf frequency also affects ion energy. At a high frequency, such as 13.56 MHz, electrons pick up the majority of the energy, and ions are left "cold." At a much lower frequency, such as 350 kHz, although electrons still acquire most of the energy, ions have a better chance to pick up some energy from the rf power due to the slower change rate of the electric field. By using the earlier analogy of the two vehicles, this would be akin to increasing the distance between stop signs from one every block to one every kilometer. In this case, the motorcycle (electron) still has the advantage at an average speed because of its fast acceleration and quick stops, but the average speed of the truck (ion) will be significantly increased because this scenario enables the big truck to reach and sustain a high speed.

Question: Why is 13.56 MHz the most commonly used frequency for rf systems?

Answer: Because the government regulates applications of rfs under an international treaty, it is necessary to regulate the applications of rfs to avoid interference with different applications. If some rf interferes with an air traffic control radio signal, it could have disastrous consequences. Industrial manufacturing, medical applications, and scientific research have 13.56 MHz assigned to them. Such rf generators have been commercially available for a long time and are more cost effective than other rf sources such as 2 MHz, 1.8 MHz, etc.

7.6 Direct-Current Bias

In an rf system, the potential of an rf hot electrode changes rapidly. Plasma potential also changes very quickly. Because electrons move faster than ions, anything close to the plasma will be charged negatively; plasma always has a higher potential than anything close to it.

A plasma potential curve (solid curve) is shown in Fig. 7.10. When the rf potential (dashed curve) is in its positive cycle, plasma potential is higher than the rf hot electrode potential. When the rf potential goes negative, plasma potential will not follow it to the negative side. It must maintain a higher potential than the ground potential. When the rf potential returns to positive again, plasma potential rises again. Thus, the plasma potential cycles entirely in a range higher than the ground potential. Therefore, on average, there is a dc potential difference maintained between bulk plasma and electrodes. This difference is called the dc bias. The energy of a bombarding ion is determined by the dc bias, which is about 10 to 20 V in a PECVD chamber with parallel plate electrodes. The dc bias mainly depends on rf power; it is also related to the chamber pressure and processing gas

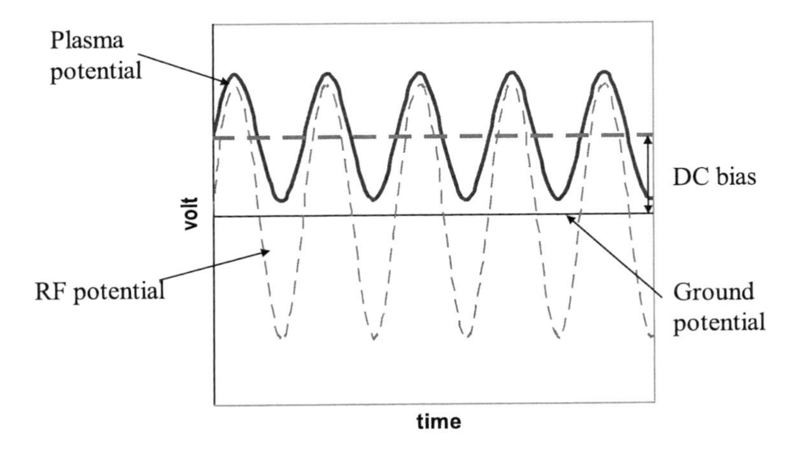

Figure 7.10 dc bias and its relation to rf power.

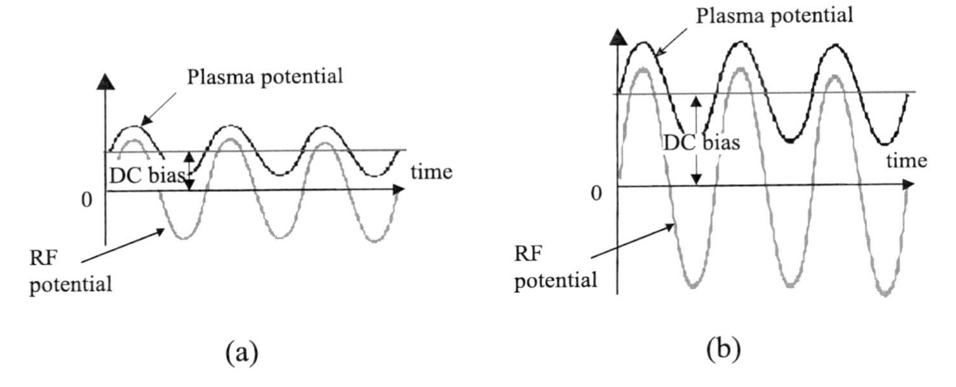

Figure 7.11 dc biases with (a) lower and (b) higher rf power.

species. When rf power increases, the amplitude of the rf potential will increase correspondingly, and the plasma potential and dc bias increases (see Fig. 7.11).

Plasma potential depends on the rf power, pressure, and spacing between the electrodes. For a symmetrical system with two electrodes having the same area (shown in Fig. 7.12), the dc bias is about 10 to 20 V. Most PECVD systems have this type of structure. Since rf power also affects plasma density, a capacitively coupled (parallel plate) plasma source cannot independently control ion energy and flux.

Figure 7.13 shows the voltage in an asymmetric electrode plasma source. In this case, the two electrodes have different areas. Current continuity induces a self-bias when the negative bias voltage builds on the smaller electrode. Sheath voltages depend on the ratio of the electrode areas, as shown in Fig. 7.14. In an ideal case (no collisions in the sheath region), voltages and electrode areas have the following relation:

$$V_1/V_2 = (A_2/A_1)^4, \tag{7.5}$$

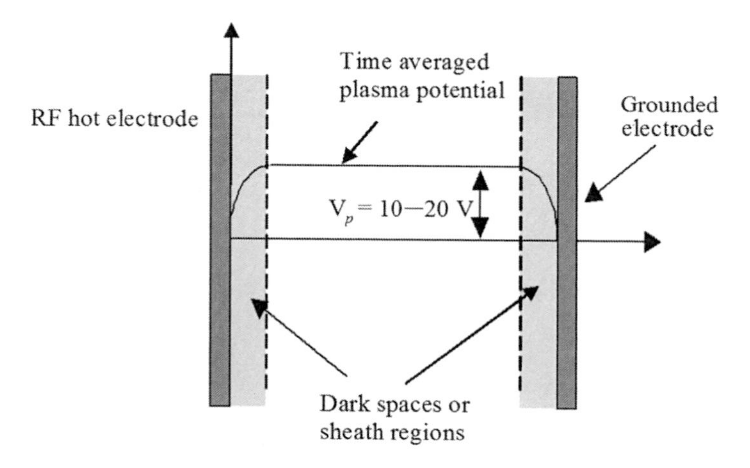

Figure 7.12 Plasma potential in a symmetric electrode system.

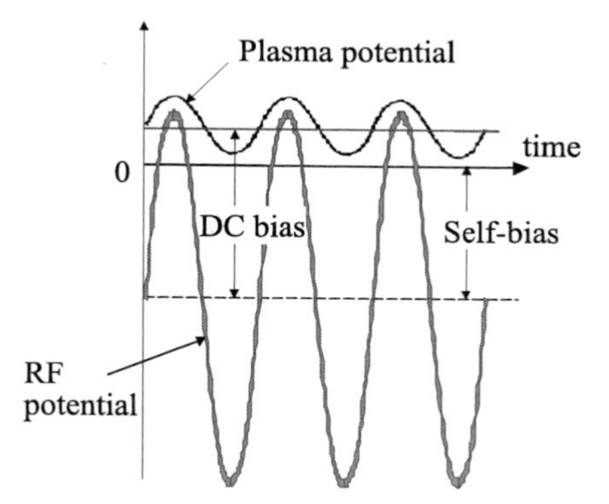

Figure 7.13 dc potential in an asymmetric electrode rf system.

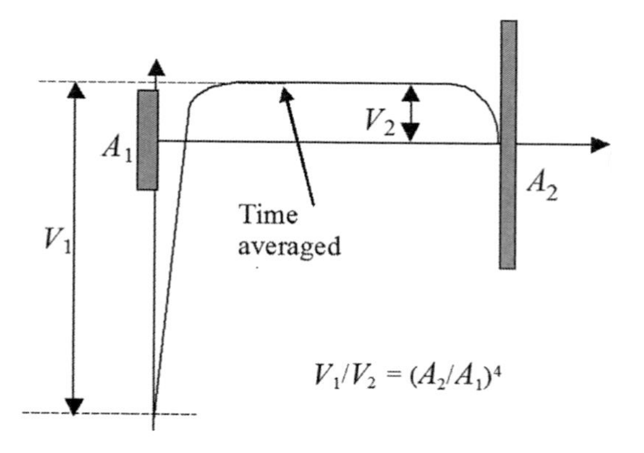

Figure 7.14 Plasma potential in an asymmetric electrode system.

where V_1 is the dc bias, which is the potential difference between the bulk plasma and the rf hot electrode; and V_2 is the plasma potential (plasma to ground). The self-bias equals $V_1 - V_2$ (rf hot to ground). The smaller electrode has a larger sheath voltage and thus receives more energetic ion bombardment.

Most etch chambers use asymmetric electrodes, and a wafer is always placed on the smaller rf hot electrode to receive more energetic ion bombardment during the entire rf cycle. Since etch processes require more ion bombardment, self-bias on the smaller electrode can help increase ion bombardment energy. For an rf system with symmetric parallel electrodes, the amount of ion bombardment on both electrodes is about the same. Therefore, if the wafer receives heavy ion bombardment, the other electrode will receive the same amount of bombardment, which is very undesirable, as this can cause particle contamination and short lifetimes of the processing chamber parts.

Question: If an electrode area ratio is 1:3, what is the difference between the dc bias and the self-bias, compared to the dc bias?

Answer: The dc bias is V_1, and the self-bias is $V_1 - V_2$; therefore, the difference is:

$$[V_1 - (V_1 - V_2)]/V_1 = V_2/V_1 = (A_1/A_2)^4 = (1/3)^4 = 1/81 = 1.23\%.$$

In this case, dc bias V_1 is significantly larger than the plasma potential V_2, and it is very close to the self-bias. Many people measure the potential difference between the rf hot electrode and the grounded electrode and call the result the dc bias, when in fact it is the self-bias. Of course, they are close enough to ignore plasma potential V_2, which is very hard to measure. Ion bombardment energy is determined by dc bias.

Question: Can a fine metal probe be inserted into the plasma to measure plasma potential V_2?

Answer: Yes, it can. However, the metal probe will be negatively charged by the fast-moving electrons and form a sheath potential between its surface and the bulk plasma when it comes close to the plasma. Therefore, measurement results are determined by theoretical models of the sheath potential, which have not been fully developed yet.

7.7 Advantage of Plasma Processes

Several qualities of plasma are very helpful in semiconductor processes. Ion bombardment is crucial for sputtering deposition, etch, and CVD film stress control. Free radicals from disassociated electron molecules greatly enhance chemical reaction rates in both CVD and etch processes. The glows emitted from plasma due to their excitation and relaxation reliably measure optical endpoints for plasma etch and plasma chamber cleaning processes.

7.7.1 Chemical vapor deposition

The major benefits of using plasma in CVD processes are:

- high deposition rates at relatively lower temperatures
- control of deposited film stress with ion bombardment
- process-chamber dry cleaning with fluorine-based plasma
- excellent gap-fill capabilities with a high-density plasma source.

Free radicals generated in plasma by dissociation processes can dramatically increase chemical reaction rates, thus significantly increasing deposition rates, especially at the relatively low temperatures required for interconnection processes after the first aluminum metallization.

7.7.1.1 Plasma-enhanced chemical vapor deposition

Comparison of silane-based silicon oxide plasma-enhanced CVD (PECVD) and low-pressure CVD (LPCVD) processes is a good example for demonstrating the advantages of using PECVD in lower-temperature (<450 °C) processes. Interlayer dielectric (ILD) deposition requires lower temperatures after the first layer of metallization.

Some dissociation collisions in the silane oxide PECVD process are

$$e^- + SiH_4 \rightarrow SiH_2 + 2H + e^-$$
$$e^- + N_2O \rightarrow N_2 + O + e^-.$$

Both SiH_2 and O are free radicals with two unpaired electrons; they are very reactive. On a heated wafer surface, they react very quickly and form silicon dioxide:

$$SiH_2 + 2O \rightarrow SiO_2 + \text{other volatiles.}$$

In comparison, the LPCVD silane oxide deposition process uses SiH_4 and O_2. SiH_4 thermally dissociates to SiH_2 when it comes close to a heated wafer surface. SiH_2 then chemisorbs on the wafer surface, reacts with oxygen, and forms silicon dioxide:

$$SiH_4 \rightarrow SiH_2 + H_2,$$
$$SiH_2 + O_2 \rightarrow SiO_2 + \text{other volatiles.}$$

Without plasma, the chemical reaction rate of this CVD process at ~400 °C is very low, which causes a very low deposition rate. Thus, low-temperature LPCVD processes for ILD applications must be performed in a batch system to achieve reasonable throughput. Table 7.2 compares the properties of PECVD and LPCVD processes for ILD silane oxide deposition at 400 °C.

Table 7.2 Comparison of PECVD and LPCVD silane oxide processes.

Processes	LPCVD (150 mm)	PECVD (150 mm)
Chemical reaction	$SiH_4 + O_2 \rightarrow SiO_2 + \cdots$.	$SiH_4 + N_2O \rightarrow SiO_2 + \cdots$.
Process parameters	p = 3 torr, T = 400 °C	p = 3 torr, T = 400 °C, and **rf** = **180 W**
Deposition rate	**100 to 200 Å/min**	**≥8000 Å/min**
Processing systems	Batch system	Single-wafer system
Wafer-to-wafer uniformity	Difficult to control	Easier to control

7.7.1.2 Stress control

Film stress is the force at the interface of two layers caused by a mismatch of the two materials. Ion bombardment of PECVD can be used to control CVD film stress. For dielectric thin films, especially silicon oxide films, compressive stress is favored. When heated, silicon expands faster than silicon oxide. If the film stress is compressive at room temperature, when the wafer is heated up for the next process step, the substrate will expand faster and therefore relieve compressive stress in the oxide film. If the oxide film has tensile stress at room temperature, it becomes even more tensile when heated. High tensile stress can cause film cracking, and in extreme cases it can even break a wafer.

Ion bombardment can hammer molecules and pack them densely into the film, which tends to make film stress become more compressive. Increasing rf power can increase both ion bombardment energy and ion flux, an increase that normally makes PECVD film more compressive. One advantage of PECVD is that by controlling rf power, the film stress can be independently controlled without major effects on other deposition characteristics, such as deposition rate and film uniformity. This is discussed in detail in Chapter 10.

7.7.1.3 Chamber cleaning

During CVD processes, deposition occurs not only on the wafer surface, but also on the processing kits, chamber walls, etc. The film deposited on these parts must be periodically cleaned to maintain stable processing conditions and prevent particle contamination of the wafers. Most CVD chambers use fluorine-based chemicals for cleaning processes.

Fluorocarbon gases such as CF_4, C_2F_6, and C_3F_8 are commonly used to clean silicon oxide CVD chambers. In plasma, these gases are dissociated and release free fluorine radicals. The chemical reaction can be expressed as

$$e^- + CF_4 \rightarrow CF_3 + F + e^-$$
$$e^- + C_2F_6 \rightarrow C_2F_5 + F + e^-.$$

Atomic fluorine is one of the most reactive free radicals; with ion bombardment from plasma, it can react with silicon oxide rapidly and form the gaseous compound SiF_4, which can be easily pumped out of the processing chamber:

$$F + SiO_2 \rightarrow SiF_4 + O + \text{volatile}.$$

SF_6 and NF_3 are commonly used fluorine-source gases for tungsten CVD chamber cleaning. Fluorine radicals can react with tungsten to form volatile tungsten hexafluoride, WF_6, which can be evacuated from the chamber with a vacuum pump.

Plasma chamber cleaning processes can be automatically endpointed (to prevent overcleaning) by monitoring the characteristic fluorine light emissions from the plasma. This is further discussed in Chapter 10.

7.7.1.4 Gap fill

When gaps between metal lines shrink to 0.25 μm with an aspect ratio of 4:1, most CVD methods cannot fill them without voids. The best known method for filling such narrow gaps without voids is the HDP-CVD process, shown in Fig. 7.15. The HDP-CVD process is covered in Chapter 10.

7.7.2 Plasma etch

Compared to wet etch, the advantages of plasma etch are anisotropic etch profile, an automatic endpoint, and less chemical consumption. Plasma etch also has a reasonably high etch rate, good selectivity, and good etch uniformity.

7.7.2.1 Etch profile control

Before plasma etch processes were widely used in semiconductor manufacturing, most wafer fabs used wet chemical etching to achieve pattern transfer. However, a wet etch is an isotropic etch (it etches in all directions at the same rate). Undercutting caused by an isotropic etch limits its application when feature size is smaller than 3 μm.

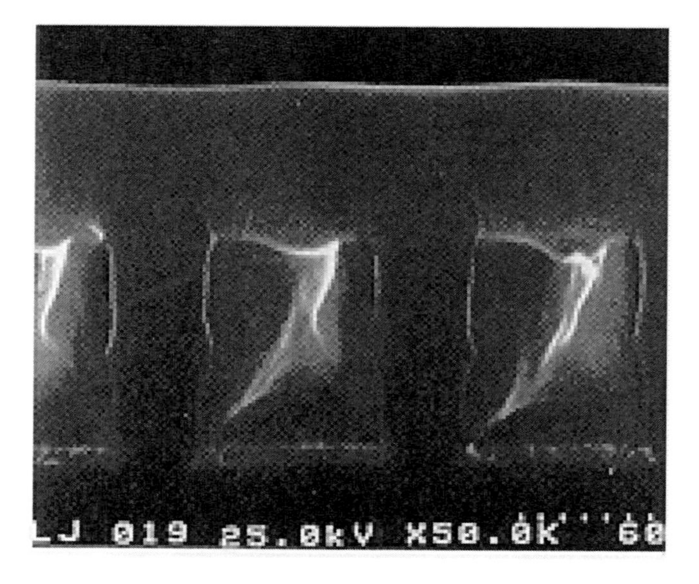

Figure 7.15 HDP-CVD oxide fills a 0.25-μm (aspect ratio 4:1) gap without voids.

Ions constantly bombard a wafer surface in plasma processes. With ion bombardment, by either a lattice-damaging mechanism or sidewall blocking mechanism, plasma etch processes can achieve an anisotropic etch profile. By reducing the etch processing pressure, ion MFP can be increased. This, in turn reduces ion collisional scattering and helps achieve better profile control.

7.7.2.2 Etch rate and selectivity

Ion bombardment associated with plasma helps break chemical bonds between surface atoms. These atoms are exposed to free radicals generated in the plasma. The combination of physical ion bombardment and chemical free radicals dramatically enhances the chemical reactions needed for etch processes. The etch rate and selectivity of the process is determined by the processing requirement. Since both ion bombardment and free radicals play an important role in etch processes, and rf power controls both ion bombardment and free radicals, rf power is the most important variable controlling etch rate. Increasing rf power can significantly increase etch rate. However, as discussed in detail in Chapter 9, this increase can affect etch selectivity.

7.7.2.3 Endpoint

Without plasma, etch endpoint is determined by time, or by an operator's visual inspection. In a plasma process, when the etch process has removed the material that requires etching and starts to remove the material underneath (the endpoint), chemical components in the plasma change due to the change in etch byproducts, signified by a variation in light emission. By detecting the change in light emission with an optical sensor, the endpoint can be automatically determined. This is a very useful tool in plasma etching for IC production.

7.7.2.4 Chemical use

Compared to wet etching, plasma etching uses significantly smaller amounts of chemicals, thus reducing the costs for chemicals and waste disposal.

7.7.3 Sputtering deposition

Compared with evaporation-deposited metal thin films, metal thin films deposited by plasma sputtering have a higher film quality, fewer impurities, and higher conductivity. They also have better uniformity, process control, and process integration capabilities. It is easier to deposit metal alloy films with sputtering deposition than with evaporation processes.

7.8 Plasma-Enhanced Chemical Vapor Deposition and Plasma Etch Chambers

7.8.1 Processing differences

CVD processes add materials to substrate surfaces, and etch processes remove materials from them. Therefore, etch processes operate at a lower pressure. Low pressure and high pumping rates help increase ion bombardment and remove etch byproducts from an etch chamber. PECVD processes normally operate at a higher pressure (from 1 to 10 torr) than etch processes (from 30 to 300 mtorr).

7.8.2 Chemical vapor deposition chamber design

PECVD processes deposit thin films on wafer surfaces. Ion bombardment is very useful in controlling film stress. For PECVD chambers, an rf hot electrode (called a face plate, showerhead, etc.) has about the same area as the grounded electrode where the wafer is placed, as illustrated in Fig. 7.16. The electrode has very little self-bias. Ion bombardment energy is about 10 to 20 eV, mainly determined by the rf power.

7.8.3 Etch chamber design

If an etch system has both rf hot and grounded electrodes of the same size, both electrodes will receive approximately the same amount of ion bombardment. Etch processes depend heavily on ion bombardment to remove materials from the wafer surface. Ion bombardment can physically dislodge material off of the substrate surface and, more importantly, break chemical bonds so molecules on the surface react easily with etchant radicals. The easiest way to increase ion bombardment on a wafer is to increase the rf power. This increases both ion bombardment energy and flux. Unfortunately, it also increases ion bombardment on other electrodes and shortens lifetimes because of particle contamination.

By designing an rf hot electrode (called a chuck or cathode) to be smaller than a grounded electrode (chamber lid), and taking advantage of the self-bias, the plasma potential of the wafer side will be much higher than the chamber lid side (see Fig. 7.17). Thus, the most energetic ion bombardment occurs on the wafer side, while the chamber lid receives minimal bombardment. Ion bombardment energy on the wafer side varies from 200 to 1000 eV, and on the lid side is about 10 to 20 eV. Ion bombardment energy is mainly determined by rf power. It also relates

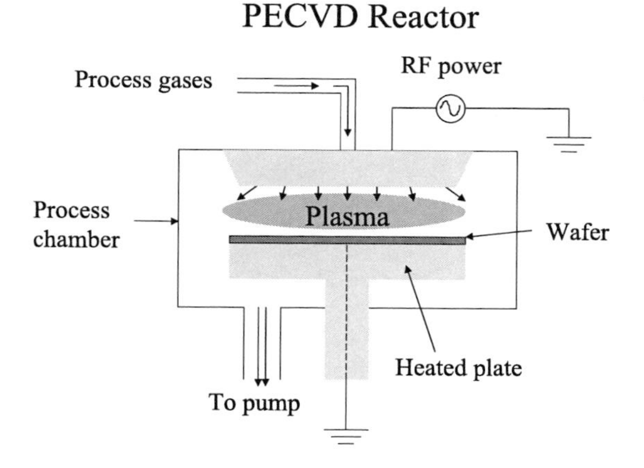

Figure 7.16 Schematic of a PECVD chamber.

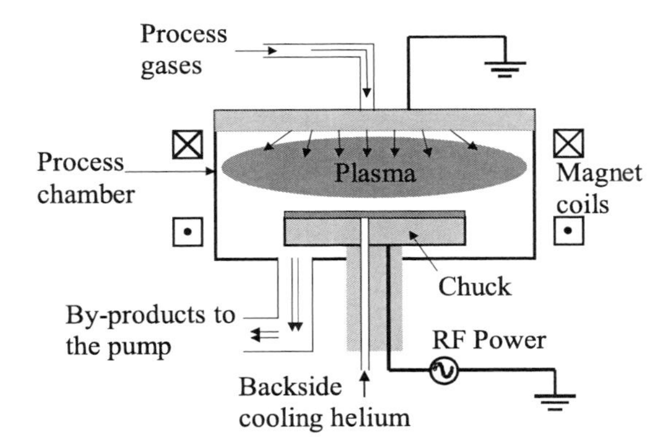

Figure 7.17 Schematic of a plasma etch chamber.

to chamber pressure, electrode spacing, processing gases, and applied magnetic field.

Plasma etch chambers operate at much lower pressure than PECVD chambers. At low pressure, the electron MFP is long. If the MFP is at the same order of magnitude as the electrode spacing or chamber dimension (~10 cm), it will not take many collisions before electrons are lost by hitting electrodes or chamber walls. Since ionization collisions are necessary for generating and maintaining plasma, it is difficult to generate plasma when pressure is so low.

Magnetic fields cause electrons to move in a spiral path. This helical path forces electrons to travel much longer distances and increases the chance of ionization collisions. A magnetic field can help generate and maintain plasma at lower pressure (<100 mtorr). Increasing the magnetic field can effectively increase plasma density, especially at low pressure. Since it increases electron density near the electrode surface, increasing the magnetic field can reduce dc bias.

Heavy ion bombardment generates a large amount of heat and can raise wafer temperature if the wafer is not properly cooled. For patterned etch, wafers are coated with a thin layer of photoresist as a pattern mask. The photoresist will reticulate if wafer temperature is above 150 °C. Therefore, a chamber designed for patterned etch requires a cooling system to prevent photoresist reticulation. Since chemical etch rates are sensitive to wafer temperature, even some blanket etch processing chambers [such as spin-on-glass (SOG) etchback chambers] need a wafer cooling system to regulate wafer temperature and control etch rate. Because etch processes require low operation pressure, and low pressure is not conducive for heat transfer, pressurized helium is normally introduced on the backside of a wafer to transfer heat from the wafer to the water-cooled pedestal (also called a chuck, cathode, etc.). An electrostatic chuck (E-chuck) is needed to prevent the wafer from being blown away from the pedestal due to the high backside pressure. Helium, which has very high thermal conductivity (second only to hydrogen), can provide a conductive path for heat transfer from the backside of the wafer to the water-cooled pedestal. For some old etch chambers, clamp rings were used to hold the wafer onto the pedestal. Because an E-chuck has much better performance with plasma uniformity and particle contamination, it replaced the clamp ring in most advanced plasma systems.

Argon-sputtering etch chambers were widely used in dielectric thin-film processes for tapering the corners of gaps prior to gap filling and thin-film surface planarization. Since the sputtering etch rate is relatively insensitive to wafer temperature, it does not need the helium backside cooling system with a clamp ring or E-chuck.

7.9 Remote Plasma Processes

Some processes require free radicals to enhance chemical reactions but do not benefit from ion bombardment because it can cause plasma-induced damage. Remote plasma systems have been developed to meet these processing demands.

Figure 7.18 shows a remote plasma system. Plasma is created in a remote chamber by either microwave or rf power. Free radicals generated in the plasma flow into the processing chamber for either etch or deposition processes.

7.9.1 Photoresist strip

A remote plasma photoresist strip process uses O_2 and H_2O to remove photoresist right after etching. A remote plasma photoresist strip system (as shown in Fig. 7.19) can be easily integrated within an etch system. A wafer can remain in the same mainframe to perform in-situ etch/strip processing in sequence. This can improve both processing throughput and yield, since photoresist and residue etchants are stripped before the wafer is exposed to the atmosphere, where moisture in the air can react with residue etchants and cause corrosion on the wafer.

7.9.2 Remote plasma etch

Some etch processes, such as a local oxidation of silicon (LOCOS) nitride strip, wine glass contact hole, etc., do not require an anisotropic etch; therefore, ion

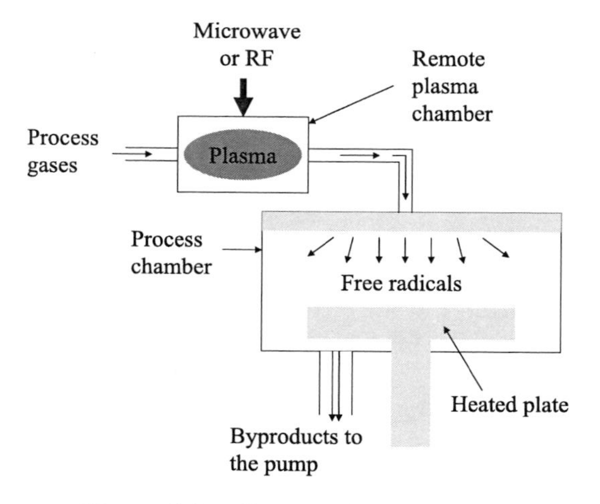

Figure 7.18 Remote plasma system.

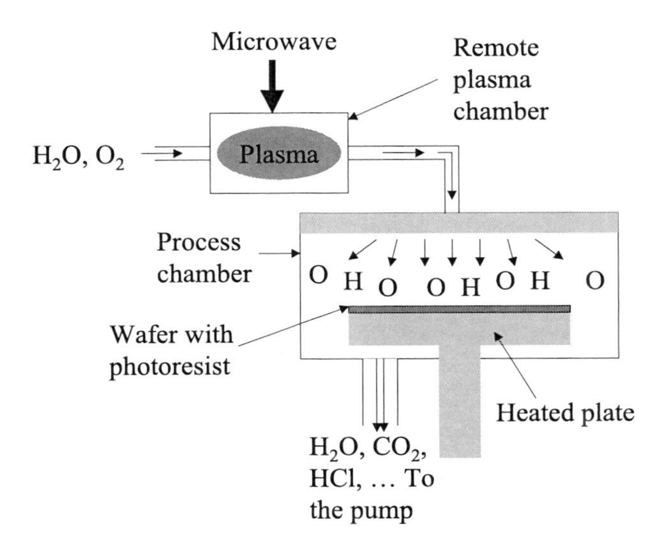

Figure 7.19 Remote plasma photoresist strip system.

bombardment is not necessary in these cases. Remote plasma etch systems are designed as dry chemical etching systems that can compete with wet etching processes for these applications. There was a trend toward replacing all wet processes in IC fabs with dry processes; however, this improvement proved to be impossible because of the wide application of CMP in advanced IC chip manufacturing processes.

7.9.3 Remote plasma cleaning

Since plasma in a processing chamber always generates ion bombardment along with free radicals, it can damage the parts inside a chamber, thus increasing cost of ownership. Another issue is that the most commonly used gases for CVD chamber cleaning processes are perfluorocarbon (PFC) gases, such as CF_4, C_2F_6, and C_3F_8. Because PFC gases contribute to global warming and ozone depletion, there is a strong political will to limit the use of these gases. Remote plasma cleaning was developed to solve these problems.

A remote plasma source uses microwave power to create a stable high-density plasma in a small cavity upstream from the processing chamber. Free radicals generated by the plasma flow into the processing chamber and react with deposited films, cleaning the chamber (shown in Fig. 7.20).

The most commonly used gas for remote plasma cleaning is NF_3. In microwave plasma, more than 99% of NF_3 dissociates. In comparison, less than 10% of carbon tetrafluoride or tetrafluoromethane (CF_4) dissociates in rf plasma. By using NF_3 microwave remote plasma cleaning processes, global warming PFC gases released from the semiconductor industry could be cut by more than 50%, and processing kit lifetimes could be significantly prolonged.

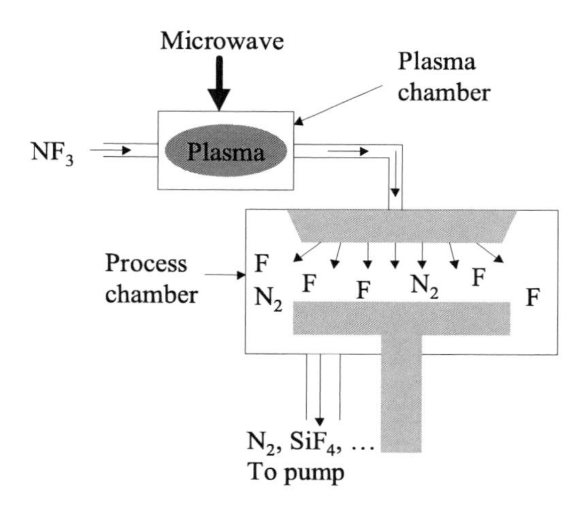

Figure 7.20 Remote plasma cleaning process.

7.9.4 Remote plasma chemical vapor deposition

Much research and development effort has been made to apply remote plasma CVD (RPCVD) processes for deposition of epitaxial silicon and epitaxial silicon germanium as semiconductor substrates, and silicon dioxide, silicon oxynitride, and silicon nitride for gate dielectric materials. Integrated with rapid thermal processes, RPCVD could also be used to deposit high-κ dielectrics such as TiO_2 and Ta_2O_5 for deep submicron devices in the future. RPCVD could also find applications for ILD0 barrier nitride deposition for sub-0.13-μm devices, since thermal budget limitations rule out the application of LPCVD nitride, and plasma-induced damage would limit the application of PECVD nitride, especially for large wafers.

7.10 High-Density Plasma

A plasma source that can generate HDP at low pressure (approximately a few millitorr) is highly desired for both etch and CVD processes. For etch processes, lower pressure gives longer ion MFP and less ion collisional scattering, enhancing etch profile control. HDP also provides more free radicals, which accelerate etching processes. For CVD processes, HDP can help achieve excellent gap-fill capabilities by in-situ, simultaneous deposition/etchback/deposition processing.

A conventional capacitively coupled plasma source cannot generate HDP. In fact, it is very difficult to generate plasma, even in a magnetic field when chamber pressure is a few millitorr. At that low pressure, electron MFP would be about the same or even longer than the electrode spacing, so there would not be enough ionization collisions. Therefore, different mechanisms are needed to generate HDP at very low pressure.

Another drawback of a capacitively coupled plasma source is that both ion flux and energy are directly related to rf power; therefore, this source cannot independently control them. To achieve better etch and CVD processing control as feature size continuously shrinks, a plasma source capable of independently controlling both ion flux and energy is needed.

There are two types of HDP sources commonly used in the semiconductor industry: the ICP (also called transformer-coupled plasma, or TCP) source, and the ECR plasma source. Both can generate HDP at a few millitorr, with the added ability of independently controlling ion bombardment flux and energy.

7.10.1 Inductively coupled plasma

The mechanism of an ICP source is very similar to a transformer, explaining why it is also called a TCP source. The inductive coils shown in Fig. 7.21(b) perform like the initial coils of a transformer. When an rf current flows into the coils, a changing magnetic field is generated, which in turn generates a changing electric field through inductive coupling, as shown in Fig. 7.21(a). The inductively coupled electric field accelerates electrons and causes ionization collisions. Since the electric field is in an angular direction, electrons are accelerated in an angular direction. This allows electrons to travel a long distance without collision with the chamber wall or electrode. That is why an ICP system can generate HDP at the low pressure of a few millitorr.

The ICP design is very popular in the semiconductor industry. Systems using ICP include HDP dielectric CVD systems; silicon, metal, and dielectric HDP etching systems; native oxide sputtering cleaning systems; and ionized metal plasma PVD systems.

A bias rf system is added to the ICP chamber to generate self-bias and control ion bombardment energy. Since ion bombardment from HDP generates a considerable amount of heat, a helium backside cooling system with an E-chuck is needed for better wafer temperature control. Figure 7.21(b) illustrates an ICP chamber. In the ICP system, ion flux, which is mainly determined by plasma density, is controlled by the rf power source; ion bombardment energy is mainly controlled by the bias rf power.

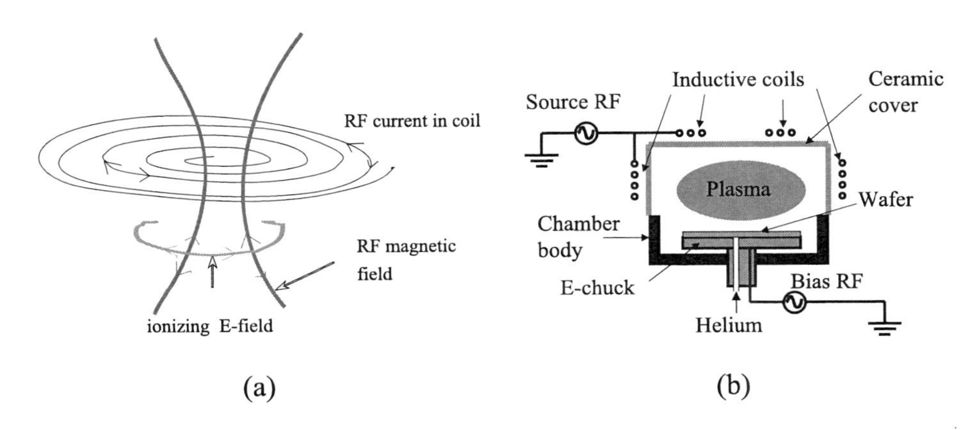

Figure 7.21 (a) Inductive coupling and (b) an ICP chamber.

7.10.2 Electron cyclotron resonance

Charged particles tend to rotate in a magnetic field. The frequency of the rotation, called gyrofrequency or cyclotron frequency, is determined by magnetic field strength. From Eq. (7.4), electron gyrofrequency is

$$\Omega_e \text{ (MHz)} = 2.8B \text{ (gauss)}.$$

In a magnetic field, when the frequency of the imposed microwave power equals the gyrofrequency of the electron, $\omega_{MW} = \Omega_e$, ECR occurs. Electrons can acquire energy from microwaves, growing more and more energized. Electrons then collide with other atoms or molecules, and the ionization collisions generate more electrons. These electrons are also resonant with the microwaves, gaining energy and creating even more electrons by ionization collisions. Since electrons are spiraling around the magnetic field line, as shown in Fig. 7.22(a), they cannot escape to the chamber wall or to the electrodes without many collisions, even if the MFP is longer than the chamber dimensions. This is how an ECR system can generate high-density plasma at low pressure.

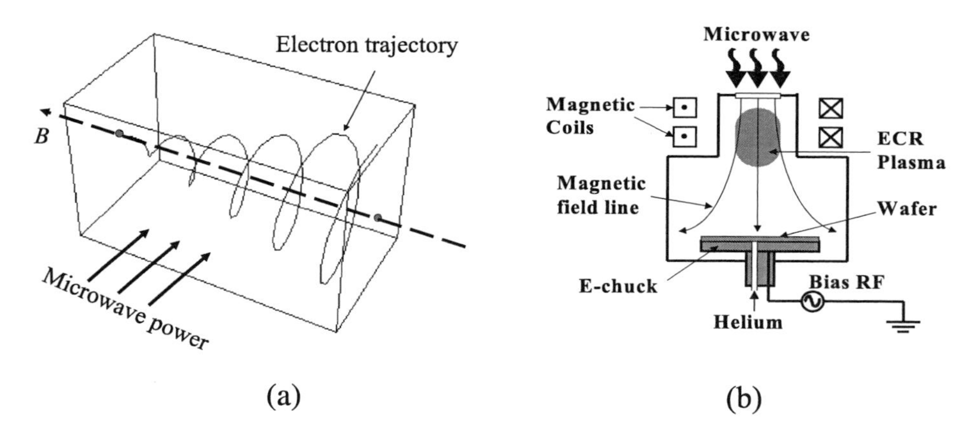

Figure 7.22 (a) ECR and (b) ECR chamber.

Like an ICP system, the ECR system also has a bias rf system to control ion bombardment energy, and an E-chuck with a helium backside cooling system to control wafer temperature [as shown in Fig. 7.22(b)]. Ion bombardment flux is mainly controlled by microwave power. One advantage of an ECR system is that, by changing the electric current in the magnetic coils, the position of resonance can be adjusted. Thus, process uniformity can be optimized by adjusting the current in the magnetic coils, which control plasma position.

7.11 Summary

- Plasma consists of ions, electrons, and neutrals.
- Three important collisions are ionization, excitation-relaxation, and disassociation.

- MFP is the average distance a particle can travel before it collides with another particle. It is inversely proportional to pressure.
- Free radicals from dissociation collisions enhance chemical reactions in CVD, etch, and dry cleaning processes.
- Plasma always has higher potential than the electrodes. Higher plasma potential causes ion bombardment.
- Increasing rf power increases both ion bombardment flux and ion bombardment energy in capacitively coupled plasmas.
- Low-frequency rf power gives ions more energy, leading to heavier ion bombardment.
- Etch processes need much more ion bombardment than PECVD processes. Etch chambers use a magnetic field to increase plasma density at low pressure.
- Capacitively coupled plasma sources cannot generate HDP.
- HDP at low pressure is desirable for both etch and CVD processes.
- ICP and ECR are the two most commonly used HDP sources.
- Both ICP and ECR plasma sources can independently control ion bombarding flux and energy.

7.12 Bibliography

B. Chapman, *Glow Discharge Process*, John Wiley & Sons, New York (1980).

F. F. Chen, *Introduction to Plasma Physics and Controlled Fusion, Volume 1: Plasma Physics*, 2nd ed., Plenum Press, New York (1984).

S. Dushman, *Scientific Foundations of Vacuum Technique*, J. M. Lafferty, Ed., John Wiley and Sons, New York (1962).

M. A. Lieberman and A. J. Lichtenberg *Principles of Plasma Discharges and Materials Processing*, John Wiley & Sons, New York (1994).

7.13 Review Questions

1. List the three components of plasma.
2. Which component of plasma moves the fastest?
3. True or false: the ionization rate of a conventional PECVD chamber is ~ 100%.
4. List three important collisions in plasma and explain why they are important.
5. How can a PECVD process achieve a high deposition rate at a lower temperature?
6. What is MFP? How does it change with pressure?
7. How does the dc bias change when rf power increases?
8. List the importance of plasma ion bombardment for etching, PECVD, and sputtering PVD processes.
9. What are the major differences between plasma etch and PECVD processing chambers?
10. In an etch chamber, the wafer normally sits on which electrode? Why?

11. Why does an etch chamber need a helium backside cooling system with an E-chuck?
12. What are the advantages of an E-chuck over a clamp ring?
13. When the etch rate goes awry in a plasma etch system, why do fabs always check the rf system first?
14. Can a capacitively coupled plasma source generate HDP?
15. List the two most commonly used HDP systems.

Chapter 8
Ion Implantation

Objectives

After finishing this chapter, the reader will be able to:

- list at least three dopants commonly used in IC chip fabrication
- identify at least three doped areas from a CMOS chip cross section
- describe the advantages of ion implantation over diffusion
- describe the major components of an ion implanter
- explain the channeling effect and list at least two ways to minimize this effect
- explain the relation of ion range to ion species and energy
- explain the requirements of postimplantation annealing
- identify safety hazards related to ion implantation processes.

8.1 Introduction

One of the most important properties of semiconductor materials is that their conductivity can be controlled by adding dopants. Semiconductor materials such as silicon, germanium, and III–VI compounds (e.g., gallium arsenate) are doped with either n- or p-type dopants in IC fabrication. There are two methods used to dope semiconductors: diffusion or ion implantation. Before 1970, a diffusion process was used in IC fabrication. Currently, doping processes are mainly performed by ion implantation.

Ion implantation is an adding process in which dopant atoms are forcefully added into a semiconductor substrate by means of energetic ion beam injection. It is the dominant doping method in the semiconductor industry and is commonly used for various doping processes in IC fabrication. Figure 8.1 shows the relationship of the ion implantation process to other processes in IC manufacturing flow.

8.1.1 Brief history

Pure single-crystal silicon has a high resistivity; the purer the crystal, the higher the resistivity. Its conductivity can be improved by adding dopants such as boron (B), phosphorus (P), arsenic (As), and antimony (Sb). Boron is a p-type dopant with only three electrons in the outermost orbit (the valence shell). When a boron atom replaces a silicon atom in a single-crystal silicon lattice, a hole is created

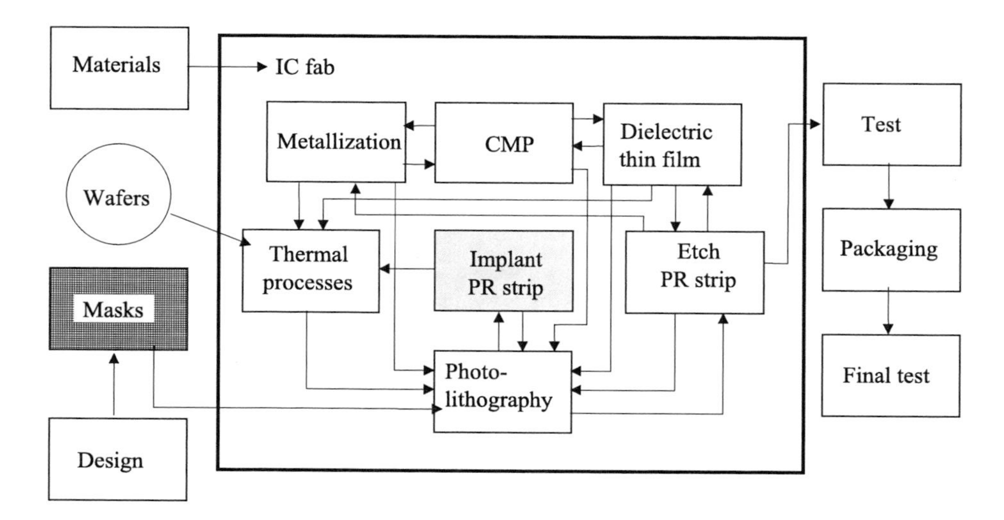

Figure 8.1 An advanced IC fabrication process flow.

in that single-crystal silicon. Holes can carry electric currents as positive charges. Phosphorus, arsenic, and antimony atoms have five electrons in their valence shells, thus they can cause electrons in single-crystal silicon to conduct electrical current. Because an electron has a negative charge, P, As, or Sb are called n-type dopants, and semiconductors with these dopants are called n-type semiconductors.

Before the mid-1970s, doping was accomplished with a diffusion process in high-temperature furnaces. The area where the furnaces were located was called the diffusion bay, and the furnaces were called diffusion furnaces, whether or not they were used for diffusion or other processes such as oxidation or annealing. Currently, few diffusion doping processes are performed in advanced IC fabs; furnaces are mainly used for oxidation and annealing processes. However, as mentioned in Chapter 2, the furnace area in an IC fab is still called the diffusion bay, and furnaces are still called diffusion furnaces for historical reasons.

The diffusion process usually takes several steps. Normally, a dopant oxide layer is deposited on the wafer surface in a predeposition process. Then an oxidation process converts the dopant oxide into silicon dioxide and forms a high-concentration dopant in the silicon substrate near the silicon–silicon dioxide interface. A high-temperature drive-in process diffuses dopant atoms into the substrate, to a depth required by the device design. All three processes, predeposition, oxidation, and drive-in, are high-temperature processes that take place in furnaces. After dopant drive-in, the oxide layer is stripped by a wet etching process. Figure 8.2 illustrates the diffusion doping process.

The physics of thermal diffusion are well known, and the processing tool is relatively simple and inexpensive. However, the diffusion process has some major limitations. For instance, dopant concentration and junction depth cannot be independently controlled, since both of them are strongly related to diffusion temperature. Another major disadvantage is that the dopant profile is always

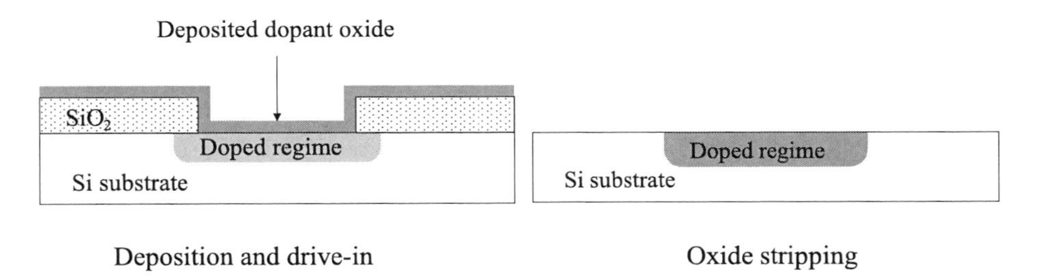

Figure 8.2 The diffusion process.

isotropic due to the nature of the diffusion process. These problems drove the search for an alternative doping method.

The idea of using ion implantation to dope semiconductors was first proposed by one of the three inventors of the first transistor, William Shockley, at Bell Laboratories in 1954. Shockley also held the patent for ion implantation (U.S. Patent 2787564). The physics and technology of a high-energy ion beam was developed during World War II, driven by research and manufacture of the first nuclear bomb. Accelerator and isotope separation technologies were directly employed in the design of the ion implanter. The introduction of ion implantation in the mid-1970s revolutionized the IC fabrication process.

Prior to the mid-1970s, semiconductor doping always used a diffusion process, which required silicon dioxide hard masks. In that time, bipolar transistors dominated IC production; when a metal-oxide semiconductor (MOS) transistor was made, it was the slower pMOS instead of the faster nMOS. This is because the p-type dopant, boron, diffuses much faster than n-type dopants, such as phosphorus and arsenic, in single-crystal silicon, making it easier to form a heavily doped p-type source/drain (S/D) than to form a heavily doped n-type S/D.

For pMOS made with a diffusion doping process, S/Ds were formed by boron diffusion with a silicon dioxide mask because boron diffuses much slower in silicon dioxide than it does in silicon. After S/D diffusion, the gate area was etched and cleaned, the thin gate oxide was grown, and the metal gate was formed. If the metal gate was misaligned with the S/D, as illustrated in Fig. 8.3, the transistor did not work properly. Oversized gates were required to ensure complete gate coverage over the S/D. The gate alignment issue presented a great challenge for shrinking feature size with this type of MOSFET manufacturing technology.

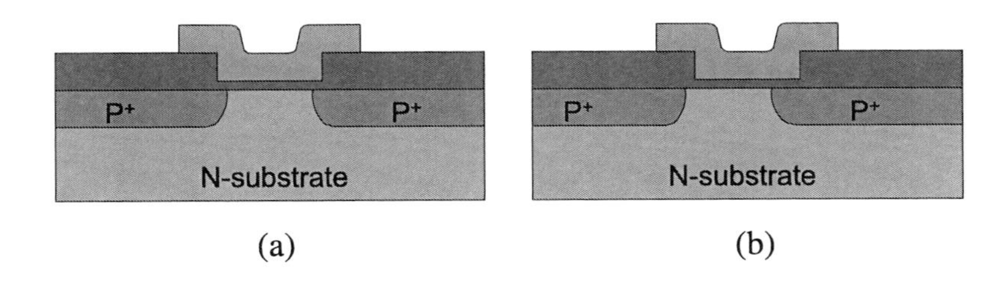

Figure 8.3 Gate alignment with S/D: (a) well aligned and (b) misaligned.

The application of ion implantation solved the gate alignment problem by using a self-aligned S/D doping process. In this case, the gate oxide is grown, and polysilicon is deposited, patterned, and etched. After photoresist stripping, high-current ion implantation is used to form the S/D. Because the polysilicon gate structure and field oxide block the ions, the S/D is always aligned with the polysilicon gate, as shown in Fig. 8.4.

It is not difficult to form heavy n-type junctions with ion implantation processes. Therefore, nMOS quickly replaced the slower pMOS after introduction of the ion implantation process. Because energetic dopant ion bombardment damages the single-crystal structure of the silicon substrate after ion implantation, a high-temperature (>1000 °C) annealing process is required to repair the damage and activate dopant atoms. The implantation annealing temperature is higher than the melting point of aluminum; therefore, different gate material must be used. Polysilicon and polysilicon-silicide stack (called polycide) have been used as gate materials for a long time, since the introduction of ion implantation. However, the transistor is still called an MOS, and no one calls it a POS (polysilicon-oxide semiconductor).

8.1.2 Advantages of implantation

Ion implantation has much better control over the doping process than the diffusion process. For example, ion implantation can independently control both dopant concentration and junction depth. Dopant concentration can be controlled by the combination of ion beam current and implantation time, and junction depth can be controlled by ion energy. Ion implantation processes can dope with a wide range of dopant concentrations, from 10^{11} to 10^{17} atoms/cm^2. The diffusion process is a high-temperature process and requires silicon dioxide hard masks. Before the diffusion process, a thick oxide layer needs to grow, then be patterned and etched to define the areas needing to be doped. The ion implantation process is a room-temperature process, and a layer of photoresist can block the energetic dopant ions. The thickness of the photoresist can be determined by the ion species and ion energy. Therefore, compared with diffusion processes, ion implantation has a lower cost because it does not require oxide growth or etch processes. Of course, the wafer holder of an ion implanter must have a cooling system that removes heat generated by energetic ions to prevent photoresist reticulation.

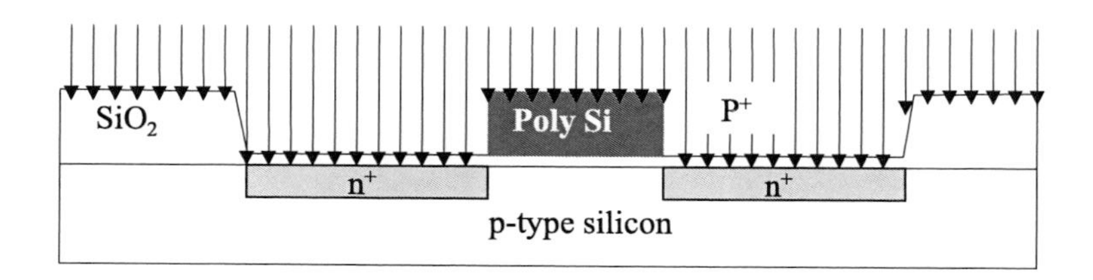

Figure 8.4 The self-aligned S/D with the ion implantation process.

The mass analyzer of an implanter selects exactly the ion species that needs to be implanted and generates a very pure ion beam, thus ion implantation has less possibility of elementary contamination. The ion implantation process always operates in a high vacuum, which is an inherently clean environment. Ion implantation is an anisotropic process, and dopant ions are implanted into the silicon mainly in the vertical direction. The doped region very closely reflects the area defined by a photoresist mask. Diffusion is an isotropic process; dopant always diffuses laterally underneath the silicon dioxide hard mask. For small feature sizes, it is very difficult to use diffusion processes to form a dopant junction. Figure 8.5 gives a comparison of the diffusion and ion implantation doping processes. Table 8.1 summarizes the advantage of ion implantation over diffusion in the doping processes.

8.1.3 Applications

The main application of ion implantation is doping the semiconductor substrate. A silicon wafer needs to be doped in order to change its conductivity in designated areas to form junctions such as wells and S/Ds for a CMOS IC. For a bipolar IC, doped junctions are needed to form buried layers, emitters, collectors, and bases.

Other applications of ion implantation are preamorphous implantation and buried layer implantation. Preamorphous implantation with silicon or germanium creates an amorphous layer on the substrate surface. It allows easier junction depth and profile control in the subsequent dopant implantation process. Germanium is a heavier atom than silicon, thus it has better damaging effects and is more favored for this application. There are two methods to make silicon-on-insulator (SOI)

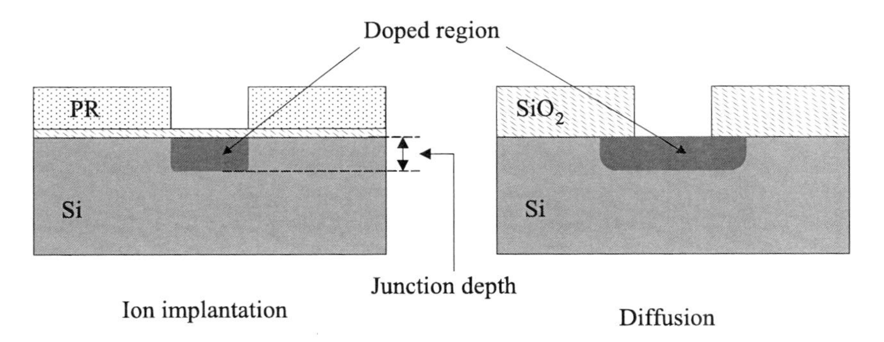

Figure 8.5 Illustration comparing of ion implantation and diffusion doping profiles.

Table 8.1 Comparison of ion implantation and diffusion.

Diffusion	Ion implantation
High temperature, hard mask	Low temperature, photoresist mask
Isotropic dopant profile	Anisotropic dopant profile
Cannot independently control the dopant concentration and junction depth	Can independently control dopant concentration and junction depth
Batch process	Both batch and single-wafer processes

wafers for high-end electronic devices; both involve ion implantation processes. One involves the implantation of high-dose oxygen ions into a silicon wafer and annealing the wafer to form a buried oxide layer underneath a thin single-crystal silicon layer. Another way is to use hydrogen implantation of the first wafer and bond it with the second wafer with a silicon dioxide layer on the surface. Because of the complete isolation between neighboring transistors, IC chips made on SOI wafers have lower leakage, higher noise immunity, higher radiation resistance, and better system reliability than IC chips made on bulk silicon wafers.

In advanced CMOS IC chips, the polysilicon gate for the nMOS is heavily n-type doped, and the polysilicon gate for the pMOS is heavily p-type doped. Silicide on top of a polysilicon structure shorts the p–n junction of the polysilicon and forms a local interconnection. Nitrogen can be implanted into the n-type doped polysilicon to form a barrier and prevent p-type dopant boron from diffusing into n-type doped polysilicon and causing device malfunction. Table 8.2 summarizes ion implantation applications, and Tables 8.3 through 8.7 list facts about phosphorus, arsenic, antimony, boron, and germanium.

8.2 Ion Implantation Basics

8.2.1 Stopping mechanisms

When ions bombard and penetrate the silicon substrate, they collide with lattice atoms. The ions then gradually lose their energy and eventually stop inside the silicon. There are two stopping mechanisms. Nuclear stopping occurs when the implanted ions collide with the nuclei of lattice atoms, are scattered significantly by the collision, and then transfer energy to the lattice atoms. In this hard collision, lattice atoms absorb enough energy to break free from the lattice binding energy, causing disorder and damage of the crystal structure. Another stopping mechanism occurs when incident ions collide with electrons of the lattice atoms. The incident ion path in the electronic collision is almost unchanged, energy transfer is very minimal, and crystal structure damage is negligible. This soft collision is called electronic stopping. The total stopping power, which is the energy loss of the ion per unit distance it travels inside the substrate, can be expressed as

$$S_{\text{total}} = S_{\text{n}} + S_{\text{e}},$$

where S_{n} is the nuclear stopping power, and S_{e} is the electronic stopping power. Figure 8.6 illustrates the stopping mechanisms, and Fig. 8.7 shows the relation between stopping power and ion velocity.

Table 8.2 Summary of ion implantation applications.

Applications	Doping	Preamorphous	Buried oxide	Poly barrier
Ions	n-type: P, As, Sb p-type: B	Si or Ge	O	N

Table 8.3 Facts about phosphorus.

Symbol	P
Atomic number	15
Atomic weight	30.973762
Discoverer	Hennig Brand
Place of discovery	Germany
Discovery date	1669
Origin of name	From the Greek word phosphoros, meaning bringer of light (an ancient name for the planet Venus)
Density of solid	1.823 g/cm^3
Molar volume	17.02 cm^3
Velocity of sound in substance	N/A
Electrical resistivity	10 $\mu\Omega \cdot$ cm
Refractivity	1.001212
Reflectivity	N/A
Melting point	44.3 °C
Boiling point	277 °C
Thermal conductivity	0.236 W m^{-1} K^{-1}
Coefficient of linear thermal expansion	N/A
Applications	n-type dopant in diffusion, ion implantation, epitaxial growth, and polysilicon deposition. Dopant of CVD silicate glass (PSG and BPSG).
Main sources	P (red), PH_3, $POCl_3$

Table 8.4 Facts about arsenic.

Symbol	As
Atomic number	33
Atomic weight	74.9216
Discoverer	Known since ancient times
Place of discovery	Not known
Discovery date	Not known
Origin of name	From the Greek word arsenikon, meaning yellow orpiment
Density of solid	5.727 g/cm^3
Molar volume	12.95 cm^3
Velocity of sound in substance	N/A
Electrical resistivity	30.03 $\mu\Omega \cdot$ cm
Refractivity	1.001552
Reflectivity	N/A
Melting point	614 °C
Boiling point	817 °C
Thermal conductivity	50.2 W m^{-1} K^{-1}
Coefficient of linear thermal expansion	N/A
Applications	n-type dopant in diffusion, ion implantation, epitaxial growth, and polysilicon deposition.
Main sources	As, AsH_3

Ion energy of the ion implantation process ranges from ultralow energy (0.1 keV) for ultrashallow junction (USJ), to high energy (1 MeV) for deep well implantation, which is located in region 1 of Fig. 8.7. From the far-left side of

Table 8.5 Facts about antimony.

Symbol	Sb
Atomic number	51
Atomic weight	121.760
Discoverer	Known since ancient times
Place discovered	Not known
Discovery date	Not known
Origin of name	From the Greek words anti and monos, meaning not alone (the origin of the symbol Sb comes from the Latin word stibium)
Density of solid	6.697 g/cm^3
Molar volume	18.19 cm^3
Velocity of sound in substance	3420 m/sec
Electrical resistivity	40 $\mu\Omega \cdot$ cm
Refractivity	1.001212
Reflectivity	55%
Melting point	630.78 °C
Boiling point	1587 °C
Thermal conductivity	24 W m^{-1} K^{-1}
Coefficient of linear thermal expansion	11×10^{-6}K^{-1}
Applications	n-type dopant in ion implantation
Main sources	Sb

Table 8.6 Facts about boron.

Symbol	B
Atomic number	5
Atomic weight	10.811
Discoverer	Sir Humphrey Davy, Joseph-Louis Gay-Lussac, and Louis Jaques Thénard
Place discovered	England, France
Discovery date	1808
Origin of name	From the Arabic word buraq and the Persian word burah
Density of solid	2.460 g/cm^3
Molar volume	4.39 cm^3
Velocity of sound in substance	16200 m/sec
Electrical resistivity	$>10^{12}$ $\mu\Omega \cdot$ cm
Refractivity	N/A
Reflectivity	N/A
Melting point	2076 °C
Boiling point	3927 °C
Thermal conductivity	27 W m^{-1} K^{-1}
Coefficient of linear thermal expansion	6×10^{-6}K^{-1}
Applications	p-type dopant in diffusion, ion implantation, epitaxial growth, and polysilicon deposition. Dopant of CVD silicate glass (BPSG).
Main sources	B, B_2H_6, BF_3

the figure, it is clear that the main stopping mechanism is nuclear stopping for low-energy and high-atomic-number ion implantation processes. The electronic

Table 8.7 Facts about germanium.

Symbol	Ge
Atomic number	32
Atomic weight	72.61
Discoverer	Clemens Winkler
Place discovered	Germany
Discovery date	1886
Origin of name	From the Latin word Germania, meaning Germany
Density of solid	5.323 g/cm^3
Molar volume	13.63 cm^3
Velocity of sound in substance	5400 m/sec
Electrical resistivity	~50000 μΩ · cm
Refractivity	N/A
Reflectivity	N/A
Melting point	938.25 °C
Boiling point	2819.85 °C
Thermal conductivity	60 W m^{-1} K^{-1}
Coefficient of linear thermal expansion	6×10^{-6} K^{-1}
Applications	Ge, GeSi, and semiconductor substrate; Ge ions for preamorphous implantation.
Main sources	Ge, GeH_4

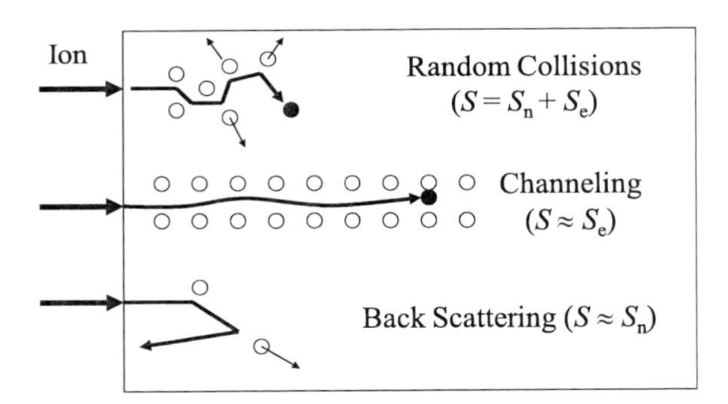

Figure 8.6 Different stopping mechanisms.

stopping mechanism is more important for high-energy, low-atomic-number ion implantation.

8.2.2 Ion range

When an energetic ion penetrates a target, it gradually loses its energy through collisions with electrons and nuclei of the atoms in the substrate, and eventually stops inside the substrate. Figure 8.8 shows the trajectory of an ion inside a substrate and the definition of an ion's projected range.

Generally, the higher the ion energy, the deeper the ion can penetrate into the substrate. However, even with the same implantation energy, ions do not stop exactly at the same depth in the substrate because each ion has different collisions

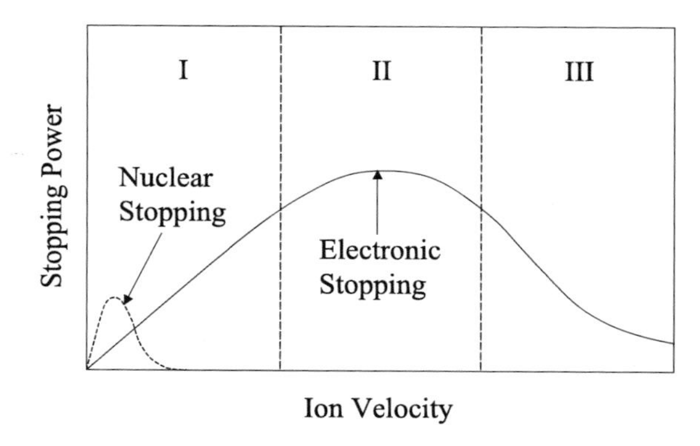

Figure 8.7 Relationship between stopping power and ion velocity (S. A. Cruz).

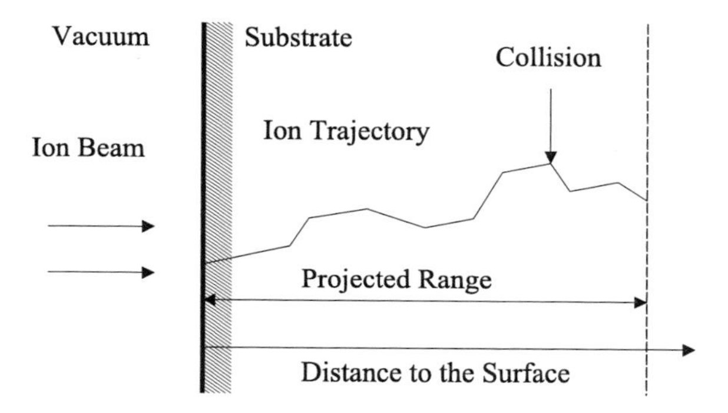

Figure 8.8 Ion trajectory and projected range.

with different atoms, causing each ion to stop at a different place. The projected ion range always has a distribution, as illustrated in Fig. 8.9.

Higher-energy ion beams can penetrate deeper into the substrate and therefore have longer projected ion range. Smaller, lighter ions have smaller collision cross sections; therefore, with the same ion energy, they can penetrate deeper into substrate and mask materials than larger, heavier ions. Figure 8.10 illustrates the projected range at different energy levels for boron, phosphorus, arsenic, and antimony ions in silicon substrate.

Ion projection range is an important parameter for ion implantation because it indicates the necessary ion energy for certain dopant junction depths. It also gives information on the required implantation barrier thickness for the ion implantation process. Figure 8.11 shows the required thickness of different barrier materials for 200-keV dopant ions. The figure shows that a boron ion requires the thickest masking layer because for the four dopant ions, boron has the lowest atomic number, smallest atom size, and longest projection range, and can penetrate deeper into the materials than any of the other three dopant ions. At high energy levels, for atoms with low atomic numbers such as boron, the main stopping mechanism is electronic stopping. Nuclear stopping is the main stopping mechanism for high-

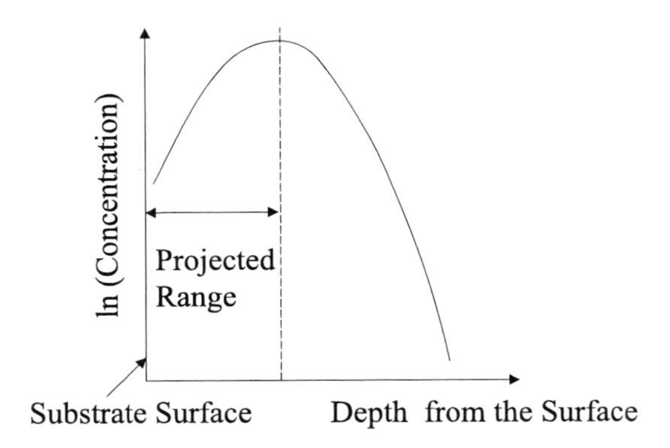

Figure 8.9 Distribution of projected ion range.

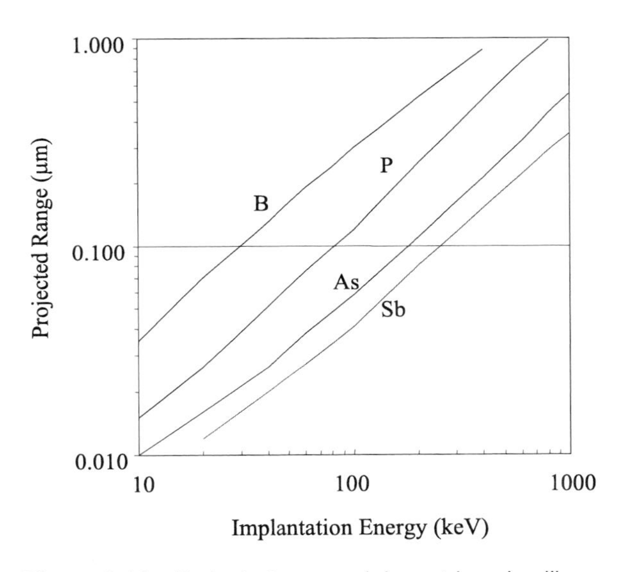

Figure 8.10 Projected range of dopant ions in silicon.

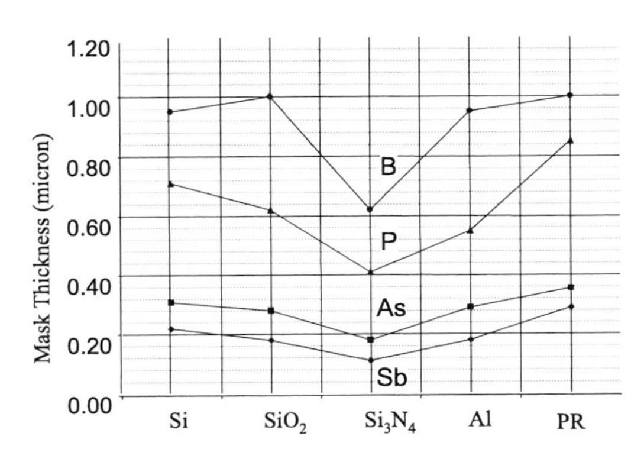

Figure 8.11 Required barrier thicknesses for 200-keV dopant ions.

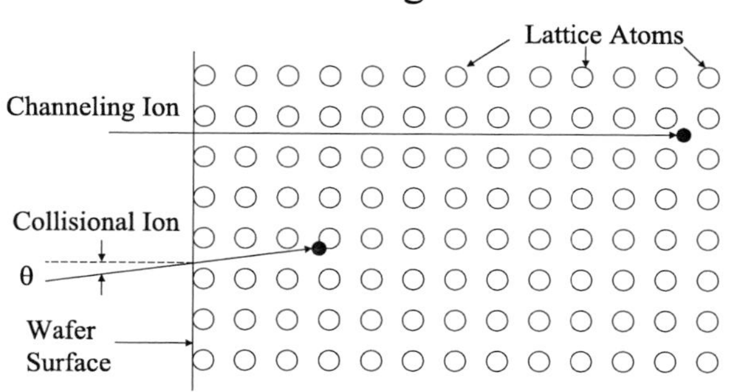

Figure 8.12 The channeling effect.

atomic-number dopant atoms. Similarly, the dopant ion with the highest atomic number, antimony, has the highest stopping power and shortest projection range, and therefore requires the thinnest masking materials.

8.2.3 Channeling effect

The projected range of an ion in an amorphous material always follows Gaussian distribution, also called normal distribution. In single-crystal silicon, lattice atoms are arranged in an orderly fashion, and many channels can be seen at certain angles. If an ion is implanted at those same angles into single-crystal silicon, it can travel a long distance along the channel, as illustrated in Fig. 8.12. This effect is called the channeling effect.

The channeling effect allows some ions to penetrate very deeply into a single-crystal substrate and causes a "tail" on the normal dopant distribution curve, as shown in Fig. 8.13. It is an undesired dopant profile, which could affect performance of the microelectronic device. Therefore, several methods have been developed to minimize this effect.

Question: The channeling effect allows an ion to penetrate deep into a single-crystal silicon substrate with relatively low energy. Why is this effect not used by fabs to create deep dope junctions?

Answer: If all ions of the ion beam are vertically implanted perfectly into the substrate, the channeling effect could be quite useful to form deep junctions with relatively low ion energy. However, ions always repel each other due to Coulombic forces; thus, an ion beam cannot be perfectly parallel. This means that many ions impact the wafer surface at small tilt angles and begin to have numerous nuclear collisions with

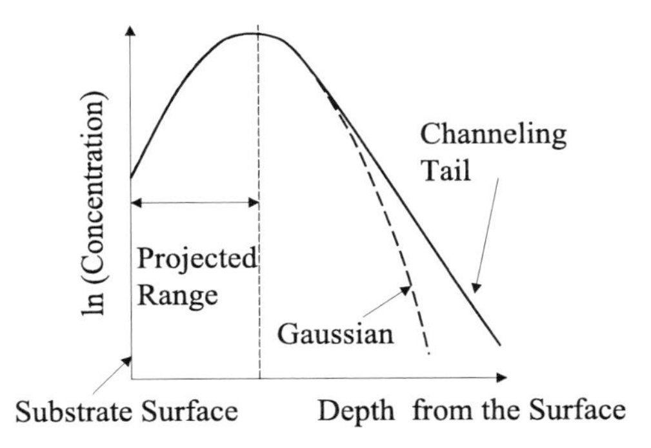

Figure 8.13 Channeling effect on the projection range distribution.

lattice atoms immediately after penetrating the substrate. The result is that only some ions channel deep into the substrate, but many ions are stopped due to normal Gaussian distribution.

One way to minimize the channeling effect is ion implantation on a tilted wafer, typically with a tilt angle of $\theta = 7$ deg. By tilting the wafer, ions impact the wafer at an angle that prevents them from channeling, as shown in Fig. 8.12. Incident ions have nuclear collisions immediately after entering the silicon substrate, effectively reducing the channeling effect. Most ion implantation processes use this technique, and most wafer holders used by ion implanters have the ability to adjust the tilting angle of the wafers.

Tilting a wafer can cause the photoresist to create a shadowing effect made by the photoresist, as shown in Fig. 8.14. This problem can be solved by rotating the wafer and/or a small amount of dopant diffusion during postimplantation annealing. If the tilt angle is too small, the dopant concentration in silicon can form a hump distribution due to the channeling effect, as shown in Fig. 8.15.

Another method widely used to handle the channeling problem is implantation through a thin layer of screen silicon dioxide. Thermally grown silicon dioxide is an amorphous material. Passing implantation ions will collide and scatter with silicon and oxygen atoms in the screen layer before they enter the single-crystal silicon substrate. Due to the collisional scattering, the pitch angle of the ions to the silicon crystal will be distributed in a wider range, thus reducing their chances of channeling. The screen oxide also can prevent the silicon substrate from touching and being contaminated by the masking photoresist. In some cases, both screen oxide and wafer tilting are used to minimize the channeling effect in ion implantation processes.

The problem with a screen layer is that some atoms in that layer can attain enough energy from energetic ions to implant themselves in the silicon. This is called the recoil effect. For a silicon dioxide screen layer, recoiling oxygen atoms can become implanted in the silicon substrate and form a high-oxygen-concentration region in the substrate near the silicon oxide interface, this oxygen-rich region can degrade carrier mobility and introduce deep-level traps. Therefore, in some implantation processes, a screen oxide cannot be used. In some cases,

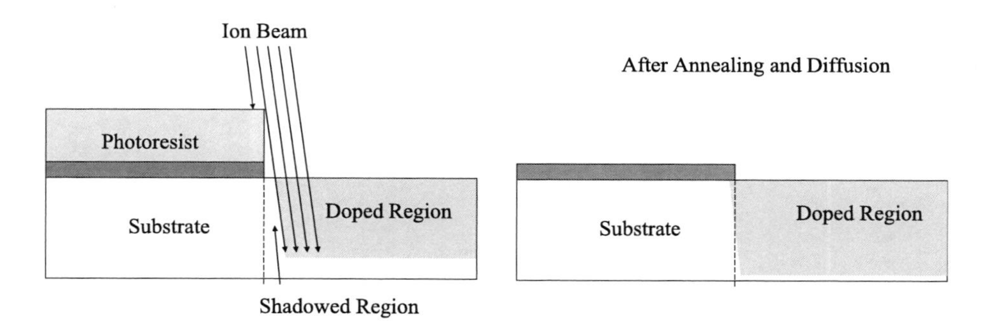

Figure 8.14 Shadowing effect and diffusion treatment.

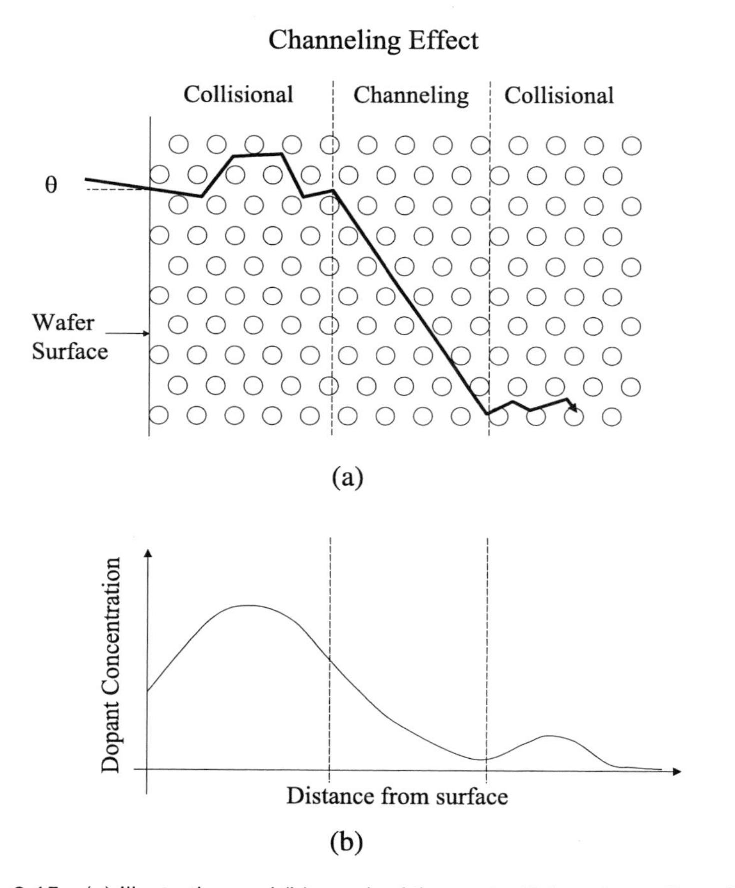

Figure 8.15 (a) Illustration and (b) graph of the postcollision channeling effect.

postimplantation oxidation and sacrificial oxide strips are needed to remove the oxygen-rich silicon layer. During the oxidation process, crystal damage caused by implantation can be annealed, and the oxide layer grows into the silicon substrate to consume the oxygen-rich region. Oxide stripping helps remove surface defects and the high-oxygen-concentration layer. However, this technique is impracticable for USJ implantation because it consumes doped silicon substrate, and there will be no extra silicon for USJ to consume.

High-current silicon or germanium ion implantation can heavily damage a single-crystal lattice structure and create an amorphous layer near the wafer surface. By using Si or Ge amorphous implantation processes, the channeling effect can be completely eliminated because, in an amorphous substrate, a dopant junction profile formed by ion implantation always follows Gaussian distribution, which is very predictable, repeatable, and controllable. The tradeoff for this preamorphous implantation solution is that it increases production costs by adding an extra ion implantation step. Also, as feature size shrinks, the thermal budget for annealing also decreases. For advanced nanometer nodes, the thermal budget may not fully anneal and recover the crystal damage caused by the preamorphous implant, and the end-of-range defect could cause junction leakage.

8.2.4 Damage and annealing

In an ion implantation process, ions gradually lose their energy by colliding with lattice atoms, transferring their energy to these atoms during the process. The transferred energy is high enough (typically about 25 eV) for these atoms to break free from the lattice binding energy. These freed atoms will also collide with other lattice atoms while traveling inside the substrate, knocking them free from the lattice by transferring enough energy to them. These processes continue until none of the freed atoms have enough energy to free other lattice atoms. One energetic ion can cause thousands of displacements of lattice atoms. Damage caused by one energetic implanted ion is shown in Fig. 8.16.

Damage created by one ion can be quickly self-annealed due to thermal movement of atoms inside the substrate at room temperature. However, in an ion implantation process, the number of ions is so great that substantial lattice damage always occurs in a single-crystal substrate. This causes the layer close to the surface to become amorphous, and the self-annealing process cannot repair the crystal damage in a short period of time. Damage effect is related to the dose, energy, and mass of the ion species. Damage increases when dose and ion energy increase. If the implantation dose is high enough, the substrate crystal structure will be completely destroyed and become amorphous near the substrate surface within ion range.

To achieve device requirements, lattice damage must be repaired in an annealing process to restore the single-crystal structure and activate the dopant. Only when dopant atoms are close to the single-crystal lattices can they effectively provide electrons or holes as majority carriers to help carry electric current. In a high-temperature process, atoms have random thermal movement until they move into single-crystal lattice locations, where the lowest free energy is found. Since the undamaged substrate underneath is single-crystal silicon, silicon and dopant atoms in the damaged amorphous layer reconstruct the single-crystal structure by falling into the lattice grids and becoming bonded by the lattice energy.

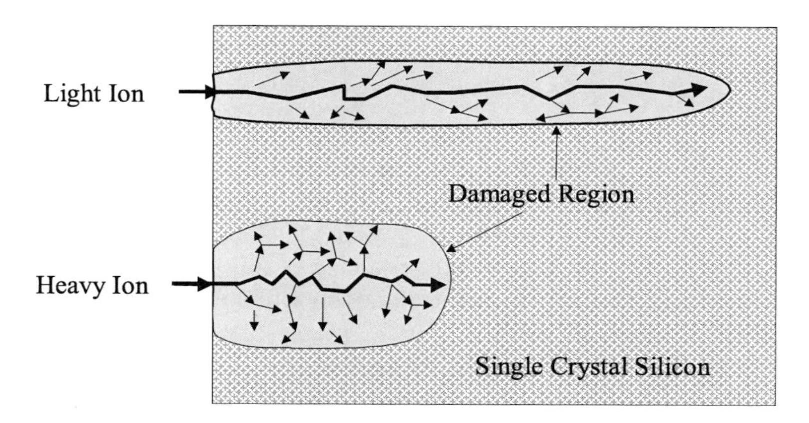

Figure 8.16 Damage caused by one ion.

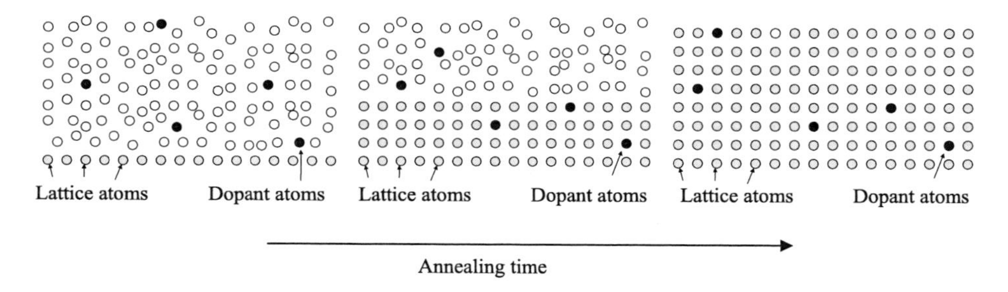

Figure 8.17 Postimplantation annealing process.

Figure 8.17 illustrates crystal recovery and dopant activation in a thermal annealing process.

During high-temperature annealing, single-crystal lattice recovery, dopant atom activation, and dopant atom diffusion happen simultaneously. When IC feature size shrinks deep into the submicron level, there is less room for dopant atoms to diffuse; thus, minimizing diffusion during the thermal annealing process becomes very important. In amorphous silicon, dopant atoms have free thermal movement without any restrictions and thus diffuse much faster than in single-crystal lattices because the lattice binding energy severely restricts movement of the dopant atoms. It has been found that at a lower temperatures, the diffusion process outpaces the annealing process. When the temperature is high enough, such as >1000 °C, the annealing process is faster than the diffusion process. This is because annealing activation energy (~5 eV) is higher than diffusion activation energy (3 to 4 eV).

Furnaces have been used for postimplantation annealing processes. A furnace is considered a batch system, which can process more than 100 wafers with temperature ranges from 850 to 1000 °C and a processing time of about 30 min. Since the furnace annealing process takes a long time, the diffusion of dopant atoms can be quite substantial; thus, for devices with small feature sizes,

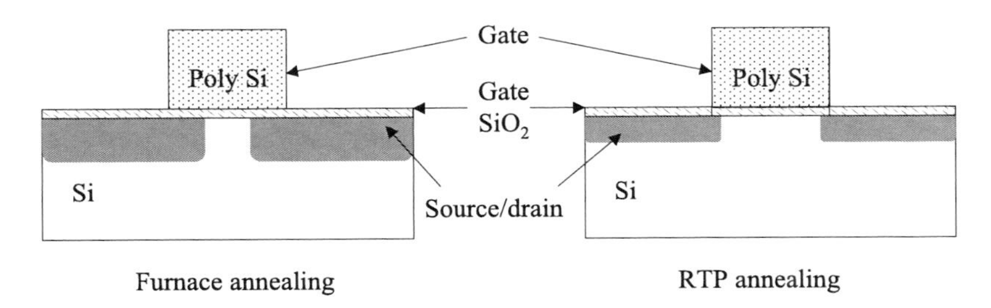

Figure 8.18 Dopant diffusion in furnace and rapid thermal annealing processes.

it becomes unacceptable. Only for some noncritical implantations with large feature sizes, such as well implantation processes, are furnaces still used for postimplantation annealing and dopant drive-in. For critical doping processes such as S/D implantation annealing, furnaces induce too much dopant diffusion and cause intolerable performance degradation for nanometer microelectronic devices.

Rapid thermal processing (RTP) has been developed to anneal the damages caused by implantation while minimizing dopant diffusion to meet the requirements of dwindling IC devices. An RTP system is a single-wafer system with a temperature ramp rate of up to 250 °C/sec, and good temperature and temperature uniformity control at about 1100 °C. With an RTP system, the rapid thermal annealing (RTA) process can operate up to 1150 °C, at which temperature implantation damage can be annealed in less than 20 sec. RTA systems can process a wafer in about a minute—from chamber entry to temperature ramp up, annealing, wafer cool-down, and wafer exit.

Question: Can a furnace temperature be ramped up and cooled down as quickly as an RTP system?

Answer: No. A furnace has very large thermal capacity, and a great deal of heating power is required to rapidly ramp up its temperature. This is very difficult to achieve without large temperature oscillation due to temperature overshoot and undershoot.

Because of the slow temperature ramp rate (usually less than 10 °C/min), it takes a long time to ramp up the processing tube from an idle temperature (normally between 650 to 850 °C) to a required annealing temperature, such as 1000 °C. While the temperature ramps up, some damage begins to anneal. Wafers must be pushed in and pulled out of a furnace very slowly to prevent wafer warping due to thermal stress caused by sudden temperature changes. Because a furnace is always at a high temperature when idling, wafers at both ends of a wafer boat have different annealing times due to the slow pushing and pulling, which causes wafer-to-wafer (WTW) nonuniformity.

RTA systems can ramp up wafer temperature from room temperature to 1100 °C very quickly (within 10 sec), while maintaining precise control of wafer temperature and temperature uniformity within the wafer. At about 1100 °C, a single-crystal lattice can be recovered in 10 sec with minimum dopant diffusion.

The RTA process has better WTW uniformity control than the furnace annealing process.

With the continued decrease in device feature size, even RTA processes are not fast enough to achieve dopant activation while keeping dopant diffusion in a tolerable range. Other annealing techniques such as spike, laser, and flush annealing have been developed and applied in IC manufacturing.

8.3 Ion Implantation Hardware

An ion implanter is a very large piece of equipment. It is one of the largest processing systems found in a semiconductor fab. It has several subsystems, including gas, vacuum, electrical, and control systems, and most importantly, the beam line system, shown in Fig. 8.19.

8.3.1 Gas system

Ion implanters use many dangerous gases and vapors to generate dopant ions. Examples are: flammable and poisonous gases such as arsine, phosphine, and diborane; corrosive gases such as boron trifluoride; and harmful vapor-form solid materials such as boron and phosphorus. To reduce the risk of these hazardous gases leaking into a fab, a special gas cabinet is placed inside the enclosure of an ion implanter to store these chemicals close to the ion source where they will be used. For some poisonous gases, subatmospheric cylinders are required for safety.

8.3.2 Electrical system

To generate high-energy ions, high-voltage dc power is needed to accelerate the ions. Normally, up to 200-kV dc power supply systems are equipped in an implanter. To generate ions in an ion source, either a hot filament or rf plasma system is used. A hot filament system requires a large current and a few hundred volts of bias power supply, and a rf ion source requires about a thousand watts

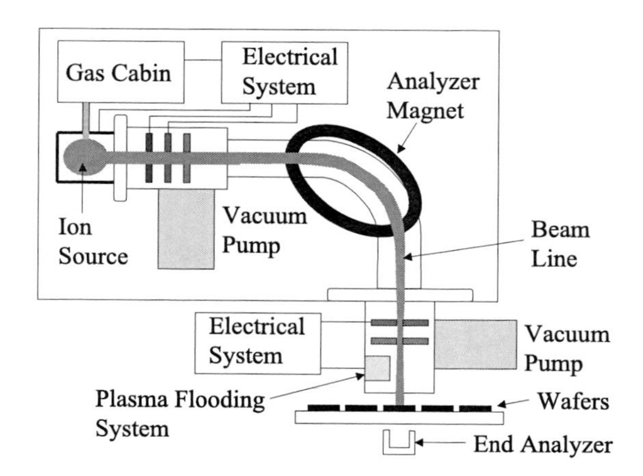

Figure 8.19 An ion implanter.

of rf power supply. Analyzer magnets need a high current to generate magnetic fields strong enough to bend the trajectories of energetic ions, and help select the correct ion species to create an ultrapure ion beam. Power supply systems need to be calibrated accurately. The voltage and current of the power supply must be very stable to ensure high and consistent process yields.

8.3.3 Vacuum system

A beam line must operate entirely in high-vacuum conditions to minimize collisions between energetic ions and neutral gas molecules along the ion trajectory. Collisions can cause ion scattering and loss, as well as implantation of unwanted species due to charge exchange collisions between ions and neutral atoms that cause elementary contamination. The pressure of an ion beam line should be low enough that the MFP of the ion is much longer than the length of the ion trajectory from the ion source to the wafer surface. Combinations of cryo-, turbo, and dry pumps are used to achieve 10^{-7}-torr high vacuum in a beam line system.

Because of the dangerous gases used in ion implantation processes, the exhaust of an implanter vacuum system must be kept separate from other process exhaust systems. Exhaust gases need to pass through a burn box and a scrubber before they can be safely released into the atmosphere. In a burn box, flammable and explosive gases are neutralized by oxygen in a high-temperature flame. In a scrubber, flushing water dissolves the corrosive gases and the burnt dust that pass through it.

8.3.4 Control system

To achieve design requirements, an ion implantation process needs to precisely control ion beam energy, current, and ion species. The process also needs to control mechanical parts, such as the robot for wafer loading and unloading, and control wafer movement to achieve uniform implantation across a wafer. To maintain system pressure, throttle valves are controlled according to a pressure setting point.

The center of the control system is a CPU board. Different control boards collect signals from the systems within the implanter and send them to that CPU board, which processes the data and sends instructions to the implanter's systems via the control board.

8.3.5 Beam line system

The ion beam line system is the most important part of an ion implanter. It consists of an ion source, extraction electrodes, mass analyzer, postacceleration electrodes, plasma flooding system, and end analyzer. Figure 8.20 illustrates a beam line system of an ion implanter.

8.3.5.1 Ion source

Dopant ions are generated in an ion source through ionization discharge of the atoms (or molecules) of dopant vapors or gaseous dopant chemical compounds. A hot filament ion source is one of the most commonly used ion sources. In this

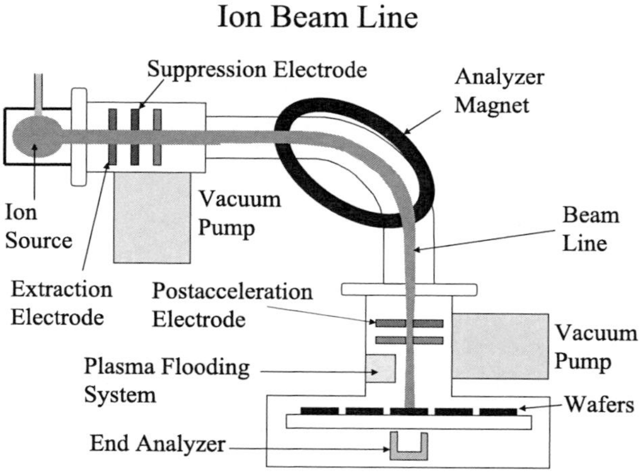

Ion Beam Line

Figure 8.20 The beam line system of an ion implanter.

type of ion source, large electrical currents flow through the filament, heat the tungsten filament, and cause thermal electron emission from the red-hot filament surface. Thermal electrons are accelerated by the arc power supply voltage with enough energy to dissociate and ionize the dopant gas molecules and atoms. Figure 8.21 shows a hot filament ion source. The magnetic field in the ion source forces electrons into gyromotion, which helps the electrons travel longer distances and increases their probability of colliding with dopant molecules, generating more dopant ions. The negatively biased anticathode plate expels electrons from its vicinity and reduces electron loss by collision with the wall along the magnetic field line.

Other types of ion sources such as rf and microwave sources have also been developed and applied to ion implantation processes. The rf ion source uses inductively coupled rf power to ionize dopant ions. The microwave ion source uses ECR to generate plasma and ionize dopant ions. Figure 8.22 shows the schematics of rf and microwave ion sources.

8.3.5.2 Extraction system

An extraction electrode with negative bias draws ions out of the plasma in an ion source and accelerates them to sufficiently high energy, about 50 keV. It is necessary for ions to have high enough energy for the analyzer magnetic field to select the correct ion species. Figure 8.23 shows an extraction system. When dopant ions accelerate toward extraction electrodes, some of the ions pass through the slit and continue to travel along the beam line. Some ions hit the extraction electrode surface, which generates x rays and excites some secondary electrons. A suppression electrode with sufficiently lower electrical potential (up to 10 kV) than an extraction electrode is used to prevent these electrons from being accelerated back toward the ion source, causing damage there. All electrodes are shaped with a narrow slit, through which ions are extracted as a collimated ion flux forming an ion beam.

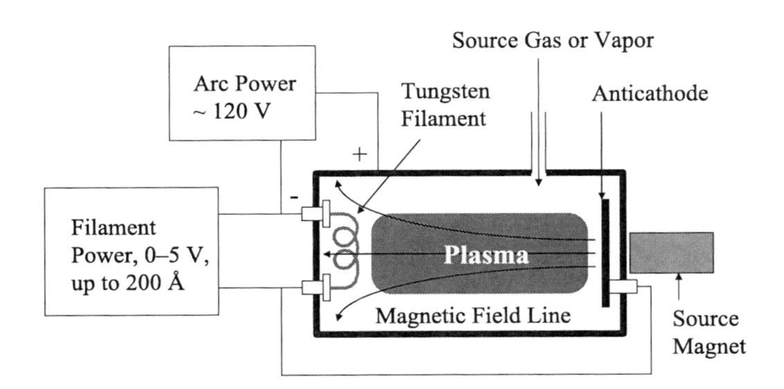

Figure 8.21 Hot filament ion source.

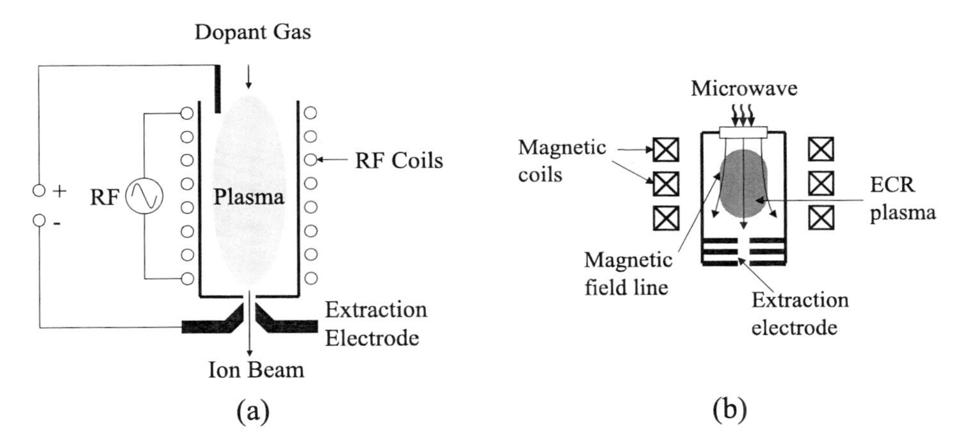

Figure 8.22 Schematics of (a) the rf and (b) microwave ion source.

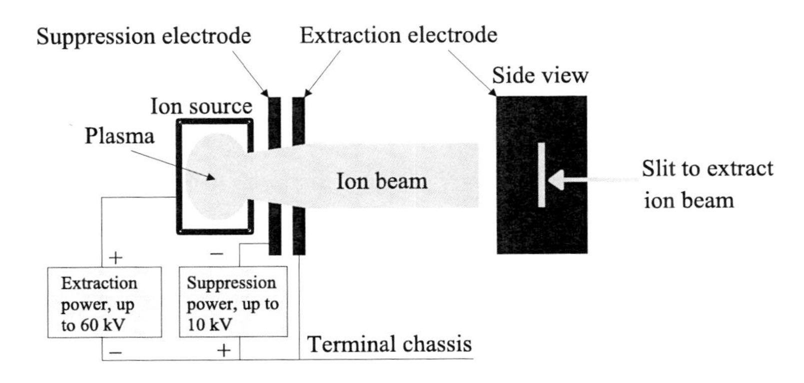

Figure 8.23 Schematics of an extraction system.

Ion beam energy after extraction is determined by the difference between the ion source and the extraction electrode. The extraction electrode potential is the same as the terminal frame and is often called the system ground potential. Notice that the potential difference between system ground and real ground (implanter cover plates) can be as high as −50 kV, which can cause a lethal electrical shock through arcing discharge without direct contact!

8.3.5.3 Mass analyzer

In a magnetic field, moving charged particles start rotating due to magnetic force, which is always perpendicular to the movement direction of a charged particle. For fixed magnetic field strength and ion energy, the gyroradius of the charged particle is only related to its mass/charge ratio, or m/q. This property was used to separate isotopes to generate enriched uranium-235 (^{235}U) from uranium-238 (^{238}U) for making the atomic bomb. In almost every ion implanter, a mass analyzer is applied to precisely select the desired ion species and weed out the unwanted ion species. Figure 8.24 illustrates the mass analyzer of an ion beam.

For example, BF_3 is commonly used as a boron dopant source. In plasma, combinations of dissociation and ionization collisions create a variety of ions. Because boron has two isotopes—^{10}B (19.9%) and ^{11}B (80.1%)—it also has several ionization states, which further increase the number of ion species, as shown in Table 8.8.

For p-well implantation, $^{11}B^+$ is preferred because, at the same energy level, it can penetrate deeper into silicon substrate due to its lighter weight. For shallow junction implantation, the $^{11}BF_2^+$ ion is the most likely species to be selected due to its large size and heavy weight. At the lowest energy level an implanter can remain stable, the $^{11}BF_2^+$ ion has the shortest ion range of these boron-containing ions; therefore, it can create the shallowest p-type junction. A small amount of fluorine integrated into the silicon substrate can bind with dangling silicon bonds at the silicon–silicon dioxide interface, which can help minimize the interface state charge and improve device performance.

Before ions enter the analyzer, their energy is determined by the potential difference between the ion source and the extraction electrode, the potential difference normally being set at approximately 50 kV. The energy of the single

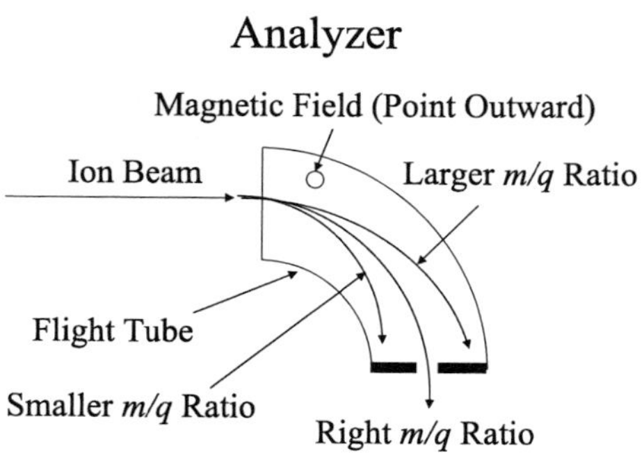

Figure 8.24 Mass analyzer of an ion implanter.

Table 8.8 Possible ion species from BF_3 plasma.

Ions	Atomic or molecular weight
^{10}B	10
^{11}B	11
^{10}BF	29
^{11}BF	30
F_2	38
$^{10}BF_2$	48
$^{11}BF_2$	49

charged ions after extraction is 50 keV. With knowledge of the ion m/q ratio and ion energy, the required magnetic field strength for ion trajectory passing through the narrow slit can be calculated. By adjusting electrical current in the magnetic coils, a mass analyzer can precisely select the desired dopant ions.

Question: $^{10}B^+$ is lighter than $^{11}B^+$, thus it can penetrate deeper than $^{11}B^+$ when both ions have the same energy. Why is $^{10}B^+$ not selected for p-well implantation?

Answer: Since only one in five boron atoms are ^{10}B and the rest are ^{11}B, in the plasma the $^{10}B^+$ ion concentration is only about one quarter of the $^{11}B^+$ ion. If a $^{10}B^+$ ion is selected, the ion beam current will be about one-fourth of the ion current of the $^{11}B^+$ beam. To achieve the same level of dopant concentration would take four times the amount of time to implant using a $^{10}B^+$ ion beam as with a $^{11}B^+$ ion beam, which causes lower throughput and is not cost effective.

Question: In plasma, phosphorus vapor can be ionized and form different ions. P^+ and P_2^{++} are two of them. Can a mass analyzer separate these two?

Answer: If P^+ and P_2^{++} ions have the same energy, a mass analyzer cannot separate them because they have exactly the same m/q ratio; therefore, they will have the same ion trajectories. When they implant into the substrate, P_2^{++} will not penetrate as deeply as P^+ because it has shorter ion range due to its larger size and heavier mass. This could cause energy contamination, which induces unwanted dopant concentration profiles and affects device performance. Fortunately, most P_2^{++} ions have about twice the amount of energy of P^+ ions after preacceleration by the extraction potential because they are doubly charged. With the same m/q ratio, higher energy ions have larger gyroradii and therefore will hit the outer wall of the flight tube of the mass analyzer. Their trajectory is similar to the ion beam with the larger m/q ratio shown in Fig. 8.24.

8.3.5.4 Postacceleration

After the mass analyzer selects the correct ion species, ions pass into the postacceleration section, where beam current and final ion energy are controlled.

Ion beam current is controlled by a pair of adjustable vanes, and ion energy is controlled by postacceleration electrode potential. Ion beam focus and beam shapes are also controlled in this section by the defining apertures and electrodes. Figure 8.25 shows a postacceleration assembly.

For the high-energy ion implanters used in well and buried layer implantation processes, several stages of high-voltage acceleration electrodes are required, connected in a series along the beam line to accelerate ions to several megaelectron volts. For ion implanters used for USJ applications (especially for p-type with boron implantation), an electrode in postacceleration is connected in a reverse way so that the ion beam is decelerated instead of accelerated when passing the electrode. This can generate a pure ion beam with energy as low as 0.1 keV.

In some implanters, after postacceleration, an electrode is used to bend the ion beam at a small angle, such as 10 deg, which helps dispose of energetic neutral particles. Neutral particles will not respond to the electric field generated by the electrode; they will keep moving straight forward while the ions' trajectories bend and move toward the wafer, as shown in Fig. 8.26. Some implantation systems bend the ion beam twice, in an S-shaped trajectory, to achieve even higher energy purity.

8.3.5.5 Charge neutralization system

When ions implant into the silicon substrate, they carry a positive charge to the wafer surface. If positive charges accumulate on the wafer, they can cause a wafer charging effect; positively charged wafer surfaces tend to expel positive ions and

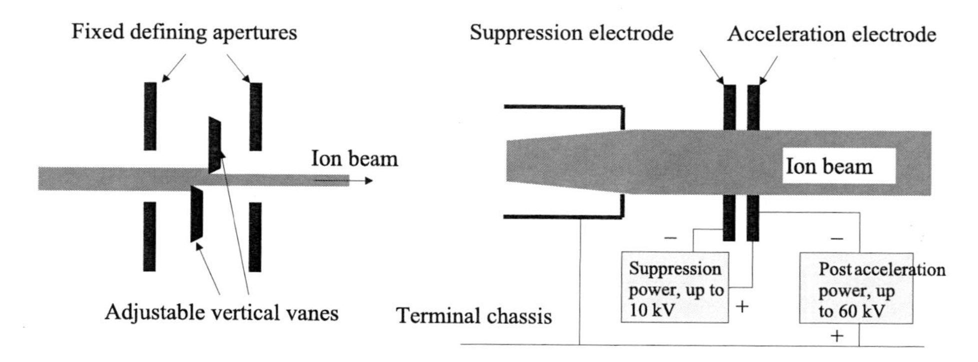

Figure 8.25 Beam current control and postacceleration assembly.

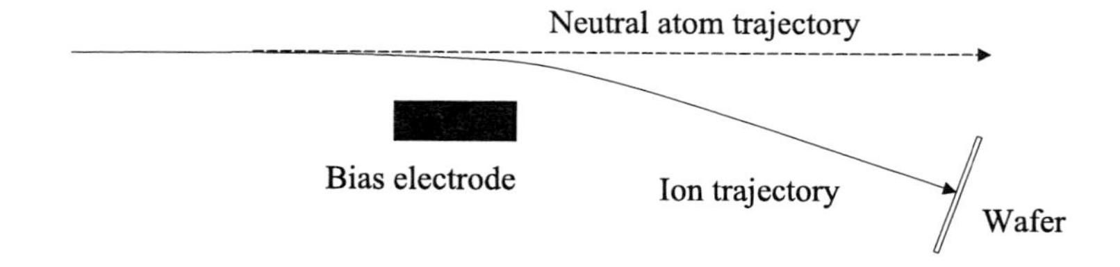

Figure 8.26 Bent ion beam trajectory.

cause beam blowup. This results in nonuniform ion implantation and nonuniform dopant distribution across the wafer, as shown in Fig. 8.27.

When the surface charge concentration is too high, charge-induced electric fields can be high enough to break down the thin gate-oxide layer and significantly impact IC chip yield. When positive charges accumulate to a certain level, they discharge in the form of arcing and cause defects on the wafer surface.

To manage the wafer charging problem, large numbers of electrons, which have negative charges, are required to neutralize positive ions on the wafer surface. There are several approaches for wafer neutralization: plasma flooding systems, electron guns, and electron showers are all used for providing electrons to neutralize ions and minimize the wafer charging effect. Figure 8.28 shows a plasma flooding system.

In a plasma flooding system, thermal electrons are emitted from a hot tungsten filament surface and accelerated by a dc power supply. They collide with neutral atoms in the chamber and generate plasma by means of many ionization collisions with electrons and ions. Electrons in the plasma are attracted into the ion beam and

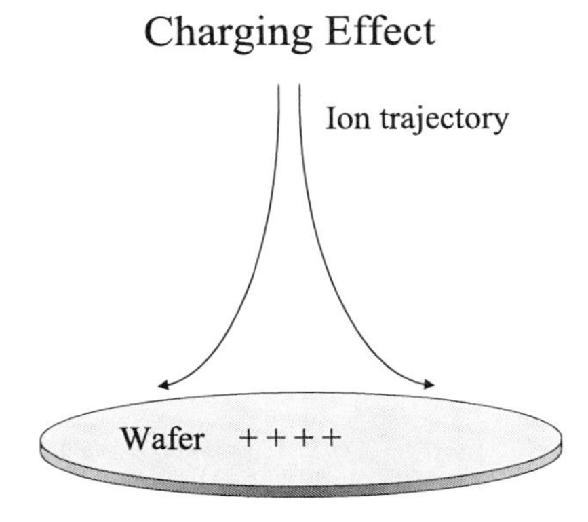

Figure 8.27 Nonuniform implantation caused by the wafer charging effect.

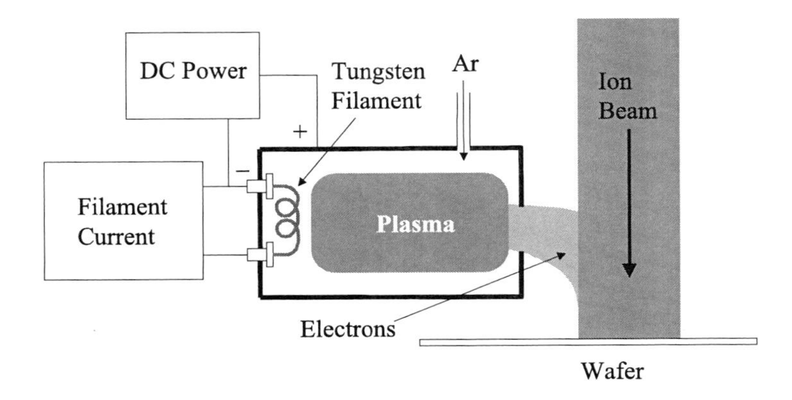

Figure 8.28 A plasma flooding system.

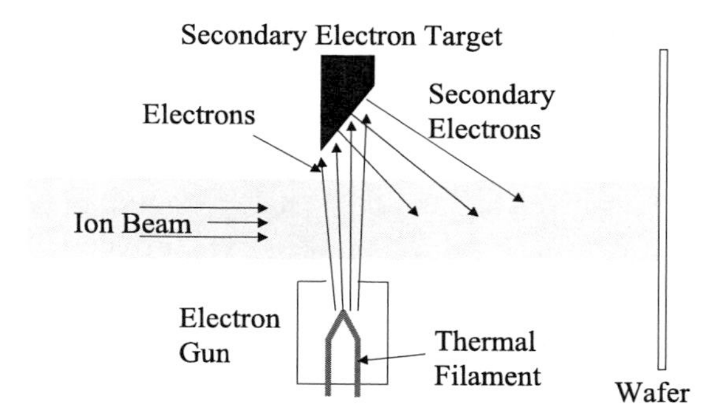

Figure 8.29 Electron gun system.

flow to the wafer surface with ions to neutralize the wafer and minimize the wafer charging effect.

Figure 8.29 illustrates an electron gun system. Thermal electrons emitted from the hot filament accelerate to a secondary electron target and collide into that target with high enough energy to generate large amounts of secondary electrons, which are knocked loose from the target surface. These electrons flow with the ions in the beam line and neutralize ions on the wafer surface.

8.3.5.6 Wafer handler

The most important task for a wafer handler is to achieve uniform implantation across a wafer. The diameter of an ion beam is about 25 mm (~1 in.); it needs to move either the ion beam or wafer—or for some implanters, both—to scan the ion beam uniformly across the wafer, which can have a diameter up to 300 mm.

In a spin wheel system, the wheel spins at a very high rate. An arc band with the ion beam diameter as its bandwidth is implanted with ions when a wafer passes through the ion beam. The center of the wheel swings back and forth, allowing the ion beam to scan uniformly across the entire wafer for all wafers on the spin wheel. Figure 8.30 illustrates a spin wheel wafer-handling system.

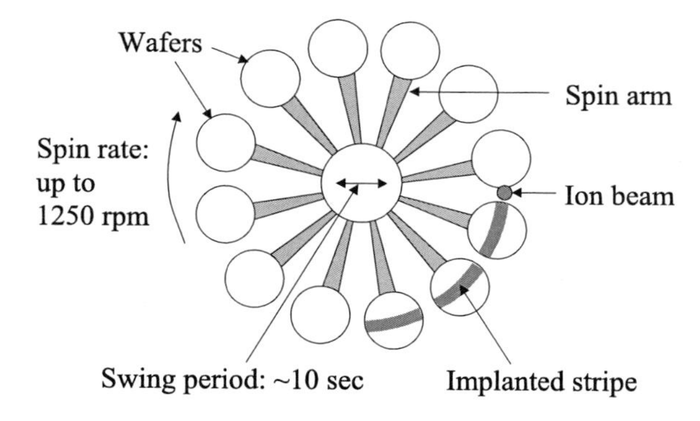

Figure 8.30 Schematic of a spin wheel wafer-handling system.

A spin disk is very similar to a spin wheel, the difference being that instead of swinging the entire disk, the spin disk scans the ion beam to achieve uniform ion implantation across the wafer. Figure 8.31 illustrates a spin disk system.

Another type of wafer-handling system used in ion implanters, which combines ion beam scanning and wafer movement, is shown in Fig. 8.32(a). By varying the applied bias potential between the scanning electrodes, the ion beam can scan back and forth in the x direction, while the wafer moves up and down in the y direction, driven by a step motor. The entire wafer can be uniformly implanted in this way. This type of scanning technique can be used in single-wafer implantation systems.

Some single-wafer implantation systems use wide-band or ribbon beams instead of scanning beams in the x direction while moving the wafer up and down for uniform ion implantation, as shown in Fig. 8.32(b). Some other systems use wide-band beams placed in the y direction and swing the wafer back and forth in the x direction to achieve uniform implantation. Single-wafer ion implantation systems have become mainstream in advanced nanometer IC manufacturing for critical ion implantation processes such as USJ S/D formation, including S/D extension (SDE) implantation and S/D implantations.

A wafer holder must have a cooling system to remove heat generated by the energetic ion bombardment and control wafer temperature. Otherwise, wafer

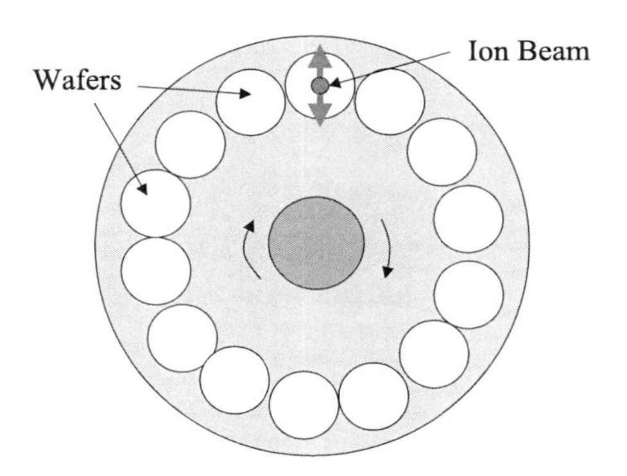

Figure 8.31 Schematic of spin disk wafer-handling system.

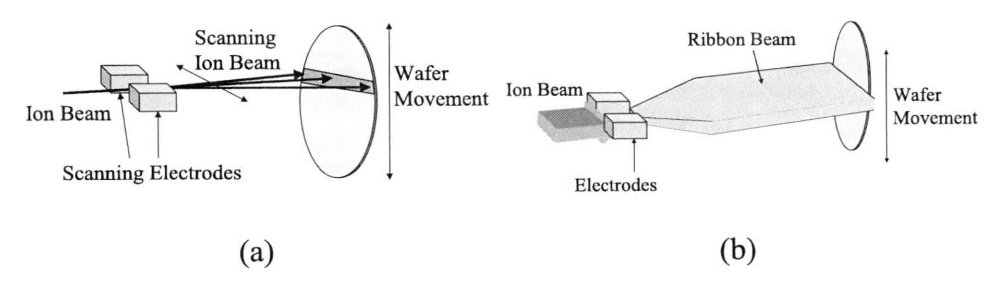

Figure 8.32 Single-wafer ion implantation systems: (a) scanning ion beam and (b) ribbon beam.

temperature can be high enough to cause photoresist reticulation. Wafer holders are usually water cooled, and temperatures are controlled below 100 °C.

8.3.5.7 Beam stop

At the end of a beam line, a beam stop or end station is needed to absorb ion beam energy; it also serves as an ion beam detector for the beam current, beam energy, and beam shape measurement. A water-cooled metal plate is used to remove heat generated by energetic ion bombardment and to block the x-ray radiation generated by the sudden stopping of the energetic ions on the target surface.

Figure 8.33 illustrates a beam stop. At the bottom of the ion beam stop, there is an array of ion detectors, which can be used to measure ion energy and energy spectrum, beam current, and beam shape. In an ion beam, many electrons travel with the ion beam. The major source of these electrons is a charge neutralization system, such as an electron flooding system, electron gun, or some other electron source, that generates a large number of electrons to neutralize the wafer surface and minimize the wafer charging effect. If these electrons enter the ion beam stop and hit the Faraday detector, current reading can be reduced and the accuracy of the beam current measurement can be affected. Permanent magnets are used to generate a magnetic field to prevent electrons from entering the beam stop due to their small gyroradii. The magnetic field can also prevent secondary electrons emitted from the graphite surface from backstreaming to the postacceleration electrode and causing damage there.

8.4 Ion Implantation Process

The three main issues in ion implantation processing are: dopant type, which is determined by ion species; junction depth, which is determined by ion energy; and dopant concentration, which is controlled by the combination of ion current and implantation time.

8.4.1 Device applications

In IC chip manufacturing, many ion implantation processes are involved to make billions of tiny transistors on a silicon wafer surface. Due to the different requirements of dopant concentration and junction depth, ion energy and beam currents for these implantation processes are quite different. In an advanced semiconductor fab, different types of implanters are employed to meet these requirements.

Table 8.9 lists a few examples of low-energy, high-current ion implantation applications in 32-nm CMOS ICs. Symbols and numbers in the right column are dopant isotopes or molecules, ion energy in keV, and the dosage in ions/cm^2. For example, pMOS S/D implantation, B11/0.4/1E15, indicates that it uses ions of boron 11 (^{11}B) at 0.4 keV with 1×10^{15} ions/cm^2 dosage. nMOS SDE, As2_150/2/5E14, indicates that it uses ions of a gaseous arsenic molecule (As_2) of atomic weight 150, or two ^{75}As, with 2 keV and 5×10^{14} ions/cm^2 dosage.

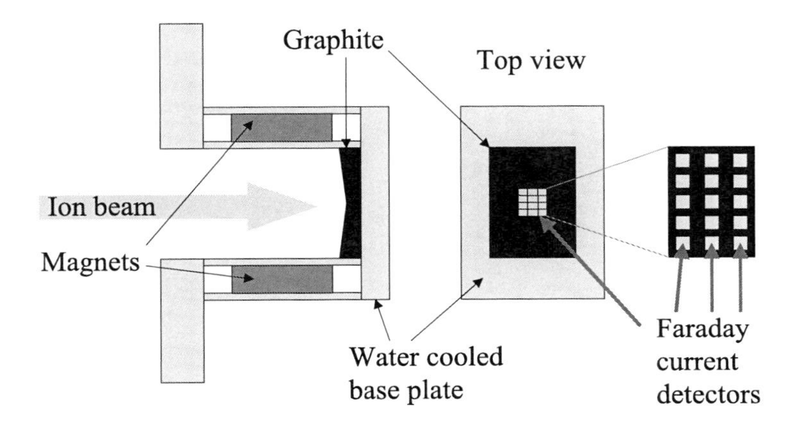

Figure 8.33 Schematic of a beam stop.

Table 8.9 Examples of 32-nm CMOS ion implantation.

Implantation step	Implantation conditions
pMOS S/D extension	B11/0.4/1E15
pMOS S/D	B11/2/3E15
nMOS S/D extension	As2_150/2/5E14
nMOS S/D	P31/4/3.5E15

Fig. 8.34 illustrates a well implantation process, which is a high-energy ion implantation process because it needs to form deep junctions called wells or tubs on which MOSFETs can be built. An nMOS is built on p-wells, and a pMOS is built on n-wells. Well implantation usually requires a high-energy ion implanter.

Antipunch-through implantation is also called mid-well implantation and is used to suppress the punch-through effect, which can cause transistor breakdown. Large angle tilt (LAT, normally 35 to 45 deg) implantation, or halo implantation, is also used for punch-through suppression for many IC devices.

Threshold implantation is also called V_T adjustment implantation, which is a low-energy, low-dosage implantation process. Threshold implantation determines at what voltage a MOSFET can be turned on or off. This is called threshold voltage, or V_T, which can be expressed as

$$V_T = \Phi_{ms} - Q_i/C_{ox} - Q_d/C_{ox} + 2\Phi_f \tag{8.1}$$

where Φ_{ms} is the potential difference between the gate material and the semiconductor substrate. In the case of polysilicon gates, the potential difference is controlled by poly-dope implantation. Q_i is the interface charge, which is determined by preoxidation clean and gate oxidation processes; $Q_d = -2(k_{Si}eN_c\Phi_j)^{1/2}$ is the depletion charge and is determined by V_T adjustment implantation, which controls majority carrier concentration N_c. $C_{ox} = k_{ox}/t_{ox}$ is the unit gate capacitance and is determined by gate dielectric material k_{ox} and gate dielectric thickness t_{ox}. Φ_f is the

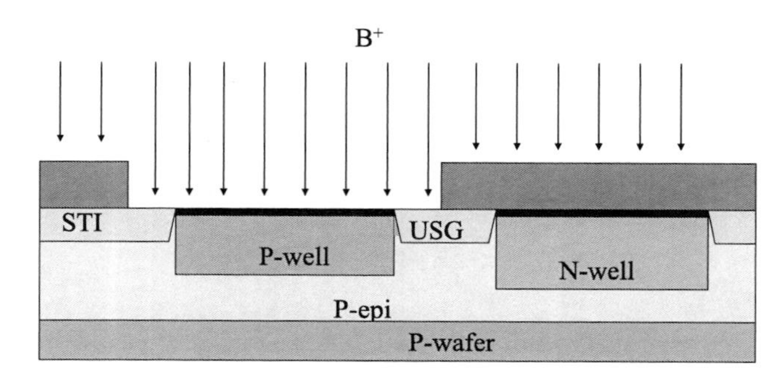

Figure 8.34 Well implantation.

Fermi potential of the substrate (silicon for most cases). Threshold voltage is one of the most important characteristics of MOSFETs, and V_T adjustment implantation is one of the most critical ion implantation processes.

Some old electronic devices require a 12-V dc power supply, many electronic circuits need 5 or 3.3 V, and advanced IC chips operate at 1.0 V. Ultralow-power IC chips can even operate at 0.4 V. These operation voltages must be higher than threshold voltages to guarantee that these transistors can be turned on or off; however, they cannot be so high that they break down the gate oxide and destroy the transistors. Figure 8.35 shows the threshold implantation of a CMOS IC chip. V_T adjustment implantations are performed with well implantations using the same implanter in the same processing sequence with significantly lower energy, as shown in Fig. 8.35.

Polysilicon needs to be heavily doped to reduce resistivity. This can be achieved either by introducing dopant gases to the CVD reactor with the silicon precursor for in-situ doping during the deposition process, or by high-current ion implantation. For advanced CMOS chips, ion implantation is preferred because it can dope pMOS and nMOS polysilicon gates separately. Polysilicon gates of pMOS are heavily p-type doped, and polysilicon gates of nMOS are heavily n-type doped. This allows better device property control. It also generates p–n junctions on the polysilicon lines that form local interconnections for connecting neighboring gates at the junctions of the p- and n-type MOSFET of the CMOS circuit. These p–n junctions must be shorted by the silicide formed on the top of the polysilicon lines in a later salicide process. Otherwise, they can cause intolerably high resistance between neighboring gates.

Poly-dope implantations need two masks, one for nMOS and another for pMOS. To reduce production cost, a poly-counter doping process has been developed and applied in IC production. It uses an implanter to heavily dope n-type dopants on polysilicon layers of entire wafers without a mask. It then patterns the wafer to expose the pMOS polysilicon layer and heavily dope it with p-type dopant. The p-type dopant concentration is so high that the polysilicon reverses from n- to p-type. Plasma doping systems, which can achieve much higher dopant concentrations than high-current ion implanters, have been developed and used

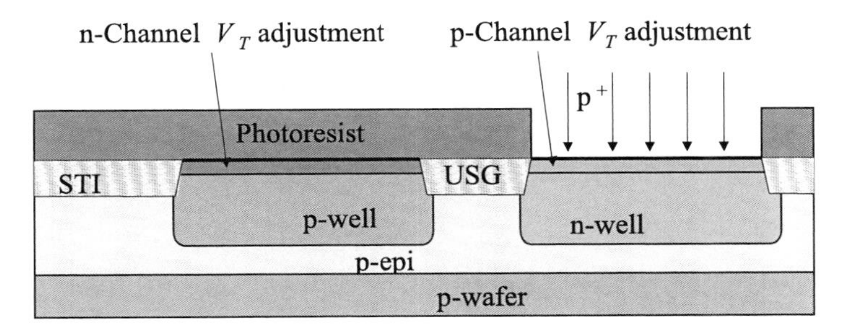

Figure 8.35 V_T adjustment implantation.

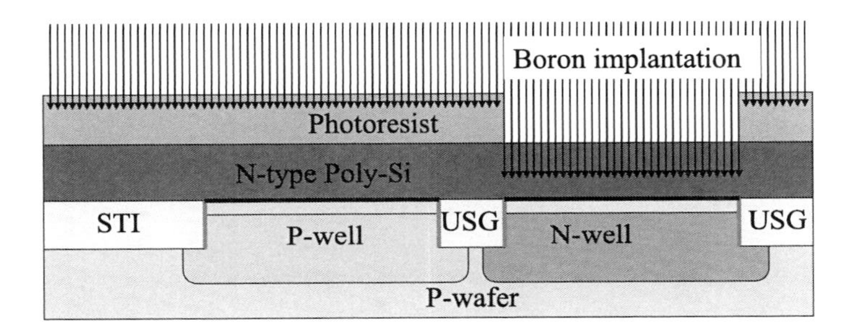

Figure 8.36 Poly-dope implantation.

for this application. Figure 8.36 shows a heavy p-type (boron) poly-counter dope process.

When transistor size shrinks, the effects of dopant diffusion in polysilicon gates can become evident and affect performance of microelectronic devices. It is very important to suppress dopant diffusion, especially to prevent boron in pMOS polysilicon gates from diffusing into nMOS polysilicon gates; otherwise, transistor characteristics can change. A diffusion blocking implantation is introduced, where a high dosage of nitrogen is implanted into the polysilicon to trap boron atoms and prevent them from diffusing too far from the polysilicon.

Lightly doped drain (LDD) is a low-energy, low-current implantation process by which a shallow junction is formed near the gate to reduce vertical electric field near the drain. LLD is required for submicron MOSFETs in order to suppress the hot electron effect, which can cause device degradation and affect chip reliability. The hot electron effect (or hot carrier effect) occurs when electrons tunnel through ultrathin gate oxide from drain to gate due to acceleration of the vertical component of the electric field induced by S/D bias and short channel. Because the dopant concentration of this implantation has increased with diminishing feature size, for sub-0.25-μm devices, the dosage has become high enough that the process can no longer be called LDD. It has been renamed SDE implantation, which provides a buffer for high-concentration S/D dopants. SDE has the shallowest junction depth in an IC chip and requires a low-energy implanter. Figure 8.37 illustrates the SDE implantation process for CMOS IC chip processing.

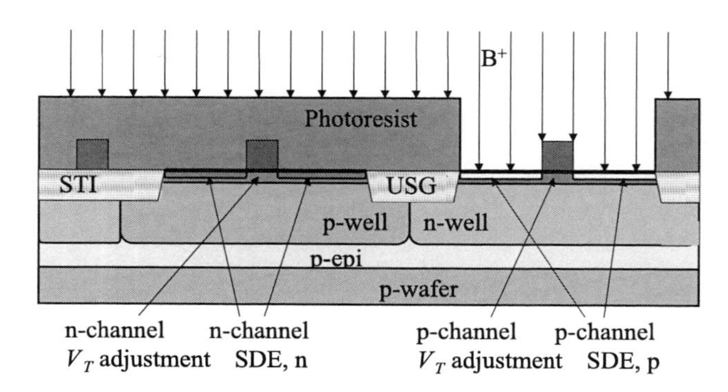

Figure 8.37 SDE implantation.

S/D implantation is a high-current, low-energy implantation process. It is the last implantation process of a MOSFET in CMOS IC chip fabrication processes. One significant difference between S/D and SDE implantation is that the dosage of S/D implantation is higher, and S/D is performed after sidewall spacer formation. Sidewall spacers keep the heavily doped S/D slightly apart from the channeling area directly underneath the gate and help suppress the hot electron effect. Figure 8.38 illustrates the S/D implantation process.

S/D implantation uses a high-current ion beam to heavily dope silicon. The masking photoresist receives a high concentration of dopant atoms after S/D implantation. This can create difficulties for the photoresist dry strip process. Oxygen radicals can oxidize and remove the photoresist, which is mainly hydrocarbon polymer. However, most dopant oxides, such as phosphorus and boron oxides, are solids instead of gases. They tend to stay on the wafer surface and cause defective residue called scum. A wet process is required after dry strip to remove these residues. This process is called descuming. In some fabs, the masking photoresist for S/D implantation is stripped in a wet process with a strong oxidizer solution, such as H_2O_2.

For the older technology with LOCOS isolation, ion implantation is normally used to form a p-type doped isolation region before the thick field oxide growth. This process is called isolation implantation or channel stop implantation. It is also used to form guarding rings around the active regions; these rings help to electrically isolate neighboring transistors.

For CMOS processes, almost every type of implantation process needs to be performed twice, once for pMOS and once for nMOS. A state-of-the-art CMOS IC device requires up to 20 implantation processes to create the multiple tiny transistors. For bipolar and BiCMOS IC chips, ion implantation is widely applied for buried layer doping, isolation formation, and base, emitter, and collector formations.

Unique DRAM ion implantation applications are contact implantations that reduce contact resistance between the polysilicon plugs and the silicon substrate, and between two polysilicon plugs. It uses high-current p-type implantation to heavily dope the silicon or polysilicon at the bottom of a contact hole after contact

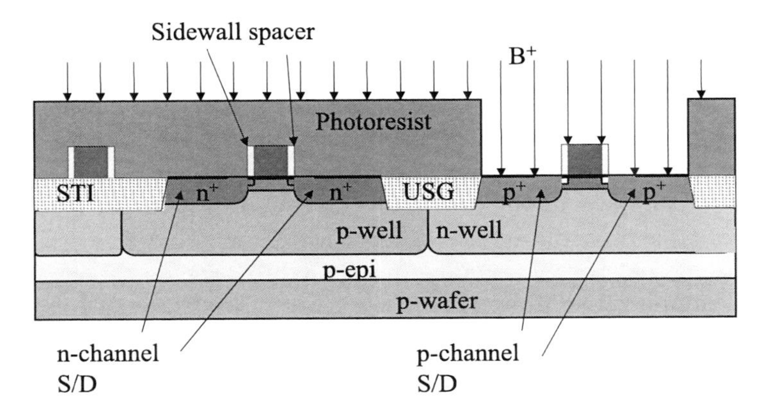

Figure 8.38 S/D implantation.

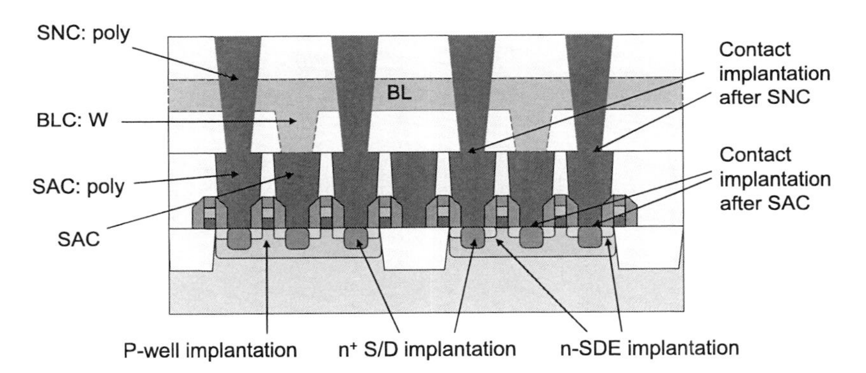

Figure 8.39 Ion implantation applications in DRAM cell array transistors and connections.

etches. Figure 8.39 illustrates the applications of ion implantations in cell array MOSFET formation and contact implantations for connection.

Since the ion implantation process is directly related to microelectronic devices, some background in device physics is helpful to better understand the process. Readers are strongly recommended to acquire information about semiconductor devices from the books by Streetman and Banerjee (entry level) and *Physics of Semiconductor Devices* by Sze (advanced level) given in the Bibliography section at the end of this chapter.

8.4.2 Other ion implantation applications

As device dimensions continue to shrink, soft error problems (due to trace amounts of natural background α-decay) become more and more pressing, especially for memory chips. Every α-particle can generate more than one million electron–hole pairs in the silicon substrate. Capacitance of a memory capacitor or transistor of a memory circuit must be large enough that stored data will not be overwritten by the surging electrons coming from the α-particle-induced electron–hole pairs when α-decay happens. Figure 8.40 illustrates the process of electron–hole pair generation by an α-particle.

Capacitance can also decrease when device dimension shrinks. One approach to solve this problem is by using a silicon-on-insulator (SOI) substrate. Figure 8.41 shows a MOSFET circuit built on a SOI substrate.

Figure 8.41 shows that each transistor is created on its own silicon substrate tub and is completely isolated from neighboring transistors and the bulk silicon substrate. Therefore, the possibility of crosstalk, latch up, and soft error are completely eliminated. SOI-based IC chips can work in extreme conditions, such as high-radiation environments, in which regular IC chips cannot work.

One approach for making a SOI substrate is to use high-energy, high-current ion oxygen implantation (up to 10^{18} ions/cm^2) to form an oxygen-rich layer in the silicon substrate. A high-temperature annealing process causes a chemical reaction between silicon and oxygen atoms and forms a buried silicon dioxide layer, while the implant-damaged silicon lattices near the surface recover the single-crystal structure. This process is called separation by implantation of oxygen, or SIMOX (see Fig. 4.26).

Another technique for making SOI wafers also involves high-current ion implantation. The Smart Cut™ (trademarked by Soitec) uses hydrogen ion implantation to generate a hydrogen-rich layer in an initial wafer with oxide growth on its surface, which bonds with a second wafer in a thermal process. During the

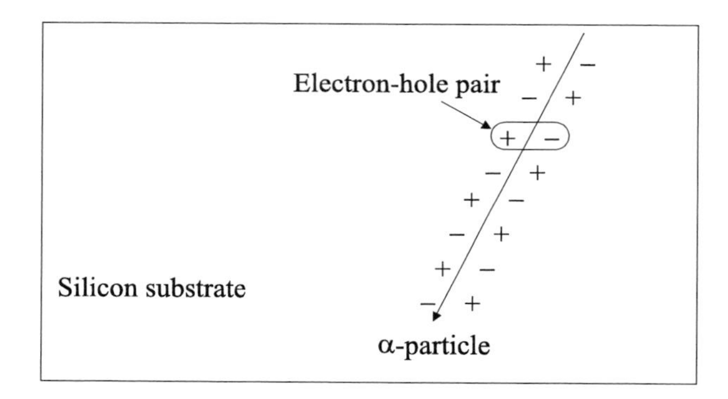

Figure 8.40 α-particle-induced electron–hole pairs.

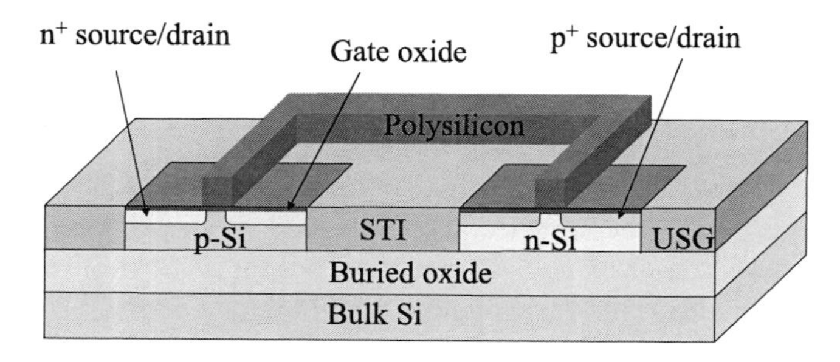

Figure 8.41 MOSFET on the SOI substrate.

bonding process, the hydrogen-rich layer splits the first wafer, leaving thin layers of silicon and oxide on the surface of the second wafer (see Fig. 4.27). Because SOI wafers made by Smart Cut^{TM} technology have a lower cost, especially for the thick buried oxide (BOX) layer, they dominate the SOI wafer market.

Ion implantation has been used to harden photoresist to improve its resistance during etch processes in IC fabs. Ion implantation is also used as one possible way to freeze the first photoresist pattern in a litho-freeze-litho-etch (LFLE) style of double-patterning technology (see Fig. 6.59).

In pattern media research and development in the hard disk drive (HDD) industry, ion implantation is one candidate for isolating magnetism patterns; other candidates include metal etch, nonmagnetic material deposition, and planarization. Figure 8.42(a) illustrates the patterning process based on etch, and Fig. 8.42(b) shows the patterning process based on ion implantation. It can be seen that the ion-implantation-based process potentially could have a lower cost than the etch-based process; however, there are still plenty of issues that need to be solved.

Another possible application of ion implantation is EUV lithography mask manufacturing. It is known that ion implantation can cause degradation of reflectivity of the multilayer. By using ion implantation damaging patterns instead of TaBN absorber patterns, shadowing effects from the TaBN absorber can be avoided during EUV lithography due to the reflective nature of EUV masks. If EUV lithography can be implemented in mass production for IC manufacturing, the technology node most likely will become 10 nm, which means the clear gap

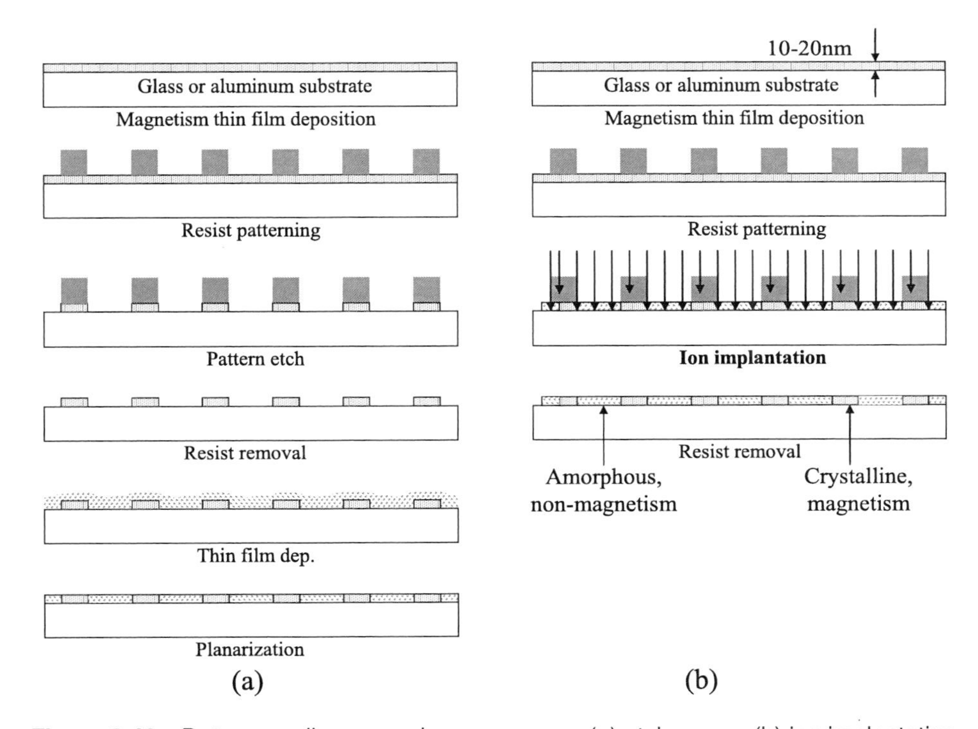

Figure 8.42 Pattern media processing sequences: (a) etch versus (b) ion implantation.

would be only 40 nm on the mask, the shadowing effect from the 6-deg incident angle would become very significant for EUV masks with absorber patterns, and ion implantation would be needed. Figure 8.43 shows these two EUV masks.

Ion implantation can also be used in solar panel manufacturing. By using external hard masks, dopant ions can be implanted into a designated area without photolithography processes, potentially saving solar cell manufacturing costs and driving down solar panel prices.

8.4.3 Processing issues

There are many processing challenges for ion implantation processes, such as wafer charging, contamination control, and process integration.

8.4.3.1 Wafer charging

Wafer charging can cause gate oxide breakdown. For example, the dielectric strength of silicon dioxide is about 10 MV/cm. Theoretically, when the surface charge is 2.2×10^{13} ions/cm^2, the electric field inside the oxide can be high enough to break down the 100-Å gate oxide if there is absolutely no electron leaking. Since electrons can always tunnel through thin dielectric layers due to the quantum effect, 100-Å oxide could survive dosages of up to 6.2×10^{18} ions/cm^2 as long as the gate voltage is lower than the breakdown voltage. For 100 Å, gate voltage is 10 V.

As transistor geometry continues to shrink, the thickness of the gate oxide becomes thinner and thinner. It can be as thin as 12 to 15 Å for 65-nm generation devices, which require even better charge neutralization techniques. Wafer charging can be monitored by several mechanisms: test wafers with capacitors, erasable programmable read-only memory (EPROM), and transistor structures are the most commonly used tools. These structures can be built on wafers specially made for the charging test. In-situ charging sensors are developed and applied in implanters next to the wafer to monitor charging conditions on the wafer surface. Figure 8.44 shows an antenna capacitor charging test structure—the ratio of the polysilicon pad area to the thin oxide area is called the antenna ratio, which can be as high

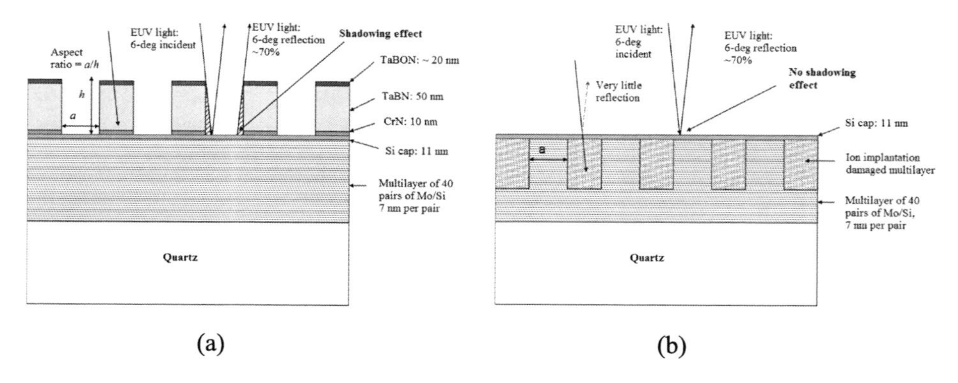

Figure 8.43 (a) A normal EUV mask with absorber patterns that have shadowing effects, and (b) an EUV mask with ion implantation patterns that do not have shadowing effects.

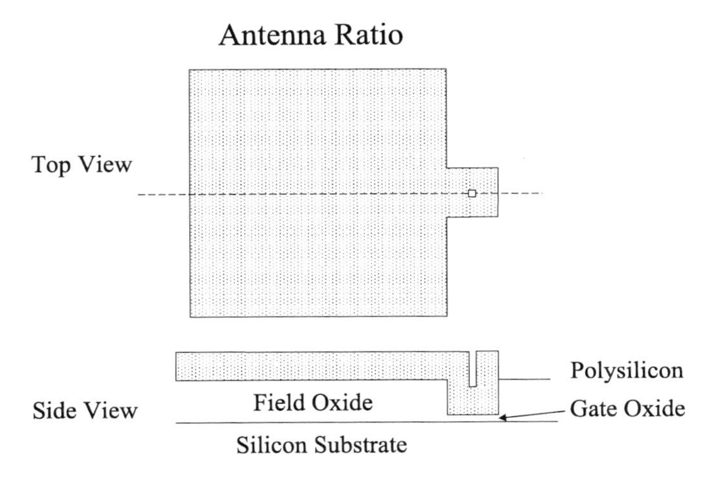

Figure 8.44 An antenna-style capacitor.

as 100,000:1. The larger the antenna ratio, the easier it is to breakdown the thin gate oxide.

The main factors that affect wafer changing are the beam current, beam scan width, disk or spin wheel radius, and spin rate. Reducing the ion beam current obviously will reduce the charging effect. The tradeoff is that it will also reduce wafer throughput. Increasing the beam scan width, disk or spin wheel radius, and spin rate effectively reduces the dosage per wafer per pass, thus reducing the charging effect. By using a large beam scan width, large disk or spin wheel radius, high spin rate, and an electron flooding system (which flows electrons to the wafer surface with the ion beam), advanced implanters have successfully addressed the charging issue for several generations of nanometer IC devices.

8.4.3.2 Particle contamination

Particle contamination is always an area of concern in almost every IC processing step. Large particles on a wafer surface can block ion beams, especially for low-energy ion implantation processes such as threshold, SDE, and S/D implantations, and cause incomplete dopant junction, which can be very harmful to IC yield. When device dimensions decrease, the size of killer particles also decreases. Figure 8.45 illustrates a killer particle causing incomplete junction implantation by shielding part of a substrate from the incident ion beam.

For many implantations still using spin wheel batch implanters, large particles can destroy patterns on a wafer's surface, similar to when a high-speed missile collides with the wall of a building. Because wafers require 15 to 20 different ion implantation processes in advanced IC chip fabs, the effects of particle-induced yield loss is accumulative. Therefore, it is extremely important to minimize the addition of killer particles in each implantation process.

Particles can be measured by scanning the wafer surface with a laser beam, and then using a photomultiplier tube to collect, transfer, and amplify the particle-induced scattered light signal. The number and location of particles of a certain size

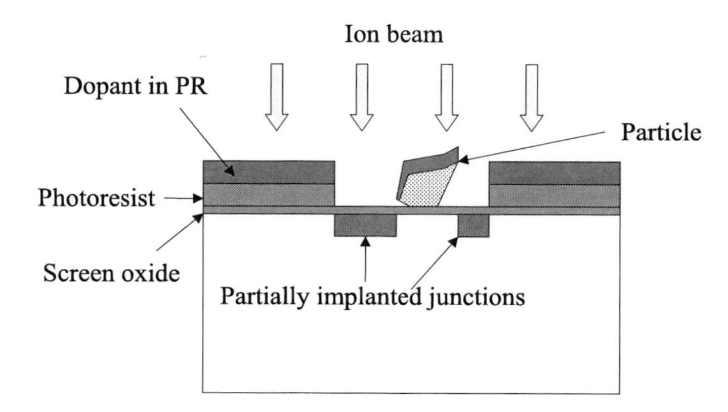

Figure 8.45 Effect of particle implantation.

or larger are measured immediately before and immediately after the implantation process; the difference in particle count is called particle adding. Locations of the added particles are also recorded, as the signatures of particles of different sizes are very useful in determining the source or sources of the particles.

Particles can be mechanically introduced by the wearing down of moving parts, typically valves and seals, clamps, and loading robots. They can also be introduced by processing. For example, arsenic, phosphorus, and antimony vapors can recondense along the beam line, and the flaking residues can reach the wafer surface during the pump-down cycle. Energetic ion sputtering is also a major particle source. Aluminum and carbon sputtered from the beam line and beam stop also can contribute to particle adding in implantation processes. A silicon wafer itself can be a particle source whenever there is wafer chipping or breakage. Photoresist film is brittle, and improper EBR in photolithography processes can leave photoresist on the edges of wafers. During wafer transfer and handling, robot fingers and wafer holder clamps can crack the photoresist on the edge, chip it off, and generate particles.

Both implanter design improvement and routine PM contribute to the minimization of particle adding in ion implantation processes. Statistic process control (SPC) methods can be applied to identify major contamination sources and help improve process control.

8.4.3.3 Elemental contamination

Elemental contamination is the coimplantation of other elements within an intended dopant. For example, both double-charged molybdenum $^{94}Mo^{++}$ and boron fluoride ions $^{11}BF_2^+$ have the same mass/charge ratio (AMU/e = 49); therefore, a mass analyzer cannot separate them, and $^{94}Mo^{++}$ can implant into a silicon wafer along with $^{11}BF_2^+$ to cause heavy metal contamination. This is the reason ion sources cannot use standard stainless steel, which contains molybdenum. Other materials such as graphite and tantalum are usually used.

If there is a small air leak, nitrogen can slip into an ion source chamber; $^{28}N_2^+$ ions have the same mass/charge ratio as silicon ions $^{28}Si^+$, which are used for preamorphous implantation. Similarly, outgassing of an ion source chamber wall can release carbon monoxide; when it becomes ionized, its ion also has the same mass/charge ratio (AMU/e = 28).

A mass analyzer also has difficulties separating ions with very close mass/charge ratios, due to insufficient resolution. $^{75}As^+$ ions can contaminate $^{74}Ge^+$ or $^{76}Ge^+$ ions in germanium preamorphous implantations, and $^{30}BF^+$ ions can contaminate $^{31}P^+$ ion implantation.

Other elemental contamination can be caused by the sputtering of beam line and wafer holder materials, such as aluminum and carbon, which can implant these ions into the wafer. Aluminum and carbon in the silicon substrate can cause device degradation.

8.4.4 Process evaluation

Dopant species, junction depth, and dopant concentration are the most important factors for ion implantation. Dopant species can be determined by the mass analyzer of an implanter, and dopant concentration is determined by ion beam current multiplied by implantation time. In ion implantation process monitoring, a four-point probe is commonly used to measure sheet resistance on a silicon surface. Sheet resistance R_s is defined as $R_s = \rho/t$ for ion implantation processes, where resistivity ρ is determined by dopant concentration, and thickness t is determined by the doping junction depth, which is controlled by dopant ion energy. Therefore, sheet resistance measurements can give information about dopant concentration, since junction depth can be estimated by knowing the ion energy, ion species, and substrate materials.

8.4.4.1 Secondary ion mass spectroscopy

By using a primary heavy ion beam to bombard a sample surface and collect a mass spectrum of sputtered secondary ions at different times, fabs can measure the dopant species, dopant concentrations, and the depth profile of the dopant concentration. Secondary ion mass spectroscopy (SIMS) is a standard measurement of ion implantation because of its ability to measure all factors critical to the ion implantation process. However, it is destructive; the sputtered spot size is large and very slow. SIMS is widely used in laboratories as the golden tool for off-line ion implanter tuning and early ion implantation process development. It is not used as an in-line monitoring system for ion implantation process control. Figure 8.46(a) illustrates how SIMS works; the depth of the profile can be calibrated with sputtering time using profilometer measurements. Figure 8.46(b) shows a SIMS measurement result of 1-keV ^{11}B ion implantation on a silicon wafer.

8.4.4.2 Four-point probe

A four-point probe (see Fig. 5.37) is the most commonly used tool employed to measure sheet resistance. Since sheet resistance is related to dopant concentration, this measurement provides useful information for ion implantation process control. By applying a certain amount of current between two of the pins and measuring the

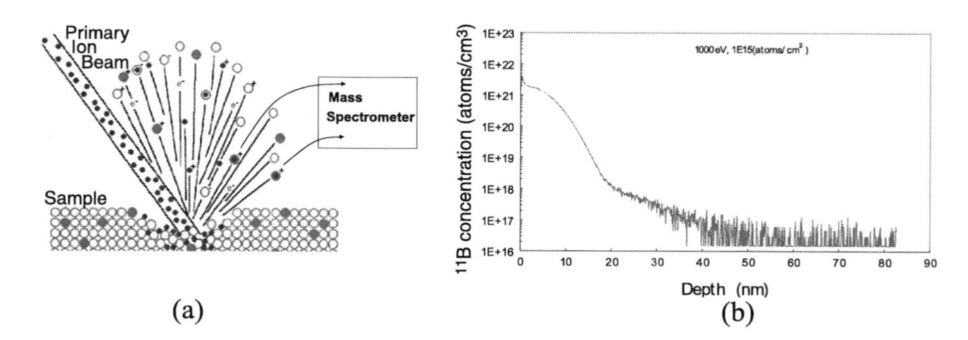

Figure 8.46 (a) SIMS and (b) an example of SIMS results for ^{11}B implantation (Advanced Ion Beam Technology, Inc.).

voltage difference between the other two pins, sheet resistance can be calculated. Four-point probe measurement is performed after the annealing process activates the dopants. Since the four-point probe makes direct contact with the wafer surface, it is only used on test wafers for process development, qualification, and control. During the measurement, the probes must come in contact with the silicon surface with enough force that the pins break through the thin (~10 to 20 Å) native oxide and make solid contact with the silicon substrate. Figure 8.47 shows an example of a four-point probe measurement of an ion implantation wafer after annealing.

8.4.4.3 Thermal wave

Another common ion implantation process monitoring technique is the thermal wave probe system. In a thermal wave system, an argon pump laser generates thermal pulses on the wafer surface, and an He-Ne probe laser measures the dc reflectivity R and reflectivity modulation ΔR induced by the pump laser at the same spot. The ratio of the two, $\Delta R/R$, is called a thermal wave (TW) signal, which is related to crystal damage, and the crystal damage in turn is a function of the implant dose. Figure 8.48 illustrates a thermal wave system.

A thermal wave measures the degree of crystal damage inflicted by ion implantation. The measurement is performed immediately after ion implantation, prior to the annealing process. This is one advantage of a thermal wave over a four-point probe, which needs annealing to active dopants prior to the measurement. Unlike a four-point probe, which requires firm contact with a wafer surface for reliable measurement, a thermal wave probe measures the wafer without physical contact. Therefore, thermal wave measurements are nondestructive and can be used to measure production wafers. One disadvantage of thermal wave measurements is that they have low sensitivity at low dosages; for example, with arsenic and phosphorus implantation dosage around 10^{12} ion/cm^2, 10% dosage change only causes a 2% change of TW signal. Another disadvantage is the drift of the TW signal over time due to room temperature annealing (or ambient annealing). Thus, a thermal wave measurement needs to be taken as soon as implantation is finished. Wafer heating caused by the laser beam during measurement also accelerates relaxation of the damage, which can change the reflectivity of the substrate. This

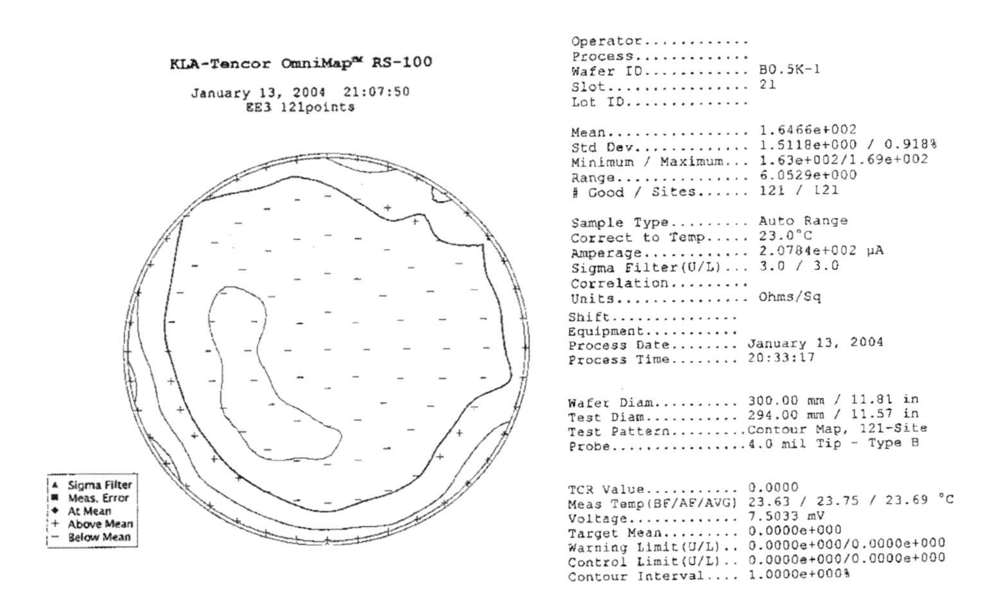

Figure 8.47 Example of a four-point probe measurement result (Advance Ion Beam Technology, Incorporated).

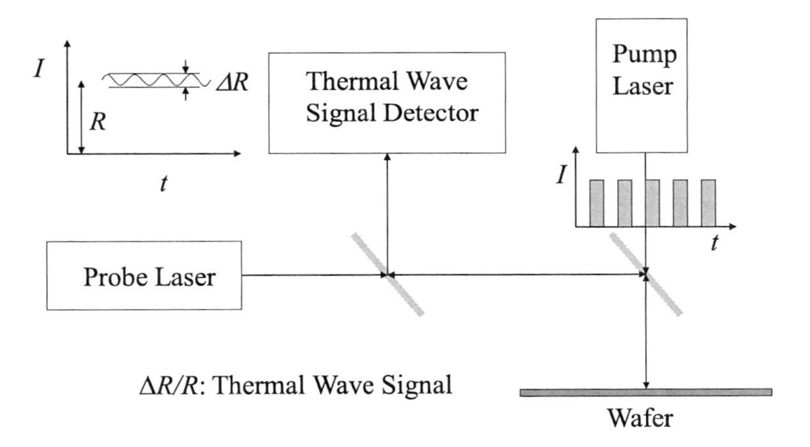

Figure 8.48 Thermal wave system.

is a standard case: the measurement process disturbs what is supposed to be measured. Therefore, inherently the process does not have very high measurement accuracy. Many factors can affect thermal wave measurement, such as ion beam current, ion beam energy, wafer patterns, and thickness of the screen oxide. The primary advantage of the thermal wave is that it can measure production wafers. It gives process engineers a very useful tool for better process control by measuring production wafers immediately after the implantation process without the long wait that other process monitors require.

8.4.4.4 Optical-electrical sheet resistance measurement

Optical-electrical sheet resistance measurement uses a pulse laser to illuminate semiconductor substrates and generate electron–hole pairs. The electron–hole pairs disperse to sensor electrodes that can detect voltage change caused by diffusion of the carriers. Diffusion rate is related to sheet resistance, and the measured voltage ratio V_1/V_2 is almost linearly related to sheet resistance. Figure 8.49 illustrates the processes of laser pulse and carrier diffusion, which allow measurement of the voltage ratio of the first and second sheet resistance R_s using this optical-electrical system.

8.5 Safety

All ion implanters use hazardous solids and gases, which are toxic, poisonous, flammable, explosive, or caustic. Very high electrical voltage (normally up to 250,000 V) is applied to produce the desired processing results.

8.5.1 Chemical hazards

Both solid and gaseous dopant sources are used in ion implantation processes. Antimony, arsenic, and phosphorus are the common solid sources, and arsine, phosphine, and boron trifluoride are the commonly used gases.

Antimony (Sb) is a brittle, silvery-white metallic element used as an n-type dopant in implantation processes. It is poisonous; direct contact with solid antimony can cause irritation of the skin and eyes. Antimony dust is extremely toxic; direct contact with it can cause severe irritation of the skin, eyes, and lungs. It can also cause heart, liver, and kidney damage.

Arsenic (As) is poisonous; direct contact with solid arsenic can cause irritation of the skin and eyes, and it also can cause discoloration of the skin. Arsenic dust is very toxic; direct contact can cause irritation of the skin and lungs. It can also cause nose and liver damage, and possibly lung and skin cancer.

Red phosphorus (P) is a commonly used solid material in implantation processes as an n-type dopant and can also be found on the side of matchboxes in the form of a dark stripe. It is flammable and can ignite with friction. Direct contact can be irritating to the skin, eyes, and lungs.

Arsine (AsH_3) is commonly used as an arsenic source gas; it is one of the most poisonous gases used in the semiconductor industry. At 0.5 to 4 ppm, a garlic-like

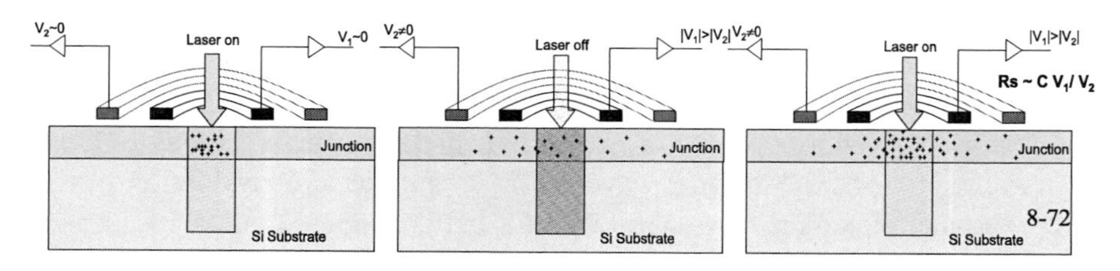

Figure 8.49 Optical-electrical measurement.

odor can be detected; at 3 ppm it can cause an immediate danger to life and health (IDLH). This is the reason that the smell of garlic is the most feared smell inside an IC fab. It can cause irritation to the nose and eyes at low concentrations. Exposure at 500 ppm for a few minutes can be fatal. Arsine is also flammable and becomes explosive when its concentration in air is 4 to 10%.

Phosphine (PH_3) is commonly used as a phosphorus source gas. It is flammable and becomes explosive when its concentration in air is higher than 1.6%. Phosphine is a poisonous gas with a rotten fish odor, which is detectable at 0.01 to 5 ppm; IDLH limit is 50 ppm. With exposure at low concentration, it can be irritating to the eyes, nose, and lungs. Exposure at 10 ppm can cause headache, breathing difficulties, cough and chest tightness, loss of appetite, stomach pain, vomiting, and diarrhea.

Boron trifluoride (BF_3) is commonly used as a boron source gas. It is corrosive because it forms hydrogen fluoride acid when it comes in contact with water. Exposure to it can cause severe skin, eye, nose, throat, and lung irritation; it also can cause a build up of fluid on the lungs.

Decaborane ($B_{10}H_{14}$) is used for molecular ion implantation for USJ formation. It is a solid with low vapor pressure at room temperature. It is toxic and when absorbed through skin can affect the central nervous system. The Occupational Safety and Health Administration (OSHA) permissible exposure limit (PEL) for decaborane is 0.05 ppm (0.3mg/m^3).

Octadecaborane ($B_{18}H_{22}$) and carborane ($C_2B_{10}H_{12}$ or CBH) are two other promising dopants for forming boron USJ with high throughput materials for molecular ion implantation. Both of them are solid materials with very low vapor pressure at room temperature.

8.5.2 Electrical hazards

Contact with high voltage and current flow can cause electrical shock, skin burns, muscle and nerve damage, heart failure, and death. About 1 mA of current across the heart can be fatal. Statistics show that death rates from contact with a 250-V ac power line is about 3%. Predictably, this percentage increases dramatically when the voltage of the electrical power supply is higher than 10 kV.

The spark breakdown voltage of air is about 8 kV/cm. For an implanter with an acceleration electrode powered at 250 kV, the breakdown distance is about 31 cm. With a sharp point, breakdown distance can be even longer. Therefore, safety interlocks are required for ion implanters to prevent electrical power from ramping up when the implanter is not fully shielded.

Because high voltage can generate large amounts of electrostatic charge, which can cause a dangerous electrical shock if contacted without full discharge, grounding bars are required for making contact and discharging all parts before any contact is made.

Ion implanters are fully enclosed systems and are so big that workers can easily be hidden inside them without others noticing. Therefore, before entering a system, it is very important to work with a partner and tag out of the system, to make sure others know someone is working inside and not to start the equipment and raise the voltage. Employees must always take a door key when entering an implanter so that other employees cannot lock the door for system start up.

8.5.3 Radiation hazards

When an energetic ion beam hits a wafer, slits, beam stop, or anything along the beam line, energy loss from the ion emits out in the form of x-ray radiation. A safety interlock is needed to prevent systems from being turned before the system is completely shielded.

Electrons generated by ion collisions with neutral atoms along the beam line and by secondary electron emission from the solid surface can also be accelerated by the electric field due to dc voltage of the accelerating electrodes. Suppression electrodes are used to prevent these electrons from accelerating to a high energy, becoming backstreamed to bombard the ion source and other beam line parts, and causing x-ray radiation and parts damage.

8.5.4 Mechanical hazards

A spin wheel and spin disk can rotate up to 1250 rpm; the velocity of the wafer can be as fast as 90 m/sec (~220 mph) while in full-speed spin. In the event of malfunction, these systems can release huge amounts of energy and cause extensive damage. Constant monitoring of the vibration magnitude of the spin wheel or disk is necessary to make certain that the spin wheel or disk can be stopped before failure happens. Spin and scan motors are very powerful; any attempt to adjust something when these motors are still active could cost fingers and arms.

8.6 Recent Developments and Applications

As the minimum feature size of devices continues to reduce, the junction depth of the channel and S/D of the MOSFET will become shallower and shallower. USJ ($x_j \leq 0.05$ μm) formation raises a big challenge for both ion implantation and annealing, especially for p-type USJ, since it requires low- to high-current, pure boron ion beams at very low energy. Requirements for the USJ are shallow junctions, low sheet resistance, low contact resistance, compatibility with silicide with low diode leakage current, minimal impact on channel profile, and compatibility with polysilicon or high-κ/metal gates. Other requirements are low cost, good within-wafer and wafer-to-wafer uniformity, low added particle count, and reliable transistors and contacts.

In addition to developing single-wafer implanters with high-current and ultralow energy (as low as 0.1 keV) with high energy purity, scientists and engineers have also searched for other approaches for USJ formation. Molecular ion implantation is one of them. In regular ion implantation, BF_3 is commonly used for p-type shallow junction implantation (not B, because the BF_2^+ ion is larger and heavier). Decaborane ($B_{10}H_{14}$), octadecaborane ($B_{18}H_{22}$), and carborane ($C_2B_{10}H_{12}$ or CBH) are the large molecules used for these studies. There are several benefits to using large molecules for USJ formation. It is very difficult to achieve high energy purity for 0.1-keV ion beams because the ion beam must be accelerated to about 5 keV for a magnetic mass spectrum analyzer to effectively isolate the desired ion species, then be decelerated by 4.9 keV. The power supply voltages for acceleration and deceleration must be very accurate and stable to achieve energy purity. Because large molecules are significantly bigger and heavier, they require a higher energy to achieve the same junction depth as BF_2^+ or B^+ ions. It is much easier to achieve high energy purity for a 1-keV (or higher) energy ion beam than it is for a 0.1-keV ion beam. These larger molecules also consist of many boron atoms; for example, CBH has ten boron atoms, thus CBH molecular ion implantation can achieve ten times higher throughput than BF_2^+ or B^+ ion implantation at the same beam current. Large-molecule implantation also causes more lattice damage, thus it has less tunneling effect and better junction profile control.

A plasma immersion ion implantation (PIII) or plasma doping (PLAD) system has been developed for targeting low-energy, high-dosage applications such as USJ and deep trench. The PIII system usually has a plasma source power for generating high-density plasma to ionize dopant gas, and a bias power to accelerate dopant ions into a wafer surface, as shown in Fig. 8.50. The most commonly used dopant gas in PLAD is B_2H_6, used for boron doping. Source power can be rf or MW. B_2H_6 can quickly dope a wafer with a very high dose; that dosage can be so high that even at the highest beam current, a beam-line-based implanter would require such a long implantation time that throughput requirements would not be met. The tradeoff of PLAD is that it cannot select ion species or precisely control ion flex or dosage. Therefore, its main applications are in high-dose, noncritical layers. It has been widely used in DRAM chip manufacturing for polysilicon counter-doping of periphery devices. It also could be used in contact implantations of array devices of DRAM.

In plasma immersion systems, dopant ions bombard the wafer and are implanted into the substrate, dopant ion flux is mainly controlled by MW power, and ion

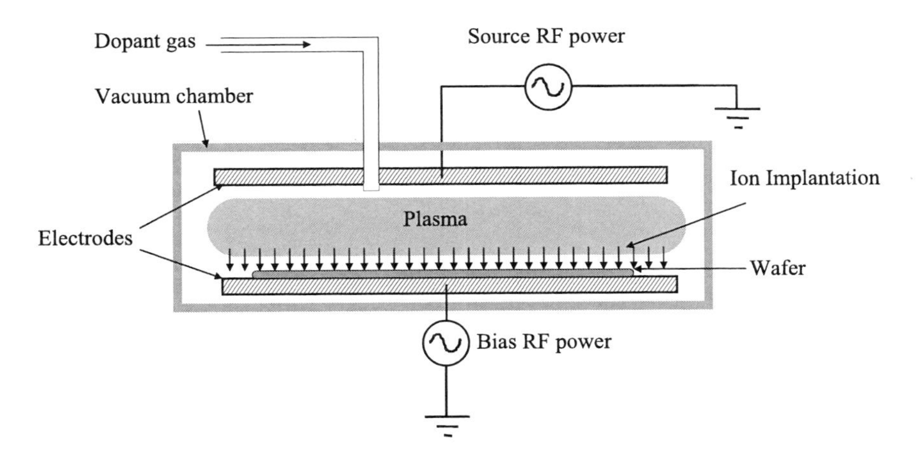

Figure 8.50 Plasma immersion ion implantation (or plasma doping) system.

energy is mainly determined by the bias rf power. Current through the magnets can affect resonance position, thus it can be used to control the position of plasma, which in turn controls dopant uniformity.

PIII is a low-energy process; its ion energy is less than 1 keV. Therefore, it can be applied for USJ formation for sub-0.1-μm device applications. A plasma immersion system cannot select specific ion species; this is a disadvantage compared with standard ion implantation. Other disadvantages are that ion flux can be affected by plasma position and chamber pressure, and that ion energy is distributed in a wide range instead of in a sharp narrow distribution like the ion implanter. Thus, PIII has difficulties precisely controlling dopant concentration and junction depth.

8.7 Summary

- Dopants commonly used in IC chip fabrication are boron for p-type and phosphorus, arsenic, and antimony for n-type.
- CMOS processes require many ion implantations, such as well/threshold implantations, LDD or SDE implantations, poly-dope implantations, and S/D implantation processes, one each for pMOS and nMOS.
- In addition to the above implantations, DRAM chip manufacturing has unique contact ion implantation.
- Ion implantation processes can precisely and independently control dopant junction depth and dopant concentration. Ion implantation is an anisotropic process, which is required for submicron processes.
- An ion implanter consists of gas, electrical, vacuum, control, and beam line systems.
- A beam line system consists of an ion source, extraction electrodes, mass analyzer, postacceleration assembly, charge neutralization system, wafer handler, and beam stop.
- Formation of abnormally deep junctions caused by long ion travel ranges in single-crystal channels is called the channeling effect. Wafer tilt and screen oxide are two commonly used methods to minimize this effect.
- For the same ion species, the higher the ion energy, the longer the ion range. At the same energy level, the lighter the ion, the longer the ion range.
- Implantation causes crystal structure damage; postimplantation thermal annealing is needed to recover the single-crystal structure and activate the dopant.
- Ion implantation processes use many hazardous solids and gases, most of which are poisonous, flammable, explosive, or corrosive. Other safety hazards are high electrical voltage, x-ray radiation, and moving parts.

8.8 Bibliography

D. G. Baldwin, M. E. Williams, and P. L. Murphy, *Chemical Safety Handbook for the Semiconductor/Electronics Industry*, 2nd ed., OME Press, Beverly, MA (1996).

S. A. Cruz, "On the energy loss of heavy ions in amorphous materials," *Radiat. Eff.* **88**, 159 (1986).

M. I. Current, *Basics of Ion Implantation*, Ion Beam Press, Austin, TX (1997).

T. Roming, J. McManus, K. Olander, and R. Kirk, "Advanced in ion implanter productivity and safety," *Solid State Technol.* **39**(12), 69-74 (1996).

W. Shockley, *Forming Semiconductor Devices by Ion Bombardment*, U.S. Patent No. 2787564 (1954).

B. G. Streetman and S. Banerjee, *Solid State Electronic Devices*, Prentice Hall, Upper Saddle River, NJ (1999).

S. M. Sze, *Physics of Semiconductor Devices*, 2nd ed., John Wiley & Sons, Inc., New York (1981).

S. M. Sze, *VLSI Technology*, 2nd ed., McGraw-Hill Companies, Inc., New York (1988).

J. F. Ziegler, *Ion Implantation - Science and Technology*, Ion Implantation Technology Co., Yorktown, NY (1996).

See http://www.webelements.com for further information on the elements in this chapter.

8.9 Review Questions

1. List the advantages of ion implantation over diffusion in doping processes.
2. Name at least three implantation processes for CMOS IC fabrication.
3. What is unique DRAM implantation? What is the purpose of this implantation?
4. What major changes did ion implantation bring to IC fabrication processes?
5. What are the two ion stopping mechanisms?
6. The most important factors for a doping process are dopant concentration and dopant junction depth. Which systems in an ion implanter can be used to control these two factors?
7. When two ions with the same energy and incident angle implant into single-crystal silicon, will they stop at the same depth in the silicon? Explain your answer.
8. Describe the relationships between the ion projection range, ion energy, and ion species.
9. Why does a wafer need annealing after ion implantation?
10. List the advantages of RTA over furnace annealing.
11. What ion implantations require spike and laser annealing?
12. Why does an ion implanter beam line require high vacuum during processing?
13. What is the most poisonous gas used in IC fabrication? What does it smell like? Is it a p-type or n-type dopant?

14. Why does an ion implanter need a high-voltage power supply?
15. Before entering an ion implanter, why is the use of a grounding bar required?
16. Explain the reason for wearing double-layer gloves when wet cleaning a beam line tube.
17. Why are wafers tilted during the ion implantation process?
18. List the energy and current requirements of well implantation and S/D implantation processes. Explain your answer.
19. If the dc current in analyzer magnets is not correct, what could be a possible processing problem?
20. Compare four-point probe and thermal wave measurements and list their advantages and disadvantages.
21. What are the main differences between ion implantation and PIII?
22. What other industry or industries could use ion implantation besides the semiconductor industry?

Chapter 9
Etch

Objectives

After finishing this chapter, the reader will be able to:

- list at least three materials that require etching in IC chip fabrication
- identify at least three etch processes from a CMOS chip cross section
- describe both wet and dry etch processes and identify their differences
- describe the plasma etch processing sequence
- name the most commonly used etchants and identify their safety hazards.

9.1 Introduction

Etch is a process that removes materials from a wafer surface to achieve the requirements of IC design. There are two types of etch processes: pattern and blanket. Pattern etch selectively removes materials from designated areas and transfers the patterns of the photoresist or hard mask on the wafer surface to films underneath. Blanket etch removes all or part of the film on the surface to achieve the desired processing results. This chapter covers both etch processes, with an emphasis on pattern etch. Figure 9.1 illustrates a MOSFET gate patterning process. First, a photolithography process with gate mask defines the photoresist pattern on the polysilicon film on the wafer surface, as shown in Fig. 9.1(a). A pattern etch process transfers the pattern of the photoresist to polysilicon film underneath [see Fig. 9.1(b)]. Figure 9.1(c) shows the photoresist after being stripped by either a dry or wet process (or a combination of both), which finishes the gate patterning.

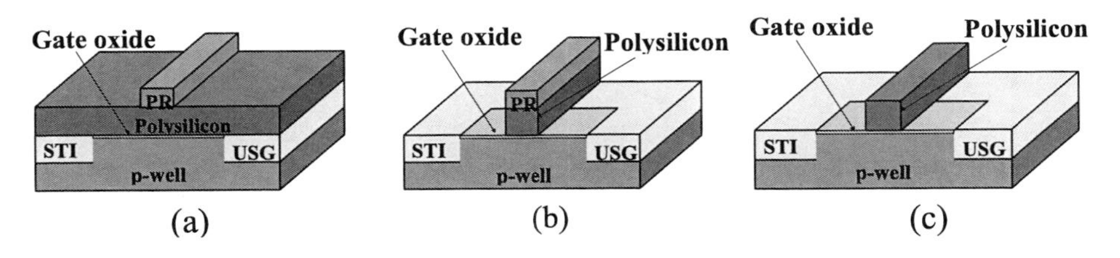

Figure 9.1 Pattern etch processes of a polysilicon gate: (a) photolithography, (b) etch polysilicon, and (c) strip photoresist.

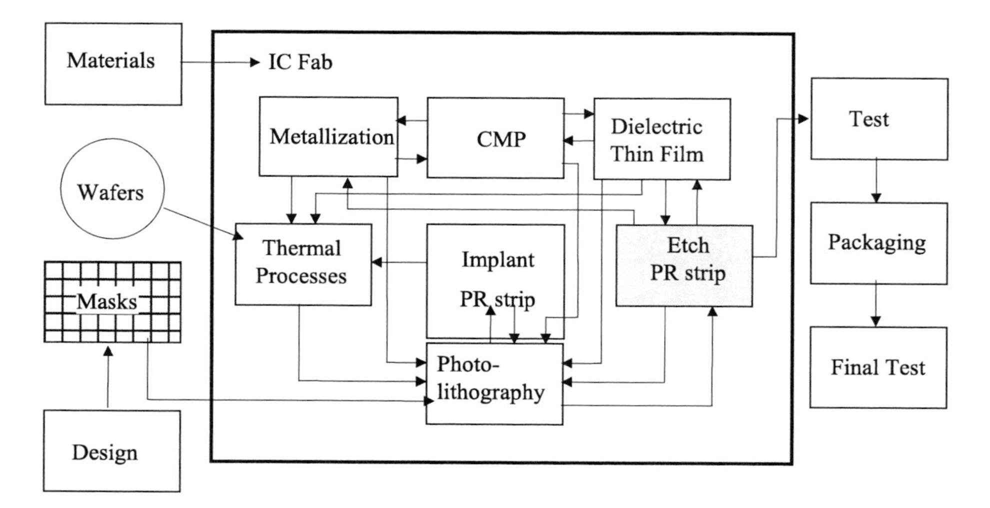

Figure 9.2 Advanced IC fabrication process flow.

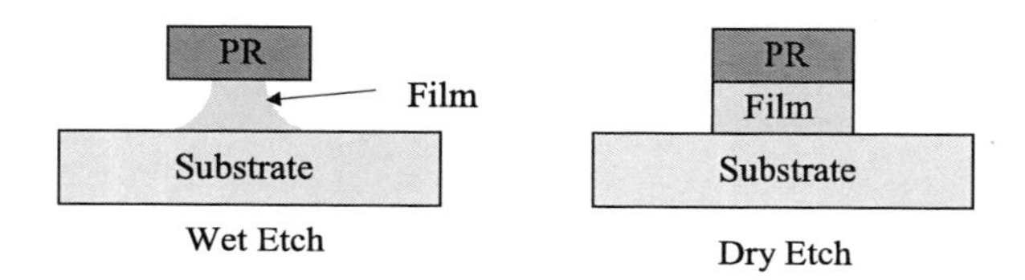

Figure 9.3 Wet and dry etch profiles.

A combination of photolithography and wet etch processes has been used by the printing industry for a long time and is still used to make printed circuit boards (PCBs). These technologies were adopted by the semiconductor industry for transistor and IC manufacture in the 1950s. Device and circuit designs are transferred to a wafer, by either etch or ion implantation, through the pattern defined on the photoresist on a wafer surface by means of the photolithography process. Figure 9.2 shows pattern etch in an IC chip manufacturing flow.

Before 1980, wet processes dominated semiconductor fabs. Chemical solutions were used to dissolve the materials not covered by photoresist to achieve pattern transfer. After 1980, when the minimum feature size became smaller than 3 μm, wet etch processes were gradually replaced by dry (plasma) etch processes. The reason for this is that wet etch always has an isotropic etch profile, which can cause CD loss, as shown in Fig. 9.3.

In advanced semiconductor fabs, almost all pattern etch processes are plasma etch processes. However, wet etch processes are still widely used for film stripping and thin-film quality control. Figure 9.4 illustrates the layers of some etch processes from the cross section of a CMOS IC chip with aluminum metallization technology.

Many etch processes (including both pattern and blanket etches) are involved in IC chip processing. For example, single-crystal silicon etch is needed to

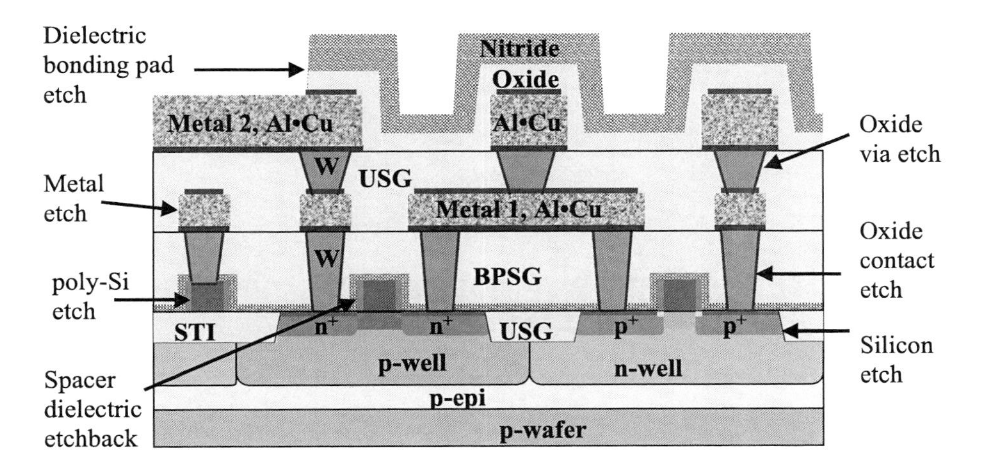

Figure 9.4 Etch processes in a CMOS IC chip with polysilicon gate and aluminum metal interconnection.

form shallow trench isolation (STI), and polysilicon etch defines gates and local interconnections. Oxide etch defines contact and via holes, and metal etch forms interconnections. There are also some blanket etch processes: the nitride stripping process after oxide chemical mechanical polishing (CMP) stops on the nitride layer (not shown in Fig. 9.4) during STI formation; dielectric anisotropic etchback to form sidewall spacers; and titanium stripping after titanium silicide formation (not shown in Fig. 9.4).

Figure 9.5 shows the cross section of an advanced CMOS IC with selective epitaxial S/D, gate-last high-κ and metal gates (HKMG), and copper/low-κ interconnection. Plasma etching of single-crystal silicon is required to form STI, and wet etching of single-crystal silicon is required to form the selective epitaxial S/D. Also, a dry etch of polysilicon is required to form the dummy polysilicon gate, and a wet etch of polysilicon is required to strip the dummy gate to create space for the HKMG. Note that there is no metal etch process for copper/low-κ interconnection. This connection uses a dielectric trench etch instead.

Topics covered in this chapter are etch basics; wet and dry etch; chemical, physical, and reactive ion etch; and etch processes for silicon, polysilicon, dielectric, and metal. Future trends of etch processes are also discussed in this chapter.

9.2 Etch Basics

9.2.1 Etch rate

Etch rate is the measure of how fast material is removed during the etch process. It is a very important characteristic, since it directly affects throughput of the etch process. Etch rate is defined as thickness change caused by the etch process divided by etch time. To calculate etch rate, film thickness must be measured before and

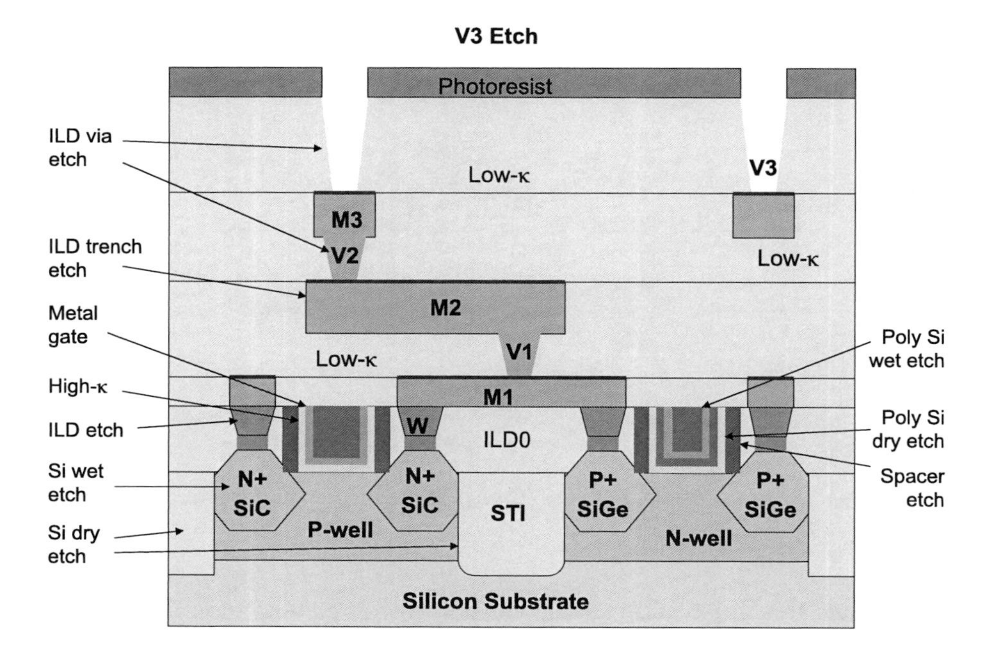

Figure 9.5 Etch processes of an advanced CMOS IC with selective epitaxial S/D, high-κ metal gate, and copper metal interconnection.

after the etch process, and etch time needs to be recorded.

$$\text{etch rate} = \frac{\text{thickness before etch} - \text{thickness after etch}}{\text{etch time}}.$$

Because etch rate is measured on blanket films, it is also called blanket etch rate. For pattern etch, the etch rate can be determined by a cross-section scanning electron microscope (SEM) measurement, which can directly measure the removed film thickness.

Question: A thermal oxide film thickness is 5000 Å; after a 30-sec plasma etch, the thickness becomes 2400 Å. What is the etch rate?

Answer: Etch rate = (5000 Å − 2400 Å)/0.5 min = 2600 Å/0.5 min = 5200 Å/min.

Question: A borophosphosilicate glass (BPSG) contact hole etch profile is shown in Fig. 9.6. What is the etch rate?

Answer: Etch rate = 4500 Å/(45/60) min = 4500Å/0.75 min = 6000 Å/min.

9.2.2 Uniformity

It is very important to have uniform etch rate across a wafer or a good within-wafer (WIW) uniformity, and high repeatability or good wafer-to-wafer (WTW) uniformity. Etch-rate uniformity is calculated from etch rates measured at certain points of a wafer. If these values are $x_1, x_2, x_3, x_4, \ldots, x_N$, N is the total number of

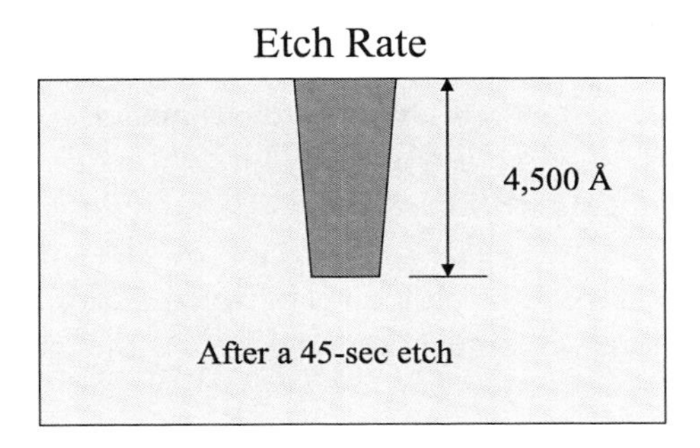

Figure 9.6 A contact profile.

data points. The mean value (or average value) of the measurement is

$$\bar{x} = \frac{x_1 + x_2 + x_3 + \cdots + x_N}{N}.$$

The standard deviation of the measurement is

$$\sigma = \sqrt{\frac{(x_1 - \bar{x})^2 + (x_2 - \bar{x})^2 + (x_3 - \bar{x})^2 + \cdots + (x_N - \bar{x})^2}{N - 1}}.$$

The standard deviation nonuniformity (in percentages) is defined as

$$NU(\%) = (\sigma/\bar{x}) \times 100.$$

Maximum-minus-minimum nonuniformity is defined as

$$NU_M = \frac{x_{max} - x_{min}}{\bar{x}} \times 100.$$

Question: What is the maximum-minus-minimum NU_M for the five-point measurement?

Before etch:
3500, 3510, 3499, 3501, 3493 Å.

After a 60-sec etch:
1500, 1455, 1524, 1451, 1563 Å.

Answer: Etch rates are 2000, 2055, 1975, 2055, and 1930 Å/min. The average etch rate is: $\bar{x}$ = 2003 Å/min, and the maximum-minus-minimum nonuniformity: $NU_M = (2055 - 1930)/(2 \times 2003) = 3.12\%$.

It is very important to have a clear definition of uniformity in the tool specifications, especially when working with the vendors or customers, because different definitions give different results.

9.2.3 Selectivity

The pattern etch process involves three kinds of materials: photoresist, the film that needs to be etched, and the film underneath. During the etch process, all three can be etched by chemical reaction of the etchants and by ion bombardment in a plasma etch process. The difference in the etch rates is characterized as selectivity.

Selectivity S is the ratio of the etch rates between different materials, especially the material that requires etching compared to the material that should not be removed.

$$S = \frac{ER_1}{ER_2}.$$

For example, in the gate etch illustrated in Fig. 9.3, photoresist provides the etch mask, and polysilicon is the material that needs to be etched. In a plasma etch process, inevitably photoresist will be etched; therefore, it is necessary to have high enough polysilicon-to-photoresist selectivity to prevent excessive photoresist loss before the etch process is finished. Underneath the polysilicon is the ultrathin gate oxide (15 to 100 Å, depending on device requirement); the process must have a very high polysilicon-to-oxide selectivity to prevent etching through the gate oxide layer during the polysilicon overetch step.

9.2.4 Profile

One of the most important characteristics of etch is the etch profile, which affects all of the deposition processes that follow. Etch profile is measured by a cross-section SEM. Figure 9.7 shows different types of etch profiles.

A perfect vertical profile is a theoretically ideal profile, since it can transfer images from the photoresist to the underside of the film without any CD loss. However, in many cases (especially for contact and via etch), an anisotropic tapered profile is preferred because the tapered contact and via hole create larger incoming angles and are easier for the following tungsten CVD processes to fill without voids. A pure chemical etch process has an isotropic profile, which causes undercuts beneath the photoresist as well as CD loss. Reactive ion etch (RIE)

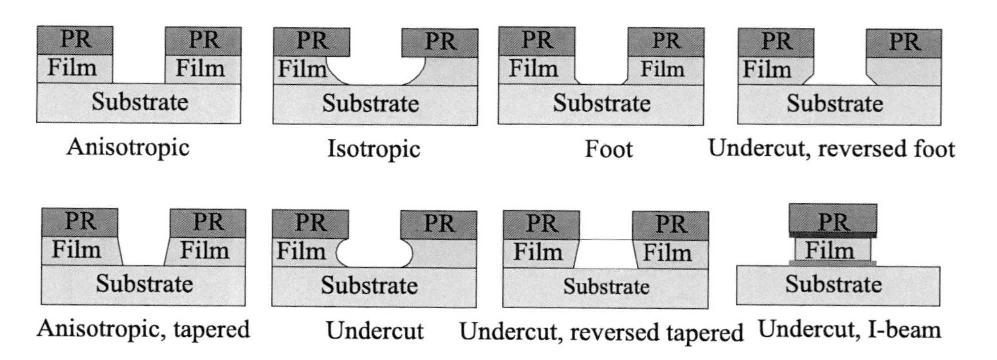

Figure 9.7 Different etch profiles.

combines both physical and chemical etches. Undercut profiles are caused by too many chemical etch components in the RIE process, or excessive ion scattering to the sidewalls. Undercut profiles are undesirable, since it is extremely difficult for the following deposition processes to fill the gaps or holes without voids. An I-beam profile is caused by incorrect etch chemistry for the middle layer of sandwiched film.

9.2.5 Etch bias

Etch bias is the difference between the CD of an etched pattern and the CD of a photoresist pattern. In Fig. 9.1, the anisotropic etch profile has the same CD as the photoresist, or zero etch bias. The undercut I-beam etch profile has CD loss, or negative etch bias. It is possible to have positive etch bias, which occurs when the CD of a photoresist pattern is smaller than the CD of an etched pattern.

9.2.6 Loading effects

In plasma pattern etch processes, etch rate and etch profile are related to the etched pattern. This phenomena is called the etch loading effect. There are two types of loading effects: macroloading and microloading.

9.2.6.1 Macroloading

The etch rate of a wafer with a larger open area is different from a wafer with a smaller open area; this WTW etch-rate difference is called the macroloading effect. Macroloading mainly affects the batch etch process and has minimal effect on the single-wafer process.

9.2.6.2 Microloading

For contact and via hole etch processes, smaller holes usually have a lower etch rate compared to larger holes, as shown in Fig. 9.8. This is called the microloading effect. Etchants have a difficult time passing through the smaller holes to reach the film that needs to be etched, and etch byproducts are harder to diffuse out. These two factors slow down the etch process.

Reducing process pressure can minimize the microloading effect. At lower pressure, the MFP is longer. Therefore, it is easier for etchants to pass through the tiny holes to reach the film. It is also easier for etch byproducts to escape from the tiny holes and be removed.

The etch pattern profile in an isolated pattern area is usually wider than the densely packed area due to photoresist sputtering deposition on the sidewall. Lack of sidewall ion bombardment from the ion scattering from neighboring patterns causes buildup of photoresist on the pattern sidewall and results in a wider profile. Figure 9.9 illustrates the profile microloading effect.

9.2.7 Overetch

During thin-film etching—including polysilicon, dielectric, and metal etch—the etch rate within the wafer is not perfectly uniform, and neither is the film thickness. Therefore, while some parts of the film have already been etched away, other parts

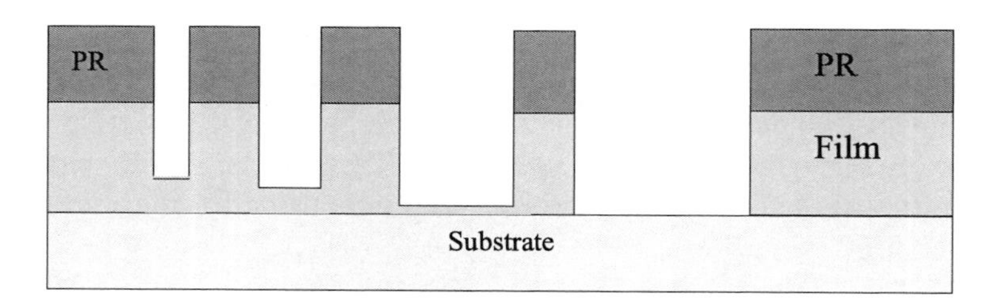

Figure 9.8 The microloading effect.

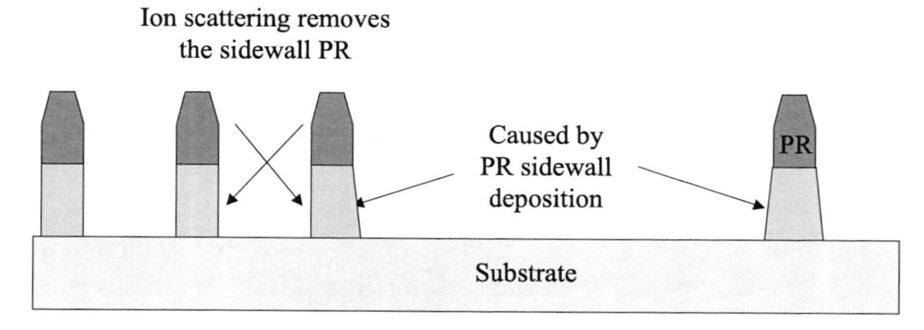

Figure 9.9 Profile microloading effect.

still have some film left behind that needs to be removed. The process that removes the leftover film is called the overetch, and the etch process that removes the bulk of the film is called the main etch.

During the overetch process, selectivity between the etched film and the substrate material must be high enough to prevent excessive substrate material loss. By using different etch conditions from the main etch process, etch selectivity between the etched film and substrate can be significantly improved during the overetch process. For a plasma etch process, overetch can be automatically triggered by an optical endpoint detector, since the chemical composition of the plasma changes when etchants in the main etch process start to etch the substrate film. For example, during the polysilicon gate etch process illustrated in Fig. 9.1, the main etch step removes the bulk polysilicon film and does not bother with selectivity over silicon dioxide. When the polysilicon has been etched away and the etchants in plasma start to etch the silicon dioxide, the radiation intensity of the oxygen line in the plasma increases. This can trigger an electrical signal to end the main etch and switch to overetch, which the high selectivity of polysilicon to silicon dioxide requires.

Figure 9.10 illustrates the main and overetch processing steps. Δd is the film thickness variation due to thickness nonuniformity. $\Delta d'$ is the maximum substrate thickness loss. The minimum requirement of film-to-substrate selectivity during the overetch processing step is: $S > \Delta d/\Delta d'$ if the etch rate is uniform in the illustrated area.

Question: For an IC chip, polysilicon film thickness is 3000 Å, and the thickness nonuniformity is 1.5%. The polysilicon etch rate is 5000 Å/min, and the

Main Etch

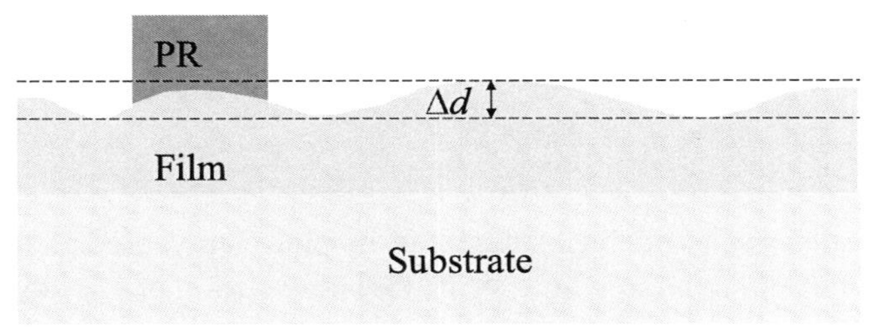

Before Main Etch

Main Etch

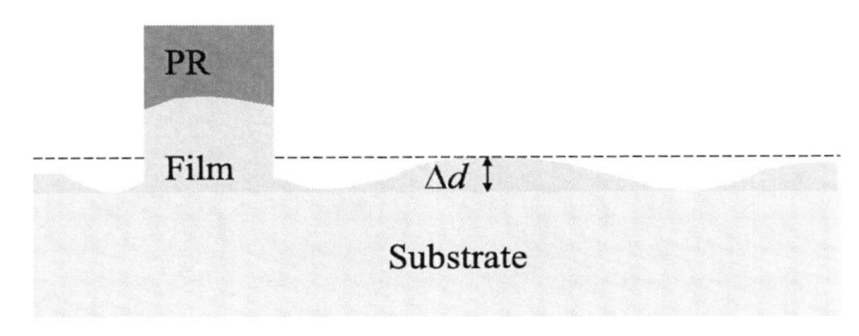

Before Overetch

Overetch

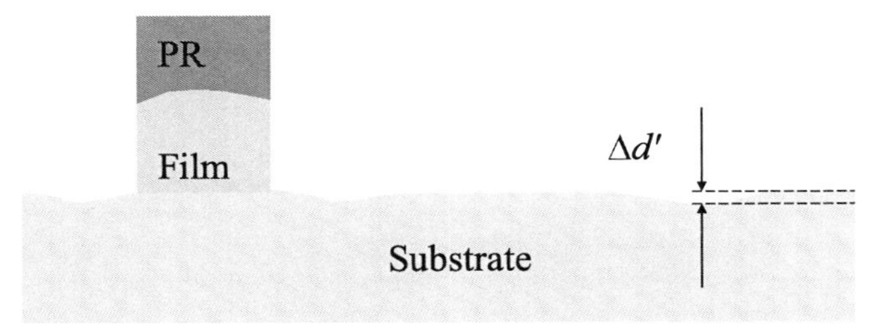

After Overetch

Figure 9.10 Main etch and overetch.

etch-rate nonuniformity within the wafer is 5%. If only 5 Å of gate oxide (t_{ox} ~ 40 Å) is allowed to be lost, what is the minimum poly-to-oxide selectivity during the overetch step?

Answer: Consider the worst possible scenario: the highest etch rate (5000 + 5000 × 5% = 5250 Å/min) on the thinnest film (3000 – 3000 × 1.5% = 2955 Å), and the lowest etch rate (5000 – 5000 × 5% = 4750 Å/min) on the thickest film (3000 + 3000 × 1.5% = 3045 Å). The etch time difference between the two (3045/4750 – 2955/5250 = 0.0782 min) is the overetch time. Since 5 Å of oxide cannot be lost during overetch, the maximum oxide etch rate is 5/0.0782 = 64 Å/min. The minimum poly-to-oxide selectivity is 5000/64 = 78.2.

9.2.8 Residue

After the etch process is finished, there can be some unwanted leftover materials either on the sidewall or the wafer surface. These unwanted leftovers are called residues. Residue can be caused by insufficient overetch of a film with complex surface topography or nonvolatile etch byproducts. Figure 9.11 shows a formation of residue called a stringer on the sidewall due to insufficient overetch of a film with step-like topography. In polysilicon etch processes, stringers are killer defects because they can cause short-circuiting between the polysilicon lines.

Adequate overetching removes most stringers. Sufficient ion bombardment can help remove surface residue, or the right amount of chemical etch can scoop out nonvolatile etch byproducts, such as copper chloride generated during the Al-Cu alloy etch process (illustrated in Fig. 9.12). Organic residue can be cleaned up by an oxygen plasma ashing process, which is also used to strip the remaining photoresist. Wet chemical clean processing can help remove the inorganic residue.

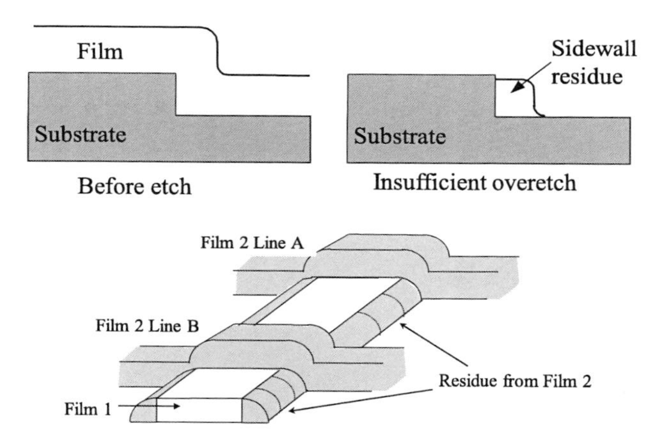

Figure 9.11 Residue caused by insufficient overetching and step topography.

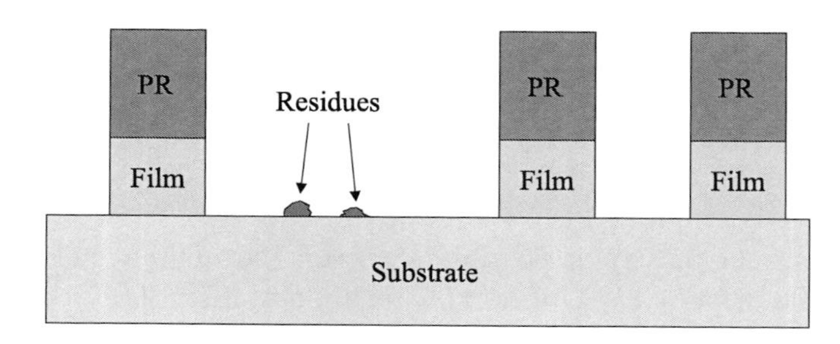

Figure 9.12 Surface residue caused by nonvolatile etch byproducts.

9.3 Wet Etch Process

9.3.1 Introduction

Wet etch is a process that uses chemical solutions to dissolve materials on the wafer surface and achieve device and circuit requirements. The byproducts of wet etch chemical reactions are gases, liquids, or materials that are soluble in the etchant solution. It has three basic steps—etch, rinse, and dry—as shown in Fig. 9.13.

Wet etching was widely used for patterning processes in semiconductor fabs before 1980, when feature size was larger than 3 μm. Wet etching normally has a high etch rate, which is controlled by etchant temperature and concentration. Wet etch processes have very good selectivity. For instance, hydrofluoric acid (HF) etches silicon dioxide very quickly, while it hardly etches silicon at all if used alone. Therefore, using HF to etch a silicon dioxide layer grown on a silicon wafer can achieve very high selectivity of silicon substrate. Compared to dry etch, wet etch uses much less expensive processing equipment because it does not require the vacuum, rf power, and complex gas delivery systems. However, after feature size shrank to less than 3 μm, it became very difficult to continue using wet etch for pattern etch processing because of its isotropic etch profile (shown in Fig. 9.14). It is impossible to wet etch a densely packed pattern with a feature size less than 3 μm; therefore, plasma etch has gradually been replacing wet etch for patterning since the 1980s because it can etch patterns with anisotropic profiles.

Fabs tried very hard to eliminate all wet processes; however, when CMP and electrochemical deposition became widely used in advanced IC fabrications, it became impossible to get rid of all wet processes. Wet etch processes are still widely used in IC fabs for thin-film stripping due to their high selectivity. Wet etch is also applied in blanket thin-film etch processing to determine deposited thin film quality.

9.3.2 Oxide wet etch

Hydrofluoric acid (HF) is commonly used for silicon dioxide wet etch. Since 1:1 HF (49% HF in H_2O) etches oxide too quickly at room temperature, it is very difficult to control the oxide etch process with 1:1 HF. HF is further diluted either

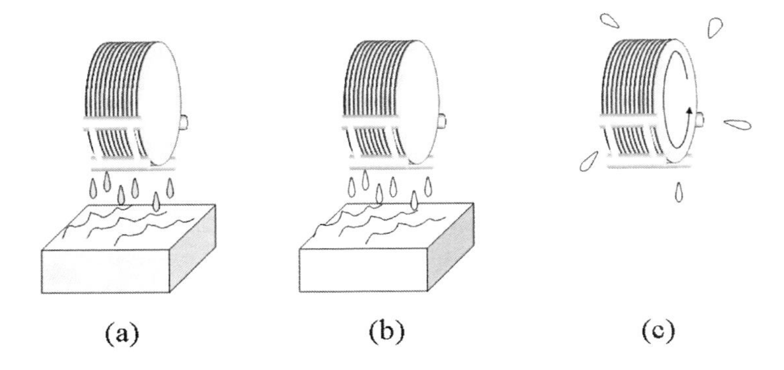

Figure 9.13 Wet etch processing steps: (a) wet etch, (b) rinse, and (c) dry.

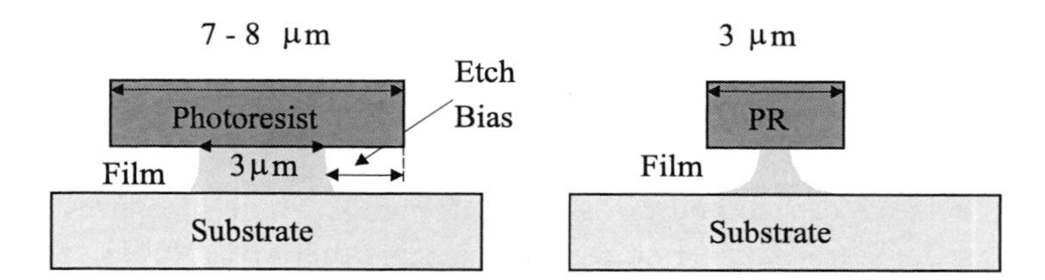

Figure 9.14 Wet etch profiles.

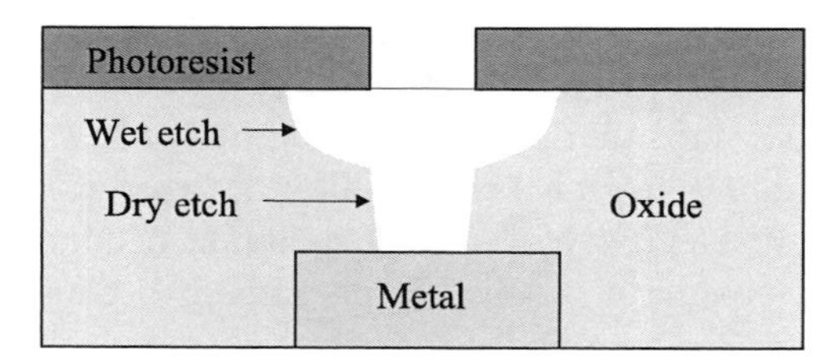

Figure 9.15 Wine-glass contact.

in water or in some buffer solution, such as ammonium fluoride (NH_4F), to slow down the oxide etch for better control of etch rate and uniformity. 6:1 buffered solution of HF, 10:1 HF, and 100:1 in H_2O are commonly used solutions for oxide wet etch.

The chemical reaction of oxide wet etch is

$$SiO_2 + 6\ HF \rightarrow H_2SiF_6 + 2\ H_2O.$$

H_2SiF_6 is water soluble, so an HF solution can easily etch away silicon dioxide. This is why HF solution cannot be kept in a glass container, and HF etch experiments cannot be performed in glass beakers or test tubes.

Some IC fabs still use combined HF oxide wet etching and plasma oxide etching to create so-called wine-glass contact holes (as shown in Fig. 9.15), which are easier for PVD aluminum to fill.

The 6:1 buffered oxide etch (BOE) and 100:1 diluted HF (DHF) etch are used daily in semiconductor fabs, even those with the most advanced technology. They are used to monitor CVD oxide film quality by comparing the wet etch rate of CVD oxide to thermally grown silicon dioxide, called the wet etch rate ratio (WERR). 10:1 HF is also used to strip native oxide from the silicon wafer surface before the thermal oxidation process.

HF is corrosive, and the unique hazard of HF is that skin or eye contact may not be immediately noticed, until up to 24 h later when severe pain can be felt as HF starts to attack the bone. HF reacts with calcium in the bone to form calcium fluoride (which is eventually neutralized). Calcium-containing solution injections can be used to treat HF attacks to prevent or reduce bone loss. Common sense requires that employees treat all clear liquid in fabs as HF—never assume it is water! If direct contact with HF is suspected, the area affected should be washed thoroughly and immediately, and medical attention should be sought.

9.3.3 Silicon etch

Single-crystal silicon etch is used for forming an isolation block between neighboring transistors, and polysilicon etch forms gates and local interconnections. Single-crystal and polycrystalline silicon can be etched isotropically with a mixture of nitric acid (HNO_3) and hydrofluoric acid (HF) through a complex chemical reaction. At first, HNO_3 oxidizes silicon on the surface to form a thin layer of silicon dioxide, which blocks the oxidation process. HF then reacts with the silicon dioxide and dissolves it to expose the silicon underneath, which is oxidized again by HNO_3. Then the oxide is etched away by the HF, and the process repeats again and again. The overall chemical reaction can be expressed as

$$Si + 2HNO_3 + 6\ HF \rightarrow H_2SiF_6 + 2HNO_2 + 2\ H_2O.$$

The mixture of potassium hydroxide (KOH), isopropyl alcohol (C_3H_8O), and water can selectively etch to the different orientations of single-crystal silicon. With 23.4 wt% KOH, 13.3 wt% C_3H_8O, and 63.3 wt% H_2O at 80 to 82 °C, the etch rate along $\langle 100 \rangle$ planes is about 100 times higher than along $\langle 111 \rangle$ planes. The hexagon-shaped groove shown in Fig. 9.16 can be wet etched by this type of anisotropic single-crystal silicon mechanism.

Nitric acid is very corrosive, and it becomes an oxidizer at concentrations higher than 40%. Direct contact can cause severe burns to the skin and eyes and can leave a bright yellow stain on the skin. Nitric acid vapor has a sharp odor and can be irritating to the throat when exposure is at low concentrations. If nitric acid vapor is inhaled at high concentrations, it can cause choking, coughing, and chest pain. In more serious cases, it can cause severe breathing difficulty, bluish skin color, and even death within 24 h due to accumulation of fluid in the lungs.

KOH is corrosive and can cause serious burns. It is harmful if ingested, inhaled, or touched. If the solid or solution comes into contact with the eyes, serious eye damage can result.

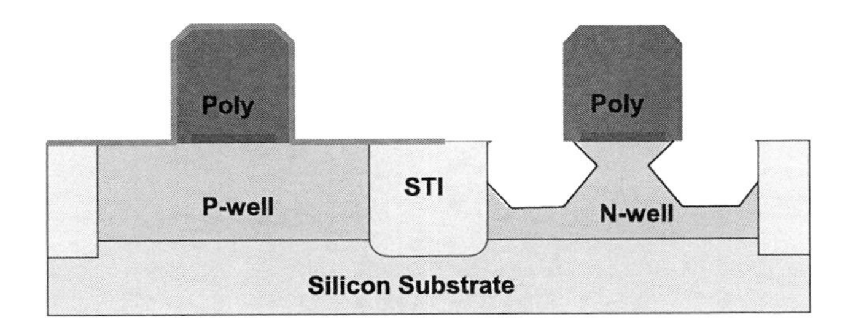

Figure 9.16 The anisotropic KOH silicon etch forms cavities for selective epitaxial SiGe pMOS S/D.

9.3.4 Nitride etch

Nitride is widely used for isolation formation processes. Figure 9.16 shows an isolation process used in the 1970s for bipolar transistor-based IC chip manufacturing, which involved silicon nitride, silicon oxide, and single-crystal silicon etches.

Hot phosphoric acid (H_3PO_4) is most commonly used to etch silicon nitride. At 180 °C with 91.5% H_3PO_4 concentration, the nitride etch rate is about 100 Å/min. This silicon nitride etch process has very good selectivity to thermally grown silicon dioxide (> 10:1) and to silicon (> 33:1). Increasing H_3PO_4 concentration to 94.5% and the temperature to 200 °C increases the nitride etch rate to 200 Å/min, while the selectivity to silicon dioxide drops to about 5:1, and selectivity to silicon reduces to around 20:1.

Question: HF can also be used to etch silicon nitride. However, in the isolation formation process (as shown in Fig. 9.17), both patterned nitride etch and the nitride strip cannot use HF. Why not?

Answer: HF etches silicon nitride at a much lower rate than it etches silicon dioxide. Therefore, using HF for nitride pattern etching can cause excessive pad oxide loss and severe undercuts. If HF is used for the nitride strip, it will very quickly etch away the pad oxide and the isolation oxide!

The chemical reaction of the silicon nitride etch is

$$Si_3N_4 + 4\,H_3PO_4 \rightarrow Si_3(PO_4)_4 + 4NH_3.$$

Both byproducts, silicon phosphate [$Si_3(PO_4)_4$] and ammonia (NH_3), are water soluble. This process is still used in isolation formation processes to strip the nitride—either after the field oxide is grown in the local oxidation of silicon (LOCOS) process, or after undoped silicate glass (USG) polishing and annealing in the STI process.

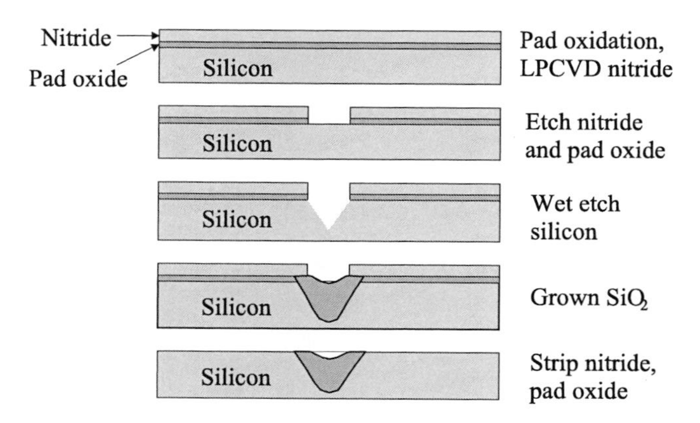

Figure 9.17 Isolation formation processing steps.

Phosphoric acid is an odorless liquid; it is corrosive and can severely burn the skin and eyes by direct contact. Exposure to its mist can cause irritation to the eyes, nose, and throat at a low concentration, while it can cause coughing and burning of skin, eyes, and lungs at a high concentration. Long-term exposure can also cause tooth erosion.

9.3.5 Metal etch

Aluminum can be etched by a wide variety of acidic formulations. One of the most commonly used formulations is the mixture of phosphoric acid (H_3PO_4, 80%), acetic acid (CH_3COOH, 5%), nitric acid (HNO_3, 5%), and water (H_2O, 10%). The etch rate for pure aluminum is about 3000 Å/min at 45 °C. The etch mechanism of aluminum etching is very similar to silicon etching: HNO_3 oxidizes the aluminum to form aluminum oxide, and H_3PO_4 dissolves Al_2O_3. Both oxidization and oxide dissolution proceed simultaneously.

In advanced IC fabs, wet processing is no longer used for aluminum pattern etching; however, some small fabs and research laboratories still use this process.

The most commonly used metal wet etching process in advanced semiconductor fabs is nickel stripping after nickel silicide formation, as shown in Fig. 9.18. The 1:1 mixture of hydrogen peroxide (H_2O_2) and sulfuric acid (H_2SO_4) is commonly used to selectively etch away nickel (Ni) while keeping silicon oxide and nickel silicide intact. The etching mechanism is similar to other metal wet etching processes. Hydrogen peroxide oxidizes nickel to form nickel oxide, while sulfuric acid simultaneously reacts with nickel oxide to form water-soluble $NiSO_4$.

Acetic acid (CH_3COOH, solutions of 4 to 10% in water make up vinegar) is a corrosive and flammable liquid with a strong vinegar odor. Direct contact with acetic acid can cause a chemical burn. High concentrations of acetic acid vapors can cause coughing, chest pain, nausea, and vomiting. Hydrogen peroxide (H_2O_2) is an oxidizer. Contact can cause irritation and burns to the skin and eyes. High concentrations of its vapor can cause severe irritation of the nose and throat, and fluid in the lungs. H_2O_2 is unstable; it self-decomposes while in storage. Sulfuric acid (H_2SO_4) is corrosive and can cause the skin to burn with direct contact. Even

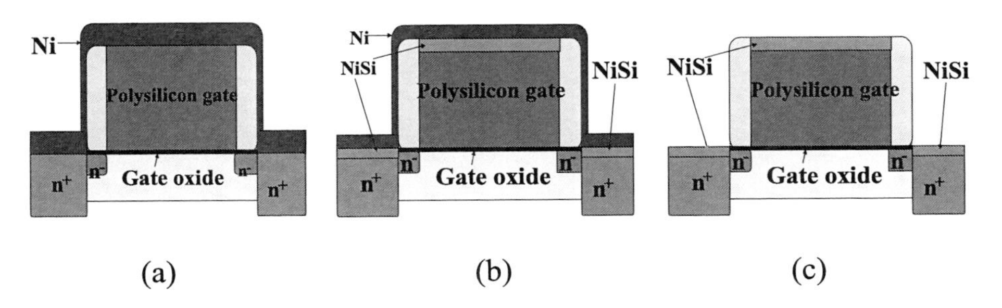

Figure 9.18 Self-aligned silicide (salicide) processes: (a) nickel deposition, (b) nickel silicide annealing, and (c) nickel wet stripping.

the diluted solution can cause a skin rash. Highly concentrated mist can cause severe chemical burns to the skin, eyes, and lungs.

9.4 Plasma (Dry) Etch

9.4.1 Introduction

Dry etch processes use gaseous chemical etchants to react with materials to be etched to form volatile byproducts, which will be removed from the substrate surface. Plasma generates chemically reactive free radicals that can significantly increase the chemical reaction rate and enhance the chemical etch. Plasma also causes ion bombardment to the wafer surface. Ion bombardment can either physically remove materials from the surface or break the chemical bonds between atoms on the surface, significantly accelerating the chemical reaction rate for the etch process. This is why most dry etch processes are plasma etch processes.

After the 1980s when feature size became smaller than 3 μm, plasma etch processes gradually replaced wet etching for pattern etch processes. The reason for this was that isotropic etch profiles of wet etch processes could no longer meet the requirements of such a small geometry. Plasma etch is anisotropic due to ion bombardment associated with the presence of plasma. Therefore, it has a much smaller etch bias and CD loss compared to the wet etch process. Table 9.1 compares wet and dry etch processes.

9.4.2 Plasma review

Plasma is ionized gas with an equal number of negative and positive charges. It also consists of ions, electrons, and neutral atoms or molecules. The three important collisions in plasma are ionization, excitation-relaxation, and disassociation. These collisions generate and sustain the plasma, cause the glow of gas discharge, and create chemically reactive free radicals that enhance chemical reactions, respectively.

MFP is the average distance a particle can travel before it collides with another particle. Reducing pressure increases MFP, increases ion bombardment energy, and reduces collision-induced ion scattering, helping to achieve vertical etch profiles.

Plasma always has a higher potential than electrodes because the faster moving electrons always charge the electrodes negatively at the start of the plasma. Higher

Table 9.1 Comparison of wet and dry etch processes.

	Wet Etch	Dry Etch
Etch bias	Unacceptable for < 3 μm	Minimum
Etch profile	Isotropic	Anisotropic to isotropic, controllable
Etch rate	High	Acceptable, controllable
Selectivity	High	Acceptable, controllable
Equipment cost	Low	High
Throughput	High (batch)	Acceptable, controllable
Chemical usage	High	Low

plasma potential causes ion bombardment because positively charged ions are accelerated to the lower potential electrodes by the sheath potential. Increasing rf power increases both ion bombardment flux and ion bombardment energy in capacitively coupled plasmas. Increasing rf power also increases free radical concentration.

Since etch is a removal process, low operation pressure is desired. The MFP is longer at a lower pressure, which is good for both ion bombardment and byproduct removal. Some etch chambers also use magnetic coils to generate a magnetic field for increasing plasma density at low pressure (<100 mtorr). As a removal process, the plasma etch process requires much more ion bombardment than does a PECVD process. Therefore, in most etch processes, a wafer is put on the smaller electrode to take advantage of the self-bias and receive more energetic ion bombardment.

High-density plasma at a low pressure is desired for both etch and CVD processes. However, commonly used capacitively coupled plasma sources cannot generate high-density plasma. ICP and ECR systems are the two most commonly used high-density plasma sources. By using a separated bias rf system, both ICP and ECR plasma sources can independently control both flux and energy of ion bombardment.

9.4.3 Chemical, physical, and reactive ion etches

There are three kinds of etch processes: pure chemical etch, pure physical etch, and RIE, which is something in-between.

Wet etch is a pure chemical etch process. Another example of pure chemical etch is the remote plasma photoresist strip. For pure chemical etch, there is no physical bombardment; materials are solely removed by chemical reaction. The etch rate of a pure chemical etch can be very high or fairly low, depending on the process. Pure chemical etch always has an isotropic etch profile, which prevents it from being applied to pattern etching when the feature size is smaller than 3 μm. Since it normally has very good etch selectivity, the pure chemical etch process is used for strip processes, such as photoresist, silicon nitride, pad oxide, screen oxide, and sacrificial oxide strips. Remote plasma (RP) etch processes, which use plasma to generate free radicals in a remote chamber and flow them into the processing chamber to react with the wafer, are pure chemical etch processes.

Argon sputtering is a purely physical etch. It is widely used in dielectric sputtering etchback to taper the openings for easier gap fill in the subsequent deposition. It is also used in precleaning to remove native oxide before metal PVD processes to reduce contact resistance. Argon is an inert gas. Therefore, there is no chemical reaction during the process. The material is physically dislodged from the surface by energetic argon ions, similar to being hit with a hammer. Etch rate of a pure physical etch is very low and mainly depends on the flux and energy of the ion bombardment. Since ion bombardment hits and removes anything in its way, selectivity for a pure physical etch is very poor. The direction of ion bombardment is mostly perpendicular to the wafer surface in plasma etch processes. Therefore, a pure physical etch is an anisotropic etch process, which etches mainly in the vertical direction.

The name RIE can be slightly misleading. The proper name of this type of etch process should be an ion-assisted etch, since the ions in this etch process are not necessarily reactive. For example, in many cases argon ions are used to increase ion bombardment. As an inert atom, argon is not chemically reactive at all. Reactive species in most etch processes are neutral free radicals, which have much higher concentrations than ions in semiconductor etch processing plasmas. This is because the activation energy of ionization is significantly higher than the activation energy of dissociation, and the species concentration is exponentially related to the activation energy. However, the term RIE has been used in the semiconductor industry for such a long time that no one will likely change it. Figure 9.19 shows a schematic of an early experiment with ion-assisted etch, along with the results.

The experiment starts with a XeF_2 gas flow, achieved by opening the shutoff valve. XeF_2 is an unstable gas because xenon is a noble gas and does not like to bond with any other atom. It is always used as a free fluorine carrier for dry chemical etch processes. When XeF_2 hits the heated single-crystal silicon sample, it disassociates and releases two free fluorine radicals. A free fluorine radical has only one unpaired electron, which makes it seek out an electron from other atoms; this makes it chemically very reactive. Fluorine reacts with silicon on the sample surface, forming volatile silicon tetrafluoride (SiF_4) and chemically dry etching the silicon. Measurement results in Fig. 9.19 show that the etch rate is very low for this purely chemical etch process.

Next, the argon ion gun is turned on. The silicon etch rate increases dramatically with the combination of the physical ion beam bombardment and the free fluorine radical chemical etch. After termination of the XeF_2 gas flow, silicon is etched solely by argon ion sputtering. This is a pure physical etching process with very low etch rate, even lower than a pure chemical etch with only XeF_2 gas flow.

From Fig. 9.19, it can be seen that the combination of both XeF_2 gas flow and argon ion bombardment has the highest etch rate, which is much higher than the sum of the two individual etch rates. This is because the argon ion bombardment breaks the chemical bonds of the silicon atoms on the surface. Silicon atoms

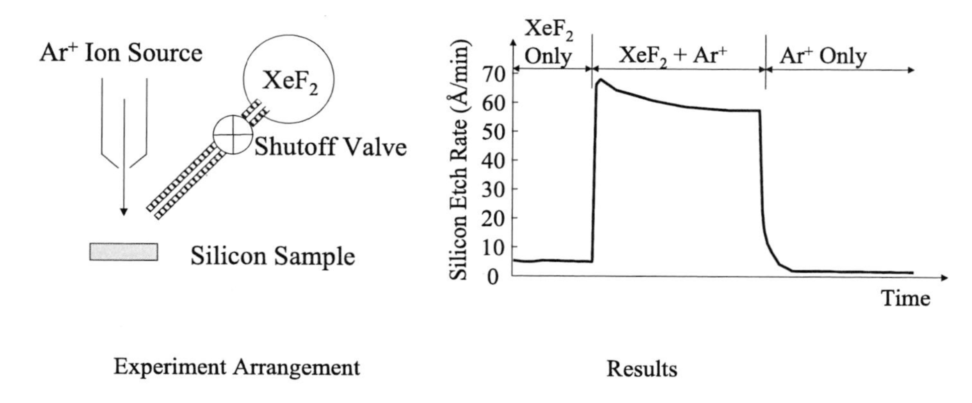

Figure 9.19 The ion-assisted etch experiment (Coburn and Winters, *J. Applied Physics*).

with dangling bonds on the surface bond with free fluorine radicals and form silicon tetrafluoride much more easily than do silicon atoms without broken bonds. Because ion bombardment is mainly perpendicular to the wafer surface, vertical etch rate is significantly higher than horizontal etch rate. This is the reason RIE has an anisotropic etch profile.

Almost all pattern etch processes in advanced semiconductor fabs are RIE processes. RIE has a reasonable and controllable etch rate and etch selectivity. It also has an anisotropic and controllable etch profile. Table 9.2 summarizes these three types of etch processes.

9.4.4 Etch mechanisms

In plasma etch processes, first the etchants are introduced to the vacuum chamber. After pressure is stabilized, rf power is used to strike discharge plasma glow. Some etchant molecules are dissociated in the plasma by the impact of collisions with electrons, which generate free radicals. The free radicals then diffuse across the boundary layer, reach the wafer surface, and are adsorbed on the surface. With the help of ion bombardment, these free radicals react with surface atoms or molecules very quickly and form gaseous byproducts. The volatile byproducts desorb from the surface, diffuse across the boundary layer, move into the convection flow, and are pumped out of the chamber. This plasma etch process sequence is illustrated in Fig. 9.20.

Plasma etch can achieve an anisotropic etch profile due to ion bombardment associated with the plasma. There are two anisotropic mechanisms: damaging and blocking; both are related to ion bombardment.

For damaging mechanisms, energetic ion bombardment breaks the chemical bonds between the atoms on a wafer surface. Atoms with dangling bonds on the surface are vulnerable to the etchant free radicals. They bond easily with the etchant radicals to form volatile byproducts, which can be removed from the surface. Because ions mainly bombard in a perpendicular direction, etch rate in the vertical direction is much higher than etch rate in the horizontal direction. Therefore, ion bombardment can achieve an anisotropic etch profile. Etch processes with a damaging mechanism are RIE processes that are closer to the physical etch side. The damaging mechanism for anisotropic etch is illustrated in Fig. 9.21.

Dielectric etch processes (mainly silicon oxide, silicon nitride, and low-κ dielectric etch processes) normally use a damaging mechanism. They are the RIE

Table 9.2 Three types of etch processes.

	Chemical etch	RIE	Physical etch
Examples	Wet etch, strip, PR etch	Plasma pattern etches	Argon sputtering
Etch rate	High to low	High, controllable	Low
Selectivity	Very good	Reasonable, controllable	Very poor
Etch profile	Isotropic	Anisotropic, controllable	Anisotropic
Endpoint	By time or visual	Optical	By time

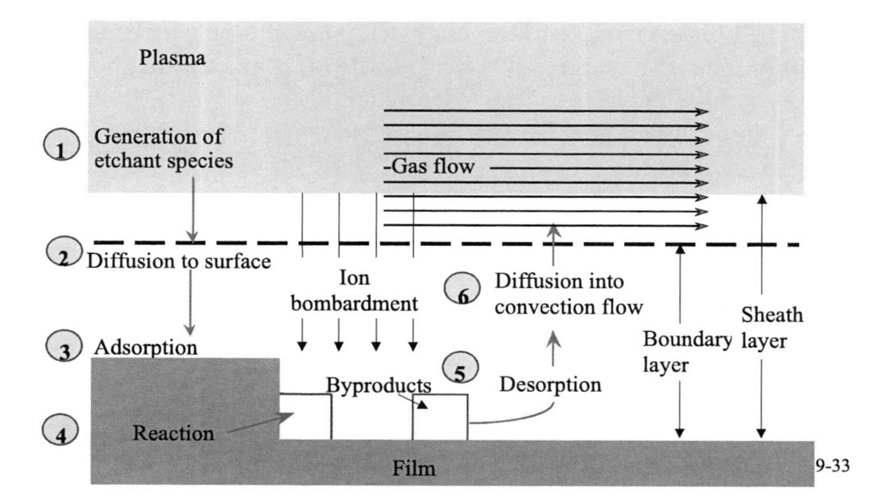

Figure 9.20 Plasma etch sequence.

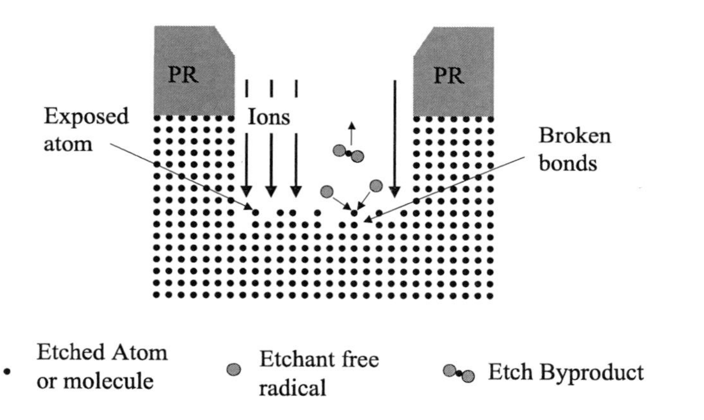

Figure 9.21 Damaging mechanism for anisotropic etch.

processes that are closer to the physical etch side. To improve the anisotropic profile with a damaging mechanism, ion bombardment needs to increase. By using heavy ion bombardment at high rf power and low pressure, a nearly vertical etch profile can be achieved. However, this can also cause plasma-induced device damage, especially for a polysilicon gate etch. Therefore, another anisotropic etch mechanism with less ion bombardment is needed to meet the requirements of these etch processes.

During the development of the single-crystal silicon etch process, someone forgot to strip the photoresist used for the silicon dioxide hard mask patterning before the silicon etch process. (The normal procedure requires stripping of the photoresist before silicon etch to prevent contamination). What happened next led to the discovery of another anisotropic etching mechanism, the blocking mechanism. During the plasma etch process, the ion bombardment sputtered some

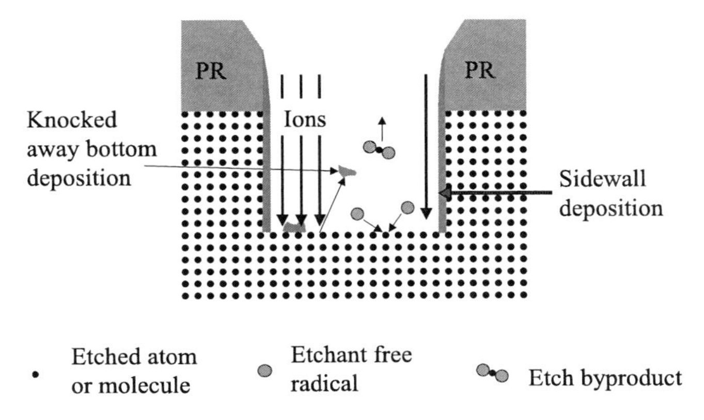

Figure 9.22 Blocking mechanism for anisotropic etch.

photoresist into the gap. The photoresist coating on the sidewall blocked the etch in the horizontal direction, while photoresist deposited on the bottom was constantly removed by ion bombardment from the plasma and exposed the surface to the etchants. Thus, the etch process is now mostly in the vertical direction, as shown in Fig. 9.22.

This mechanism has been used to develop a variety of anisotropic etch processes, in which chemical deposition during etch processing protects the sidewall and blocks etching in the horizontal direction. Etch processes using a blocking mechanism generally require less ion bombardment than etch processes using a damaging mechanism. Single-crystal silicon etch, polysilicon etch, and metal etch processes use this mechanism and are the RIE processes that are closer to the chemical etch side. The sidewall deposition needs to be cleaned by either (or both) dry and wet cleaning processes. Table 9.3 summarizes the two anisotropic etch mechanisms and their applications in the IC processes.

9.4.5 Plasma etch chamber

The plasma process was first applied to etch carbonaceous materials, such as photoresist, in an oxygen plasma, a process also known as plasma stripping or plasma ashing. In this case, oxygen free radicals generated in the plasma by electron dissociation collisions react rapidly with carbon and hydrogen in carbonaceous material to form volatile CO, CO_2, and H_2O. This process effectively removes the carbonaceous material from the surface. A barrel system with an etching tunnel used for this application is illustrated in Fig. 9.23.

This concept was extended in the late 1960s to silicon etch and silicon compound etch that used fluorine-containing gas such as CF_4 as the etchants and SiF_4 as the gaseous etch byproducts.

Another type of dry etch system is a downstream or remote plasma system, which generates plasma in a remote chamber. Etchant gases are allowed to flow through the plasma chamber and dissociate in the plasma, and the free radicals are allowed to flow into the processing chamber to react and etch the materials on the wafer. Figure 9.24 shows the schematics of a downstream etch processing system.

Table 9.3 The two anisotropic etch mechanisms and their applications.

	RIE		
Pure chemical etch	**Blocking mechanism**	**Damaging mechanism**	**Pure physical etch**
No ion bombardment	Light ion bombardment	Heavy ion bombardment	Only ion bombardment
Photoresist strip	Single-crystal silicon etch	Oxide etch	
Silicide metal strip	Polysilicon etch	Nitride etch	Sputtering etch
Nitride strip	Metal etch	Low-κ dielectric etch	

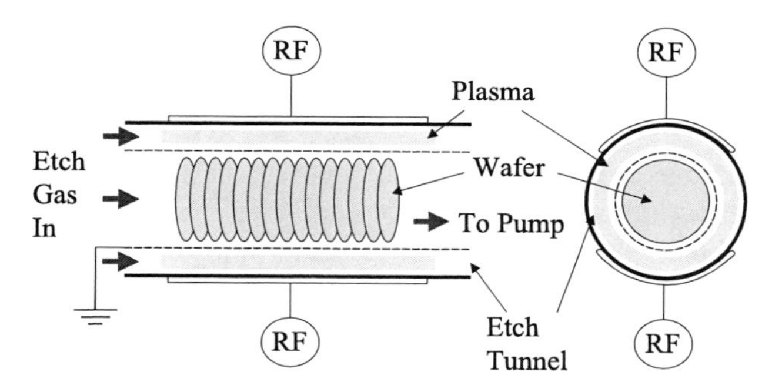

Figure 9.23 Barrel etch system.

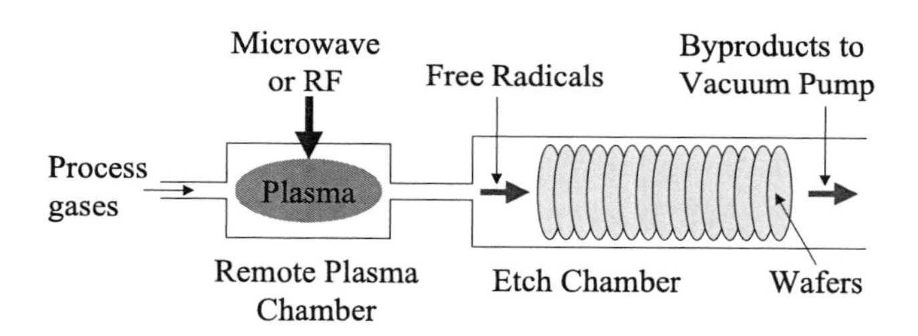

Figure 9.24 Downstream plasma etch system.

Barrel and downstream systems are designed as isotropic etch systems. To achieve a directional etch profile, different etch systems were developed. The parallel plate plasma etch system is one of those approaches. It operates at about 0.1 to 10 Torr, and the wafer sits on the grounded electrode, as shown in Fig. 9.25. Since both rf hot electrodes and grounded electrodes have similar areas, there is very little self-bias on the wafer. Both electrodes receive about the same amount of ion bombardment due to the dc bias of the rf plasma.

By increasing ion bombardment, etch rate can be increased, and more importantly, the directional etch profile can be improved. To increase ion bombardment, rf power needs to increase, and pressure must be reduced. For a parallel plate plasma etch system with equal-area electrodes, increasing rf power increases ion bombardment and etch rate on both the wafer surface and the other

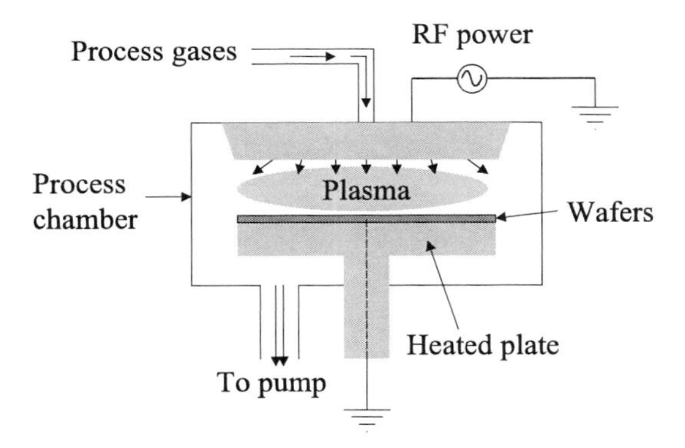

Figure 9.25 Parallel plate plasma etch system.

electrode, the chamber lid. This is very undesirable, since it causes shorter lifetimes for parts inside the chamber and increases particle contamination.

By designing an etch system with a wafer on the smaller, rf hot electrode, and taking advantage of the self-bias, the wafer surface can receive significantly high-energy ion bombardment, while the ion-bombardment grounded energy on the chamber lid is much lower. Ion energy to bombard the wafer is the sum of the plasma dc bias and the self-bias. Ion energy to bombard the chamber lid is the dc bias, which is much lower than the self-bias when the area of the wafer-side electrode is less than one half of the chamber lid area. This type of etch system is the batch RIE system and has been the most popular etch system since the 1980s. Figure 9.26 illustrates this system.

With device dimensions shrinking, requirements of etch uniformity (especially WTW uniformity) become higher. Single-wafer processing tools, which have better WTW process control, started dominating the etch process. Figure 9.27 gives a schematic of a single-wafer magnetically enhanced RIE (MERIE) system.

Reducing process pressure increases the MFP. Therefore, ions have higher energy for bombardment and less scattering due to collisions. Both are good for anisotropic pattern etch profiles. However, at low pressure, the MFP of an electron can be too long to have enough ionization collisions to generate and sustain the plasma. A magnetic field is commonly used to force electrons into a helical movement. This causes electrons to travel much longer distances and increases their chances of ionization collision. Increasing the magnetic field increases plasma density, especially at low pressure. However, it also increases electron density near the wafer surface, which reduces plasma sheath potential (or dc bias) and ion bombardment energy. Without a magnetic field, electrons in the sheath region are repelled to the bulk plasma because the wafer is negatively charged by the fast moving electrons as soon as the plasma begins. With a magnetic field, electrons are not easily expelled due to the helical motion, as shown in Fig. 9.28; this increases electron density and reduces dc bias.

Heavy ion bombardment generates a large amount of heat and raises wafer temperature if the wafer is not properly cooled. For pattern etch, a wafer is coated with a thin layer of photoresist as the pattern mask, which can be burned if the

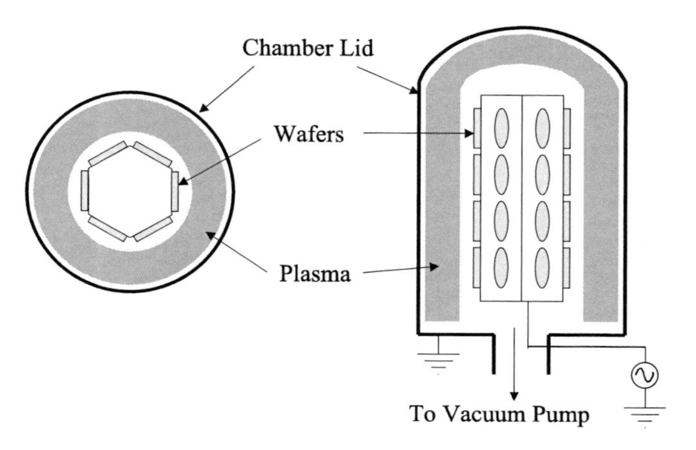

Figure 9.26 Batch RIE system.

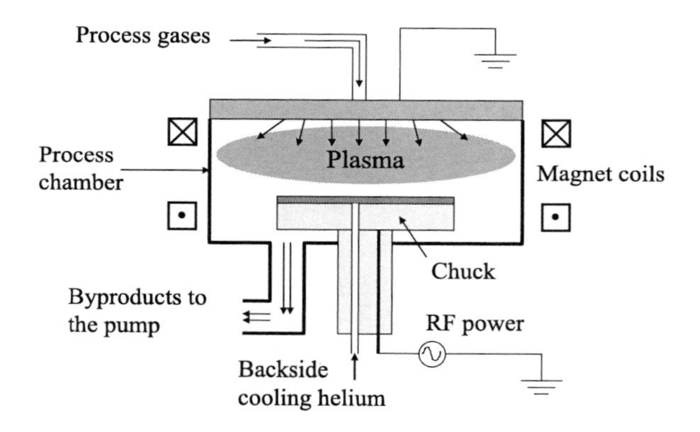

Figure 9.27 Single-wafer MERIE system.

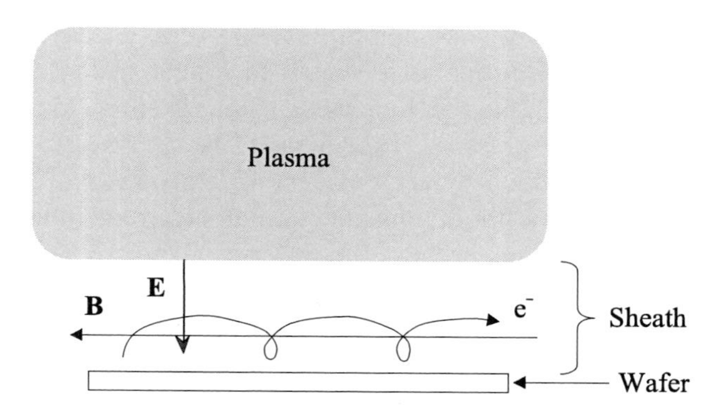

Figure 9.28 Electron helical movement in the sheath layer.

wafer temperature is more than 150 °C. Also, the chemical etch rate is sensitive to the wafer temperature. Therefore, a cooling system is required for the pattern etch chamber to prevent photoresist reticulation, and control wafer temperature and etch rate. Since an etch process needs low pressure and low pressure is not very good for heat transfer, pressurized helium at the backside of the wafer is used to remove the heat from the wafer. To prevent high-pressure backside helium from blowing the wafer off the pedestal, either a clamp ring (which mechanically holds the wafer on the pedestal) or an electrostatic chuck (E-chuck, which uses electrostatic force to hold the wafer) is required. Figure 9.29 gives the schematics of both a clamp ring and E-chuck.

The E-chuck became more popular in the 1990s because it had better etch uniformity and less particle contamination. It has better etch uniformity because it does not have the shadowing effect of the clamp ring on the wafer edge. It also has better temperature uniformity due to uniform cooling across the wafer, and no center bowing effect associated with a clamp ring. Since the E-chuck does not have mechanical contact (as the clamp ring does), it can contribute to the reduction of particles in the etch process. Helium, which has very high thermal conductivity, is commonly used to provide a conductive path for heat transfer from the wafer to a water-cooled pedestal. Some facts of helium are listed in Table 9.4.

Question: Hydrogen has a higher thermal conductivity than helium. Why not use hydrogen for wafer cooling in etch processes?

Answer: For safety reasons. Because hydrogen is explosive and flammable, whereas helium is inert, helium is much safer. As the second-best thermal conducting gas, helium can meet the cooling requirements of plasma etch systems.

Since there are always some depositions in the plasma etch process, a routine plasma dry cleaning process is used to remove most of the deposited materials left inside the chamber. However, after several thousand micron film etches, the deposition film eventually builds up, increasing the risk of particle contamination. Therefore, scheduled preventive maintenance (PM) chamber wet cleanings are needed to manually remove deposited materials from the surface of parts inside the chamber and chamber wall. Some etch chambers are designed with a liner

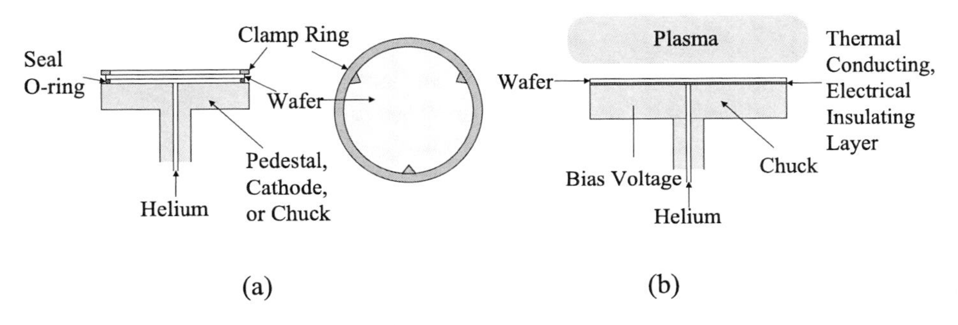

Figure 9.29 (a) Clamp ring and (b) E-chuck.

Table 9.4 Facts about helium.

Symbol	He
Atomic number	2
Atomic weight	4.002602
Discoverer	Sir William Ramsay, and independently by N. A. Langley and P. T. Cleve
Place of discovery	London, England, and Uppsala, Sweden
Discovery date	1895
Origin of name	From the Greek word helios, meaning sun. Its line radiation (a yellow line at 587.49 nm) was first detected from the solar spectrum during the solar eclipse of 1868 in India by French astronomer Pierre-Jules-César Janssen.
Molar volume	21.0 cm^3
Velocity of sound in substance	970 m/sec
Refractive index	1.000035
Melting point	0.95 K or −272.05 °C
Boiling point	4.22 K or −268.78 °C
Thermal conductivity	0.1513 $Wm^{-1}K^{-1}$
Applications	Cooling gas and carrier gas in CVD and etch processes

inside the chamber. During PM, technicians only need to replace the liner and send the dirty one out to a specialized shop for cleaning and preparation for reuse. This practice can sharply reduce system downtime for wet cleaning and improve production throughput.

As the minimum feature size continues to shrink, pattern etch processes are required to operate at even lower pressures to reduce ion scattering for better etch profile and tighter CD control. A capacitively coupled plasma source cannot generate and sustain plasma at a low pressure of a few mtorr, because the electron MFP is too long to receive enough ionization collisions. ICP and ECR are two types of plasma sources commonly used in the semiconductor industry to generate high-density plasma at low pressure for deep submicron feature pattern etch. Figure 9.30 illustrates the two systems.

One of the most important advantages of these high-density plasma systems is that ion bombardment flux and energy can be independently controlled by the source and bias rf power, respectively. In capacitively coupled plasma, both ion flux and energy are related to rf power.

9.4.6 Endpoint

For wet etch, in most cases the endpoint of the etch process is determined by time, which is calculated the by premeasured etch rate and required etch thickness. Visual inspection by an operator is also used for determining the endpoint, since there is no adequate solution for automatic endpoint determination. Wet etch rate is very sensitive to etchant temperature and concentration, which can be slightly different from station to station and lot to lot. Therefore, postetch visual inspection by an operator is always recommended.

One of the advantages of plasma etch is that it can use an optical system to automatically determine the endpoint of an etch process. At the end of an

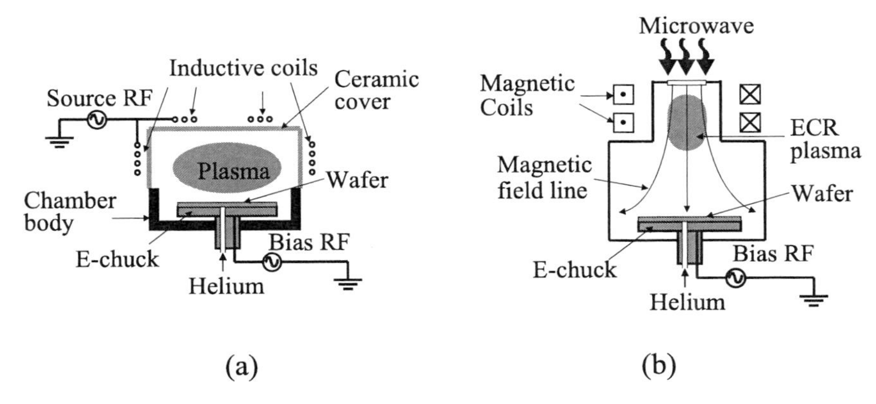

Figure 9.30 (a) ICP and (b) ECR high-density plasma etch systems.

etch process, chemical components change in the plasma, leading to a change in color and intensity of light emission from the plasma. By monitoring a certain wavelength of light emission with a spectroscope and detecting the signal change (due to the chemical changes at the end of the etch process), an electrical signal can be sent to a computer controling the etch system to automatically stop the etch process. Table 9.5 lists some of the wavelengths of chemicals used for etch endpoint detection.

For example, at the end of an aluminum etch process, the AlCl line intensity reduces due to a lack of etch byproduct $AlCl_3$ because most of the aluminum has been already etched out. This change in line intensity can be the signal to end the etch process.

Some other methods, such as pressure change, bias voltage change, mass spectrometry, etc., can also be used for endpoint detection. However, changes in pressure and bias voltage during the process are very unfavorable from a processing point of view, since they can affect process repeatability. Mass spectrometry measures the concentration of different chemical species in the processing chamber that change at the end of the etch process. Therefore, it can be used for endpoint detection. Since mass spectrometry requires a vacuum chamber and pump system, it is not cost effective compared to an optical endpoint system. It is rarely used in RIE process endpoint systems. For downstream or remote plasma etch systems, there is no plasma inside the processing chamber, therefore there is no light emission glow to provide an optical system endpoint, and mass spectrometry becomes the system of choice.

9.5 Plasma Etch Processes

9.5.1 Dielectric etch

Silicon compound-based dielectrics, such as silicon oxide, silicon nitride, and silicon oxynitride, have been widely used in IC chip manufacturing since the early 1960s. Dielectric etch is used mainly to create contact and via holes for interconnections between different layers of conductors. The etch process

Table 9.5 Wavelengths of the chemicals used for etch endpoint.

Film	Etchant	Wavelength (Å)	Emitter
Al	Cl_2, BCl_3	2614 3962	AlCl Al
Poly Si	Cl_2	2882 6156	Si O
Si_3N_4	CF_4/O_2	3370 3862 7037 6740	N_2 CN F N
SiO_2	CF_4 and CHF_3	7037 4835 6156	F CO O
PSG, BPSG	CF_4 and CHF_3	2535	P
W	SF_6	7037	F

that creates the contact holes for connection between the first metal layer and silicon S/D and polysilicon gates is called a contact etch. The contact etch needs to etch the ILD0 layer, which normally is doped silicate glass, either PSG or BPSG.

Via etch is very similar to contact etch. It etches intermetal dielectric (IMD) or interlayer dielectric (ILD), which is mainly USG, fluorinated silicate glass (FSG), a low-κ dielectric such as SiCOH, or an even lower-κ dielectric of porous SiCOH, depending on the technology nodes. Via etch ends on a metal surface, while contact etch ends on a silicon or silicide surface.

Other dielectric etch processes are hard mask etch and bonding pad etch. Both LOCOS and STI need to etch nitride and pad oxide hard masks. For the LOCOS process, a silicon nitride layer serves as the oxidation mask, and in the STI process, it serves as a hard mask of silicon etch and the stop layer of the oxide CMP process. There are also silicon oxide hard masks for silicon alignment mark etching, and hard mask etching for copper, gold, and platinum etches. Bonding pad etch removes passivation nitride and oxide stacked film that opens the metal pad for wire bonding or contact bump formation.

Dielectric etch processes generally use fluorine chemistry with heavy ion bombardment. A damaging mechanism is employed to achieve an anisotropic etch profile. The most commonly used gases for dielectric etches are carbon fluorides, such as CF_4, CHF_3, C_2F_6, and C_3F_8. Some oxide etch systems also use SF_6 as the fluorine source gas. In normal conditions, carbon fluorides are very stable and will not react with silicon dioxide or silicon nitride. In plasma, carbon fluorides dissociate and generate very reactive free fluorine radicals. These radicals chemically react with silicon dioxide or silicon nitride to form volatile silicon tetrafluoride, which can be readily pumped away from the surface. Some facts about fluorine are listed in Table 9.6.

Table 9.6 Facts about fluorine.

Symbol	F
Atomic number	9
Atomic weight	18.9984032
Discoverer	Henri Moissan
Place of discovery	France
Discovery date	1886
Origin of name	From the Latin word fluere, meaning to flow
Molar volume	11.20 cm^3
Velocity of sound in substance	N/A
Refractive index	1.000195
Melting point	53.53 K or −219.47 °C
Boiling point	85.03 K or −187.97 °C
Thermal conductivity	0.0277 W m^{-1} K^{-1}
Applications	Free fluorine is the main etchant for silicon oxide and silicon nitride etching processes. WF_6 is widely used as a W precursor for WCVD. FSG is widely used in 90-nm technology node logic IC. F_2 excimer lasers and CaF_2 have been used in 157-nm lithography research and development.

Chemical reactions for a plasma etch of silicon dioxide and silicon nitride are

$$CF_4 \overset{\text{plasma}}{\rightarrow} CF_3 + F,$$

$$F + SiO_2 \overset{\text{plasma}}{\rightarrow} SiF_4 + O,$$

$$F + Si_3N_4 \overset{\text{plasma}}{\rightarrow} SiF_4 + N.$$

Argon is commonly used in dielectric etch to increase ion bombardment. This helps increase etch rate and achieves the anisotropic etch profile by breaking the strong Si–O and Si–N bonds. Oxygen can be added to react with carbon and free more fluorine radicals, increasing the etch rate. However, introducing oxygen can affect the dielectric etch selectivity to the silicon and photoresist. Hydrogen can be added to improve selectivity over silicon.

In BPSG contact etching, when a contact hole reaches the silicide gate and local interconnection, contact with the S/D is only about halfway complete. While the BPSG etch process continues to reach the silicide S/D contact, overetch on the gate/local interconnection must be minimized. Therefore, the contact etch requires a very high oxide-to-silicide selectivity to prevent excessive loss of silicide during metal polycide contact. The required selectivity is $S \geq t/\Delta t$, as shown in Fig. 9.31.

For dielectric etch processes, the fluorine-to-carbon (F/C) ratio plays a very important role in etch selectivity. When $F/C < 2$, the polymerization process can occur, and a layer of Teflon®-like polymer can be deposited inside the processing chamber. For CF_4, the F/C ratio starts with 4. In plasma, CF_4 dissociates to CF_3 and F; F is consumed in the etching process, while CF_3 can continue to dissociate into CF_2. This sequence reduces the F/C ratio in the processing chamber. The polymerization process starts when many CF_2 radicals link into a

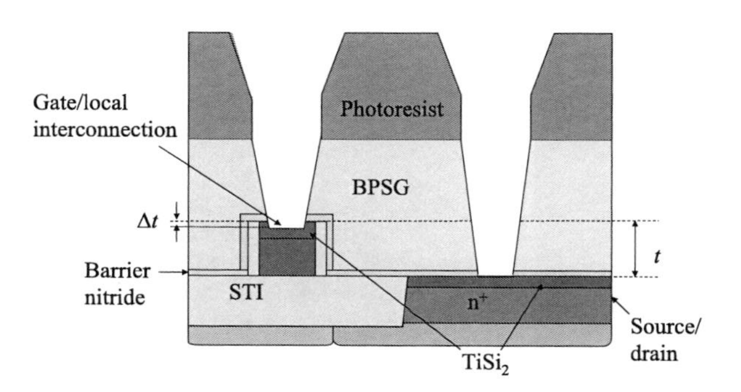

Figure 9.31 Contact etch.

long chain. Ion bombardment associated with dc bias can suppress polymerization by physically removing the deposited polymer before it builds up and forms a continuous film. Figure 9.32 shows the relationship between F/C ratio, dc bias, and the polymerization process.

For oxide etching, especially contact etch, the F/C ratio is chosen in an etching region close to the boundary of the polymerization region. While fluorine etches oxide, oxygen comes into play when fluorine replaces oxygen to bond with silicon. Oxygen reacts with carbon in carbon fluorides to form CO and CO_2. This frees more fluorine radicals to maintain the F/C ratio in the etching regime. When the etch process reaches the silicon or silicide surface, fluorine is consumed while carbon is not because there is no oxide as an etch byproduct, since the etched film does not contain oxygen. The F/C ratio decreases at the silicon or silicide surface, and the process quickly moves into the polymerization regime and begins depositing polymer, which stalls the etch process on the silicon or silicide surface. This provides a mechanism for achieving very high oxide-to-silicon/silicide etch selectivity. It allows oxide etch to continue for another few thousand angstroms to reach the S/D, while contact to the silicide gate/local interconnection is already reached (shown Fig. 9.31). Polymer deposited on the silicon or silicide surface can be removed by either an oxygen plasma ashing process (which also removes photoresist), or by a wet cleaning process.

Since dielectric etch mainly uses the damaging mechanism, this process is very physical, with heavy ion bombardment to the wafer surface when the wafer is in the etching region. The wafer must be very close to the boundary of the polymerization regime to achieve high selectivity to silicon or silicide. At the same time, the larger grounded electrode (the chamber lid) is in the polymerization region. It encounters much less energetic ion bombardment than the wafer surface. Thus, polymerization is always happening in the dielectric etch chamber during contact and via etch processes. Polymers deposited in the plasma etch chamber need to be removed with a routine O_2/CF_4 plasma cleaning process to prevent particle contamination caused by cracking of the polymer film. For etch chamber cleaning, it can be necessary to put a dummy wafer on the chuck to protect it from ion bombardment damage. A seasoning step always follows the cleaning process to intentionally deposit a thin layer of polymer inside the chamber to prevent loose residue falling from the

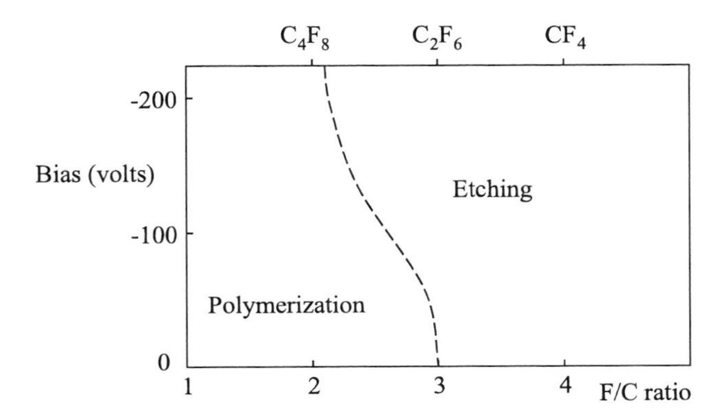

Figure 9.32 Relationships between F/C ratio, dc bias, and polymerization (From *Plasma Etching and Reactive Ion Etching*, J. W. Coburn).

chamber wall, and to keep the process in similar processing condition and prevent "first-wafer" effects.

Densities of wordlines (WLs) and contact holes are quite different from SRAM in logic devices, DRAM cell arrays, and NAND flash cell arrays, as shown in Fig. 9.33. The line/space ratio of a NAND flash WL is about 1:1, DRAM WL is about 1:2, and SRAM gate CD and gap is about 1:3. Figure 9.33 shows that NAND flash does not require a contact hole between WLs. In fact, it only needs contact holes to connect to bit or source lines for a string of 32 (32-bit) or 64 (64-bit) WLs. However, DRAM and SRAM need a contact between gates.

For DRAM applications, because spacing between the array transistor gates or WLs is significantly smaller than that of logic devices at the same technology nodes, there is little room for contact holes between WLs. To avoid shorting between the contact and WL in such a dense WL pattern, a self-aligned contact (SAC) process has been developed. As shown in Fig. 9.34, the SAC process etches BPSG, with a high selectivity to silicon nitride, which forms a hard mask of the WL stack and sidewall spacer that fully covers WLs. The SAC process allows high contact hole density of DRAM applications. After polysilicon deposition and CMP, it provides landing pads for bitline contact (BLC) and storage node contact (SNC). Therefore, SAC is also called landing pad contact (LPC) after etching, and landing pad poly (LPP) after polysilicon CMP.

For copper metallization, more dielectric etches are added in IC manufacturing processes. Besides via etch, there are trench etches. Two approaches are used: via first and via last. The via first process is illustrated in Fig. 9.35. It starts with via mask lithography to define the via holes, then etches the via holes and stops in the middle of the etch-stop layer (ESL). After being stripped, the photoresist is used to fill the via holes to protect them during trench etching. Trench mask lithography defines the photoresist patterns, and trench etch transfers the patterns to a low-κ dielectric ILD film. The ESL is broken through during the cleaning process and is called the cap layer because it is on top of any previous metal layer. ESL is formed with a thin layer of silicon nitride (SiN), silicon oxynitride (SiON), or silicon carbon nitride (SiCN).

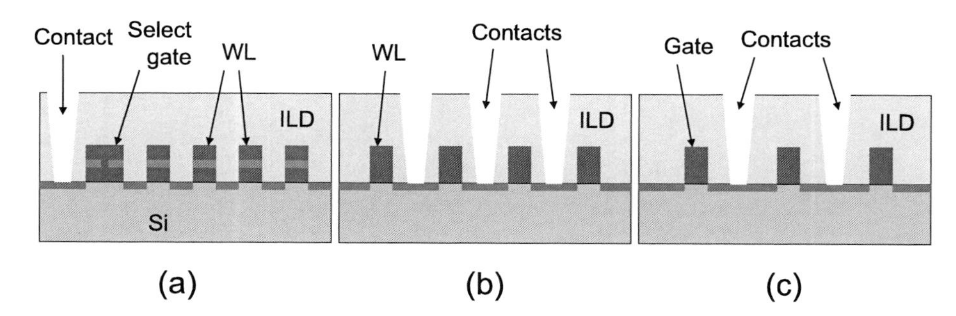

Figure 9.33 Wordlines and contact hole densities of different memory devices: (a) NAND flash, (b) DRAM, and (c) SRAM.

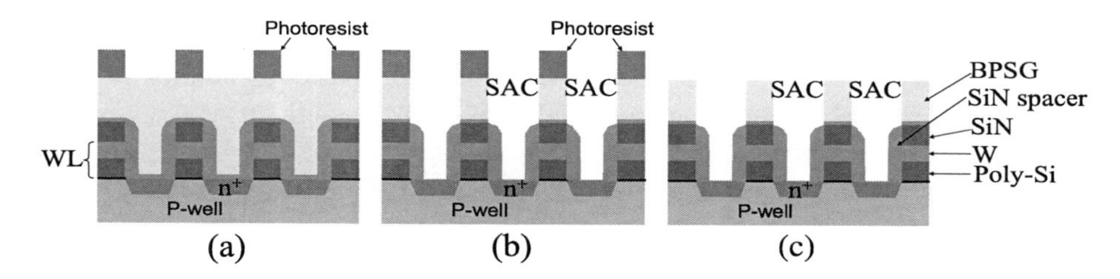

Figure 9.34 DRAM SAC process: (a) SAC mask lithography, (b) SAC etch, and (c) photoresist strip.

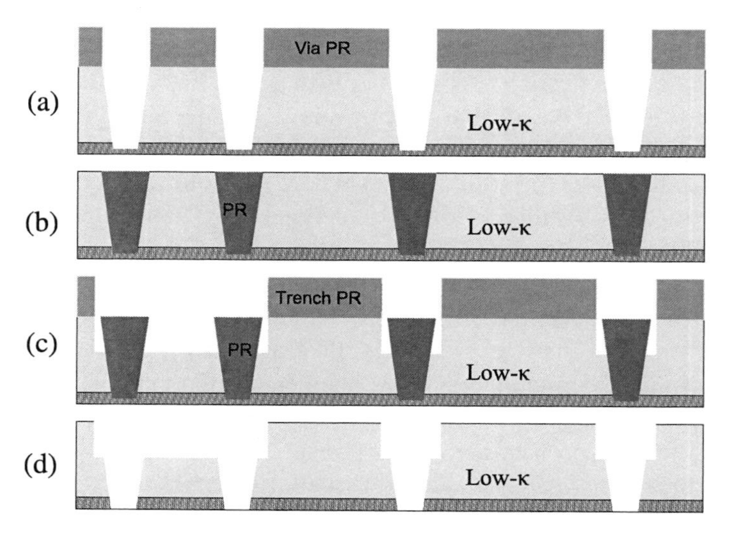

Figure 9.35 A via-first dual damascene process: (a) via etch, (b) photoresist fill and etchback, (c) trench etch, and (d) ESL breakthrough.

Figure 9.36 illustrates a trench-first dual damascene process. After trench mask lithography, trenches are etched onto ILD film, as shown in Fig. 9.36(a). For advanced nanometer technology node devices, ILD film is a low-κ dielectric film. After via mask lithography [illustrated in Fig. 9.36(b)], via holes are etched on the ESL, which is broken through after the photoresist cleaning step.

Because many low-κ and ultralow-κ (ULK) dielectric films are organosilicate glass (OSG) or porous OSG with large amounts of carbon and hydrogen integrated into the films, they can be damaged by oxygen radicals in photoresist ashing and stripping processes. To protect low-κ or ULK films from damage during photoresist striping processes, a metal hard mask such as titanium nitride (TiN) is commonly used for ILD etch. After ILD deposition, the metal hard mask layer is deposited.

Figure 9.37 illustrates a copper metallization process with TiN and TEOS oxide hard masks. BARC is a spin-on material coated onto the wafer surface before photoresist spin-coating. Tetraethoxysilane or tetraethylorthosilicate (TEOS) is a widely used precursor for silicon oxide CVD processes. TEOS is also commonly known in IC fabs as undoped silicon oxide deposited by a PECVD process using TEOS as precursor. The PECVD TEOS process is described in detail in Chapter 10. Trench masks are used to pattern photoresist first, then BARC and hard masks are etched with the photoresist pattern, as shown in Fig. 9.37(a). TiN and TEOS hard masks can prevent low-κ or ULK films from being damaged by oxygen radicals during the two photoresist stripping processes. After photoresist ashing, a via mask is patterned, and via holes are etched more than halfway, as shown in Fig. 9.37(b). A metal hard mask is used to etch the low-κ dielectric trenches and via holes simultaneously (with high selectivity to ESL), so the via holes stop on the ESL, as illustrated in Fig. 9.37 (c). The TiN hard mask stays on the wafer surface after copper deposition, shown in Fig. 9.37(d), and eventually is removed by the metal CMP of copper metallization, shown in Fig. 9.37(e).

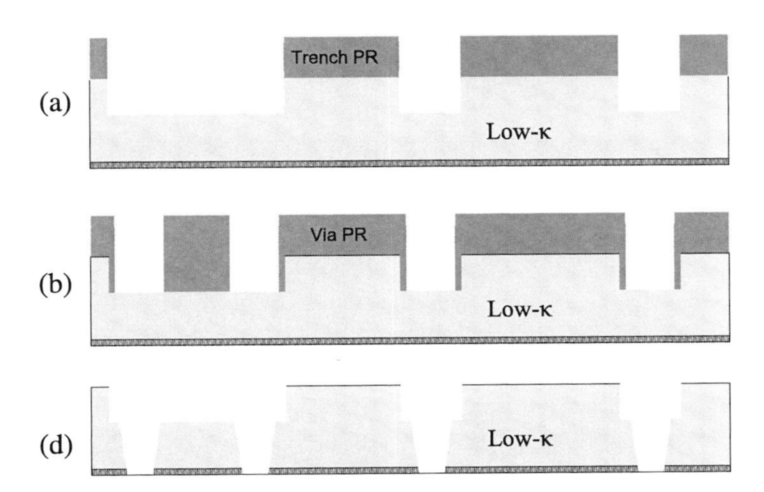

Figure 9.36 A trench-first dual damascene process: (a) trench etch, (b) via mask lithography, (c) via etch, and (d) ESL breakthrough.

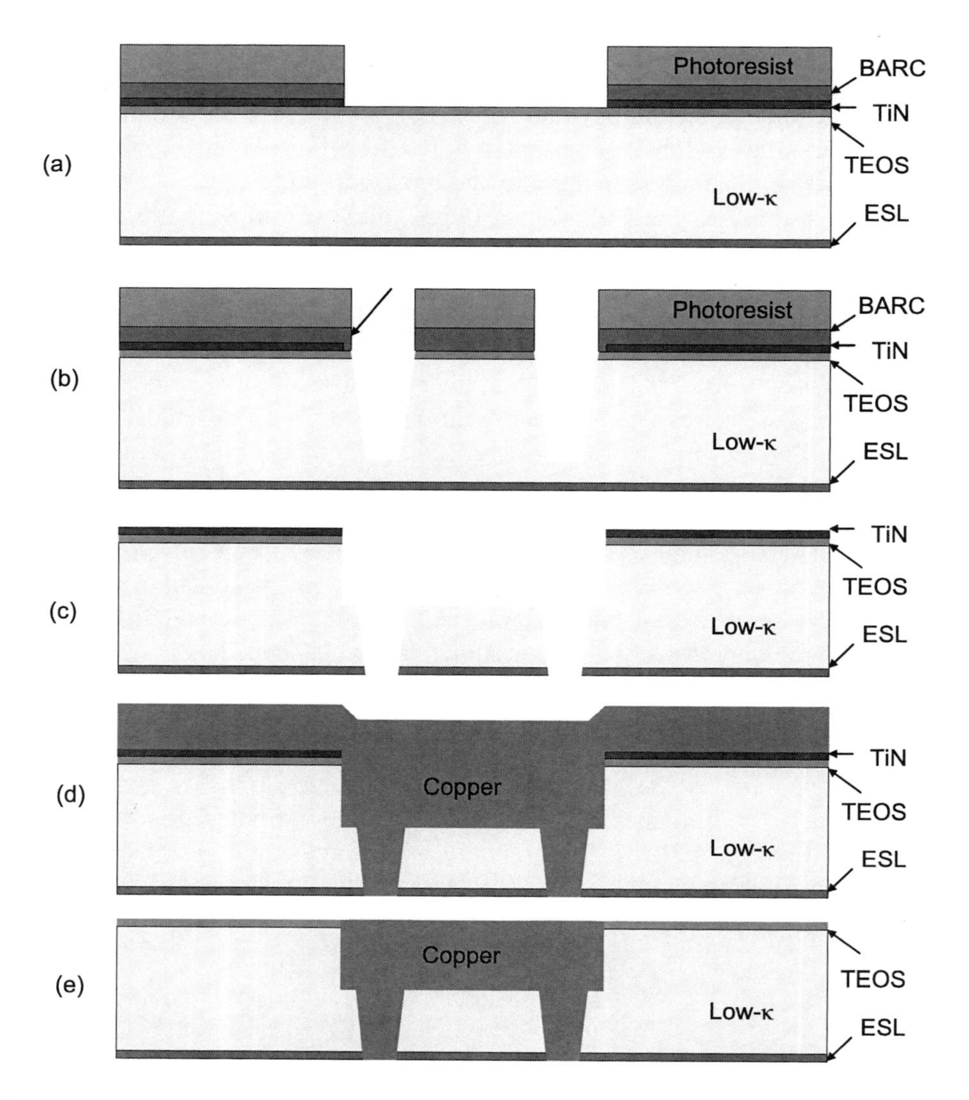

Figure 9.37 Copper metallization with a metal hard mask: (a) etch bottom antireflective coating (BARC) and metal hard mask with trench mask, (b) via etch halfway with a via mask, (c) trench and via etch with metal hard mask and ESL breakthrough, (d) copper deposition, and (e) metal CMP to remove the metal hard mask.

The main etch processes of low-κ or ULK dielectrics can use CF_4, CHF_3, or C_4F_8 chemistry, with argon for ion bombardment. Other fluorocarbon gases, such as C_4F_6 (hexafluorobutadiene) and c-C_5F_8 (octafluorocyclopentene), can also be used with O_2 in low-κ and ULK dielectric etching. Carbon monoxide (CO) can also be used to control the F/C ratio. Reducing the F/C ratio in the etchant gas can also help improve etch selectivity of the low-κ dielectric to the photoresist. Table 9.7 summarizes dielectric etch processes.

Table 9.7 Summary of dielectric etch processes.

Name of etch	Hard mask	Contact	Via (Al-Cu)	Via/trench (Cu/low-κ)	Bonding pad
Materials	Si_3N_4 or SiO_2	PSG or BPSG	USG or FSG	Low-κ or ULK	Nitride and oxide
Etchants	CF_4, CHF_3, etc.	CF_4, CHF_3, etc.	CF_4, HF_3, etc.	CF_4, CHF_3, CO, etc.	CF_4, CHF_3, etc.
Underlying layer	Si, Cu, Au,	Polysilicon or silicide	TiN/Al-Cu	ESL/Cu	Metal
Endpoints	CN, N, or O	P, O, and F	O, Al, and F	CN, O, and F	O, Al, and F

9.5.2 Single-crystal silicon etch

Single-crystal silicon etch is required for forming STI, which has been used in advanced IC chips since the mid-1990s, when IC technology nodes went to a half-micron. STI has replaced LOCOS field oxide as the isolation between neighboring microelectronic devices. This is because STI does not have "bird's beak," and the surface topography of STI is much flatter than that of LOCOS.

Single-crystal silicon etch is also needed for deep trench capacitor formation in DRAM chip manufacturing. The capacitor packing density is increased by forming the capacitor in a vertical direction deep in the silicon substrate, as shown in Fig. 9.38. Deep trench capacitor DRAM is widely used as embedded DRAM in system-on-chip (SOC) IC chips because it is more compatible with standard CMOS processes.

Figure 9.39(a) shows a stack capacitor DRAM with planar array nMOS, which is commonly used for DRAM with technology nodes larger than 100 nm. Figure 9.39(b) shows a stack capacitor DRAM recessed gate array nMOS, which is widely used in DRAM with sub-100-nm technology. In Fig. 9.39, BL and BLC in the cell array area are marked with a dashed outline because they are not in the cross section. Instead, they are behind and in front of the cross section, sandwiched between the SNC. To avoid shorting between the SNC polysilicon plug and tungsten BL, a silicon nitride liner is deposited on the sidewall of the SNC, as shown in Fig. 9.39(b). SN stands for storage node.

By using single-crystal silicon etch with a reversed cell gate mask, recessed gate electrodes of cell array nMOS can be created, which can help reduce the short channel effect due to shrinking of the gate feature size. It can be seen that the stack capacitor DRAM with a planar array transistor only needs one single-crystal silicon etch that forms the STI, while the recessed gate array transistor DRAM needs two single-crystal silicon etches, one for STI and another for recessed gate.

Buried wordline (bWL) DRAM technology (illustrated in Fig. 9.40) has been developed based on deep trench DRAM and recessed gate transistor technology. For recessed gate array transistors, the gate electrode is buried under the wafer surface. Conducting wires with polysilicon/metal/nitride stacks above the wafer surface serve as WLs of the DRAM cells. By placing the cell array nMOS and WL under the wafer surface, the bWL DRAM reduces a masking step because

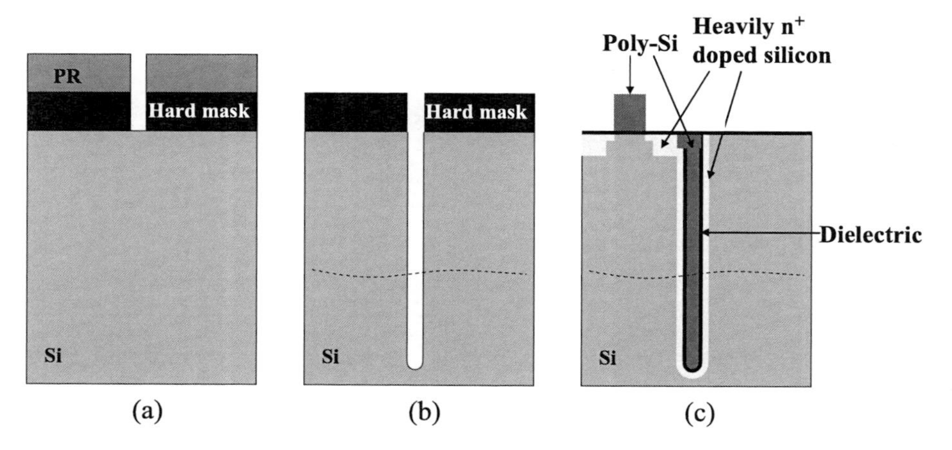

Figure 9.38 Deep trench capacitor DRAM formation: (a) hard mask etch, (b) single-crystal silicon etch, and (c) deep trench capacitor DRAM formation.

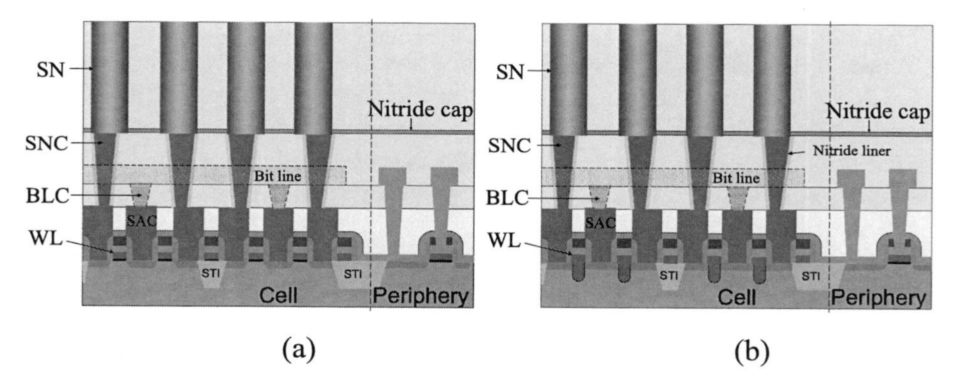

Figure 9.39 (a) Cross section of a stack capacitor DRAM with planar array nMOS and (b) with recessed gate array nMOS.

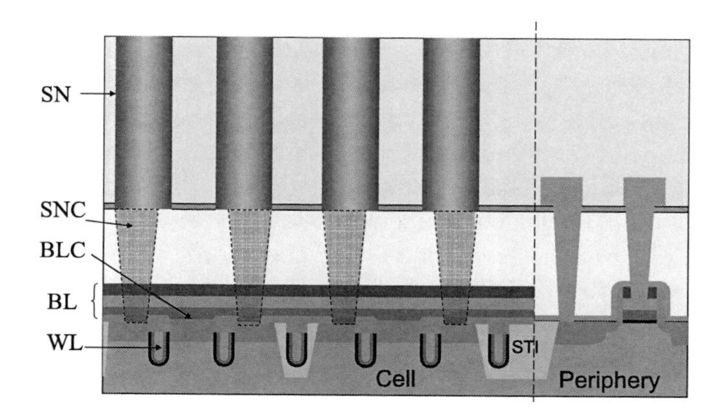

Figure 9.40 Buried WL DRAM.

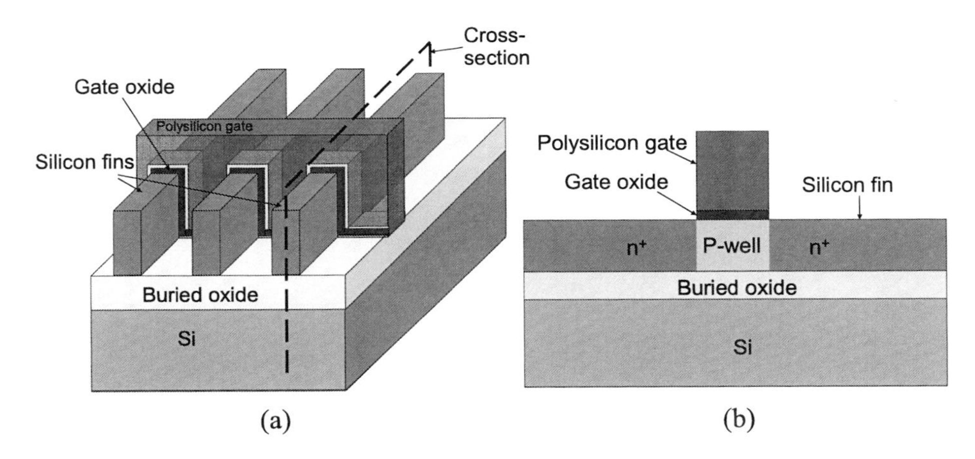

Figure 9.41 (a) Three FinFETs and (b) the cross section of one of them.

WL patterning is no longer required. By removing the topography of the wafer surface due to the dense WL pattern, another masking layer is reduced, since the SAC layer is not needed. The tradeoff is the challenge of WL trench etching, which must be well controlled for CD, depth, and profile. It also must achieve nearly the same etch rate and etch profile in both single-crystal silicon and in the STI oxide. Both deposition and etchback of the metal layers, TiN and W, in the trench also need to be well controlled. Another challenge is that bWL DRAM array transistors are nMOS with metal gates, usually titanium nitride (TiN), not the polysilicon gates commonly used in both planar and recessed gate array nMOS. Also, for bWL DRAM, polysilicon and metal stacks (usually polysilicon/WN/W stack) are used as gate electrodes for the periphery MOSFETs, which are also used as bitline stacks. Therefore, it is very challenging to form bitline contacts that connect bitlines to the array nMOS while still maintaining the integrity of the gate oxide of the periphery MOSFETs.

Single-crystal silicon etch is also needed for making 3D devices such as FinFET, which is regarded by many as one of the most promising candidates for next-generation devices when the widely used planar MOSFETs can no longer scale. Figure 9.41(a) illustrates three FinFETs with a shared gate electrode built on an SOI substrate. Figure 9.41(b) shows the cross section of the FinFET along the dashed surface of Fig. 9.41(a). It can be seen from the cross section that the FinFET looks just like a planar MOSFET made with an SOI substrate. Figure 9.41(a) shows that the silicon fins can be formed with a single-crystal silicon etch. It is significantly easier to form silicon fins with SOI wafers than with bulk wafers because the buried oxide can be used as an etch endpoint, and the height of the fin is easier to control because it is the thickness of the silicon on the buried oxide.

FinFET can also be made from a bulk silicon wafer, which requires better control of the single-crystal silicon etch process, such as CD, depth, and profile. The height of the silicon fin is controlled by the amount of recess of the STI oxide, as shown in Fig. 9.42. The CD and height of the silicon fins can be measured by atomic force microscope (AFM) and scatterometry.

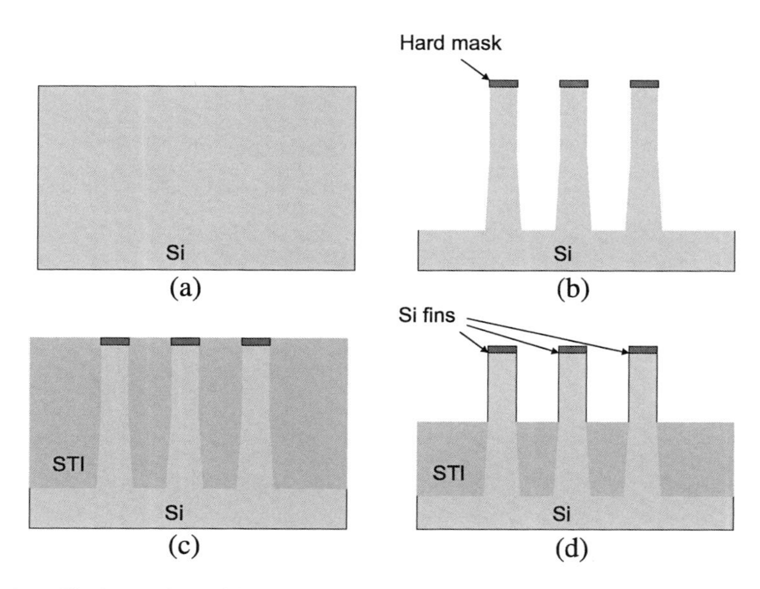

Figure 9.42 Fin formation with a bulk silicon wafer: (a) bulk silicon substrate, (b) trench etch, (c) STI oxide fill and CMP, and (d) STI oxide recess.

Single-crystal silicon etch normally employs silicon dioxide (or in many cases, both silicon dioxide and silicon nitride) hard masks instead of photoresists to avoid contamination, as illustrated in Fig. 9.42(b). It uses a blocking mechanism, with HBr as the main etchant and O_2 as a sidewall passivation agent. In plasma, HBr dissociates and releases free bromine radicals, which can react with silicon to form volatile silicon tetrabromide ($SiBr_4$). Oxygen oxidizes silicon on the sidewalls to form silicon dioxide, which protects silicon from the free bromine radicals. At the bottom of the trench, ion bombardment keeps the oxide from growing, thus the etch process continues only in the perpendicular direction. Some facts about bromine are listed in Table 9.8.

The main chemical reaction of single-crystal silicon plasma etching processes is

$$\mathrm{HBr} \overset{\text{plasma}}{\rightarrow} \mathrm{H} + \mathrm{Br},$$

$$\mathrm{Br} + \mathrm{Si} \overset{\text{plasma}}{\rightarrow} \mathrm{SiBr_4}.$$

Oxygen is used to improve the selectivity over oxide hard masks. It is also used to react with etch byproduct $SiBr_x$ to form and deposit $SiBr_xO_y$ on the trench sidewall. This protects the sidewall and restricts the etch to mainly the vertical direction because $SiBr_xO_y$ deposited on the trench bottom is constantly removed by ion bombardment. Fluorine source gases such as SiF_4 and NF_3 can be used to improve etch profile control on the trench sidewall and bottom. Fluorine is also needed to achieve an etch rate similar to single-crystal silicon and silicon oxide, required by bWL DRAM.

Single-crystal silicon etch takes two steps—breakthrough and the main etch. A brief breakthrough step removes the thin native oxide on the silicon surface

Table 9.8 Facts about bromine.

Symbol	Br
Atomic number	35
Atomic weight	79.904
Discoverer	Antoine J. Balard
Place of discovery	France
Discovery date	1826
Origin of name	From the Greek word bromos, meaning stench
Molar volume	19.78 cm^3
Velocity of sound in substance	No data
Resistivity	$>10^{18} \mu\Omega \cdot cm$
Refractive index	1.001132
Melting point	−7.2 °C
Boiling point	59 °C
Thermal conductivity	0.12 $W\ m^{-1}\ K^{-1}$
Applications	Free bromine as the main etchant for single-crystal silicon etching processes
Source	HBr

with heavy ion bombardment and fluorine chemistry. The main etch step etches silicon with HBr and O_2 (diluted to 30% in He) chemistry. After the etch process is finished, the wafer requires wet cleaning processes to remove sidewall deposition. A significant difference in single-crystal silicon etch compared with other plasma etch processes is that it does not have an underlying layer. Therefore, it cannot be endpointed by an optical signal and usually endpoints by time.

Single-crystal silicon etch chambers always acquire some depositions of complex compounds of silicon, bromine, oxygen, and fluorine on their walls. These must be removed with a routine plasma cleaning process with fluorine chemistry to control particle contamination. Similar to other cleaning processes, a seasoning step is needed after the cleaning process.

9.5.3 Polysilicon etch

Polysilicon gate etch is the most important etch process because it defines the gates of MOSFET. Polysilicon gate etch has the smallest CD of all the etching processes. In the good old days, when people talked about N-micron technology, it indicated that the gate CD was N μm.

When feature size shrinks deep into the nanometer range, gate CD and technology nodes no longer remain the same number. Technology node is mainly determined by the pitch of gate patterns or portion of the gate pitch. The definition of a technology node is also different for different devices. The NAND flash technology node is half-pitch: 20-nm NAND flash has 40-nm WL pitch, usually with a 20-nm gate CD and 20-nm gap CD. DRAM technology nodes are about one-third WL pitch; a 33-nm DRAM has about a 99-nm WL pitch with ~30-nm gate CD. Technology nodes of CMOS logic devices are defined as about one-quarter of SRAM gate pitch because there is contact between the gates. For example, a 22-nm SRAM device (published in 2008 by Haran) has a gate pitch of 90 nm and gate CD of 25 nm. Figure 9.43 illustrates cross sections of array transistors of NAND flash, DRAM, and SRAM.

Figure 9.44 shows Intel's six-transistor (6T) SRAM size scaling timeline and top-view SEM images of unit cells of the 6T SRAM from 90- to 22-nm technology nodes after polysilicon gate etch. The SRAM layout had a revolutionary change starting with the 65-nm node, which was quite different from the layout of the 90-nm node. After the 45-nm node, a double-patterning technique must be applied in gate patterning. As the technology node continued to shrink, the CDs of MOSFET gates had trouble continuing to shrink, and IC designers started to reduce the gaps between the gates.

MOSFETs with polysilicon gates require polysilicon etch to form gate patterns. MOSFETs with HKMG must etch polysilicon, whether it is used as a gate-first approach or gate-last technique. In fact, the polysilicon gates of the Intel SRAMs with 45-, 32-, and 22-nm technology nodes shown in Fig. 9.43 will be removed after ILD0 CMP and replaced by metal stacks to form the metal gates in a gate-last processes. Because the gaps between gates are so narrow for 32- and 22-nm SRAMs, trench-style contact, with etch processes similar to the SAC of DRAM, have been developed and applied to form the contact plug.

Figure 9.45 illustrates a polysilicon etch process that forms CMOS gates and local interconnections. This process uses photoresist as the etch mask, and the etch stops on gate and STI oxides.

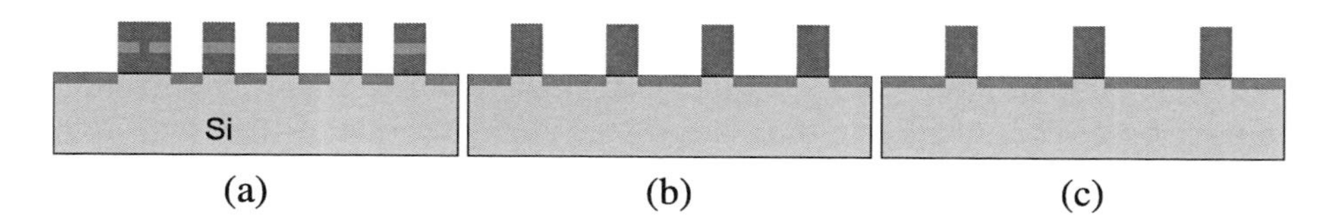

Figure 9.43 Cross sections of array transistors for different memory devices with the same technology node: (a) NAND flash, (b) DRAM, and (c) SRAM.

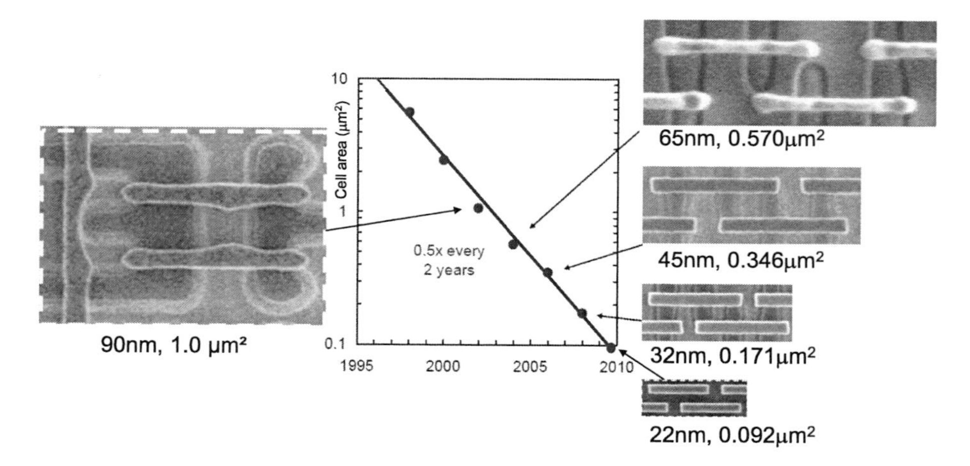

Figure 9.44 Intel SRAM size scaling (Intel).

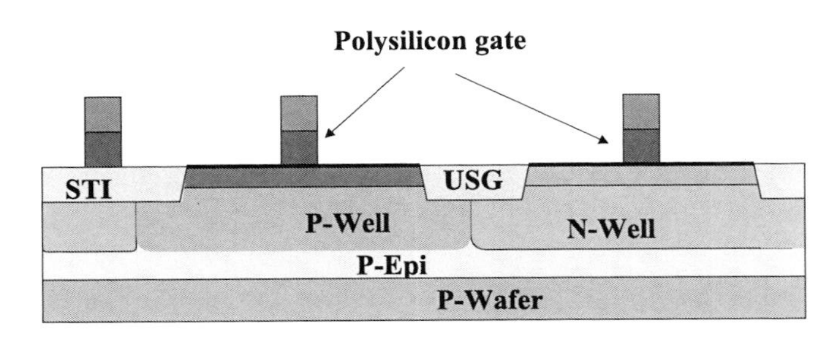

Figure 9.45 Polysilicon gates and local interconnection etches.

Cl_2 is the most popular main etchant for polysilicon etch processes. In plasma, Cl_2 molecules are dissociated to generate free chlorine radicals, which are very reactive and can react with silicon, forming gaseous silicon tetrachloride ($SiCl_4$). Table 9.9 lists some facts about chlorine.

Chlorine tends to combine with photoresist materials and deposit a thin polymer layer on the sidewall. This can help achieve an anisotropic etch profile and reduce CD bias (polysilicon CD loss or gain compared to the CD of photoresist). HBr can also be used as a secondary etchant and sidewall passivation promoter. O_2 can be used to improve selectivity over oxide.

One of the biggest challenges for polysilicon gate etching is the high selectivity to silicon dioxide, since underneath the polysilicon is the ultrathin gate oxide. For a 45-nm device, thickness of the gate oxide is about 12 Å, which is equivalent to only two layers of a silicon dioxide molecule! Since neither the etch rate nor polysilicon film thickness is perfectly uniform, some parts of the polysilicon may already be etched away, while for some other parts the etch is still going on (shown in Fig. 9.46). The thin gate oxide cannot afford to be lost, otherwise the etchant that removes polysilicon will also remove the single-crystal silicon underneath the

Table 9.9 Facts about chlorine.

Symbol	Cl
Atomic number	17
Atomic weight	35.4527
Discoverer	Carl William Scheele
Place of discovery	Sweden
Discovery date	1774
Origin of name	From the Greek word chloros, meaning pale green
Molar volume	17.39 cm^3
Velocity of sound	206 m/sec
Resistivity	$>10^{10}\mu\Omega\cdot cm$
Refractive index	1.000773
Melting point	−101.4 °C or 171.6 K
Boiling point	−33.89 °C or 239.11 K
Thermal conductivity	0.0089 W m^{-1} K^{-1}
Applications	Used as the main etchant for polysilicon and metal etching processes; and for polysilicon, epitaxy silicon deposition chamber cleaning.
Sources	Cl_2, HCl

gate oxide, causing silicon pit defects. Therefore, the selectivity to oxide must be high enough during the polysilicon overetch process.

Figure 9.46 shows requirements of overetch for a polysilicon etch process. The etch process in the higher-etch-rate portions of the wafer have already reached the gate oxide [left side of Fig. 9.46(a)], while in the lower-etch-rate portion there is still a remaining thin layer of polysilicon needing to be removed [right side of Fig. 9.46(a)]. Assume the etch rate has 3% nonuniformity. Thus, for a perfectly uniform 50-nm polysilicon film, the remaining polysilicon thickness is about 1.5 nm. By only allowing ~0.5 Å loss of gate oxide, the overetch selectivity should be higher than 30:1. For the LOCOS case illustrated in Fig. 9.46(b), residue polysilicon on the sidewall of the field oxide needs to be etched away, while in the active area the polysilicon is already etched out. Selectivity is determined by the ratio of the sidewall residue thickness and the tolerated gate oxide loss. If the sidewall residue thickness is 1500 Å and the maximum gate oxide loss is 30 Å, selectivity of the polysilicon-to-silicon dioxide during the overetch needs to be at least 1500:30 = 50:1.

Polysilicon etch has three steps: breakthrough, main etch, and overetch. The breakthrough step removes the thin native oxide (~10 to 20 Å) with high ion bombardment and sometimes uses fluorine chemistry. The main etch step removes polysilicon from designated areas and forms the gate and local interconnect patterns. In the overetch step, the etch condition is changed to remove the polysilicon residue while minimizing the gate oxide loss. The main step etches polysilicon at a high rate. The selectivity over silicon dioxide is not a concern, since the etch process has not reached the oxide yet. As soon as the etchants start to etch the gate oxide, oxygen will be released from the film and penetrate the plasma. Optical spectroscopic sensors monitoring oxygen line radiation detect an intensity increase of the oxygen radiation line and trigger the endpoint to stop the main etch process and start the overetch process. The overetch process flows oxygen, reduces rf power, and reduces Cl_2 flow to improve polysilicon-to-oxide selectivity. Some fabs have a phororesist trimming step before breakthrough to remove any residue photoresist in the bottom of the gap between the two photoresist patterns. It also can reduce gate CD and improve device speed.

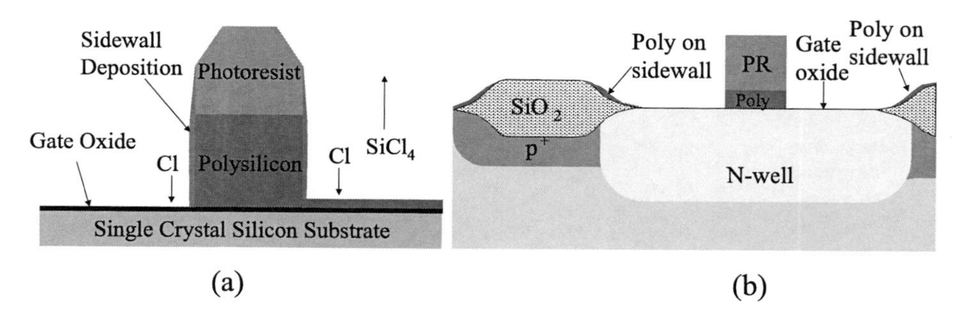

Figure 9.46 Polysilicon overetch requirement.

For advanced technology node IC chips, in order to pattern the fine patterns with 193-nm ArF optical lithography, photoresist patterns alone cannot meet the requirements of etch bias. Dielectric hard masks on top of the polysilicon are commonly used. This dielectric layer sometimes serves as an ARC. Above the dielectric hard mask, a spin-on BARC layer is added prior to photoresist coating. These are commonly called trilayers. After photoresist patterning by a photolithography process, BARC can be etched with oxygen plasma, and the dielectric hard mask can be etched with fluorine chemistry and argon ion bombardment, which is similar to a breakthrough step. The main etch of polysilicon can use fluorine chemistry to endpoint (by means of an optical system monitor) the thickness variation of the polysilicon. The overetch chemistry is usually HBr with O_2, which allows for very high selectivity of the gate oxide.

Fluorine can be used to etch polysilicon. Some polysilicon etch processes use SF_6 and O_2 chemistry. Because fluorine etches silicon dioxide faster than chlorine does, it has a lower selectivity to oxide; therefore most fabs prefer to use chlorine chemistry for the main etch step.

In some DRAM gate (WL) processes, tungsten (W) or tungsten silicide (WSi_2) is deposited on top of the polysilicon to reduce the resistance of the WL and local interconnections. The etch process for this silicide–polysilicon stack film requires one more step to etch W or WSi_2, first with fluorine chemistry, then switching to chlorine chemistry to etch the polysilicon.

9.5.4 Metal etch

Metal etch has been used to form the interconnection lines for an IC to connect transistors and circuit units. For some CMOS IC chips with mature technology (and even advanced DRAM and flash chips), the metal layer consists of three layers: TiN ARC, aluminum copper alloy, and titanium nitride plus titanium or a titanium tungsten welding layer. The TiN ARC layer reduces reflection from the aluminum surface and helps improve lithography resolution. The Al–Cu alloy is the main metal material for carrying the electric current and forming global interconnections. The Ti, TiN/Ti, or TiW layer can help reduce contact resistance between the Al–Cu and tungsten plug. This layer also prevents the copper in aluminum from diffusing into the silicate glass, reaching the silicon substrate, and causing device damage.

The most commonly used chemical for metal etch is chlorine. In plasma, Cl_2 dissociates and generates the free radicals Cl, which react to TiN, Al, and Ti and form the volatile byproducts of $TiCl_4$ and $AlCl_3$:

$$\mathrm{Cl_2} \overset{\text{plasma}}{\rightarrow} \mathrm{Cl} + \mathrm{Cl},$$

$$\mathrm{Cl} + \mathrm{Al} \overset{\text{plasma}}{\rightarrow} \mathrm{AlCl_3},$$

$$\mathrm{Cl} + \mathrm{TiN} \overset{\text{plasma}}{\rightarrow} \mathrm{TiCl_4} + \mathrm{N},$$

$$\mathrm{Cl} + \mathrm{Ti} \overset{\text{plasma}}{\rightarrow} \mathrm{TiCl_4}.$$

Metal etch normally uses Cl_2 as the main etchant, and BCl_3 is commonly used for sidewall passivation. It is also used as a secondary Cl source and a heavy ion source to provide the BCl_3^+ ion with ion bombardment. In some cases, argon is also used to increase ion bombardment. N_2 and CF_4 can be employed to improve sidewall passivation.

Question: Why does no one use fluorine to etch a TiN/Al-Cu/Ti metal stack?

Answer: The main byproduct of an aluminum-fluorine reaction is AlF_3, which has very low volatility. At normal etch conditions, ~100 mtorr, <60 °C, AlF_3 is a solid. Therefore, fluorine cannot be used in a plasma etch process to pattern the TiN/Al-Cu/Ti metal stack.

It is very important that metal etch processes have good profile control, residue control, and metal corrosion prevention. During the metal etch process, a small amount of copper in the aluminum can cause a residue problem because $CuCl_2$ has very low volatility and will stay on the wafer surface. It can be removed by physical ion bombardment that dislodges it from the surface. It can also be removed by chemical etching, which cuts $CuCl_2$ off of the surface. Since both $CuCl_2$ particles and the wafer surface are negatively charged by the plasma, $CuCl_2$ can be pushed away from the surface by electrostatic force. It is very important for the metal etch process to strip the photoresist in situ before the wafer is exposed to moisture in the atmosphere. Otherwise, chlorine residue in the photoresist and sidewall deposition will react with H_2O to form HCl and cause metal corrosion problems.

For the gate-first approach of a HKMG process, the stack film of a dielectric hard mask, the polysilicon, and TiN must be gate etched, as shown in Fig. 9.47. The etching process is similar to a normal polysilicon gate etching process: first a breakthrough etch is required to pattern the dielectric hard mask, then the main etch process removes bulk polysilicon with fluorine plasma, and then the metal etch step uses Cl or HBr to etch TiN with high selectivity to the capping layers and high-κ dielectric layer. For polysilicon gate etch, oxygen can be added to improve etch selectivity to silicon oxide. However, for a metal gate etch, adding oxygen to the plasma can result in the oxidation of TiN, which forms TiO_2 and causes metal gate loss.

TiN etch is also needed to pattern hard masks of ULK dielectric in copper low-κ interconnection formation, as shown in Fig. 9.37(a).

9.5.5 Photoresist strip

After the etch process, the photoresist has served its purpose and needs to be stripped. Photoresist can be stripped with either a wet or dry process. Oxygen is most commonly used for dry stripping. Water vapor (H_2O) is usually added to plasma to provide an extra oxidation agent (HO) to remove photoresist, and hydrogen radicals (H) to remove the chlorine on the sidewall and in the photoresist. For the metal etch process, it is very important to strip the photoresist right after the etch process, before the wafer is exposed to moisture in the atmosphere. This is

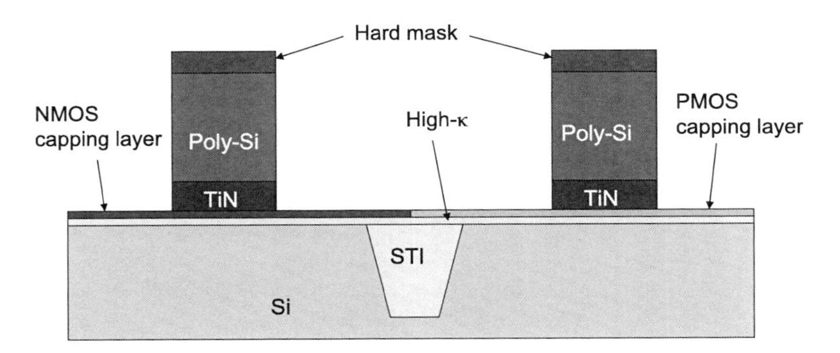

Figure 9.47 Gate etch of the gate-first approach for a HKMG device.

because the chlorine absorbed on the sidewall deposition and photoresist will react with moisture and form hydrochloric acid, which can etch aluminum and cause metal corrosion. The basic chemistry involved in the photoresist stripping process is

$$\begin{aligned} O_2 &\overset{\text{plasma}}{\rightarrow} O + O \\ H_2O &\overset{\text{plasma}}{\rightarrow} 2H + O, \\ H + Cl &\rightarrow HCl, \\ O + PR &\rightarrow H_2O + CO + CO_2 + \cdots. \end{aligned}$$

Figure 9.48 shows a schematic of a photoresist stripping chamber with a remote plasma source. It can be put in line with the etch chambers to strip the photoresist in situ in the same mainframe.

9.5.6 Dry chemical etch processes

Dry chemical etch can be achieved by either using thermally unstable chemical gases, such as XeF_2 and ozonator-generated O_3, or by using a remote plasma source, which generates free radicals in a remote plasma chamber and flows them into the processing chamber. Since these unstable gases are expensive and difficult to store, remote plasma processes are more commonly used in IC fabrication.

By using a remote plasma source and taking advantage of plasma-generated chemically active free radicals that are free from ion bombardment on the wafer surface, a dry chemical etch can be applied to thin-film stripping and wine-glass contact etch processes. The advantage of a dry chemical etch over a wet one is that it can be placed in line with another RIE chamber on one system. It can process wafers in situ and improve production throughput. One example is the wine-glass contact shown in Fig. 9.15. The remote plasma etch chamber can be placed in line with the RIE chamber in the same mainframe. First, the wafer is in the remote plasma etching chamber for isotropic etch, then it is transferred

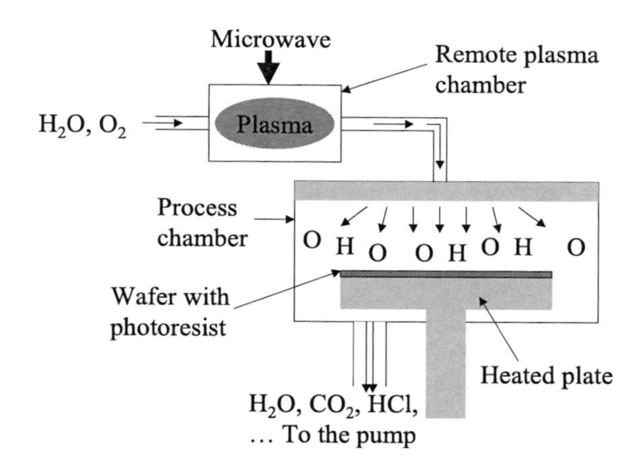

Figure 9.48 A remote plasma photoresist stripping system.

to the RIE chamber for anisotropic etch. Another application for remote plasma etch processes is silicon nitride layer stripping in the LOCOS process, and both silicon nitride and polysilicon layer stripping in poly-buffered LOCOS (PBL) processes.

9.5.7 Blanket dry etch processes

Blanket plasma etch removes material from the entire wafer surface. It does not have a photoresist pattern on the wafer surface. The main applications of blanket etch processes are etchback, spacer formation, and film strip.

Argon sputtering etchback is a pure physical etch process; it removes materials by physically dislodging tiny pieces of the material from the surface with energetic ion bombardment. It is widely used in dielectric thin-film applications to taper gap openings. It increases accessibility to CVD precursor molecules and improves gap-fill abilities. Ar sputtering is also widely used to remove native oxide from wafer surfaces prior to metal deposition.

RIE etchback combines both the physical and chemical etch processes. It can be used in line with a dielectric CVD tool for sidewall spacer formation so that the CVD chamber can deposit a conformal dielectric film on the pattern, as shown in Fig. 9.49(a). The RIE etch chamber can etch back the conformal dielectric film from the sidewall spacer, as shown in Fig. 9.49(b). RIE etch can also be applied with a tungsten CVD tool to form tungsten plugs. In this case, a WCVD chamber deposits tungsten into contact holes or via holes and all across the wafer surface. An RIE etch chamber with fluorine plasma had been used to remove bulk tungsten from the surface. The RIE process has also been used in photoresist or SOG etchback processing for dielectric planarization.

9.5.8 Plasma etch safety

There are some safety issues in plasma etch processes, as they involve certain chemicals that are corrosive and toxic, such as Cl_2, BCl_3, SiF_4, and HBr. Inhalation

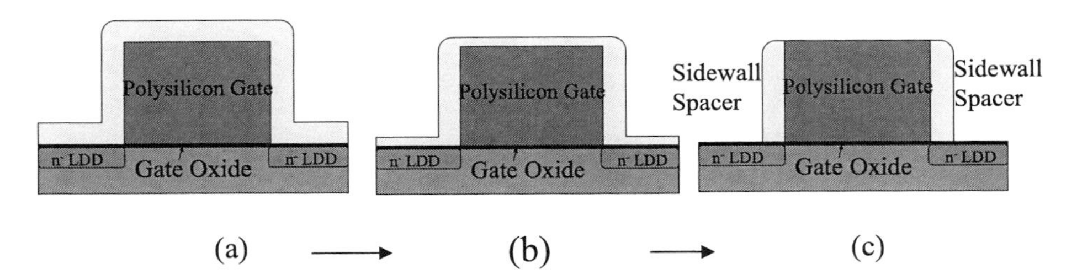

Figure 9.49 RIE etchback for sidewall spacer formation: (a) CVD conformal dielectric film, (b) etchback dielectric film, and (c) formation of sidewall spacer.

of these gases at a high concentration (>1000 ppm) can be fatal. Carbon monoxide (CO) is a colorless and odorless gas. It is flammable and can cause flash fires. It is toxic and harmful if inhaled because it combines with hemoglobin in the blood and decreases the amount of oxygen delivered to tissues, causing blood damage, difficulty breathing, and even death.

Radio frequency power can cause electric shock, and at high power can be lethal. All moving parts, including robot and vacuum valves, are mechanical hazards and can cause injury. Lock out and tag out are important for preventing someone from turning on a system from the cleanroom side while others are still working on the system in a gray area.

9.6 Process Trends

For wet etch, etch rate is mainly determined by temperature and etchant concentration. Increasing temperature increases etch rate due to the higher chemical reaction rate and higher etchant and etch byproduct diffusion rates. Increasing etchant concentration also increases etch rate. Selectivity is mainly determined by the chemistry of the etchant and the materials involved in the etching process. Wet etch normally has very good selectivity. For wet etch, the etch profile is always isotropic, and there is no way to control it. Etch-rate uniformity is mainly determined by the uniformity of the etchant solution temperature and concentration. Solution and wafer agitation can help to improve etch uniformity. The endpoints of wet etch processes are determined by time and detected by the visual inspection of an operator.

For plasma etch, especially the commonly used parallel plate electrode rf plasma, etch rate is most sensitive to rf power. Increasing rf power increases both ion bombardment flux and ion bombardment energy. This significantly enhances physical etch rate and damaging effects. Increasing rf power also increases free-radical concentration, an increase that enhances the chemical etch effect. Therefore, if the etch rate of a plasma etch system is outside of specifications, the rf system should be checked first. The rf system includes the source, cables and connections, and the rf match. Increasing rf power causes the RIE process to become more physical. It normally reduces etch selectivity, especially the selectivity to photoresist, due to physical sputtering. The relationship between etch rate and selectivity to rf power is illustrated in Fig. 9.50.

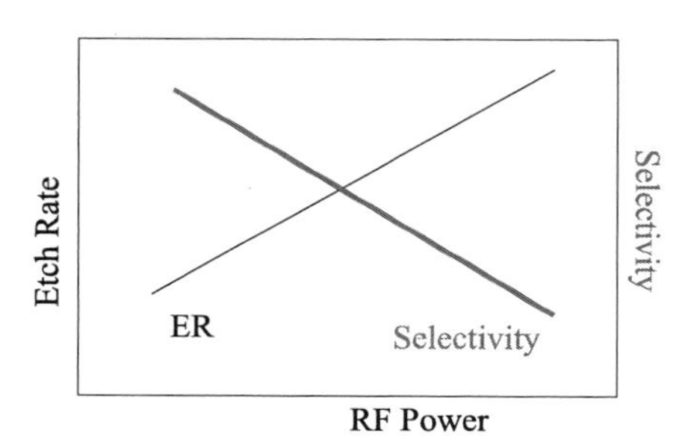

Figure 9.50 Etch process trends with rf power.

Pressure mainly controls etch uniformity and profile. It can also affect etch rate and selectivity. Changing pressure changes the MFP of electrons and ions. It affects the uniformity of the plasma and etch rate. By increasing pressure, the MFP becomes shorter, causing more collisions for the ions. Ion energy decreases, and ion collisional scattering increases, increasing the chemical etch components in the RIE process. Normally, increasing the pressure increases etch rate if the etch process is chemically dominant and reduces etch rate if the etch is physically dominant.

Increasing the magnetic field increases plasma density. Therefore, this increase also increases the ion bombardment flux, which increases the physical etch component. However, it decreases ion energy by reducing the sheath bias, and increases free radical concentration, an increase that can cause etch to be more chemical. Normally, at low pressure, when the magnetic field is weak, physical etch improvement outweighs chemical etch. Increasing the strength of the magnetic field causes the etch to become more physical. When magnetic field strength attains a certain value, the etch becomes more chemical due to the lesser ion energy caused by the reduction of dc bias. Figure 9.51 summarizes etch trends when rf power, pressure, and magnetic field strength B increases.

If the etch chamber is leaking, the etching rate to the photoresist increases due to the presence of oxygen in the plasma. The selectivity to photoresist degrades, and particle count increases. Insufficient photoresist hard bake during the photolithography process can also cause high photoresist etch rates and excessive photoresist loss during the etch process.

Since each etch process requires different chamber designs, uses different chemicals, and operates at different conditions, processing trends can be quite different from one another. Normally, a tool vender provides the basic processing information, including process trends and troubleshooting guides for the tools.

9.7 Recent Developments

To create better anisotropic etch profiles and reduce CD loss, lower pressure is desired because it can increase MFP and reduce ion scattering. Increasing plasma density can increase ion bombardment flux; for a certain amount of ion bombardment requirements, increasing ion bombardment flux can reduce the

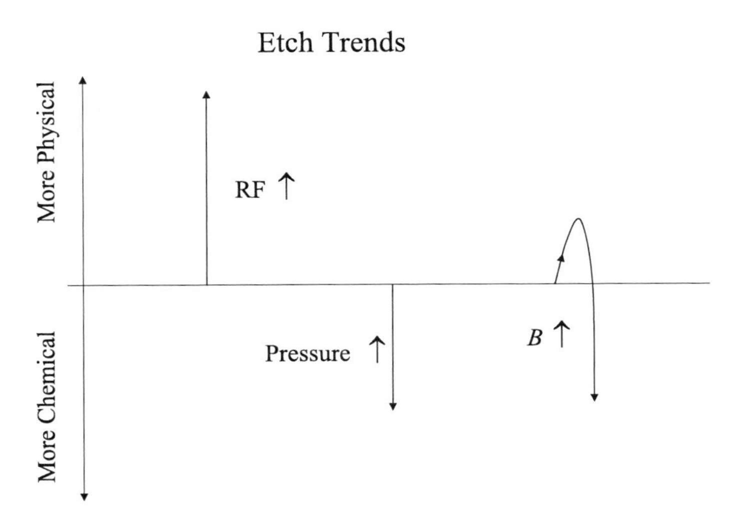

Figure 9.51 Summary of etch trends.

required ion energy, a method that is a favorite for reducing device damage. Plasma chambers with higher density at lower pressures are the desired chamber designs for achieving higher etch rates and better profile control. ICP and ECR etch chambers have been developed to meet these requirements. Both can generate high-density plasma at a low pressure and can independently control plasma density and ion bombardment energy, a feature that is very important for etch process control. For both ICP and ECR plasma sources, the ionization rate is not very high, about 1 to 5%. In advanced 300-mm fabs, ICP etch systems are widely used to etch conductive materials such as silicon, polysilicon, and metal layers, while capacitively coupled plasma etch systems are commonly used for dielectric etch processes. A helicon plasma source (illustrated in Fig. 9.52) can achieve almost 100% ionization rate at a low pressure of a few millitorr. It could very well be one of the candidates for future etch chamber designs.

Another concern is plasma uniformity control, especially for larger wafer sizes. Plasma uniformity and position control will become more important in the future. For the mature technology node, high throughput and low cost of ownership are key factors to competing with the existing in-production systems. If a reliable etch system can be manufactured with a low cost of ownership that can save money for customers in the long run, it would be possible for IC manufacturers to replace the existing systems and introduce the new low-cost systems. The key is that yield and up-time should be the same or higher than existing systems, throughput should be higher, and the consumable lower, so that customers can justify replacement if one year of operation savings can pay off the cost of the equipment.

New materials have recently been added to IC chip manufacturing processes, such as HKMG and ULK dielectrics. Etch of these new materials is one of the challenges for etch process developers. New device structures have also been applied in advanced IC chip manufacturing, such as 3D FinFET, trigate devices,

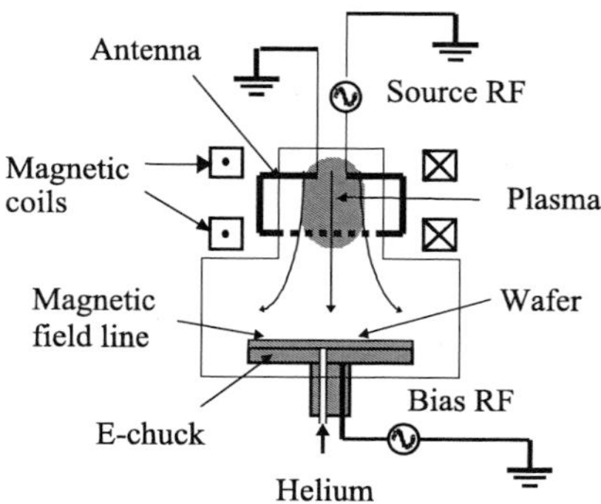

Figure 9.52 Etch chamber with a helicon plasma source.

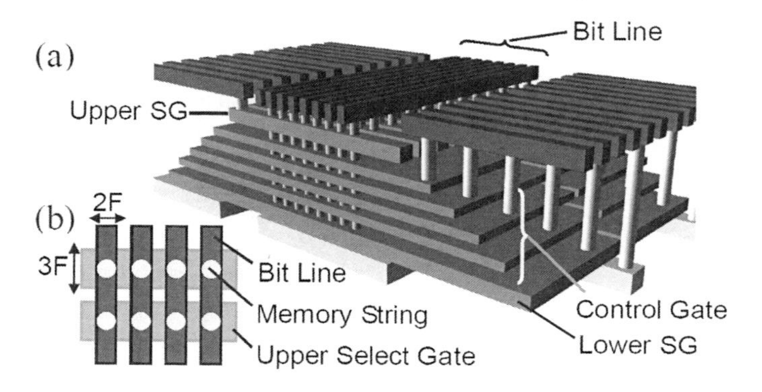

Figure 9.53 (a) Birds-eye view and (b) top view of a 3D flash memory array (*IEDM Technical Digest*).

vertical devices, etc. For 3D FinFET, a single-crystal silicon etch process becomes more challenging, especially for a FinFET built on bulk silicon wafers. For advanced bWL DRAM, an etch process with about 1:1 etch selectivity of single-crystal silicon and silicon oxide is required. One of the approaches for increasing the storage density of NAND flash is to adapt a 3D stacked cell structure, as shown in Fig. 9.53, which illustrates four layers of NAND flash cells used to form a string of 4-bit flash memory. Real applications probably require 16-bit strings. The etch process of holes for memory strings can be very challenging, and an etch process for contact holes of WLs that have different depths would be even more challenging. 3D packaging such as through-silicon via (TSV) processes provide more challenges and opportunities for etch processes. Different from submicron and nanometer pattern etch processes in front-end-of-line (FEoL) processes, TSV etch processes have significantly larger CDs (50 to 5 μm), and require significantly higher etch rates to achieve the required reasonable throughput.

9.8 Summary

- The four materials that need to be etched in IC chip fabrication are single-crystal silicon, polysilicon, dielectric (silicon oxide and nitride), and metals (TiN, Al–Cu, Ti, W, and WSi_2).
- Four major etch processes are silicon, poly, dielectric, and metal etches.
- Wet etch uses a chemical solution to dissolve material that needs to be etched.
- Dry etch uses chemical gases to remove material from the substrate surface by physical etch, chemical etch, or a combination of both.
- Wet etch has high selectivity, high etch rate, and low equipment cost. The isotropic etch profile limits its applications in pattern etch to feature sizes larger than 3 μm.
- Wet etch is still widely used in advanced semiconductor fabs for thin-film stripping and dielectric thin-film quality control.
- In plasma etch processes, etchant flows into the chamber and dissociates in the plasma, and free radicals diffuse across the boundary layer and are adsorbed on the surface. The free radicals react with an atom or molecule on the surface with the help of ion bombardment. Volatile byproducts desorb from the surface, diffuse across the boundary layer, and are pumped out of the chamber in the convection flow.
- There are two anisotropic etch mechanisms—damaging and blocking. Dielectric etches normally use a damaging mechanism. Silicon, polysilicon, and metal etches use a blocking mechanism.
- Dielectric etch uses fluorine chemistry. CF_4, CHF_3, and Ar are commonly used for the dielectric etch processes. CF_4 is the main etchant, and CHF_3 is a polymer precursor used to improve etch selectivity to photoresist and silicon. Ar is used to enhance ion bombardment. O_2 can be used to increase etch rate, and H_2 can be used to improve etch selectivity to photoresist and silicon.
- For low-κ and ULK dielectric etching, CO can be added to improve etch process control.
- Silicon etch uses HBr as a main etchant, and O_2 and fluorine compounds are used for sidewall deposition.
- Polysilicon can be etched with Cl_2 or SF_6. O_2 is normally used to improve the selectivity to oxide, and HBr is used to help with sidewall protection.
- Metal etch uses Cl_2, BCl_3 and N_2 to improve sidewall passivation.
- Copper metallization does not require metal etching; it requires dielectric trench etches.
- Photoresist stripping after ULK dielectric etch becomes very challenging, and a metal hard mask (TiN) is commonly used.

9.9 Bibliography

D. G. Baldwin, M. E. Williams, and P. L. Murphy, *Chemical Safety Handbook for the Semiconductor/Electronics Industry*, 2nd ed., OME Press, Beverly, MA (1996).

R. Bowman, G. Fry, J. Griffin, D. Potter, and R. Skinner, *Practical VLSI Fabrication for the 90s*, Integrated Circuit Engineering Corp., Raleigh, NC (1990).

J. W. Coburn, *Plasma Etching and Reactive Ion Etching*, AVS Monograph Series, M-4, American Institute of Physics, Inc., New York (1982).

J. W. Coburn, H. F. Winters, "Ion- and electron-assisted gas-surface chemistry: an important effect in plasma etching," *J. Appl. Phys.* **50**, 3189–3196 (1979).

Y. Fukuzumi et al., "Optimal integration and characteristics of vertical array devices for ultrahigh-density, bit-cost scalable flash memory," *IEDM Tech. Dig.*, 449–452 (2007).

S. K. Ghandhi, *VLSI Fabrication Principles*, 2nd ed., Wiley-Interscience Publication, John Wiley & Sons, Inc., New York (1994).

B. S. Haran et al., "22-nm technology compatible fully functional 0.1 μm^2," *IEDM Tech. Dig.*, 625–628 (2008).

D. M. Manos and D. L. Flamm, *Plasma Etching, An Introduction*, Academic Press, San Diego, CA (1989).

I. Morey and A. Asthana, "Etch challenges of low-κ dielectric," *Solid State Technol.* **42**(6), 71–78 (1999).

T. Schloesser et al., "A 6F2 buried wordline DRAM cell for 40 nm and beyond," *IEDM Tech. Dig.*, 809–812 (2008).

See http://www.webelements.com for further information on the elements in this chapter.

9.10 Review Questions

1. What is the pattern etch process?
2. Define etch selectivity.
3. What are the differences between wet etch and RIE?
4. What is the chemical most commonly used to wet etch silicon dioxide? What is the special safety concern of that particular chemical?
5. Why is wet etch preferred for thin-film stripping processes?
6. Explain the two anisotropic etch mechanisms.
7. Name the etch process that has the smallest critical dimension.
8. Why does the polysilicon gate etch process require high polysilicon-to-oxide selectivity?
9. Why does polysilicon etch prefer chlorine over fluorine as the main etchant?
10. How does the F/C ratio affect oxide etch processes?
11. Why isn't fluorine used as the main etchant for Al–Cu metal etches?

12. Explain the benefit of a low-pressure, high-density plasma source in pattern etch processes.
13. What metallic materials have been used for hard masks in low-κ and ULK dielectric etch?
14. After low-κ or ULK dielectric etch, is a metal etch process required to remove the metal hard mask after photoresist stripping?

Chapter 10
Chemical Vapor Deposition and Dielectric Thin Films

In the semiconductor industry, electrically insulating materials are commonly called dielectrics. Dielectric thin-film processes are adding processes because they add a thin layer of dielectric materials to the wafer surface. While the majority of dielectric thin-film processing is CVD processing, spin-on dielectric processes are also used in IC manufacturing. The main issues with dielectric thin-film processes are filling gaps without voids, deposition of uniform film with high throughput, and making the final surface as planarized as possible.

Objectives

After finishing this chapter, the reader will be able to:

- list two dielectric thin-film materials used in IC chip fabrication
- identify at least four applications of dielectric thin films from a CMOS chip cross section
- describe the CVD processing sequence
- list the two deposition regimes and describe their relation to temperature
- name the two most commonly used silicon precursors for dielectric CVD.

10.1 Introduction

There are two types of dielectric thin films widely used in the semiconductor industry: thermally grown and deposited thin films. Thermally grown dielectric thin films are discussed in Chapter 5. In this chapter, we cover deposited dielectric thin films. The fundamental difference between thermally grown and deposited thin films is that grown film consumes silicon from the substrate, while deposited film does not. Figure 10.1 shows the difference between silicon dioxide films. For thermally grown silicon dioxide, oxygen comes from the gas phase, and silicon comes from the silicon substrate. This process consumes silicon from the substrate, and the film grows into the substrate. For CVD oxide, both silicon and oxygen come from the gas phase, so there is no consumption of silicon substrate.

The quality of thermally grown oxide film is much better than that of deposited film. Therefore, the gate oxide of a MOSFET always uses thermally grown silicon

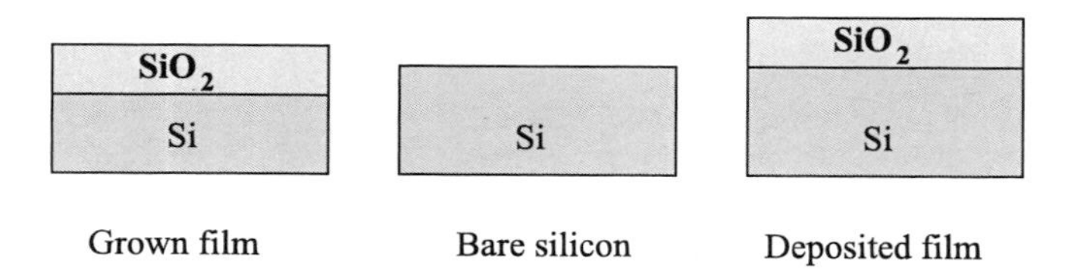

Figure 10.1 Thermally grown and deposited thin films.

dioxide or nitridated thermal oxide. There are many applications of deposited dielectric thin films in IC chip fabrication, including electrical isolation, ion implantation blocking, dopant source, ARC, hard mask, capping layer, ESL, and circuit passivation.

The main role of dielectric thin film is as a dielectric layer for electrical insulation in multilevel metal interconnections. Both CVD dielectrics and the combination of spin-on and CVD dielectrics are used for this application. It is also used in STI as electrical insulation between neighboring transistors. It is used to form spacers on the sidewalls of polysilicon or polycide (polysilicon/silicide stack) gate that are needed to form an LDD or SDE and the diffusion buffer. A passivation dielectric seals the IC chip to protect the tiny circuit from chemical damage due to moisture and mobile ions. It is used as an ESL for dual damascene copper metallization. It also can be used as a cap barrier layer for low-κ or ULK dielectrics. It is also used as a passivation layer to protect the IC chip from mechanical damage during testing and packaging. The CVD or spin-on dielectric ARC is used to minimize reflection from the wafer surface to meet the resolution requirements of photolithography processes. The commonly used metallic ARC can no longer meet the resolution requirements once pattern feature size becomes smaller than 0.25 μm.

Applications of dielectric thin films in a CMOS IC with Al–Cu interconnection are illustrated in Fig. 10.2. Different companies use different acronyms for dielectric layers in the interconnection applications. Many companies use ILD, which stands for interlayer dielectric. For the dielectric between the polysilicon and first metal layer, some fabs call it a premetal dielectric (PMD), while some others call it ILD0. Some fabs call the dielectric layers between the metal layers an intermetal dielectric (IMD), while others call them ILD-*X* (*X* from 1 to whatever the number of metal layers minus one).

In the first edition of this book, PMD and IMD were the only acronyms used for dielectrics because at that time most interconnections in IC chips were aluminum–copper alloys (Al–Cu). For IC chips with Al-Cu interconnections, the deposition conditions of PMD and IMD are quite different. IMD layers used USG and were deposited at around 400 °C due to the limitations of deposition temperature of Al–Cu alloys. PMD normally used a doped oxide, either PSG or BPSG. Deposition temperature was limited by the thermal budget, which was the diffusion tolerance of the dopants in the activation area defined by the device design. However, device structure has since been changed. For example, fabs are manufacturing IC chips with metal gates with both gate-first and gate-last approaches,

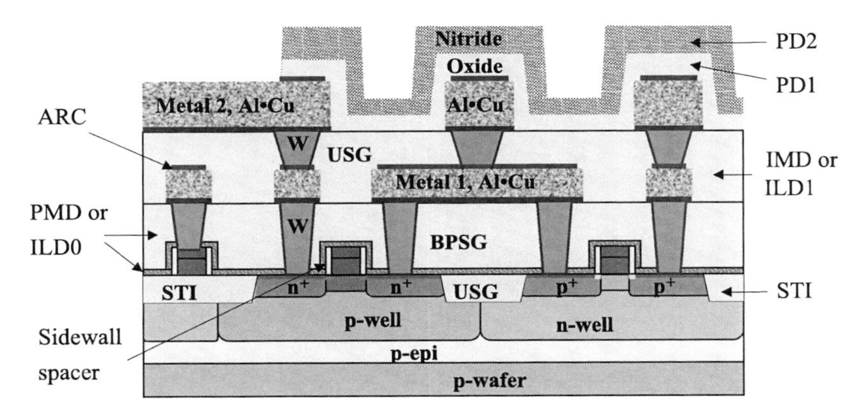

Figure 10.2 Applications of dielectric thin film in a CMOS circuit with Al–Cu interconnection.

thus it no longer makes sense to call the dielectric between a gate and the first metal wire a PMD. Also, for the bWL DRAM, metals (TiN and W) are used for array transistor gate electrodes, thus PMD is no longer a correct description of the dielectric. Therefore, in this updated edition, the acronym ILD is used to describe the dielectric layers for the interconnection, either with Cu or with Al–Cu alloys.

Figure 10.3 illustrates a cross section of a CMOS chip with five layers of Cu/low-κ interconnection. More advanced IC chips can have up to 11 metal layers, which need 12 ILD layers. Thus, for an advanced IC fab, many dielectric CVD tools are dedicated to ILD deposition. For an N-layer metal interconnection IC chip with STI, the minimum number of dielectric layers is:

$$\text{dielectric layer} = \underset{\text{STI}}{1} \quad \underset{\text{spacer}}{+1} \quad \underset{\text{ILD}}{+N+1} \quad \underset{\text{PD}}{+1} = N+4 \qquad (10.1)$$

Some advanced IC chip devices use double or even triple spacers, and embedded DRAM in SOC devices requires dielectric deposition between the two capacitor electrodes. For these chips, there are more dielectric deposition steps than $N + 4$.

10.2 Chemical Vapor Deposition

CVD is a process in which a gaseous chemical precursor (or precursors) has a chemical reaction on the wafer surface and deposits a solid byproduct as a layer of thin film. Other byproducts are gases and leave the surface.

CVD processes are widely used in the semiconductor industry for various thin-film depositions, such as epitaxial silicon, polysilicon, dielectric thin film, and metal thin-film depositions.

Epitaxial and polysilicon depositions are the high-temperature CVD processes discussed in Chapter 5, and the metal thin-film CVD process is covered in Chapter 11. Dielectric CVDs are covered in detail in this chapter. Table 10.1 summarizes some of the important CVD thin films used in IC manufacturing.

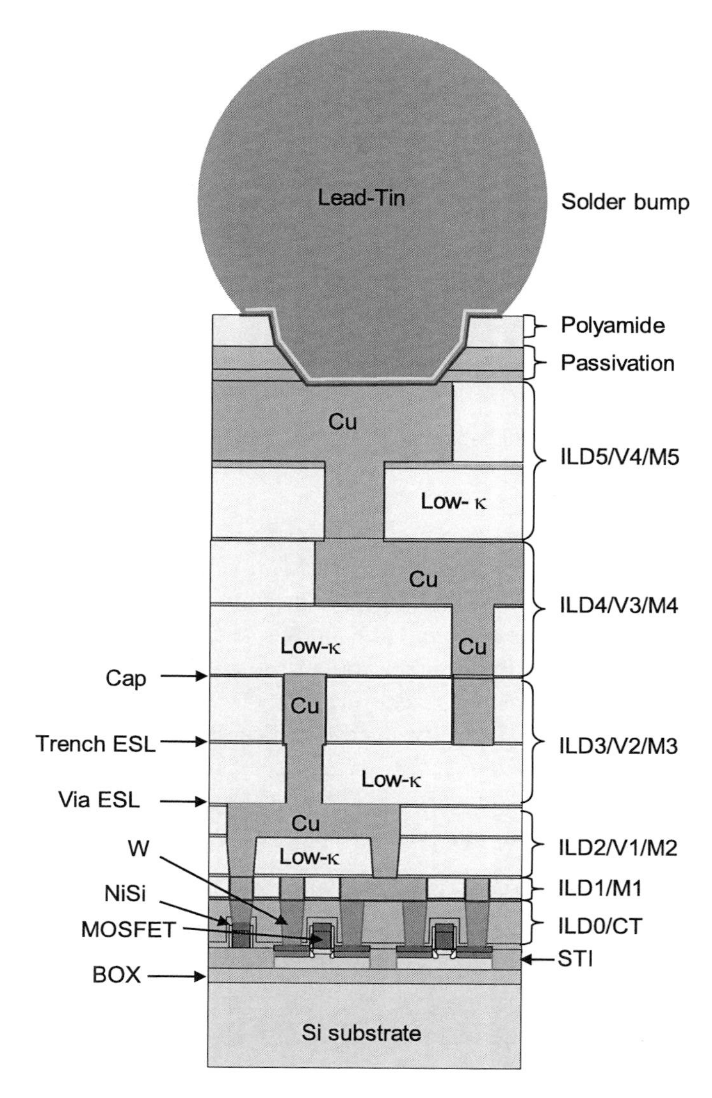

Figure 10.3 Dielectric layers in a copper/low-κ interconnection application.

10.2.1 Chemical vapor deposition process description

CVD is a sequential process, which includes several steps:

- gas or vapor phase precursors are introduced into the reactor
- precursors diffuse across the boundary layer and reach the substrate surface
- precursors adsorb on the substrate surface
- adsorbed precursors migrate on the substrate surface
- chemical reaction occurs on the substrate surface
- solid byproducts form nuclei on the substrate surface
- nuclei grow into islands
- islands merge into the continuous thin film

Table 10.1 CVD thin films and precursors used in the IC industry. Note that Cu (hfac) (vtms) stands for hexaflouroacetyl-acetonato-copper-vinyl-trimethylsilane.

	FILMS	PRECURSORS
Semiconductor	Si (poly) Si (epi)	SiH_4 (silane) $SiCl_2H_2$ (DCS) $SiCl_3H$ (TCS) $SiCl_4$ (Siltet)
Dielectrics	Oxide	SiH_4, O_2 SiH_4, N_2O $Si(OC_2H_5)_4$ (TEOS), O_2 TEOS TEOS, O_3 (ozone)
	Oxynitride	SiH_4, N_2O, N_2, NH_3 SiH_4, N_2, NH_3
	Si_3N_4	SiH_4, N_2, NH_3 $C_8H_{22}N_2Si$ (BTBAS)
	Low-κ	3MS (trimethylsilane), 4MS (tetramethylsilane), etc., and O_2
	ULK	DEMS (diethoxymethylsilane) and $C_6H_{10}O$ (cyclohexene oxide)
	W (Tungsten)	WF_6 (tungsten hexafluoride), SiH_4, H_2
Conductors	WSi_2	WF_6 (tungsten hexafluoride), SiH_4, H_2
	TiN	$Ti[N(CH_3)_2]_4$ (TDMAT)
	Ti	$TiCl_4$
	Cu	Cu (hfac) (vtms)

- other gaseous byproducts desorb from the substrate surface
- gaseous byproducts diffuse across the boundary layer
- gaseous byproducts flow out of the reactor.

Figure 10.4 illustrates the sequential process, which includes the precursor gases being introduced, precursors diffusing across the boundary layer, byproducts diffusing across the boundary layer, and byproducts flowing out of the reactor.

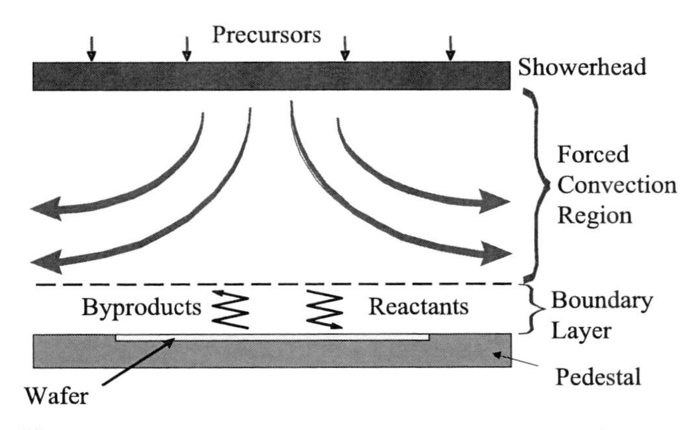

Figure 10.4 CVD process sequence.

Figure 10.5 shows that precursors reach the substrate surface by diffusing across the boundary layer, adsorbing, and migrating on the substrate surface. The ability of the precursor to migrate on the surface is called surface mobility, which is very important for thin-film step coverage and gap-fill properties. This is covered in a later section of this chapter. When precursors have chemical reactions on the surface, they form into a solid material and release gaseous byproducts. The first few molecules of the solid material form nuclei on the surface, and additional chemical reactions cause the nuclei to grow into islands. The islands grow, merge, and finally form a continuous thin film on the wafer surface.

Figure 10.5 can also be used to illustrate the physical vapor deposition (PVD) process. The major difference between CVD and PVD is at the nucleation step. CVD has a chemical reaction on the substrate surface, while PVD does not.

10.2.2 Chemical vapor deposition reactor types

There are three types of CVD reactors commonly used in the semiconductor industry: atmospheric pressure (APCVD), low-pressue (LPCVD), and plasma-enhanced (PECVD).

10.2.2.1 Atmospheric pressure chemical vapor deposition

Atmospheric pressure equals 760 torr at sea level at 0 °C. Figure 10.6 illustrates an APCVD system. It has three zones: two nitrogen buffer zones and a processing zone in between. Nitrogen curtains in both buffer zones separate the processing gases and prevent them from leaking into the atmosphere. Heater elements heat the wafer carriers or disks, which in turn heat the wafers. A conveyor belt constantly transports wafers into the processing zone. Precursor chemical gases react on the heated surface and deposit a thin film. The deposited wafers are removed from the carrier. After several depositions, the thin-film-coated carriers are sent to a clean shop to remove the coated thin film before being reused. The conveyor belt is also coated with a layer of deposited thin film as it passes through the processing zone. Therefore, it needs constant cleaning to maintain a stable processing condition. The APCVD process is controlled by temperature, processing gas flow rate, and belt speed.

The APCVD process has been used to deposit silicon oxide and silicon nitride. The APCVD O_3-TEOS oxide process is widely used in the semiconductor industry, especially in STI and ILD0 applications.

Question: A semiconductor manufacturer has its research and development laboratory on the coast near sea level and one of its manufacturing fabs on a high-altitude plateau. It was discovered that APCVD processes developed in the laboratory could not be directly transferred to that particular fab. Why?

Answer: Atmospheric pressure is related to altitude. On a high-altitude plateau, the atmospheric pressure is significantly lower than that at sea level. The earlier APCVD reactor did not have a pressure-control system, so

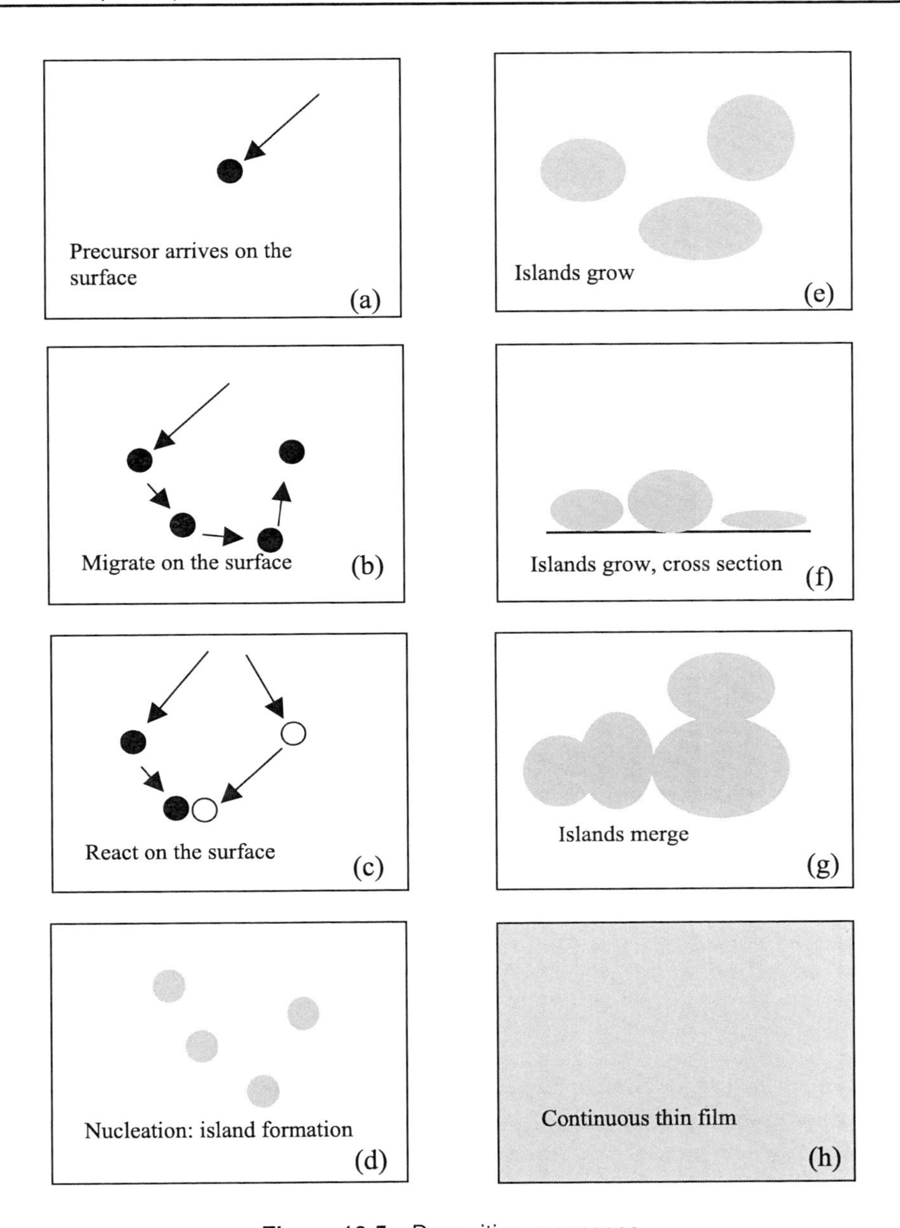

Figure 10.5 Deposition processes.

processing conditions were quite different between the laboratory and the manufacturing fab. A process that worked fine in one location might not work well in the other because of the pressure difference.

10.2.2.2 Low-pressure chemical vapor deposition

LPCVD operates at a low pressure, from 0.1 to 1 torr. The LPCVD reactor is very similar to an oxidation furnace. It has three heating zones, and wafers are

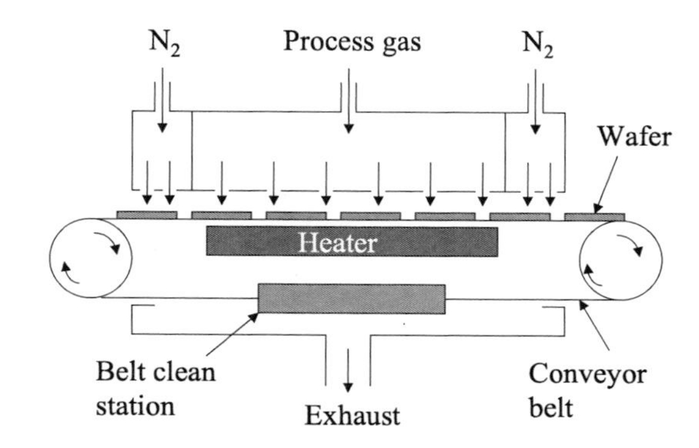

Figure 10.6 The APCVD system.

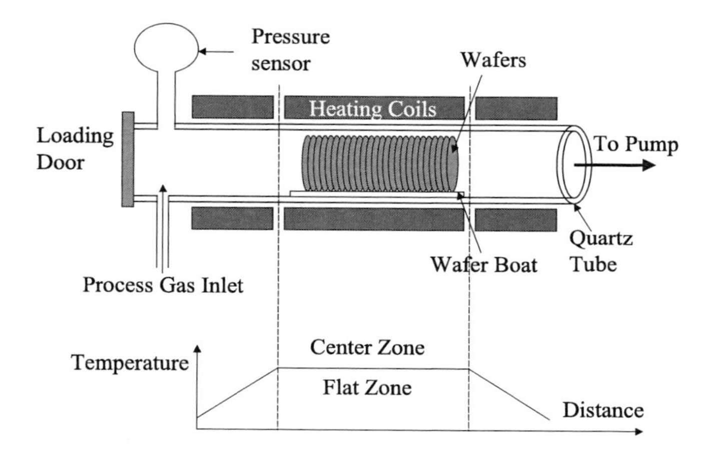

Figure 10.7 The LPCVD system.

processed in the center zone or flat zone. LPCVD systems require vacuum systems to control pressure inside the reactors. LPCVD reactors operate in surface-reaction-limited regions, which are covered in a later section. The deposition process is mainly controlled by wafer temperature and is relatively insensitive to gas flow rate. Therefore, wafers can be loaded vertically with very little spacing required. The large wafer load improves throughput and reduces wafer processing costs compared with APCVD processes. Figure 10.7 illustrates the LPCVD system.

LPCVD processes have been used to deposit oxide, nitride, and polysilicon. The majority of polysilicon and amorphous silicon are deposited in LPCVD reactors. Silicon nitride layers used for oxidation masks in LOCOS processes and for CMP hard stop layers in STI processes are deposited in LPCVD processes. LPCVD is also commonly used to deposit diffusion barrier silicon nitride layers to block diffusion of the dopant atoms in doped oxide from diffusing through the thin gate oxide layer and into activation areas.

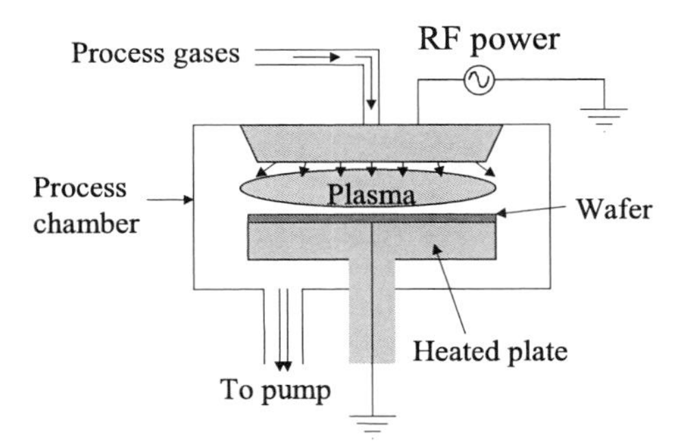

Figure 10.8 The PECVD reactor.

The LPCVD reactor operates at a high temperature, usually higher than 650 °C. Therefore, it cannot be used for the IMD depositions after the first metal layer deposition.

10.2.2.3 Plasma-enhanced chemical vapor deposition

The operating pressure of PECVD usually is from 1 to 10 torr. Because free radicals generated from the plasma dramatically increase the chemical reaction rate, PECVD can achieve high deposition rates at relatively low temperatures, which is very important for ILD depositions. Figure 10.8 gives an illustration of a PECVD reactor.

Another advantage of PECVD is that deposited film stress can be controlled by rf power without having a major effect on deposition rate. PECVD processes are widely used for oxide, nitride, low-κ, ESL, and other dielectric thin-film depositions.

10.2.3 Chemical vapor deposition basics

10.2.3.1 Step coverage

Step coverage is a measurement of the deposited film reproducing the slope of a step on a substrate surface. It is one of the most important specifications of the CVD process. Definitions of sidewall step coverage, bottom step coverage, conformality, and overhang are given in Fig. 10.9.

Step coverage is determined by both the arriving angle and precursor surface mobility. The arriving angle is illustrated in Fig. 10.10. It can be seen that corner A has the largest arriving angle (270 deg), and corner C has the smallest. Therefore, corner A will have more precursor atoms or molecules when they diffuse across the boundary layer. If these precursors react immediately after being adsorbed on the wafer surface without migration, corner A will have more deposition than corner C and will form overhangs as shown in Fig. 10.10.

The arriving angle can also be modified by controlling etch processes. For example, most fabs etch contact holes with a tapered angle instead of a straight hole (shown in Fig. 10.11) to create a larger arriving angle, and to make it easier for tungsten CVD processes to fill the contact hole.

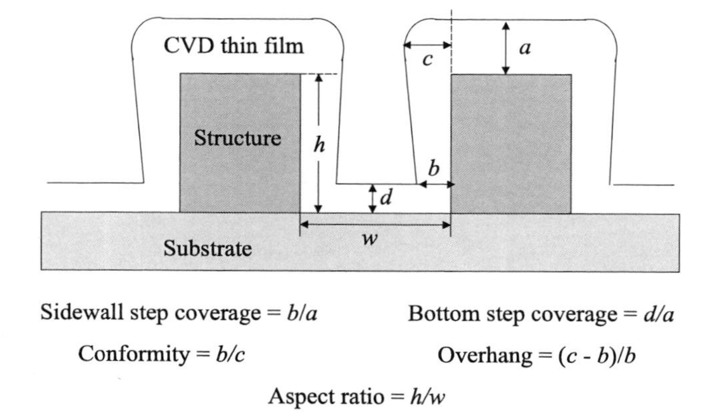

Figure 10.9 Step coverage and conformality.

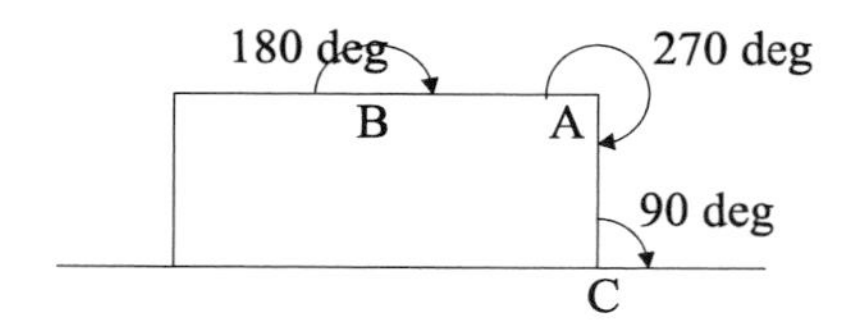

Figure 10.10 Arriving angles.

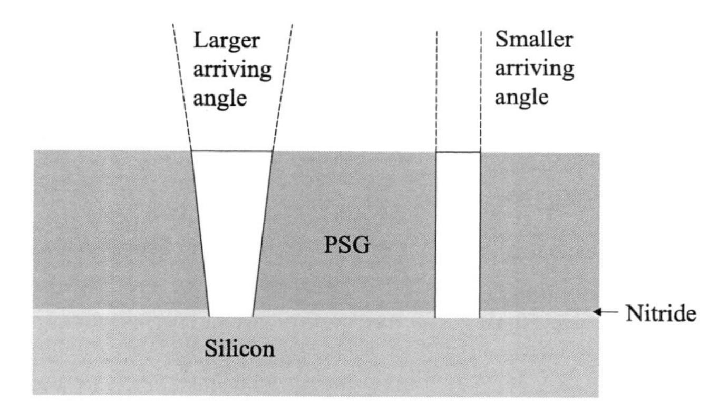

Figure 10.11 Arriving angles for tapered and straight contact holes.

Overhangs are very undesirable. If a deposited film starts with overhangs caused by the arriving angle effect and low surface mobility, as the film thickness increases, the overhang will grow faster than the film due to the larger arriving angle. Very soon the overhangs seal the top of the gap to form the void (also known as the key hole) between the polysilicon pattern, as shown in Fig. 10.12.

These voids can have processing gases sealed inside, and the diffusion of these gases inside the IC chip can cause problems in later processes, causing yield issues

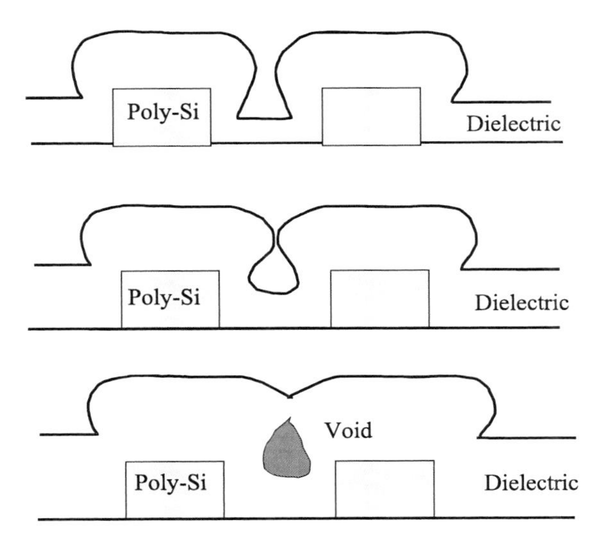

Figure 10.12 Void formation process.

or reliability issues during chip operation in an electronic system. Therefore, for most CVD processes, voids are unwanted, and void-free gap fill is required to ensure high yield and reliability of IC chips.

By reducing process pressure, the MFP of precursors increases. When the MFP is longer than gap depth h (shown in Fig. 10.9), there are very few collisions inside the gap. Thus, the precursors have little chance to go backward and reach corner A (Fig. 10.11) from inside the gap. Therefore, the MFP being longer than the gap depth effectively reduces the arriving angle and improves step coverage. This is another advantage of the LPCVD process over the APCVD process, especially for the silane-based silicon oxide CVD process, since silane has very low surface mobility.

Figure 10.13 shows that surface mobility can significantly affect step coverage. After precursors adsorb on the surface, and if the precursors have enough energy to break the adsorption bond with the surface, they can leave the surface and hop along it. If they move along the surface rapidly, migration of the precursor can smooth out the arriving angle effect. Thus, precursors with high surface mobility can achieve very good step coverage and good conformality.

Surface mobility is mainly determined by precursor chemistry, which is discussed in a later section. It is also related to wafer temperature, since heat can provide energy to the precursors that is necessary for them to break free and migrate on the wafer surface. Increasing wafer temperature can improve deposited film step coverage. For PECVD processes, ion bombardment on the wafer surface can also provide energy for the precursors to break free from the substrate surface and increase surface mobility. Therefore, increasing the rf power in a PECVD reactor can improve deposited film step coverage with an effect similar to raising the temperature.

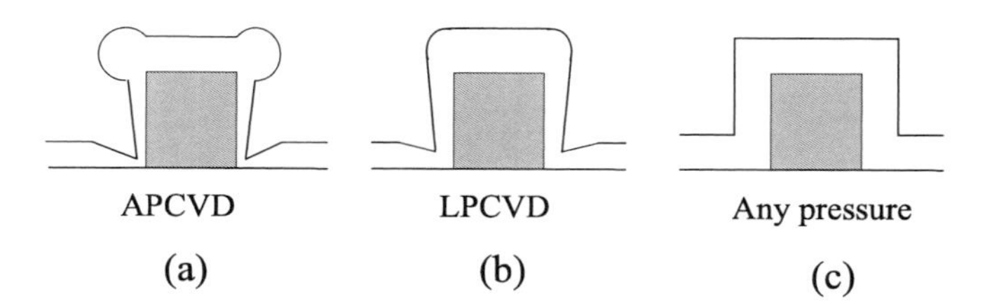

Figure 10.13 Relationship of step coverage to pressure and surface mobility: (a) high pressure, low mobility; (b) low pressure, low mobility; and (c) high mobility.

10.2.3.2 Gap fill

It is very important for CVD processes to fill gaps without voids. For example, in CMOS and DRAM ILD0 deposition processes, there is zero tolerance for any voids between the gates. This is because the voids in ILD0 can cause electrical shorts between the contact plugs, as shown in Fig. 10.14.

For tungsten CVD or copper deposition processes, voids inside contact or via plugs are also undesired; they can cause high interconnect electrical resistance. They can also induce chip reliability issues due to void migration and out-diffusion of the trapped processing gases and byproducts, such as WF_6, H_2, F, and HF, that can corrode metal wires or damage devices.

For the ILD deposition process, voids above the metal surface are intolerable. They can cause trouble for later processes, especially for the dielectric CMP process. However, if there are no contact plugs between the metal lines, voids can be tolerated when they are located inside the gap below the top metal surface. Voids can effectively reduce the dielectric constant between the metal lines, in turn reducing parasitic capacitance. Because voids always trap some processing gases and byproducts, most fabs prefer void-free gap fill (or hole fill) for CVD processes.

Question: How can voids help reduce κ-value between metal lines?

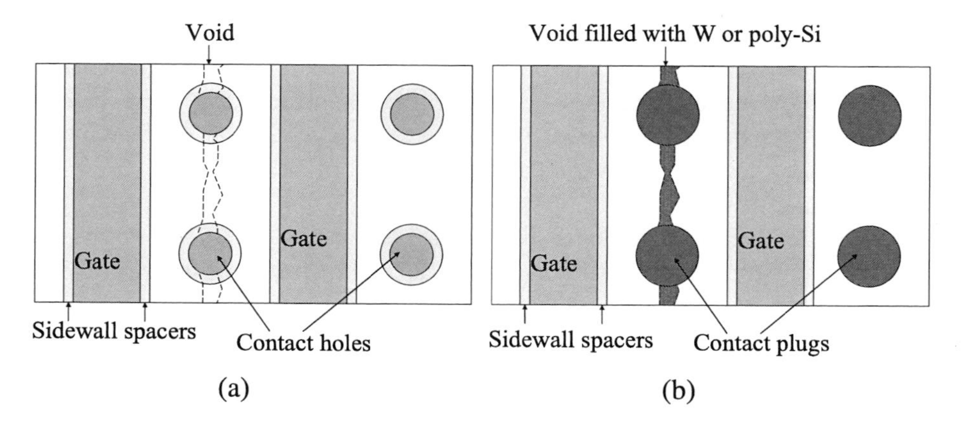

Figure 10.14 Voids that caused plug-to-plug shorts: (a) after contact hole etch and (b) after contact plug formation.

Answer: The dielectric constant (or κ-value) in a void is close to 1, which is significantly lower than the 4.0 to 4.2 of CVD silicon oxide, or 2.5 of ULK dielectric film. Therefore, a gap sealed with dielectric film with voids in the gap will have a lower effective dielectric constant than a gap filled with dielectric film without voids. The purposely formed void is called an air gap, which has the lowest achievable κ-value.

When a deposited film hangs over a pattern, it can cause voids if the film keeps growing. Different methods have been developed to alleviate this issue. One of them is argon ion sputtering etch, which chips the corner of overhangs and tapers the gap opening to increase the arriving angle, so that the subsequent deposition process can fill the gap without voids. This approach is called dep/etch/dep, short for deposition, etchback, and deposition, as illustrated in Fig. 10.15. This process is widely used for IMD processes in IC manufacturing.

If CVD precursors have very high surface mobility, the CVD film will have good step coverage and conformality, as shown in Fig. 10.13(c). The film can also grow up conformally to fill the gap without voids, as shown in Fig. 10.16. The O_3-TEOS oxide CVD and the tungsten CVD processes belong to this case.

In high-density-plasma (HDP) CVD processes, deposition and sputtering etch happen simultaneously inside the same processing chamber. Heavy ion bombardments due to low pressure (a few millitorr) and the high plasma density

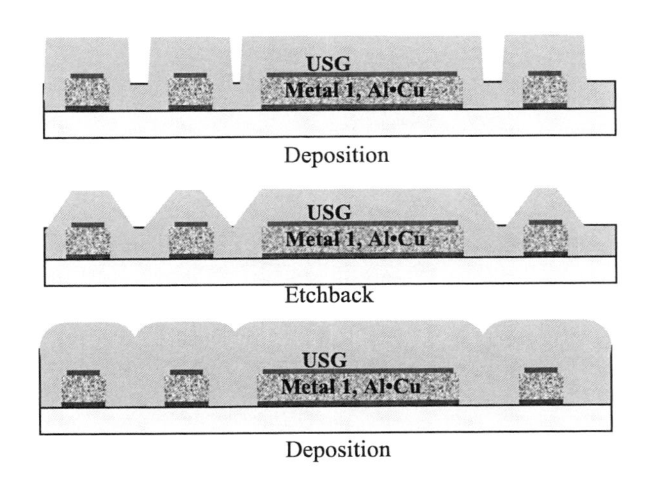

Figure 10.15 Dep/etch/dep gap-fill processes.

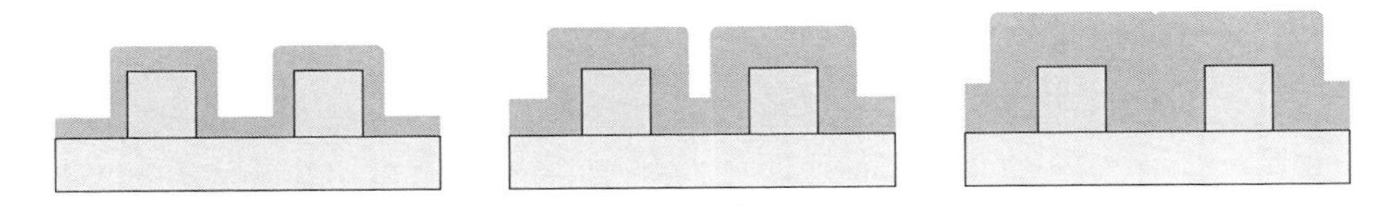

Figure 10.16 Conformal film deposition and gap fill.

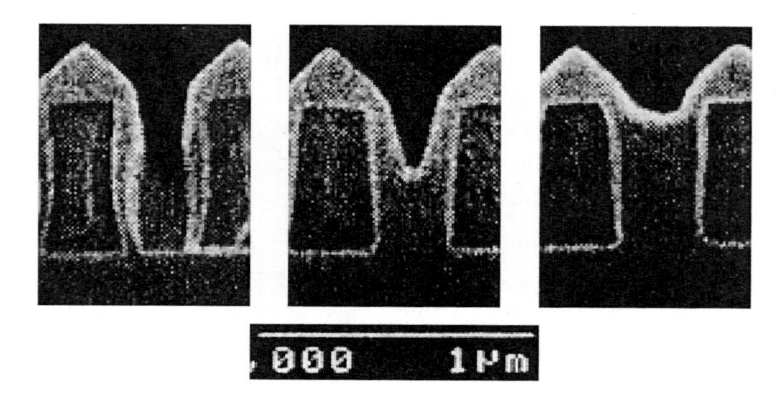

Figure 10.17 Gap fill of HDP-CVD. (Applied Materials, Incorporated).

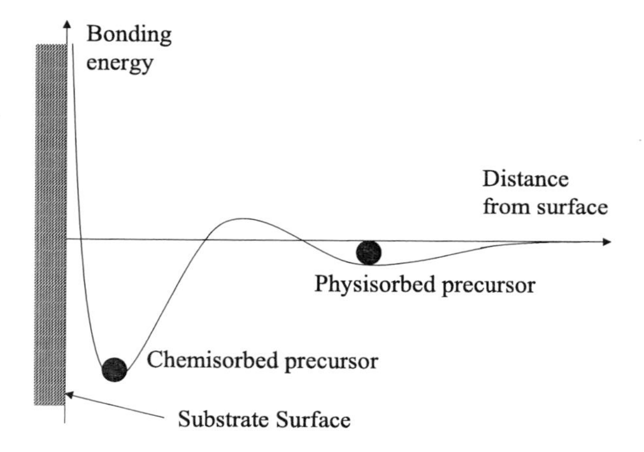

Figure 10.18 Relationship of bonding energy to chemical and physical adsorption.

constantly chip off the corner to keep the gap open and tapered, allowing a larger arriving angle and bottom-up deposition. Figure 10.17 shows a gap fill with the HDP-CVD process.

10.2.4 Surface adsorption

When precursors reach the substrate surface after diffusion across the boundary layer, they are adsorbed by the surface. There are two kinds of adsorption—chemisorption and physisorption (see Fig. 10.18).

10.2.4.1 Chemisorption

Chemisorption is the shortened term for chemical adsorption. In this case, an actual chemical bond is formed between an atom on the surface and an atom in the adsorbed precursor molecule. The chemisorbed atoms or molecules are held to the surface with energy that exceeds 2 eV. Because of the strong chemical bonds, the chemisorbed precursors have very low surface mobility.

For most dielectric CVD processes (especially ILD and passivation dielectric processes), deposition temperature cannot be higher than 450 °C because at this temperature aluminum begins to react with silicon dioxide. Even with copper interconnections, ILD deposition temperature is still limited to around 400 °C. At 400 °C (~0.06 eV), the heat alone cannot provide enough energy to cause the chemisorbed precursors to break free from their chemical bond and leave the substrate surface. In PECVD processes, ion bombardment provides enough energy (10 to 20 eV) to break the chemical bonds and free chemisorbed precursors from the surface, causing surface migration.

10.2.4.2 Physisorption

Physisorption is the shortened term for physical adsorption. In this case, the adsorbed molecules are held to the surface with forces much weaker than a chemical bonding force. Physisorption involves energies less than 0.5 eV per molecule. The nature of forces involved in physisorption varies from long-range Van der Waals forces to dipole–dipole forces (of which hydrogen bonding is a special case).

Both thermal energy at 400 °C and ion bombardment provide enough energy to cause significant amounts of physisorbed precursors to break free and leave the surface. The physisorbed precursors can move about the surface, so they have much higher surface mobility than that of chemisorbed molecules.

10.2.5 Chemical vapor deposition precursors and their adsorption

The most commonly used silicon source gases for dielectric CVD processes are silane (SiH_4) and TEOS [$Si(OC_2H_5)_4$]. For low-κ dielectrics, trimethylsilane [$(CH_3)_3SiH$] or 3MS is a commonly used precursor. Diethoxymethylsilane ($C_5H_{14}Si$) or DEMS can be used as a precursor as well, and cyclohexene oxide ($C_6H_{10}O$) or CHO can be used as a porogen for ULK dielectric deposition.

Silane is one of the most commonly used silicon source gases for dielectric thin-film deposition. The main purpose of the silane PECVD process is passivation dielectric depositions. It is used for the ILD0 barrier nitride layer and dielectric ARC layer depositions. This process is also used in HDP-CVD oxide processes. Silane is widely used in LPCVD polysilicon and epitaxial silicon deposition processes, as well as tungsten CVD processes for the tungsten nucleation step. It is also employed as a silicon source for tungsten silicide deposition.

Silane is very reactive, pyrophoric (ignites itself), explosive, and toxic. If a silane gas line is opened without thoroughly purging it beforehand, oxygen and moisture will react with the residual silane inside the gas line. This can cause fire or a minor explosion and forms fine silicon dioxide particulates that dust the gas line. This can induce significant processing downtime and increase production cost for replacement of the dusted double-layer silane line.

Since the silane molecule is perfectly symmetrical in a tetrahedral structure (shown in Fig. 10.19), it can neither chemisorb nor physisorb to the substrate surface. However, since silane is chemically very reactive, it can easily dissociate from the heat and plasma. The molecular fragments formed by pyrolysis or plasma dissociation—SiH_3, SiH_2, or SiH—are chemically very reactive free radicals and can easily form chemical bonds with surface atoms and readily chemisorb on the substrate surface. Silane-based precursors have very low surface mobility, explaining why silane-based dielectric CVD films always have overhangs at the upper corners of patterns and normally have poor step coverage, especially for the APCVD process.

TEOS is a large organic molecule with an ethoxy group (OC_2H_5) bonded to the silicon atom at each corner of its tetrahedron [Fig. 10.20(a)]. Different from a silane molecule, the TEOS molecule is no longer perfectly symmetric due to the large size of its ethyl group. It can form hydrogen bonds with surface atoms and physisorb on the substrate surface. Therefore, the TEOS precursor has high surface mobility, and TEOS-based CVD film generally has good step coverage and

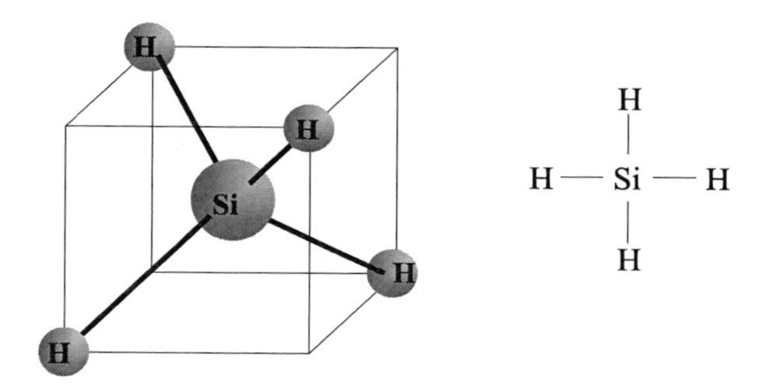

Figure 10.19 Silane molecular structure.

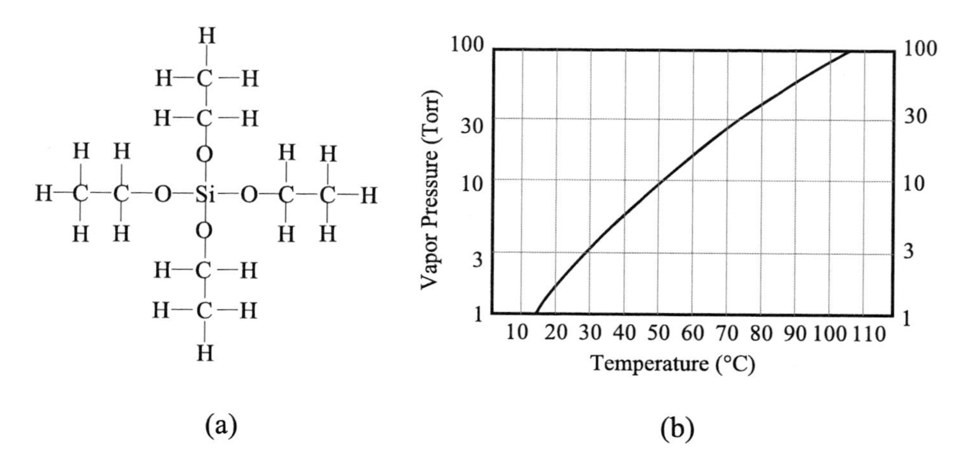

Figure 10.20 (a) TEOS molecule and (b) vapor pressure.

conformality. TEOS is widely used for oxide deposition. Its applications include STI, sidewall spacers, and ILD. Some fabs even use PE-TEOS (PECVD silicon oxide with TEOS and O_2) film for passivation oxide deposition. In IC fabs using Al–Cu interconnections, most dielectric CVD processes are TEOS-based oxide processes. For advanced IC fabs with copper low-κ interconnections, PE-TEOS film is commonly used as the capping layer for low-κ or porous low-κ dielectric films. Because PE-TEOS film is widely used in fabs, some call it TEOS for short.

TEOS is a liquid at room temperature and has a boiling point at the sea level of 168 °C. As a reference, the boiling point of water (H_2O) at sea level is 100 °C. TEOS vapor pressure is shown in Fig. 10.20(b).

To use TEOS or any liquid chemical in CVD processes, special delivery systems to vaporize the chemical and deliver the vapor into the processing reactor are required. Figure 10.21 shows three of those systems: boiler, bubbler, and injection.

All three vapor delivery systems are used in IC manufacturing for liquid chemical vapor delivery. For TEOS processes, the injection system is receiving more attention, since it can precisely and independently control the TEOS flow rate into the processing chamber.

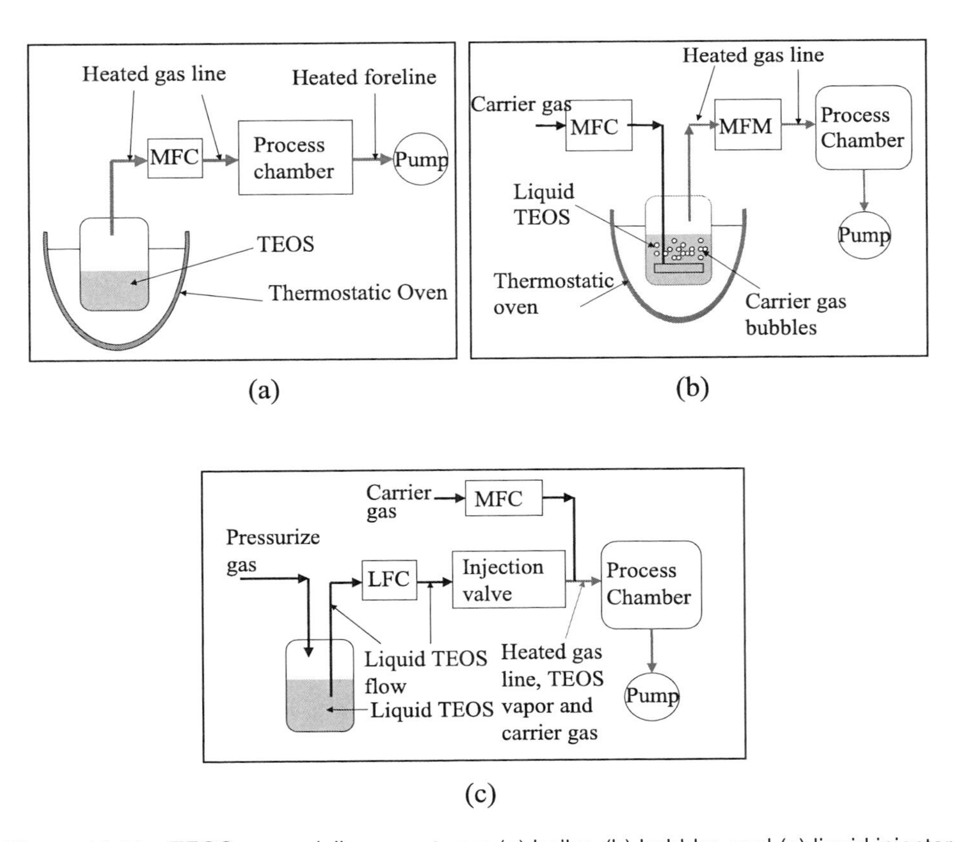

Figure 10.21 TEOS vapor delivery systems: (a) boiler, (b) bubbler, and (c) liquid injector.

3MS is a commonly used silicon precursor for low-κ dielectric film deposition. DEMS can be used as precursors, and CHO can be used as porogens. DEMS and CHO can be used to deposit porous ULK dielectric film for the ILD application of cutting-edge IC chips. Figure 10.22 shows a 3MS molecule, a DEMS molecule, and a sketch of the CHO molecule.

10.2.5.1 Sticking coefficient

The sticking coefficient is the probability that an atom or a molecule will form a chemical bond with surface atoms and chemisorb on the substrate surface in one surface collision. It can be calculated by comparing the results of the theoretically calculated deposition rate on a flat surface with 100% sticking coefficient, and the measured actual deposition rate on that surface.

The lower the sticking coefficient, the higher the surface mobility. Silane has a very low sticking coefficient due to its symmetric molecular structure. To the fragments of silane, SiH_3 is unstable, and SiH_2 and SiH have very high sticking coefficients. Both TEOS and WF_6 have very low sticking coefficients; therefore, they do not easily chemisorb on the surface and usually have to hop along the substrate surface. They have high surface mobility. This is the reason that TEOS and WF_6 CVD films always have very good step coverage and conformality.

From the SEM images in Fig. 10.23, it can be seen that the TEOS oxide film has better step coverage than the silane oxide film. That is why most silicon oxide depositions are TEOS-based processes in advanced semiconductor fabs: because they have much better gap-fill capability.

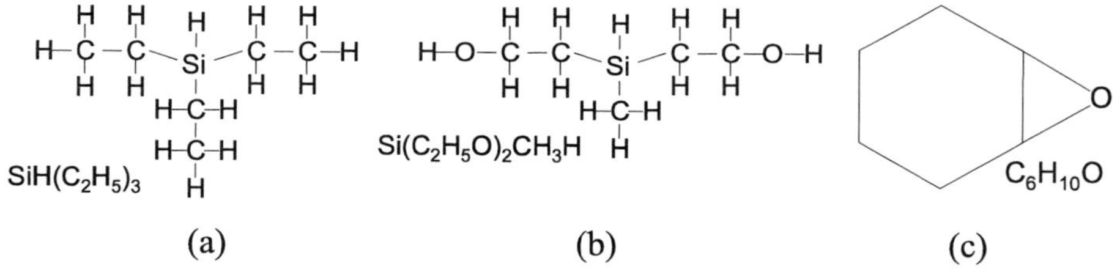

Figure 10.22 Molecular structures of (a) 3MS, (b) DEMS, and (c) CHO.

Table 10.2 Sticking coefficients for various precursors.

Precursors	Sticking coefficient
SiH_4	3×10^{-4} to 3×10^{-5}
SiH_3	0.04 to 0.08
SiH_2	0.15
SiH	0.94
TEOS	10^{-3}
WF_6	10^{-4}

Question: TEOS cannot be used as the silicon source gas for silicon nitride deposition. Why?

Answer: In the TEOS molecule, a silicon atom is bonded with four oxygen atoms. It is almost impossible to strip all of the oxygen atoms and have silicon bond only with nitrogen. Therefore, TEOS is mainly used for oxide deposition, and nitride deposition uses silane as the silicon source gas.

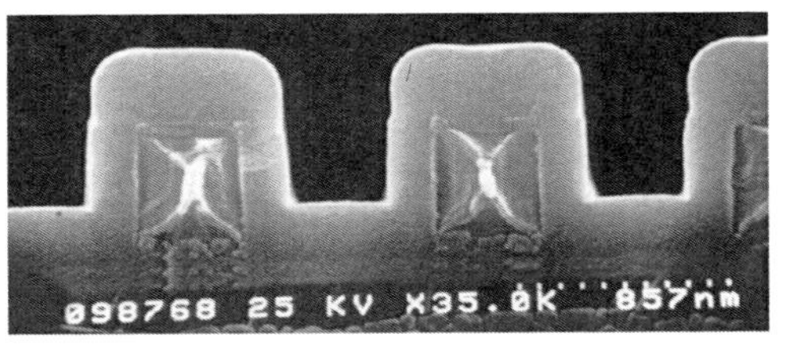

Figure 10.23 Step coverage of TEOS oxide and silane oxide (Applied Materials, Incorporated).

10.2.6 Chemical vapor deposition kinetics

10.2.6.1 Chemical reaction rate

The chemical reaction rate can be expressed as the Arrhenius equation:

$$CR = A\exp(-E_a/k_B T), \tag{10.2}$$

where A is a constant, E_a is the activation energy (as illustrated in Fig. 10.24), k_B is the Boltzmann constant, and T is the substrate temperature. Low activation energy making E_a means a low chemical reaction barrier, which makes it easier for a reaction to happen.

External energy sources such as heat, rf power, or UV radiation are needed for chemical precursors to overcome activation energy barriers and achieve a chemical reaction.

Because the chemical reaction rate is exponentially related to temperature, it is very sensitive to changes in temperature. Changing temperatures can dramatically alter chemical reaction rates. For a CVD process, the deposition rate (DR) is related to the chemical reaction rate (CR), the precursor diffusion rate (D) in the boundary layer, and the precursor adsorption rate (AR) on the substrate surface.

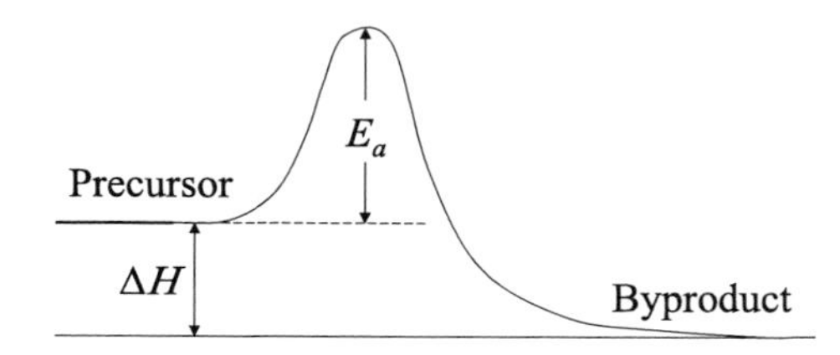

Figure 10.24 Chemical activation energy.

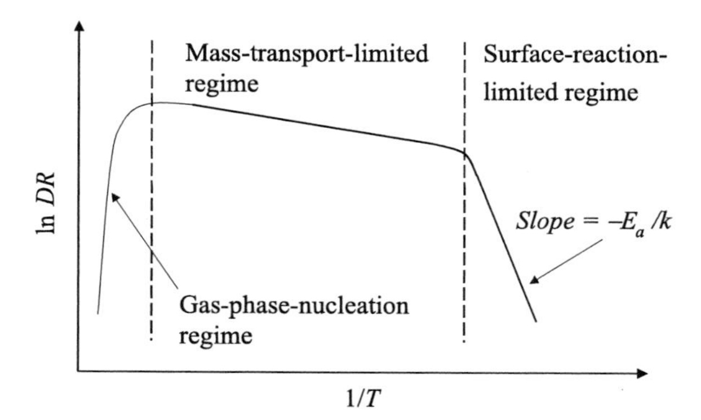

Figure 10.25 Deposition rate regimes.

From Fig. 10.25, it can be seen that the deposition rate has three regimes when the temperature changes. At low temperature, the chemical reaction rate is low, and the deposition rate is very sensitive to temperature. This is called a surface-reaction-limited regime. At higher temperature, the deposition is much less sensitive to temperature. This is the mass-transport-limited regime. When the temperature increases further, the deposition rate sharply decreases due to gas phase nucleation. This is a very undesirable deposition regime, since chemical precursors react in mid-air, generating huge amounts of particles and contaminating the wafer and reactor. Gas phase nucleation (or the homogenous nucleation regime) can be used to generate nanoparticles, but it must be avoided for all CVD processes in IC production. For silane-based dielectric PECVD processes, if the pressure is too high (>10 torr), the gas-phase-nucleation regime can begin and create particle problems.

10.2.6.2 Surface-reaction-limited regime

In a surface-reaction-limited regime, the chemical reaction rate cannot match precursor diffusion and adsorption rates; precursors pile up on the substrate surface and wait their turn to react. In this case, the deposition rate is mainly determined by the chemical reaction rate on the substrate surface:

$$DR = CR[B][C][]\ldots,$$

where CR is the chemical reaction rate defined in Eq. (10.2), and $[B]$, $[C]$, etc., are the concentrations of the adsorbed precursors.

In a surface-reaction-limited regime, deposition rate is very sensitive to temperature because it is mainly determined by the chemical reaction rate.

Some LPCVD processes, such as polysilicon and amorphous silicon deposition processes, operate in the surface-reaction-limited regime due to low deposition temperatures. It is vital to precisely control the processing temperature, since a one-percent difference in temperature can result in a five- to ten-percent difference in deposition film thickness and cause WTW uniformity problems.

10.2.6.3 Mass-transport-limited regime

When the surface chemical reaction rate is high enough, the chemical precursors react immediately when they adsorb on the substrate surface. In this case, the deposition rate is no longer determined by the surface reaction rate but by how fast chemical precursors can diffuse across the boundary layer, and reach and adsorb on the surface:

$$\text{deposition rate} = D\frac{dn}{dx}[B][C][\,]\ldots,$$

where D is the diffusion rate of the precursors in the boundary layer, dn/dx is the gradient of the precursor concentration in the boundary layer, and $[B][C]$, etc., are precursor concentrations on the substrate surface determined by precursor adsorption rates.

In a mass-transport-limited regime, deposition rate is not very sensitive to temperature, and deposition is mainly controlled by gas flow rates. The relationship between deposition rate and temperature is illustrated in Fig. 10.26, which also indicates temperature sensitivity of the two regimes.

10.2.6.4 Chemical vapor deposition reactor deposition regime

In a surface-reaction-limited regime, precursors wait on the substrate surface to react, and the deposition rate is mainly determined by the chemical reaction rate,

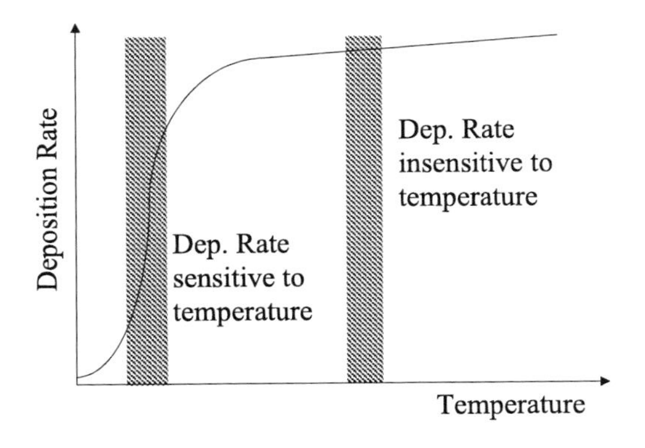

Figure 10.26 Relationship of deposition rate and temperature.

which is very sensitive to temperature. In a mass-transport-limited regime, the surface chemical reaction rate is high enough that the surface waits for chemical precursors to diffuse across the boundary layer and chemisorb on the surface. In this regime, deposition does not depend on the chemical reaction rate but is determined by the precursor diffusion and chemisorption rates.

Most single-wafer process reactors are designed to operate in mass-transport-limited regimes because it is easier to control the gas flow rate than wafer temperature. Plasma processes are commonly employed to generate chemically reactive free radicals, which can dramatically increase the chemical reaction rate and help processes achieve mass-transport-limited regimes at relatively low temperatures. Unstable chemicals such as ozone are also used to achieve high chemical reaction rates at lower temperatures and make the processes less sensitive to temperature.

10.3 Applications of Dielectric Thin Films

Commonly used dielectric thin films in the semiconductor industry are silicon-based compounds such as silicon oxide, silicon nitride, and silicon oxynitride.

Table 10.3 Properties of silicon dioxide and silicon nitride.

Oxide (SiO_2)	**Nitride** (Si_3N_4)
High dielectric strength, $> 1 \times 10^7$ V/cm	High dielectric strength, $> 1 \times 10^7$ V/cm
Lower dielectric constant, $\kappa = 3.9$	Higher dielectric constant, $\kappa = 7.0$
Not a good barrier for moisture and mobile ions (Na^+)	Good barrier for moisture and mobile ions (Na^+)
Transparent to UV	Conventional PECVD nitride opaque to UV
Can be doped with P and B	–

Both silicon oxide and silicon nitride are good electrical insulators with very high dielectric strength (breakdown voltage). Dielectric constants of CVD oxide and nitride are slightly higher than their stoichiometric values listed in Table 10.3. Silicon oxide has a lower dielectric constant than nitride. Therefore, interconnects with silicon oxide between the metal lines have a smaller parasitic capacitance and shorter RC delay time than those with silicon nitride. This is the reason that ILD mainly uses silicon oxide instead of nitride. Because nitride is a much better mobile-ion and moisture barrier layer than oxide, it is generally used for the final layer of passivation dielectric applications, and a barrier layer for doped oxide in ILD0 applications. Some devices such as those using EPROM require UV transparent passivation layers. Silicon oxynitride, which has properties somewhat in between silicon oxide and silicon nitride, is commonly used for this application. For advanced IC chips in nanometer technology, low-κ dielectrics such as PECVD OSG or SiCOH are commonly used as ILD material. To further reduce RC delay, ULK dielectrics, mainly porous SiCOH, were developed and used in ILD applications.

10.3.1 Shallow trench isolation

As IC device dimensions shrank to deep submicron levels, STI gradually replaced LOCOS as the isolation between neighboring transistors. Undoped silicate glass (USG) is used for STI trench fill. PE-TEOS, O_3-TEOS, HDP-CVD oxide, and spin-on dielectrics are used, depending on the trench geometry. Figure 10.27 shows a cross-section SEM image of a STI after CVD oxide trench fill, and Fig. 10.28 illustrates the STI formation processes.

Requirements of dielectric CVD thin films for STI applications are void-free gap fill and a dense film with little shrinkage. It uses higher deposition temperatures to help achieve better step coverage and higher film quality for STI applications.

10.3.2 Sidewall spacer

Sidewall spacers can be found in most MOS transistor-based IC chips when gate dimensions are smaller than 2 μm. They are mainly used to form the LDD or SDE to suppress the hot carrier effects. They also can provide a diffusion buffer for the dopant atoms in a S/D. A sidewall spacer is needed in self-aligned silicide formation to prevent shorting between the S/D and the gate. It is also needed for selective epitaxial growth (SEG) of SiGe for pMOS S/D and SiC for nMOS S/D. If

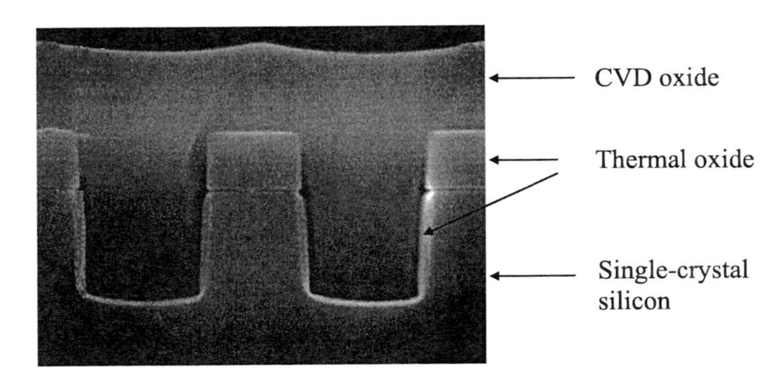

Figure 10.27 CVD oxide trench-fill application in STI process (Applied Materials, Incorporated).

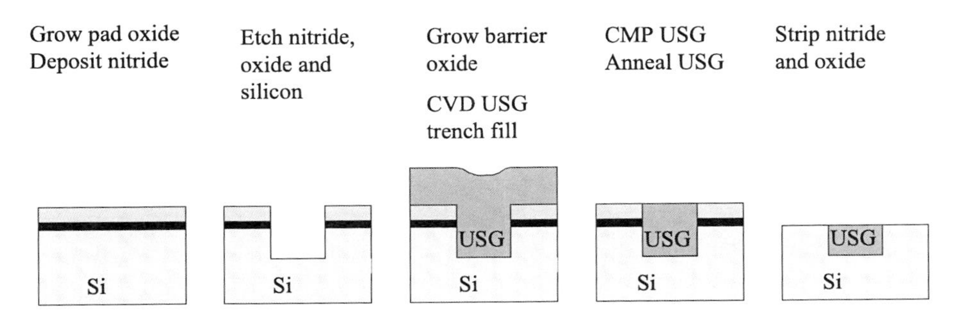

Figure 10.28 Processing steps of STI.

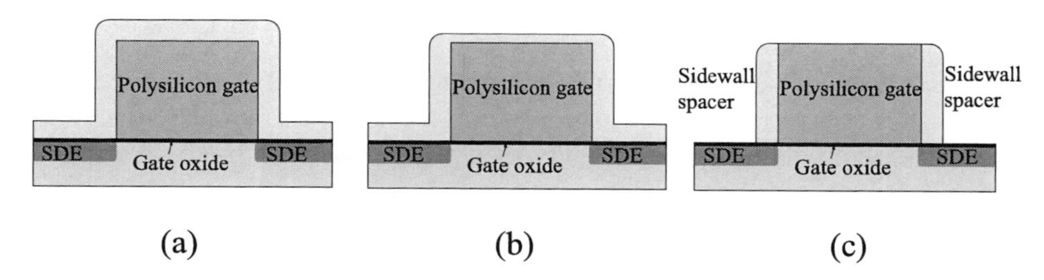

Figure 10.29 Formation of a sidewall spacer: (a) conformal dielectric film deposition, (b) dielectric etch back, and (c) spacer formation.

the spacer layer cracks, SEG film will grow on the polysilicon gate. Figure 10.29 illustrates sidewall spacer formation processes.

A sidewall spacer can use LPCVD nitride with thermal growth oxide as an etch endpoint. It can also use a LPCVD nitride liner as an etch stop to etchback TEOS-based CVD oxide film. In some devices, trilayer spacers are used.

O_3-TEOS USG film is used for oxide deposition due to its almost perfect conformality and sidewall step coverage. Requirements for spacer dielectric film are good conformality and high selectivity to polysilicon during etchback. Deposition temperature limitation for spacer dielectrics is the thermal budget, which is determined by device design.

10.3.3 Interlayer dielectric-0

ILD0 is the first dielectric layer deposited on a wafer surface after transistors have been created. The requirements of ILD0 are low dielectric constant, mobile-ion barriers, void-free gap fill, and planarization of the surface. Temperature limitations for ILD0 deposition and thermal flow are determined by the thermal budget of the device.

ILD0 is normally a phosphorus-doped silicate glass (phosphosilicate glass, PSG) or phosphorus- and boron-doped silicon oxide (borophosphosilicate glass, BPSG). To prevent phosphorus and boron from diffusing into the activation region (S/D), it is necessary to deposit a barrier layer before PSG or BPSG deposition. USG (~1000 Å) was used as a barrier layer when device geometry was still in the micron range. CVD nitride (<100 Å) is used now.

There are two important reasons to dope phosphorus with silicon oxide: gettering mobile sodium ions (Na^+) and reducing silicate glass thermal flow temperature.

Sodium is located in column IA of the periodic table, which means that it has only one electron in the outermost shell. It is very easy for sodium to lose that electron and become an ion. Sodium ions are very small and mobile. When they accumulate in the gate oxide of a MOSFET, they can change the threshold voltage of the MOSFET and cause it to turn on or off randomly. Figure 10.30 illustrates sodium ions turning on a normally off nMOS. It takes only a tiny amount of sodium to substantially damage a MOSFET and cause circuit malfunction. Therefore, it is extremely important to control mobile ion contamination. Since sodium is almost

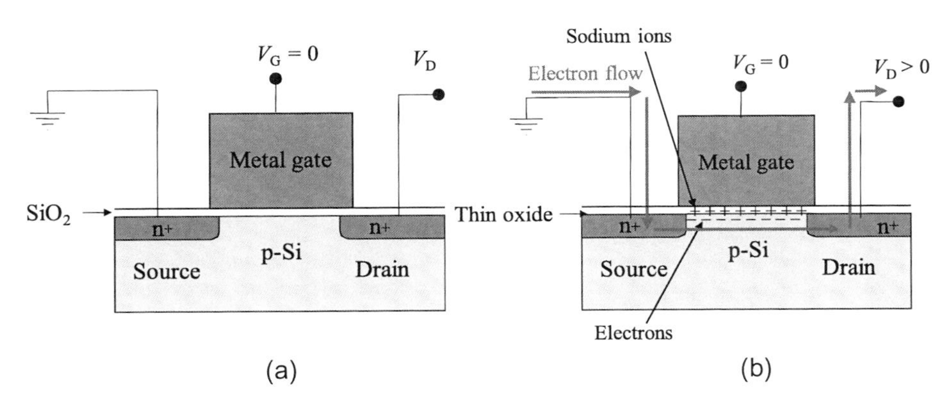

Figure 10.30 Effect of sodium ions on an nMOS: (a) normal off mode with $V_G = 0$ V and (b) turned on by Na^+, even at $V_G = 0$ V.

everywhere, and it is very difficult to completely eliminate sodium, a sodium barrier layer right above the gate of the MOSFET is required. PSG and BPSG are commonly used because they can trap sodium ions and prevent them from diffusing into gate oxide and causing device damage. Some facts about sodium are summarized in Table 10.4.

Question: Silicon nitride is a better sodium barrier layer than silicon oxide. Why isn't nitride used for ILD0?

Answer: Because silicon nitride has a higher dielectric constant than oxide, using nitride as the dielectric between the conducting lines can cause a much longer RC time delay and significantly affect circuit speed. A thin layer of nitride (to <100 Å) is used as a diffusion barrier layer in ILD0 applications to prevent diffusion of phosphorus and boron from PSG or BPSG diffusing into the S/D.

When USG is heated to >1500 °C, it becomes soft and starts to thermally flow. Since the melting point of silicon is 1414 °C, the wafer will melt before the USG starts to flow. It is a known fact from the glass industry that PSG can flow at significantly lower temperatures. As deposited, the silicate glass surface is rough and full of hills and valleys, which can cause resolution problems in photolithography processes due to DoF, and serious step coverage problems for the subsequent metal deposition process. At high temperature, glass becomes soft and viscous. It will respond to surface tension force, which creates a smoother topography on the glass surface.

As shown in Fig. 10.31, the higher the phosphorus concentration in PSG, the better the reflow results that can be achieved. The θ in the figure is the reflow angle. The smaller the reflow angle, the better the surface planarization.

If the phosphorous concentration is too high (>7 wt%), the PSG surface will be highly hygroscopic. P_2O_5 will react with moisture (H_2O) and form phosphoric acid (H_3PO_4) on the PSG surface, possibly leading to aluminum corrosion. It can also cause photoresist adhesion problems during the photolithography process

Table 10.4 Facts about sodium.

Symbol	Na
Atomic number	11
Atomic weight	22.989770
Discoverer	Sir Humphrey Davy
Place of discovery	England
Discovery date	1807
Origin of name	From the English word soda (the origin of the symbol Na comes from the Latin word natrium)
Density of solid	0.968 g/cm^3
Molar volume	23.78 cm^3
Velocity of sound in substance	3200 m/sec
Electrical resistivity	4.7 $\mu\Omega \cdot$ cm
Reflectivity	N/A
Melting point	97.72 °C
Boiling point	882.85 °C
Thermal conductivity	140 W m^{-1} K^{-1}
Coefficient of linear thermal expansion	71×10^{-6}K^{-1}
Applications	Major contaminant, needs to be strictly controlled
Main removal agent	HCl
Barrier materials used	Silicon nitride and PSG

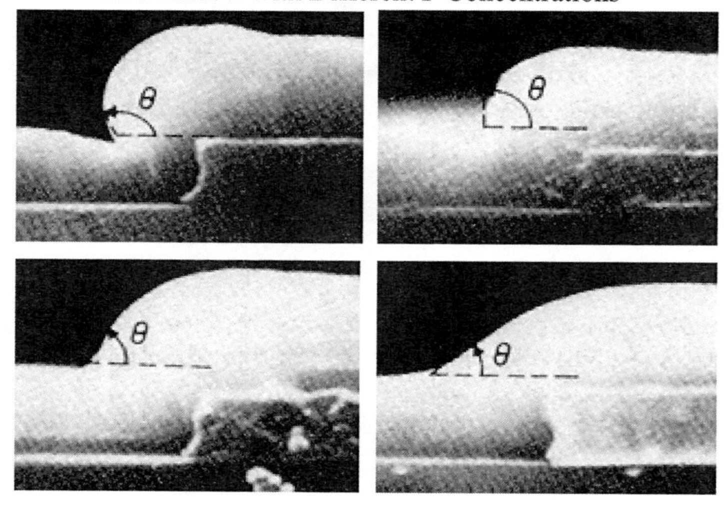

Figure 10.31 PSG reflow at 1100 °C in N_2 ambient for 20 min with different phosphorus concentrations (A.C. Adams).

for contact hole masking because photoresist will not stick well to hygroscopic surfaces.

When device dimensions shrink, thermal budget limitations require lower reflow temperatures. Boron has also been used along with phosphorus to dope silicate glass to further reduce reflow temperatures without too much phosphorus. BPSG can flow at 850 °C, as shown in Fig. 10.32. The first four of 4 × 4 indicates 4-wt% boron, and the second four indicates 4-wt% phosphorus. BPSG is widely used in IC chips with feature sizes from 2 to 0.25 μm. If the boron concentration in BPSG is too high, B_2O_3 can react with moisture (H_2O) and form boric acid (H_3BO_3) crystals on the BPSG surface, possibly causing defects such as particle contamination. The upper limit of BPSG dopant concentration is about 5 × 5.

As feature size continues to reduce, thermal reflow can no longer meet the planarization requirements of nanometer photolithography, and there is no more thermal budget for thermal reflow. CMP processing has been used to replace reflow for ILD0 planarization. Since no reflow is required, boron is no longer needed in the film, so PSG is used again for ILD0 application. Table 10.5 summarizes the development of ILD0 deposition and planarization processes.

LPCVD nitride had been used as a ILD0 barrier layer. Due to thermal budget concerns, PECVD nitride has been more widely used because PECVD nitride can be deposited at significantly lower temperatures. ILD0 was deposited with

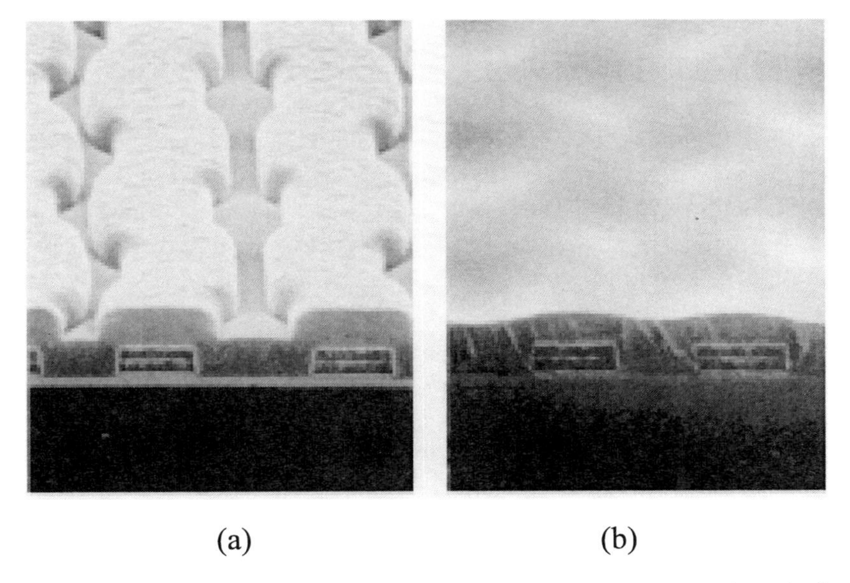

Figure 10.32 4 × 4 BPSG reflow at 850 °C, 30 min in N_2 ambient: (a) as deposited and (b) after reflow (Applied Materials, Inc.).

Table 10.5 Development of ILD0 processes.

Dimension	PMD	Planarization	Reflow temperature
>2 μm	PSG	Reflow	1100 °C
2 to 0.35 μm	BPSG	Reflow	850 to 900 °C
0.25 μm	BPSG	Reflow + CMP	750 °C
180 nm and beyond	PSG or USG	CMP	-

silane-based CVD processes long ago. O_3-TEOS-based CVD processes began to be commonly used for ILD0 deposition in semiconductor fabs after the 1990s.

10.3.4 Interlayer dielectric-1 and up

For advanced IC chips with multilayer interconnections, more than half of the dielectric thin-film processes are ILD processes. ILD normally uses USG, low-κ, or ULK dielectrics, depending on the technology node. Different processing sequences can be used for gap filling and planarization, depending on gap geometry and the manufacturer's preference. The requirements of ILD are a low dielectric constant, void-free gap fill, and planarization of the surface. ILD deposition temperature does not exceed 450 °C because of the presence of metal interconnections. Normally, the ILD deposition temperature is about 400 °C.

For metal line spacing >0.6 μm, deposition/sputter etchback/deposition is commonly used for gap filling. TEOS-based oxide deposition and argon sputter etch are employed for this process.

Spin-on glass (SOG) processes, including PECVD liner deposition, liquid glass spin-on, SOG cure, SOG etchback, and PECVD cap layer deposition, are used in semiconductor manufacturing. Figure 10.33 shows the application of the SOG process for ILD gap fill (SOG1) and planarization (SOG2) applications. Since the curing temperature is limited by the presence of metal lines and cannot exceed 450 °C, SOG film quality is not good and needs to be etched away as much as possible from the wafer surface, leaving only SOG in the gap for gap fill and planarization. Spin-on dielectrics (SOD) with processes similar to SOG were also intensively developed to compete with CVD-based dielectric thin films for low-κ ILD applications. Due to reliability issues, SOD lost the competition.

An ozone-TEOS-based process can deposit nearly perfect conformal oxide film with excellent gap-fill ability. This process is used to fill >0.35-μm gaps and >2:1 aspect ratio. Since ozone-TEOS oxide film has poor quality and unfavorable tensile stress, it needs a PECVD barrier layer and cap layer, similar to a SOG process. It does not require etchback, and all three layers can be deposited in the same processing chamber in one processing sequence.

As device dimensions continue to shrink, metal gaps will become smaller and smaller. At the same time, the thickness of the metal cannot be reduced accordingly. Otherwise, resistance of the metal line will increase accordingly. Void-free filling of narrow (<0.25-μm) and high-aspect-ratio (>3:1) gaps poses a great challenge to dielectric thin-film deposition processes. By doing dep/etch/dep simultaneously in the same processing chamber, the HDP-CVD process can fill gaps 0.20-μm wide with a 4:1 aspect ratio without voids. Figure 10.34 shows an SEM cross section of HDP-CVD USG filling a 0.25-μm, 4:1 aspect-ratio gap. Before SEM analysis, a diluted HF decoration etch is performed. Because HDP-CVD USG and PECVD USG have different etch rates, the boundary between the two layers shows up after the wet etch from the chip cross section.

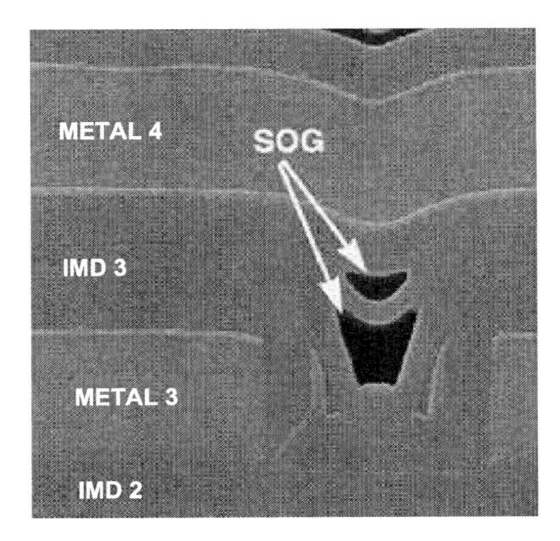

Figure 10.33 Spin-on glass for ILD gap fill (lower SOG) and planarization (upper SOG) applications (Integrated Circuit Engineering Corp.).

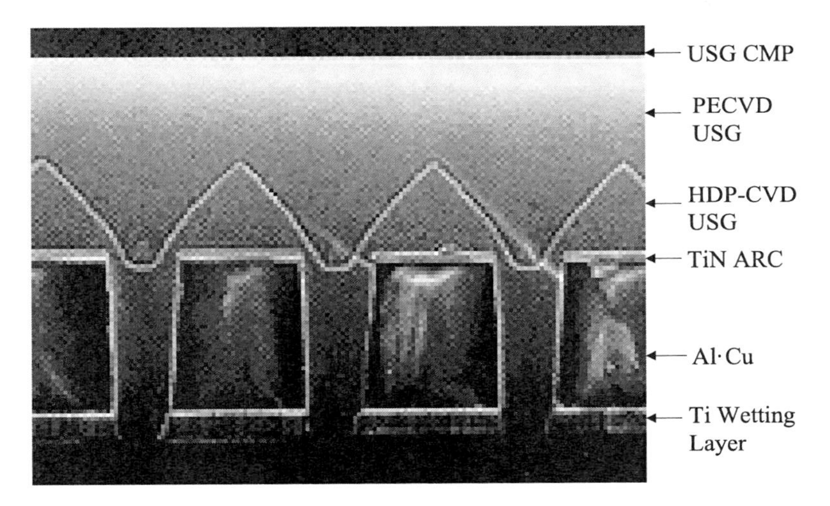

Figure 10.34 HDP-CVD, PECVD, and CMP for 0.25-μm, 4:1 aspect-ratio IMD gap fill (Applied Materials, Inc.).

Since HDP-CVD deposits and sputter etches thin film at the same time, the net deposition rate is not very high. Thus, it is mainly used for gap fill, and a PECVD film with a higher deposition rate is used to cap it. Dielectric CMP planarizes the PECVD cap layer to meet the planarization requirement for photolithography processes.

With low-κ copper interconnections, there is no metal gap to fill, and film is deposited on a flat surface after the metal CMP, thus there is no planarization issue. The requirements of dielectric thin films become simpler: high throughput (which means high deposition rate), good uniformity, and stable low-κ dielectric properties, including dielectric constant, thermal conductivity, dielectric strength, and mechanical strength.

10.3.5 Passivation dielectrics

Most fabs use plastic packaging instead of ceramic packaging to reduce costs for the back end of IC chip processing. Plastic is not a good barrier for moisture and mobile ions. Therefore, a good barrier layer deposited at a low temperature (~400 °C) with high dielectric and mechanical strength is required for the final passivation layer of an IC chip. PECVD silicon nitride satisfies all of these requirements and is commonly used for passivation layers. Since the passivation dielectric is the last layer of a chip, the higher dielectric constant of the silicon nitride does not affect the speed of the circuits. For chips using ceramic packaging, oxide passivation is adequate, because ceramic (aluminum oxide) is a very good barrier against moisture and mobile ions.

Nitride does not stick very well to aluminum lines due to stress mismatch. Therefore, an oxide layer is always deposited prior to nitride deposition to buffer the stress and help silicon nitride adhesion. In-situ PECVD silane oxide and nitride processes are commonly used for passivation dielectric depositions. The deposition temperature of passivation dielectrics is also limited by the presence of metal interconnection wires; therefore, 400 °C is the most commonly used deposition temperature.

10.4 Dielectric thin-film characteristics

In this section, dielectric thin-film characteristics, including refractive index, thickness, and stress and metrology tools for their measurement, are discussed.

10.4.1 Refractive index

The definition of refractive index is

$$\text{refractive index}, n = \frac{\text{speed of light in vacuum}}{\text{speed of light in film}}. \tag{10.3}$$

For SiO_2, $n = 1.46$, and for Si_3N_4, $n = 2.01$. Refractive index is related to the wavelength of the light used for measurement. The refractive indexes mentioned in this chapter are measured at 633 nm, the wavelength of red light emitted from a He-Ne laser, which is commonly used in laser pointers. A prism separates sunlight into a colorful spectrum. This occurs because the refractive index of the prism material—quartz—is a function of the wavelength of the light. Light with different wavelengths has different refractive angles entering and leaving the prism, giving the light a colorful spectrum.

The refractive index and refractive angle illustrated in Fig. 10.35 can be described by the law of refraction:

$$n_1 \sin\theta_1 = n_2 \sin\theta_2,$$

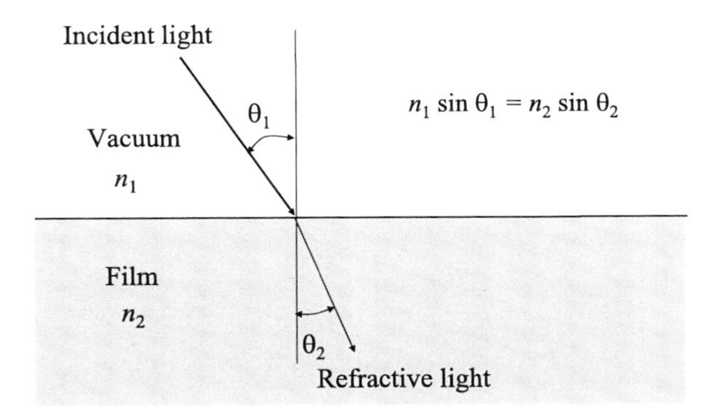

Figure 10.35 Refractive index and refractive angle.

where n_1 is the refractive index of the first dielectric material, usually air, with a dielectric index very close to 1. θ_1 is the incident angle, n_2 is the refractive index of the second dielectric material, and θ_2 is the refractive angle. This equation can be used to measure dielectric index by shining a laser beam into the dielectric material and measuring the refractive angle. However, it is impossible to measure the refractive index of a dielectric film by applying this equation when film thickness is too thin.

For silicon compound dielectric thin films, refractive index measurements can provide useful information on the chemical composition and physical conditions of the film. For a silicon- or nitrogen-rich oxide, the refractive index will be higher than 1.46 of the SiO_2, and it will be lower than 1.46 when it is oxygen rich. For nitride, silicon-rich film will have a higher refractive index than 2.01, and nitrogen- or oxygen-rich film will have a lower value, as shown in Fig. 10.36.

Question: If a process reactor is leaking during the silicon nitride deposition process, how will the nitride refractive index change?

Answer: The chamber leak introduces oxygen from the atmosphere into the processing chamber. Oxygen is fairly reactive and will react with silicon precursor gases, becoming integrated into the deposited film. This will create oxygen-rich nitride or even oxynitride. From Fig. 10.36, a decrease in refractive index can be predicted.

In dielectric thin-film thickness measurements, refractive index and film thickness are always considered together. Therefore, it is necessary to know the refractive index first before measuring film thickness. Ellipsometry and prism couplers are two of the most common metrology tools used to measure refractive index in semiconductor processing fabs.

10.4.1.1 Ellipsometry

The polarization state of light changes when a light beam is reflected from a film surface, as illustrated in Fig. 10.37. By monitoring this change, information can be attained about the refractive index and thickness of dielectric thin films. The

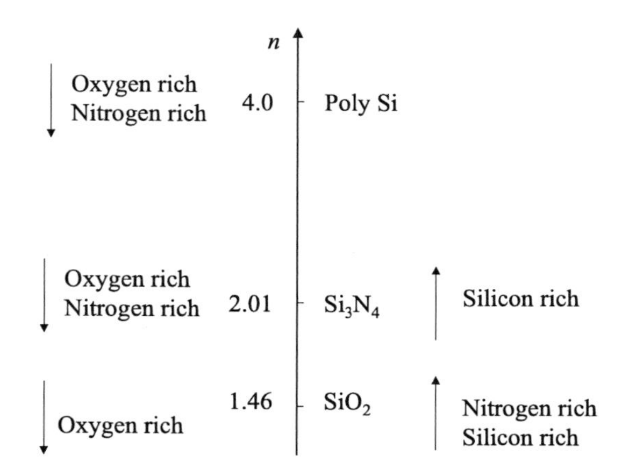

Figure 10.36 Refractive index and its importance for dielectric thin films.

change in polarization of the **p** and **s** components of a light beam during reflection can be determined. The fundamental equation of ellipsometry can be expressed as

$$\rho = r_p/r_s = \tan \Psi e^{I\Delta},$$

where ρ is a complex amplitude reflection ratio, and r_p and r_s are the Fresnel reflection coefficients. Both thickness and refractive index can be calculated from the parameters Ψ and Δ. An approximate value of film thickness is needed in advance because the measured ellipsometric quantities are the periodic functions of the thickness. Ellipsometry can also be used to measure dielectric thin-film thickness if its refractive index is known.

Question: Is a 0.98 measurement result possible for a thin-film refractive index?

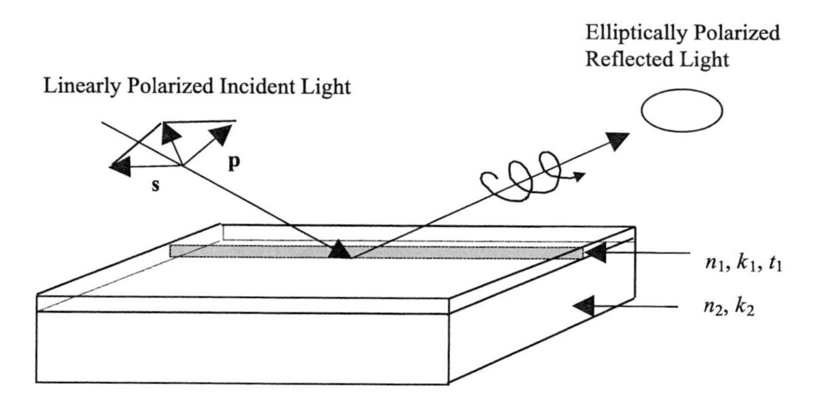

Figure 10.37 Ellipsometry system.

Answer: Based on the definition in Eq. (10.3), a refractive index cannot be less than 1, which is the refractive index of vacuum. The measurement must be incorrect.

10.4.1.2 Prism coupler

In a prism coupler, a wafer with a dielectric thin film is brought into contact with the base of a prism by a pneumatically operated coupling head. There is a small air gap between the film and prism base. A laser beam shining through the prism is totally reflected at the prism base and deflected to a photodetector, as shown in Fig. 10.38.

At certain discrete values of incident angle θ (called mode angles), photons can tunnel through the small air gap into the dielectric thin film and enter into a guided optical propagation mode. The loss of photons at the mode angles causes sudden light intensity drops on the photodetector, as illustrated in Fig. 10.39.

The first mode angle approximately determines the thin-film refractive index, and the difference between the mode angles determines film thickness, allowing an independent measurement of the dielectric thin-film refractive index and thickness.

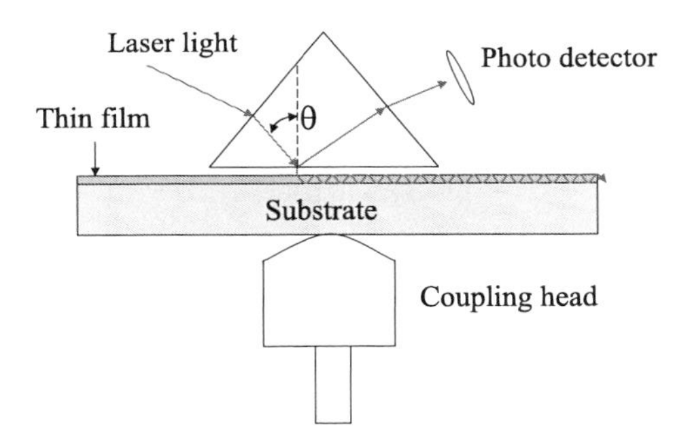

Figure 10.38 Prism coupler.

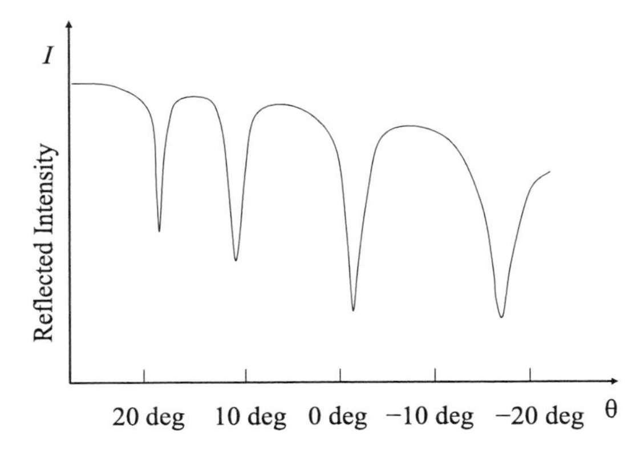

Figure 10.39 Intensity (I) of reflected light versus incidence angle (θ).

Prism coupler measurement does not require advanced knowledge of film thickness. However, it does require a minimum film thickness (typically 3000 to 4800 Å) to support at least two modes to achieve good results. The thicker the film, the more angular modes it can support, and the better the measurement results are. Prism couplers can also be used to measure refractive indices and thicknesses of stacked films.

10.4.2 Thickness

Thickness measurement is one of the most important measurements for dielectric thin-film processes. Film deposition rate, wet etch rate, and shrinkage all depend on thin-film thickness measurements.

10.4.2.1 Color chart

After deposition of the dielectric thin film, a wafer will have different colors on its surface, depending on film thickness, refractive index, and the angle of light.

Reflected light from a dielectric thin-film surface (light 1), and the reflected light from the interface of a dielectric thin film and the substrate (light 2), have the same frequencies but different phases, as shown in Fig. 10.40. There is interference between the two reflected light rays that can cause both constructive and destructive interference at different wavelengths, since the refractive index is a function of wavelength. Table 10.6 lists the color chart of oxides with different thicknesses.

The color seen on the wafer is determined by constructive interference frequency, which relates to the phase difference of the two reflected lights. The thicker the film, the larger the phase shift:

$$\Delta\phi = 2tn(\lambda)/\cos\theta, \tag{10.4}$$

where t is the thin-film thickness, $n(\lambda)$ is the thin-film refractive index, and θ is the incident angle. When the phase shift $\Delta\phi$ is larger than 2π, the color pattern repeats itself.

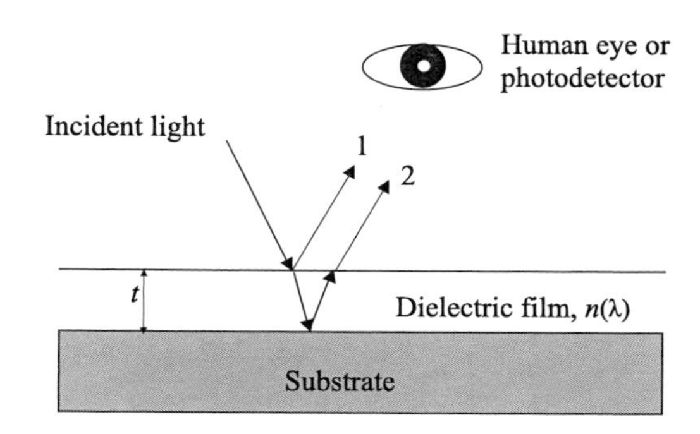

Figure 10.40 Reflection light rays and phase difference.

Table 10.6 Color chart of silicon oxide.

Thickness (Å)	Color	Thickness (μm)	Color
500	Tan	1.0	Carnation pink
700	Brown	1.02	Violet-red
1000	Dark violet to red-violet	1.05	Red-violet
1200	Royal blue	1.06	Violet
1500	Light blue to metallic blue	1.07	Blue-violet
1700	Metallic to very light yellow-green	1.10	Green
2000	Light gold or yellow slightly metallic	1.11	Yellow-green
2200	Gold with slight yellow-orange	1.12	Green
2500	Orange to melon	1.18	Violet
2700	Red-violet	1.19	Red-violet
3000	Blue to violet-blue	1.21	Violet-red
3100	Blue	1.24	Carnation pink to salmon
3200	Blue to blue-green	1.25	Orange
3400	Light green	1.28	Yellowish
3500	Green to yellow-green	1.32	Sky blue to green-blue
3600	Yellow-green	1.40	Orange
3700	Green-yellow	1.45	Violet
3900	Yellow	1.46	Blue-violet
4100	Light orange	1.50	Blue
4200	Carnation pink	1.54	Dull yellow-green
4400	Violet-red		
4600	Red-violet		
4700	Violet		
4800	Blue-violet		
4900	Blue		
5000	Blue-green		
5200	Green		
5400	Yellow-green		
5600	Green-yellow		
5700	Yellow to yellowish (at times appears light gray or metallic)		
5800	Light orange or yellow to pink		
6000	Carnation pink		
6300	Violet-red		
6800	Bluish (appears between violet-red and blue-green, overall looks grayish)		
7200	Blue-green to green		
7700	Yellowish		
8000	Orange		
8200	Salmon		
8500	Dull, light red-violet		
8600	Violet		
8700	Blue-violet		
8900	Blue		
9200	Blue-green		
9500	Dull yellow-green		
9700	Yellow to yellowish		
9900	Orange		

Using a color chart is a convenient way to measure film thickness. Although it is no longer used for thickness measurements in advanced IC fabs, it is still a useful tool for detecting obvious nonuniformity problems of deposited film.

Question: If some beautiful color rings appear on a wafer with a CVD dielectric layer, what is the conclusion?

Answer: Color change indicates dielectric thin-film thickness change; thus, the film with the color rings must have a problem with thickness uniformity, which is most likely caused by a nonuniform thin-film deposition process.

Question: Why does the color of a thin film change when seen at a different angle on a wafer?

Answer: As is seen from Fig. 10.29, when looking at a wafer from a different angle, incident angle θ changes. Equation (10.4) shows that the phase shift also changes; thus, the wavelength for constructive interference changes, which causes color change. It is important to hold a wafer straight when using a color chart to measure thin-film thickness. Tilting the wafer can cause the film thickness to seem thicker than it actually is.

10.4.2.2 Spectroreflectometry

Spectroreflectometry measures the reflected light intensity at different wavelengths. Thin-film thicknesses can be calculated from the relation of the reflected light intensity and wavelength of the light, as shown in Fig. 10.41. A photodetector is much more sensitive than human eyes in detecting the spectrum of intensity and wavelength; therefore, spectroreflectometry can obtain much higher resolution and accuracy for thickness measurements.

Thickness can be calculated by the following equation:

$$\frac{1}{\lambda_m} - \frac{1}{\lambda_{m+1}} = \frac{1}{2nt}. \qquad (10.5)$$

Here, λ_m and λ_{m+1} are the m'th and $(m + 1)$'th constructive interference wavelengths, respectively; n is the film refractive index; and t is the film thickness. From Eq. (10.5) it can be seen that using the wrong refractive index can cause an incorrect thickness measurement result.

Figure 10.42 shows a schematic of a spectroreflectometry system. UV light is used for the measurement because, for dielectric thin films, the refractive index is insensitive to wavelengths in the UV range.

Question: Many advanced thin-film thickness measurement tools allow users to choose the refractive index of a film. If the PE-TEOS USG film refractive index is chosen to measure O_3-TEOS USG film thickness, what will be the effect on the measurement result?

Answer: Since nt is always coupled together, as shown in Eq. (10.5), inputting an incorrect refractive index n will cause the wrong thickness measurement result. Because O_3-TEOS USG is a porous film with many tiny holes ($n_{hole} \sim 1$) in it, a O_3-TEOS USG film normally has a lower refractive index, about 1.44, which is slightly lower than the 1.46 of PE-TEOS USG. Therefore, the measured O_3-TEOS film thickness will be slightly thinner than its actual value.

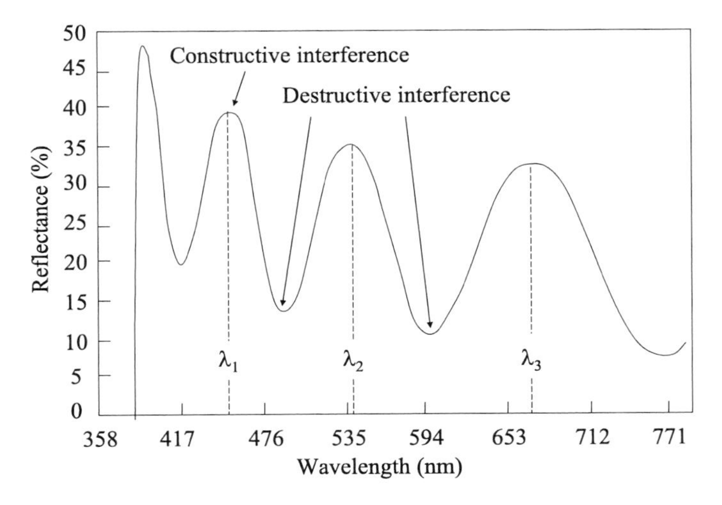

Figure 10.41 Relationship of reflectance and wavelength.

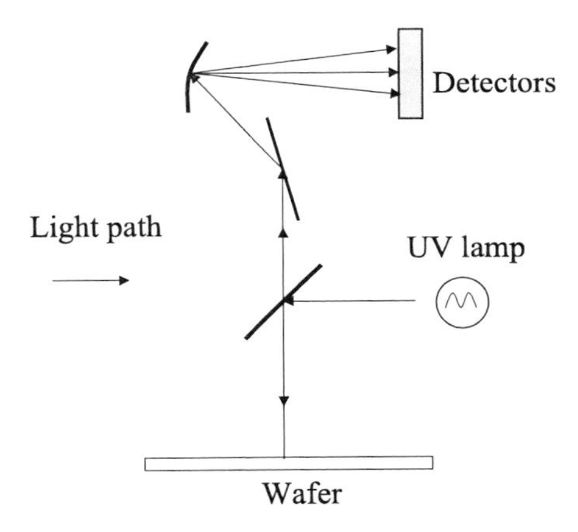

Figure 10.42 Schematic of a spectroreflectometry system.

10.4.2.3 Deposition rate

Deposition rate is defined as

$$\text{deposition rate} = \frac{\text{deposited film thickness}}{\text{deposition time}}.$$

Deposition rate is one of the most important characteristics of any deposition process. It affects process throughput, while directly relating to cost of ownership. The most commonly used unit for deposition rate is Å/min. The higher the deposition rate, the higher the throughput, and the lower the cost of ownership. For example, assume a fab has 20 CVD systems, and chamber cleaning and other overhead take 50% of the total deposition process. If there were a way to double the deposition rate while keeping all of the other factors unchanged, the fab could reduce down to five CVD systems.

Question: An LPCVD silane process deposits 10,000-Å oxide film in 50 min. What is the deposition rate?

Answer: Deposition rate = 10,000/50 = 200 Å/min.

10.4.2.4 Wet etch rate

Wet etch rate and wet etch rate ratio (WERR) are commonly used for dielectric thin-film quality control. A film with better quality has a lower WERR.

Hydrofluoric acid (HF) can etch silicon oxide and silicon nitride. Since 1:1 HF (49% HF in H_2O) etches oxide too quickly to control, HF is diluted to 6:1 and buffered with ammonium fluoride (NH_4F) to slow down the wet etch rate. This procedure is called a 6:1 BOE, and it is commonly used to qualify dielectric thin-film processes. Another popular oxide wet etch uses 100:1 HF solution.

The definition of wet etch rate is:

$$\text{wet etch rate} = \frac{\text{thickness before wet etch} - \text{thickness after wet etch}}{\text{wet etch time}}.$$

The wet etch rate of 6:1 BOE at 22 °C for thermally grown silicon dioxide is about 1000 Å/min. CVD oxide has a higher wet etch rate because it is not as dense. The wet etch rate is very sensitive to HF solution concentration and temperature. To eliminate measurement uncertainties caused by the variations of temperature and concentration, a wafer with a thermally grown silicon dioxide film is etched alongside wafers with CVD thin films as the reference. This technique is called the WERR to thermal oxide:

$$\text{WERR} = \frac{\text{thickness change of the CVD film}}{\text{thickness change of the thermal oxide film}}. \tag{10.6}$$

The WERR is preferred over the wet etch rate itself because using the ratio has eliminated the major measurement uncertainties associated with exact concentration and temperature.

Question: A O_3-TEOS CVD oxide film thickness changed from 4500 Å before 6:1 BOE, to 2400 Å after it, while the reference thermal oxide thickness changed from 2000 to 1550 Å. What is the WERR?

Answer: WERR = (4500 – 2400)/(2000 – 1550) = 2100/450 = 4.67.

10.4.2.5 Shrinkage

Heating up a wafer and cooling it back to room temperature is called a thermal cycle. During a thermal cycle, atoms inside a thin film have some thermal movement, which finishes at a lower potential position and causes the film to be denser. Therefore, after a thermal cycle, film thickness decreases. The amount of shrinkage indicates the original film quality. Film with better quality has less shrinkage:

$$\text{shrinkage} = \frac{\text{thickness change after thermal cycle}}{\text{thickness before thermal cycle}}. \quad (10.7)$$

Shrinkage is one of the major concerns for oxide films in STI applications because the trench fill oxide requires high-temperature annealing. Too much oxide shrinkage can cause stress-induced defects in silicon substrates, especially at trench corners.

Question: After 1100 °C anneal, HDP-CVD USG film thickness changes from 3500 to 3460 Å. What is the shrinkage of the film?

Answer: Shrinkage = (3500 – 3460)/3000 = 40/3500 = 1.14%.

10.4.2.6 Uniformity

Uniformity, or to be precise, nonuniformity, is a very important factor of the thin-film deposition because it can affect product yield. If a CVD tool has a uniformity issue and deposits a film with a thickness variation larger than the control limit, it can cause underetch where the film is too thick, and overetch where the film is too thin. Uniformity can be calculated by multipoint thickness measurements. A clear definition of uniformity is important, since different definitions give different results on uniformity, even with the same sets of measurement data.

Measurement results are $x_1, x_2, x_3, x_4, \ldots, x_N$, where N equals the total number of data points. In an IC fab, $N = 9$, 25, 49, or 121 are typical sample numbers. $N = 49$ is used for 200-mm wafers and $N = 121$ for 300-mm wafer. The mean value (or average value) of the measurement is

$$\bar{x} = \frac{x_1 + x_2 + x_3 + \cdots + x_N}{N}.$$

The standard deviation of the measurement is

$$\sigma = \sqrt{\frac{(x_1 - \bar{x})^2 + (x_2 - \bar{x})^2 + (x_3 - \bar{x})^2 + \cdots + (x_N - \bar{x})^2}{N - 1}}.$$

The standard deviation nonuniformity (in percentages) is defined as

$$NU(\%) = (\sigma/\bar{x}) \times 100. \quad (10.8)$$

The definition of maximum-minus-minimum nonuniformity (in percentages) is:

$$NU_{\text{max-min}} = [(x_{\text{max}} - x_{\text{min}})/(2\bar{x})] \times 100 \quad (10.9)$$

Question: Calculate the NU and $NU_{\max-\min}$ for a five-point thickness (in Å) measurement.

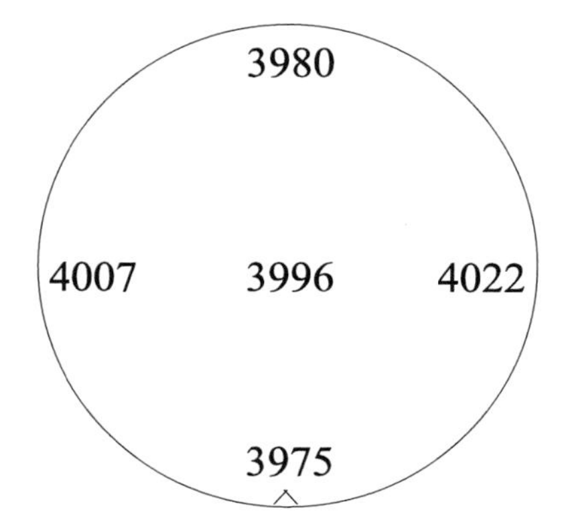

Answer: The average thickness is: $\bar{x}$ = 3996 Å, and the standard deviation is σ = 18.8 Å.

The standard deviation nonuniformity is:

$$NU = 18.8/3966 = 0.47\%,$$

and the maximum-minus-minimum nonuniformity is

$$NU_{\max-\min} = (4022 - 3975)/(2 \times 3996) = 0.59\%.$$

10.4.3 Stress

This section defines stress and details the cause and nature of stress in dielectric films. Stress arises due to a mismatch between different materials, such as between the substrate (wafer) and the film (dielectrics, metal, etc.). There are two kinds of stresses, intrinsic and extrinsic. Intrinsic stress develops during the film nucleation and growth process. Extrinsic stress results from differences in the coefficients of thermal expansion between the film and substrate. Stress can be either tensile (noted as positive) or compressive (noted as negative), as shown in Fig. 10.43. High stress on a dielectric film, whether it is compressive or tensile, can cause film cracking, metal line spiking, or void formation. In some cases, extremely high tensile stress can even break the wafer.

For PECVD dielectric thin films, intrinsic film stress can be controlled by rf power. Film stress is kept at a compressive 10^9 dyne/cm^2 (−100 MPa). Ion bombardment from the plasma can pack molecules densely inside the film, thus increasing film density and compressive stress. It is important to control dielectric thin-film stress for dielectric CVD processes because this stress can affect defect densities of both dielectrics and metals, and it can affect any dielectric CMP processes that follow.

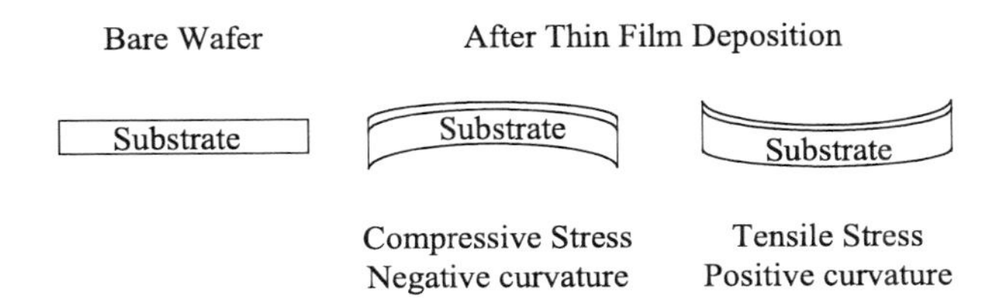

Figure 10.43 Illustration of compressive and tensile stress.

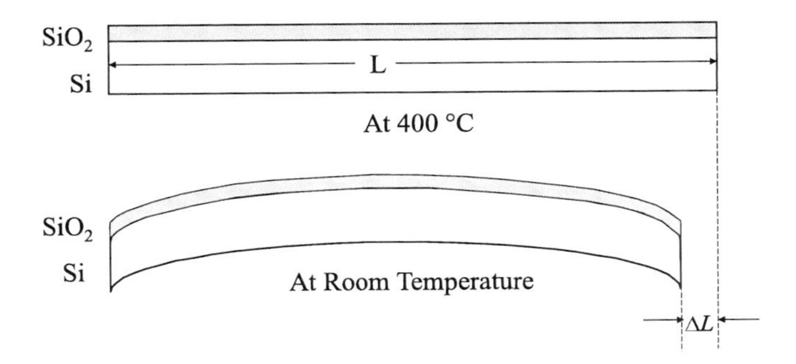

Figure 10.44 Illustration of thermal stress.

Figure 10.44 shows thermal stress on an oxide film due to the different thermal expansion rates between the oxide and silicon substrate. The thermal expansion of a material can be expressed as

$$\Delta L = \alpha \Delta T L \tag{10.10}$$

where ΔL is the change of dimension, ΔT is the change of temperature, and α is the coefficient of thermal expansion. Table 10.7 lists the coefficients of thermal expansion of some commonly used materials in IC processing. All of them are measured at 300 K (or 22.85 °C), which is close to room temperature.

Assume that there is no stress between an oxide film and a silicon substrate right after oxide film deposition at 400 °C. When the wafer has cooled down to room temperature, the silicon shrinks more than the oxide, and the substrate compresses the film. Due to a higher thermal expansion rate of the silicon, the wafer curvature becomes more concave, as shown in Fig. 10.43. Because silicon contracts more than oxide, oxide film has a compressive stress.

Question: Why does an oxide film prefer compressive stress, while metal films prefer tensile stress?

Answer: If an oxide film has tensile stress at room temperature, when the wafer is heated up for the next process, the silicon substrate expands more and causes the film stress to become more tensile. This can crack the oxide film. If an oxide film stress starts as compressive, when wafer temperature increases, the film stress becomes less compressive.

Table 10.7 Coefficients of thermal expansion (in 10^{-6} °C^{-1}).

$\alpha(SiO_2) = 0.5 \times 10^{-6}$ °C^{-1}	$\alpha(W) = 4.5 \times 10^{-6}$ °C^{-1}
$\alpha(Si) = 2.5 \times 10^{-6}$ °C^{-1}	$\alpha(Al) = 23.2 \times 10^{-6}$ °C^{-1}
$\alpha(Si_3N_4) = 2.8 \times 10^{-6}$ °C^{-1}	$\alpha(Cu) = 17 \times 10^{-6}$ °C^{-1}

Therefore, oxide films always favor compressive stress. To the contrary, metal films such as tungsten and aluminum favor tensile stress at room temperature because they have higher thermal expansion rates than silicon and silicon oxide, and expand more when temperature increases, which can make the film stress less tensile.

Stress is calculated by measuring the change of the wafer curvature before and after thin-film deposition. The most commonly used unit for film stress is MPa = 10^6 Pa (1 MPa = 10^7 dynes/cm^2):

$$\sigma = \frac{E}{1-\nu}\frac{h^2}{6t}\left(\frac{1}{R_2} - \frac{1}{R_1}\right) \tag{10.11}$$

where σ is the film stress (Pa), E is Young's modulus of the substrate (Pa), ν is Poisson's ratio of the substrate, h is the substrate thickness (μm), and t is the film thickness (μm). R_1 is the wafer curvature radius before deposition (μm), and R_2 is the wafer radius curvature after deposition (μm). Figure 10.45 shows a schematic of a stress measurement system.

Question: When using recycled dummy wafers to test thin-film deposition processes, what factors need to be taken into consideration?

Answer: Recycled wafers normally require a polishing process in which their thickness is reduced and becomes thinner than the standard thickness (725 μm for the 200-mm wafer, and 775 μm for the 300-mm wafer). Deposition rate, nonuniformity, and WERR measurements are not

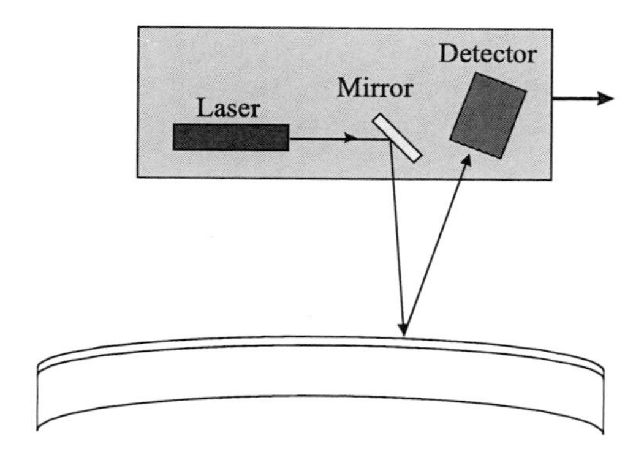

Figure 10.45 Laser scanning stress measurement tool.

affected by using recycled wafers. It is evident from Eq. (10.11) that the stress measurement is very sensitive to wafer thickness h. Therefore, it is necessary to adjust the system constant regarding the wafer thickness to achieve an accurate reading. It is very important to change that constant back to the standard value, since other people using the same stress gauge may not be aware of the change and acquire false readings on their measurements when using a normal wafer, possibly causing unnecessary process downtimes.

10.5 Dielectric Chemical Vapor Deposition Processes

Two processes are widely used for dielectric thin-film deposition: thermal CVD and PECVD. In thermal processes, heat alone provides the energy needed for chemical precursors to overcome the activation energy barrier and achieve chemical reactions. Both APCVD and LPCVD are thermal CVD processes. In PECVD processes, both heat and rf provide the energy required for chemical reactions.

The commonly used silicon precursors are silane, TEOS, and 3MS. Silane is used to deposit both nitride and oxide; TEOS is used to deposit oxide, either doped or undoped; and 3MS is used to deposit low-κ dielectric SiCOH and porous SiCOH. 3MS can also be used to deposit low-κ barrier SiCN layers.

10.5.1 Thermal silane chemical vapor definition process

Silane is one of the most commonly used silicon source gases for dielectric CVD processes. It has been used for silicon dioxide deposition with both the APCVD and LPCVD process:

$$SiH_4 + 2O_2 \underset{\text{heat}}{\rightarrow} SiO_2 + 2H_2O.$$

APCVD uses diluted silane (3% in nitrogen), and LPCVD uses pure silane. Compared with APCVD silane oxide, LPCVD silane oxide has higher throughput due to larger wafer load and better step coverage from the longer MFP of the precursors. Both APCVD and LPCVD silane oxide processes have been replaced by the TEOS-based oxide process because of the latter's superior gap-filling abilities.

LPCVD silane nitride is commonly employed as an oxidation mask for LOCOS formation. It is also used as an oxide CMP hard stop for STI processes. When thermal budget allows, fabs use LPCVD nitride as the PMD barrier layer to prevent dopants from PSG or BPSG diffusing into the activation areas:

$$3\,SiH_4 + 4\,NH_3 \underset{\text{heat}}{\rightarrow} Si_3N_4 + 12\,H_2.$$

The LPCVD nitride process is covered in detail in Chapter 5.

10.5.2 Thermal tetraethoxysilane chemical vapor deposition process

TEOS oxide film has better step coverage and conformality due to the high surface mobility of the physisorption of a TEOS molecule on a substrate surface. At high temperatures, TEOS will dissociate and form silicon dioxide:

$$Si(OC_2H_5)_4 \underset{\text{heat}}{\rightarrow} SiO_2 + \text{volatile organics.}$$

This thermal TEOS oxide deposits at temperatures higher than 700 °C in a LPCVD reactor. Deposition rate is very low, about 20 Å/min. This process has been used to deposit TEOS for sidewall spacers. By using TEOS along with TMB [trimethylborate, $B(OCH_3)_3$] and TMP [trimethylphosphite, $P(OCH_3)_3$], the reaction was used to deposit BPSG for PMD applications. This reaction temperature is too high to deposit USG for ILD applications.

10.5.3 Plasma-enhanced chemical vapor deposition silane process

For PECVD silane oxide, most fabs use nitrous oxide (N_2O, laughing gas) instead of O_2 as an oxygen source gas because it is compatible with silane at high pressure. By flowing silane and N_2O into a vacuum chamber (less than 10 torr) and striking plasma with rf power, the gases react and deposit silicon oxide on the heated wafer surface. Both the heat from the wafer surface and the energetic electrons inside the rf plasma provide the necessary energy for chemical reactions to occur. PECVD oxide is not a pure silicon dioxide SiO_2; it always has a small amount of hydrogen in the deposited film:

$$SiH_4 + N_2O \overset{\text{plasma}}{\underset{\text{heat}}{\rightarrow}} SiO_xH_y + H_2O + N_2 + NH_3 + \cdots.$$

In plasma, electrons dissociate both silane and nitrous oxide via dissociation collisions:

$$e^- + SiH_4 \rightarrow e^- + SiH_2 + H_2,$$
$$e^- + N_2O \rightarrow e^- + O + N_2.$$

Both O and SiH_2 are free radicals with two unpaired electrons and are chemically very reactive. They are easily chemisorbed on a wafer surface and can chemically react. Plasma-generated free radicals take the deposition process into the mass-transport-limited regime at a relatively low temperature. Deposition rate is high and mainly depends on the gas flow rate. Normally, N_2O is overflowed; therefore, the deposition rate is controlled by the silane flow rate. The PECVD silane oxide process has only two precursor gases—silane as the silicon source and N_2O as the oxygen source—so it is easier to explain the deposition process.

Question: Can silane be overflowed and the nitrous oxide flow rate be used to control the deposition rate?

Answer: Theoretically it can be done, but practically no one should try this because it is very dangerous and not cost effective. Overflowing silane will create a big safety hazard of fire and explosions, and silane is more expensive than nitrous oxide.

The PECVD silane oxide process has three main steps. They are stabilization (~5 sec), deposition (depending on the required film thickness), and chamber pump down. In the stabilization step, both pressure and gas flow rate are stabilized so that they do not change during the deposition process. During the deposition step, rf power is turned on to strike the plasma and start deposition. After deposition is finished, both rf power and process gases are turned off, which stops deposition. The chamber is pumped down to prepare for the next process. Sometimes a plasma purge step with rf power and N_2O flow is added to consume all of the residue silane before chamber pump down. Table 10.8 gives a sample process recipe for PECVD silane oxide (step 2 of the processing sequence).

In the table, sccm stands for standard cubic centimeter per minute. The expected deposition rate is 1 μm/min (10,000 Å/min), refractive index is 1.46, and film stress is compressive 100 MPa (10E9 dyne/cm^2).

Question: Increasing the silane flow rate alone increases the deposition rate. How do the refractive index and film stress change in this case?

Answer: By increasing silane flow, the SiH_4 to N_2O ratio increases, which causes more silicon to be deposited into the film. Figure 10.36 shows that the refractive index increases. While the deposition rate increases, rf power and pressure are kept constant. This counts as the same amount of ion bombardment on a thicker film, meaning that the film will receive less ion bombardment per unit thickness. It will make the film less dense and cause less compressive film stress.

Therefore, to increase deposition rate while keeping the same stress and refractive index, silane flow rate, N_2O flow rate, and rf power must increase at the same time.

10.5.3.1 Passivation

Silicon nitride is a very good barrier layer for moisture and mobile ions and is widely employed as the final passivation layer. PECVD nitride is used in this process because passivation dielectric deposition requires a relatively low deposition temperature (<450 °C) at a reasonably high deposition rate. In the

Table 10.8 PECVD silane oxide process.

Step 2	Deposition
Pressure (torr)	3.0
Temperature (°C)	400
rf (W)	250
SiH_4 flow (sccm)	120
N_2O flow (sccm)	2400

PECVD nitride process, silane is commonly used as the silicon source, ammonia is used as the main nitrogen source, and nitrogen is used as the carrier gas and secondary nitrogen source:

$$SiH_4 + N_2 + NH_3 \xrightarrow[\text{heat}]{\text{plasma}} SiN_xH_y + H_2 + N_2 + NH_3 + \cdots.$$

The passivation dielectric nitride requires good step coverage, high deposition rate, good uniformity, and stress control.

PECVD silane passivation dielectric deposition processes have seven steps, including stabilization 1, oxide deposition, pumping, stabilization 2, nitride deposition, plasma purging, and chamber pumping.

PECVD nitride has about 20 atomic percent (different from weight percent) of hydrogen. Hydrogen bonds with both nitrogen and silicon. Since Si–H bonds absorb UV light, conventional nitride is opaque to UV transmission.

Question: Where does the hydrogen in PECVD passivation dielectric nitride come from?

Answer: Both silicon precursor SiH_4 and nitrogen precursor NH_3 have hydrogen atoms, so naturally many of them are integrated into the film due to the low deposition temperature of passivation dielectric nitride.

By using silane and nitrogen without ammonia, nitride can be deposited with much less hydrogen concentration and high UV transmission rate. However, the deposition rate will be significantly lower due to the lack of nitrogen free radicals. A nitrogen molecule is very stable; it is much harder to dissociate nitrogen molecules than to dissociate ammonia molecules in plasma.

EPROM requires a UV transparent passivation layer to allow UV light to reach the floating gate. UV light can erase memory by exciting stored electrons in the floating gate to tunnel across the interpoly dielectric and drain out of the grounded upper gate. Oxynitride (SiO_xN_y) is commonly used for the EPROM passivation dielectric. By using silane, nitrogen, ammonia, and nitrous oxide, silicon oxynitride can be deposited:

$$SiH_4 + N_2 + NH_3 + N_2O \xrightarrow[\text{heat}]{\text{plasma}} SiO_xN_y + H_2O + N_2 + \cdots.$$

Oxynitride has properties between oxide and nitride. Its refractive index is about 1.7 to 1.8, it is UV transparent, and it is a fairly good barrier layer for moisture and mobile ions. The oxynitride process requires monitoring of the refractive index more frequently than other dielectric CVD processes. This is because the barrier property and UV transmission rate of oxynitride film are sensitive to the silicon, nitrogen, and oxygen concentrations in the film. Refractive index measurements can provide this information.

10.5.3.2 Interlayer dielectric-0 barrier layer

Most fabs use TEOS-based PSG or BPSG for ILD0 applications. A diffusion barrier layer is needed to prevent dopant atoms (phosphorus and boron) from diffusing into activation areas and damaging the transistors. Both USG (1000 Å) and nitride (<300 Å) can be used for this application. As device dimensions have continued to shrink, most fabs prefer to use thinner LPCVD nitride as the barrier layer. When device dimensions shrink even farther, thermal budget limitations will rule out the application of LPCVD for ILD0 deposition; thus, PECVD nitride will start to be used for this application.

The PECVD ILD0 nitride barrier layer usually is deposited at around 550 °C, a deposition temperature significantly lower than the 700 °C required by LPCVD nitride processes. Therefore, the PECVD nitride process has a much lower thermal budget. It is higher than the 400 °C used for PECVD passivation nitride. At higher temperatures, PECVD nitride film has higher deposition rates, lower hydrogen concentration, and better film quality.

ILD0 nitride has also been applied as a stress liner to stress the gate and form a favorite strain in the channel of a MOSFET. An nMOS channel requires a tensile strain, and a pMOS channel requires a compressive strain. Both strains can be achieved in PECVD nitride by adjusting the rf power of the deposition process. A dual stress liner can be formed by depositing the first tensile strain nitride liner all over the wafer. Then, a mask defines the pMOS area to selectively remove the first nitride liner from the pMOS area. After photoresist is removed and the wafer is cleaned, a compressive stress liner is deposited. Depending on the device requirement, the second stress liner can either stay on top of the tensile stress liner in nMOS regions, or the compressive stress liner can be removed from the nMOS region with a masking step.

10.5.3.3 Dielectric antireflective coating

To achieve high resolution for photolithography processes, an ARC layer is required to reduce reflection from the shiny aluminum and polysilicon surfaces. With reflectivity of about 30 to 40%, titanium nitride (which can be sputter deposited with aluminum–copper alloy in a cluster PVD tool) is a commonly adapted ARC layer for IC chips with Al–Cu interconnections. For polysilicon patterning, a dielectric ARC layer is used to meet the resolution requirements of photolithography.

Figure 10.46 illustrates the dielectric ARC. When the phase difference of the reflected light on the photoresist–ARC interface (light 1), and the reflected light on ARC–metal interface (light 2) equals half of the wavelength,

$$\Delta\phi = 2nt = \lambda/2,$$

the two reflection rays will create destructive interference with each other and sharply reduce the reflection light intensity inside the photoresist. In the figure,

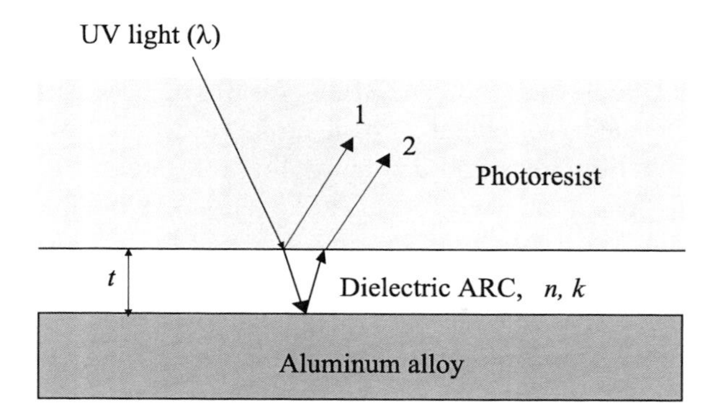

Figure 10.46 Dielectric ARC.

n is the refractive index of the ARC film, t is thickness of the ARC film, and λ is the lithography light wavelength. By controlling absorption coefficient k, the intensities of the two reflection rays can be exactly the same, so there will be complete destructive interference between the two and no reflected light in the photoresist at all. This significantly improves lithography resolution, especially when feature size is smaller than a quarter-micron.

Silane PECVD silicon-rich oxynitride film with a high refractive index ($n \sim 2.2$) has been developed for this application:

$$SiH_4 + N_2O + He \underset{\text{heat}}{\overset{\text{plasma}}{\rightarrow}} SiO_xN_y + H_2O + N_2 + NH_3 + He + \cdots.$$

SiH_4 flow rate controls the deposition rate, SiH_4/N_2O ratio determines the refractive index n, pressure and rf power control absorption rate k, and helium flow controls film uniformity. Normally, the film thickness is about 300 Å. For different wavelengths, i-line (365 nm), DUV (248 nm), and 193 nm, requirements for dielectric ARC film are different; therefore, different process recipes are needed for these applications.

10.5.4 Plasma-enhanced chemical vapor deposition tetraethoxysilane process

A thermal TEOS process can deposit oxide film with excellent step coverage and gap-filling ability. However, the deposition temperature is so high (>700 °C) that it is limited to ILD0 applications. In the 1980s, requirements of multilevel metal interconnections strongly drove the development of low-temperature TEOS oxide deposition processes for ILD applications. The PECVD TEOS oxide process uses plasma to dissociate oxygen molecules to generate free oxygen radicals, which dramatically enhance TEOS oxidation rate and help to achieve a high oxide deposition rate at relatively low temperatures (~400 °C). Because most TEOS precursors are physisorbed on the oxide surface with high surface mobility,

PE-TEOS oxide film has very good step coverage and conformality:

$$Si(OC_2H_5)_4 + O_2 \underset{\text{heat}}{\overset{\text{plasma}}{\rightarrow}} SiO_2 + \text{other volatiles}.$$

PE-TEOS USG was the dielectric most used for ILD applications in the 1990s. In a cluster tool, PE-TEOS chambers can be used with argon sputter etch chambers to complete an in-situ PE-TEOS dep/etch/dep processing sequence for gap fill and planarization. PE-TEOS USG is also widely used as barrier and cap layers for spin-on glass and O_3-TEOS USG processes in ILD applications.

PE-TEOS can also be used to deposit PSG and BPSG for ILD0 applications. By reacting phosphorus source gases such as TMP or TEPO [$PO(OCH_3)_3$], PSG can be deposited:

$$PO(OCH_3)_3 + Si(OC_2H_5)_4 + O_2 \underset{\text{heat}}{\overset{\text{plasma}}{\rightarrow}} SiO_2 + \text{other volatiles}.$$

With both phosphorus and boron (TMB or TEB) sources, a similar process can deposit PE-TEOS BPSG for PMD application. However, most fabs prefer nonplasma CVD processes such as O_3-TEOS CVD BPSG or PSG for ILD0 applications due to the concern of plasma-induced gate damage.

The PE-TEOS process is also employed to deposit fluorinated silicate glass (FSG). By using TEOS and oxygen along with a fluorine precursor gas such as SiF_4 or FTES [$FSi(OC_2H_5)_3$, fluorotriethyloxysilane], FSG can be deposited on a heated wafer surface in the rf plasma chamber. FSG has a lower dielectric constant (κ = 3.5 to 3.8) compared to USG's κ = 4.0 to 4.2 because the electronegative nature of fluorine decreases the polarizability of the overall SiO_xF_y network and lowers the dielectric constant. By using FSG, the parasitic capacitance between metal lines can be reduced, since the capacitance is proportional to the dielectric constant. Therefore, using FSG can reduce the RC time delay, minimize crosstalk, and reduce power consumption. It was used for ILD applications before maturity of the low-κ dielectrics. The chemical reaction of the FSG CVD process can be expressed as:

$$\underset{\text{(FTES)}}{FSi(OC_2H_5)_3} + \underset{\text{(TEOS)}}{Si(OC_2H_5)_4} + O_2 \underset{\text{heat}}{\overset{\text{plasma}}{\rightarrow}} \underset{\text{(FSG)}}{SiO_xF_y} + \text{other volatiles}.$$

The requirements for FSG films are stability, device compatibility, cost effectiveness, and easy process integration.

Figure 10.47 compares silicon dioxide, FSG, and silicon tetrafluoride. When fluorine is integrated into the silicate glass, the dielectric constant is reduced. The more that fluorine is integrated into the film, the lower the dielectric constant will be. However, the film will become more unstable, and fluorine tends to outgas from the film. If the fluorine concentration is too high, this can cause outgassing during the next thermal cycle when the wafer is heated up for deposition or metal

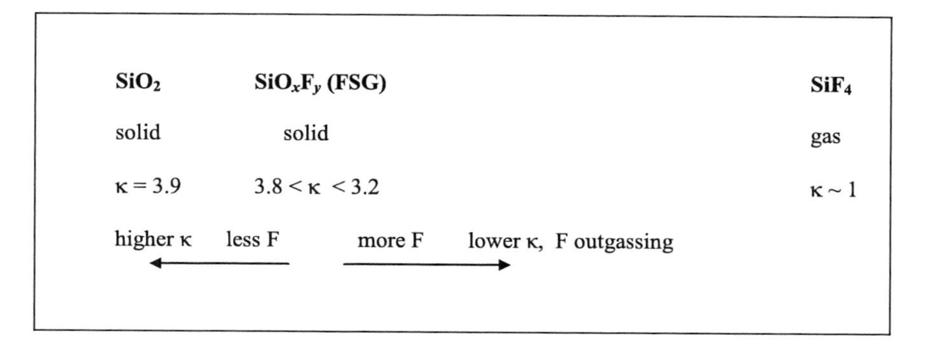

Figure 10.47 Demonstration of FSG processing trends.

annealing. HF and F_2 being released from the film can cause device or circuit damage due to corrosion.

FSG film has better gap-fill abilities than USG film due to the etch effect of the combination of plasma-induced free fluorine radicals and ion bombardment, especially on the upper corner of the gap. Because of the larger arriving angle at the upper corner, this area receives more ion bombardment and more free fluorine radicals, which keep an overhang from forming. By using PE-TEOS SiF_4 chemistry, a void-free 0.65-μm gap fill without an etchback can be achieved, as can a 0.35-μm void-free gap fill with an etchback.

10.5.5 Dielectric etchback process

Sputtering and planarization etchback processes are widely used with dielectric CVD to achieve gap fill and planarization. Unlike the deposition process, etchback is a removal process. In many IC fabs, dielectric etchback processes are performed in the thin-film area, along with the dielectric CVD process. Therefore, etchback processes are covered in detail in this chapter instead of the etch chapter (Chapter 9).

In many cases, sputtering etch chambers are placed on the same mainframe with the dielectric CVD chambers of a cluster tool to operate the in-situ integrated dep/etch/dep process, as shown in Fig. 10.48.

The sputtering etch process is a purely physical etch process. It operates at low pressure (~30 mtorr) in argon plasma. Argon ions bombard the wafer surface and dislodge the dielectric thin film on the surface by breaking chemical bonds by means of energy and momentum transfer from the bombarding ions. The dielectric fragments are negatively charged by the electrons from the plasma, and expelled from the wafer surface, since the wafer is also charged negatively by the plasma. The dislodged dielectric fragments then cross the boundary layer into the convection flow and are pumped out from the chamber.

Since the sputtering etch is a purely physical etch (which solely depends on argon ion bombardment, with no chemical reaction with dielectric thin films), it has a very low etch rate. It chips off the step corners much faster than it removes materials from the surface. The step corner facets are about 45 deg in slope,

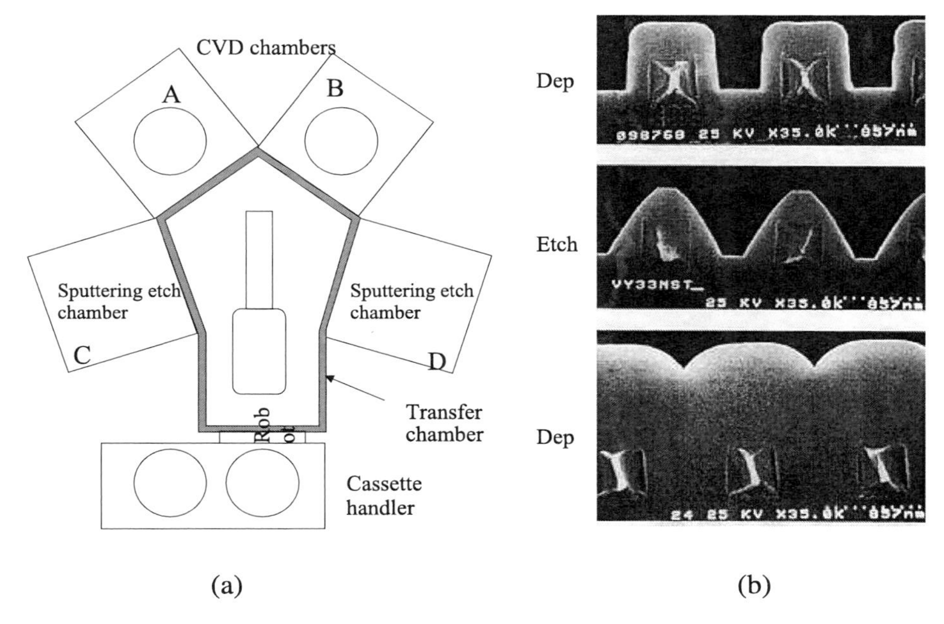

Figure 10.48 (a) Cluster tool and (b) SEM images of the PE-TEOS dep/etch/dep process (Applied Materials, Inc.).

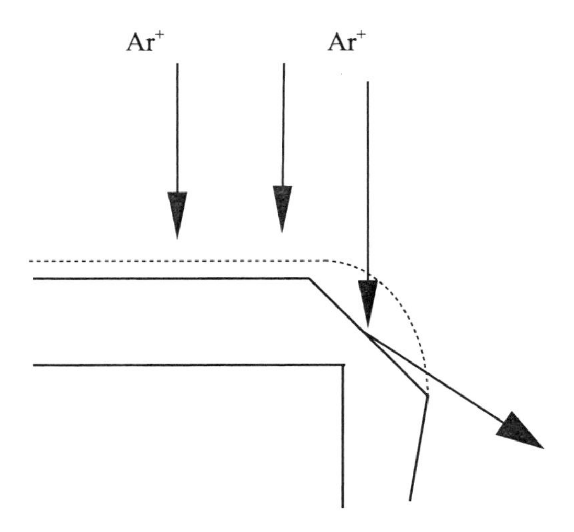

Figure 10.49 Argon sputtering chops the corner of a gap.

tapering the gaps and making it easier for the ensuing deposition step to fill the gap without voids.

Question: Why do sputtering etch processes usually use argon as a processing gas?
Answer: Gap-fill sputtering etch processes need to chip off the step corners of the gaps with purely physical bombardment. Therefore, an inert

gas is required. Since argon is a heavy and inexpensive noble gas, it became the gas of choice for almost all physical sputtering processes. The atomic weight of argon is 40, compared with silicon's 28 and helium's 4. Heavy ions are favored for achieving better bombardment results. Argon is the third most abundant component in the earth's atmosphere (~1%) [only less abundant than nitrogen (78%) and oxygen (20%)]. It can be purified directly from condensed air, making it more affordable than other heavier inert gases such as krypton, xenon, and radon. Radon is a radioactive gas and cannot be used in semiconductor processes.

A sample of an argon sputter etch process recipe is listed in Table 10.9. The expected blanket etch rate on thermally grown silicon dioxide is about 250 Å/min for this process. Thermally grown oxide is used for process qualification because of its WIW and WTW uniformity. For CVD oxide, the sputter etch rate is slightly higher. Like most plasma etch processes, the etch rate of the sputter etch process is mainly determined by rf power because both ion bombardment energy and ion density will increase when rf power increases. Increasing rf power can significantly increase the etch rate. Etch uniformity is very sensitive to the pressure.

A schematic of an etch chamber is given in Fig. 10.50. Since sputter etch operates at a low pressure and the electron MFP is long, it is difficult to generate and maintain the plasma. An external magnetic field is introduced to force electrons to spiral around the magnetic field lines and travel much longer distances. This increases the probability of ionization collisions between electrons and the neutral argon atoms. The sputtering etch process is a blanket etch process, and there is no need to protect the photoresist. The sputtering etch process is a purely physical etch process, and the sputter etch rate is not sensitive to wafer temperature, so the sputtering etch chamber does not need a wafer cooling system.

By using CF_4 and O_2, a sputter etch chamber can also be used for dielectric planarization etchback, as shown in Fig. 10.51. In this case, oxide is etched by a combination of both fluorine chemical etch and ion bombardment sputtering. Because chemical etch tends to etch isotropically, the dielectric thin-film surface after etchback will be smoother than when it is deposited. To enhance chemical etch effect, the planarization etchback process normally operates at a much higher pressure than does a sputtering etchback process. Increasing pressure can reduce MFP, increase ion collisions and scattering, and reduce ion energy. This reduces ion bombardment and the physical sputtering etch effect.

Table 10.9 Sample argon sputter etch recipe.

Pressure (mtorr)	30
rf power (W)	300
B-field (G)	50
Ar (sccm)	50

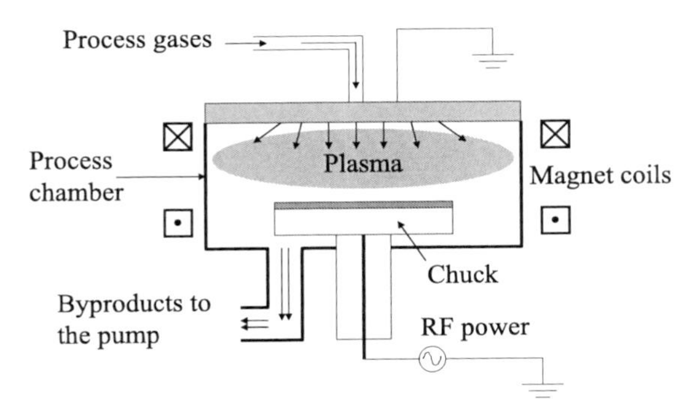

Figure 10.50 A sputtering etch chamber.

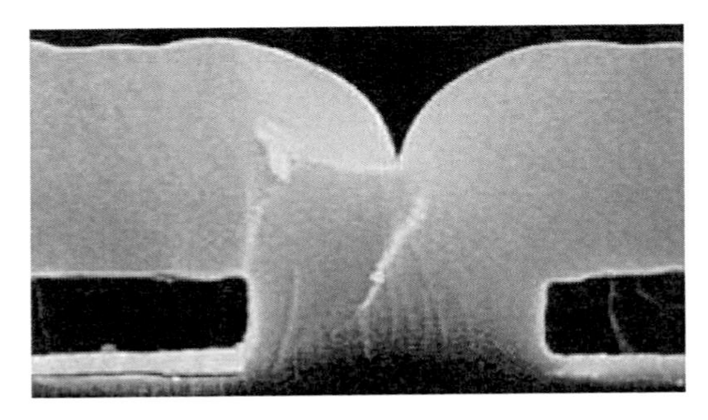

PE-TEOS oxide deposition

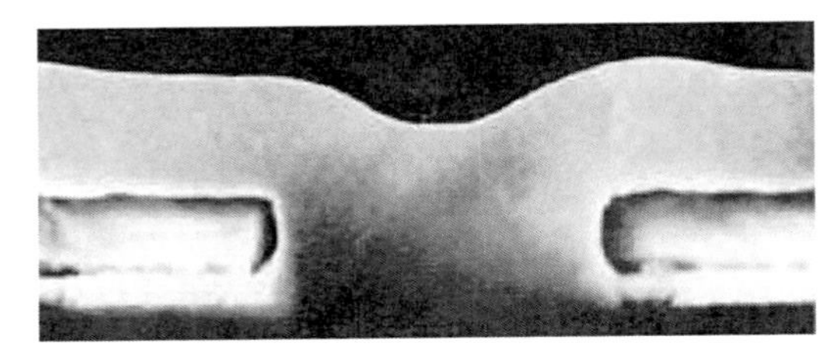

Planarization etchback

Figure 10.51 SEM images of the planarization etchback process (Applied Materials, Inc.).

10.5.6 Ozone-tetraethoxysilane process

Ozone is a very unstable molecule, so thermal energy can easily dissociate it and release chemically reactive free oxygen from the reaction:

$$O_3 \rightarrow O_2 + O.$$

At 25 °C, the half-life of ozone is about 86 h, which means that after 86 h, half of the ozone molecules will be dissociated and gone. Because chemical reaction rates increase exponentially with temperature [see Eq. (10.2)], the dissociation rate of ozone is much faster, and its half-life is much shorter at higher temperatures. At

400 °C, the half-life of ozone is less than a millisecond. Therefore, ozone can be used as a carrier of oxygen free radicals at room temperature and can generate free oxygen at higher temperatures at the wafer surface to enhance chemical reactions without the presence of plasma inside the processing chamber. Using ozone to react with TEOS oxidizes the TEOS and deposits silicon oxide. O_3-TEOS oxide film has excellent conformality and gap-fill abilities, which can fill very narrow gaps, and it is commonly employed for deep submicron IC chip dielectric thin-film deposition. O_3-TEOS oxide can be deposited at both atmospheric (APCVD) and subatmospheric (SA-CVD) pressures.

10.5.6.1 Ozonator

Since ozone has a very limited lifetime, it must be generated on site and consumed right away. In nature, ozone is generated in lightning during a thunderstorm. An ozonator generates ozone in a very similar way. In a high-pressure corona discharge, oxygen molecules are dissociated by energetic electrons, and released as free oxygen radicals:

$$\mathrm{O_2} \overset{\text{plasma}}{\rightarrow} \mathrm{O + O}.$$

The free oxygen radicals react with an oxygen molecule to form ozone in a three-body collision:

$$\mathrm{O + O_2 + M \rightarrow O_3 + M\ (M = O_2, N_2, Ar, He, etc.)}.$$

The third party in the collision is necessary to meet the requirements for conservation of both energy and momentum.

Figure 10.52 shows an ozone generation process. Mixing approximately one-percent nitrogen in the oxygen flow into the ozonator can improve and stabilize ozone yield. Several ozone cells are connected in a series to repeat the process and maximize ozone concentration. Since ozone is unstable and tends to dissociate, reducing ozone cell temperature can also help improve ozone yield by reducing ozone dissociation.

After passing several ozone cells, ozone concentration can reach up to 14 wt%, depending on the oxygen flow rate, rf power, and nitrogen concentration. Generally, at the same rf power and nitrogen concentration, a higher flow rate will have lower ozone concentration. At the same flow rate and nitrogen concentration, higher rf power will have higher ozone yield.

Ozone concentration can be monitored by its UV absorption property. From Beer–Lambert's law,

$$I = I_0 \exp(-ACL),$$

where I is the UV intensity measured with ozone, I_0 is the UV intensity measured without ozone flow, A is the UV absorption coefficient of ozone, L is the absorption

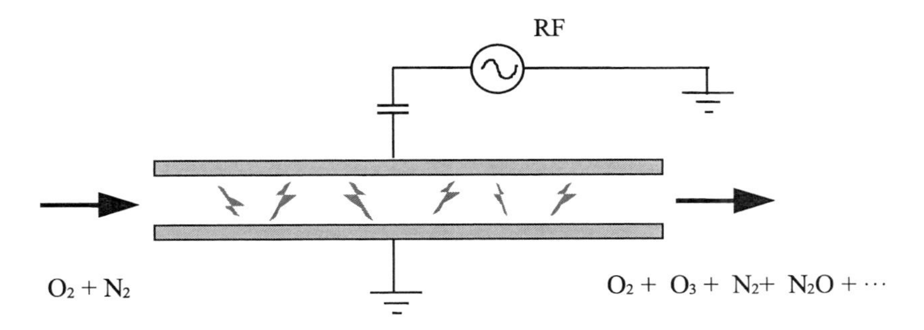

Figure 10.52 Ozone generation in an ozone cell.

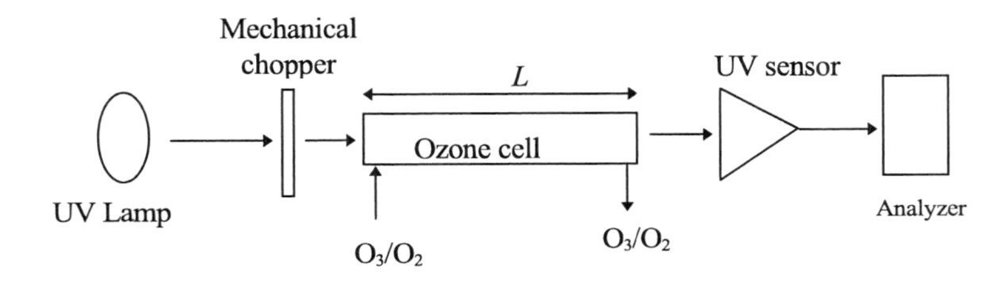

Figure 10.53 Zone concentration monitoring system.

cell length, and C is the ozone concentration. Figure 10.53 shows an ozone monitoring system.

The main goal for an ozonator is to generate stable, repeatable, and high-concentration ozone flow at high flow rates to maintain maximum ozone output.

10.5.6.2 Ozone-tetraethoxysilane undoped silicate glass process

The main applications of O_3-TEOS USG films are STI trench fill and ILD layering of IC chips with Al–Cu interconnections. Using ozone to react with TEOS in the vicinity of 400 °C deposits silicon oxide at reasonably high deposition rates, 1500 to 2000 Å/min, depending on gas flow rate, temperature, and pressure:

$$\text{TEOS} + O_3 \underset{\text{heat}}{\rightarrow} SiO_2 + \text{volatile organics}\,.$$

Because it has almost perfect step coverage and conformality, O_3-TEOS USG film can fill very small gaps without voids, as shown in Fig. 10.54. It is often used for ILD applications in multimetal layer interconnections.

In the O_3-TEOS deposition process, higher O_3:TEOS ratio means more free oxygen to react and oxidize TEOS. Therefore, the higher the O_3:TEOS ratio, the better the deposited oxide film quality. To increase the O_3:TEOS ratio, both ozone concentration and ozone flow rate need to be increased (both of which are limited by ozonator capability); another way is to reduce TEOS flow rate, causing a lower deposition rate.

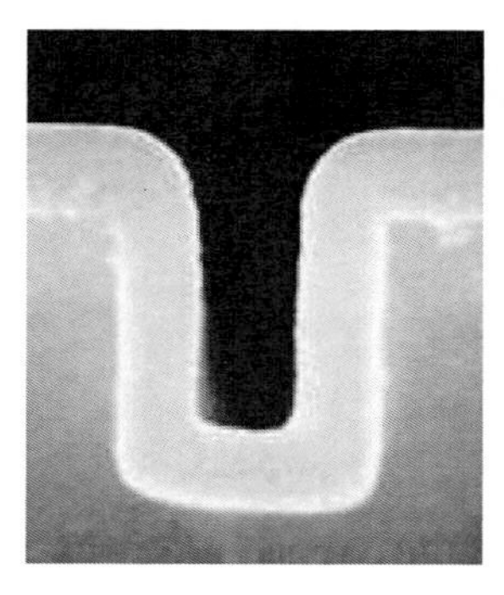

Figure 10.54 O_3-TEOS USG step coverage and gap fill (Applied Materials, Inc.).

Because O_3-TEOS oxide film is porous and absorbs moisture, it needs a dense PECVD cap film to seal it from the atmosphere. Since it has unfavorable tensile stress, it requires a compressive PECVD film to buffer the stress to avoid metal hillocks. O_3-TEOS USG film for ILD applications is always sandwiched between two PECVD USG films.

For subatmospheric pressure O_3-TEOS processes, all three layers are deposited in the same processing chamber, with the same processing steps. First, a PE-TEOS USG thin film (about 1000 Å) is deposited as a stress buffer layer, then pressure is gradually ramped up to prevent deposited film on the wall of the process reactor from flaking and causing particle contamination. After stabilization of TEOS flow and pressure, O_3-TEOS USG film is deposited to fill the gap by turning on the ozonator and flowing ozone into the processing chamber. After gradually pumping down the chamber and stabilizing the pressure, oxygen flow, and TEOS flow, rf power is turned on to strike the plasma, and a thick layer of PE-TEOS USG cap film is deposited to seal the porous O_3-TEOS USG film. Either CMP or etchback is employed to planarize the PE-TEOS USG cap layer, depending on the IC processing requirements.

10.5.6.3 Ozone-tetraethoxysilane phosphosilicate glass and borophosphosilicate glass processes

Both O_3-TEOS BPSG and PSG are commonly used for ILD0 applications. Figure 10.55 shows that O_3-TEOS BPSG fills 0.25-μm gaps with aspect ratios of 4:1 in ILD0 applications. Reacting ozone with TEOS, TEB, and TEPO will form BPSG. The reaction can be expressed as:

$$O_3 \rightarrow O_2 + O,$$
$$O + \text{TEB} + \text{TEPO} + \text{TEOS} \rightarrow \text{BPSG} + \text{volatile organics.}$$

10.6 Spin-On Glass

The SOG process and photoresist coating and baking processes are very similar, which is one important reason that fabs adopted this technology, since

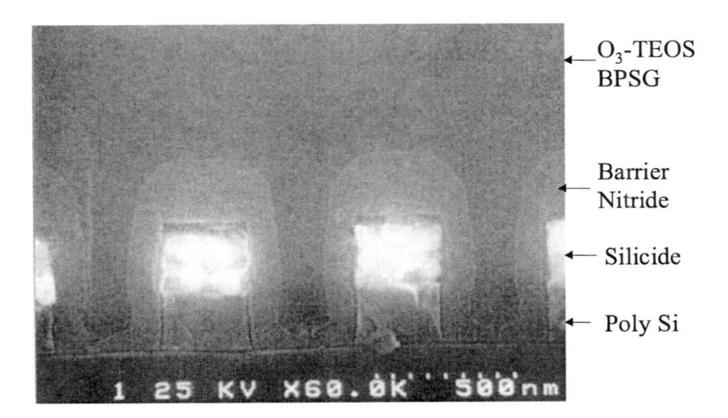

Figure 10.55 O_3-TEOS BPSG fills 0.25-μm gaps with aspect ratios of 4:1 (Applied Materials, Inc.).

manufacturers like familiar technology. Some IC producers have used a SOG process for ILD gap fill and planarization of IC chips with Al–Cu interconnections for a long time.

There are two kinds of SOGs commonly used in the semiconductor industry: one is silicate, the other is siloxane. Both of them have Si-O bonds, as shown in Fig. 10.56. Solvents used for SOG are alcohol, ketone, etc.

SOG processes require several processing tools and different process steps, such as PECVD/SOG spin coating/SOG curing/SOG etchback/PECVD. SOG is always sandwiched between two PECVD layers, as shown in Fig. 10.57.

To begin, a thin layer of PECVD film is deposited in a PECVD reactor as a liner or barrier layer; then liquid SOG is uniformly spun onto the wafer surface in a spinner and forms a thin film a few thousand angstroms thick. Film thickness is determined by the spin rate and viscosity of the liquid, just as with the spin coating of a photoresist. The liquid may need to be spun twice to achieve the required uniformity. Surface tension of the liquid forces SOG to flow into the narrow gaps and fills them without voids. After prebake on a hotplate, some solvents are driven out of the SOG, and Si–O bonds start to crosslink. Then the wafers are placed into a baking furnace to bake at 400 to 450 °C, which drives out most solvents, making it solid silicate glass. The thickness of the film shrinks about 5 to 15% after SOG curing processes. Since the quality of the glass is not very good, in most cases etchback is required to remove the majority of SOG from the surface and leave SOG only in the gaps. A PECVD cap layer is deposited to prevent both outgassing and moisture absorption of the SOG.

In some cases, fabs prefer to deposit the cap PECVD oxide without etchback to save a major processing step in order to improve throughput and reduce production cost. This is called SOG without etchback, and the possibility of SOG outgassing is much higher than that of SOG with etchback.

The disadvantages of SOG are its inherently complicated process integration, vulnerability to particulate contamination, film cracking or peeling, and residue

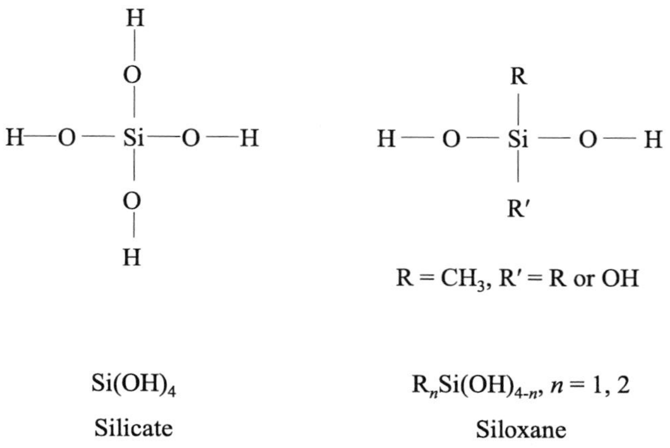

H—O—Si(—O—H)(—O—H)—O—H

H—O—Si(R)(R′)—O—H

$R = CH_3$, $R' = R$ or OH

$Si(OH)_4$

Silicate

$R_nSi(OH)_{4-n}$, $n = 1, 2$

Siloxane

Figure 10.56 Two common SOGs (without solvent): (a) silicate [$Si(OH)_4$] and (b) siloxane [$R_nSi(OH)_{4-n}$, $n = 1, 2$].

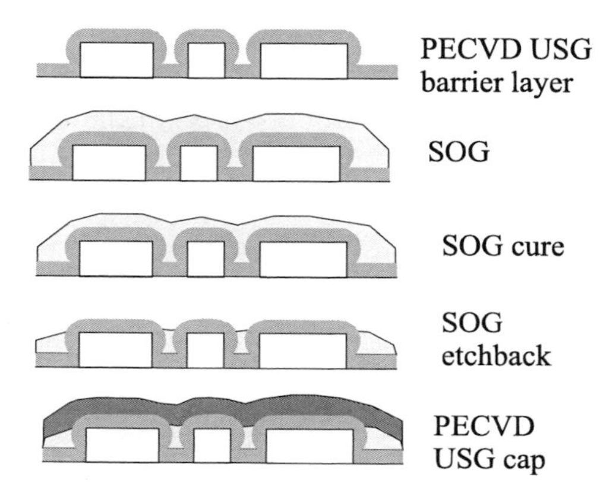

Figure 10.57 SOG processing sequence.

solvent outgassing. The last three problems can be solved by carefully controlling the process.

10.7 High-Density Plasma Chemical Vapor Deposition

While the dep/etch/dep processing sequence can fill small gaps, it requires two chambers: a CVD chamber and an etch chamber. Wafers need to be transferred back and forth between these two processing chambers. When a gap becomes narrower due to feature size, it may need an additional dep/etch cycle to become filled without voids. When the gaps become even smaller, and aspect ratios become even larger, many dep/etch cycles will be needed to fill the gaps, making it impractical for mass production because throughput will be too low. Therefore, a tool that can deposit and sputter etch simultaneously would be very helpful for deep submicron gap filling.

Question: With smaller feature sizes, the metal line width and gap between the metal lines become smaller. However, the metal line height does not shrink accordingly, causing a larger gap aspect ratio. Why does the metal height not shrink accordingly to keep the same aspect ratio and make it easier for dielectric gap fill?

Answer: Metal line resistance $R = \rho l/wh$. With smaller feature size, metal line length l and width w shrink accordingly. If metal line height also reduces accordingly, metal line resistance increases accordingly, which, of course, is unacceptable. Therefore, the metal interconnection line must retain the same height. By reducing feature size, more chips can be placed on a wafer. However, this also raises more challenges for dielectric deposition processes.

To achieve a high sputtering etch rate, the processing chamber must operate at a low pressure (<30 mtorr, the lower the better) to provide the long MFP required for ions to accelerate. At that low pressure, PECVD deposition rate is too low to match the sputtering etch rate in a standard capacitively coupled plasma chamber with two parallel plate electrodes, due to low plasma density. Keep in mind that the deposition rate must be higher than the etch rate to achieve net deposition. Therefore, different HDP sources are needed for these in-situ dep/etch/dep process reactors. Two types of HDP sources have been developed and applied in semiconductor processing: ICP and ECR, as illustrated in Fig. 10.58.

Since HDP-CVD is an in-situ dep/etch/dep process, the net deposition rate is not very high. Therefore, HDP-CVD is normally only used for gap fill, while the cap layer is deposited by the PECVD process, which has significantly higher deposition rates.

Since some of the dislodged oxide fragments that sputtered off the gap corner redeposit at the gap bottom, deposition rate of the film on the bottom is usually about three times higher than that on the sidewall in HDP-CVD processes. The result is that the sidewall films never meet, and the gap is filled from the bottom up without a seam, as shown in Fig. 10.59. This is an attractive feature with respect to subsequent integration with CMP because in CMP processes a seam provides a weak point and polishes faster than the other areas, creating a defect on the surface.

HDP-CVD oxide processes use silane as a silicon precursor and oxygen as an oxygen precursor. Argon is added for enhanced sputtering etch effect. It can also deposit PSG and FSG by reacting with phosphine and silicon tetrafluoride:

$$\text{USG}\ SiH_4 + O_2 + Ar \rightarrow USG + H_2O + Ar + \cdots,$$

$$\text{PSG}\ SiH_4 + PH_3 + O_2 + Ar \rightarrow PSG + \text{volatiles},$$

$$\text{FSG}\ SiH_4 + SiF_4 + O_2 + Ar \rightarrow FSG + \text{volatiles}.$$

Question: Why is silane used instead of TEOS as the silicon source gas in HDP-CVD oxide processes?

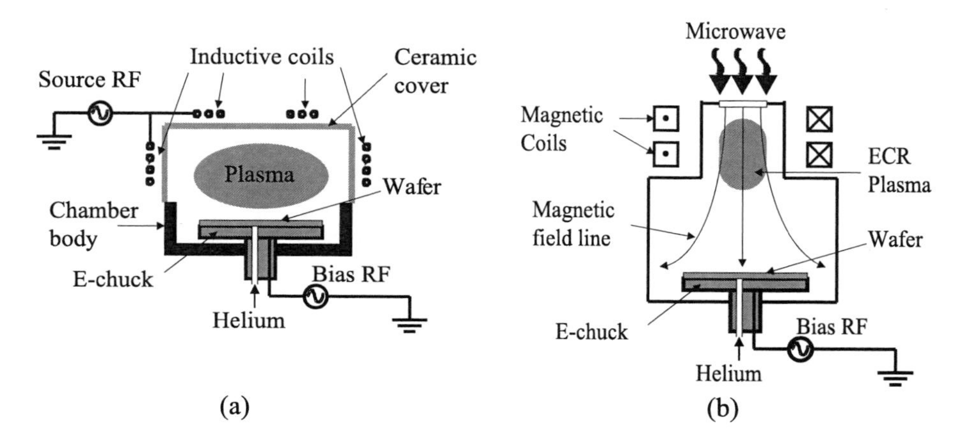

Figure 10.58 Schematics of (a) ICP and (b) ECR chambers.

Answer: For HDP-CVD, step coverage is no longer an important factor for gap fill because heavy ion bombardment always keeps the gap tapered, and deposition is from the bottom up. Using silane can save the cost and hassle related to the vapor delivery system of liquid TEOS sources.

In HDP-CVD, the deposition rate is controlled by silane flow, and the refractive index is determined by the SiH_4/O_2 flow ratio. Film stress is mainly controlled by the bias rf power, source rf, and helium backside pressure, which controls the wafer temperature. The pressure and source rf power can also affect film uniformity.

10.8 Dielectric Chemical Vapor Deposition Chamber Cleaning

During the dielectric CVD process, a dielectric thin film can be deposited not only on the wafer surface but also on everything inside the chamber, particularly the wafer chuck, showerhead, and chamber wall. It is very important to clean the chamber routinely to prevent the thin film on those surfaces from cracking and causing particulate contamination. For dielectric CVD tools, it is not unusual for chambers to be undergoing cleaning processes more than half of the time, rather than running deposition processes!

For PECVD tools, rf plasma cleaning is a popular way to clean the chamber, and remote plasma cleaning processes are receiving more attention.

10.8.1 Radio-frequency plasma cleaning

One of the advantages of dielectric PECVD over LPCVD and APCVD is that it can use plasma-generated free fluorine radicals for chamber dry cleaning. A plasma cleaning process is a plasma etch process in which both chemical and physical processes help remove dielectric film on the processing kits and chamber walls. Both silicon oxide and silicon nitride processes use carbon fluoride compounds such as CF_4, C_2F_6, and C_3F_8 as a fluorine precursor gas because they are stable and easy to handle. In some cases, NF_3 is used to generate more free fluorine radicals.

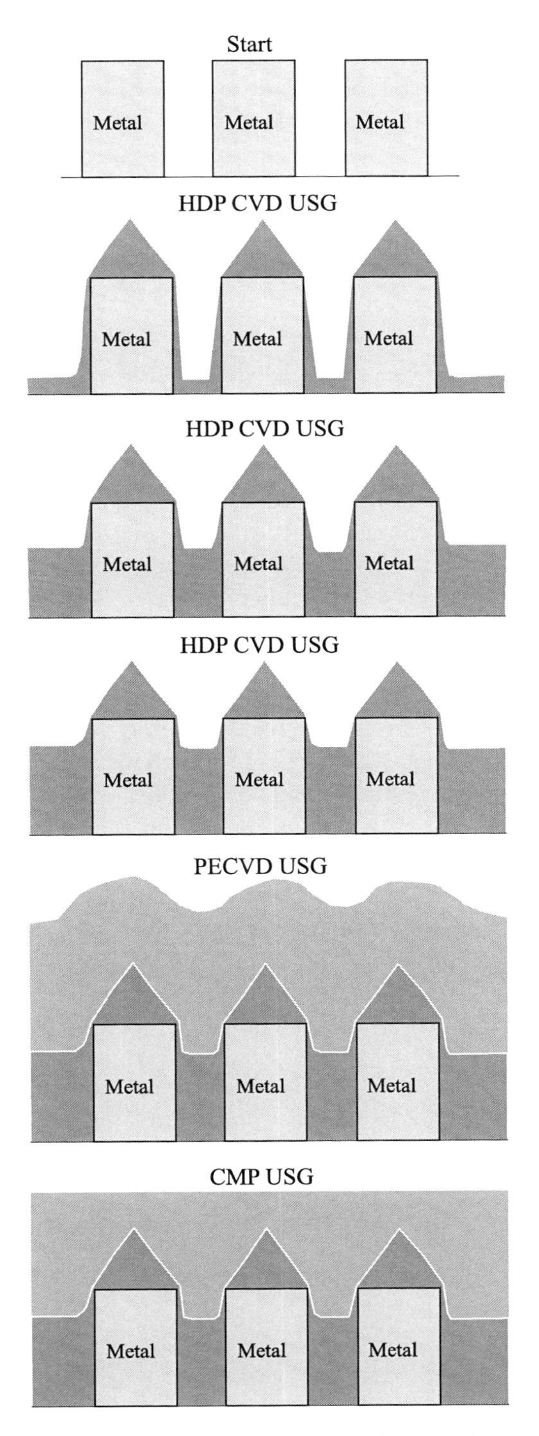

Figure 10.59 IMD HDP-CVD gap fill and planarization processes.

In plasma, carbon fluoride is dissociated to release the free fluorine needed for removing silicon oxide and silicon nitride:

$$CF_4 \xrightarrow{\text{plasma}} CF_3 + F,$$

$$F + SiO_2 \xrightarrow[\text{heat}]{\text{plasma}} SiF_4 + O,$$

$$F + Si_3N_4 \xrightarrow[\text{heat}]{\text{plasma}} SiF_4 + N.$$

In plasma cleaning processes, oxygen source gases such as N_2O and O_2 are used with carbon fluoride to react with carbon to form CO and CO_2, freeing more fluorine radicals and increasing the F/C ratio. This can prevent carbon fluoride polymerization, and increases cleaning efficiency.

It is very important to have a F/C ratio above 2, otherwise polymerization can occur. In plasma, CF_4 dissociates into CF_3 and F. CF_3 will continue to dissociate into CF_2 and F. The linking of CF_2 radicals into a long chain is called polymerization (as shown in Fig. 10.60). In this case, white Teflon®-like polymers are deposited inside the processing chamber, and it is necessary to open the chamber and do a wet clean to mechanically scrub the polymer off the chamber. This tool downtime can significantly affect throughput.

Chamber cleaning processes require six steps. Pressure stabilization, plasma cleaning, pump down, pressure set up for deposition, seasoning, and pump down. In the seasoning step, about 1000-Å oxide film is intentionally deposited on the wafer chuck and everywhere else inside the chamber. This is a very important step that can help keep the same deposition condition for each wafer, achieve good WTW uniformity, and improve production repeatability. It also can help reduce chemical contamination by covering fluorine residue, as well as reduce particulate contamination by sealing the loose fragments of the thin-film residue.

A sample recipe for 200-mm-wafer silane PECVD chamber cleaning is given in Table 10.10.

It takes about 90 sec to clean the chamber after 1-μm USG deposition. In the silicon oxide cleaning process, an etch chemical reaction releases oxygen from the

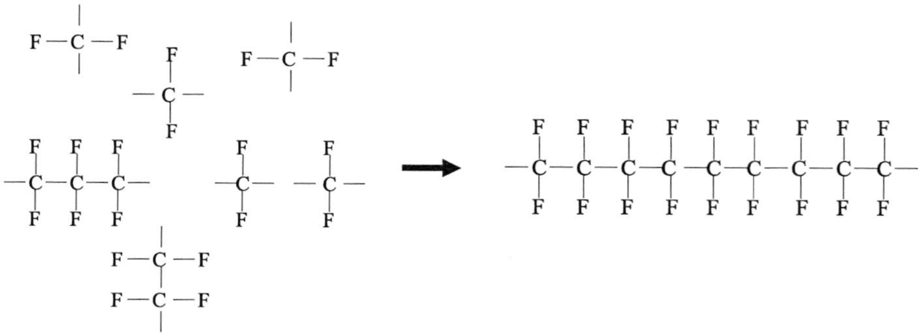

Figure 10.60 Carbon fluoride polymerization process.

Table 10.10 200-mm-wafer silane PECVD chamber cleaning recipe.

Step 2	Clean
Pressure	5 torr
rf	1000 W
CF_4	1200 sccm
N_2O	400 sccm

film and causes extra carbon oxidation. Since there is no extra oxygen released from the film, the silicon nitride cleaning needs a higher N_2O flow rate (about 600 sccm) to react with carbon, free more fluorine, and keep the right F/C ratio to prevent polymerization. It takes about 2 min to clean a nitride chamber after 1-μm nitride film deposition. By comparison, it takes about 1 min to deposit 1-μm oxide, and about 85 sec to deposit 1-μm nitride.

Question: For the 1200-sccm CF_4 flow into the processing chamber, what percentage is actually used for the oxide cleaning process?

Answer: Less than 3%! More than 97% of CF_4 simply flows into the chamber and is pumped away. CF_4 is a stable gas and cannot be removed by a scrubber; it is considered safe to release into the atmosphere. However, carbon fluoride gases are classified as global warming gases, and demands for limiting their use are increasing.

In the plasma, excitation-relaxation collisions between electrons, atoms, and molecules cause glow. Since different gases have different atomic structures, the colors of light emission in plasma can provide information about the chemical components inside the chamber and can be used to monitor the cleaning process.

When the cleaning process begins, the plasma generates free fluorine atoms, and initially the fluorine increases. During the main cleaning process, fluorine atoms are busy etching silicon oxide or silicon nitride, so the free fluorine concentration in the plasma remains low, and the intensity of the fluorine line (704 nm) also stays low. When approaching the end of the cleaning process, silicon oxide is gradually removed, and the free fluorine concentration in plasma increases; the fluorine line intensity also increases. When oxide is completely removed, the fluorine concentration becomes a constant, which indicates the endpoint, as shown in Fig. 10.61.

10.8.2 Remote plasma cleaning

Since the rf plasma cleaning process uses high rf power to generate plasma inside a processing chamber, ion bombardment associated with plasma causes processing kit damage and increases the cost of ownership for IC manufacturers. In rf plasma cleaning, most carbon fluoride gases are not dissociated and are released into the atmosphere. Carbon fluoride gases are considered global warming gases. Global warming is mainly caused by increased carbon dioxide in the atmosphere due to fossil fuel consumption, which always accompanies development and industrialization. Carbon dioxide and carbon fluoride gases absorb the inferred

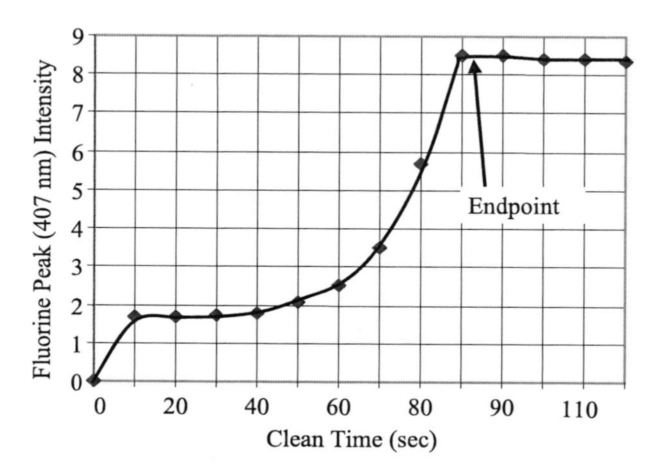

Figure 10.61 Endpoint trace of a cleaning for 1.8-μm PE-TEOS film (Applied Materials, Inc.).

radiation from sunlight more efficiently than nitrogen and oxygen. Increasing the amounts of carbon dioxide and carbon fluoride gases in the atmosphere can increase the heat absorption and raise the temperature of the earth.

Carbon fluoride gases are also believed to be responsible for holes in the ozone layer of the upper atmosphere. The ozone layer blocks UV radiation from solar rays. Areas under the holes have abnormally high UV radiation intensity, which can cause sunburn and skin cancer. Due to the long lifetimes of carbon fluoride gases in the atmosphere (half-life >10,000 years), they can cause long-term environmental damage. Therefore, there are strong demands to reduce the release and use of carbon fluoride by the semiconductor industry. Remote plasma cleaning fills this demand.

By using MW power, which has a much higher frequency than rf, a stable plasma can be generated at high pressure. Normally, NF_3 is used as a fluorine source gas. In high-density MW plasma, more than 99% of NF_3 is dissociated and releases three free fluorine atoms. The free fluorine atoms then flow into the CVD processing chamber and react with the deposited film on the processing kits and chamber wall to form gaseous SiF_4 and other gaseous byproducts, and eventually are pumped out of the chamber. Since there is no plasma inside the processing chamber, there is no ion bombardment on the chamber body or parts inside the processing chamber, prolonging their lifetime. Figure 10.62 illustrates a chamber cleaning process with a MW remote plasma source.

The advantages of MW remote plasma cleaning are longer lifetimes of the processing chamber parts, lower cost of ownership, and much less fluoride gas release. Disadvantages are a less-mature technology compared to the production-proven rf plasma cleaning process, higher cost for extra equipment, and the use of hazardous, expensive NF_3 gas. Another disadvantage is that the existing optical endpoint system for rf plasma cleaning cannot be used for remote plasma cleaning because there is no plasma-induced light emission in the processing chamber.

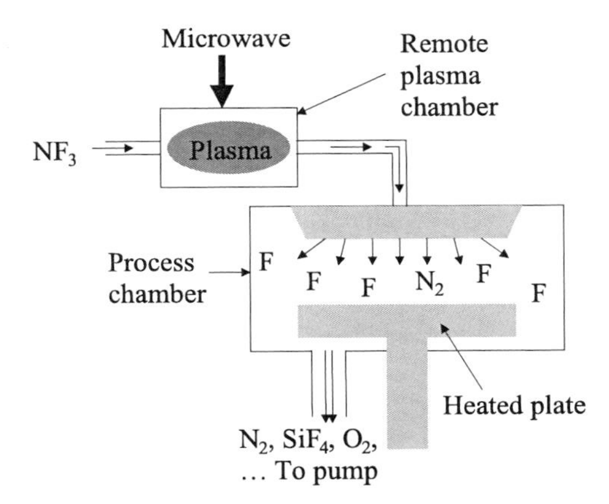

Figure 10.62 Remote plasma CVD chamber cleaning process.

Remote plasma cleaning requires different types of measurement methods, such as Fourier transform infrared (FTIR) spectroscopy, to achieve an automatic processing endpoint. FTIR systems measure chemical bonding concentrations to detect chemical component changes and determine the cleaning endpoint.

10.9 Process Trends and Troubleshooting

10.9.1 Silane plasma-enhanced chemical vapor deposition processing trends

Many silane-based thin-film processes are PECVD processes. For silane processes, due to the higher diffusion rate of the precursor in the boundary layer, increasing the process temperature always increases the deposition rate [see Fig. 10.63(a)]. Increasing temperature will also improve deposited film step coverage and quality.

For PECVD films, increasing rf power always increases compressive stress [see Fig. 10.63(b)] because increasing rf power will increase both ion density and energy of ion bombardment. Increasing rf power increases the deposition rate when rf power is low and reduces the deposition rate when rf power is high. At very low rf power, the deposition rate can be very low at relatively low temperatures (~400 °C) because there are not enough free radicals to enhance the chemical reaction rate. When rf power increases, the deposition rate increases dramatically due to the enhancement of the surface chemical reaction from the free radicals generated by the plasma. When rf power further increases, deposition will be well into the mass-transport-limited regime, and the deposition rate will no longer be related to the chemical reaction rate on the surface. Deposition rate stops increasing and starts to decrease because the heavier ion bombardment associated with increasing rf power reduces the precursor adsorption rate on the surface.

From Fig. 10.64, it can be seen why most PECVD processes operate at the high-rf-power side of the curve because in that regime, deposition rate is not very sensitive to the rf power. Therefore, rf power can be used to control the film stress and WERR without having a major effect on deposition rate. Deposition rate is too sensitive to the rf power if the processes operate on a low-rf power regime.

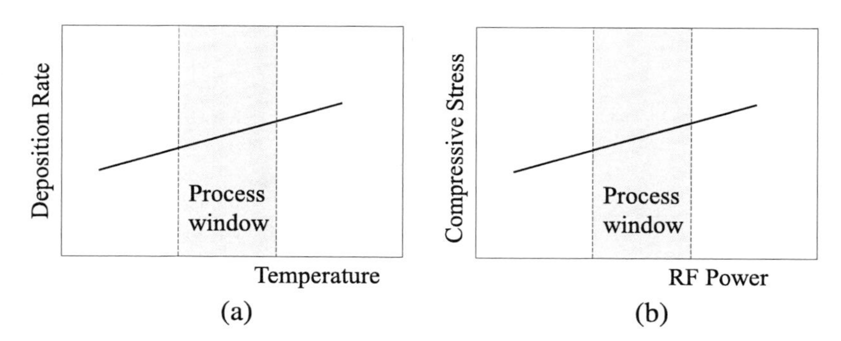

Figure 10.63 Silane PECVD processing trends: (a) temperature-to-deposition rate, and (b) rf power to film stress.

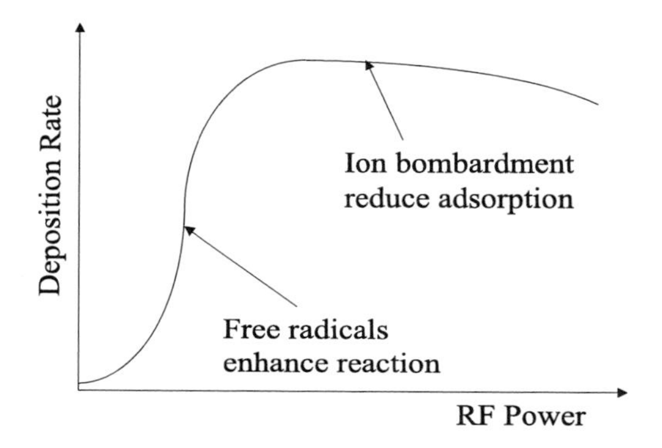

Figure 10.64 Relationship of deposition rate to rf power.

For the silane process, refractive index is mainly determined by the chemical composition of the film. Silicon-rich film has a higher refractive index, and oxygen-rich film has a lower refractive index, as shown in Fig. 10.65.

Since silane PECVD processes operate in the mass-transport-limited regime, the deposition rate is mainly determined by the gas flow rate, especially the silane flow rate. By increasing silane flow alone, the deposition rate increases while film compressive stress is reduced due to less ion bombardment per unit of film thickness. Refractive index increases because the film becomes more silicon rich; the WERR increases as well because the film is less dense due to less ion bombardment per unit thickness film.

10.9.2 Plasma-enhanced tetraethylorthosilicate trends

There are some similarities in processing trends between PE-TEOS and PE-silane processes. When increasing rf power from a very low level, the deposition rate increases very rapidly at first, then slows down and eventually starts to decrease when rf power is high enough (shown in Fig. 10.64). At the same time, film stress becomes more compressive due to increasing ion bombardment.

Similar to the PE-silane process, the PE-TEOS process also operates in the mass-transport-limited regime, and deposition rate is mainly controlled by TEOS

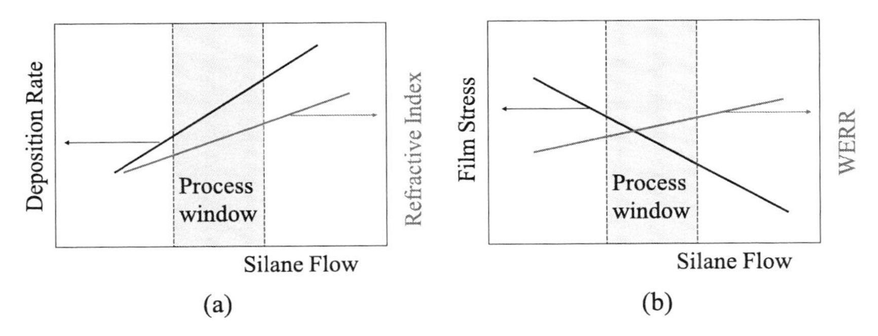

Figure 10.65 PECVD silane processing trends: (a) silane flow rate versus deposition rate and refractive index, and (b) silane flow rate versus compressive stress and WERR.

flow rate. Since TEOS vapor always flows into the processing chamber with a carrier gas such as helium, it is important to have a precise and independent control of the net TEOS flow rate. By increasing TEOS flow rate alone, the deposition rate will increase linearly, stress will become less compressive, and the wet etch rate will increase, as shown in Fig. 10.66.

Question: For silane processes, increasing silane flow rate increases the refractive index significantly because the deposited film becomes richer in silicon. Why does the refractive index almost not change when the TEOS flow rate increases, as shown in Fig. 10.66?

Answer: In a TEOS molecule, the silicon atom is already bonded with four oxygen atoms. Therefore, increasing TEOS flow does not make the deposited film silicon rich. For TEOS oxide film, the refractive index is very stable and only changes slightly due to film density change. Denser film has a slightly higher refractive index.

Different from the silane process, increasing temperature for a TEOS process at 400 °C always reduces the deposition rate, no matter whether it is a PE-TEOS or an O_3-TEOS process. The difference in this trend is caused by the different adsorption mechanisms of TEOS and silane. Silane uses chemisorption, and TEOS uses physisorption. For a chemisorption precursor such as silane, the bonding energy between the precursor and surface atom is very strong, and a temperature change near 400 °C has no significant effect on the adsorption rate. For a physisorption precursor like TEOS, the bonding energy is low, and varying the temperature near 400 °C has some strong effects on the adsorption rate. At higher temperatures, more physisorbed precursors can receive enough energy to break the weak bonds and leave the surface, thereby decreasing the adsorption rate and causing lower deposition rates, as shown in Fig. 10.67.

10.9.3 Ozone-tetraethoxysilane trends

O_3-TEOS film always has tensile stress, and there is no way to control the film stress, since its deposition is a thermal process. The O_3-TEOS film refractive index is about 1.44, due to the porosity of the film.

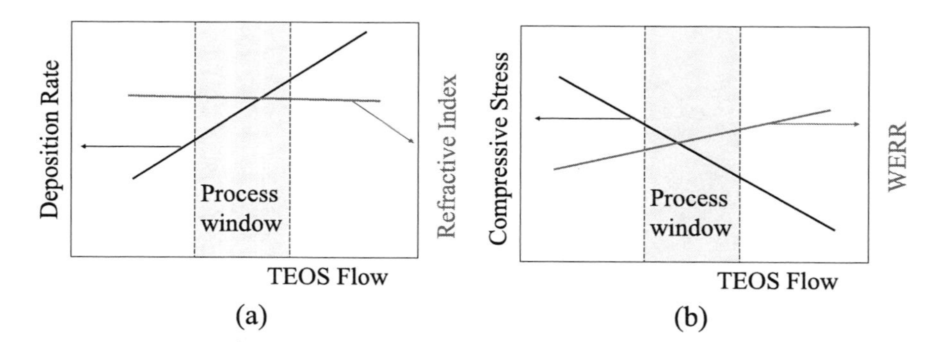

Figure 10.66 PE-TEOS processing trends: (a) TEOS flow rate versus deposition rate and refractive index, and (b) TEOS flow rate versus compressive stress and WERR.

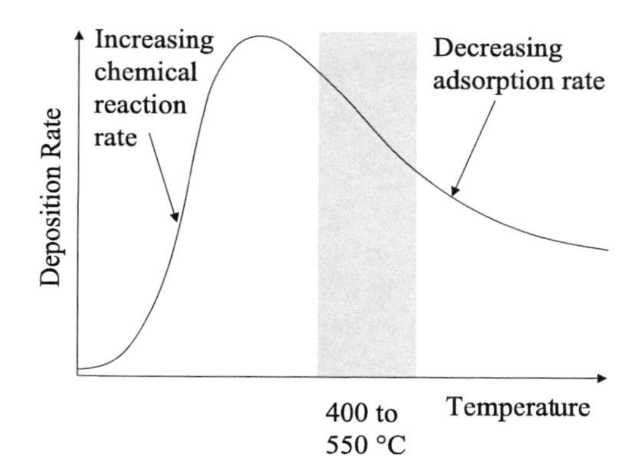

Figure 10.67 TEOS processing trend: deposition rate versus temperature.

In the O_3-TEOS deposition process, a higher O_3:TEOS ratio is always desired, since a higher ratio means more free oxygen to react and oxidize the TEOS; the higher the O_3:TEOS ratio, the better the oxide film quality. O_3-TEOS USG film quality can be monitored by measuring the WERR related to the thermally grown SiO_2, a measurement that normally is around 4 to 5. Film quality can also be monitored by the shrinkage measurement. O_3-TEOS film shrinkage is normally about 4 to 5% for the 450 °C thermal cycle. Shrinkage can reach 7% for >1000 °C STI annealing. Increasing the O_3:TEOS ratio can reduce both WERR and film shrinkage. One way to increase the O_3:TEOS ratio is to increase both ozone concentration and ozone flow rate, which is limited by the ozonator capability; another way is to reduce TEOS flow rate, causing a lower deposition rate.

At high pressure, the O_3-TEOS process has better step coverage and gap-fill abilities. This is because higher pressure reduces the MFP, allowing TEOS molecules to have more collisions in the boundary layer, causing them to hop along the substrate surface more frequently to eliminate the arriving angle effect.

Increasing the deposition temperature for the O_3-TEOS process increases chemical reaction rates and improves film quality. Increasing deposition temperature can also increase surface mobility, thus improving film step coverage and gap-fill abilities. Since a TEOS molecule is physisorbed on the substrate surface, increasing temperature reduces the adsorption rate and increases desorption rate, reducing deposition rate. The relation of temperature and deposition rate is illustrated in Fig. 10.63(a).

Question: For both PE- and O_3-TEOS processes, the maximum deposition rate can be achieved at about 250 °C. Why do ILD TEOS processes normally operate about 400 °C, and ILD0 and STI processes deposit at even higher temperatures (~550 °C)?

Answer: At higher deposition temperatures, higher film quality and better step coverage can be achieved, as well as void-free gap fill. In these cases, the concern of yield outweighs the concern of throughput.

10.9.4 Troubleshooting guide

It is impossible to learn troubleshooting skills solely from classroom training and reading books. Troubleshooting skills must be learned by hands-on experience. However, a troubleshooting guide that outlines previous fab experiences can be very helpful to engineers and technicians on the job, and can also benefit those who plan on working in the semiconductor industry in the future.

Processing problems can happen when something in the process goes wrong either suddenly or gradually; often a process can fail, then return to normal, then gradually go wrong again. In those cases, an investigation should be performed to find out what has been changed between the good process and the bad process, or vice versa. For instance, a PE-TEOS process operated normally for a while, then suddenly the deposition rate seemed decreased because measurement results showed the film thickness to be thinner than the normal value. In this case, the metrology tool should be checked first to make sure that the system constants used for measurements are the correct ones. If someone measured the nitride film thickness with that tool and did not change the refractive index back to PE-TEOS oxide, the measured oxide film thickness would be significantly thinner than it actually is, since refractive index and film thickness are always coupled together. If nothing is wrong with the metrology tool, then check whether the process recipe has been changed. Inappropriate changes in processing recipes, such as shorter deposition time, lower TEOS flow rate, and higher wafer temperature, will cause the same kind of problem in thickness measurement.

If a process always has problems at the end of each shift, someone should work cross-shifts to find out what is being changed during the shift change. Some part of the process must be different between the shifts; that change puts the process back to normal at the start of the shift. Most likely that difference is also the source of the problem, as the process gradually goes wrong at end of the shift.

If a process always has problems at the beginning of each shift and gradually returns back to normal after a few wafer depositions, check whether the tool idle time is too long. Long idle times can cause the first-wafer effect, especially for

TEOS processes. The first-wafer effect is mainly due to the cooler processing kit temperature, which causes higher viscosity of the TEOS vapor flow and lower deposition rates. Thus, running a seasoning deposition and plasma cleaning cycle process during wafer loading and transfer can heat up the processing kits and effectively reduce the first-wafer effect for the TEOS process.

Since most dielectric CVD processes operate in the mass-transport-limited regime, deposition rate is mainly determined by gas flow rate, especially the flow rate of silicon source gases such as silane and TEOS. Therefore, it is likely that the deposition rate problems are related with silane or TEOS flow.

For PECVD film, stress is mainly determined by ion bombardment. Therefore, if the film stress is not of the desired value ($\sim$100 MPa = 10^9 dyne/cm^2, compressive), anything related to ion bombardment should be on the checklist, most notably rf power system and wafer grounding. Gas flow rate and pressure also can affect film stress.

Refractive indices for PE-TEOS and O_3-TEOS processes are very stable, 1.46 and 1.44, respectively. For silane processes, the refractive index can vary in a wide range, depending mainly on the chemical composition of the film. If the refractive index is out of specification range, the gas flow ratio (such as SiH_4/N_2O or SiH_4/NH_3) should be checked first. For the nitride process, a chamber leak can cause low refractive index.

Uniformity is determined by flow pattern. If a nonuniformity pattern is thin or thick at the center, it can by adjusted by varying the spacing of the wafer and showerhead, or by changing the carrier gas flow rate, such as helium flow for TEOS processes and nitrogen flow for nitride process. The relationship between spacing and profile is illustrated in Fig. 10.68. If the nonuniformity is side to side, it could be a wafer leveling or centering problem. Valve leaks can also cause side-to-side nonuniformity. Tests with different wafer orientations can help confirm or eliminate this possibility.

If the particle counts are gradually increasing and out of specification range, check whether the chamber needs a scheduled wet clean. For particle

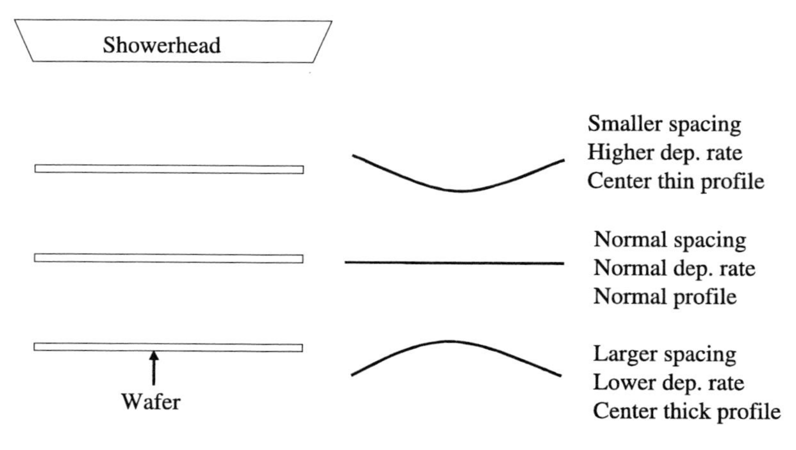

Figure 10.68 Wafer spacing and film thickness profiles.

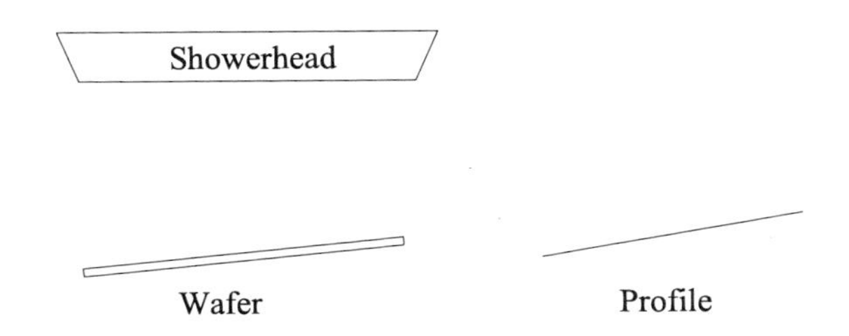

Figure 10.69 Wafer position and film thickness profile.

contamination problems, a partition method should be employed. At first, wafers should be run through the loading, transfer, and processing chamber one by one to check whether it is a mechanical particle problem. Then gases should be flowed into each processing chamber individually to check whether there is a dusted gas line. After eliminating all possible mechanical particle sources, the deposition process can be run to see whether there are process-induced particles. Improper cleaning processes, incorrect pressure, rf power changes, or a leaking process chamber can all cause particle contamination during deposition processes. The possible sources of particles should be eliminated one by one.

Troubleshooting procedures are very tool dependent, since different tools run different processes. The equipment providers normally have hardware and process troubleshooting guides for to each type of tool.

10.10 Recent Developments

In the first decade of the twenty-first century, IC device development placed new demands on dielectric thin films. The isolation trench became narrower, and aspect ratio became so high (up to 8) that it was no longer shallow. HDP-CVD oxide, O_3-TEOS oxide, flowable CVD, and spin-on oxide were used for void-free STI trench fill. Sidewall spacers are still mainly formed with the nitride/TEOS oxide combination.

As device dimensions continue to grow smaller, the thickness of gate oxides will also decrease. Eventually, the thickness will become too thin to work reliably, even with 1 V of power supply (see Example 3.6 in Chapter 3). Therefore, high-κ dielectric materials have been developed for gate dielectrics starting with the 45-nm technology node. Hafnium oxide (HfO_x) is the most commonly used high-κ gate dielectric, and atomic layer deposition (ALD) has been used to form it.

While gate dielectrics are requiring high-κ materials, the κ-value of ILD layers has gone in the other direction. To further reduce κ-value, PECVD porous low-κ dielectric deposition processes have been developed. The combination of low-κ dielectric and copper metallization reduces the RC delay to meet the demands of increasing IC speed.

The most significant difference between the dual damascene Cu process and the Al–Cu process (Fig. 10.70) is that the dual damascene process does not need to etch metal. Because copper is very difficult to dry etch, dual damascene has become the process of choice for copper metallization. In advanced CMOS ICs,

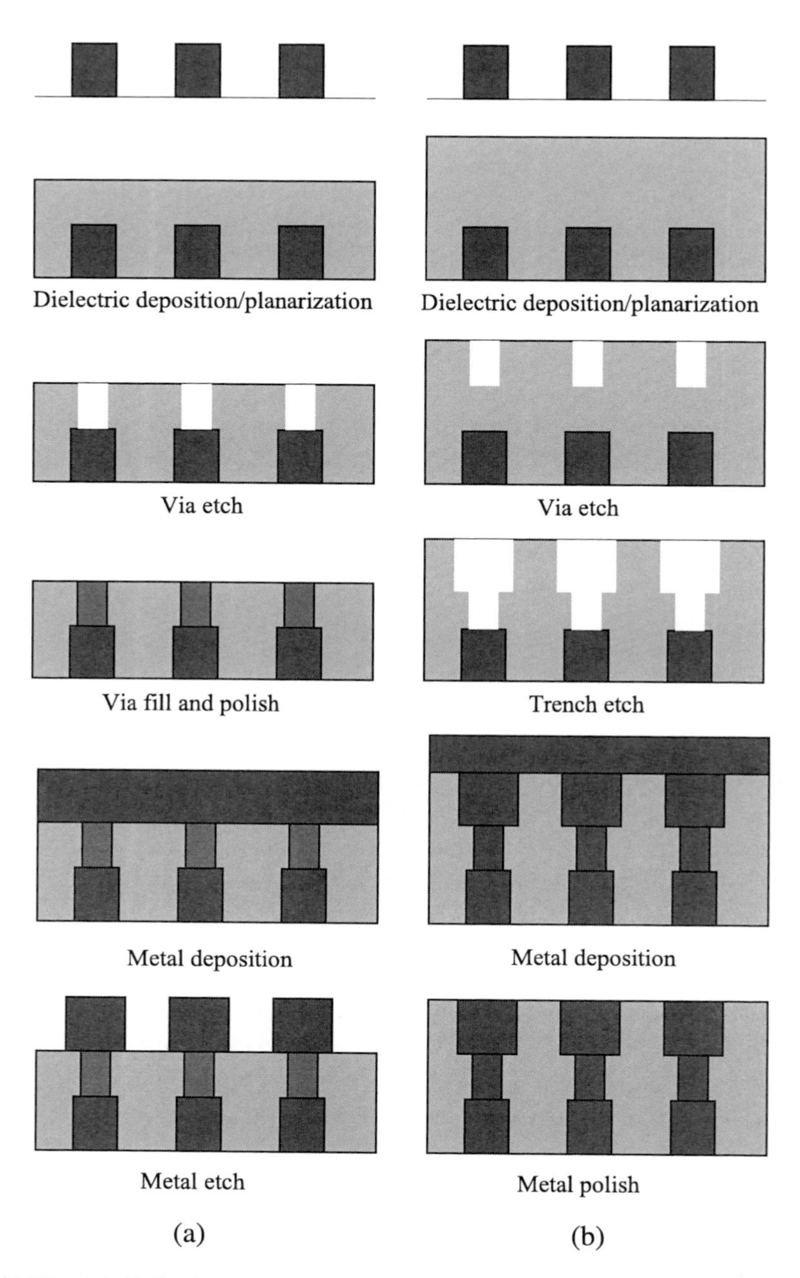

Figure 10.70 (a) Al–Cu interconnection and (b) dual damascene copper interconnection processes.

copper interconnects have already replaced Al–Cu interconnects and become mainstream. Even in memory chips, such as DRAM and NAND flash, copper interconnects are being applied in high-speed products. For Cu interconnects, ILD and passivation dielectric thin-film depositions have become relatively simple, with no more gaps to fill and no more planarization issues. Requirements for those thin-film depositions are high deposition rate, good uniformity, stable low-κ value, and capability of process integration. Of course, for STI and ILD0, there are still the great challenges of filling even narrower gaps and planarizing dielectric layer surfaces.

10.10.1 Low-κ dielectrics

3MS and other silicon-containing organic compounds are used as organosilicate precursors in PECVD processes to deposit low-κ (~2.9 to 2.7) SiCOH film for ILD applications. To further reduce RC delay and improve circuit clock speed, ULK dielectrics based on porous SiCOH (κ = 2.5 to 2.2) have been developed. Using precursors such as DEMS (diethoxymethylsilane) for the base film and a porogen such as CHO (cyclohexene oxide or $C_6H_{10}O$), composite OSG-organic film can be deposited with organic (C_xH_y) pockets. With UV and visible light post-treatment, which outgases the organic and transforms the organic pockets into air pockets, porous ULK dielectric film can be formed for ILD applications in cutting-edge IC chips, as shown in Fig. 10.71. Pore size is approximately a few nanometers.

Dielectric barrier materials with lower κ-values have also been developed for copper/low-κ interconnect applications to replace silicon nitride barrier materials, which have a dielectric constant of κ = 7. Silicon carbon nitride (SiCN) with a κ-value of approximately 4.8 has been developed and used for this application. Amorphous silicon carbide (α-SiC:H) with a κ-value of 2 to 3.6 has also gained much attention as a strong candidate for this application. The α-SiC:H can be deposited with the PECVD process, with tetramethylsilane [4MS or $Si(CH_3)_4$] or trimethylsilane [3MS or $SiH(CH_3)_3$] as a precursor. It can help to further reduce the overall dielectric constant of the ILD layer and improve IC speed.

10.10.2 Air gap

The lowest dielectric constant is 1.0, which can only be achieved in vacuum. Gases usually have dielectric constants just slightly higher than 1. At an atmospheric pressure of 1, the dielectric constant of air is 1.00059, which is the lowest κ-value achievable in ILD for metal interconnection of an IC chip.

There are two approaches to forming air gaps, one is void formation with nonconformal PECVD film, and the other uses sacrificial material between metal wires. Void formation is discussed in earlier sections of this chapter.

Figure 10.72 is an example of void-formed air gaps in a 25-nm NAND flash device. Figure 10.72(a) shows the air gaps between WLs of the 25-nm NAND flash, and Fig. 10.72(b) shows the air gaps between BLs of the same device. By using air gaps, the parasitic capacitance between WLs and BLs have been reduced by 25 and 30%, respectively.

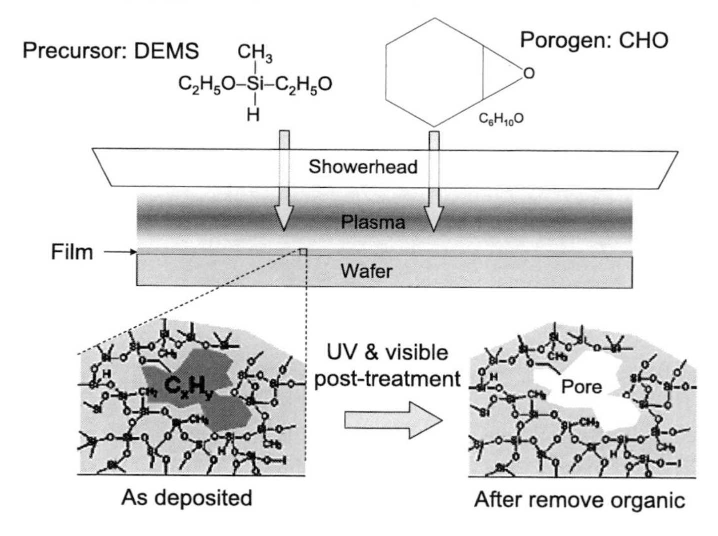

Figure 10.71 Deposition and post-treatment for forming porous low-κ dielectric film.

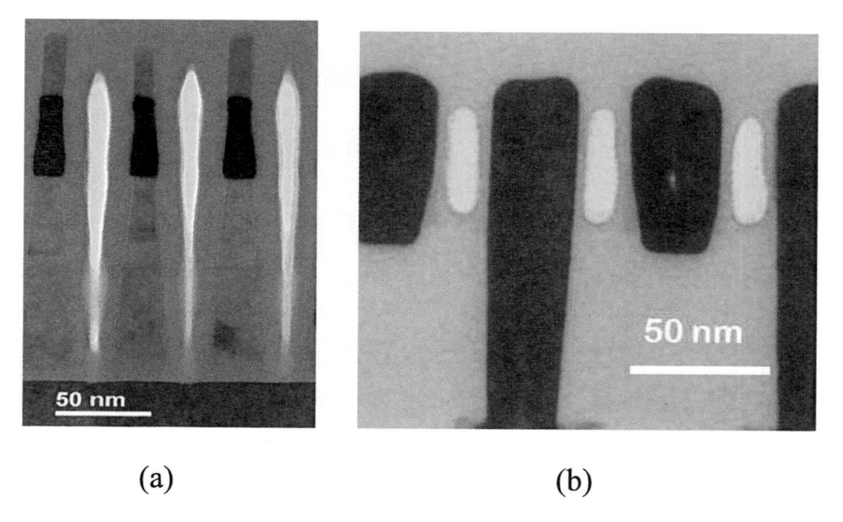

Figure 10.72 Cross-section TEM images of Intel/Micron 25-nm node NAND flash: (a) air gaps between WL and (b) BL (K. Prall and K. Parat, *IEDM*, 102–105, 2010).

The sacrificial method also has several approaches. Figure 10.73 shows one of them. It starts with an ILD stack of ESL/low-κ/ESL/TEOS. After copper CMP, TEOS can be stripped with BOE, as shown in Fig. 10.73(a). A protective liner is deposited, followed by organic coating, curing, and etchback [see Fig. 10.73(b)]. A porous cap layer is deposited, and organic outgassing with UV or visible radiation creates air gaps between the copper lines.

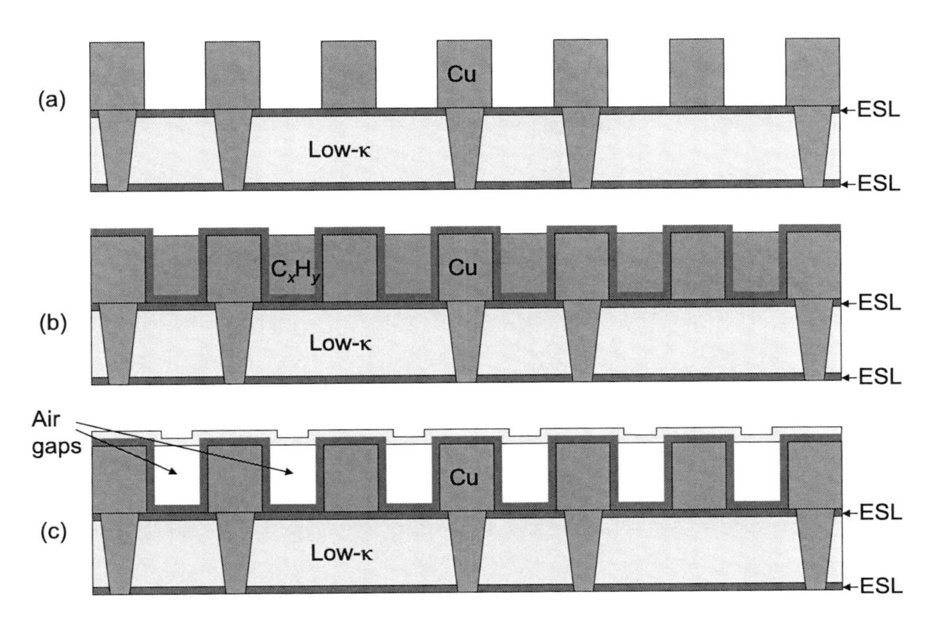

Figure 10.73 Air gap formation with sacrificial layers: (a) after oxide strip, (b) after organic etchback, and (c) after organic outgassing.

10.10.3 Atomic layer deposition

ALD is a deposition process that allows users to deposit a very thin film in the form of atomic layers in multiple deposition circles. ALD can be used to deposit compound semiconductors such as GaAs, InP, GaP, AlN, GaN, InN, etc. It can also be used to deposit high-κ dielectrics, such as Al_2O_3, TiO_2, ZrO_2, HfO_2, Ta_2O_5, La_2O_3, etc. Metallic nitrides, including those used in the metal gates of MOSFETs such as TiN, TaN, Ta_3N_5, NbN, and MoN, also can be deposited with the ALD process.

The ALD process is performed in a sealed reactor. The first processing gas flows into the chamber, and the precursor molecules adsorb on the substrate surface. The chamber is then purged, and the first processing gas is removed from the chamber, leaving behind only those molecules adsorbed on the wafer surface. The second processing gas then flows into the chamber to react with the first process molecules on the surface to form a molecular layer of compound material on the surface. After all of the first precursor molecules are consumed, the chemical reaction is self-terminated, and a purge process removes the second processing gas and byproducts of the chemical reaction from the chamber. The ALD cycle, illustrated (clockwise) in Fig. 10.74 (first precursor, purge, second precursor, and purge), deposits a molecular layer of the compound material onto the wafer surface. Multiple circles can be performed until the required compound film thickness is reached.

HfO_2 is one of the candidates for the high-κ gate dielectrics of MOSFETs. In this case, $HfCl_4$ can be the first precursor, H_2O can be the second precursor, and HCl is the byproduct of the chemical reaction that deposits the HfO_2 on the surface.

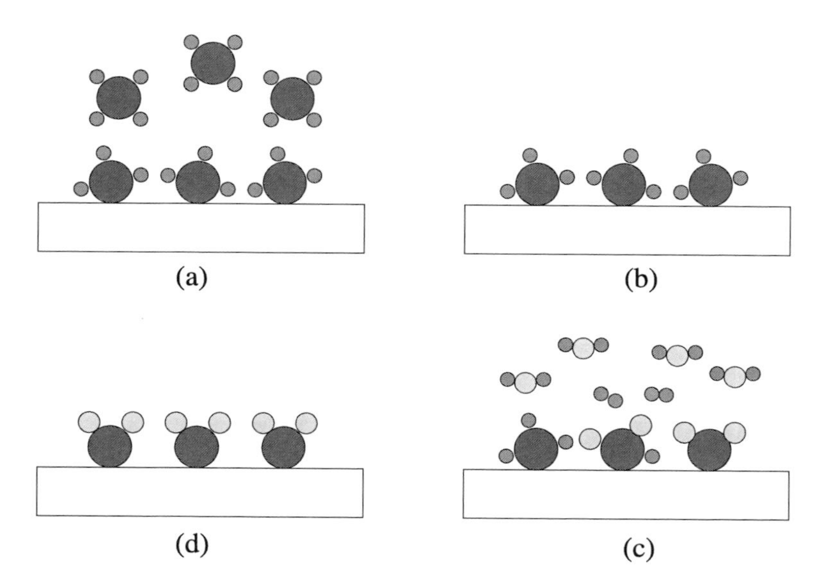

Figure 10.74 An ALD circle: (a) first precursor pulse, (b) purge, (c) second precursor pulse and reaction, and (d) purge.

The chemical reaction can be expressed as

$$HfCl_4 + 2H_2O \rightarrow HfO_2 + 4HCl.$$

The advantages of ALD processes are formation of stoichiometric films with large area uniformity and 3D conformality, and precise control of film thickness. It is possible for ALD to deposit a film at low temperatures, and the deposition process is gentle enough for sensitive substrates. A disadvantage of the ALD process is low throughput and low use of precursors.

Because of low deposition rates (typically less than 10 nm/min compared with ~800 nm/min of PECVD oxide), ALD processes are mainly applied to thin films with thicknesses less than 10 nm. ALD processes can be used to deposit high-κ gate dielectrics of MOSFETs, high-κ dielectrics of DRAM storage capacitors, metallic metal compounds of MOSFET metal gates, and some barrier layers of copper/low-κ interconnection.

The high-κ gate dielectric HfO_2 of a MOSFET can be deposited with hafnium tetrachloride ($HfCl_4$), TEMAHf $\{Hf[N(C_2H_5)(CH_3)]\}_4$, or tetrakisethylmethyl-aminohafnium as hafnium precursors, and O_3 or H_2O as the oxygen precursors. High-κ dielectric Al_2O_3 of DRAM storage node capacitors can be deposited with TMA [$Al(CH)_3$ or trimethylaluminium] as an aluminum precursor, and H_2O or O_3 as an oxygen precursor.

10.10.4 High-κ dielectrics

The basic components of a DRAM cell are a transistor and a capacitor. DRAM fabrication processing requires a dielectric material with an even higher dielectric constant for the capacitor to shrink. Al_2O_3 (κ = 11.5) and ZrO_2 (κ ~ 25) have been used in this application in advanced DRAM manufacturing. TiO_2 (κ up to 85), Ta_2O_5 (κ ~ 25), and $Ba_{0.6}Sr_{0.4}TiO_3$ or BST (κ up to 600) can also be employed for this application. ALD and CVD processes can be used to deposit these high-κ dielectrics for DRAM capacitor applications.

The thickness of gate oxide in 65-μm devices is about 15 Å; this is too thin to withstand any future scaling down without significantly increasing gate leakage due the quantum tunneling effect. By switching to materials with dielectric constants significantly larger than the 3.9 of silicon dioxide, gate dielectric thickness can be increased, while the effective oxide thickness (EOT) can be further reduced without excessive gate leakage. Gate dielectrics of MOSFETs made a dramatic change starting at the 45-nm technology node when Intel introduced high-κ dielectric gate oxides and metal gate electrodes in 2007. The industry grew away from thermally grown silicon dioxide, which had been used for about 50 years as a gate dielectric, and adapted hafnium-based high-κ gate dielectrics. ALD processes are used to deposit HfO_2 for this application.

To conduct electric current between the S/D, a conductive channel must be formed underneath the gate by reversion of the minority carrier there. Minority carrier reversion can be achieved by charging a metal–oxide interface with the applied gate voltage. Therefore, capacitance of an MOS gate capacitor must be large enough to hold the charges. When feature size reduces, the area of the capacitor electrode reduces quickly, as does the gate capacitance. To retain enough capacitance, the distance between the two electrodes (which is the thickness of the gate dioxide layer) must be reduced. For 0.18-μm IC chips, the gate oxide is about 35 Å. It will be approximately 25 Å for 0.13-μm minimum-feature-size IC chips and about 15 Å for 90-nm devices. A problem will occur when the gate oxide thickness further reduces, as gate-to-substrate leakage will significantly increase, and IC chip reliability will be affected. Nitridated silicon oxide with κ ~ 5 has been used in 90-, 65-, 45-, and even 32/28-nm devices. High-κ dielectrics such as hafnium dioxide (HfO_2, κ ~ 25) have been developed and implemented in HKMG CMOS manufacturing, starting from the 45-nm technology node. With HfO_2 as the gate dielectric, gate dielectric thickness can be increased significantly, and gate leakage can be reduced exponentially. Gate capacitance is determined by EOT, which can be expressed as:

$$\mathrm{EOT} = \frac{\kappa_{SiO_2}}{\kappa_x} t_{ox}.$$

Here, $\kappa_{SiO_2} = 3.9$ is the dielectric constant of silicon dioxide, which is used as the standard κ-value in the IC industry to determine high- or low-κ of dielectric materials. X can be HfO_2 or another of the high-κ dielectrics listed in Table 10.11. Gate leakage is determined by the gate dielectric thickness, t_{ox}. The high-κ gate dielectric allows fabs to use thicker gate dielectric layers to reduce gate leakage while still keeping the gate capacitance large enough for device performance. It allows fabs to continue traditional scaling, which requires reducing EOT.

Many other high-κ dielectrics such as titanium dioxide (TiO_2, κ ~ 80), tantalum pentoxide (Ta_2O_5, κ ~ 26), and zirconium dioxide (ZrO_2, κ ~ 25) have also been researched and could be used in IC chip manufacturing in the future. Figure 10.75 shows suitable elements with oxides that could be used for gate dielectrics of silicon MOSFETs. Table 10.11 lists some dielectrics and their κ-values.

10.11 Summary

- Applications of dielectric thin film are STI, high-κ gate dielectrics, sidewall spacers, ILD, and passivation dielectrics; ILD applications are most prevalent.

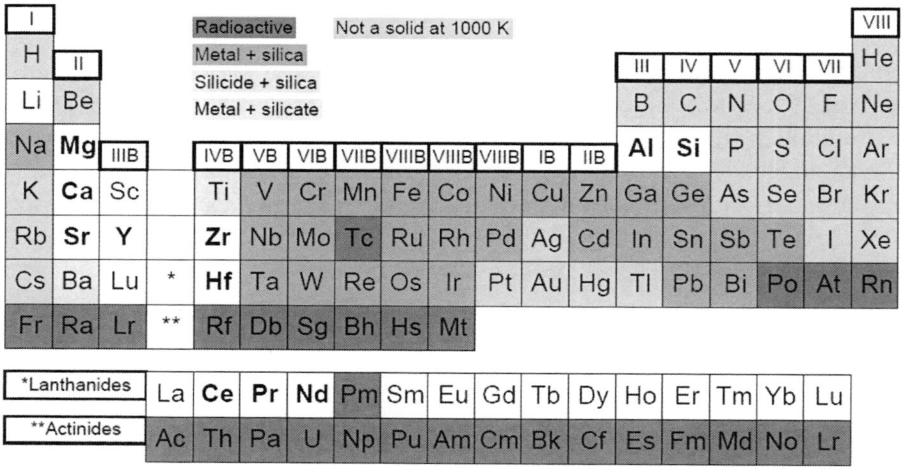

Figure 10.75 Elements in the table with bold letters have oxides that are suitable for gate dielectric applications of silicon MOSFETs. Elements in a shaded box are not suitable due to radioactivity or instability of their oxides on silicon substrates (Osten et al.).

Table 10.11 Dielectric materials and their κ-values.

Material	*k*
SiO_2	3.9
Si_3N_4	7.8
Y_2O_3	15
La_2O_3	30
Pr_2O_3	31
Dy_2O_3	14
Ta_2O_5	26
TiO_2	80
HfO_2	25
ZrO_2	25
Gd_2O_3	12–13
CeO_2	26
Sc_2O_3	>10
BST	>200

- Silicon oxide and silicon nitride are the two most commonly used dielectric materials in ILD applications.
- Low-κ and porous ULK dielectrics are widely used in advanced Cu interconnections.
- Basic CVD processing sequence is as follows: introduction of precursor, precursor diffusion and adsorption, chemical reaction, gaseous byproducts desorption, and diffusion.
- There are two deposition regimes: surface-reaction-limited and mass-transport-limited. Most dielectric CVD reactors are designed to operate in the mass-transport-limited regime because it is less sensitive to temperature variations.
- Film step coverage and conformality are determined by the arriving angle and surface mobility of precursors.
- ILD0 uses PSG or BPSG; deposition and reflow temperatures are determined by the thermal budget.
- ILD mainly uses USG, FSG, SiCOH, and porous SiCOH, depending on the process technology nodes. Deposition temperature is limited by the presence of metal interconnections.
- A passivation dielectric is mainly formed with silicon oxide/nitride stack films.
- Silane and TEOS are the two silicon sources most commonly used for dielectric CVD processes.
- 3MS is widely used for low-κ OSG deposition.
- O_2, N_2O, and O_3 are the most frequently used oxygen precursors for silicon oxide depositions.
- NH_3 and N_2 are the most common nitrogen source gases for nitride depositions.
- Fluorine chemistry is used for dielectric CVD chamber dry cleaning. CF_4, C_2F_6, C_3F_8, and NF_3 are commonly used fluorine source gases.
- The argon sputtering etch process is widely used along with the dielectric CVD process for gap-fill applications. The CF_4/O_2 etchback process is used for dielectric layer planarization.
- Both ellipsometry and prism couplers can be used for measurement of a film's refractive index, which can provide information about the film's chemical composition and physical condition.
- Both ellipsometry and spectroreflectometry can be used for dielectric thin-film thickness measurement. A correct refractive index must be used for an accurate thickness measurement.
- Compressive stress (~100 MPa) is favored for dielectric thin films.
- Most dielectric CVD processes are operated in the mass-transport-limited regime, and the deposition rate is mainly controlled by the silicon source gas flow rate.
- Radio frequency power can be used to control film stress and WERR for PECVD dielectric thin films. The higher the rf power, the more compressive the film stress, and the lower the WERR.
- Film refractive indexes are mainly controlled by the flow rate ratio for silane processes and are determined by the film density for TEOS processes.

- HDP-CVD processes uses silane and oxygen to deposit oxide, and use argon for in-situ sputtering to achieve high-aspect-ratio gap fill.
- ICP and ECR are the two most commonly used HDP sources in the semiconductor industry.
- HfO_2 has been used as high-κ gate dielectric for advanced HKMG MOSFETs.
- High-κ gate dielectrics and metal gate electrodes of HKMG MOSFETs usually are deposited by ALD processes.
- ALD has excellent film uniformity and step coverage. It can control film composition and film thickness very well. However, its deposition rate is very low, thus it is limited to ultrathin-film applications.

10.12 Bibliography

A. C. Adams and C. D. Capio, "Planarization of phosphorus-doped silicon dioxide," *J. Electrochem. Soc.* **128**, pp. 423–429 (1981).

C. Bencher, C. Ngai, B. Roman, S. Lian, and T. Vuong, "Dielectric antireflective coatings for DUV lithography," *Solid State Technol.* **40**(3), pp. 109–114 (1997).

K. Cherkaoui, M. Modreanu, A. Negara, P. Hurley, A. Groenland, and S. H. Lo, "High dielectric constant (high-κ) materials for future CMOS processes," *Electron Dev. Lett.* **18**, 209 (1997).

C. C. Chiang, M. C. Chen, C. C. Ko, Z. C. Wu, S. M. Jang, and M. S. Liang, "Physical and barrier properties of plasma-enhanced chemical vapor deposition α-SiC:H films from trimethylsilane and tetramethylsilane," *Japan. J. Appl. Phys.* **42**, pp. 4273–4277 (2003).

"Dielectric PE- and SA-CVD Processes," training manual of Applied Materials, Inc. Santa Clara, CA (1997).

L. Favennec, V. Jousseaume, V. Rouessac, J. Durand and G. Passemard, "Ultra low κ PECVD porogen approach: matrix precursors comparison and porogen removal treatment study," *Mater. Res. Soc. Symp. Proc.* **863**, B3.2.1 (2005).

H. D. B. Gottlob, T. Echtermeyer, M. Schmidt, T. Mollenhauer, J. K. Efavi, T. Wahlbrink, M. C. Lemme, M. Czernohorsky, E. Bugiel, A. Fissel, H. J. Osten, and H. Kurz, "0.86-nm CET gate stacks with epitaxial Gd2O3 high-κ dielectrics and FUSI NiSi metal electrodes," *IEEE Electron Dev. Lett.* **27**(10), pp. 814–816 (2006).

E. Korczynski, "Low-k dielectric costs for dual-damascene integration," *Solid State Technol.* **42**(5), pp. 43–51 (1999).

H. J. Osten, J. P. Liu, P. Gaworzewski, E. Bugiel, P. Zaumseil, "High-κ gate dielectrics with ultra-low leakage current based on praseodymium oxide," *IEDM Tech. Dig.*, pp. 653–656 (2000).

H. J. Osten, J. Dabrowski, H. J. Müssig, A. Fissel, V. Zavodinsky, "High-κ dielectrics: the example of Pr_2O_3," in *Challenges in Process Simulation*, J. Dabrowski and E. R. Weber, Eds., pp. 259–293, Springer Verlag, Berlin (2004).

J. E. J. Schmitz, *Chemical Vapor Deposition of Tungsten and Tungsten Silicides For VLSI/ULSI Applications*, Noyes Publications, Park Ridge, NJ (1992).

S. Sivaram, *Chemical Vapor Deposition Thermal and Plasma Deposition of Electronic Materials*, Van Nostrand Reinhold, New York (1995).

S. M. Sze, *VLSI Technology*, 2nd ed., McCraw-Hill, New York (1988).

D. Wang, S. M. Chandrashekar, S. P. Beaudoin and T. S. Cale, "Nonuniformity in CMP Processes Due to Stress," 2nd Intl. Conf. Chem.-Mech. Polish Plan. ULSI Multi. Intercon., Santa Clara, CA (1997).

D. L. W. Yen and G. K. Rao, "SOG without etchback," *Proc. VMIC*, pp. 85 (1988).

See http://www.webelements.com for further information on the elements in this chapter.

10.13 Review Questions

1. Describe the CVD process sequence. What are the major differences between CVD and PVD?
2. List at least three CVD dielectric thin-film layers in an IC chip.
3. What is the basic difference between a thermally grown oxide and a CVD oxide?
4. Of APCVD, LPCVD, PECVD, and HDP-CVD processes, which one has the highest processing pressure? Which has the lowest?
5. Describe the three deposition regimes at different temperatures. What is the relationship between deposition rate and temperature in each regime?
6. List the most commonly used silicon source gases for dielectric CVD.
7. List three commonly used oxygen precursor gases.
8. List three precursor gases used for silicon nitride deposition.
9. What type of chemistry is used for dielectric CVD chamber plasma cleaning?
10. Explain why compressive stress is favored for ILD films.
11. For PECVD processes, how does the dielectric film stress change when rf power increases?
12. List two reasons that PSG is used for ILD0. What is the reason for using BPSG?
13. List at least two applications of PE-TEOS USG film.
14. Why are both silicon oxide and nitride used as a passivation dielectric?
15. What are the differences between HDP-CVD and PECVD processes?
16. List two commonly used HDP sources.
17. If the silane flow rate is increased in a PECVD silane oxide deposition process, what are the effects on deposition rate, refractive index, and film stress?

18. Increasing temperature has different effects on deposition rates for PECVD silane USG and PE-TEOS USG processes. Explain.
19. Using the philosophy behind Fig. 10.47, predict the changes for OSG (SiCOH) film properties (such as dielectric constant κ and film stability) after increasing carbon concentration in the film.
20. What is the purpose of porogen in porous low-κ dielectric deposition?
21. What are the main advantages and disadvantage of ALD compared to CVD?
22. What materials can be used as precursors for HfO_2 deposition?
23. What is the standard κ-value when discussing high- and low-κ dielectrics?
24. Low-κ dielectrics are mainly applied in ILDs of metal interconnects. What about high-κ dielectrics?
25. How is stress controlled in an ILD0 stress liner? Which MOSFET needs tensile stress and which one needs compressive stress?

Chapter 11
Metallization

Various types of conductors are applied in IC chip manufacturing. Metals with high conductivity are widely used for the interconnections that form microelectronic circuits. Metallization is an adding process that deposits metal layers on a wafer surface.

Objectives

After finishing this chapter, the reader will be able to:

- explain the device application of metallization processes
- list at least five metals commonly used in IC chip manufacturing
- describe the advantages of copper interconnections over Al–Cu interconnections
- list three different methods used in metal deposition
- describe the sputtering deposition process
- explain the purpose of high vacuum in metal deposition processes
- list the metals used in high-κ, metal gate MOSFETs.

11.1 Introduction

Metals such as copper and aluminum are very good conductors. They are widely used to make conducting lines for transporting electrical power and electric signals. On an IC chip, miniature metal lines are used to connect billions of transistors made on the surface of a semiconductor substrate.

Requirements for metallization are: low resistivity for low power consumption and high IC speed, smooth surfaces for high-resolution patterning processes, high electromigration resistance for achieving high chip reliability, and low film stress for good adhesion to underlying substrates. Other requirements are stable mechanical and electrical properties during subsequent processing, good corrosion resistance, and relatively easy deposition, etch, or CMP.

It is very important to reduce resistance of the interconnect lines because IC device speed is closely related to the RC delay time, which is proportional to the resistivity of the conductor used to form the metal wires. Metal lines with low resistivity have shorter RC times and faster clock speed.

Although copper (Cu) has lower resistivity than aluminum (Al), technical difficulties such as adhesion, diffusion problems, and difficulties with dry etching hampered its applications in IC chip manufacturing. Al interconnections dominated metallization applications for a long time. In the 1960s and 1970s, pure Al and Al–Si alloys were used as metal interconnection materials. In the 1980s, when device dimensions shrank, one layer of metal interconnection was no longer enough to route all of the transistors, and multilayer interconnections became popular. Soon there was no room for widely tapered contact and via holes when device density increased. Metallization on tapered contact holes is not good for multilayer interconnections because it leaves a very rough surface topography. It is difficult to uniformly deposit a thin-film layer over a rough surface, and it is difficult to precisely pattern the rough layer. To increase packing density, near-vertical contact and via holes are needed; these holes are too narrow and have an aspect ratio that is too high for PVD Al alloys to fill without voids. Tungsten (W) was introduced and became widely used to fill the contact and via holes, and served as a plug to connect the different metal layers. Titanium (Ti) and titanium nitride (TiN) barrier/adhesion layers are required prior to tungsten deposition to prevent diffusion and film peeling of the tungsten. Figure 11.1 illustrates a cross section of a CMOS IC chip with an Al–Cu alloy interconnection and tungsten plug.

Question: Can all dimensions of a metal line be reduced according to the shrinkage of minimum feature size to connect every transistor in one layer of metal?

Answer: Because metal line resistance $R = \rho l/wh$, if a metal line shrinks in every dimension (length l, width w, and height h) according to the shrinkage of the device dimensions by a factor of 1.4 (on the technology node), metal line resistance R will increase by factor of 1.4. This shrinkage will affect device performance and slow down circuit speed due to increasing power consumption and heat generation. Line resistance will be the same if only the width and length of the metal line are reduced while keeping the height unchanged. However, keeping height unchanged will cause high-aspect-ratio metal stacks, which are difficult to etch. It also forms narrow gaps, which are hard to fill without voids for dielectric deposition processes.

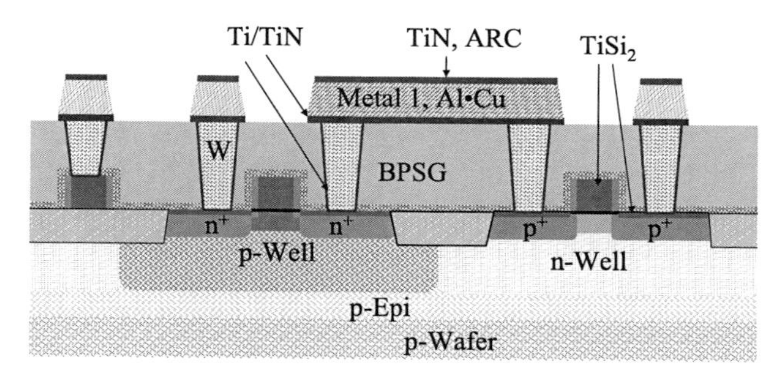

Figure 11.1 Cross section of an IC chip with Al–Cu interconnection.

In the late 1990s, development of the CMP process paved the way for using copper in IC interconnection applications with damascene or dual damascene processes. Tantalum (Ta) or tantalum nitride (TaN) is used as the barrier layer, and silicon nitride (SiN) or silicon carbon nitride (SiCN) has been used as barrier and cap layers to isolate the copper layer and prevent it from diffusing through the low-κ dielectric layer into the silicon substrate, causing interference with the normal operations of the transistors. Silicon nitride is also used as an etch-stop layer for the dual damascene dielectric etching process. Figure 11.2 illustrates a CMOS chip cross section with a copper interconnection.

Figure 11.3 illustrates the position of metallization in IC manufacturing processes. This chapter covers the metals used in IC chip manufacturing, their applications, and their deposition processes.

Ti/TiN
SiN
SiCN
NiSi
Ta or TaN
M1
Cu
Low-κ
Cu
Cu
PSG
W
STI
n+
n+
USG
p+
p+
p-Well
n-Well
p-Epi
p-Wafer

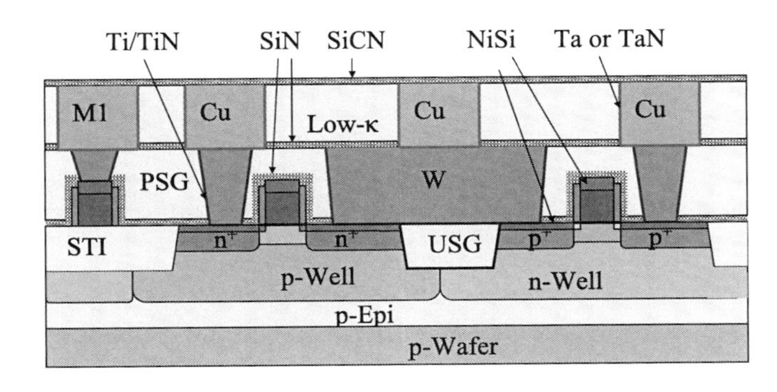

Figure 11.2 Cross-section of an IC chip with Cu/low-κ interconnection.

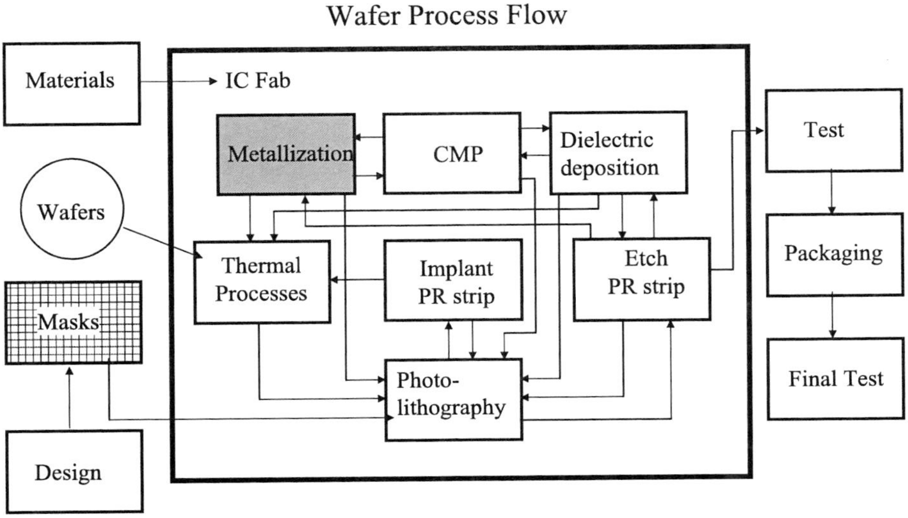

Wafer Process Flow
Materials
IC Fab
Metallization
CMP
Dielectric deposition
Test
Wafers
Thermal Processes
Implant PR strip
Etch PR strip
Packaging
Masks
Photo-lithography
Final Test
Design

Figure 11.3 Metallization in an IC fabrication process flow.

11.2 Conducting Thin Films

Polysilicon, silicides, aluminum alloys, titanium, titanium nitride, and tungsten are the most commonly used conductors in IC chip manufacturing for gate materials, and for location and global interconnections. Copper has been used for interconnections in IC production since the late 1990s.

11.2.1 Polysilicon

Polysilicon is the most common material used to form gates and local interconnections. Polysilicon has replaced aluminum as gate material since the introduction of ion implantation technology in the mid-1970s. Polysilicon has high temperature stability, which is necessary for self-aligned S/D implantation and postimplantation high-temperature annealing processes. An aluminum gate cannot sustain the high temperature (>1000 °C) required for postimplantation annealing processes.

Polysilicon normally is deposited with an LPCVD process; either SiH_4 or SiH_2Cl_2 can be used as a silicon precursor. Deposition temperatures range from 550 to 750 °C, and the polysilicon can be heavily doped with boron, phosphorus, or arsenic, either *in situ* during deposition or *ex situ* by an ion implantation process. The polysilicon deposition process is discussed in detail in Chapter 5.

11.2.2 Silicides

Even heavily doped polysilicon has a high resistivity of approximately a few hundred μΩ·cm. When device dimensions shrink, resistance of the polysilicon local interconnection will also increase, which causes more power consumption and longer RC time delay. To reduce the resistance and increase device speed, silicides (which have much lower resistivity than polysilicon) have been developed and used along with polysilicon to form a polycide stack. Titanium silicide ($TiSi_2$) and tungsten silicide (WSi_2) are the two commonly used silicides in IC devices that use mature technologies.

Titanium silicide is formed in a self-aligned silicide (salicide) process. First, the wafer surface is cleaned in chemical solutions to remove contaminants and particles. Then, argon sputtering in a vacuum chamber removes the native oxide from the wafer surface. A layer of titanium is deposited by a sputtering process on the wafer surface, with silicon exposed at the S/D and on top of the polysilicon gate. A thermal annealing process, performed in a rapid thermal processing (RTP) system, allows titanium to chemically react with silicon to form titanium silicide on the top of polysilicon and on the surface of the S/D. Because titanium does not react with silicon dioxide and silicon nitride, silicide is only formed at the places where silicon directly comes into contact with titanium. After a wet etch process strips the unreacted titanium with a mixture of hydrogen peroxide (H_2O_2) and sulfuric acid (H_2SO_4), the wafer is annealed once more to fully silicide the titanium and increase its grain size, improving conductivity and reducing contact resistance.

Tungsten silicide processing is quite a different process. First, a thin film of tungsten silicide is deposited on a polysilicon surface by a thermal CVD process

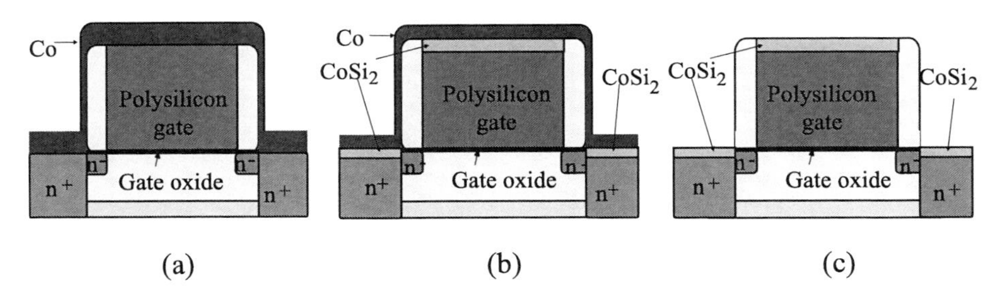

Figure 11.4 Self-alignment cobalt silicide formation process steps: (a) cobalt deposition, (b) silicide annealing, and (c) cobalt wet striping and second annealing.

with WF_6 as the tungsten precursor and SiH_4 as the silicon precursor. Then, a polycide stack is etched in a multistep process in which fluorine chemistry is used to etch the tungsten silicide and chlorine chemistry is used to etch the polysilicon. After photoresist stripping, a RTA process follows to increase tungsten silicide grain size and improve conductivity.

As device dimensions grow smaller, the feature size of a gate can become so small that it is narrower than a titanium silicide grain size, which is about 0.2 μm. Cobalt silicide ($CoSi_2$) has been used in gate and local interconnection applications for CMOS devices from 180- to 90-nm technology nodes. Figure 11.4 illustrates the process steps of self-aligned cobalt silicide formation. Titanium silicide was used before cobalt silicide in earlier technology nodes. Replacing the cobalt (Co) with titanium (Ti), the $TiSi_2$ formation process is almost the same as the steps shown Fig. 11.4.

With shrinking device dimensions, the annealing temperature of $CoSi_2$ (typically ~750 °C) and its annealing time of ~30 sec take away too much of the precious thermal budget. A nickel (Ni) silicide process has been developed and applied to CMOS devices since the 65-nm technology node. Nickel silicide (NiSi) processing is also very similar to the process shown in Fig. 11.4. The main advantage of NiSi processing is that it requires a significantly lower annealing temperature, typically less than 500 °C. NiSi can be used for CMOS devices down to 10-nm technology node.

Because NiSi is thermally unstable, it tends to further react with silicon to form $NiSi_2$, which can cause nickel silicide to grow into the silicon substrate, commonly called NiSi piping or NiSi encroachment, inducing junction leakage and causing yield loss. Many methods have been studied to solve this killer defect. A popular solution is to add platinum (Pt) to a nickel target that can be used to sputter NiPt alloy on the wafer surface to form NiPt silicide.

11.2.3 Aluminum

At one time, aluminum was the most common metal used for interconnections, which route thousands or millions of transistors built on a single wafer surface. Copper interconnections became mainstream technology in the first decade of the twenty-first century. Aluminum is the fourth-best metal for electrical conduction

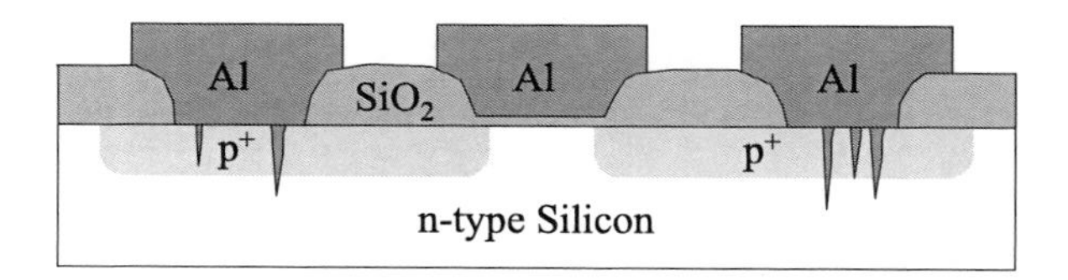

Figure 11.5 Junction spiking.

(2.65 μΩ·cm), following silver (Ag, 1.6 μΩ·cm), copper (Cu, 1.7 μΩ·cm), and gold (Au, 2.2 μΩ·cm). Of these four, it is the only metal that can be easily dry etched to form tiny metal interconnection lines. It was also used as a gate material before the introduction of ion implantation in the mid-1970s.

Silicon will dissolve in aluminum. At the S/D area, where aluminum metal lines come into direct contact with silicon, silicon can dissolve into the aluminum, and the aluminum will diffuse into the silicon to form aluminum spikes. Aluminum spikes can punch through the doped junction and short the S/D junction with the substrate, causing excessive device leakage and reliability issues. This effect is called junction spiking and is illustrated in Fig. 11.5.

The saturation concentration of dissolved silicon in aluminum is about 1%. Therefore, adding about 1% of Si into Al can effectively prevent further dissolving of the silicon. Thermal annealing at 400 °C forms Si–Al alloys at the silicon-aluminum interface that also can help prevent aluminum-silicon interdiffusion and junction spiking.

Metallic aluminum is a polycrystalline material that consists of many small monocrystalline grains. When electric current flows through an aluminum wire, a stream of electrons constantly bombards the grains. The smaller grains will start to move after a certain amount of time, similar to small rocks on the bottom of a creek being flushed down a stream during flood season. This effect is called electromigration. Figure 11.6 shows the electromigration process.

Electromigration can cause serious problems in aluminum wires. When a few grains start to move due to electron bombardment, the metal line begins to tear apart at certain points and causes higher current density in the line at these points. This aggravates the bombardment and causes further aluminum grain migration. The high current density and high resistance generate a large amount of heat and eventually break down the metal line. Electromigration can affect IC chip reliability because it can break the aluminum line and causes an open loop in the tiny circuit after the chip is employed in an electronic system. For old houses with aluminum wiring, this can become a safety hazard because it can cause fire due to the intense heat generated at the tearing point of the aluminum line, normally at the contact point.

It was discovered that when a small amount of copper is added to aluminum to form an Al–Cu alloy, electromigration resistance can be significantly improved, since copper can act like glue between the aluminum grains and prevent them from migrating due to electron bombardment. An Al–Si–Cu alloy was used after this discovery, and it could still be used by some older fabs because IC manufacturers normally are very reluctant to replace a production-proven process.

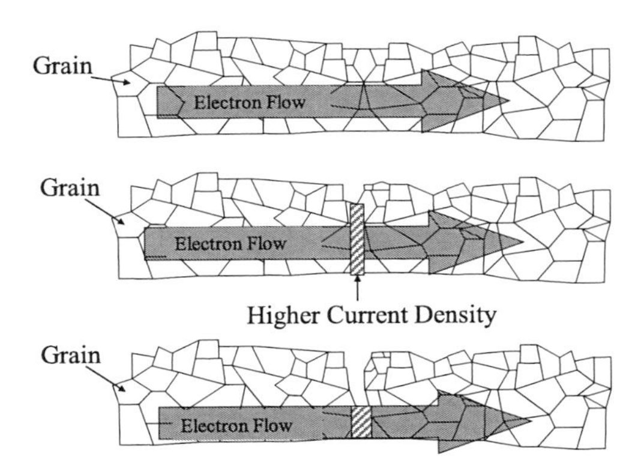

Figure 11.6 Electromigration.

By then end of the 1990s, the most commonly used metal for interconnection was the Al–Cu alloy. Because the aluminum no longer made direct contact with silicon, silicon was no longer required in the aluminum alloy. Copper concentration varied from 0.5 to 4%, depending on the process requirements and manufacturer's specifications. High copper concentration improves electromigration resistance. However, higher copper concentrations can make metal etch processes more difficult. Chapter 9 revealed that the main etchant for Al–Cu alloy etch is chlorine. One of the byproducts of Al–Cu alloy etch is copper chloride, which has a very low volatility and tends to stay on the wafer surface to form unwanted residues, commonly called scum. High copper concentrations in Al–Cu alloy can worsen the scum issue. Heavy ion bombardment in the plasma etch process can help dislodge the nonvolatile copper chloride, or a little isotropic chemical etch could scoop it out from the surface and prevent scum from forming. The etch residues can cause defects on the wafers and affect IC chip yield. An extra wet descumming process is often required to remove these residues.

Both CVD and PVD processes can be used to deposit aluminum. Because PVD aluminum has a higher quality and lower resistivity, it is the most popular method used in the IC industry. Thermal evaporation, electron beam evaporation, and plasma sputtering can be used for Al PVD. Magnetron sputtering deposition is the most popular PVD process for aluminum alloy deposition in semiconductor fabs. Electron beam evaporation and thermal evaporators are no longer used in advanced IC fabs. They are still used by universities and research laboratories for education and research applications. Some less-advanced fabs could still use them for IC chip manufacturing that does not require state-of-the-art technology.

Aluminum CVD is a thermal CVD process with aluminum organic compounds such as dimethylaluminum hydride [DMAH, $Al(CH_3)_2H$] as precursors. Compared with PVD aluminum, CVD aluminum has much better step coverage and gap-fill abilities, making it a very attractive candidate to replace tungsten CVD processes for contact/via hole filling. The quality of CVD aluminum film is poorer and resistivity is higher than with PVD films. It is not very difficult to deposit aluminum in a CVD process. However, it is not easy to deposit Al–Cu alloy film in a CVD

Table 11.1 Facts about aluminum.

Symbol	Al
Atomic number	13
Atomic weight	26.981538
Discoverer	Hans Christian Oersted
Place of discovery	Denmark
Discovery date	1825
Origin of name	From the Latin word alumen, meaning alum
Density of solid	2.70 g/cm^3
Molar volume	10.00 cm^3
Velocity of sound in substance	5100 m/sec
Hardness	2.75
Electrical resistivity	2.65 μΩ·cm
Reflectivity	71%
Melting point	660 °C
Boiling point	2519 °C
Thermal conductivity	235 W m^{-1} K^{-1}
Coefficient of linear thermal expansion	23.1 ×10^{-6} K^{-1}
Etchants (wet)	H_3PO_4, HNO_4, CH_3COOH
Etchants (dry)	Cl_2, BCl_3
CVD precursor	$Al(CH_3)_2H$
Main applications in IC manufacturing	To form Al–Cu alloy in metal interconnections. Used as a bulk gate material in gate-last HKMG of MOSFET.

process, so the applications of aluminum CVD processes are limited. Without copper in the aluminum, electromigration can cause serious problems in device reliability. Some facts about aluminum are summarized in Table 11.1.

11.2.4 Titanium

There are several applications that use titanium: silicide formation, titanium nitridation, wetting layers, welding layer, metal gates, and metal hard masks.

Titanium silicide is commonly used in IC fabrication for chips with 250nm or larger technology modes. It has low resistivity and can be formed on the polysilicon gate and the S/D in a self-aligned silicide (salicide) process. In the salicide process, titanium is deposited on the wafer surface by a sputtering deposition process. It is then annealed in a thermal process to form titanium silicide ($TiSi_2$). Titanium can also be deposited by a CVD process, which uses $TiCl_4$ and H_2 to react and deposit titanium at 650 °C, and can form titanium silicide in situ in the regions where Ti directly contacts with Si.

Titanium is also widely used as a welding layer for tungsten and aluminum to reduce contact resistance. This is because titanium can scavenge oxygen atoms and prevent them from bonding with tungsten and aluminum atoms to form high-resistivity WO_4 and Al_2O_3. It is also used with titanium nitride as a diffusion barrier layer for tungsten plugs and for local interconnection to prevent tungsten from diffusing into the silicon substrate. Figure 11.7 shows an application of titanium and titanium nitride in the interconnection application of an IC chip with Al–Cu interconnection.

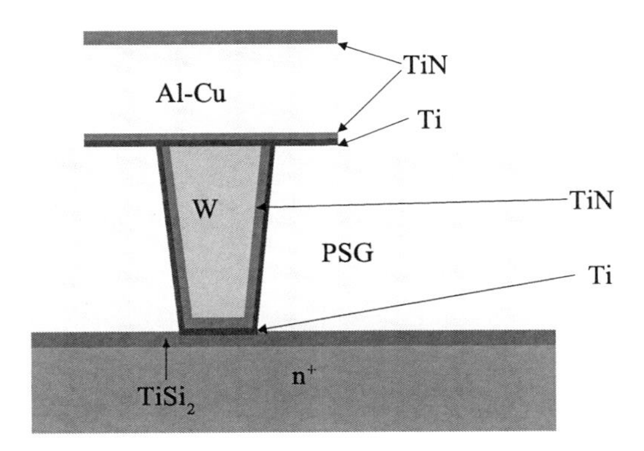

Figure 11.7 Titanium applications of IC chips with Al–Cu interconnection.

Titanium can also be used in titanium nitride formation; by depositing titanium and annealing it at a high temperature in the nitrogen/ammonia ambient, TiN can be formed through a chemical reaction on the surface. Of course, most people prefer PVD titanium nitride for its easy process control and lower temperature requirement.

Titanium normally is deposited with a magnetron plasma sputtering process. It also can be deposited by a CVD process with $TiCl_4$ as the precursor and reacts with hydrogen at high temperatures. Some facts of titanium are listed in Table 11.2.

11.2.5 Titanium nitride

Titanium nitride is widely used in IC manufacturing for barrier layers, adhesion layers, and antireflection coating (ARC). A thin layer (50 to 200 Å) of TiN is also needed for barrier and adhesion layers for tungsten. It prevents tungsten from diffusing into the oxide and silicon, and helps tungsten stick to silicon oxide surfaces. TiN also has a much lower reflectivity than the Al–Cu alloy and has been used as an ARC to improve photolithography resolution in metal patterning processes. A TiN layer on top of the aluminum alloy also can prevent hillocks and help control electromigration.

In advanced HKMG MOSFET, TiN is used as a metal gate electrode. Gate-first HMMG MOSFET uses TiN as a gate electrode for both nMOS and pMOS, as shown in Fig. 11.8(a). Gate-last HKMG MOSFET uses TiN for nMOS and to react with TiAl to form TiAlN for nMOS gate electrodes [see Fig. 11.8(b)]. TiN is also used as gate electrodes of cell array transistors of bWL DRAM.

For copper/low-κ and copper/ULK interconnections (especially for porous ULK dielectrics), TiN has been used as metal hard masks for low-κ dielectric dual damascene etch processes to protect the OSG film from attacking oxygen radicals during photoresist ashing. Otherwise, oxygen radicals can react with methyl or ethyl groups in the OSG film and increase the dielectric constant of the film.

TiN has also been used as the electrodes of storage node capacitors in DRAM chips, which require very good film conformality and excellent step coverage of both sidewalls and bottoms of high-aspect-ratio storage node holes.

Table 11.2 Facts about titanium.

Symbol	Ti
Atomic number	22
Atomic weight	47.867
Discoverer	William Gregor
Place of discovery	England
Discovery date	1791
Origin of name	Named after the Titans, the sons of the earth goddess in Greek mythology
Density of solid	4.507 g/cm^3
Molar volume	10.64 cm^3
Velocity of sound	4140 m/sec
Hardness	6.0
Electrical resistivity	40 μΩ·cm
Melting point	1668 °C
Boiling point	3287 °C
Thermal conductivity	22 W m^{-1} K^{-1}
Coefficient of linear thermal expansion	8.6 ×10^{-6} K^{-1}
Etchants (wet)	H_2O_2, H_2SO_4
Etchants (dry)	Cl_2, NF_3
CVD Precursor	$TiCl_4$
Main applications in IC manufacturing	Wetting layers for contact and Al–Cu metallization. Barrier layer for W. Reacts with nitrogen to form TiN in reactive sputtering.

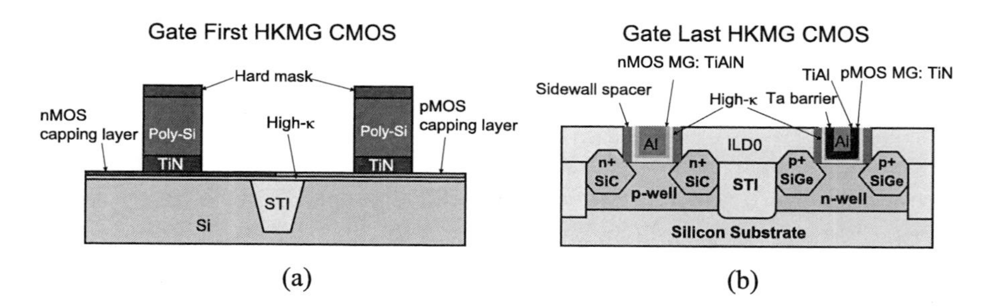

Figure 11.8 Applications of different metal layers in HKMG CMOS: (a) gate-first and (b) gate-last. (MG stands for metal gate.)

Titanium nitride can be deposited by PVD, CVD, and ALD processes. PVD TiN is deposited in a sputtering chamber with a titanium target, which is capable of depositing titanium in situ with TiN. Since a thin layer of titanium (50 to 100 Å) is always required for trapping oxygen and reduces contact resistance before titanium nitride deposition, in-situ Ti/TiN deposition can improve process throughput.

Reactive sputtering is the most commonly used method to deposit TiN in PVD processes. It uses argon and nitrogen as processing gases. In plasma, parts of both gases are ionized, and some nitrogen molecules are dissociated; these molecules generate chemically reactive free nitrogen radicals. Titanium atoms sputtered off from the target surface by the argon ions react with nitrogen when they pass through the argon-nitrogen plasma, and titanium nitride is formed and deposited on the wafer surface. Some titanium atoms can pass through the plasma and be deposited on the wafer surface, where they react with nitrogen radicals and form a titanium nitride there. Nitrogen radicals can also react with the titanium target and form a titanium nitride layer on the target surface. Ar^+ ions sputter the TiN molecules off the target surface and deposit them on the wafer surface.

CVD processes have been developed and applied to deposit conformal TiN films in IC chip manufacturing. They can be either high-temperature (~700 °C) CVD with precursors of $TiCl_4$ and NH_3, or low-temperature (~350 °C) metal organic CVD (MOCVD) with precursors of tetrakis-dimethylamino-titanium {TDMAT or $Ti[N(CH_3)_2]_4$}. TiN film deposited at a high temperature has better quality, lower resistivity, and better step coverage. However, it cannot be used for via applications, since the processing temperature is so high that it will melt the aluminum lines already on the wafer. MOCVD TiN has much better film conformality and step coverage compared with PVD TiN and has been widely used in advanced CMOS logic IC fabs for contact barrier/glue layer deposition. As deposited, MOCVD TiN has many organic groups integrated in the film. To densify the film, plasma treatment usually follows. N_2-H_2 plasma bombards the TiN film, helping to fully nitridize the film and remove residual methyl groups. MOCVD TiN usually needs several deposition-treatment cycles.

For HKMG applications, especially gate-last approaches, the ALD process has been used to deposit TiN due to its excellent step coverage. The ALD process has also been used for DRAM storage node TiN depositions.

In some old fabs, titanium nitride is still formed by the nitridation of the titanium surface with ammonia in a thermal annealing process.

11.2.6 Tungsten

Tungsten is the metal of choice for filling contact holes and forming the plugs that connect the metal layer and silicon surface. It is also applied to fill via holes between the different metal layers for Al–Cu interconnections, which are used in memory chips and mature (0.25 μm or earlier technology nodes) CMOS logic chips. When IC device dimension shrank to the submicron level, contact holes between the interconnection layers became smaller with higher aspect ratio, and it was impossible for PVD aluminum to fill them without voids. Hence, tungsten CVD (WCVD) processes were developed. With excellent step coverage and gap-fill abilities, it became the film most used to fill high-aspect-ratio contact or via holes, as illustrated in Fig. 11.9. Because WCVD has higher resistivity (8.0 to 12 μΩ·cm) than PVD aluminum–copper alloys (2.9 to 3.3 μΩ·cm), it is only used for local interconnections and plugs for connection between different layers.

Question: Why is a void in a contact hole unacceptable?

Answer: Because a void in a contact hole will cause high contact resistance and high current density due to the small metal cross section in the contact hole. Excessive heat generated there can damage the IC device very quickly.

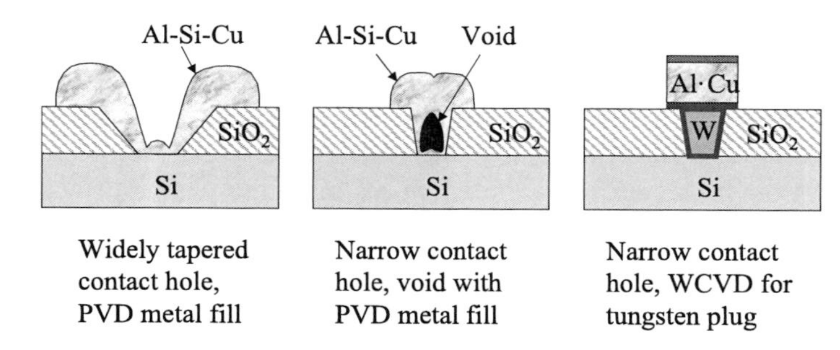

Figure 11.9 Evolution of the contact metallization processes.

Tungsten has been widely used in DRAM chips to form BLs and BL contact plugs. Tungsten or tungsten silicide is also added to polysilicon to form WLs in DRAM. Polysilicon plugs have long been used for DRAM FEoL contacts, such as self-aligned contact (SAC) and storage node contact (SNC). Due to smaller feature sizes, tungsten is being employed to form DRAM FEoL contact plugs. Because most DRAM chip do not need to operate at high clock speeds (2 GHz or higher) as CMOS logic ICs do, some advanced DRAM chips still use Al–Cu interconnections instead of Cu interconnections to reduce fabrication costs. Tungsten plugs are widely used in DRAM metal contact plugs between the metal layer and devices in the periphery area, and in via plugs between the metal layers, which in most cases are limited to two or three layers. For bWL DRAM, tungsten is added to TiN to form WLs and gate electrodes of the array transistors. It is also used with polysilicon to form the BLs and gate electrodes of periphery MOSFETs. NAND flash memory devices use tungsten as well to form BLs and source lines.

Tungsten is deposited during the CVD process, with WF_6 as the tungsten precursor. WF_6 can react with SiH_4 at 400 °C to deposit tungsten, which is used as the nucleation layer. A WF_6 and H_2 reaction at about 400 °C is applied for bulk tungsten deposition, which fills the narrow contact/via holes. Since tungsten does not stick very well to silicon dioxide surfaces, a layer of TiN is always needed to help tungsten adhere to the oxide surface. TiN and Ti stacks can prevent tungsten from diffusing into the silicon dioxide layer and cause heavy metal contamination; it can also block tungsten from contacting the silicon and reacting with it to form tungsten silicide and cause junction loss. Figure 11.10 shows an SEM photograph of a tungsten plug and TiN/Ti barrier/adhesion layer. Some facts of tungsten are summarized in Table 11.3.

11.2.7 Copper

Copper has much lower resistivity (1.7 μΩ·cm) than an aluminum–copper alloy (2.9 to 3.3 μΩ·cm). It also has significantly higher resistance to electromigration. It is highly desirable to use copper in an IC chip as the metal interconnection wire because copper can reduce power consumption and increase IC speed. However, copper has poor adhesion to silicon dioxide. It also has a very high diffusion rate in silicon and silicon dioxide, and copper diffusion can cause heavy metal

Table 11.3 Facts about tungsten.

Symbol	W
Atomic number	74
Atomic weight	183.84
Discoverer	Fausto and Juan Jose de Elhuyar
Place of discovery	Spain
Discovery date	1783
Origin of name	From the Swedish words tung sten, meaning heavy stone. W comes from wolfram, named after the tungsten mineral wolframite.
Density of solid	19.25 g/cm^3
Molar volume	9.47 cm^3
Velocity of sound in substance	5174 m/sec
Hardness	7.5
Reflectivity	62%
Electrical resistivity	5 $\mu\Omega\cdot$cm
Melting point	3422 °C
Boiling point	5555 °C
Thermal conductivity	170 W m^{-1} K^{-1}
Coefficient of linear thermal expansion	4.5 $\times 10^{-6}$ K^{-1}
Etchants (wet)	KH_2PO_4, KOH, and $K_3Fe(CN)_6$; boiling H_2O
Etchants (dry)	SF_6, NF_3, CF_4, etc.
CVD precursor	WF_6
Main applications	Contact plugs for CMOS logic devices. WLs, BLs, and contact plugs for DRAM and flash memory chips.

Figure 11.10 Tungsten plug and TiN/Ti barrier/adhesion layer (Applied Materials, Inc.).

contamination and device malfunction. Copper is very difficult to dry etch because copper–halogen compounds have very low volatility. After feature sizes dipped below 3 μm, all patterned etching processes have been dry etch (RIE) processes. Lack of an effective method to anisotropically etch copper became one of the important reasons that copper was not as popular as aluminum before the 0.18-μm technology node. Some basic facts about copper are listed in Table 11.4.

Technology developments, including high-aspect-ratio dielectric contact/via and trench etch processes, barrier layer deposition, copper deposition, and more importantly copper chemical mechanical polishing (CMP), led to the maturity of

Table 11.4 Facts about copper. Note that Cu(hfac)(vtms) stands for hexaflouroacetyl-acetonato-copper-vinyl-trimethylsilane.

Symbol	Cu
Atomic number	29
Atomic weight	63.546
Discoverer	
Place of discovery	Copper has been used by human beings since ancient times, long before any written history.
Discovery date	
Origin of name	From the Latin word cuprum, meaning the island of Cyprus
Density of solid	8.92 g/cm^3
Molar volume	7.11 cm^3
Velocity of sound in substance	3570 m/sec
Hardness	3.0
Reflectivity	90%
Electrical resistivity	1.7 μΩ·cm
Melting point	1084.77 °C
Boiling point	5555 °C
Thermal conductivity	400 W m^{-1} K^{-1}
Coefficient of linear thermal expansion	16.5×10^{-6} K^{-1}
Etchants (wet)	HNO_4, HCl, H_2SO_4
Etchants (dry)	Cl_2, needs low pressure and high temperature
CVD precursor	Cu(hfac)(vtms)
Main applications in IC manufacturing	Main metal for interconnections

dual damascene metallization technology. The fact that the dual damascene process does not require a metal etch step paved the way for copper metallization of IC interconnections in the late 1990s.

Copper deposition requires two steps: deposition of barrier and seed layers that cover the sidewall and bottom of the trenches and via holes, and bulk layer deposition by electrochemical plating (ECP) that completely fills the via holes and trenches. An annealing process follows bulk copper deposition to increase grain size and improve electrical conductivity before the metal CMP process, which removes copper and the barrier metal layer from the wafer surface, leaving the copper inside the trenches and via holes to form metal interconnects.

11.2.8 Tantalum

Tantalum is used as a barrier layer prior to copper deposition to prevent copper from diffusing across silicon oxide into the silicon substrate and causing device damage. Tantalum is a better barrier material for copper than other barrier materials such as titanium and titanium nitride. Tantalum is deposited with a sputtering process. Some facts about tantalum are listed in Table 11.5.

11.2.9 Cobalt

Cobalt is mainly used to form cobalt silicide ($CoSi_2$) for CMOS logic devices with 180- to 90-nm technology nodes. Cobalt silicide is still used for advanced flash memory devices and is deposited with a sputtering process. Some facts about cobalt are listed in Table 11.6.

Table 11.5 Facts about tantalum.

Symbol	Ta
Atomic number	73
Atomic weight	180.9479
Discoverer	Anders Ekeberg
Place of discovery	Sweden
Discovery date	1802
Origin of name	From the Greek word Tantalos, meaning father of Niobe due to its close relation to niobium in the periodic table.
Density of solid	$16.654 g/cm^3$
Molar volume	$7.11\ cm^3$
Velocity of sound in substance	3400 m/sec
Hardness	3.0
Reflectivity	90%
Electrical resistivity	$12.45\ \mu\Omega\cdot cm$
Melting point	2996 °C
Boiling point	5425 °C
Thermal conductivity	$57.5\ W\ m^{-1}\ K^{-1}$
Coefficient of linear thermal expansion	$6.3 \times 10^{-6}\ K^{-1}$
Etchants (wet)	2:2:5 mixture of HNO_3, HF, and H_2O
Etchants (dry)	Cl_2
Main applications in IC manufacturing	Used as a copper barrier layer. Also used as a barrier layer in pMOS of gate-last HKMG MOSFETs to block TiAl from reacting with TiN. TaN is also used as a copper barrier layer. TaBN is widely used as an absorber of EUV masks.

11.2.10 Nickel

Nickel (see Table 11.7) is required to form nickel silicide (NiSi) for CMOS logic devices with 65-nm and smaller technology nodes because NiSi formation needs significantly lower temperatures than other salicide formations. This makes it suitable for devices with smaller feature sizes and lower thermal budgets. It is deposited by means of a sputtering process.

Nickel is a magnetic material. Nickel–iron (NiFe) alloys can be used to form a magnetic memory cell in magnetoresistive random access memory (MRAM).

11.3 Metal Thin-Film Characteristics

Thickness measurements of metal thin films are different from dielectric thin-film measurements. It has been very difficult to directly and precisely measure the thickness of opaque thin films, such as metal films; the measurement needed to be performed on a test wafer in a destructive way. Recently, some nondestructive methods, such as laser acoustic measurements and x-ray reflectometry, have been developed and applied in metal thin-film metrology and processing controls.

Conducting films have polycrystalline structures. A metal's conductivity and reflectivity are related to grain size; metal films with larger grain sizes have higher conductivity and lower reflectivity. These properties are related to the deposition process. For instance, a higher deposition temperature provides higher mobility to deposited atoms on the substrate surface and forms larger grains in the deposited

Table 11.6 Facts about cobalt.

Symbol	Co
Atomic number	27
Atomic weight	180.9479
Discoverer	Georg Brandt
Place of discovery	Sweden
Discovery date	1735
Origin of name	From the German word kobald, meaning goblin or evil spirit
Density of solid	8.900 g/cm^3
Molar volume	6.67 cm^3
Velocity of sound in substance	4720 m/sec
Hardness	6.5
Reflectivity	67%
Electrical resistivity	13 μΩ·cm
Melting point	1768 K or 1495 °C
Boiling point	3200 K or 2927 °C
Thermal conductivity	100 W m^{-1} K^{-1}
Coefficient of linear thermal expansion	13.0 ×10^{-6} K^{-1}
Etchants (wet)	H_2O_2 and H_2SO_4
Etchants (dry)	-
Main applications in IC manufacturing	To form cobalt silicide for salicide applications.

Table 11.7 Facts about nickel.

Symbol	Ni
Atomic number	28
Atomic weight	58.693
Discoverer	Axel Fredrik Cronstedt
Place of discovery	Sweden
Discovery date	1751
Origin of name	From the German word kupfernickel, meaning devil's copper or Saint Nicholas' (old Nick's) copper
Density of solid	8.908 g/cm^3
Molar volume	6.59 cm^3
Velocity of sound in substance	4970 m/sec
Hardness	4.0
Reflectivity	72%
Electrical resistivity	7.2 μΩ·cm
Melting point	1728 K or 1455 °C
Boiling point	3186 K or 2913 °C
Thermal conductivity	91 W m^{-1} K^{-1}
Coefficient of linear thermal expansion	13.4 ×10^{-6} K^{-1}
Etchants (wet)	Mixture of H_2O_2 and H_2SO_4
Etchants (dry)	Cl_2
Main applications in IC manufacturing	To react with silicon to form NiSi for reducing the resistance of S/D contacts and local gate interconnections.

film. Sheet resistance is routinely measured and monitored for metal deposition process control in IC fabs.

11.3.1 Thickness and deposition rate

Metal films such as aluminum, titanium, titanium nitride, and copper are opaque films. Optically based techniques such as reflectospectrometry and ellipsometry commonly used in dielectric thin-film measurements cannot be used to measure metal film thicknesses. A destructive process is required to precisely measure an actual metal film thickness by cross-section SEM, or by measuring the step height with a profilometer after removing part of the deposited film.

For SEM measurement, a test wafer needs to be cut after metal thin-film deposition, and the cut samples are placed on the stage of the SEM system. An energetic electron beam scans across the sample, and the bombardment causes a secondary electron emission from the sample. Since different materials have different yield rates of secondary electron emissions, SEM can precisely measure the metal film thickness by measuring the intensity of the secondary electron emission. For example, a TiN film thickness can be measured on the top and sidewall of the contact hole from the SEM image shown in Fig. 11.10. Cross-section SEM can also detect whether there are voids in the plugs, and whether the tungsten plugs have good contact with the substrate. However, the process is expensive, destructive, and very time consuming; it is difficult to measure the uniformity of film across an entire wafer.

Question: Why are SEM images always in black and white?
Answer: SEM photographs are taken from the intensity of a secondary electron emission, which only emits strong or weak signals, and these transfer to bright and dim in a photographic image. SEM images with beautifully colored patterns are artificially painted after the photograph has been taken and the image has been analyzed.

Profilometer measurements can provide information about the thickness and uniformity of thick films (>1000 Å). A profilometer requires a patterned etch process prior to the measurement. First, a layer of metal film is deposited on the wafer surface, then a photolithography process patterns the photoresist in certain areas. After a wet etch process, most of the metal film on the wafer is removed, and the photoresist is stripped, leaving metal steps on certain parts of the wafer. Then the wafer is put in a stylus profilometer, which can sense and record the microscopic surface profile with a stylus probe (see Fig. 11.11).

Deposition rate is determined by the measured film thickness divided by the deposition time. In a metal CVD process, deposition rate can be affected by gas flow rate and processing temperature. In a magnetron sputtering deposition process, deposition rate is mainly controlled by the bias power. For an evaporator process, deposition rate is mainly determined by the filament current.

Ultrathin (50 to 100 Å) titanium nitride film is almost transparent. Therefore, reflectospectrometry and ellipsometry can be used to measure its thickness.

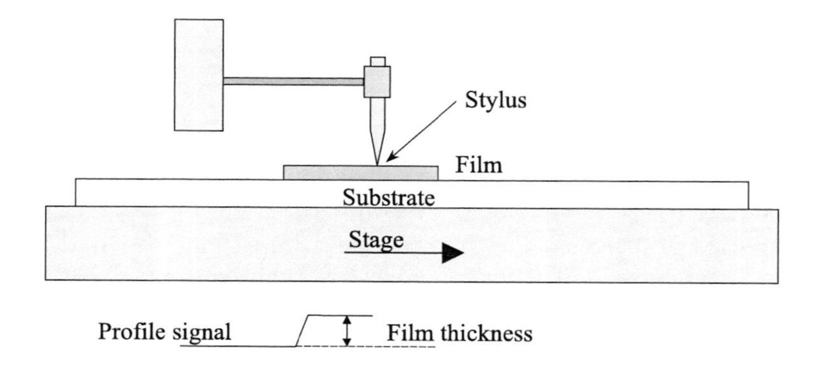

Figure 11.11 Schematic of a profilometer.

A four-point probe is commonly used to indirectly monitor metal film thickness by assuming that the resistivity of a metal film is constant over the entire wafer surface.

The acoustic method is a technique that can directly measure opaque thin-film thicknesses without direct contact with the film. Therefore, it is capable of measuring metal film thickness and uniformity on production wafers, an advantage that is very helpful for metal deposition process control.

The basic principle of the acoustic measurement method can be illustrated in Fig. 11.12. A laser beam is shot onto the thin-film surface; a photodetector is then used to measure the intensity of the reflected light. A very short laser pulse (~0.1 ps or 10^{-13} sec) from a pump laser is focused and shot onto the same spot, about 10×10 μm, which heats up the surface from 5 to 10 °C in that short time. Thermal expansion of the material at that spot causes a sound wave, which propagates through the film at the speed of sound of that particular material. When the acoustic wave reaches the interface of different materials, part of the wave is reflected from that interface, while the other part continues to propagate in the material underneath. The reflected sound wave, or echo, can cause reflectivity changes when it reaches the thin-film surface. The acoustic wave will echo back and forth in the film until it is damped out. The time between peaks Δt of the reflectivity change indicate the time the sound wave traveled back and forth in the films. When the speed of sound (V_s) of the material is known, the film thickness can be calculated by

$$d = V_s \Delta t / 2.$$

The decay rate of the echo is related to film density. This method can also be used to measure film thickness for each film in a multilayer structure.

Figure 11.12(a) shows the measurement of a TiN film on a thick TEOS oxide layer, which is treated as the substrate. Figure 11.12(b) shows the time between peaks of reflectivity change $\Delta t \approx 25.8$ ps. The speed of sound in TiN film is $V_S = 9500$ m/s $= 95$ Å/ps; therefore, the TiN film thickness can be calculated as $d = 1225$ Å.

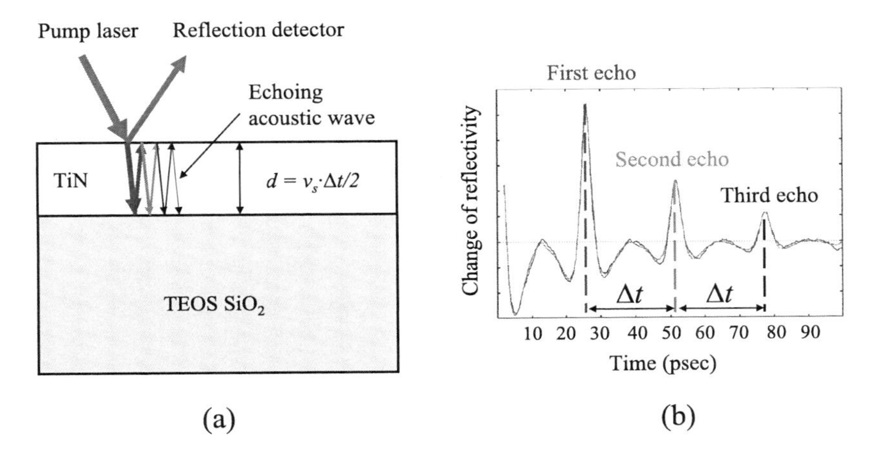

Figure 11.12 Acoustic method for metal thin-film measurement [Fig. 11.12(b) courtesy of Rudolph Technologies, Inc.].

X rays can reflect on a sample surface when the incident angle is very small. The reflectivity of an x ray is determined by the incident angle, surface roughness, film thickness, and film density. Due to interference between the reflection from the film surface and reflection from the interface of the film and substrate, reflectivity of an x ray oscillates with the changing incident angle. By varying the glancing incident angle and measuring the intensity of the reflected x ray, a spectrum of oscillating x ray intensity can be acquired as a function of incident angle. By fitting with the theoretical model, x-ray reflectometry (XRR) can be used to measure film thickness and film density, which provides information on film composition. Figure 11.13(a) illustrates the principle of the XRR measurement, and Fig. 11.13(b) gives an example of XRR measurement results of TaN/Ta film on Si substrate, where $q_z = 4\pi \sin\theta/\lambda$.

11.3.2 Uniformity

Uniformity (actually nonuniformity) of the thickness, sheet resistance, and reflectivity are routinely measured during process development and for process maintenance. It can be calculated by measuring the sheet resistance and reflectivity at multiple locations on a wafer in the patterns illustrated in Fig. 11.14.

The more measurement points are taken, the higher the precision of the measurement. However, more measurement points require longer measurement time, which means lower throughput and higher production cost because, in an IC manufacturing fab, time is money—a lot of money.

49-point measurement, 3σ standard deviation nonuniformity is the most commonly used nonuniformity definition for process qualification in 200-mm semiconductor fabs. To attain better control of the larger wafer size, more measurement points are needed. A 300-mm wafer usually needs a 121-point measurement. It is very important to clearly define nonuniformity because, for the same set of measurement data, different definitions can cause different nonuniformity results. For

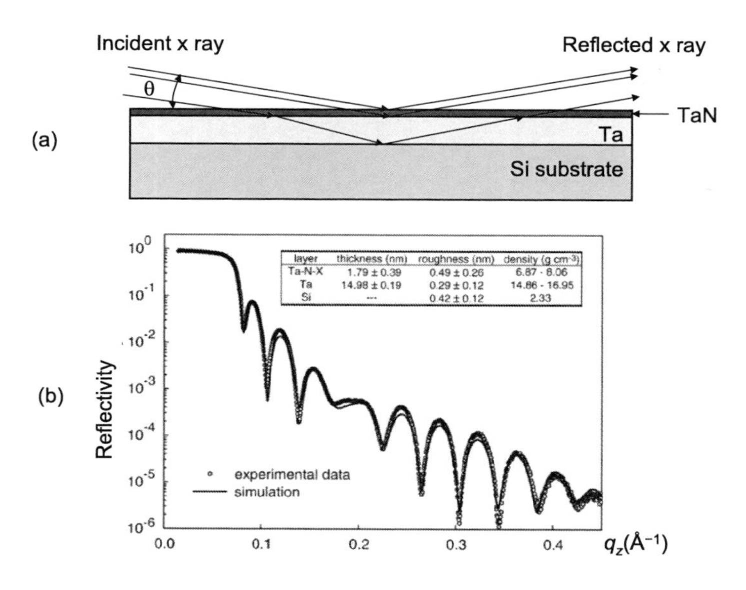

Figure 11.13 (a) XRR measurement principle and (b) an example of measurement results. [Fig. 11.13(b), from R. J. Matyi)].

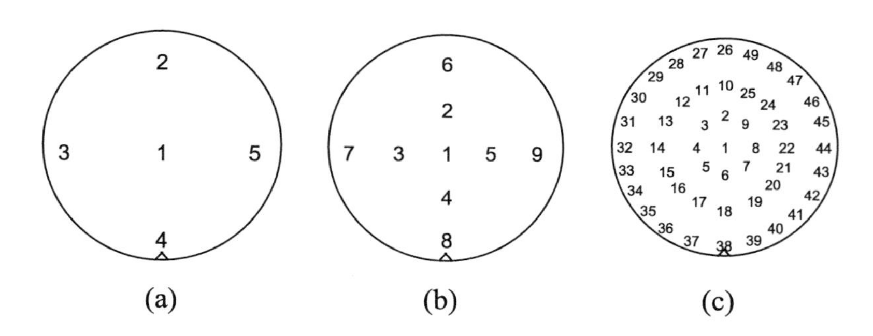

Figure 11.14 Mapping patterns for uniformity measurement: (a) 5 point, (b) 9 point, and (c) 49 point.

production wafers, sometimes less-time-consuming 5- and 9-point measurements are used for process monitoring and control, determined by processing requirements and metrology tool throughput.

11.3.3 Stress

Stress is due to material mismatch between a film and substrate. There are two types of stress, compressive and tensile. If stress is too high, whether compressive or tensile, it can cause serious problems. For metal thin films, high compressive stress can cause hillocks, which can short metal wires between different layers;

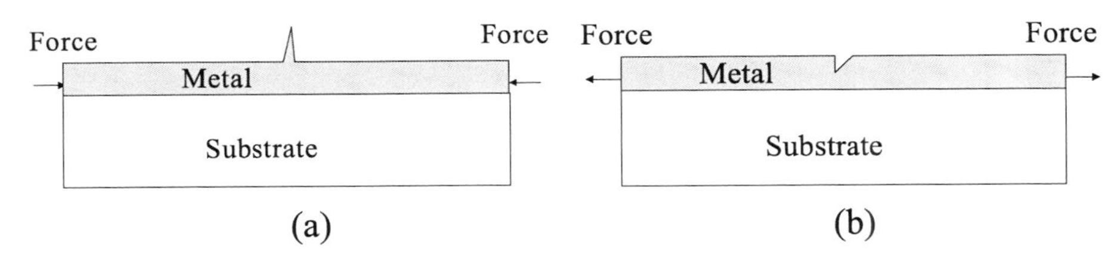

Figure 11.15 Different defects caused by high film stresses. (a) Hillock caused by compressive stress and (b) crack caused by tensile stress.

high tensile stress can cause films or interconnection lines to crack or peel. Figure 11.15 illustrates thin film defects caused by high film stresses.

There are two causes of stress: intrinsic and thermal. Intrinsic stress is caused by film density and determined by ion bombardment in the plasma sputtering deposition process. When atoms on a wafer surface are bombarded by energetic ions from plasma, they are squeezed densely together while forming the film. This type of film likes to expand, thus the film stress is compressive. Higher deposition temperature increases the mobility of the adatoms, also increasing film density and causing less tensile stress. Intrinsic stress is also related to the deposition temperature and pressure. Thermal stress is due to wafer temperature change and the different thermal expansion rates of the film and substrate. Aluminum has a very high thermal expansion rate, $\alpha_{Al} = 23.6 \times 10^{-6}\ K^{-1}$, compared to silicon, $\alpha_{Si} = 2.6 \times 10^{-6}\ K^{-1}$. For aluminum film deposited at a higher temperature such as 250 °C, when the wafer is cooling down to room temperature, the aluminum film shrinks more than the silicon substrate due to its higher thermal expansion rate. In this case, the wafer pulls the aluminum film and causes tensile stress. In fact, at room temperature, a little tensile stress is favorable for aluminum thin films, because the stress becomes less tensile when the wafer is heated during the succeeding metal annealing (~450 °C) and dielectric deposition (~400 °C) processes.

Question: Why does silicon oxide film favor compressive stress at room temperature?

Answer: Because silicon oxide has a lower thermal expansion rate ($\alpha_{SiO_2} = 0.5 \times 10^{-6}\ K^{-1}$) than the silicon substrate. If it has tensile stress at room temperature, it becomes more tensile in later processes when the wafer is heated up, which can crack the oxide film and cause hillocks on the aluminum line.

Stress can be measured from the change of wafer curvatures before and after thin-film deposition, described in detail in Chapter 10. The normal procedure for metal thin-film stress measurement is: measurement of wafer curvature, deposition of thin film with known thickness, and second wafer curvature measurement.

11.3.4 Reflectivity

Reflectivity is an important property of thin metal films. For a stable metallization process, reflectivity of the deposited film should be a constant. A change of reflectivity during the process indicates drifts of the processing conditions. Reflectivity is a function of film grain size and surface smoothness and needs to be controlled. The larger the grain size is, the lower the reflectivity. The smoother the metal surface is, the higher the reflectivity. Reflectivity measurement is an easy, quick, and nondestructive process, and it is frequently performed in the metal bays of semiconductor fabs.

Reflectivity is very important to photolithography processes because it can cause a standing wave effect due to interference between incoming light and reflecting light. This can affect photolithographic resolution by creating wavy grooves on the sidewall of the photoresist stack due to periodic overexposure and underexposure. ARC is required for metal patterning processes, especially for aluminum patterning, because aluminum has very high reflectivity—180 to 220% relative to silicon.

Reflectivity can be measured by focusing a light beam on a film surface and then measuring the intensity of the reflected beam. Reflectivity measurement results use the value relative to silicon.

11.3.5 Sheet resistance

Sheet resistance is one of the most important characteristics of conducting materials, especially for conducting films. It is commonly used to monitor the conducting thin-film deposition process and deposition chamber performance. For conducting films with known conductivity, a sheet resistance measurement is widely used to determine the film thickness, since this method is much faster than SEM and profilometer thickness measurements. Resistivity is one of the most fundamental properties of a material. For a conducting thin film, resistance can be calculated by the production of film sheet resistance and film thickness.

Sheet resistance R_s is a defined parameter. A four-point probe is a measurement tool that measures voltages and currents, and calculates sheet resistance. By measuring sheet resistance, film resistivity ρ can be calculated if the film thickness is known, or film thickness can be determined if its resistivity is known.

For the conducting line shown in Fig. 11.16(a), resistance can be calculated as

$$R = \rho \frac{L}{A},$$

where R is the resistance, ρ is the resistivity of the conductor, L is the length of the conducting line, and A is the area of the line cross section. If the wire cross-section is a rectangle, as shown in Fig. 11.16(b), its area can be simply expressed as the

Resistance of a Metal Line

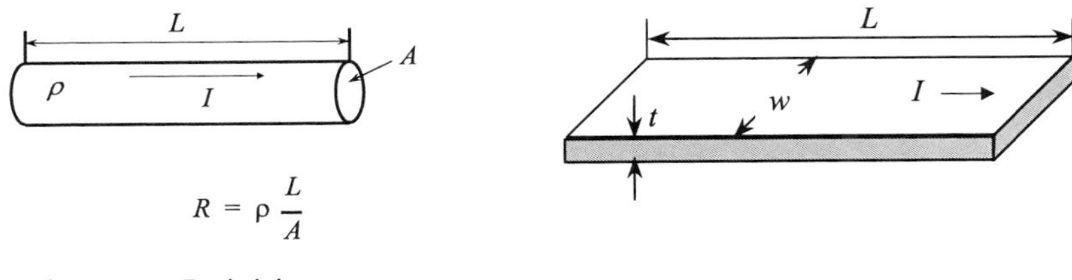

Figure 11.16 Conducting lines with (a) round and (b) rectangular cross sections.

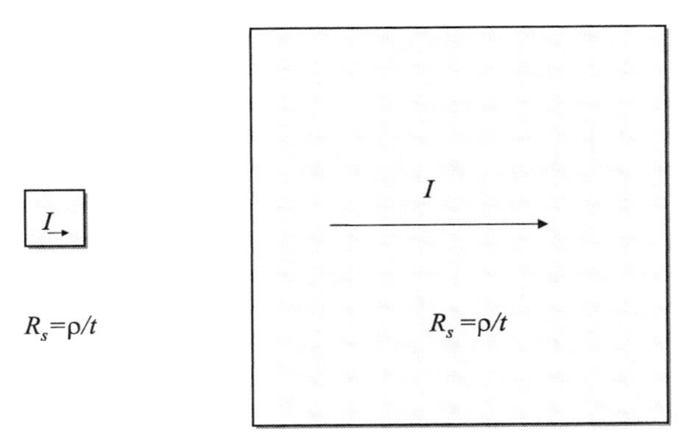

Figure 11.17 Sheet resistance is the same for these two thin-film squares.

product of the width and thickness (wt). The line resistance can be expressed as

$$R = \rho \frac{L}{wt}.$$

For a square sheet, the length is equal to the width, $L = w$, thus the two cancel each other. Therefore, resistance of a square conducting sheet, defined as sheet resistance, can be expressed as

$$R_s = \rho/t.$$

The unit of sheet resistance is ohms per square ($\Omega/\square$). The square symbol $\square$ here only denotes that the value is the resistance of a square; the size of the square does not matter. If the metal film thickness is perfectly uniform, sheet resistance of the square sheet with one micron at each side will be the same as a square sheet with one inch at each side, as shown in Fig. 11.17.

Question: For two conducting lines patterned from the same thin metal film with the same length-to-width ratios (as shown in Fig. 11.18), are their line resistances the same?

Answer: Yes. Both lines are consistent with the same number of square sheet resistors in serial connection. Since each square sheet has the same sheet resistance, their line resistances are the same.

Resistivity is related to the film material, grain size, and structure. For a particular metal film, the larger the grain size, the lower the resistivity.

The four-point probe shown in Fig. 11.19 is the most popular tool used to measure sheet resistance. A certain amount of current is applied between two of the pins, and voltage is measured between the other two pins. Sheet resistance equals the ratio of the voltage to the current and multiplies a constant, which depends on the pins being used. Typically, the distances between the probes are $S_1 = S_2 = S_3 = 1$ mm; if a current I is applied between P_1 and P_4, $R_s = 4.53\ V/I$, where V is the voltage between P_2 and P_3. If the current is applied between P_1 and P_3, $R_s = 5.75\ V/I$, where V is measured between P_2 and P_4. These two equations are derived with the assumption of a infinite film area, which is not quite true for thin-film measurements on a wafer.

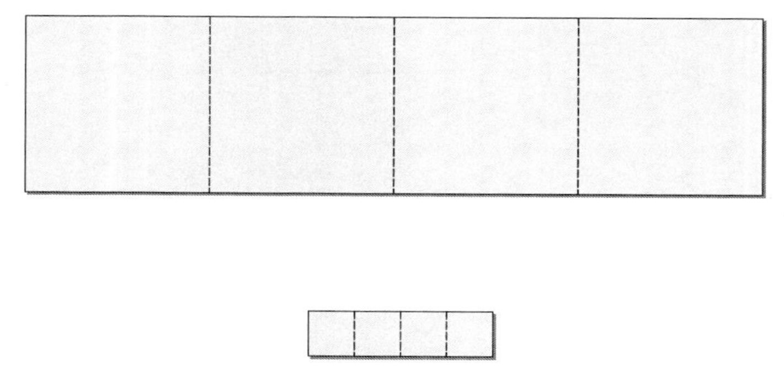

Figure 11.18 Line resistance of two conducting lines.

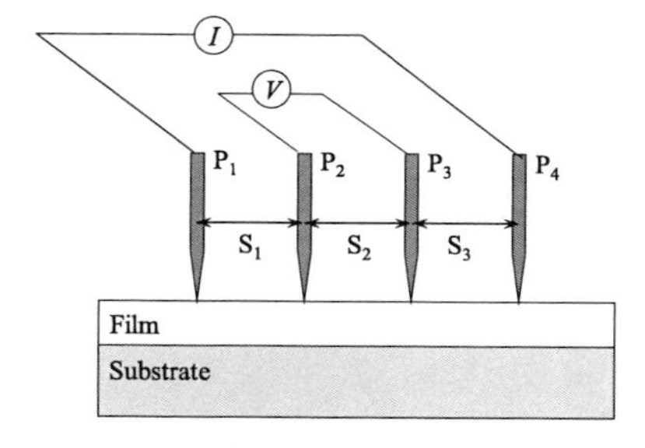

For a typical four-point probe, $S_1 = S_2 = S_3 = 1$ mm,
If current is applied between P_1 and P_4, $R_s = 4.53\ V/I$
If current is applied between P_1 and P_3, $R_s = 5.75\ V/I$

Figure 11.19 A four-point probe.

An advanced four-point measurement system runs four measurement procedures automatically in sequence, with both configurations and reverse current for each configuration to minimize the edge effect and achieve optimized results. Since four-point probes make direct contact with the film, they cannot be used on a product wafer. They are only used on test wafers for process development, tool qualification, and process control.

Sheet resistance measurement can provide important information about the film resistance, which is affected by film thickness, grain size, alloy concentration, and the incorporation of impurities, such as oxygen. Since process development usually sets up the correlation between the sheet resistances of metal layers and the product yield, these measurements are always closely monitored in the IC manufacturing process.

The resistivities of Al–Cu alloys and Cu films are well known and stable at certain deposition conductions. Therefore, sheet resistance measurements can provide a quick and convenient way to monitor film thickness and uniformity. However, this does not work with titanium nitride, because TiN resistivity is very sensitive to several process variables, such as nitrogen/argon flow ratio and processing temperature. Measured sheet resistance cannot be used to calculate TiN film thicknesses, since there is little confidence that the assumed film resistivity is the right value.

11.4 Metal Chemical Vapor Deposition

11.4.1 Introduction

Metal CVD is widely used to deposit metal in IC processing. CVD metal films have very good step coverage and gap-fill abilities and can fill tiny contact holes to create connections between metal layers. CVD metal thin films normally have poorer quality and higher resistivity than PVD metal thin films. Therefore, they are mainly used for contact and via plugs and local interconnections, not for global interconnections.

The most popular CVD metals are tungsten, tungsten silicide, and titanium nitride. This section covers these metal CVD processes. Most metal depositions are thermal processes; external heat from the heating elements provides the free energy needed for chemical reaction. In some cases, remote plasma sources are used to generate free radicals and increase the chemical reaction rate. A schematic of a metal CVD system is given in Fig. 11.20. The rf unit in the system is used for the plasma dry clean of the processing chamber.

A normal processing sequence for metal CVD deposition is as follows:

- wafer slides into the chamber
- slip valve closes
- pressure and temperature are set, with secondary process gas(es)
- all process gases flow in; deposition starts
- main processing gas is terminated (secondary processing gas remains on)
- all processing gases are terminated

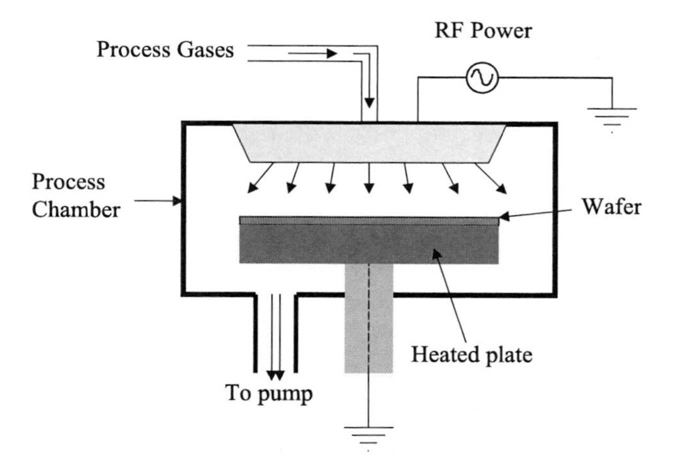

Figure 11.20 A metal CVD system.

- chamber is purged with nitrogen
- slip valve opens, and robot pulls wafer out.

During the metal deposition process, the metal thin film is deposited not only on the wafer surface, but also anywhere inside the processing chamber. A periodic cleaning process is needed to prevent the film inside the chamber from cracking and flaking, which generates particles. A plasma-enhanced chemistry is employed to routinely dry clean the chamber. For metal CVD chambers, fluorine- and chlorine-based chemicals are used in chamber plasma cleaning processes to dry etch films deposited inside the process chamber.

A processing sequence for plasma cleaning is as follows:

- chamber pumps down
- pressure and temperature are set with cleaning (etchant) gas(es)
- rf turns on; plasma and cleaning process is started
- rf turns off; chamber is purged
- set up pressure and temperature, with secondary process gas(es)
- main processing gas flows in; seasoning layer is deposited
- termination of main processing gas; secondary processing gas(es) remains on
- All processing gases are terminated
- chamber is purged with nitrogen
- chamber is ready for next deposition.

The seasoning step is very important. A thin layer of metal film intentionally deposited inside the chamber can effectively prevent loose residue from flaking and generating particles. It also can seal the remaining etchant gas and prevent it from interfering with the deposition process, as well as help prevent the first-wafer effect.

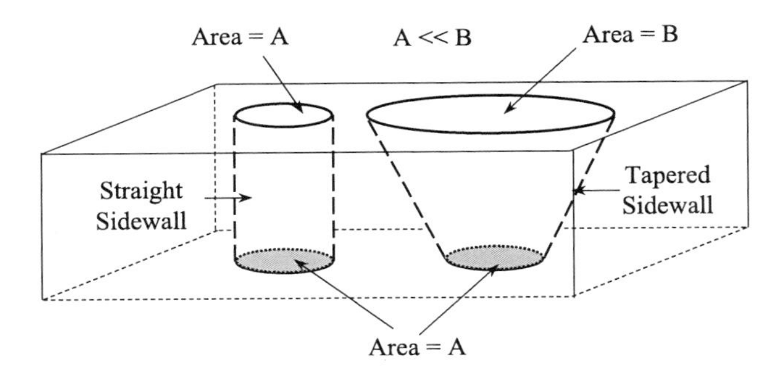

Figure 11.21 Vertical and tapered contact holes.

11.4.2 Tungsten

Tungsten has been extensively used in the semiconductor industry for metallization since the 1980s. Smaller feature sizes no longer have room for tapered contacts for the contact/via holes. Narrow and near-vertical contact/via holes must be used in metallization processes; however, PVD metal cannot fill these holes without voids. Thus, tungsten CVD processes have been developed. CVD tungsten film has almost perfect step coverage and conformity and can fill tiny holes without voids. It has been widely used as a plug to connect conducting layers in semiconductor fabrication processes. Figure 11.21 illustrates the vertical and tapered contact holes and clearly shows that using vertical contact holes can increase the packing density of IC wiring.

Because tungsten has lower resistivity than silicide and polysilicon, it can also be used for local interconnection applications. A local interconnection tungsten CVD process is different from the plug CVD process, since interconnection processing requires lower resistivity, while the main concern for tungsten plug application is gap-fill capability.

Tungsten hexafluoride (WF_6) is always used as a tungsten precursor for tungsten CVD processes. Silane (SiH_4) and hydrogen (H_2) are used to react with WF_6 and reduce fluorine to deposit tungsten. WF_6 is corrosive; when it reacts with water, it forms HF. The special safety hazard of HF is that it cannot be felt when direct contact is made with skin, and it can cause severe pain and damage once it reaches the bone and reacts with calcium. Double-layer rubber gloves are always required for tungsten CVD chamber wet cleaning. SiH_4 is pyrophoric (ignites itself), explosive, and toxic. H_2 easily leaks due to its small molecular size, and it is flammable and explosive.

For contact holes between metal and silicon, a silicon reduction process can be used to deposit tungsten:

$$2\,WF_6 + 3\,Si \rightarrow 2\,W + 3\,SiF_4.$$

It can be used as a nucleation layer to form selective tungsten. In this case, tungsten only deposits on the nucleation layer, which is on the silicon surface and not on the silicon dioxide surface during bulk deposition with H_2/WF_6 chemistry.

One of the advantages of selective tungsten is that it has good contact with silicon and thus a lower contact resistance. Tungsten can be selectively deposited in contact holes and therefore does not need etchback or polishback of bulk tungsten on the wafer surface. However, this process consumes silicon from the substrate and can cause junction loss. It also introduces fluorine into the substrate. Tungsten will react with silicon to form tungsten silicide during subsequent processes and will further consume the substrate silicon and cause junction loss. Another problem of the selective tungsten CVD process is that it is difficult to achieve perfect selectivity, since there are always some unpredictable nucleation sites on the silicon oxide surface. Selective tungsten processes are not commonly used in IC applications.

Most fabs use a blanket tungsten CVD process with a titanium/titanium nitride barrier/glue layer, which reduces contact resistance, prevents tungsten from reacting with silicon, and helps tungsten stick to silicate glass. The blanket tungsten CVD process has two steps: nucleation and main deposition. First, a thin layer of tungsten is deposited on the glue layer by using SiH_4/WF_6 chemistry:

$$2\,WF_6 + 3\,SiH_4 \rightarrow 2\,W + 3\,SiF_4 + 6\,H_2.$$

The nucleation step can strongly affect the uniformity of the subsequent film. After the nucleation step, H_2/WF_6 chemistry is used to deposit bulk tungsten to fill the contact/via holes:

$$WF_6 + 3\,H_2 \rightarrow W + 6\,HF.$$

During tungsten CVD processes, Ar is used for backside purging to prevent deposition on the edge and backside of the wafer. N_2 is used to improve film reflectivity and conductivity. Figure 11.22 illustrates the CVD tungsten seed and bulk layers.

Question: Plasma can generate free radicals and enhance the deposition rate of CVD processes. However, most tungsten CVD processes are thermal processes, not PECVD. Why?

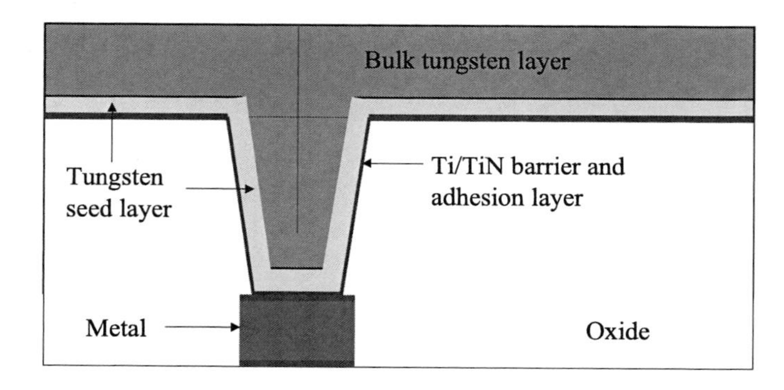

Figure 11.22 CVD tungsten seed and bulk layers.

Answer: In plasma, WF_6 dissociates and generates fluorine free radicals. With the help of ion bombardment (which always comes along with plasma), fluorine radicals can etch silicate glass during tungsten CVD and create many defects.

After each deposition, the CVD chamber must be cleaned with fluorine chemistry. NF_3 is used to provide free fluorine radicals in a plasma process, and Ar ions are used to provide heavy bombardment. Ion bombardment helps to damage bonds between tungsten atoms and significantly improves the etching effect:

$$6\,F + W \rightarrow WF_6.$$

After the cleaning process, a thin layer of tungsten is deposited in the chamber in the seasoning step to reduce particle contamination and eliminate the first-wafer effect.

The bulk tungsten CVD process fills contact/via holes and also coats a layer of tungsten over the entire wafer surface. Bulk tungsten on the wafer surface must be removed so that only tungsten is left behind in the contact/via holes. (These serve as plugs to connect different conducting lines between different layers.) Tungsten etchback processes with fluorine chemistry were extensively used in the industry until tungsten CMP processes become widely adapted by IC fabs in the late 1990s.

Question: By patterning and etching bulk tungsten film, a metal interconnection can be formed directly with tungsten wires, allowing many processing steps [such as tungsten CMP, metal stack (Ti/Al–Cu/TiN) PVD, metal wire patterning, and metal wire etch] to be skipped. Why do fabs not do this to save processing costs?

Answer: CVD tungsten has a much higher resistivity (8 to 11 $\mu\Omega\cdot$cm) than a PVD Al–Cu alloy (2.9 to 3.3 $\mu\Omega\cdot$cm). Tungsten can be used for local interconnections, not for global interconnections, because highly resistant interconnects will slow down circuit speed and increase power consumption, which are unacceptable consequences for most CMOS logic devices. However, DRAM chips are very cost sensitive, and many of them indeed use tungsten to fill BL and contact holes, then pattern and etch the tungsten film to form the BL directly. DRAM clock speed is not that high, so tungsten wire BLs are acceptable, and this low-cost approach helps drive down cost. Also, using a tungsten BL allows designers to put BLs below the storage node capacitors, and allow higher temperature processes such as LPCVD of polysilicon and silicon nitride.

Figure 11.23 shows applications of conducting layers in a recessed-gate stacked capacitor DRAM. In the figure, BLC is the bitline contact, and SN is the storage node. It is evident that the WL is formed by a polysilicon-tungsten stack. A thin layer of tungsten nitride (WN) is deposited between the polysilicon and tungsten. SAC is formed by polysilicon. Both BLCs in the array and periphery areas are

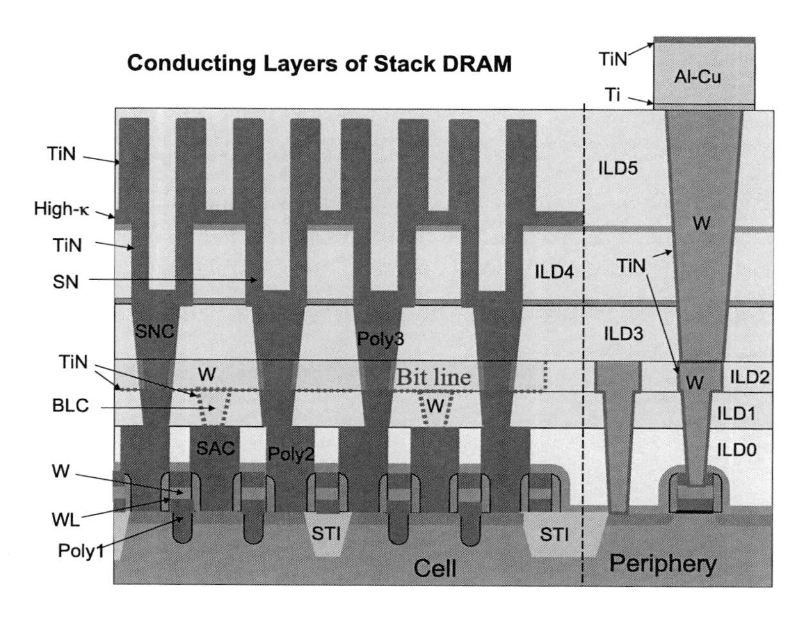

Figure 11.23 Tungsten layers in a recessed-gate stacked DRAM.

formed by tungsten with a Ti/TiN barrier/glue layer. SNC is formed by polysilicon. The SN capacitor is formed by a TiN/high-κ/TiN stacked layer. High-κ can be an Al_2O_3 film with silicon oxide on both sides. The metal contact is filled with tungsten using Ti/TiN as a barrier/glue layer.

Some DRAM manufacturers implement bWL technology for advanced DRAM chip fabrication. A bWL DRAM uses tungsten for WL and a tungsten/polysilicon stack for BL. It also uses tungsten with a TiN/Ti liner to form SNC plugs.

After tungsten deposition, bulk tungsten on a wafer surface needs to be removed by either tungsten etchback or tungsten CMP. The advantage of the tungsten etchback process is that it can be performed in situ with a tungsten CVD process in a single mainframe. However, tungsten CMP has much better process quality control and can significantly improve chip yield. It quickly replaced the etchback process when CMP technology matured and became more broadly accepted.

11.4.3 Tungsten silicide

Tungsten silicide (WSi_2) is used for gate and WL applications for DRAM chips with mature technology. Both SiH_4 and SiH_2Cl_2 (dichlorosilane, or DCS) are employed as silicon source gases, and WF_6 is employed as the tungsten precursor. SiH_4/WF_6 chemistry requires a lower processing temperature, typically at 400 °C, while DCS/WF_6 chemistry requires a higher processing temperature, about 550 to 575 °C.

For SiH_4/WF_6 chemistry,

$$WF_6 + 2\,SiH_4 \rightarrow WSi_2 + 6\,HF + H_2.$$

This process is very similar to the nucleation step in tungsten CVD processes. The main difference is the ratio of the SiH_4/WF_6 flow rate; when the ratio is lower than 3, the chemical replacement reaction deposits silicon-rich tungsten instead of tungsten silicide. To ensure tungsten silicide (WSi_x, with x between 2.2 to 2.6) deposition, the SiH_4/WF_6 flow ratio must be larger than 10.

The DCS/WF_6 chemistry can be expressed as

$$2\,WF_6 + 7\,SiH_2Cl_2 \rightarrow 2\,WSi_2 + 3\,SiF_4 + 14\,HCl.$$

DCS/WF_6-based processing requires a higher deposition temperature, has a higher tungsten silicide deposition rate, and has better film step coverage than SiH_4/WF_6-based processes. It also has much lower fluorine concentration in the film, and less film peeling and cracking problems due to its lower tensile stress. DCS/WF_6 silicide processing is gradually replacing silane-based processes.

The advantages of WSi_x over $TiSi_2$ are fewer process steps and the ease of integration with polysilicon deposition in one processing tool. However, tungsten silicide has higher resistivity than titanium silicide, and it only forms silicide on the gate, while the salicide process can form titanium silicide on the gate and S/D at the same time.

11.4.4 Titanium

There are two major applications of titanium in IC chip manufacturing. It is needed to reduce contact resistance before the titanium nitride barrier/glue layer deposition because direct contact between TiN and Si can cause high contact resistance. Titanium also reacts with silicon to form titanium silicide. For barrier layer applications, most choose PVD titanium instead of CVD titanium because PVD film has better quality and lower resistivity.

For titanium silicide processes, CVD titanium has some advantages. It has better step coverage than PVD titanium; this is important because Ti is deposited after gate etch, and the wafer surface is not planarized. At high temperatures (~600 °C), CVD Ti can react with Si to form $TiSi_2$ simultaneously during Ti deposition. The process can be expressed as

$$TiCl_4 + 2\,H_2 \rightarrow Ti + 4\,HCl,$$
$$Ti + Si \rightarrow TiSi_2.$$

11.4.5 Titanium nitride

Titanium nitride (TiN) is extensively used as a barrier/glue layer for tungsten plugs. Although the quality of CVD TiN film is not as good as PVD, and it has higher resistivity than PVD TiN, it has much better sidewall step coverage than PVD TiN films, 70 versus 15%. A thin layer of CVD TiN (~75 to 200 Å) is applied for contact/via holes as the barrier and glue layer for a tungsten plug after thin layers of PVD Ti deposition, as shown in Fig. 11.24.

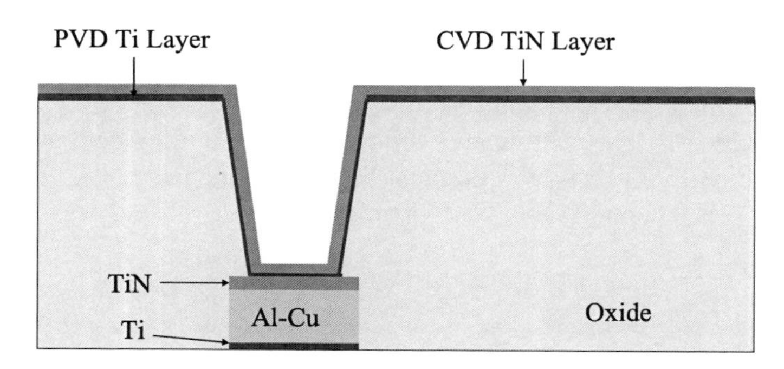

Figure 11.24 PVD and CVD titanium nitride layers.

TiN can be deposited from the reaction of $TiCl_4$ and NH_3 at 400 to 700 °C:

$$6\,TiCl_4 + 8\,NH_3 \rightarrow 6\,TiN + 24\,HCl + N_2.$$

TiN film deposited at higher temperatures has better film quality and a lower chlorine concentration. For low-temperature processes at 400 °C, Cl concentration can be as high as 5% in film; even high-temperature processes have ~0.5% Cl in film. Chlorine causes aluminum corrosion, which affects IC chip reliability. One of the possible byproducts, ammonium chloride (NH_4Cl), is a solid and can cause particle contamination.

Metal organic CVD (MOCVD) processes have been developed and widely used for TiN deposition. Metal organic compounds such as tetrakis-dimethylamido-titanium {$Ti[N(CH_3)_2]_4$, TDMAT} and tetrakis-diethylamido-titanium {$Ti[N(C_2H_5)_2]_4$, TDEAT} can be used as titanium nitride precursors, which dissociate at low temperatures (<450 °C) and deposit TiN with excellent step coverage. TDMAT is the more commonly used precursor. Its deposition temperature is about 350 °C and pressure is about 300 mtorr. The reaction can be expressed as:

$$Ti[N(CH_3)_2]_4 \rightarrow TiN + \text{organics}.$$

As deposited, a TiN thin film (~100 Å) is not as dense as a high-temperature deposited film, and has higher resistivity. It also has a high concentration of carbon and hydrogen. A postdeposition N_2-H_2 plasma treatment combined with chemical reactions from N and H free radicals can help to scavenge the C and H in the film and reduce their concentration. Ion bombardment of plasma can densify the film and reduce its resistivity. RTP annealing with N_2 ambient around 450 °C can also be used to densify the film and reduce resistivity.

TDMAT is a very poisonous liquid, and swallowing it can be fatal. $TiCl_4$ is highly irritating to the skin, eyes, and mucous membranes. Acute exposure can result in skin burns and marked congestion of the mucous membranes. It can also cause cornea damage. Ammonia is corrosive with a sharp, penetrating odor. Direct contact with liquid ammonia can cause severe chemical burns on skin and eyes. Contact and inhalation of ammonia vapors at low concentrations

(~25 ppm) can instigate irritation of the skin, eyes, nose, throat, and lungs. At high concentrations (~5000 ppm), it can cause severe eye irritation, chest pain, and fluid in the lungs.

11.4.6 Aluminum

Aluminum CVD processes to replace tungsten plugs and reduce interconnection resistance have been in research and development for a long time. Aluminum can be deposited at a relatively low temperature with aluminum organic compounds, such as dimethylaluminum hydride [$Al(CH_3)_2H$, or DMAH] and tri-isobutylaluminum [$Al(C_4H_7)_3$, or TIBA]. These compounds can decompose and deposit aluminum in a thermal process in a vacuum chamber. At about 350 °C, DMAH dissociates and deposits aluminum; the chemical reaction is expressed as

$$Al(CH_3)_2H \rightarrow Al + \text{volatile organics.}$$

DMAH is pyrophoric (igniting on contact with air), expensive, and often has high viscosity, which makes it difficult to handle. CVD aluminum film has very good trench and hole fill ability, and has lower resistivity than that of CVD tungsten film. Unlike tungsten film, which requires an etchback or a CMP process to remove the bulk tungsten deposition on the wafer surface, CVD aluminum film does not need to be removed from the surface for aluminum metallization. Because it is difficult to integrate a small amount of Cu into CVD Al, a layer of PVD Al–Cu is needed on top of CVD Al. Theoretically, CVD Al/PVD Al–Cu processing can save a process step, improve overall throughput, and reduce the processing cost for Al metal interconnections. Because copper metallization technology matured rapidly and replaced Al–Cu metallization in advanced CMOS logic ICs, the CVD Al process did not have a chance to be widely used for metal interconnections in IC manufacturing.

By using the cluster tools shown in Fig. 11.25, the entire ALD/CVD metallization process for gate-last HKMG integration can be performed in situ with a multichamber configuration. The integrated processing sequences are: wafer loading, preclean, HfO_2 ALD, TiN ALD, Ta ALD, cool down, and wafer unloading, performed in the cluster system shown in Fig. 11.25(a). After a pattern etch process removes the Ta protection layer from nNMOS, the photoresist is removed, and the wafer is cleaned, the cluster tool shown in Fig. 11.25(b) can be used for TiAl ALD, TiN ALD, and Al CVD. For planar MOSFET with gate-last HKMG process, CVD Al is commonly used to fill the gate trench after deposition of the work-function metals [see Fig. 11.8(b)]. For HKMG FinFET, CVD W is used to fill the gate trench.

11.5 Physical Vapor Deposition

11.5.1 Introduction

PVD is a deposition process that vaporizes solid materials, either by heating or by sputtering, and recondenses the vapor on the substrate surface to form a solid

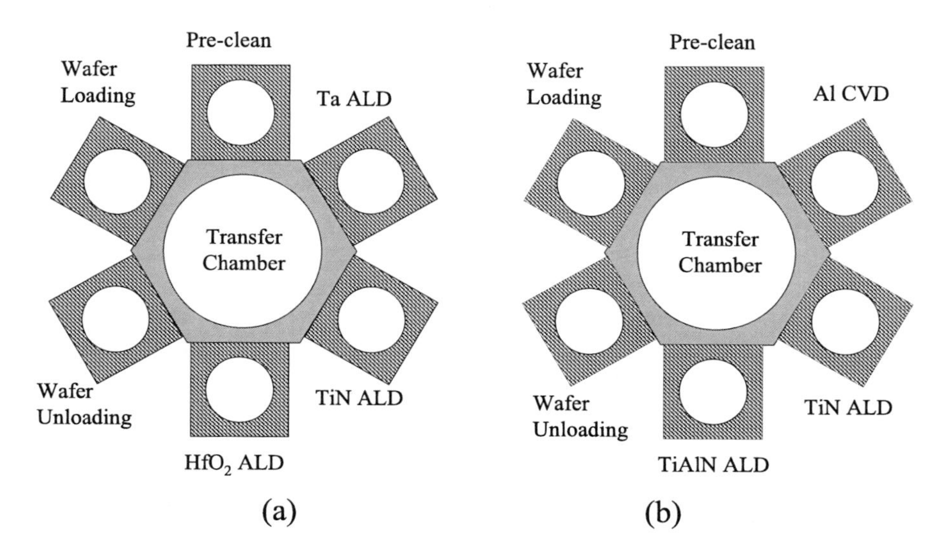

Figure 11.25 Cluster tools used for gate-last HKMG metallization.

thin film. It plays a very important role in metallization processes in semiconductor manufacturing.

Comparing CVD and PVD processes, the PVD process uses solid sources, while CVD processes use gaseous or vapor precursors. A CVD process relies on chemical reaction on the substrate surface; the PVD process does not. CVD film has better step coverage, and PVD film has better quality, lower impurity concentration, and lower resistivity. Figure 11.26 compares CVD and PVD processes.

For metallization processes in IC manufacturing, PVD processes are used to deposit Ni layers for NiSi formation, Ta barrier layers, TiN glue layers, Ta and TaN barrier layers, copper seed layers, Al–Cu layers, and TiN ARC layers. CVD processes are commonly employed to deposit TiN layers for barrier/adhesion, and tungsten for plugs.

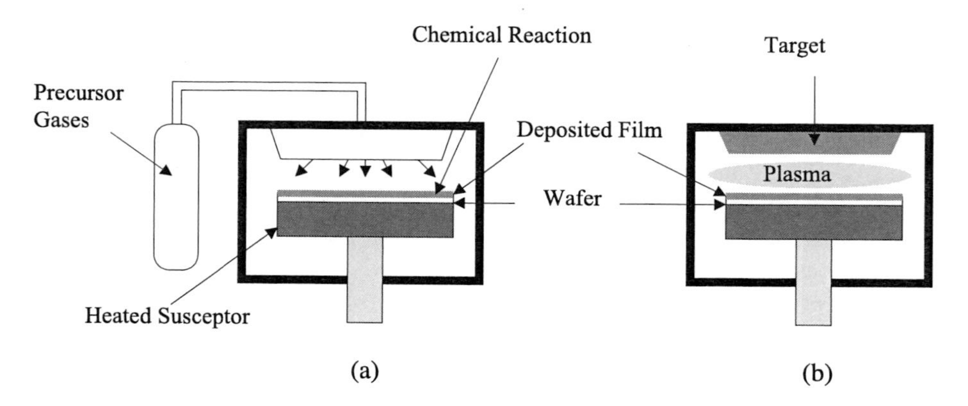

Figure 11.26 Comparison of (a) CVD and (b) PVD processes.

Two methods are used in metal PVD processes, evaporation and sputtering. Sputtering is the dominant PVD process in semiconductor fabrication processes because it can deposit high-purity, low-resistivity metal films with excellent uniformity and reliability.

11.5.2 Evaporation processes

In the early years of IC manufacturing, only aluminum was used for the metallization process, and thermal evaporation was generally used for Al deposition. Thermally evaporated Al films always have high mobile-ion concentration, which can affect transistors and circuits. An electron beam evaporator was developed to deposit high-purity Al and Al alloys.

11.5.2.1 Thermal evaporation

Aluminum has relatively low melting (660 °C) and boiling points (2519 °C), and it is relatively easy to vaporize under low pressure. In the early years of IC processing, thermal evaporators were widely used to deposit aluminum thin film to form gates and interconnections. A schematic of a thermal evaporation system is illustrated in Fig. 11.27. Throughout the process, the system needs to be under high vacuum, about 10^{-6} torr, to minimize residual oxygen and moisture. These residues can react with aluminum to form high-resistivity aluminum oxide and significantly increase the sheet resistance of the film.

Flowing a large amount of electric current through a tungsten filament heats up the filament by resistive heating, $P = RI^2$. A red-hot tungsten filament heats up the aluminum charge, melts it, and vaporizes it in the vacuum chamber. When aluminum vapor reaches the wafer surface, it recondenses and forms a thin layer of aluminum film on the surface.

In a filament evaporation system, a shutter mechanism is placed between the filament and wafers. At the beginning of the deposition process, the filament is heated to just above the metal melting point to melt all of the metal charge while the shutter is closed. After the temperature is stabilized and volatile impurities are

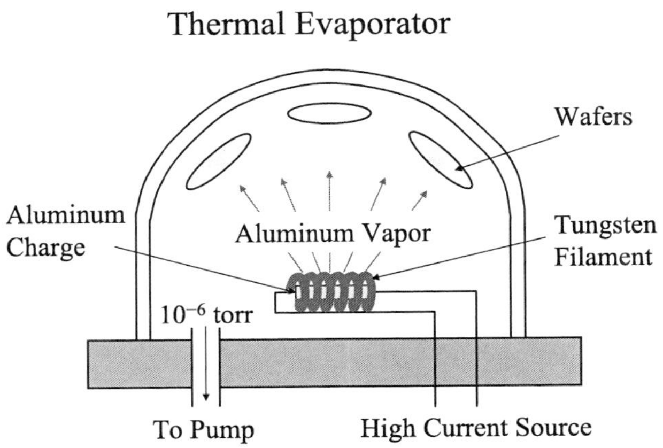

Thermal Evaporator

Wafers
Aluminum Charge
Aluminum Vapor
Tungsten Filament
10^{-6} torr
To Pump
High Current Source

Figure 11.27 A thermal evaporator.

driven away from the charge surface by the heat, the current ramps up to raise the temperature and evaporate the metal. Then the shutter is opened to allow the metal vapor to emit, reach the wafer, condense on the surface, and deposit metal thin film on the wafer.

For thermal evaporation deposition processes, the deposition rate of aluminum is related to the heating power, which is controlled by the electric current; usually, a higher current has a higher deposition rate. One important safety issue for the thermal evaporator is electrical shock. The high current (~10 A) used by an evaporator can cause a fatal electric shock in the case of direct contact. In fact, a 1-mA electric current across a human heart can be lethal.

Aluminum thin film deposited with a thermal evaporator always has a trace mount of sodium from the tungsten filament; it is high enough to shift the threshold voltage of MOSFETs and affect IC device reliability. It also has a low deposition rate and poor step coverage. It is also very difficult to precisely control the proper proportions for alloyed films such as Al-Si, Al–Cu, and Al–Cu-Si. It is no longer used for metallization processes in VLSI and ULSI chip manufacturing. It could still be used for wafer backside gold coating processes for facilitating the die attachment. Some university and college research and education laboratories still use thermal evaporation systems because they are a simple tool and are easy to operate with relatively low maintenance cost.

11.5.2.2 Electron beam evaporation

To replace filament heating, which can cause contamination and poor step coverage, electron beam (e-beam) heating technology was developed to evaporate metals for IC metallization. A beam of electrons, typically with energy of about 10 keV and currents up to several amperes, is directed at the metal in a water-cooled crucible in a vacuum chamber, and heats the metal to an evaporation temperature. During the evaporation deposition process, the outer portion of the charge does not melt and remains in a solid state, minimizing film contamination from the trace amounts of impurities inside the graphite or silicon carbide crucible. Figure 11.28 illustrates a typical e-beam evaporating system.

To improve deposition uniformity and step coverage, planetary wafer holding systems are used. By using multiple electron guns and crucibles, it is possible to simultaneously evaporate different metals, and deposit metal alloys such as Al-Si, Al–Cu, Al-Si-Cu, etc. Heating the wafers with IR lamps can raise the wafer temperature, increase adatom surface mobility, improve film step coverage, form large grains, and reduce resistivity.

Although e-beam evaporation has better processing results than thermal evaporation, it cannot match the metallization process results of sputtering deposition. It is rare to see an e-beam evaporator in an advanced semiconductor fab. The e-beam evaporator also can cause device damage due to x-ray radiation generated when the energetic electrons hit the metal charge. Some less advanced semiconductor fabs are still using this tool for metallization processes due to its high-throughput and low equipment cost.

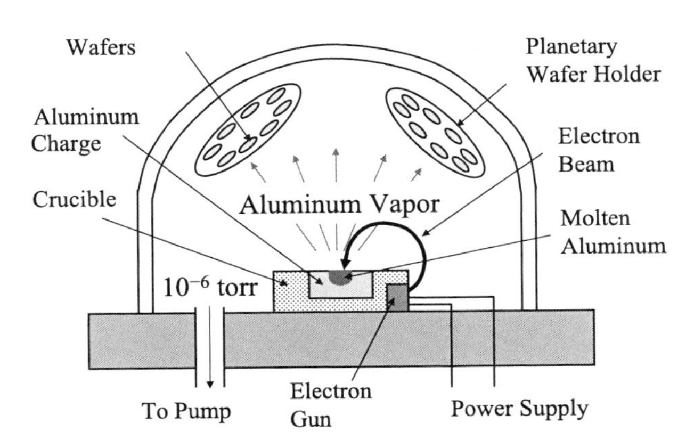

Figure 11.28 E-beam evaporator.

11.5.3 Sputtering

Sputtering deposition is the most common PVD process used for metallization in the IC industry. In fact, it has almost become a synonym for PVD in advanced semiconductor fabs. It involves energetic ion bombardments, which physically dislodge atoms or molecules from a solid metal surface and redeposit them on a substrate surface to form a metal thin film. Argon is the most frequently used gas for sputtering processes because it is inert, relatively heavy, abundant (~1% of atmosphere), and cost effective.

By applying electric power between the two electrodes under low pressure, a free electron between the electrodes is accelerated by the electric field and becomes more energetic as it gains energy from the electric field. When it collides with a neutral argon atom, it can knock one of its orbiting electrons out of the range of the attractive force of the nuclei to become a free electron. This is called an ionization collision, which generates a free electron and a positively charged argon ion. When a neutral argon atom loses a negatively charged electron in a collision, it becomes a positively charged argon ion. The two free electrons repeat this process to generate even more free electrons and ions, while electrons and ions are constantly lost due to collisions with electrodes, chamber walls, and electron ion recombinations. When their generation rate equals their loss rate, a steady state is reached, and stable plasma is generated.

While the negatively charged electrons are accelerated to a positively biased electrode called an anode, the positively charged argon ions accelerate toward a negatively biased cathode plate, called the target. The target plate is made from metal that is to be deposited on the wafer. When these energetic argon ions hit the target surface, atoms of the target material are physically dislodged from the surface by the momentum transfer from the impacting ions and are thrown into the vacuum in the form of a metal vapor. Figure 11.29 illustrates the sputtering process.

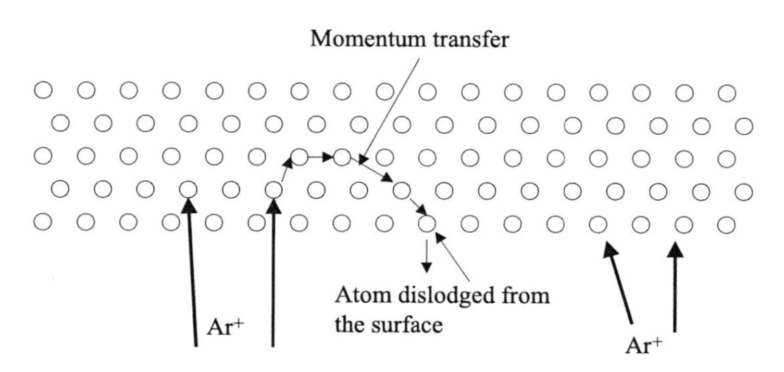

Figure 11.29 The sputtering process.

After the atoms or molecules leave the target surface, they travel inside the vacuum chamber in the form of a metal vapor. Eventually, some of them reach the wafer surface, adsorb on the surface, and become adatoms. The adatoms migrate on the wafer surface until they find nucleation sites and rest there. Other adatoms stop and recondense around the nucleation sites to form grains, which are in a single-crystal structure. When the grains grow and touch other grains, they form a continuous polycrystalline thin metal film on the wafer surface. The border between the grains is called a grain boundary, which can scatter electron flows and cause higher resistivity. Grain size is mainly determined by surface mobility, which is related to many factors, such as: wafer temperature, substrate surface condition, baseline pressure (contamination), and the final annealing temperature. Higher temperatures result in higher surface mobility and large-sized grains. Grain size has a strong impact on film reflectivity and sheet resistance. A metal film with a larger grain size has fewer grain boundaries for scattering electron flow and therefore has lower resistivity.

Question: High wafer temperatures increase grain size and improve conductivity, electromigration resistance, and film step coverage. Why do fabs not always use high wafer temperatures in PVD processes?

Answer: A metal film with large grain size is hard to etch with smooth sidewalls. Therefore, it is common practice to deposit metal films with smaller grain sizes at lower temperatures, and anneal metal films at higher temperatures after metal etch and photoresist strip. The annealing process forms larger grain sizes and reduces film resistivity. For Al–Cu alloy metal interconnects, processing temperature is limited to ~400 °C after deposition of the first layer of Al–Cu film.

The simplest sputtering system is the dc diode sputtering system, as illustrated in Fig. 11.30. In this system, a wafer is placed on a grounded electrode, and the target is a cathode, the negatively biased electrode. When a high-power dc voltage (several hundred volts) is applied to the system under low pressure, argon atoms are ionized by the electric field, and argon ions are accelerated and bombard the target, sputtering the target material from the surface.

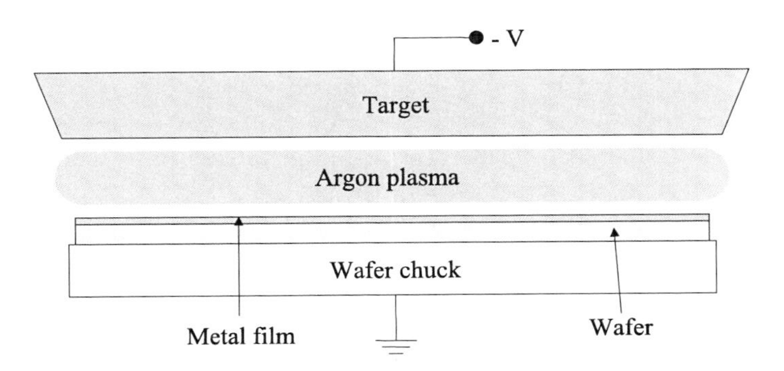

Figure 11.30 A dc sputtering system.

Other basic types of sputtering systems include dc triode, rf diode, dc magnetron, rf triode, and rf magnetron. The dc magnetron is the most popular method for PVD metallization processes because it can achieve high deposition rates, good film uniformity, good step coverage, high film quality, and easy process control. High deposition rates allow single-wafer sputtering deposition processes, which have several advantages over batch systems with planetary wafer holders. They have better WTW uniformity, higher system reliability, and lower particle contamination levels.

In a magnetic field, magnetic force causes gyromotion of charged particles. Because an electron is a very lightweight particle, it has a very small gyroradius and will be constrained near the magnetic field line. In a magnetic field, electrons have to travel much longer distances by spinning around the magnetic field lines; this gives electrons more chances for ionization collisions. Therefore, a magnetic field can help increase plasma density, especially at low pressure. In a dc magnetron sputtering system, a rotating magnet is placed on top of the metal target. The magnetic field generates higher plasma density and causes more sputtering near the magnet, where the magnetic field strength is stronger. By adjusting the location of the magnets, uniformity of the deposited film can be optimized. Magnetron sputtering creates a ring of erosion grooves on the target surface after a certain number of watt-hours of sputtering operation, as illustrated in Fig. 11.31.

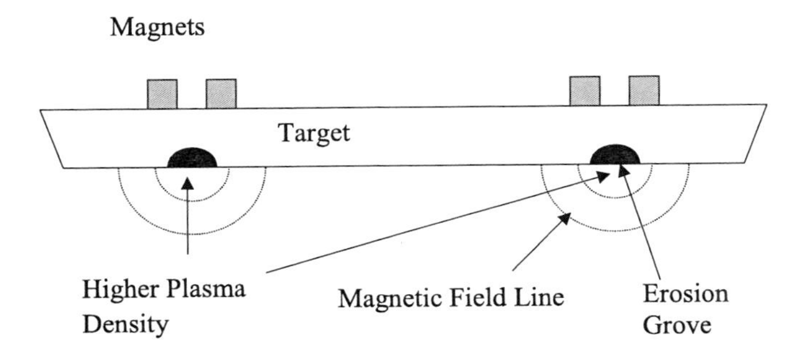

Figure 11.31 Magnetron sputtering system.

Another advantage of the sputtering process is that it can be easily used to deposit metal alloy films by using alloy targets with proper metal proportions. The sputtering deposition process can also be employed to deposit metal oxide or nitride, such as titanium nitride and tantalum nitride, in a reactive sputtering process by using oxygen or nitrogen with argon.

Normally, a shield or liner is installed inside a PVD chamber to protect the chamber wall and chamber parts from deposition of metal thin films, as shown in Fig. 11.32. The shield also serves as an anode of the dc discharge system while the wafer is placed on a floated chuck. After a certain amount of deposition, the metal-coated shield is replaced by a clean one during routine PM. The dirty one is sent out for cleaning and prepared for reuse.

With an rf power supply, a sputtering system can also be employed to deposit dielectric thin film by sputtering the nonconducting target. Radio frequency power can be applied to the backside of an insulating target and capacitively coupled to the plasma. Since electrons are much more mobile than ions in rf plasma, the surface of the insulating target accumulates net negative charges and dc voltage, up to a thousand volts. A negatively charged surface expels electrons and attracts positive ions, causing positively charged argon ions in the plasma to bombard the dielectric target surface and sputter molecules off the target surface to recondense on the wafer surface.

Argon is inert, relatively heavy, and inexpensive due to its abundance in the atmosphere. In fact, argon is the third most abundant elementary material in the atmosphere (about 1%), after only nitrogen (78%) and oxygen (20%). It is the most commonly used gas in sputtering processes, for both deposition and etch. It has also been applied in RIE and ion implantation for amorphous implantation. Table 11.8 lists some facts of argon.

11.5.4 Basic metallization processes

In an advanced wafer fab, the metallization process is normally performed in one integrated process sequence using a cluster tool with multiple chambers. Figure 11.33 shows a photograph of a cluster system used in advanced metallization processes.

Metal PVD chambers must achieve a high-baseline vacuum to minimize contamination, especially for aluminum deposition. It is necessary to achieve

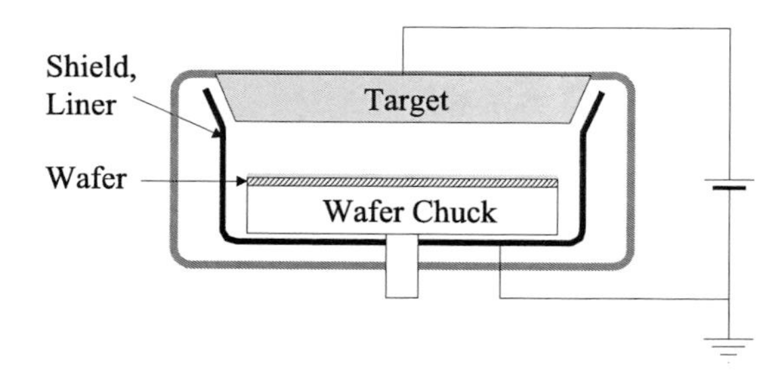

Figure 11.32 Sputtering system with a shield.

Table 11.8 Facts about argon.

Symbol	Ar
Atomic number	18
Atomic weight	39.948
Discoverer	Sir William Ramsay, Lord Rayleigh
Place of discovery	Scotland
Discovery date	1894
Origin of name	From the Greek word argos, meaning inactive
Molar volume	22.56 cm^3
Speed of sound in substance	319 m/sec
Refractive index	1.000281
Electrical resistivity	N/A
Melting point	−189.2 °C
Boiling point	−185.7 °C
Thermal conductivity	0.01772 W m^{-1} K^{-1}
Main applications in IC manufacturing	PVD, sputtering etch, dielectric etch, and ion implantation

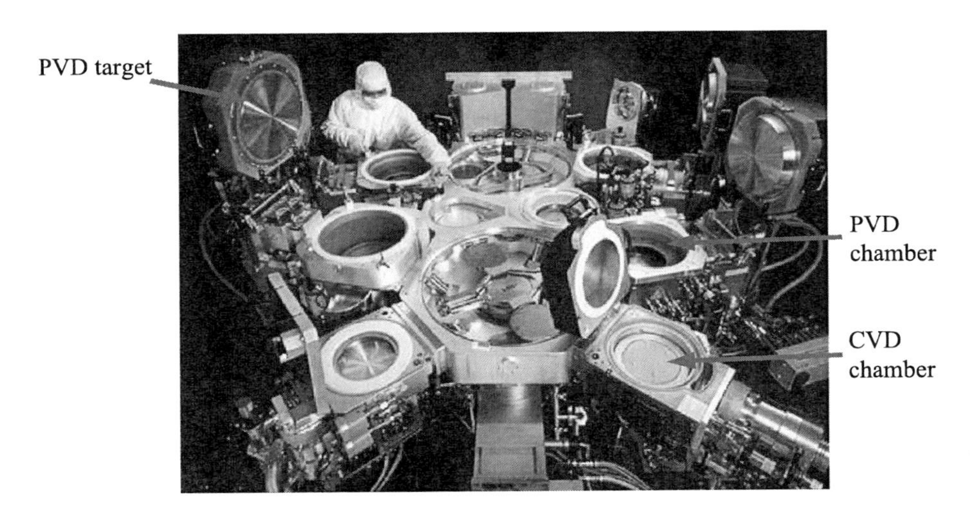

Figure 11.33 Cluster tool for integrated metallization process (Applied Materials, Inc.).

ultrahigh vacuum (UHV, $<10^{-9}$ torr) baseline pressure to minimize contamination and improve conductivity. Combinations of dry, turbo, and cryo-pumps are used to achieve the required vacuum level for PVD systems.

Since a metal target always grows a thin layer of native oxide when it is exposed to the atmosphere, a burn-in process is required to condition the target before it processes production wafers for a new target, or after opening a chamber for maintenance. By sputtering the target with argon ions, the native oxide and defects generated during the target manufacturing process can be removed. Several dummy wafers are needed to protect the wafer chuck during the burn-in process.

A PVD process for Al–Cu metallization involves several steps: degas, predeposition sputtering clean, barrier layer deposition, bulk aluminum alloy layer deposition, and ARC layer deposition. For copper metallization, the PVD process is used to deposit Ta barrier and Cu seed layers. For contact and via processes, a combination of PVD and CVD processes is used. The process starts with degassing and a predeposition sputtering clean, then Ti/TiN barrier/glue layers PVD and TiN CVD, and finally tungsten CVD to fill the holes.

11.5.4.1 Degas

Before starting the PVD process, it is very important to heat a wafer to a temperature high enough to drive out gases and moisture adsorbed on the wafer surface. Otherwise, outgassing of the adsorbed gasses and moisture during the deposition process can cause serious contamination and result in a high resistivity of the deposited metal film.

11.5.4.2 Preclean

Before metal deposition, a preclean step is required to remove the native oxide on the metal surface to reduce contact resistance. It is very important to remove this native oxide, since metal oxides such as aluminum and tungsten have very high resistivity. Without a precleaning process, contact resistance can be much higher than it should and can affect IC chip performance and reliability.

The argon sputtering process is commonly used for precleaning processes. In argon plasma, argon ions are accelerated by external power, in most cases by rf power. They bombard the wafer surface and blast off the ultrathin native oxide layer from the metal surface. This process exposes the metal under high vacuum and prepares the wafer for metal deposition. The precleaning process can also remove polymer residues from the bottom and sidewall of the contact/via holes. Carbon fluoride polymer residues are common byproducts of the silicate glass etching process when carbon fluoride etchant gases are used. Ion bombardment in the precleaning process always chops away some oxide from the upper corners of the contact/via holes and tapers the opening prior to the metal deposition process to improve step coverage and plug fill. Figure 11.34 illustrates the argon sputtering preclean process.

Both capacitively coupled plasma sources and inductively coupled plasma (ICP) sources are used for the preclean process. Advantages of an ICP plasma source are higher plasma density at lower pressure, and independent control of ion energy and flux.

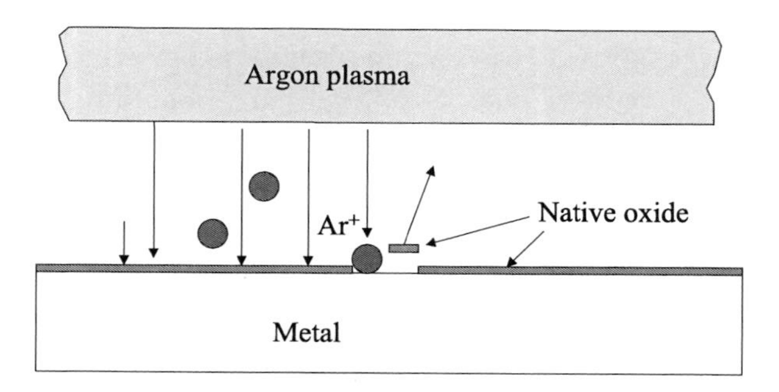

Figure 11.34 Argon sputtering preclean process.

11.5.4.3 Titanium physical vapor deposition

Before tungsten CVD and aluminum alloy PVD processes are implemented, a thin layer of titanium is always deposited to reduce contact resistance because titanium can trap oxygen and prevent it from bonding with tungsten or aluminum to form high-resistivity tungsten and aluminum oxides. For titanium deposition, larger grain size with low resistivity is desired. Therefore, a wafer is heated to approximately 350 °C during the deposition process to improve surface mobility of the titanium adatoms; this step also helps to improve step coverage.

For interconnection metal applications, titanium PVD with a standard magnetron chamber can meet the process requirements. However, for submicron contact/via applications, good bottom step coverage of the titanium film across the entire wafer is needed to reduce contact resistance. Standard magnetron systems can no longer meet this demand, since they always cause uneven step coverage from a wafer's center to its edge. Thus, collimating devices and metal ionization techniques have been developed for contact/via Ti/TiN sputtering processes.

A collimating system allows metal atoms or molecules to move mainly in a vertical direction, so that they can reach the bottom of the narrow contact/via holes. Although this slows down the deposition rate (because collimators stop many metal atoms and molecules from reaching the wafer surface), it improves bottom step coverage over the entire wafer. To compensate for the decrease in deposition rate, a higher dc power is applied to the electrodes to increase the sputtering effect. During this process, collimator holes near the center receive more deposition than those on the edge and begin to clog more quickly. Therefore, the magnetic system of a magnetron is designed to start the process with a center-thick deposition profile. The collimated sputtering system is illustrated in Fig. 11.35.

The rf current in inductively coupling coils can ionize metal atoms through an inductive coupling mechanism. The positive metal ions impact the negatively charged wafer surface mostly in a vertical direction, helping to improve bottom step coverage and reduce contact resistance. Ionized metal plasma systems have been used for copper metallization processes to deposit tantalum or tantalum nitride

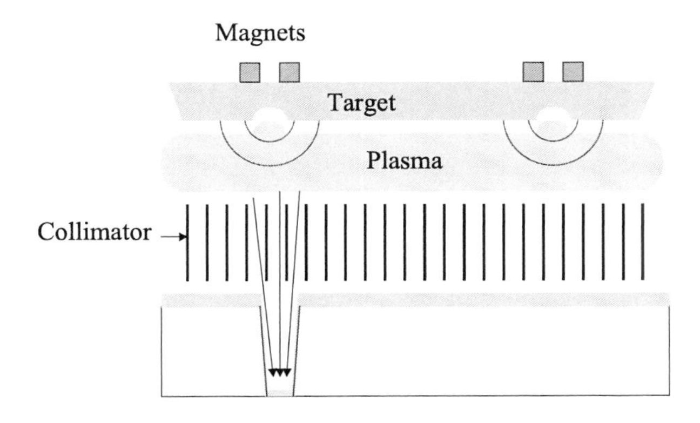

Figure 11.35 Collimated sputtering system.

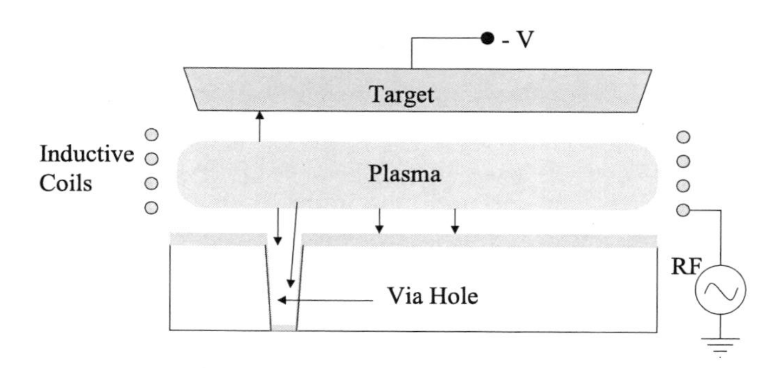

Figure 11.36 Ionized metal plasma system.

barrier layers and copper seed layers into high-aspect-ratio trench and via holes. An ionized metal plasma system is illustrated in Fig. 11.36.

The inductively coupled coils of an ionized metal plasma system are located inside the vacuum chamber and are made of the same material as the metal target because it will also be sputtered by ion bombardment.

11.5.4.4 Titanium nitride physical vapor deposition

Titanium nitride PVD uses a reactive sputtering process. By flowing nitrogen with argon into the processing chamber, some nitrogen molecules become dissociated by electron impact collisions in the plasma. Dissociated free nitrogen radicals have three unpaired electrons, which make them very reactive. Nitrogen radicals can react with sputtered Ti atoms to form and deposit TiN on the wafer surface. They also can react with a titanium target to form a thin layer of TiN on the target surface. Argon ions bombard the target surface and sputter titanium nitride from the target surface, then redeposit it on the wafer surface.

There are many applications involving titanium nitride, and all require different deposition processes. For TiN used as barrier and glue layers for tungsten, low resistivity and good step coverage (especially good bottom step coverage) are required. Therefore, collimated systems or ionized metal plasma systems are needed. For a TiN layer underneath Al–Cu, the main concern is lower resistivity. A standard magnetron system can meet the requirement. For ARC applications, low reflectivity is the major concern, and the coating can be deposited in a standard magnetron chamber. These three applications of titanium nitride are shown in Fig. 11.37.

TiN is also used for metal gate electrodes in HKMG processes. For gate-first HKMG, PVD TiN can be used. However, it is not suitable for gate-last HKMG because PVD nitride step coverage does not deposit film into narrow gate trenches very well. CVD TiN or ALD TiN can be used for gate-last HKMG.

Titanium nitride can be deposited in situ with titanium in the same PVD chamber. First, argon is used to sputter titanium from the target and deposit it onto the wafer surface. Then nitrogen flows into the processing chamber with argon for reactive sputtering, which deposits the titanium nitride. For integrated Ti and TiN in-situ deposition processes, a target cleaning process is necessary after the

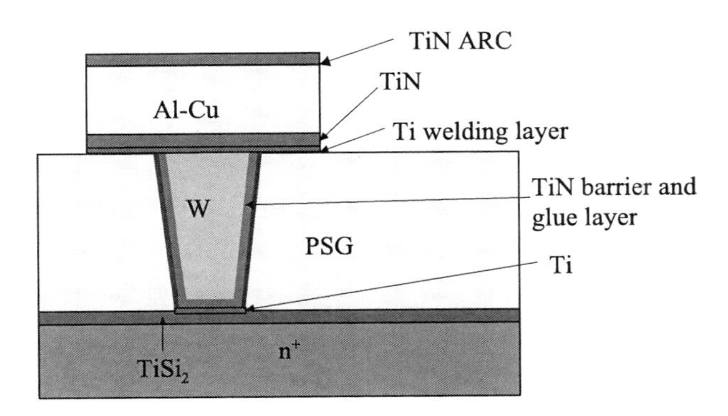

Figure 11.37 Three applications of TiN.

production wafer deposition process. Argon is used to sputter away the titanium nitride layer from the target surface to prepare it for the next titanium deposition process. A titanium dummy wafer is required during the cleaning process to protect the wafer chuck from being coated with titanium thin film. A silicon dummy wafer is not as good as a titanium dummy wafer because both titanium and titanium nitride stick well to a titanium wafer. This reduces the risk of particle contamination due to film cracking and flaking from the dummy wafer surface.

For a PVD chamber dedicated to TiN deposition, a periodic titanium coating is required to prevent flaking of the TiN layer off of the chamber parts and possibly causing particle contamination.

Nitrogen is the most commonly used gas in semiconductor fabs. It is used for nitride formation processes, such as silicon nitride and titanium nitride. It is the most abundant gas species in the earth's atmosphere (78%). Nitrogen is widely used as a purge gas because N_2 molecules are very stable, and it can be used as an inert gas as well for low-temperature (<700 °C) processes. Nitrogen is also used in pneumatics systems in processing tools. Some facts about nitrogen are listed in Table 11.9.

11.5.4.5 Aluminum–copper physical vapor deposition

Al–Cu alloy deposition requires an ultrahigh-baseline vacuum to achieve low film resistivity. At the start of the chamber pump-down process, air begins to be evacuated from the vacuum chamber. Residue gas inside the vacuum chamber, such as N_2 and O_2, is mainly from the atmosphere. When the chamber pressure reduces down to the millitorr range, the majority of residue gases are no longer the leftovers from the atmosphere, but the outgassing of adsorbed gases from the chamber wall. Moisture is one of these residual gases and is the most difficult to eliminate. If moisture is present during the aluminum sputtering deposition process, aluminum atoms will react with the residual H_2O to from aluminum oxide (Al_2O_3), which is a very good insulator. Integration of a tiny amount of oxygen in aluminum alloy film can significantly increase the film's resistivity. Therefore, to achieve high-quality, low-resistivity aluminum film deposition, the processing chamber must achieve a very high baseline vacuum to minimize the amount of moisture inside

Table 11.9 Facts about nitrogen.

Symbol	N
Atomic number	7
Atomic weight	14.007
Discoverer	Daniel Rutherford
Place of discovery	Scotland
Discovery date	1772
Origin of name	From the Greek words nitron genes, meaning nitre and forming, and from the Latin word nitrum (nitre is a common name for potassium nitrate)
Molar volume	13.54 cm^3
Speed of sound in substance	333.6 m/sec
Refractive index	1.000298
Electrical resistivity	N/A
Melting point	−209.95 °C
Boiling point	−195.64 °C
Thermal conductivity	0.02583 W m^{-1} K^{-1}
Main applications in IC manufacturing	As a purge gas for all IC manufacturing equipment, CVD precursor and carrier gas, and PVD reactive gas.

the chamber. A cluster tool is usually equipped with a staged vacuum system using a combination of pumps—including dry, turbo, and cryo-pumps—to achieve UHV in the aluminum PVD chambers. A cryo-pump can help a PVD chamber reach up to 10^{-10}-torr base pressure by freezing residual gases in a freeze trap.

There are two types of aluminum processes, standard and hot aluminum. A standard process is used in the deposition of aluminum alloy over a tungsten plug, normally after titanium and titanium nitride wetting layers. The main concerns with this process is uniformity of the deposited film and the consistency of the process during the lifetime of the target. Al–Cu alloy film deposited at a higher temperature has larger grain size, better electromigration resistance (EMR), and lower film resistivity. The problem is that films with larger grain sizes are difficult to etch with smooth pattern sidewalls. A standard aluminum process is normally operated at around 200 °C and can deposit metal films with smaller grain sizes, making it easier to achieve good interconnect line pattern etching. Larger grain sizes with high EMR and lower resistivity can be achieved later with a metal annealing process after photoresist stripping.

Some IC fabs have attempted to replace tungsten plugs with aluminum plugs for contact and via hole filling. Hot aluminum processing was developed for this application, which allows aluminum to fill contact and via holes. This reduces resistance between the conducting layers, since PVD aluminum is a much better conductor (2.9 to 3.3 μΩ·cm) than CVD tungsten (8 to 11 μΩ·cm). Hot aluminum processing consists of several steps. First, either titanium or both titanium and titanium nitride are deposited to lower contact resistance, and to prevent junction spiking for contact with silicon. A thin aluminum film is deposited at a low temperature (<200 °C) as the seed layer for the hot aluminum deposition. Then a thicker aluminum layer is deposited at a higher temperature (450 to 500 °C). Hot aluminum can diffuse into the contacts/via holes due to the high adatom surface mobility. Hot aluminum fills the holes without voids while keeping the aluminum surface relatively planarized due to thermal flow. It also forms interconnections between the conduction layers with aluminum plugs.

High-pressure aluminum reflow processes have also been developed to fill the small contact/via holes with aluminum. Another approach for aluminum metallization is the combination of aluminum CVD and PVD processes.

11.6 Copper Metallization

Copper has a lower resistivity than an aluminum–copper alloy. It also has better electron migration resistance because copper atoms are much heavier than aluminum. Copper has always been an attractive choice for metal interconnections in IC chips. By using copper to replace the aluminum alloy, interconnection resistance can be significantly reduced due to copper's lower resistivity (1.8 to 1.9 μΩ·cm for copper versus 2.9 to 3.3 μΩ·cm for aluminum–copper alloy). The high EMR of copper allows higher current density in copper interconnection lines. Both properties help increase IC speeds. However, the high diffusion rates of copper in both silicate glass and silicon, and device reliability degradation from copper contamination in silicon substrates, hampered copper applications from 1960 to 1970. After the 1980s, when plasma dry etch replaced wet etch for pattering etch processing, lack of volatile inorganic copper compounds made application of copper for IC metal interconnections even more difficult.

In the 1990s, development of the CMP process paved the way for copper interconnections by applying the dual damascene process, which does not require a metal etching step. There are several other applications for the dual damascene copper interconnection process. A barrier layer of copper can be deposited into high-aspect-ratio via holes to prevent copper diffusion. This barrier layer provides good sidewall and bottom step coverage, good adhesion to dielectrics, and low contact resistance. Other applications are the deposition of copper film with high-quality, low-resistivity, and void-free filling of high-aspect-ratio trench and via holes. Finally, there are defect-free copper polishing and post-CMP cleaning processes. A copper interconnection processing sequence is illustrated in Fig. 11.38.

11.6.1 Precleaning

All metallization processes require a precleaning process to remove thin native oxide and possible polymer depositions from dielectric etch on the metal surface and bottom of the via holes. It is a very important step because insufficient cleaning can result in high contact resistance, induced by either the presence of native oxide between the metal layers or voids in the via plugs.

Like other metallization processes, argon sputtering etch is predominantly used for predeposition cleaning to remove native oxide and any other materials on the metal by physical ion bombardment. Radio frequency power is used to generate argon plasma that is either inductively coupled, called the source rf, or capacitively coupled, called bias rf. Source rf mainly controls ion flux, whereas bias rf controls ion bombardment energy. After a sputtering clean removes native oxide at the

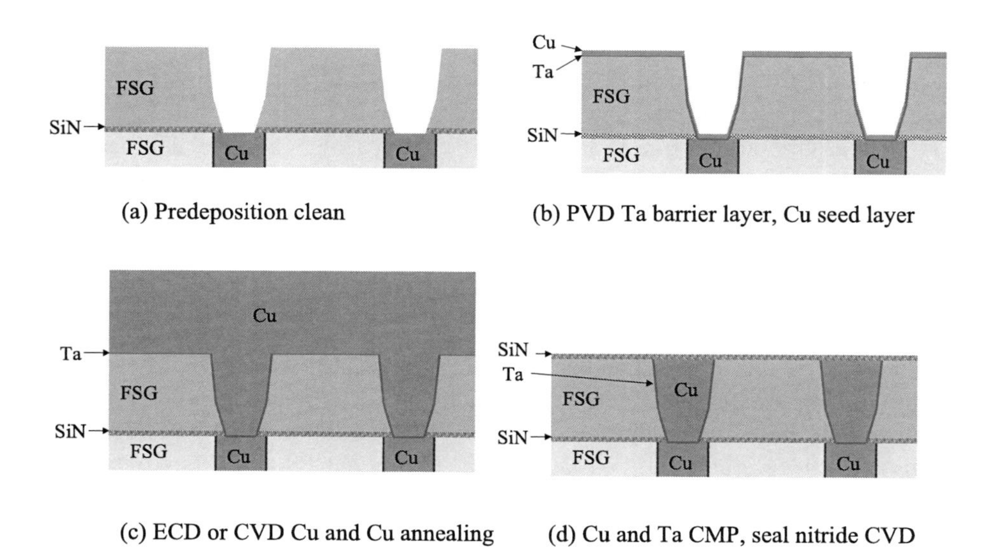

(a) Predeposition clean (b) PVD Ta barrier layer, Cu seed layer

(c) ECD or CVD Cu and Cu annealing (d) Cu and Ta CMP, seal nitride CVD

Figure 11.38 Dual damascene copper interconnection process.

bottom of the metal surface, the wafer is transferred to a PVD chamber in an UHV environment, which prevents the metal surface from reoxidizing. Because barrier and seed layer depositions require a low wafer temperature, an electrostatic chuck (E-chuck) system is desired for the sputtering etch chamber, since it can effectively cool down the wafer during the sputtering etch process; this eliminates the cooling time required for systems without E-chuck. It is difficult to transfer heat effectively in the UHV environment required for PVD processes.

At the beginning of copper metallization, an argon sputtering preclean process is used to remove copper oxide before barrier and seed layer deposition. As technology nodes continue to shrink, low-κ and porous low-κ dielectrics have been integrated with copper to form interconnections. Fabs started to worry about possible contamination issues due to Ar sputtering preclean because this process can sputter copper from the bottom to the sidewall of a via and penetrate the pores causing shorts between copper wires. Also, copper has a high diffusion rate in silicate glass; thus, it can diffuse into the silicon substrate later and cause reliability problems for the IC chip due to device malfunction from heavy metal contamination. Therefore, reactive preclean techniques were developed. By striking plasma with hydrogen (H_2) and helium (He) gases, free hydrogen radicals (H) can be generated at low pressures. Hydrogen radicals can diffuse into the via holes, react with copper oxide, and replace the copper atoms to bond with oxygen. This process forms moisture, which is driven away by the heated wafer surface, and effectively removes native oxide from the copper surface at the bottom of the via. The chemical reaction of hydrogen reduction in copper precleaning processes can be expressed as

$$4\,H + CuO_2 \rightarrow Cu + 2\,H_2O.$$

Reactive precleaning has some advantages, but it also introduces hydrogen, which is flammable and explosive, into the system. Since the wafer heats up during this process, it needs extra time to cool down after precleaning; this cooldown time can affect throughput. Most likely, engineers will continue to use the production-proven Ar sputtering etch for precleaning processes in copper metallization until it no longer works.

11.6.2 Barrier layer

Several barrier materials such as Ti, TiN, Ta, TaN, W, WN, etc., have been studied for adoption as diffusion barrier layers for copper metallization processes to prevent copper from diffusing into the silicon substrate and damaging microelectronic devices. Both tantalum and tantalum nitride can be used for this application. Most fabs use a few hundred angstroms of tantalum or tantalum nitride, or a combination of both films, as a copper barrier layer.

Requirements for the tantalum barrier layer are low contact resistance, good bottom and sidewall step coverage, and film quality integrity on the trench/via sidewalls. An ionized metal plasma chamber is customarily used to achieve high bottom step coverage and reduce contact resistance. Although high deposition temperatures improve deposited film step coverage, quality, and electrical conductivity, low temperatures are desired for barrier layer deposition because they deposit film with smoother surfaces. At lower temperatures, adatoms have lower surface mobility, causing smaller grain size and smoother film surfaces, which are important for copper seed layer deposition. To effectively remove heat generated on the wafer surface due to ion bombardment during deposition, an E-chuck system is desired for the barrier layer deposition chamber. Conductivity of the metal films can be improved by a thermal annealing process after deposition of copper seed and bulk layers.

11.6.3 Copper seed layer

Before bulk copper deposition (either by ECP or by CVD), a copper sputtering deposition process is always needed to deposit a thin seed layer (500 to 2000 Å). The process provides nucleation sites for bulk copper grain and film formation. Without this conducting surface, copper atoms will not stick to the wafer surface very well when they migrate to the surface. If this seed layer is not present, there could be either no deposition or deposition with poor uniformity.

An ionized metal plasma system is used for this application, since it has good bottom step coverage. Deposition is performed at a low temperature to achieve a smooth film surface, which is important for bulk copper deposition with ECP. A rough seed layer surface in the trench and via sidewalls can cause voids during ECP bulk copper deposition. As in the barrier layer deposition, an E-chuck system is preferred in the PVD chamber for copper seeding.

Since copper has low ionization energy, copper vapor can be easily ionized. After initial argon plasma bombardment, argon flow can be turned off, and plasma is almost solely formed with copper vapor. At low pressure, the MFP of a copper ion is much longer than the via depth (a few thousand angstroms); therefore, copper ions can be thrown to the bottom of the high-aspect-ratio trench and via holes to achieve reasonably good bottom and sidewall coverage and smooth film surface. As feature sizes shrink, via holes could become so narrow that PVD copper can no longer achieve the requirements for a seed layer due to its poor step coverage; CVD or ALD copper processes could be used for this application. This is discussed in a later section.

11.6.4 Copper electrochemical plating

ECP is an old technology still used for metal plating processes in a wide range of industries including hardware, glass, automotive, and electronics. It is extensively used to plate copper layers on both sides of fiber plastic boards for printed circuit board fabrication. ECP processes were developed for copper deposition in IC interconnection processes at the end of the 1990s and became the process of choice in the early 2000s. One of the advantages of ECP is that it is a low-temperature process. This makes it compatible with the low-κ polymeric dielectrics that could be used in future interconnection processes. Other names used for ECP are electrochemical deposition (ECD) and electroplating deposition (EPD).

The copper ECP process involves the following steps. First, a wafer is placed on a plastic wafer holder. A cathode holds the wafer with a conducting seal ring and immerses it into plating solution containing sulfuric acid (H_2SO_4), copper sulfate [$Cu(SO_4)$], and other additives. An electric current flows from the anode (which is a pure copper plate) to the cathode. In the solution, the $Cu(SO_4)$ dissociates to copper (Cu^{2+}) and sulfate (SO_4^{2-}) ions. Copper ions in the solution flow to the wafer surface carrying the electrical current that follows the applied electric field. When they reach the wafer surface, copper ions start to adsorb on the surface, nucleate, and deposit copper film on the copper seed layer (or strike layer). At the same time, some copper atoms on the anode surface are ionized, leave the plate surface, and dissolve in the ECP solution. Figure 11.39 shows the copper ECP process.

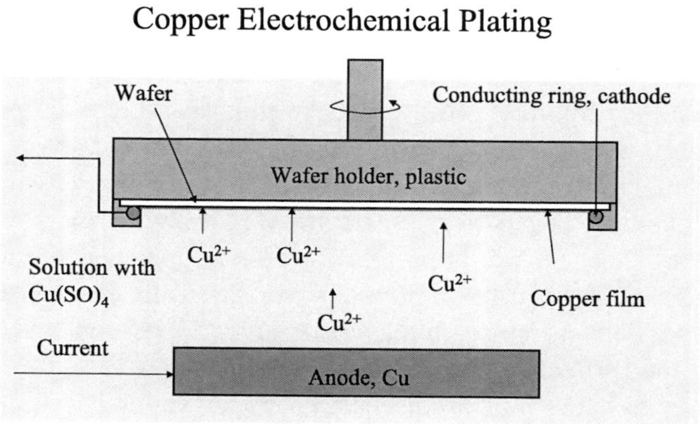

Copper Electrochemical Plating

Wafer
Conducting ring, cathode
Wafer holder, plastic
Cu^{2+}
Solution with $Cu(SO)_4$
Copper film
Current
Anode, Cu

Figure 11.39 Copper ECP process.

During the ECP process, a copper-sulfate solution flows into the high-aspect-ratio trench and via holes due to surface tension at the liquid–solid interface. Sulfuric acid in the solution dissolves the native oxide from the surface of the copper seed layer and exposes the copper underneath, similar to a pre-PVD sputtering etch. The applied electric current drives the copper ions toward the sidewalls and bottoms of the via holes. With the presence of the copper seed layer, a conformal film on the sidewalls and bottoms of the via holes is deposited. The consumed copper ions in the sulfate solution inside the trench and vias are constantly resupplied by the diffusion of copper ions. Thus, the conformal copper film eventually fills the trenches and via holes without voids (see Fig. 11.40).

Increasing the driving current between the two electrodes increases the copper deposition rate. However, high deposition inside the trench and via holes will consume all of the copper ions present there. When copper ions outside the trench and via holes diffuse into them (driven by the concentration gradient), they are more likely to deposit on the gap corners and form overhangs. This can cause trouble in via filling and contribute to void formation. To achieve better gap fill, a pulse current with large forward amperage and small reversed amperage is used. During a reversed current pulse, copper is removed from the wafer surface, reducing the overhang of the gap. Therefore, an ECP process with a pulse-driven current has effects similar to the dep/etch/dep process described in Chapter 10. Some additives such as inhibitors are used to reduce deposition on the corner of the trench and via holes, and other additives such as accelerators are used to accelerate the deposition on the bottom. Thus, the ECP process can deposit copper film from the bottom up to further improve via and trench filling abilities.

Some fabs have attempted to develop electrochemical processes which—after copper fills the via holes and trenches—can reverse the electrodes, making the copper-plated wafer the anode, and the copper plate the cathode. The majority of bulk copper film on the wafer surface can be etched away with an electrochemical etching process.

After copper ECP processing, the wafer is rinsed with deionized (DI) water to remove chemical solutions on the surface and avoid metal corrosion. Then the

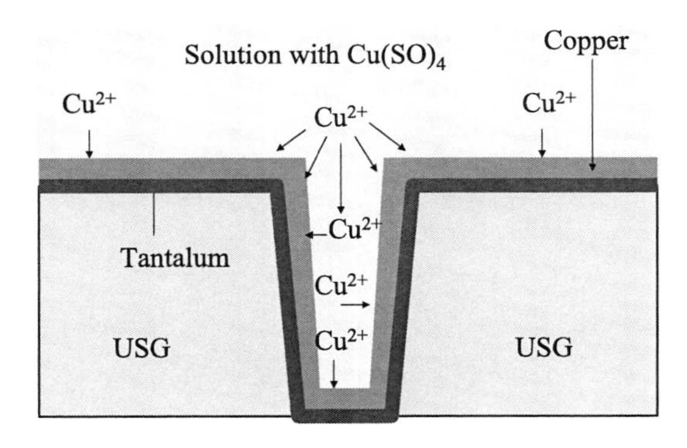

Figure 11.40 ECP via fill.

wafer is spin dried, and a robot removes it from the ECP tool (called a plater). An ECP copper wafer must be annealed in a thermal furnace to densify the film and reduce resistivity before being sent to the CMP bay to remove the bulk copper film on its surface. In an approximately 30-min thermal process at around 250 °C, copper grain size, film density, and conductivity can be increased. ECP processes can provide acceptable films with excellent trench and via fill abilities for copper dual damascene processes in deep submicron IC metallization applications.

Some challenging issues for ECP processes are determining and controlling three aspects—the copper ion concentration in the copper sulfate solution, the effects of additives, and the repeatability of trench and via filling as device dimensions continue to shrink. For small trench and via holes, the liquid surface tension tends to create bubbles inside the via holes that prevent deposition and cause voids.

11.6.5 Copper chemical vapor deposition

Another approach to filling trench and via holes is copper CVD. Frequently used copper precursors are bis-hexafluoroacetyl-acetonate copper [$Cu(hfac)_2$] and hexaflouroacetyl-acetonato-copper-vinyl-trimethylsilane [Cu(hfac)(vtms)]. $Cu(hfac)_2$ is a solid at room temperature, and it sublimes at low heat (35 to 130 °C). Its chemical structure is shown in Fig. 11.41. $Cu(hfac)_2$ can deposit copper by the hydrogen reduction reaction:

$$Cu(hfac)_2 + H_2 \rightarrow Cu + 2\ H(hfac).$$

Copper deposited in this reaction requires a temperature of 350 to 450 °C to achieve low resistivity. This is a little too high for process integration with the polymeric low-κ dielectric needed in IC interconnection applications.

Cu(hfac)(vtms) is a liquid at room temperature. It can deposit copper at a low temperature (<200 °C) with high quality, low resistivity, and good gap and via filling abilities. The chemical reaction is expressed as

$$2\ Cu(hfac)(vtms) \rightarrow Cu + Cu(hfac)_2 + 2\ (vtms).$$

This reaction is reversible; therefore, vtms can be used to dry clean the deposition chamber. By removing copper deposited on the chamber wall and parts

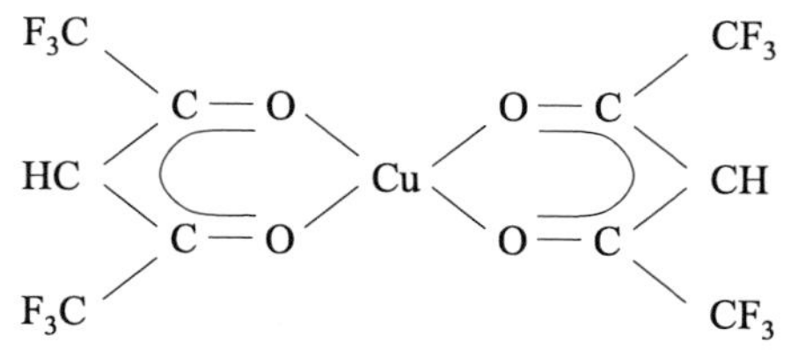

Figure 11.41 Chemical structure of $Cu(hfac)_2$.

inside the chamber, particle contamination caused by film cracking and flaking can be prevented. Because byproducts of the process are stable, they can be easily stripped and recycled for future applications. The Cu(hfac)(vtms) process is the most promising copper CVD process. However, it faces tough competition from the production-proven copper ECP process.

One possible application of copper CVD is depositing a conformal copper seed layer in narrow trenches and via holes, which is an increasingly difficult achievement using PVD processes with shrinking feature size. However, PVD engineers are still attempting to develop a copper seed PVD process that will meet next-generation copper/ULK interconnection requirements. It is quite possible that copper CVD will never become mainstream in the IC manufacturing process.

11.7 Safety

Metal sputtering PVD processes do not use any dangerous chemicals; their processing gases are argon and nitrogen, both of which are considered safe gases. However, metal CVD processes use a variety of chemicals with different chemical hazards.

WF_6, SiH_4, and H_2 are widely used in tungsten and tungsten silicide deposition. WF_6 is corrosive; silane is pyrophoric, explosive, and toxic; and hydrogen is flammable and explosive. TDMAT is commonly used for titanium nitride deposition and is very poisonous. DMAH is a precursor used for aluminum deposition; it is pyrophoric and explosive.

Other safety issues are electrical and mechanical in nature, such as electrical shock from dc and rf power sources, moving parts, and hot surfaces. Detailed information about safety issues related to the processes and processing tools should be obtained from the equipment providers.

11.8 Summary

- Metallization processes are mainly used in the formation of metal interconnections.
- For mature IC products using Al–Cu interconnections, the most commonly used metals are Al, W, Ti, and TiN.
- For advanced IC chips with copper metallization, the most frequently used metals are Cu, Ta, and/or TaN.
- Metals can be deposited with CVD, PVD, ALD, and EPD processes.
- Higher deposition temperatures can increase metal grain size, which reduces film resistivity.
- Silicides are used to reduce the resistance of local interconnections and contact resistance.
- CVD tungsten can fill high-aspect-ratio contact and via holes and is commonly used as metal plugs in interconnection processes.
- Titanium is used to reduce contact resistance.

- Titanium nitride is frequently used as a tungsten diffusion barrier layer and adhesion layer. It is also used as an ARC to improve resolution for metal patterning processes. TiN has been used as hard masks for ULK dielectric etching. It is also used as metal gates for HKMG MOSFETs.
- In a sputtering process, argon ions bombard a metal target, dislodging metal atoms or molecules from the surface as vapor. These atoms or molecules travel to the substrate surface, adsorb, and deposit on the surface as a thin film.
- A dc magnetron sputtering system is the most commonly used sputtering system in metal PVD processes.
- A high-baseline vacuum is required in metal PVD processes to reduce contamination and resistivity of deposited films. For aluminum PVD, UHV baseline pressure (10^{-9} torr) is needed to minimize moisture concentration inside the PVD chamber.
- Titanium nitride can be deposited with a reactive sputtering process by using nitrogen with argon in the plasma.
- CVD titanium nitride is commonly used in submicron IC interconnection processes due to its excellent sidewall step coverage.
- For contact and via application, either a collimated or ionized metal plasma system is used for Ti and TiN PVD to improve bottom step coverage and reduce contact resistance.
- Poor adhesion to silicon dioxide, high diffusion rates in silicon and silicon dioxide, device degradation due to deep-level acceptors caused by copper contamination, and hard-to-dry etching due to lack of simple volatile chemical compounds prevented copper applications in IC chip metallization processess before the 1990s.
- Improvement in copper barrier deposition, copper deposition, and more importantly, the copper CMP process, paved the way for copper metallization, which is the future of IC metal interconnections.

11.9 Bibliography

D. G. Baldwin, M. E. Williams and P. L. Murphy, *Chemical Safety Handbook for the Semiconductor/Electronics Industry*, 2nd ed., OME Press, Beverly, MA, 1996.

J. Baliga, "Depositing Diffusion Barriers," *Semicond. Intl.* **20**(3), pp. 76–81 (1997).

A.E. Braun, "Copper Electroplating Enter Mainstream Processing," *Semicond. Intl.* **22**(4), pp. 58–66 (1999).

C. Y. Chang and S.M. Sze, *ULSI Technologies*, McGraw-Hill, New York (1996).

A. V. Gelatos, C. J. Mogab, R. Marsh, E. T. T. Kodas, Jain, A, and M. J. Hampden-Smith, "Selective chemical vapor deposition of copper using (hfac) copper(I) vinyltrimethylsilane in the absence and presence of water," *Thin Solid Films* **262**(1–2), pp. 52–59 (1995).

D. James, "Intel pushes lithography limits, co-optimizes design/layout/process at 45 nm," *Solid State Technol.*, 30–33 (March 2007).

J. A. Kittl, W. T. Shiau, D. Miles, K. E. Violette, J. C. Hu, and Q. Z. Hong, "Salicides and alternative technologies for future ICs: Part 1," *Solid State Technol.* **42**(6), pp. 81–92 (1999).

J. A. Kittl, W. T. Shiau, D. Miles, K. E. Violette, J. C. Hu, and Q. Z. Hong, "Salicides and alternative technologies for future ICs: Part 2," *Solid State Technol.* **42**(8), pp. 55–62 (1999).

X.W. Lin and Dipu Pramanik, "Future interconnection technologies and copper metallization," *Solid State Technol.* **41**(10), pp. 63–78 (1998).

R. J. Matyi, L. E. Depero, E. Bontempi, P. Colombi, A. Gibaud and M. Jergel, "The international VAMAS project on x-ray reflectivity measurements for evaluation of thin films and multilayers: preliminary results from the second round-robin," *Thin Solid Films* **516**(22), 7962–7966 (2008).

K. A. Mistry et al., "A 45-nm logic technology with high-k & metal gate transistors, strained silicon, 9 Cu interconnect layers, 193-nm dry patterning, and 100% Pb-free packaging," *Proc. IEDM*, 247–250 (2007).

C. Ryu, H. Lee, K. W. Kwon, A. L. S. Loke, and S. Wong, "Barriers For Copper Interconnections," *Solid State Technol.* **42**(4), pp. 53–56 (1999).

L. Shon-Roy, A. Wiesnoski, and R. Zorich, *Advanced Semiconductor Fabrication Handbook*, Integrated Circuit Engineering Corp., Scottsdale, AZ (1998).

See http://www.webelements.com for further information on the elements in this chapter.

11.10 Review Questions

1. List the four metals commonly used in IC chip manufacturing.
2. Why does aluminum always alloy with copper? Why does it sometimes also alloy with silicon?
3. What is the advantage of an e-beam evaporator over a filament evaporator?
4. Why is tungsten used for metal plugs to connect different layers of conductors?
5. List at least four applications of titanium nitride.
6. What is the metrology tool used to measure sheet resistance?
7. Describe the sputtering deposition process.
8. Explain why a dc sputtering system cannot be used to deposit dielectric materials.
9. In what application does Ti and TiN PVD need a collimated sputtering system?
10. Why does an aluminum alloy sputtering chamber require a UHV?
11. What could happen to deposited film if the argon gas line has a small leak?
12. If the power in a dc magnetron sputtering system is increased, how will the deposition rate change?

13. If the deposition temperature in a dc magnetron sputtering system is increased, how will the metal film sheet resistance change?
14. List the reasons that hampered the application of copper in IC fabrication before 1990s.
15. Compare copper metallization and the standard aluminum alloy-tungsten metallization; what are the main differences?
16. List three methods used in copper deposition.
17. For HKMG processes, which approach uses CVD Al, gate-first or gate-last?
18. What are work-function metals for pMOS and nMOS?
19. Why do fabs use Ta in gate-last HKMG MOSFETs?
20. How is it that DRAM can use tungsten wires while CMOS logic ICs cannot?

Chapter 12
Chemical Mechanical Polishing

CMP is a removal process that strips small portions of film deposited on a wafer by a combination of chemical reaction and mechanical polishing, thus making the surface smoother and more planarized. It is also used to remove bulk dielectric film on the surface to form STI on the silicon substrate, and remove bulk metal film from the wafer surface to form metal interconnection lines or plugs in the dielectric film. This chapter covers CMP processes.

Objectives

After finishing this chapter, the reader will be able to:

- list the applications of CMP
- describe the necessity of dielectric planarization
- describe the basic structure of a CMP system
- list the differences between oxide CMP slurry and metal CMP slurry
- describe the oxide CMP process
- explain the metal polishing process
- describe the importance of post-CMP cleaning
- describe CMP applications in copper metallization
- list two CMP processes required in gate-last high-κ, metal gate MOSFET fabrication processes.

12.1 Introduction

After wafers have been sawed from a single-crystal silicon ingot, there are many processing steps to prepare the flat, shiny, and defect-free wafer surfaces for IC processing. Along with wafer edge rounding, lapping, and etching, there is always a CMP process used in the final steps of wafer manufacturing that planarizes the wafer and virtually eliminates all defects from the surface caused by the wafer sawing process. However, the initial reaction of the industry to CMP processing was not encouraging. ("You're doing *what* to my wafers?!")

Traditionally, direct contact with the frontal surface of a product wafer in semiconductor fabs was strictly forbidden. The reason is simple and straightforward: any direct contact can create defects and particles that reduce the yield of IC chips and cause lower profit margins for the IC fab. In the case of CMP, a wafer is not only held face down and pushed forcefully against a rotating polishing pad, but it also endures this process in the presence of alkaline or acidic polishing slurry, which contains large amounts of silica or alumina particles. To the initial surprise of skeptics, the CMP process not only planarizes wafer surfaces (as it was designed to accomplish), but it also reduces defect density and improves IC chip yield.

As the technology matured, most IC companies gradually adopted the CMP process as a standard routinely performed in semiconductor fabs. A brief overview of CMP development history, advantages of CMP, and applications of CMP processes are given in this section.

12.1.1 Overview

After the 1980s, more than two metal layers were required to connect the increasing number of transistors in IC chips. One of the great challenges was planarization of the dielectric layers between the metal layers. It is very difficult to achieve high-resolution photolithography on a rough surface due to the DOF requirements of optical systems. A rough dielectric surface can also cause problems in metallization because metal PVD processes usually do not have good sidewall step coverage. Thinner metal lines on the sidewall have a higher current density and are more vulnerable to electromigration issues.

Figure 12.1 shows a brief summary of the IC manufacturing flow and makes it clear that CMP is a very important part of the process. Wafers come to CMP from a thin-film process, either dielectric or metal thin-film deposition. In most cases, wafers are transported out of the CMP bay into a photo bay or thin-film bay. Dielectric CMP is followed by either photolithography or metal thin-film deposition, and metal CMP is only followed by dielectric thin-film deposition.

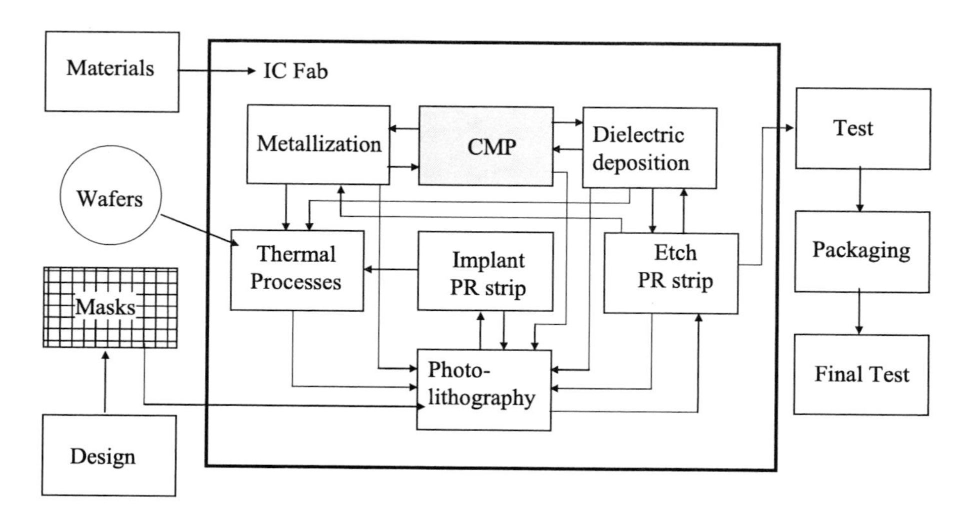

Figure 12.1 CMP applications in IC manufacturing flow.

Several methods, such as thermal flow, sputtering etchback, photoresist etchback, and SOG etchback, have been used for dielectric planarization. Dielectric CMP processing, pioneered by IBM in the mid-1980s, was developed for dielectric planarization applications. In fact, some fabs would prefer that the acronym CMP stand for chemical mechanical planarization.

Tungsten has been used to form metal plugs to connect different layers of conductors. CVD tungsten fills narrow contact and via holes and also covers the surface of the entire wafer. To remove bulk tungsten film from the wafer surface and form tungsten plugs, fluorine-based plasma etchback processes were developed and widely used in IC fabrication. Later, tungsten CMP (WCMP) processes were developed for bulk tungsten removal. WCMP replaced tungsten etchback for plug formation processes in a very short time because of its superior performance in yield improvement, giving fab managers enough justification to make the switch due to cost savings.

Figure 12.2 illustrates that many CMP processing steps are required to fabricate a CMOS IC chip with Al–Cu interconnections. It starts with two dielectric CMP processes, STI dielectric and ILD0. Then, every metal layer requires two CMP processes, a dielectric CMP and a WCMP. For the four-layer metal IC chip illustrated in Fig. 12.2, at least eight CMP processing steps are required: five dielectric and three tungsten CMP processes.

Figure 12.3 shows a cross section of an advanced CMOS chip with HKMG, selective epitaxial S/D, contact bottom silicide, and copper/ULK interconnection,

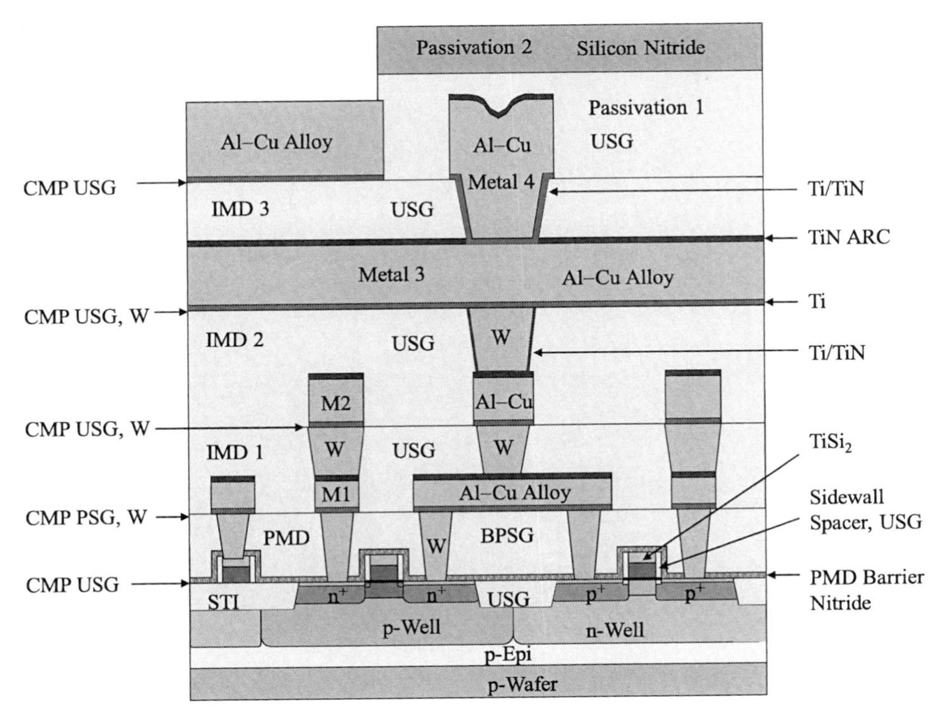

Figure 12.2 CMP applications in CMOS IC chip processing.

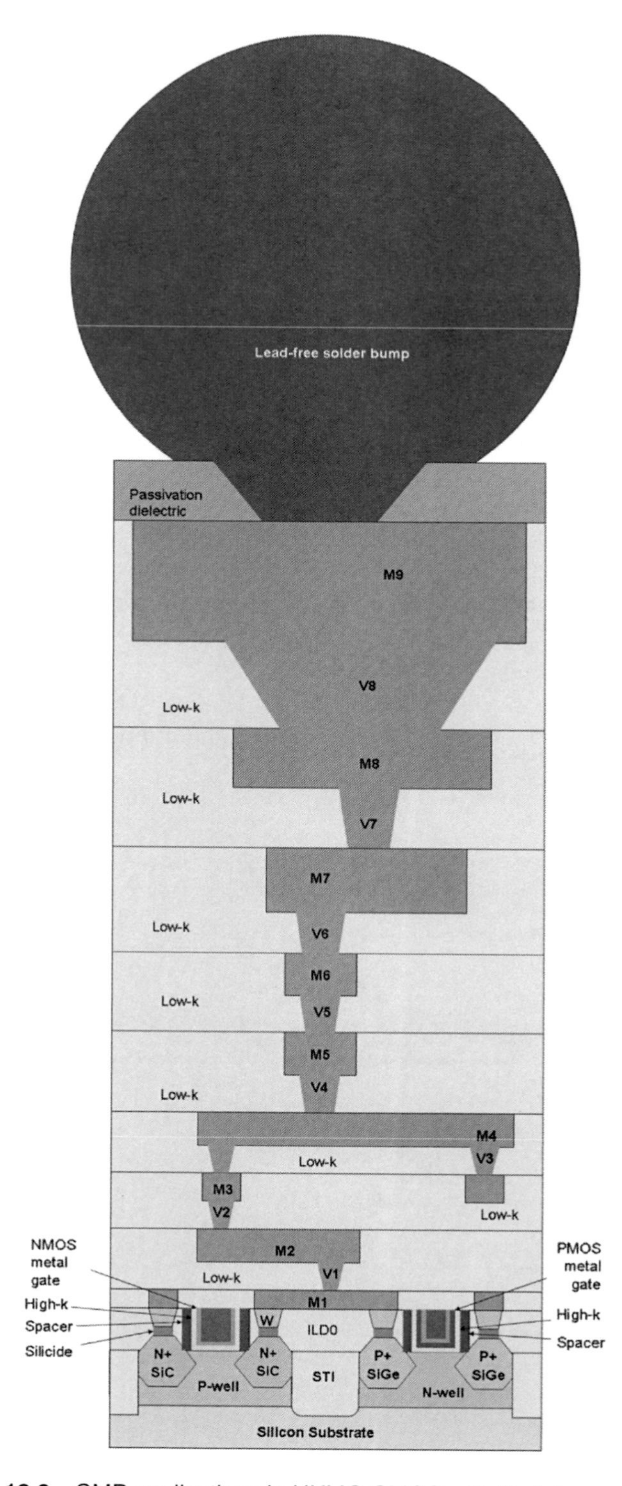

Figure 12.3 CMP applications in HKMG CMOS with Cu interconnection.

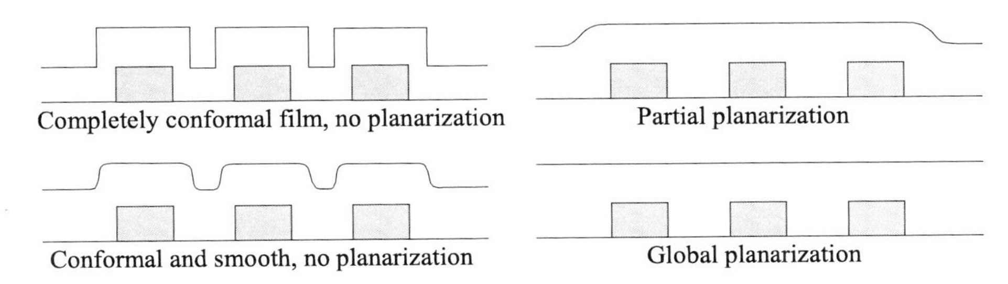

Figure 12.4 Definition of planarization.

which requires more metal CMP processes than dielectric CMP processes. This nine-metal-layer chip requires two dielectric CMP processes, STI and ILD0. The metal needs to be polished 11 times: MG CMP, WCMP, and one Cu CMP for each metal interconnection layer.

12.1.2 Definition of planarization

Planarization is a process that removes surface topological features and smoothes and flattens the surface. The degree of planarization indicates the flatness and smoothness of a wafer surface, particularly for a dielectric thin-film surface after deposition over a patterned wafer surface. The definition of planarization is illustrated in Fig. 12.4.

The degrees of planarity are defined in Table 12.1 and Fig. 12.5, where R is resolution. Smoothing and local planarization can be achieved by thermal flow and etchback processes. Feature sizes smaller than 0.35 μm require global planarization, which can only be achieved by CMP.

12.1.3 Other planarization methods

Thermal flow has long been used for ILD0 planarization. By heating a wafer to a high temperature between 800 to 1000 °C, the doped silicate glass (either PSG or

Table 12.1 Degrees of planarity.

Planarity	R (μm)	θ
Surface smoothing	0.1 to 2.0	>30 deg
Local planarization	2.0 to 100	30 to 0.5 deg
Global planarization	>100	<0.5 deg

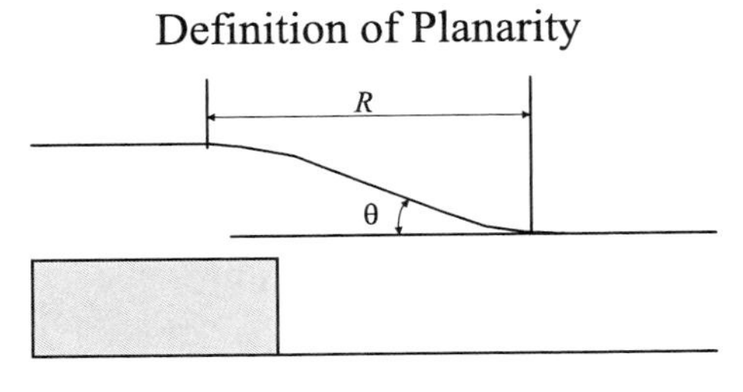

Figure 12.5 Thin-film surface topography.

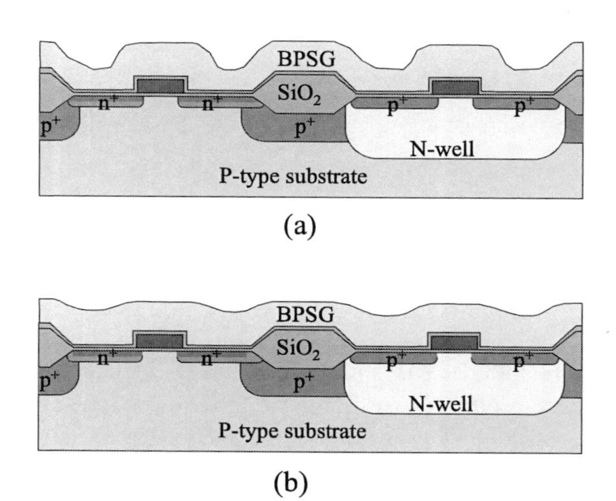

Figure 12.6 Planarization by thermal flow in CMOS IC chip processing: (a) as BPSG is deposited and (b) after reflow.

BPSG, depending on the allowed reflow temperature) becomes soft and starts to flow due to surface tension, as shown in Fig. 12.6.

There are several limitations in thermal reflow planarization. Planarization is mainly determined by reflow temperature and dopant concentration. Higher temperatures give better planarization results. However, high temperatures also cause degradation of transistor performance due to excessive dopant diffusion. To lower the reflow temperature, higher dopant concentrations are needed. These higher concentrations can cause metal corrosion issues if the phosphorus concentration is too high (>7 wt%) due to the phosphoric acid formed when phosphorus oxide (P_2O_5) reacts with moisture (H_2O). They also can cause surface defects due to crystallization of the boric acid formed when boron oxide (B_2O_3) reacts with moisture, if the boron concentration in BPSG is too high.

Because the reflow temperature is far higher than the melting point of aluminum, it cannot be used to planarize the dielectric after the first aluminum alloy layer. Another planarization method is needed for the ILD.

Argon sputtering etchback (ion mill) has been developed and used for ILD planarization. In the sputtering etch process, energetic argon ions bombard the wafer surface, chip off the corners of the gap, and taper the gap openings. This allows subsequent CVD processes to fill the gap, leaving a reasonably planarized surface. Reactive ion etchback processes with CF_4/O_2 chemistry can further planarize the dielectric surface. Figure 12.7 illustrates this etchback planarization process. A detailed description of this dep/etch/dep/etch planarization method is given in Chapter 10.

Photoresist etchback is another method employed to planarize the dielectric surface. After the deposition of the dielectric layer, a layer of photoresist is spin coated on the wafer surface. Due to surface tension, the liquid photoresist can fill the gaps and retain a very flat surface. After the baking process, the

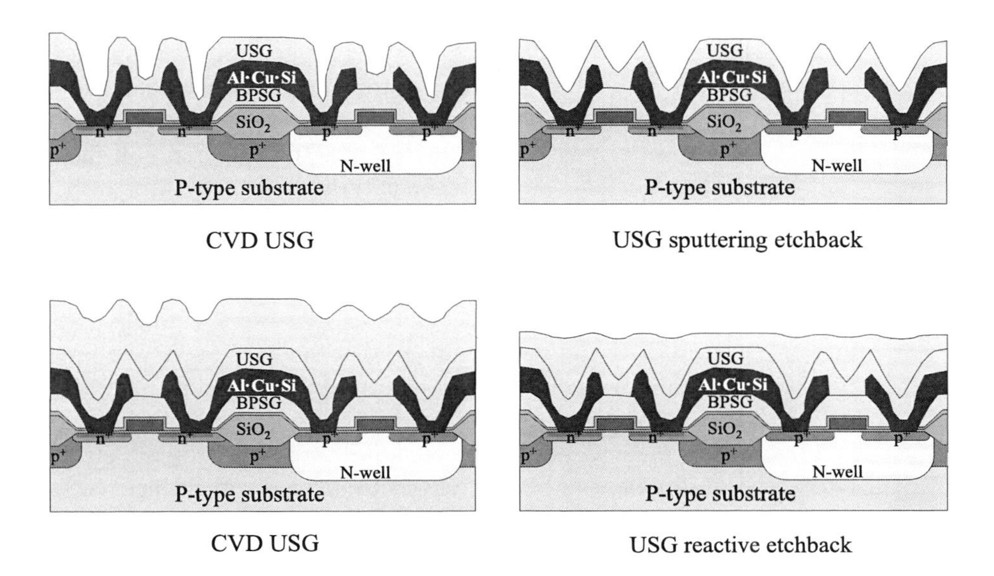

Figure 12.7 Dep/etch/dep/etch gap fill and planarization methods.

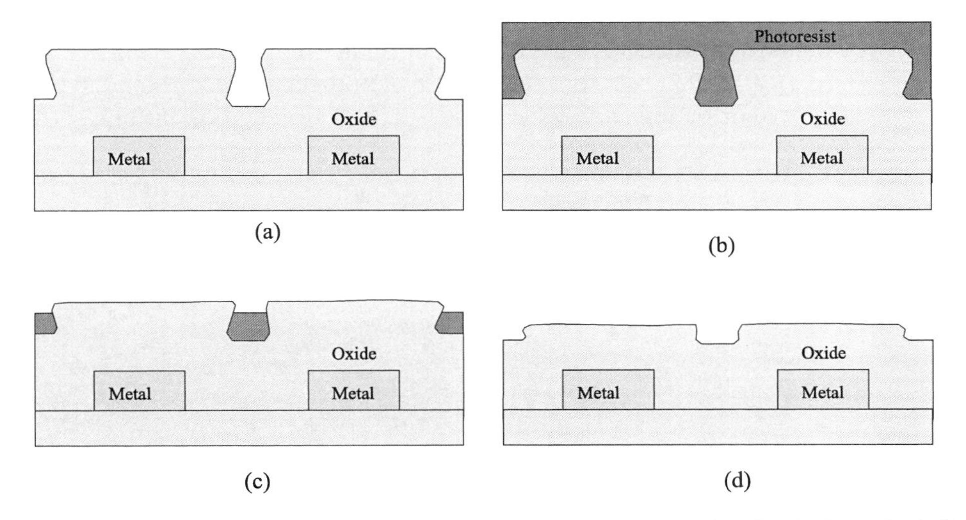

Figure 12.8 Photoresist etchback planarization process: (a) as oxide film is deposited, (b) photoresist coating, (c) photoresist and oxide etch, and (d) after etchback.

photoresist becomes a solid thin film with a planarized surface coated on the wafer surface. Using a plasma etch process with CF_4/O_2 chemistry, silicon oxide can be anisotropically etched by fluorine radicals, and the photoresist can be anisotropically etched by oxygen radicals. By adjusting the gas flow ratio of CF_4 and O_2, theoretically, close to 1:1 etch selectivity of silicon oxide and photoresist can be achieved. Therefore, after etchback, the oxide surface can be planarized. Figure 12.8 shows the photoresist etchback process.

When fluorine radicals start to etch silicon oxide, oxygen replaced by fluorine is released from the etched oxide film. These extra oxygen radicals help to etch the photoresist and cause a higher photoresist etch rate. This is the reason that photoresist etchback cannot achieve a very high degree of planarization the way it is designed. However, after photoresist etchback, the dielectric film surface is significantly flatter than when it was first deposited. In some cases, fabs repeat the photoresist etchback process one or more times to achieve the desired planarization result.

By using SOG to replace photoresist, SOG etchback processes can help both the gap fill and planarization of ILD. One advantage that the SOG etchback process has over photoresist etchback is that a small portion of SOG can stay on the wafer surface to fill narrow gaps between the metal lines. PECVD USG liner and cap layers are used in SOG processes, and ILD with a sandwich structure of USG/SOG/USG can fill gaps and has a reasonable planarized surface. In some cases, two SOG coating, curing, and etchback processes are used to meet gap-fill and planarization requirements. Figure 12.9 shows an SEM photograph of an IC chip with SOG gap fill and planarization.

12.1.4 Necessity of chemical mechanical polishing

As device dimensions reduce, the resolution of photolithography becomes higher. Using Eq. (6.1), $R = K_1\lambda/\mathrm{NA}$, resolution can be improved by increasing the NA of an optical system or by reducing the exposure light wavelength λ. From Eq. (6.2),

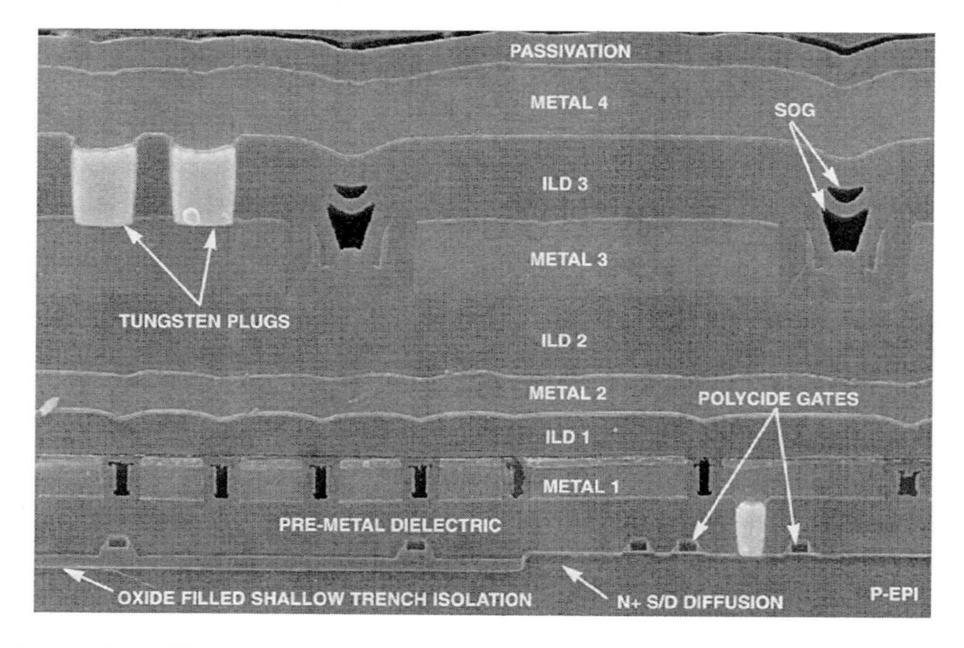

Figure 12.9 SOG gap fill and planarization of ILD. (Integrated Circuit Engineering Corporation)

DOF = $K_2\lambda/2(\text{NA})^2$, it is obvious that both approaches reduce the DOF of an optical system. With a simple calculation using Eqs. (6.1) and (6.2), the DOF is found to be about 2083 Å for 0.25-μm resolution, and 1500 Å for 0.18-μm resolution. Assume that $K_1 = K_2$, $\lambda = 248$ nm (DUV), and NA = 0.6. Therefore, when the patterning feature size is smaller than one-quarter of a micron, or 250 nm, the surface topography must be controlled to less than 2000 Å to ensure the required photolithography resolution. For feature sizes larger than 0.35 μm, other planarization methods can be used to meet the DOF requirements of photolithography. When the feature size is smaller than 0.25 μm, the required planarization level can only be achieved by using CMP processes.

12.1.5 Advantages of chemical mechanical polishing

CMP results in a planarized wafer surface, which allows high resolution of the photolithography process. A planarized surface also eliminates the high resistance and electromigration problems of metal lines caused by the sidewall thinning associated with poor step coverage of metal PVD processes (shown in Fig. 12.10).

A planarized surface can also eliminate the requirements of excessive exposure and development to clear the thicker photoresist regions due to dielectric steps. This improves the resolution of via holes and metal line patterning processes, as shown in Fig. 12.11.

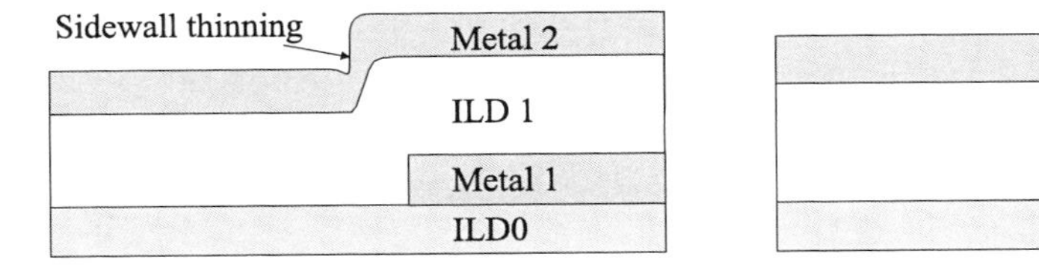

Figure 12.10 (a) Sidewall thinning of a metal line caused by a nonplanarized surface. (b) No such issue with CMP-planarized ILD1.

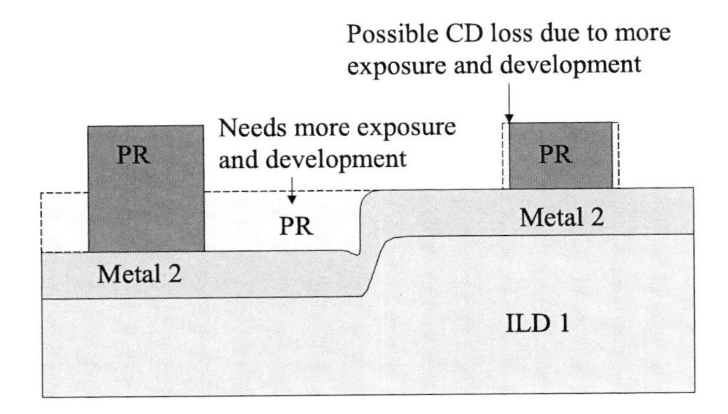

Figure 12.11 Overexposure and overdevelopment due to a nonplanarized surface.

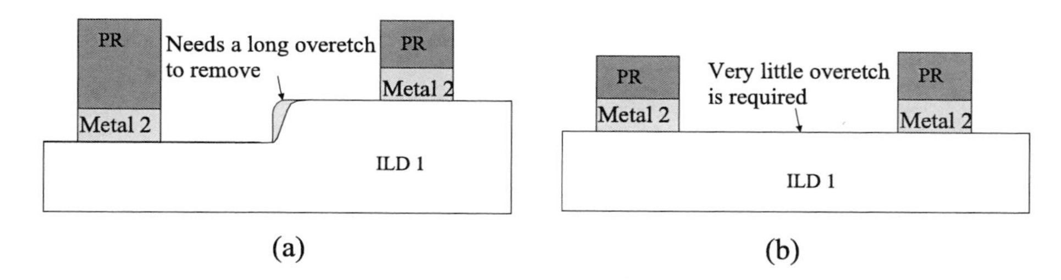

Figure 12.12 Surface planarization and overetch requirements (a) without CMP and (b) with CMP.

A planarized surface allows more uniform thin-film deposition, which helps to sharply reduce any required overetch time. This reduces the chance of undercut or substrate loss associated with long overetch steps during the etch processes, as shown in Fig. 12.12.

Planarization with CMP can minimize defects and improve yield by reducing the processing problems in thin-film deposition, photolithography, and etch processes. The application of CMP also widens IC chip design parameters.

The CMP process can effectively reduce defect density. Inherently, the CMP technique can remove asperities, stringers, and foreign particles from a wafer surface. However, CMP can introduce defects of its own, such as scratches, residues, delamination, dishing and erosion, etc. Large foreign particles can cause scratches, and too much downward force can cause delamination. If the pattern densities in the chip vary from one area to another area, dishing and erosion can happen during the CMP process. CMP systems need to be in conjunction or integrated with appropriate post-CMP cleaning processes to ensure that wafer surfaces are free of slurry residues and other impurities.

12.1.6 Applications of chemical mechanical polishing

CMP processes are routinely performed in IC chip manufacturing. They are used to remove bulk oxide film in STI formation processes. A detailed STI description and illustration is given in Chapter 13. CMP processes are also used to planarize ILD layers in Al–Cu interconnection processes. This allows higher lithography resolution in contact and metal patterning, and easier metal deposition because CMP processes can effectively remove dielectric steps. CMP processing replaced the RIE tungsten etchback process and since the late 1990s has been widely used to remove bulk tungsten and TiN/Ti glue/barrier layers from the wafer surface to form tungsten plugs. Figure 12.13 shows the applications of CMP processes from the cross section of a CMOS chip with a traditional aluminum alloy interconnection.

In DRAM manufacturing processes, polysilicon is frequently used to form contact plugs. After polysilicon deposition fills the contact holes, a polysilicon CMP process removes the polysilicon on the surface and keeps the polysilicon in the contact holes as conducting plugs in DRAM array areas. For recessed-gate

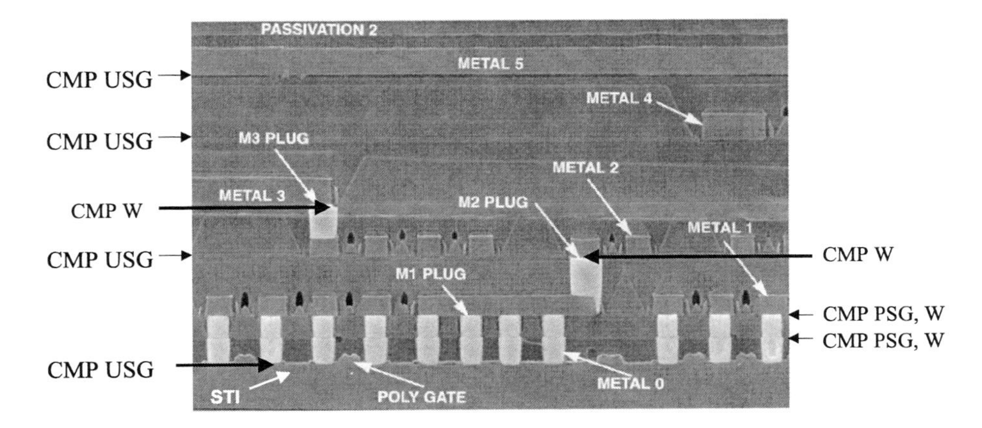

Figure 12.13 CMP applications in IC chip fabrication (Integrated Circuit Engineering Corp.).

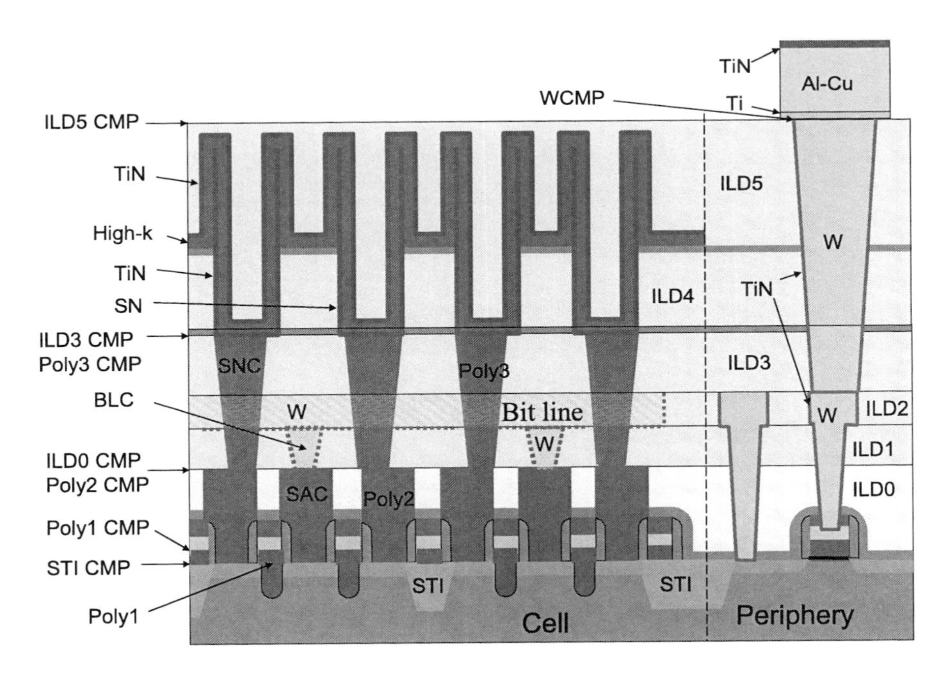

Figure 12.14 CMP applications in a recessed-gate stacked capacitor DRAM.

DRAM, after poly1 fills the recessed trench of the array transistor, a polysilicon CMP process is needed to planarize the polysilicon surface before WL metal deposition. Figure 12.14 illustrates CMP applications in recessed-gate DRAM.

One of the most important applications of CMP processes are copper interconnections. Because copper is very difficult to dry etch, dual damascene has become the process of choice for copper metallization in IC chip manufacturing. The term damascene originates from the ancient Middle East in today's Damascus, capital of Syria. People there invented a process to decorate sword surfaces with gold inscriptions by using diamond to cut trenches into the steel sword surface, polishing gold into the trenches, washing away the gold on the surface, and leaving gold in the trenches. After this process, the sword was decorated with inscriptions on its surface. This technique has been used in the jewelry industry to place gold inscriptions on gemstone surfaces and is called a damascene process. In fact, the tungsten plug formation process is a damascene process.

For copper applications, the dual damascene process is used. It uses two dielectric etch processes, one for via etch and one for trench etch. After dielectric etches, metal layers (PVD Ta/PVD Cu/ECP Cu) are deposited into via holes and trenches. After metal annealing, a metal CMP process removes the copper and tantalum barrier layer from the wafer surface and leaves copper lines and plugs embedded inside the dielectric layer. Figure 12.15 illustrates the dual damascene copper metallization process.

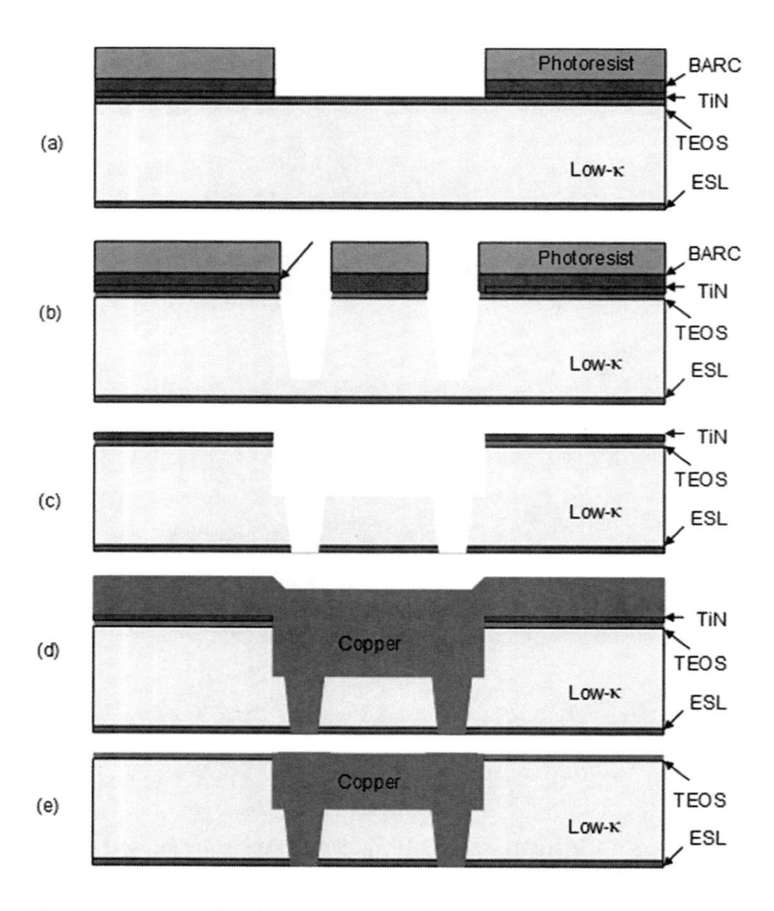

Figure 12.15 Copper metallization with a metal hard mask: (a) etch BARC and metal hard mask with trench mask, (b) etch via halfway with via mask, (c) etch trench and via with metal hard mask and ESL breakthrough, (d) metal deposition, and (e) metal CMP.

12.2 Chemical Mechanical Polishing Hardware

12.2.1 Introduction

A CMP system consists of a polishing pad, a rotating wafer carrier (which holds the wafer face down against the polishing pad), and a slurry dispenser unit. Figure 12.16 illustrates the most commonly used CMP system, with a polishing pad affixed on a rotating polishing table or platen.

CMP processes in the IC industry use water-based polishing slurries with abrasive particles and chemical additives. Slurry is dispensed onto the polishing pad surface, and the wafer is pressed face down against the polishing pad. Both the platen and wafer carrier rotate, usually in the same direction. The combined effect of mechanical abrasion and chemical etch removes material from the wafer surface. Protruding areas receive more mechanical abrasion and are removed faster than recessed areas; this helps planarize the wafer surface.

12.2.2 Polishing pad

A polishing pad is made of a porous, flexible polymer material, such as cast and sliced polyurethane or urethane-coated polyester felt. The pad properties directly affect the quality of the CMP process. The polishing pad materials must be durable, reproducible, and compressible at the processing temperature. The primary processing requirement is high topographic selectivity to achieve surface planarization.

The main qualities of polishing pads are hardness, porosity, fillers, and surface morphology. A harder polishing pad allows for higher removal rates and better within-die (WID) uniformity, while a softer pad allows for better within-wafer (WIW) uniformity. Extremely hard pads cause wafer scratches more easily. The hardness of a pad can be controlled by changing the pad chemical compositions or by changing the cellular structure. Porous cells in the pad absorb polishing slurry and help transfer slurry to the wafer surface, especially at the contact points of the pad and wafer. This is similar to holes in a bath sponge helping deliver liquid soap to the skin. Filler materials can be added to the polymer to improve

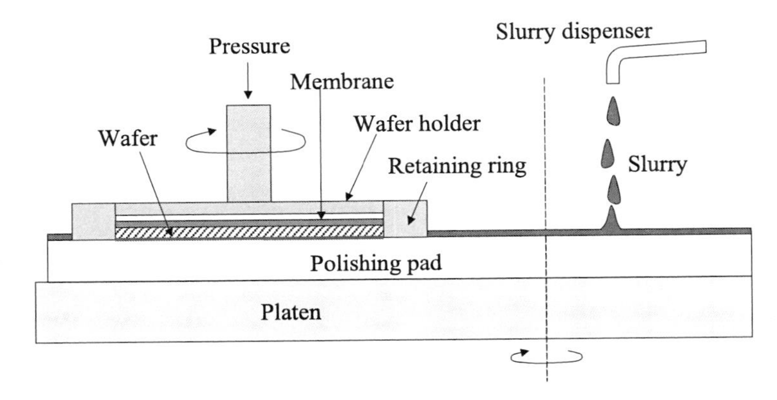

Figure 12.16 Schematic of a CMP system.

its mechanical properties and tailor the pad properties to meet specific process requirements. Polishing pad surface roughness determines the conformality range. Polishing pads with smoother surfaces have a shorter conformality range, which results in poorer topographical selectivity and causes fewer planarization polishing effects. Pads with rougher surfaces have a longer conformality range and better planarization polishing results. Figure 12.17 illustrates the pattern removal effect of rough and smooth polishing pads.

During processing, the pad becomes smoother due to polishing. Therefore, it is necessary to have a reconditioning process to recreate the rough pad surface. Most CMP tools have an in-situ pad conditioner for each polishing pad. The conditioner resurfaces the polishing pad, removes the used slurry, and supplies the surface with fresh slurry. Figure 12.18 shows the position and movement of the pad conditioner on a polishing pad.

The slurry-free polishing pad is composed of four layers: a microreplicated abrasive, a rigid layer, a resilient layer, and a self-adhesive backing layer. Since the abrasive particulates come from the pad surface, only DI water or a basic solution needs to be applied to the pad surface during the CMP process. Advantages of the slurryless pad are that it dramatically simplifies the requirements of slurry storage, delivery, and mixture processes, and it can be used in existing CMP systems with limited retrofitting. Some CMP systems are designed with a roll of slurryless polishing pads that roll during CMP processing and do not require a conditioning system.

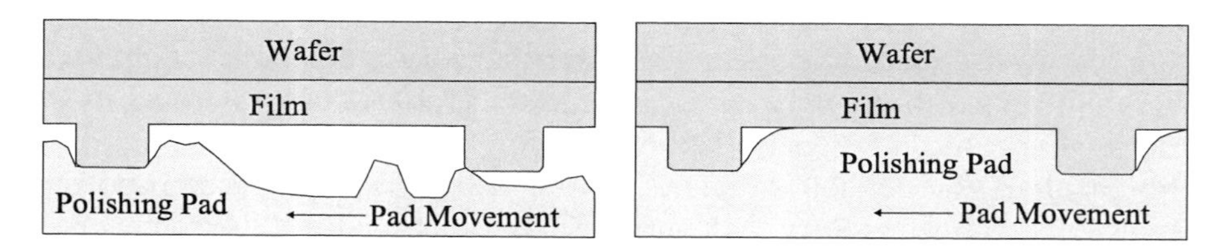

Figure 12.17 Rough pad (left) and smooth pad (right).

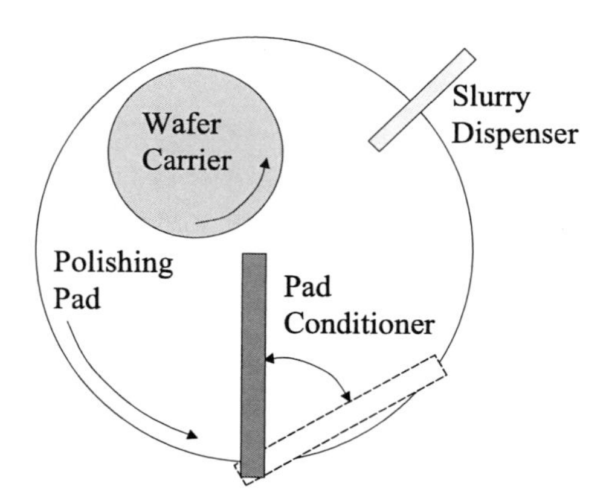

Figure 12.18 Polishing pad and pad conditioner.

12.2.3 Polishing head

A polishing head is also called a wafer carrier and consists of a polishing head body, retaining ring, carrier membrane, and a downward force driving system, as shown in Fig. 12.19. Carrier membranes or carrier films are made of polyurethane or rubber-like materials with a tubular structure. The main purpose of a carrier membrane is to support and cushion wafers in the carrier. Otherwise, the metal polishing head with its downward force can cause wafer damage. A flexible carrier membrane conforms to the wafer's backside so that pressure is applied uniformly. The membrane can compensate for wafer deformations, such as bow or warp due to thin-film or thermal stresses, and help to improve the uniformity of the CMP process. The important parameters of a carrier membrane are its porosity and compressibility. An evenly compressible membrane is important to achieve a uniform CMP process. Membranes must be kept clean because slurry particles can reside in them and cause wafer damage.

The retaining ring is made of plastic and prevents the wafer from slipping out of the wafer carrier. This ring can last for several thousand wafer polishings. The wafer is held to the wafer carrier by a vacuum chuck; the pressurized carrier chamber delivers the downward force to the wafer, and pushes the wafer to the polishing pad. A pneumatic system is used to control the position of the retaining ring to independently control the polishing rate near the wafer edge and contributes to the reduction of edge exclusion due to the shadowing effect. Figure 12.20 illustrates the schematic of a polishing head.

12.2.4 Pad conditioner

A pad conditioner uses a rotating disk with diamond coating that sweeps across the surface of a polishing pad to increase roughness of the pad surface and remove the used slurry. During this process, the polishing pad maintains the surface roughness required for achieving good planarization results. Most CMP tools have in-situ pad conditioners. Some CMP conditioners have a stainless steel plate coated with nickel-plated diamond grits on the surface. Diamond CMP conditioners have also been used and are made from stainless steel plates coated with CVD silicon, then layered with diamond grits coated by CVD diamond film. Figure 12.21 illustrates the surface of these pad conditioners.

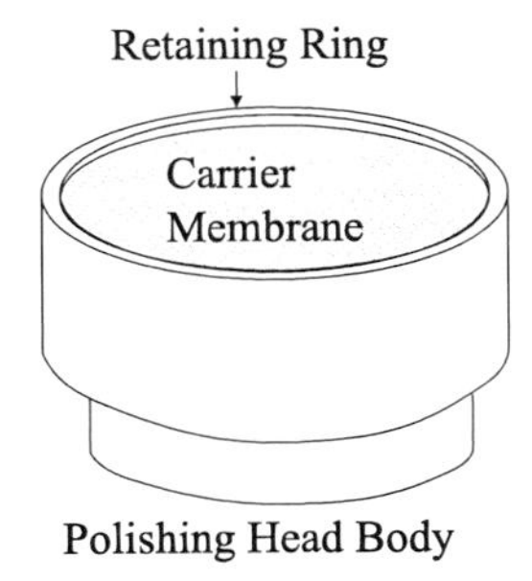

Figure 12.19 Schematic of a polishing head.

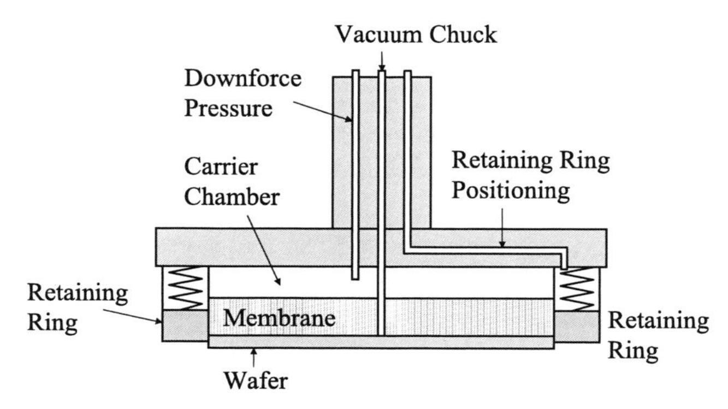

Figure 12.20 Cross section schematic of a polishing head.

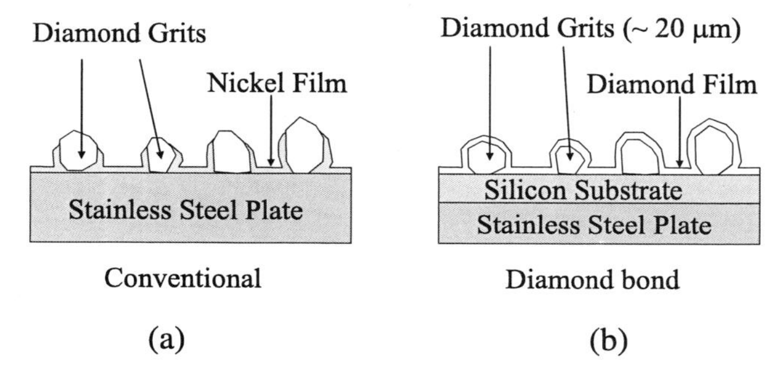

Figure 12.21 Surface of (a) nickel-plated conditioner and (b) diamond-plated conditioner.

A pad conditioner roughens the polishing pad surface while the pad is polishing the wafer. This ensures that roughness of the pad remains unchanged during the polishing process and retains consistent WTW process uniformity.

12.3 Chemical Mechanical Polishing Slurries

Slurry plays a very important role in the CMP process. Particulates in slurry mechanically abrade the wafer surface and remove surface materials. Chemicals in a slurry solution react with surface materials or particulates and dissolve the materials or form chemical compounds that can be removed by abrasive particles. Additives in CMP slurries help to achieve the desired polishing results. In fact, CMP slurries work like toothpaste. Toothpaste provides particulates to abrade the unwanted coating from the surface of the teeth during brushing, while chemical reactions from additives kill gems, remove tartar, and form a protective layer on the teeth.

Slurries used in CMP processes are water-based chemicals with abrasive particles and chemical additives. Different polishing processes require different slurries. Slurries can impact the removal rate, selectivity, planarity, and uniformity of CMP processes. Therefore, slurries always are engineered and formulated for specific applications. There are two types of slurries in CMP processes: one for oxide removal and one for metal removal. Oxide slurries are alkaline solutions with suspended silica. Metal slurries are acidic solutions with suspended alumina particles. Additives are used in the slurries to control the pH value, which affects chemical reactions in CMP processes and helps to achieve optimized processing results.

Components of slurries are stored in different bottles, such as DI water with particulates in one bottle, additives for pH control in another, and oxidants for metal oxidation in a third bottle. They are flowed into a mixer where the components of the slurries are mixed according to the ratio determined by the processing requirements. A slurry delivery system is illustrated in Fig. 12.22, where LFC stands for the liquid flow controller.

12.3.1 Oxide slurry

The most commonly used dielectric in IC processing is silicon oxide. STI formation needs oxide CMP, and ILD0 needs CMP to planarize the surface before contact photolithography. Slurry for an oxide CMP process was developed from the experiences of the optical industry, which requires a fine polish of silicate glass to make lenses and mirrors for optical equipment. Oxide slurry is a colloidal suspension of fine fumed silica (SiO_2) particles in water with base additives. KOH is used to adjust the pH, and sometimes NH_4OH is used. A diluted basic solution of KOH (usually <1%) is frequently used to control the slurry pH at 10 to 12.

The pH value represents the acidity or alkalinity of an aqueous solution, which ranges from 0 to 14. Neutral solutions such as water have pH value equals to 7. Solutions with pHs lower than 7 are acids; the lower the pH, the stronger the acid. Alkaline solution has pH higher than 7. The higher the pH, the higher the alkalinity, as shown in Fig. 12.23.

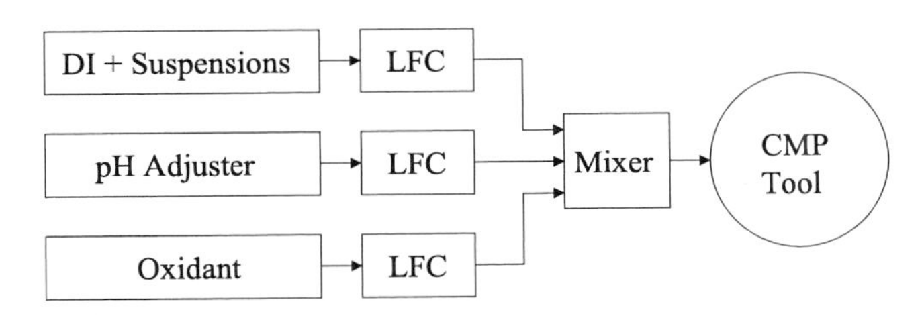

Figure 12.22 Slurry delivery system.

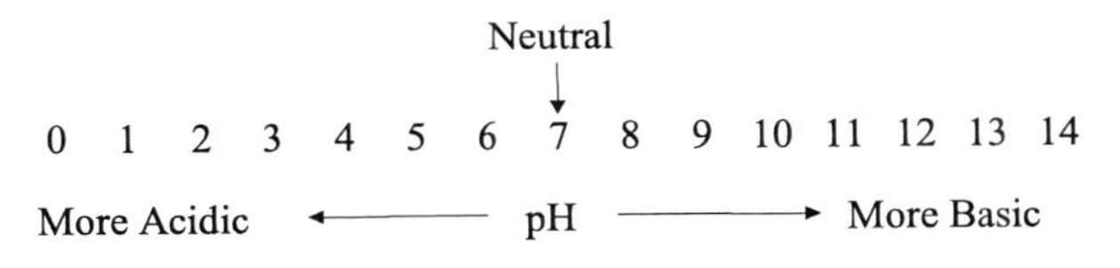

Figure 12.23 Relationship of pH values and acidity or alkalinity.

The fine fumed silica particles that are suspended in solution are abrasives for CMP. Oxide polishing slurries contain approximately 10% solids. With proper temperature control, these slurries can have a shelf lifetime of up to one year.

Fumed silica particles can be formed in a vapor phase hydrolysis of silicon tetrachloride in a hydrogen-oxygen flame. The chemical reaction can be expressed as

$$2\,H_2 + O_2 \rightarrow 2\,H_2O,$$
$$SiCl_4 + 2\,H_2O \rightarrow SiO_2 + 4\,HCl \uparrow .$$

The overall reaction can be expressed as

$$SiCl_4 + 2\,H_2 + O_2 \rightarrow SiO_2 + 4\,HCl \uparrow .$$

At about 1800 °C, these reactions form molten spheres of silicon dioxide. The size of the silica spheres varies from 5 to 20 nm, depending on process parameters. The molten spheres can collide and fuse with one another to form branched, 3D, chain-like aggregates. Figure 12.24 illustrates the fumed silica formation process.

The pH of a slurry can significantly affect dispersion of the silica particles. The particles attain a surface charge, and the sign and magnitude of the charge depend on the pH of the solution. Up to a pH of 7.5, the viscosity of silica slurry in an aqueous medium is high enough to prevent silica particles from dispersing. At a pH higher than 7.5, the silica particles attain sufficient charges to generate electrostatic repulsion, which effectively disperses the slurry. When the pH is higher than 10.7, the particles dissolve and form silicate.

Another type of silica abrasive used in oxide CMP processes is colloidal silica, which is also called precipitated silica. Colloidal silica can be prepared in an alkali silicate solution. At near-neutral pH conditions, silica nucleation forms colloidal silica particles of about 1 to 5 nm in diameter. If the pH is kept just slightly

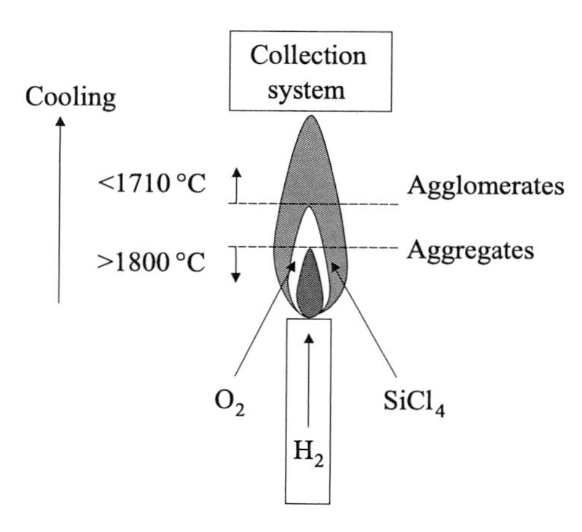

Figure 12.24 Fumed silica formation.

alkali, the colloidal silica particles will not fuse together, and they gradually grow larger in size, 100 to 300 nm. Some particles grow even bigger, and large-particle count (LPC) is an important factor of slurry. Very large particles (>1 μm) need to be filtered out because they tend to generate scratch defects in CMP processes. Figure 12.25 shows the differences between fumed silica and colloidal silica particulates.

For STI oxide CMP processes, high selectivity to nitride is required so that the process can stop on the silicon-nitride polish stop layer, which is also the hard mask layer for STI etching. Slurry with a cerium oxide (CeO_2 or ceria) abrasive has been developed and implemented in IC fabrication. This slurry contains about 5% CeO_2 abrasive in a weak alkaline aqueous media. By using CeO_2 abrasive with resin, the slurry can achieve high topographic selectivity and help to avoid the dishing effect in large field oxide areas, as illustrated in Fig. 12.26. Slurry with a CeO_2 abrasive can achieve better selectivity to nitride than slurry with silica abrasive particles. It also has less dishing effect in STI oxide CMP processes.

12.3.2 Metal polishing slurry

Metal CMP processing is similar to a metal wet etch process. At first, the oxidant in the slurry reacts with metal to form an oxide on the metal surface; then the oxide is removed, exposing the metal surface for oxidation, and the oxide is removed again. Metal polishing slurries are pH-adjusted suspensions of alumina (Al_2O_3). The slurry pH controls the two mechanisms attributed to metal removal: corrosive wet etching and oxidation passivation.

In metal CMP processes, oxidant in the slurries oxidizes metal surfaces. In varied conditions, different types of metal oxides can be formed, and each of them has a different solubility. This causes two competing removal mechanisms. If

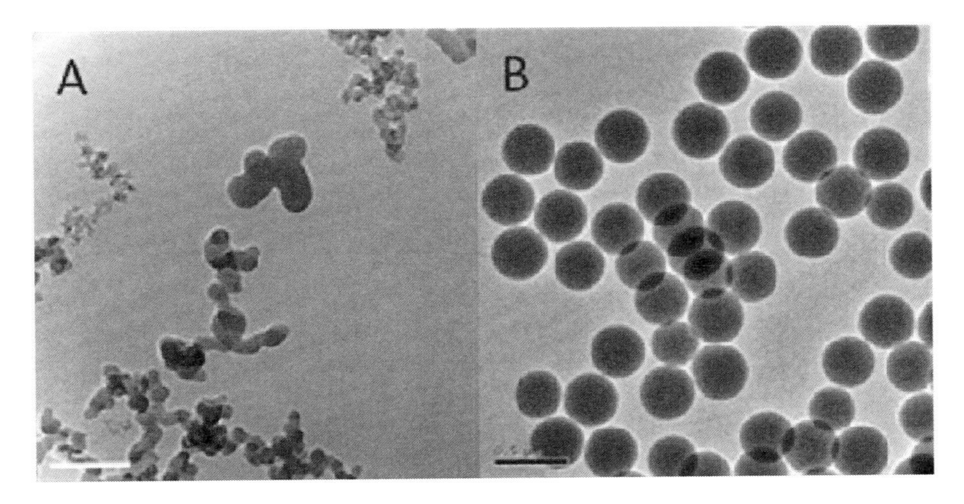

Figure 12.25 SEM images of silica abrasive particles: (a) fumed silica and (b) colloidal silica (Fujimi Corp.).

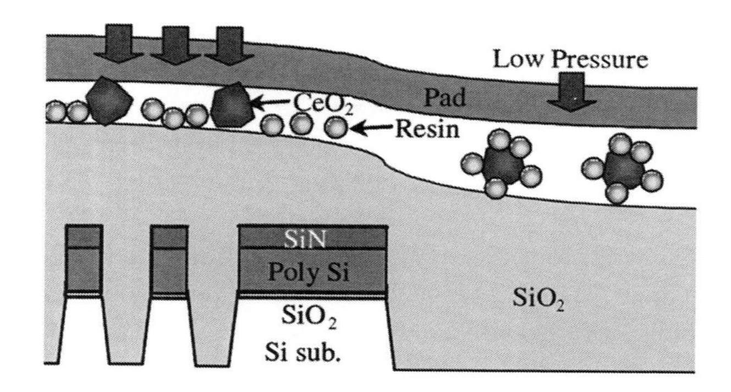

Figure 12.26 STI CMP with CeO_2 abrasive and resin additive (Y. Matsui, et al.).

an oxidation process mainly generates oxide ions that are soluble in the slurry solution, the wet etch will dominate the metal removal process. This is not a favorable process for planarization applications because wet etch is an isotropic process with no topographic selectivity. If the metal oxide is insoluble, the oxide will coat the metal surface and block any further oxidation processes. The fine alumina particles in the polishing slurry mechanically abrade the passivation oxide layer, expose the metal surface, and allow metal oxidation and oxide abrasion processes to be repeated. This chemical mechanical removal process has a high surface topographic selectivity, which is favorable for surface planarization. Additives are used in metal CMP slurries to control the pH and achieve the delicate balance between wet etch, passivation, and oxide removal to optimize the metal CMP processing results.

12.3.3 Tungsten slurry

Tungsten can be passivated by the formation of WO_3 due to chemical reaction with an acidic solution (pH lower than 4) during the tungsten CMP process. For a higher pH, soluble $W_{12}O_{41}^{10-}$, WO_4^{2-}, and $W_{12}O_{39}^{6-}$ ions are formed in the solution, and the tungsten is wet etched at a high etch rate. Figure 12.27 illustrates a potential pH diagram, called a Pourbaix diagram, for tungsten and indicates the passivation and wet etch regime at different pH values and potential. When pH < 2, tungsten is in the passivation regime.

In the presence of an oxidant, such as potassium ferricyanid [$K_3Fe(CN)_6$], ferric nitrade [$Fe(NO_3)_3$], and hydrogen peroxide (H_2O_2), the range of pH in which tungsten becomes passivated can be extended to 6.5. By adjusting the slurry pH, both low wet etch rates and high CMP removal of tungsten film can be achieved simultaneously.

Tungsten slurries are quite acidic, with pH levels from 4 to 2. Compared to oxide slurries, tungsten slurries have lower solid contents and much shorter shelf lifetimes. Alumina particles are not in colloidal suspension in a slurry at that pH value. Therefore, tungsten slurries require mechanical agitation prior to and during delivery to the CMP tools.

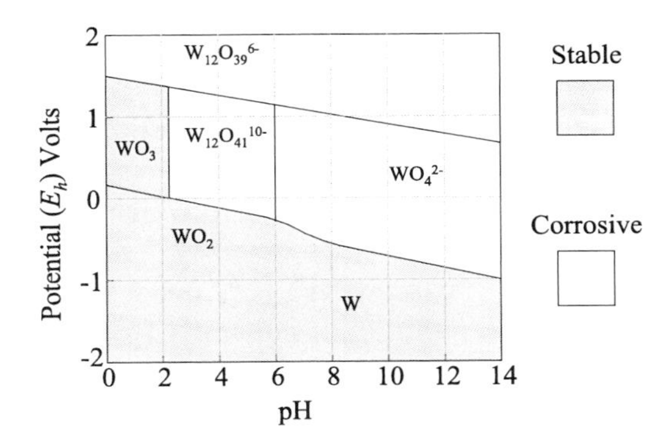

Figure 12.27 Pourbaix diagram for tungsten.

12.3.4 Aluminum and copper slurries

Aluminum slurries are water-based acidic solutions, with H_2O_2 or amines and H_2O_2 as oxidants, and alumina as abrasives. They also have very limited shelf lifetimes, since an H_2O_2 molecule is unstable and tends to dissociate to H_2O, releasing a free oxygen radical. Aluminum CMP has been used in replacement gate or gate-last processes of HKMG to form metal gate electrodes for advanced MOSFETs since the 45-nm technology node.

Figure 12.28 illustrates the potential pH diagram for copper. It can be seen that at $5 < pH < 13$, copper is in the passivation regime. To achieve consistent polishing process results, a colloidally stable slurry is needed. A colloidally stable alumina suspension can be achieved at a pH just below 7. Therefore, this suspension has only a small window in which copper slurries can achieve both electrochemical passivation and colloidally stable suspension of aqueous alumina particulates.

Copper slurries are acidic solutions with alumina as abrasives. Different oxidants can be used, such as hydrogen peroxide (H_2O_2), ethanol (HOC_2H_5) with nitric acid (HNO_4), ammonium hydroxide (NH_4OH) with potassium ferri- and ferrocyanide, or nitric acid with benzotriazole.

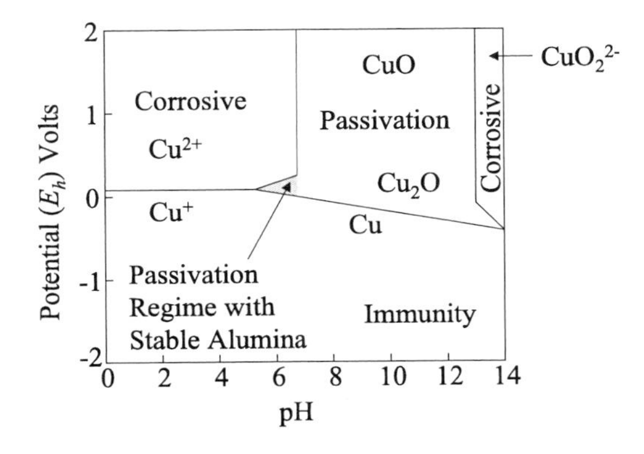

Figure 12.28 Potential pH diagram for copper.

Question: Oxide slurries use silica, while metal slurries use alumina as abrasives. Can oxide slurries use alumina and metal slurries use silica?

Answer: Silica particles form chemical bonds with atoms on the surface of silicate glass. This helps to chemically remove silicon oxide by tearing atoms or molecules from the glass surface with the silica particles, and dissolving the particulates in a high-pH solution. Alumina cannot form chemical bonds with oxide film and is not soluble in high-pH aqueous solutions. Therefore, alumina only provides mechanical abrasion for oxide removal, causing a low oxide removal rate. Both silica and alumina can be used in metal slurry as abrasives. However, using silica can cause a higher oxide removal rate and worsen metal-to-oxide polish selectivity.

By adding aluminate ions $Al(OH)_4^-$ into a slurry, colloidal stability can be significantly improved. Copper CMP using slurries with aluminate-modified colloidal silica abrasives can achieve high removal rates and good planarization and dishing performance.

12.4 Chemical Mechanical Polishing Basics

12.4.1 Removal rate

Mechanical material removal rate R was found by Preston in research related to glass polishing processes. The Preston equation can be expressed as

$$R = K_p \cdot p \cdot \Delta v,$$

where p is the polishing pressure, which is determined by downforce divided by the contact area; K_p is the Preston coefficient, which is related to a specific process and must be determined empirically; and Δv is the relative velocities of the wafer and polishing pad. The Preston equation works very well for bulk film polishing processes. Because protruding portions on a rough surface have a higher polishing pressure than other parts of the surface, the Preston equation reveals that the removal rate of the protruding parts would be higher than that of the rest of the surface. This helps remove surface topography and planarize the surface. Figure 12.29 shows the higher polishing pressure of the protruding parts of the wafer.

Since CMP processes are never purely mechanical processes, the Preston equation does not always describe the process accurately. Chemical interactions always play a crucial role in removal processes, especially in metal CMP processes.

Polish rate can be determined by measuring the film thickness change before and after CMP processing, and dividing that value by the CMP time. For dielectric CMP processes, the removal rate can be monitored in situ by optical reflective interferometry, described in Chapter 10. An optical reflective interferometry system is built into the CMP system to detect the endpoint of a polishing process.

Removal rate of a CMP process is several thousand angstroms per minute. It is mainly determined by the downforce pressure, stiffness of the polishing pad, and

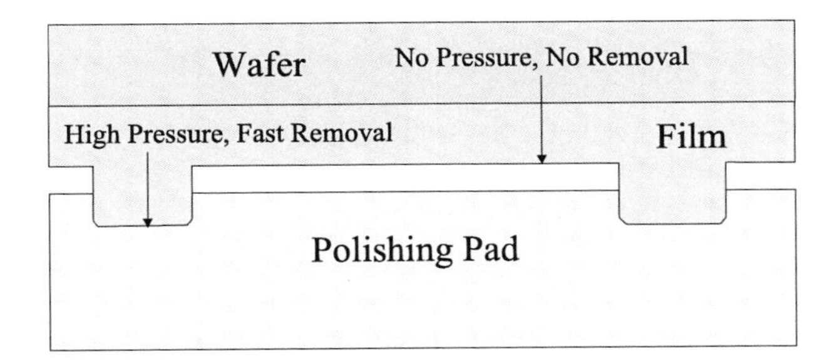

Figure 12.29 Topographic selectivity of a mechanical polish.

amount of applied polish slurry. Different films also have different polishing rates. For instance, different silicate glasses, such as SOG film, PECVD oxide film, and O_3-TEOS oxide film, have different polishing rates. Doped oxide film has different polishing rates from undoped oxide film, etc.

If the removal rate begins to drop gradually during a routine process, the most likely source of the problem is pad surface deterioration. If pad conditioning is adequate, the pad could need replacement.

12.4.2 Uniformity

For 200-mm wafers, 49-point, 3σ standard deviation measurements are used as the definition of uniformity for CMP process qualifications, and for changes in film uniformity before and after CMP processing is monitored. For 300-mm wafers, 121-point measurements might be required. For production wafers, only the uniformity after CMP processing is monitored.

Both WIW and WTW uniformity can be affected by the polishing pad condition, downforce pressure distribution, relative speeds of the wafer and the polishing pad, retaining ring position, and shape of the wafers. By using a harder pad and lower pressure (as low as 2 psi), good global uniformity within less than 3% nonuniformity can be achieved.

12.4.3 Selectivity

Removal selectivity is the ratio of removal rates of different materials. In CMP processing, a large ratio is desired for film that should be removed and material that should not be removed. Removal selectivity is a very important factor in CMP processing. It can significantly affect CMP-induced defects, such as erosion or dishing, and can also be very important for endpoint detection. Slurry chemistry is the primary factor that affects the removal selectivity of a CMP process. For oxide CMP in STI formation processes, high oxide-to-nitride selectivity is very important to ensure that the polishing process stops on the nitride surface. Oxide-to-nitride selectivity is between 3 and 100, depending on slurry type, pad stiffness, downforce pressure, and pad rotation speed. Because only the oxide layer is polished, selectivity is not important in ILD planarization processes in Al-Cu interconnections. For tungsten CMP processes, selectivity of oxide and titanium nitride is very important. Tungsten-to-TEOS oxide selectivity is very high, from 50 to 200.

For a particular metal CMP slurry, reactivity of the chemical oxidant is critical to control the removal rate and selectivity. This makes the choice of oxidant one of the most critical factors in metal slurry engineering. Selectivity is also related to pattern density. For example, the tungsten-to-oxide removal rate ratio in tungsten CMP processes can be as high as 150:1, based on blanket film polishing results. In reality, the ratio can be much smaller, depending on the pattern density of each material; the higher the pattern density, the lower the removal selectivity. This loss of selectivity can lead to erosion of the tungsten and oxide films, as illustrated in Fig. 12.30.

IC design layout can directly affect erosion problems. Designing opening areas on less than 30% of a chip's surface can help to solve erosion problems. Dummy patterns are a very important part of design for manufacturing (DFM), because dummy pattern placement can significantly affect dishing and erosion issues, and impact product yield.

12.4.4 Defects

CMP processing can remove many defects from a wafer's surface, thereby helping to improve product yield. However, it also can introduce defects such as scratches, residual slurry, particles, erosion, and dishing.

Large foreign particles and hard polishing pads can cause scratches on wafer surfaces. Tungsten can fill scratches in the oxide surface caused by oxide CMP processing and can form microscopic tungsten wires after tungsten CMP that cause short circuits or crosstalk, reducing IC yield.

Improper downforce pressure, worn polishing materials, inadequate pad conditioning, particle surface attraction, and slurry drying can cause slurry residue on wafer surfaces. This can cause defects due to contamination and reduce the IC yield. Post-CMP cleaning is very important to remove slurry residue and improve process yield.

Erosion is mainly caused by the pattern-density-induced selectivity degradation, as shown in Fig. 12.30. It can result in incomplete interconnections in the next layer of metal interconnections because it increases the depth of the via holes. This can cause incomplete via hole etch and create open loops between the different layers in the next dual damascene interconnection, as shown in Fig. 12.31.

The dishing effect happens at a bigger open area, such as at large metal pads or STI oxide in the trenches. Because more material is removed from the center of a

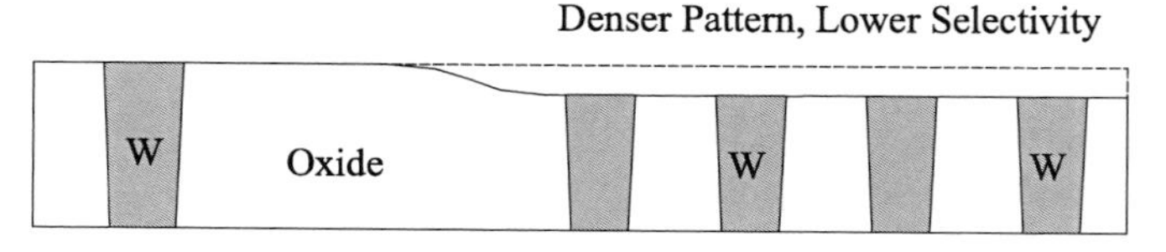

Figure 12.30 Erosion caused by high pattern density.

larger pattern, and the cross-sectional view looks like a dish (shown in Fig. 12.32), this is called the dishing effect.

Both dishing and erosion effects are related to removal selectivity. For example, in tungsten CMP processing, if tungsten-to-oxide selectivity is too high, dishing and recessing effects can occur at the tungsten plugs and pads during tungsten overpolishing, after the many tungsten layers have been removed. If selectivity is not high enough, both oxide and tungsten will be polished away during the overpolishing step, causing erosion problems. Oxide CMP with a high selectivity of oxide to nitride can cause oxide dishing during the oxide overpolishing step in STI formation, as shown in Fig. 12.33.

Dishing and erosion can be measured by an atomic force microscope (AFM). An AFM usually has a microscopic silicon or silicon nitride cantilever with a sharp-tipped probe at the end. The tip of the probe has a radius on the order of nanometers and can scan the proximity of a sample surface without directly touching it, due to the force between the tip atoms and surface material atoms. Interaction can deflect the cantilever along the sample surface. By recording the deflection variation caused by the surface variation, AFMs can measure surface roughness in the nanometer range. A silicon nanotube whisker or carbon nanotube

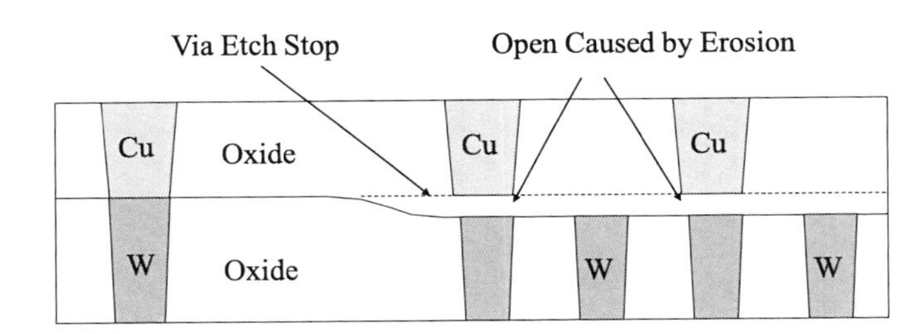

Figure 12.31 Erosion caused by open circuit problems.

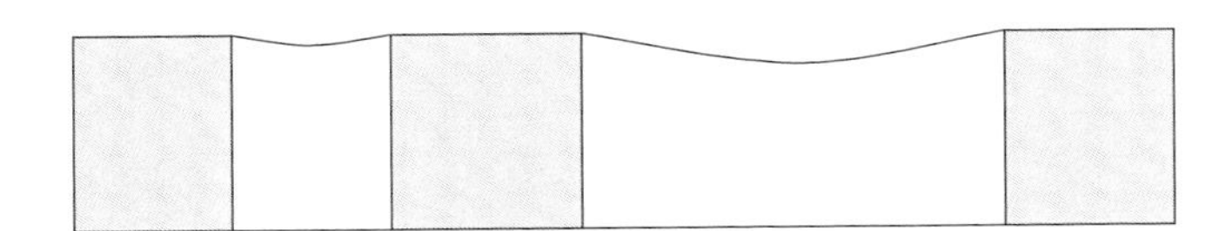

Figure 12.32 The dishing effect.

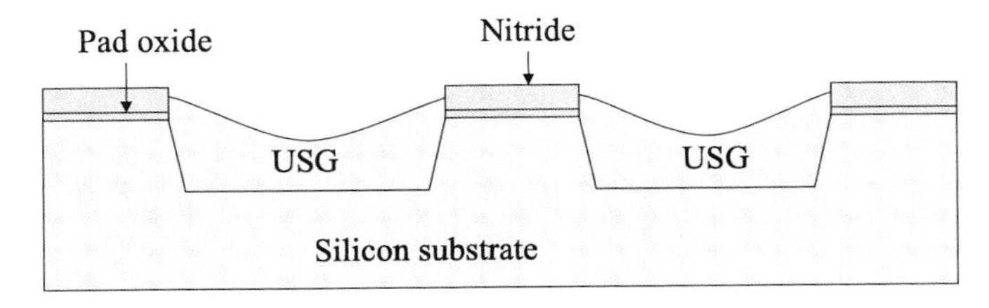

Figure 12.33 Dishing and recessing effects of the STI USG.

whisker can be grown on the AFM tip to help improve measurement resolution of the tiny patterns in the nanometer range. Figure 12.34 shows a cantilever and probe tip without and with a nanotube whisker on an AFM system, and Fig. 12.35 illustrates the AFM measurement sequence.

By scanning a smaller area with dense scanning lines, an AFM image can be formed that shows a 3D profile of a microscopic pattern. AFMs can also be used to measure pattern CD, height and profile of photoresists, and etched patterns. Because they are very slow, AFMs are mainly used for research and development, and engineering troubleshooting. They are also used as golden standard tools to calibrate CD measurement systems such as scatterometry systems.

Particles and defects can be measured by light scattering. Because particles and defects have irregular surface topographies, they scatter incident light while smooth surfaces reflect it. By detecting the scattered light, particles and defects can be monitored on the wafer surface. Figure 12.36 illustrates the idea of particle detection by scattering light.

Because the intensity of the scattered light is very weak, an elliptical mirror is used to collect it. An elliptical mirror collects all of the light scattered from one focus and reflects it to another focus. Particle detection systems are designed so that a laser beam scans the wafer surface vertically at one focus of an elliptical mirror, and a photodetector is placed at another focus. This design allows the user to collect the most scattered light by moving the wafer to detect tiny particles and defects, and then map their position on the wafer surface. Figure 12.37 illustrates this type of particle detection system.

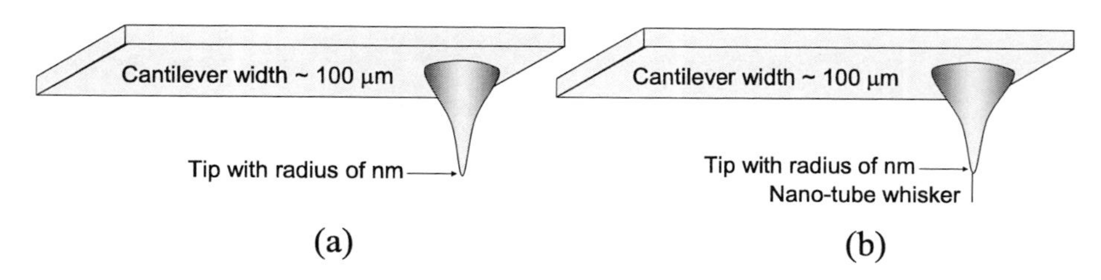

Figure 12.34 Cantilever and probe tip of the AFM: (a) conventional tip and (b) tip with a nanotube whisker.

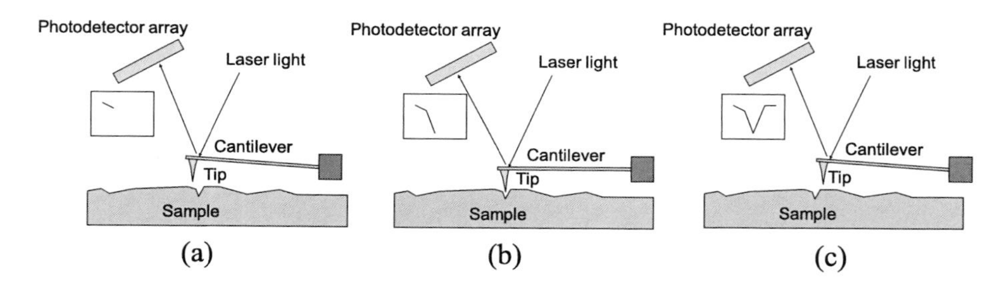

Figure 12.35 AMF measurement sequence.

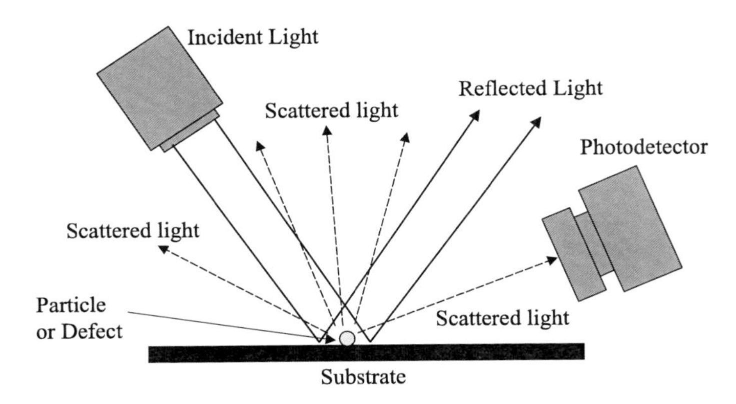

Figure 12.36 Light-scattering particle measurement.

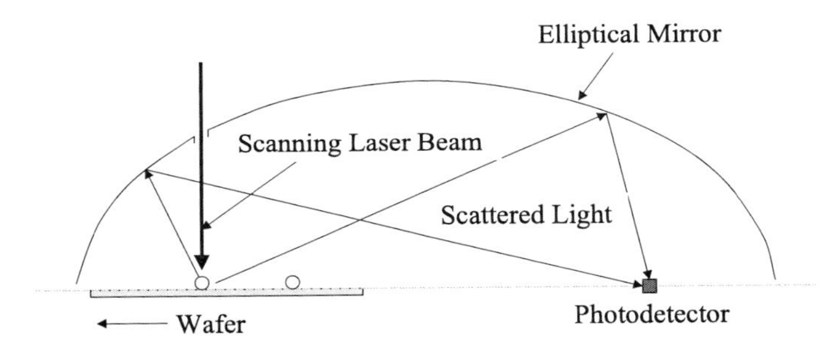

Figure 12.37 Light-scattering particle detector with an elliptical mirror.

SEMs are also widely used for defect detection. The SEM uses an energetic electron beam to scan across the wafer surface and collect the secondary electron emission signals for detection of microscopic features. It is widely used to review physical defects captured by optical inspection and helps to determine the type of defect by looking at it with a high-resolution image.

Because an electron beam charges up the surface to affect the secondary electron emission, an SEM signal of the metal contact plugs connected to the ground can have a higher contrast than a signal from those not connected to the ground. This is called voltage contrast, and it is unique to electron beam inspection (EBI), which can be used to capture electrical defects such as open circuits of contact and via plugs (Fig. 12.31) and junction leakage, which can be caused by junction spiking (see Fig. 11.5). Since the development of 90-nm technology nodes, EBI has been widely implemented after WCMP because it can capture electrical or voltage contrast defects that optical inspections cannot. Figure 12.38(a) shows an nMOS leakage captured by EBI as a bright voltage contrast defect, and Fig. 12.38(b) is the cross-section transmission electron microscope (TEM) image of that defect. The TEM confirmed that the bright tungsten plug came into contact with a leaking n+/p-well junction, and the leak was caused by nickel silicide diffusion along the dislocation fault line that shorted the nMOS S/D junction to the substrate.

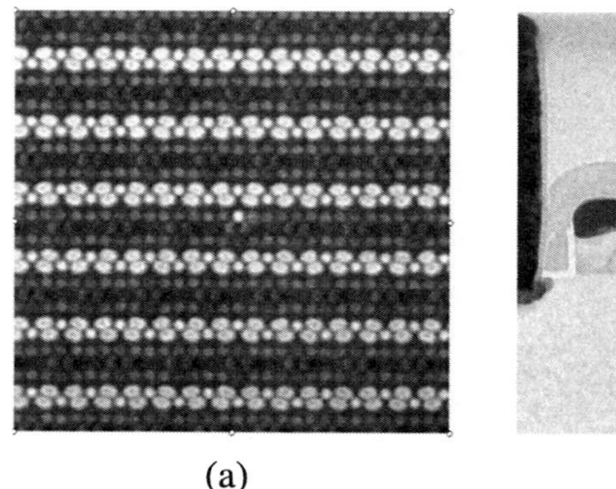

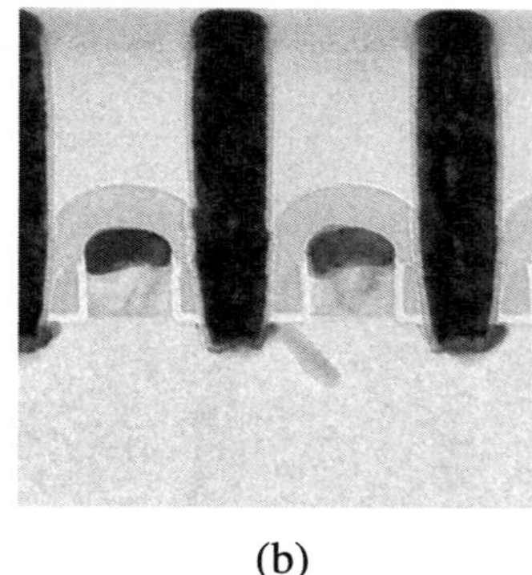

(a) (b)

Figure 12.38 (a) Top-down SEM image of an nMOS leak captured as a bright voltage contrast (BVC) defect by post-WCMP EBI. (b) Cross-section TEM image of that BVC defect.

One of the most important advantages of the CMP process is that it can reduce defect density induced by rough surface topography. The benefits of defect reduction by CMP processes far outweigh the defects introduced by them.

12.5 Chemical Mechanical Polishing Processes

There are two types of CMP processes. One is for planarization, which removes part of the film (usually about 1 μm) and planarizes the thin-film surface. Another one is a bulk removal process in which bulk film on the surface is removed by a polishing process, and only parts of the film are left behind to fill the trenches or holes.

For Al-Cu interconnections, the most commonly used CMP processes are oxide and tungsten CMP. The majority of oxide CMP processes are planarization processes, such as ILD CMP processes. Only STI oxide CMP is a bulk removal process, which removes the oxide from the wafer surface and leaves it only in the trenches to form separations between neighboring microelectronic devices. Tungsten CMP processes are bulk removal processes, which remove bulk tungsten from the wafer surface and leave small amounts of tungsten in the contact or via holes to form plugs of interconnection between different conducting layers.

Question: Dual damascene copper CMP belongs to what type of process?

Answer: Copper CMP is a bulk removal process that removes bulk copper and leaves copper only in the trenches and via holes to form metal interconnections.

12.5.1 Oxide chemical mechanical polishing

Silicon oxide CMP processes have been developed and applied in the optical industry for a long time to fine grind and polish glass surfaces for lenses and mirrors. In the mid-1980s at IBM, an early oxide CMP process was developed by combining the knowledge and experience of glass polishing and bare silicon wafer polishing.

The chemical interaction between the silica particles in slurry and the oxide thin-film surface during the silicon oxide CMP process can be described as follows. First, hydroxyls are formed on both oxide film and silica particle surfaces when contact is made with the water-based slurry. Then, hydrogen bonds are formed between hydroxyls on the oxide surface and silica particle surfaces in the slurry. With help from heat generated by mechanical polishing, molecular bonds between the surfaces can be formed. Mechanical removal of the particles bonded to the wafer surface also tears away the atoms or molecules from the wafer surface, significantly helping the removal process. Figure 12.39 illustrates this oxide polishing mechanism.

No oxide polishing effects are observed in nonaqueous solution, highlighting the importance of surface hydroxylation in silicate glass polishing processes. No oxide polishing occurs without silica abrasives, demonstrating that particles in the slurry are the primary polishing mechanism. A silica slurry has a very low oxide polishing ability except at very high pH (>10) when silica dissolves and enters the solution as a silicate anion. A high concentration of hydroxyl ions in the slurry significantly increases the removal rate of silicate glass.

12.5.2 Tungsten chemical mechanical polishing

Tungsten is widely used as a plug to connect different levels of metal lines. Since CVD tungsten films have excellent gap-fill abilities, removing bulk tungsten film

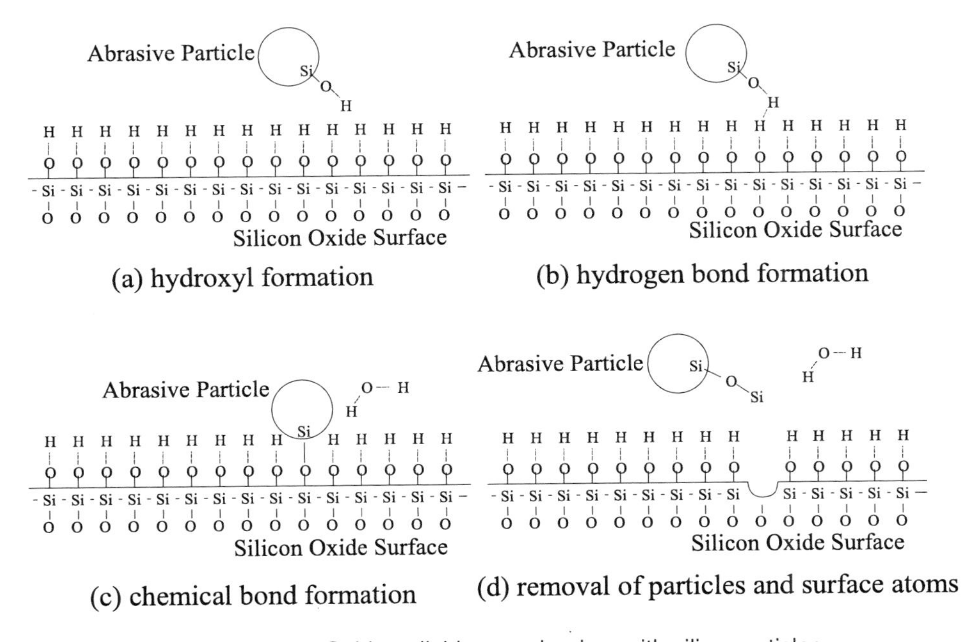

Figure 12.39 Oxide polishing mechanism with silica particles.

and leaving tungsten in the contact/via holes can form tungsten plugs. For IC fabs still using Al–Cu interconnections, the most frequently used metal CMP processes are tungsten CMP processes. These fabs include very advanced nanotechnology memory fabs of both DRAM and flash, and CMOS logic fabs with technology nodes from 0.8 to 0.25 μm. Before tungsten CMP processing was widely adapted by IC fabs, fluorine-based RIE etchback processes were commonly used to remove bulk tungsten on the wafer surface after tungsten CVD processing.

One advantage of the tungsten etchback process is that it can be performed in situ with the tungsten CVD process in a cluster tool. However, tungsten etchback processes always cause the recessing of tungsten and the Ti/TiN barrier/adhesion layer due to the aggressive fluorine chemical etch of Ti/TiN, and this affects chip yield, as shown in Fig. 12.40. The ex-situ tungsten CMP process quickly replaced the tungsten etchback process because it significantly improves yield.

There are always two competing removal mechanisms in metal CMP processes. One is wet etch, in which the oxidant oxidizes the metal and forms a metal oxide that is soluble in the slurry solution. This is a pure chemical process. The other removal mechanism is a combination of chemical and mechanical processes. In this case, the oxidant oxidizes the metal surface and forms a strong metal oxide, which protects the metal surface and stops further oxidation. Mechanical abrasions from particulates in the slurry remove the passivation metal oxide and expose the metal surface to repeated oxidation and oxide removal processes. Figure 12.41 illustrates the two removal mechanisms in the metal CMP process.

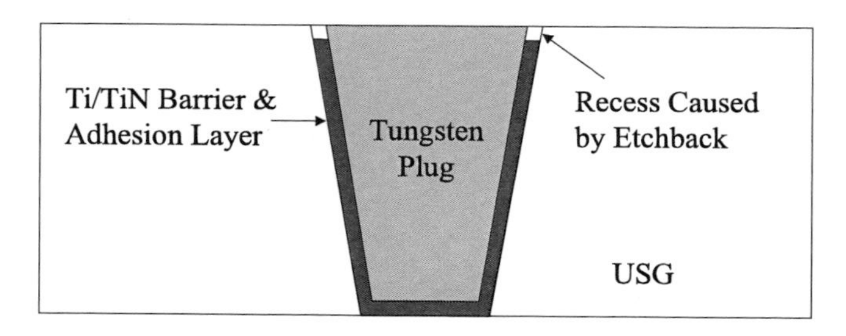

Figure 12.40 Ti/TiN recessing due to tungsten etchback.

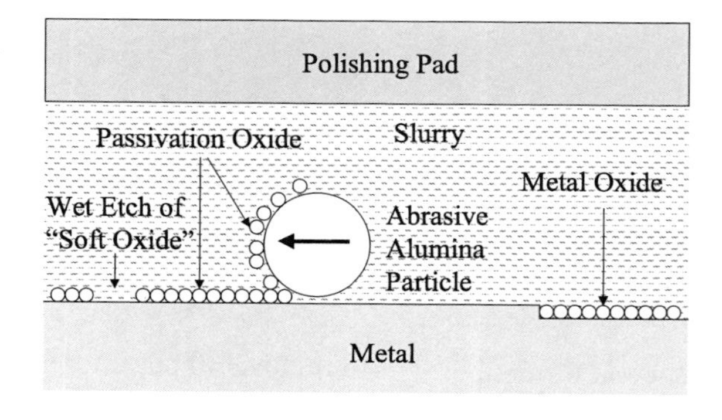

Figure 12.41 Metal polishing mechanisms.

Fine alumina powder is used in tungsten CMP slurry, and potassium ferricyanide [$K_3Fe(CN)_6$] is used as both an etchant and oxidant. The wet etch chemistry can be expressed as

$$W + 6Fe(CN)_6^{-3} + 4H_2O \rightarrow WO_4^{-2} + 6Fe(CN)_6^{-4} + 8H^+,$$

in which the ferricyanide serves as an electron sink that oxidizes and dissolves the tungsten as a WO_4^{-2} ion in the slurry solution. The competing passivation oxidation reaction can be expressed as

$$W + 6Fe(CN)_6^{-3} + 3H_2O \rightarrow WO_3 + 6Fe(CN)_6^{-3} + 6H^+.$$

This reaction forms the passivating oxide WO_3.

In tungsten CMP processes, these two competing processes are controlled by the local pH at the tungsten–slurry interface. This is accomplished by using additives, such as potassium hydrogen phosphate (KH_2PO_4), which adjusts the pH to between 5 and 6. To improve polishing planarity, a weak organic base such as ethylenediamine is used to further adjust the pH closer to 7 (a neutral value), which increases passivation and reduces wet etch.

Because potassium ferricyanide is highly toxic, and used slurry disposal causes severe environmental problems, many fabs prefer to use ferric nitrate [$Fe(NO_3)_3$] as an oxidizer. Tungsten CMP uses a two-step polishing process. The first step is removal of bulk tungsten with slurry pH < 4, and the second step is removal of the titanium nitride/titanium stacked barrier/adhesion layer with slurry pH > 9.

12.5.3 Copper chemical mechanical polishing

It is very difficult to pattern copper with plasma etching due to the lack of volatile inorganic copper compounds. Since a dual damascene process does not require a metal etch, it is the perfect candidate for copper metallization. Copper CMP is one of the most challenging processes in dual damascene interconnection processes for copper applications.

Hydrogen peroxide (H_2O_2) or nitric acid (HNO_3) can be used as an oxidant in copper polishing slurry, and alumina particulates are used for abrasion. Because copper oxide CuO_2 is porous and cannot form a passivation layer to stop further copper oxidation on the surface, an additive that can enhance the passivation effect is needed. Ammonia (NH_3) is one additive used in copper CMP slurry. Other additives such as ammonium hydroxide [NH_4OH], ethanol, or benzotriazole can also be used as complexing agents to reduce the wet etch effect.

In dual damascene copper metallization processes, both bulk copper and barrier tantalum layers need to be removed by a CMP process. Because copper slurry cannot effectively remove tantalum, the lengthy overpolishing step for tantalum removal can cause copper recessing and dishing effects, as shown in Fig. 12.42.

To solve this problem, a two-slurry polishing approach is adopted. In this approach, the first slurry is used primarily to remove the bulk copper layer, while

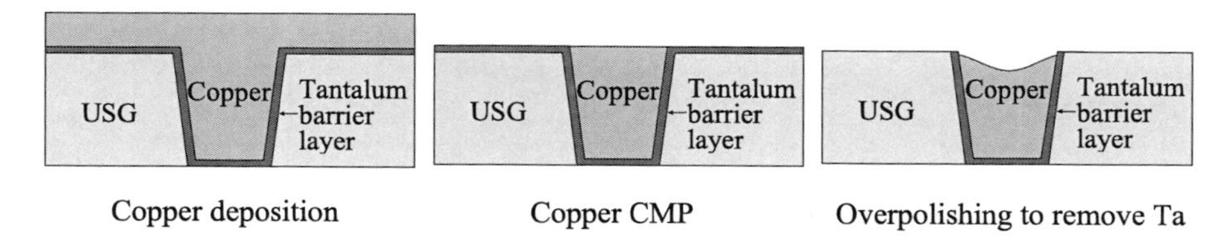

Figure 12.42 Dishing and recessing caused by overpolishing.

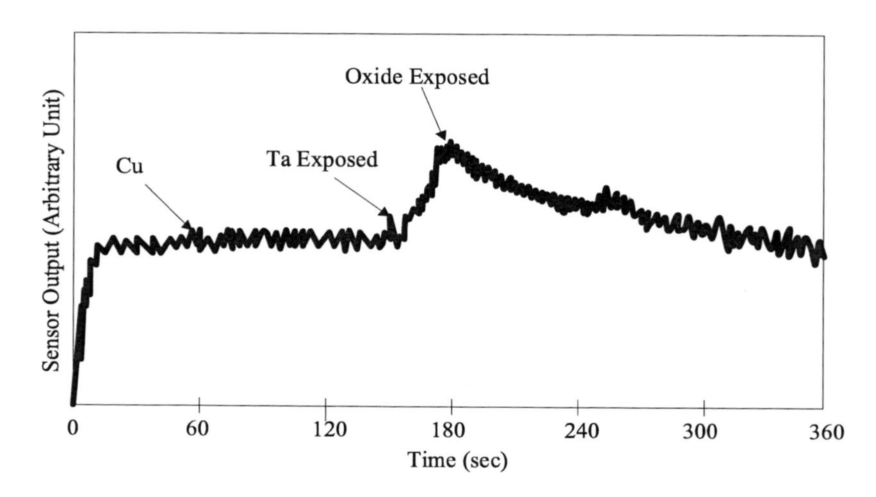

Figure 12.43 Motor current output during a copper CMP process (Aplex).

the second slurry is used to remove the tantalum barrier layer. It is very important to remove all copper from the surface with the first slurry before using the second one, since the second slurry has a very low copper removal rate. The extent of copper loss with overpolishing can be significantly reduced compared to the one-slurry approach. The two-slurry CMP process with separate copper and barrier polishing steps can help reduce copper dishing and oxide erosion effects. The use of multiple polishing platens greatly simplifies multislurry CMP processing.

12.5.4 Chemical mechanical polishing endpoint detection

The endpoints of CMP processes can be detected by monitoring the motor current or by optical measurement. When the CMP process comes to an end, the polishing pad reaches and begins to polish the underside, causing the friction force to change. To maintain a constant pad rotation rate, the current of the polishing head rotary motor changes. By monitoring the change in motor current, the endpoint of the CMP process can be detected. Figure 12.43 illustrates current change during the copper CMP process.

This endpoint detection process is described in detail as follows. During processing, the rotary motor current is close to constant while polishing copper.

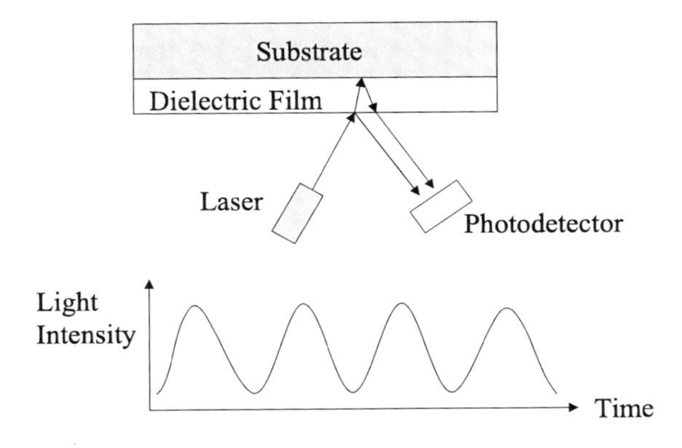

Figure 12.44 Dielectric CMP endpoint.

Noise in the signal comes from the rotation frequency of the motor. When copper is removed and the tantalum barrier layer is exposed, polishing friction increases. To keep the rotation rate, the current of the rotary motor increases. When the tantalum barrier layer is removed, the pad starts to polish oxide. The current starts to gradually drop while tantalum is gradually removed from the surface. Sensing current change as the endpoint for the copper CMP process allows for an in-situ, two-slurry polishing process. In that process, the copper CMP stops when the endpoint monitor senses the motor current increasing. Then the wafer is transferred to another pad that uses a different slurry to effectively remove tantalum.

Another commonly used endpoint technique for CMP processes is the optical endpoint. For dielectric CMP, either the film thickness itself or the film thickness change can be detected in situ with spectroreflectometry. Endpoint can be achieved by either monitoring the thickness change or measuring the film thickness. Reflected lights from dielectric surfaces and dielectric–substrate interfaces interfere with each other. Depending on the film refractive index, film thickness, and incident angle of the light, the interference status can vary from constructive to destructive. Constructive interference gives a brighter reflection, and destructive interference causes a dimmer reflection. When polishing dielectric thin films, a change in film thickness causes periodic changes of the interference states between constructive and destructive. This causes repeated high and low intensities of the detected reflected light. If a single-wavelength light source such as a laser is used, dielectric film thickness changes can be monitored by the change in reflected light. By using a light source with a wide spectrum, such as a UV lamp or multiple-wavelength lasers, the type and thickness of the dielectric film can be directly monitored. Figure 12.44 illustrates dielectric thin-film CMP endpoint detection.

For metal CMP, the change of reflectivity can be used for the processing endpoint. Metal surfaces have high reflectivity; when metal film is removed, the reflectivity significantly reduces, indicating the endpoint. Figure 12.45 illustrates a metal CMP endpoint.

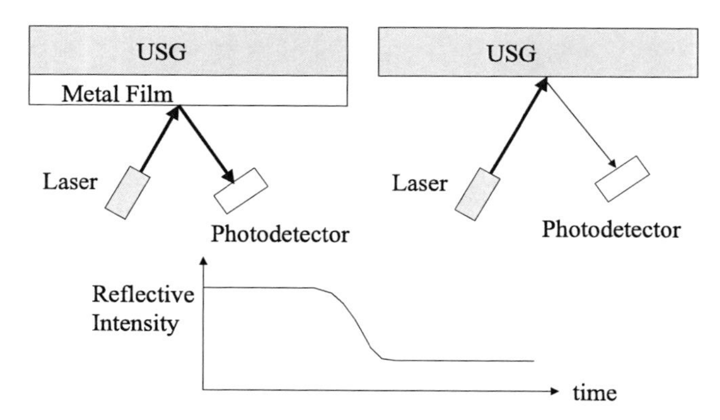

Figure 12.45 Metal CMP endpoint.

12.5.5 Post-chemical-mechanical-polishing wafer cleaning

Post-CMP wafer cleaning is an inseparable part of the CMP process. Immediately after the CMP process, wafers must be thoroughly cleaned, otherwise many defects are left on the wafer surface that are associated with the polishing process and polishing slurries. Post-CMP cleaning must effectively remove both residual slurry particles and other chemical contaminants introduced during the CMP process by slurries, pads, and conditioning tools. The post-CMP cleaning process usually includes poly-vinyl-alcohol (PVA) brush cleaners with DI water, and mechanical scrubbing to remove CMP slurry particles. Most brush cleaning involves the application of DI water through rinse nozzles. Higher cleaning efficiency can be achieved by increasing the DI water volume or brush pressure, or by application of ultrasonic waves. A PVA brush is made of porous polymers, which allows chemicals to penetrate and be delivered to the wafer surface, as illustrated in Fig. 12.46(a). Double-sided scrubbers have also been used in post-CMP clean processes [see Fig. 12.46(b)].

Some slurry particles are chemically bonded to atoms on the wafer surface, which happens when slurry dries on the wafer surface. Chemical additives, such

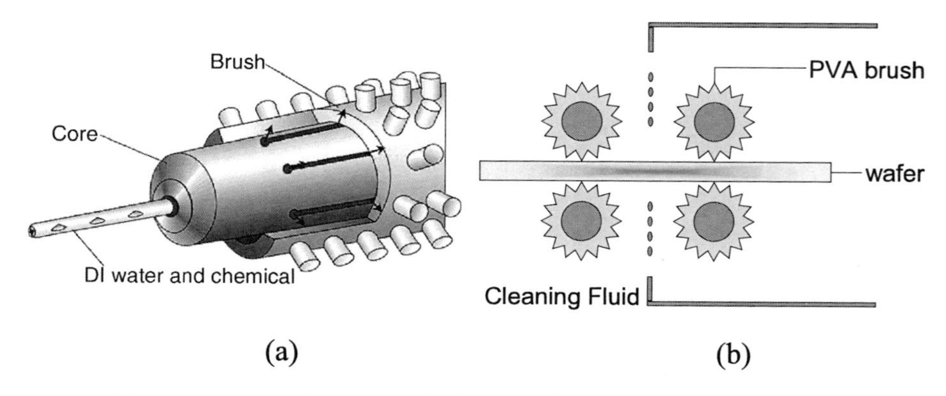

Figure 12.46 (a) A slip-on-the-core (SOTC) design PVA brush (Lam Research Corp.). (b) Illustration of a double-sided scrubber.

as ammonium hydroxide (NH_4OH), hydrofluoric acid (HF), or surfactants can be necessary to efficiently remove these bonded particles by weakening the particle-to-surface bonds or breaking the bonds. The additives also help particles diffuse away from the surface and prevent new particles from forming in the vicinity of the wafer. Residual slurry should never be allowed to dry on the wafer surface because slurry particles can develop a strong chemical bond if the wafer dries. DI water with chemical additives can be used to reduce the adhesion between the particles and wafer. A basic solution is used to adjust the wafer and particle surface charges so that electrostatic repulsion keeps particles from redepositing on the surface. Acidic solutions can be used to oxidize and dissolve organic or metal particles. Figure 12.47 illustrates particle removal processes with acidic and alkaline chemicals.

After oxide CMP processing, silica particles from slurry either adhere to the oxide surface or are embedded in it. It is very important to keep the wafer wet while transferring the wafer to the cleaning station. An alkaline chemical, usually NH_4OH, is used for postoxide CMP cleaning. The alkaline solution negatively charges both silica particles and oxide surface; thus, the electrostatic force expels particles from the surface. For particles strongly bound to the surface with molecular bonds, HF is used to remove such particles by either breaking the bonds or dissolving the silica particles and part of the oxide surface. Megasonics (ultrasound with frequency in the few hundred hertz to megahertz range) is frequently used to form microbubbles in the chemical solution that release shock waves at implosion that help to dislodge particles.

Tungsten slurries are much harder to remove than oxide slurries. DI wafers with NH_4OH additives are commonly used in post-tungsten CMP cleaning processes. The use of ferric nitrate [$Fe(NO_3)_3$] as the oxidant results in high Fe^{3+} ion concentration in the solution. The Fe^{3+} ion interacts with OH^- to form $Fe(OH)_3$ particulates during cleaning with DI water and NH_4OH, and the particles can grow to larger than 1 μm. $Fe(OH)_3$ particulates can cause high surface defect density and can contaminate the brush (called brush loading). Defects caused by $Fe(OH)_3$ particles can be reduced by using 100:1 HF cleaner.

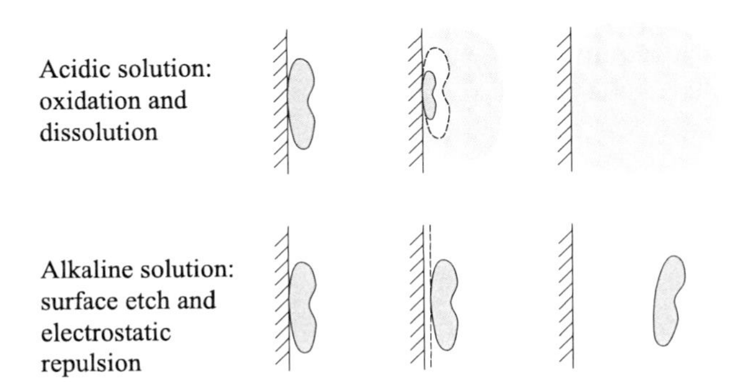

Figure 12.47 Particle removal mechanisms with acidic and basic chemical solutions.

After the cleaning process, a DI water rinsing process follows. Then, the water must be completely removed from the wafer surface without leaving any residue. Wafer drying must be a physical removal process without water evaporation, since evaporative drying is likely to leave dissolved chemicals in the DI water as contaminants.

Spin drying is the most commonly used technique in both single-wafer and batch spin dryers. The centrifugal force from spinning drives water toward the wafer edge and off of the wafer. Ultraclean drying air or nitrogen flow helps drive residual water from the center, where centrifugal force is very low.

Another drying method is vapor drying, which uses high vapor pressure of an ultrapure solvent, usually isopropyl alcohol (IPA, C_3H_8O), to displace the water from the wafer surface.

CMP tools are connected with wet cleaning tools. These integrated systems allow the dry-in dry-out CMP, post-CMP cleaning, and drying process sequence, and improve process yield.

12.5.6 Process issues

The main concerns with CMP processes are polishing rate, planarization capability, WID uniformity, WIW uniformity, WTW uniformity, removal selectivity, defects, and contamination control.

Polishing rate is determined by downforce pressure, pad hardness, pad condition, and applied slurry. Different films also have different polishing rates. Planarization capability is determined by the stiffness and surface condition of the polishing pad. Uniformities can be affected by the polishing pad condition, downforce pressure, relative speeds of the wafer and the polishing pad, and the curvature of the wafers, which is related to film stress. Downforce pressure distribution is the most important factor for controlling CMP uniformity. Removal selectivity is controlled by slurry chemistry and is related to pattern density, which is determined by design layout. There are many different types of defects that are related to many different processing parameters. For example, a copper CMP process can have corrosion on Cu pads that connect to pMOS, and dendrite growth on Cu pads that connect to nMOS. This is because in an acidic solution, copper ions tend to leave pMOS-contacting metal surfaces and move toward nMOS-connecting metal surfaces, especially if light shines on the wafer surface to create a 0.6- to 0.7-V photovoltage between the p and n junctions. Figure 12.48 illustrates the photoelectrochemical process of dendrite-corrosion defect formation.

One way to eliminate this defect is running copper CMP and post-CMP processes in the dark, or sealing the process system so that photovoltage between the p–n junctions can be avoided.

Contamination control is one of the most important issues related to CMP processes. Because slurries contain large amounts of particles and alkali ions, the CMP bay has a much higher probability of mobile-ion and particle contamination than other processing areas in an IC fab. Therefore, some fabs isolate their CMP

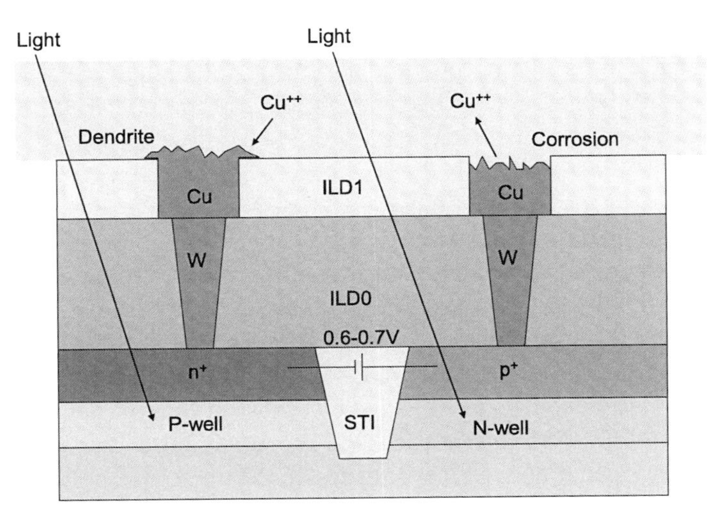

Figure 12.48 Dendrite and corrosion on p^+/n-well metal.

bay from other processing areas to avoid cross-contamination. Wafer cassettes need to be changed when product wafers are transferred in and out of the CMP bay. Personnel movement between the CMP bay and other areas without a gown change is restricted. A good rule of thumb is that anything belonging to the CMP bay stays in the CMP bay; anything not belonging to the CMP bay should be kept out.

Copper CMP tools must be dedicated tools used only for copper polishing processes to avoid copper contamination of the silicon wafer, since copper contamination can cause unstable performance of MOSFETs and ruin IC chips.

If slurry is spilled, it is very important to immediately wash and clean the area thoroughly before the slurry dries. Dried slurry leaves huge amounts of tiny particulates, which easily become airborne. Good housekeeping is needed to avoid slurry spillage and residue build up. Strict procedures and disciplines are required when changing polishing pads and carrier films, and the fresh pads usually need three to five dummy wafer polishings to season them before being ready to polish production wafers.

12.6 Recent Developments

The CMP process is one of the fundamental processes in IC chip manufacturing, along with etch and CVD. Multilayer Al–Cu interconnections need oxide and tungsten CMP processes to planarize ILD layers and form tungsten plugs, respectively. Copper interconnections have become a mainstream technology in CMOS logic IC since the development of the 130-nm technology node. Copper CMP processes have been widely used in dual damascene processes to form copper interconnections in advanced IC fabs.

Interconnections with a combination of copper metallization and low-κ dielectrics raise new challenges for CMP processes. Because low-κ dielectric materials have less mechanical strength compared to silicon oxide, it is easier for low-κ dielectrics to crack and delaminate during their interface with copper. A low-downforce CMP process has been sought for copper and barrier metal CMP to avoid damaging low-κ dielectrics.

For DRAM applications, polysilicon and oxide-nitride-oxide dielectrics have been used for storage capacitor formation. To shrink the size of the capacitors, high-κ dielectric materials such as Al_2O_3 have been used. DRAM manufacturers are also looking for copper interconnections as a cost-cutting solution. Copper CMP processes involved with these low-κ dielectric materials can also be applied in advanced DRAM fabrication.

The most significant development in CMOS IC is the implemention of HKMG in MOSFETs; these revolutionized gate dielectrics for the first time in the 50-year history of MOSFETs. The gate-last HKMG process requires two CMP processes. The first is poly-open CMP, which polishes ILD0 to expose polysilicon dummy gates. After dummy gate removal and cleaning, high-κ dielectrics and multiple metal layers are deposited to fill the gate trenches to form HKMG. Dummy gate removal can help to further increase silicon strain in the channel, and improve device speed. This is one of the advantages of gate-last over gate-first HKMG. The second process is metal CMP, which removes bulk metal layers on the wafer surface and keeps them only in the gate trenches; this finishes the replacement metal gate process. Figure 12.49 shows the CMP applications in gate-last HKMG CMOS formation.

To avoid high-downforce-induced copper delamination during copper CMP processing, low-downforce CMP processes have been developed. Electrochemical mechanical polishing (ECMP) processes have been introduced to remove copper from wafer surfaces, which are positively biased and serve as anodes. Some companies even introduced downforce-free electrochemical polishing processes to meet the requirements of copper CMP for copper/ULK interconnections.

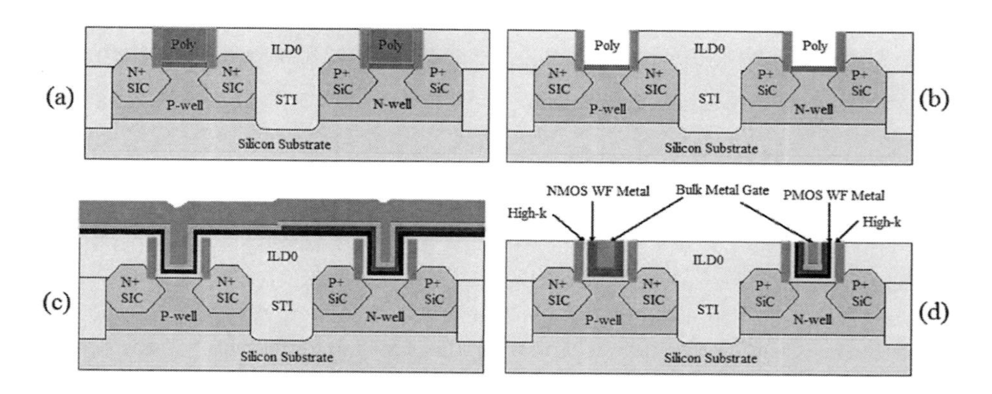

Figure 12.49 Gate-last HKMG processes: (a) ILD0 CMP or poly-open CMP, (b) polysilicon dummy gate removal, (c) high-κ and metal layer deposition, and (d) metal CMP to form HKMG MOSFETs.

12.7 Summary

- The main applications of CMP processes are dielectric planarization and bulk film removal for STI, tungsten plugs, and dual damascene copper interconnections.
- Multilevel metallization needs planarized dielectric surfaces for both high-resolution photolithography in via hole and metal line patterning, and for easier metal deposition.
- For 0.25-μm and smaller patterned features, CMP processes must prepare the planarized surface required by high-resolution photolithography processes due to the small DOF.
- The advantages of CMP processes are: possibility of high-resolution photolithography patterning due to the planarized surface; higher yield; lower defect density due to the removal of rough surface topologies; and more flexibility for IC design.
- A basic CMP system consists of a rotating wafer carrier, a polishing pad on a rotating platen, a pad conditioner, and a slurry delivery system.
- Oxide CMP slurries are alkaline solutions at $10 < pH < 12$, with colloidal suspension of silica abrasives. Metal CMP slurries are acidic solutions at $4 < pH < 7$ with alumina abrasives.
- Important factors for CMP processes are polish rate, planarization capability, selectivity, uniformities, defects, and contamination controls.
- Polishing rate is related to downforce pressure, pad stiffness, pad surface condition, relative speeds of the pad and wafer, and slurry type.
- CMP uniformity is determined by downforce pressure distribution, pad stiffness, and pad surface condition.
- Removal selectivity is determined by slurry chemistry.
- In oxide CMP processing, silica abrasive particles in slurry form chemical bonds with the silicon oxide surface, and polishing slurry mechanically abrades the bonded silica along with bonded surface materials from the surface.
- There are two metal removal mechanisms in a metal CMP process: wet etch and passivation/abrasion. In wet etch, an oxidant forms metal oxides that are soluble in slurry. In passivation/abrasion, an oxidant oxidizes metal to form a passivation oxide, which stops metal oxidation and is removed by abrasive particles in the polishing slurry.
- Post-CMP cleaning is a very important step for reducing defects and improving yield. DI water with ammonium hydroxide (NH_4OH) is commonly used for post-CMP cleaning. For oxide CMP processes, HF is used to remove silica particulates that form chemical bonds with surface molecules and cannot be removed by NH_4OH solution. For metal processes, oxidants such as surfactants and nitric acid are used to oxidize and dissolve metal particles that cannot be removed by NH_4OH solution.
- CMP-related defects are scratches, residues, delamination, metal corrosion, dielectric film cracks, etc. These defects can be captured by optical inspection.

- Post-WCMP and post-Cu-CMP electron beam inspections can effectively capture electrical defects such as contact-plug open circuits and junction leakage.
- CMP processes are required for gate-last HKMG CMOS manufacturing.

12.8 Bibliography

T. Ashizawa, "Novel cerium oxide slurry with high planarization performance for STI," *Proc. 4th CMP Symp.* **S-6** (1999).

I. Belov, J.Y. Kim, P. Watkins, M. Perry, and K. Pierce, "Polishing slurries with aluminate-modified colloidal silica abrasive," *MRS Proc.* **867**, W6.9.1 (2005).

A. E. Braun, "Slurries and pads face 2001 challenges," *Semicond. Intl.* **21**(13), 65–74 (1998).

C. Y. Chang and S. M. Sze, *ULSI Technologies*, McGraw-Hill, New York (1996).

P. Feeney, "CMP for metal-gate integration in advanced CMOS transistors," *Solid State Technol.*, 14 (Nov. 2010).

M. A. Fury, "CMP processing with low-k dielectric, *Solid State Technol.* **42**(7), 87–96 (1999).

R. R. Jin, S. H. Ko, B. A. Bonner, S. Li, T. H. Osterheld, and K. A. Perry, "Advanced front-end CMP and integration solutions," *Proc. CMP-MIC*, 119–129 (2000).

Y. Matsui, Y. Tateyama, K. Iwade, T. Mishioka, and H. Yano, "High-performance CMP slurry with CeO_2/resin abrasive for STI formation," *ECS Trans.* **11**, 277–283 (2007).

K. Mistry et al., "A 45-nm logic technology with high-k + metal gate transistors, strained silicon, 9 Cu interconnect layers, 193-nm dry patterning, and 100% Pb-free packaging," *IEDM Tech. Dig.*, 247–250 (2007).

F. W. Preston, "The theory and design of plate glass polishing machine," *J. Soc. Glass Technol.* **11**, pp. 214–256 (1927).

W. S. Rader, T. Holt, and K. Tamai, "Characterization of large particles in fumed silica-based CMP slurry," *MRS Proc.* **1249** (2010).

S. Reabke Selinidis, D. K. Watts, J. Saravia, J. Gomez, C. Dang, R. Islam, J. Klain, and J. Farkas, "Development of a copper CMP process for multilevel, dual inlaid metallization in semiconductor devices," *Proc. Electrochem. Soc.* **98**(7), 9 (1998).

C. Sainio and D. J. Duquette, "Electrochemical characterization of copper in ammonia-containing slurries for chemical mechanical planarization of interconnects," *Proc. Electrochem. Soc.* **98**(7), 126–133 (1998).

R. K. Singh, D. W. Stockbower, C. R. Wargo, V. Khosla, M. Vinogradov and N. V. Gitis, "Post-CMP cleaning applications: challenges and opportunities," *Proc. 13th Intl. CMP-MIC*, 355 (2008).

J. M. Steigerwald, S. P. Murarka, and R. J. Gutmann, *Chemical Mechanical Planarization of Microelectronic Materials*, John Wiley & Sons, Inc., New York (1997).

"Trends and future developments for diamond CMP pad conditioners," *Industrial Diamond Rev.* **1**(4), 16–21 (2004).

H. Xiao, L.E. Ma, Y. Zhao, and J. Jau, "Study of devices leakage of 45-nm node with different SRAM layouts using an advanced ebeam inspection systems," *Proc. SPIE* **7272**, 72721E-1 (2009) [doi:10.1117/12.813885].

12.9 Review Questions

1. What were the earliest applications of CMP processes in the semiconductor industry?
2. What are the two main applications of CMP processes in IC chip fabrication in Al–Cu interconnections?
3. Name the other dielectric planarization methods used before CMP processes became widely applied in IC manufacturing.
4. Why does the fabrication of IC chips with feature sizes smaller than 0.25 μm require CMP processes?
5. Compared with other planarization methods, what are the benefits of CMP processes?
6. Why do polishing pads need reconditioning?
7. What abrasive particles are commonly used in oxide slurry?
8. What kind of particle is normally used in metal CMP slurry?
9. Why do oxide slurries need a high pH?
10. Describe the two competing removal mechanisms in the metal CMP process.
11. What are erosion and dishing effects?
12. What metrology tool can be used to measure dishing and erosion?
13. How can particles and defects on the wafer surface be measured?
14. What inspection system can be used after WCMP to capture electrical defects such as contact open circuits and junction leakage? Can an optical inspection be used to capture these defects? Explain your answer.
15. Explain the importance of post-CMP cleaning.
16. Describe the two particle removal mechanisms in a wet chemical cleaning process.
17. If slurry spills and dries in the fab, what problems can it cause?
18. Why is a copper CMP tool only used for a copper polishing process?

19. When silica particulates form molecular bonds with oxide surfaces, they cannot be removed by NH_4OH. What is the chemical usually used to remove these silica particles? Try to describe the removal process.
20. What two CMP processes are needed in advanced gate-last HKMG CMOS fabrication?

Chapter 13
Process Integration

IC manufacturing involves many processing steps that have been described in the previous chapters. To fabricate a functional chip, every processing step must be well integrated with the other processing steps. This chapter describes the integration of these steps in CMOS IC chip fabrication.

Objectives

After finishing this chapter, the reader will be able to:

- list the three methods used for isolation formation
- describe the three-well formation processes
- explain the purpose of threshold adjustment implantation
- describe the sidewall spacer process and explain its applications
- explain the advantages of high-κ gate dielectrics over SiON gate dielectrics
- describe the gate-first and gate-last approaches of metal gate formation
- name at least three silicides used for gate and local interconnections
- name three metals used in a traditional aluminum interconnection process
- list the basic steps for a copper metallization process
- identify the material most commonly used as a final passivation layer for an IC chip
- describe the main differences of CMOS, DRAM, and NAND flash processes.

13.1 Introduction

It takes more than 30 masks and several hundreds of processing steps to finish the fabrication of an advanced CMOS IC chip. Every step is related to all of the other steps. For CMOS processes, the steps can be classified as front-end of line (FEoL), mid-end of line (MEoL) and back-end of line (BEoL). FEoL includes active area (AA) formation, well implantations, gate patterning, and S/D formation that create the transistors. MEoL includes self-aligned silicide, contact hole pattern and etch, and tungsten deposition and CMP that form the contacts between the devices and the metal wires. BEoL steps form the interconnections and passivation. Traditional aluminum interconnections include metal stacked (Ti/TiN/Al-Cu/TiN) PVD and etch; dielectric CVD; dielectric planarization; and via hole pattern and etch. For copper interconnections, BEoL includes via hole patterning and etch;

trench patterning and etch; barrier layer (Ta or TaN) and copper seed layer PVD; bulk copper plating and annealing; and metal (Cu/Ta) CMP.

For flash processes, FEoL includes the AA, WL, contact to BL/source line (CB1), source line, contact to BL (CB2), and BL. The BEoL includes via and metal layers.

For DRAM processes, two major competing technologies coexisted for a long time; one is stacked capacitor technology and the other is deep trench capacitor technology. Stacked capacitor DRAM dominates the DRAM market, while deep trench capacitor DRAM is widely used as embedded DRAM for system-on-chip applications because it is more compatible with CMOS and logic processes. More-advanced DRAM processes have also been developed; one of them in mass production is the buried wordline (bWL) technology.

13.2 Wafer Preparation

Single-crystal silicon wafers with ⟨100⟩ orientation are frequently used for CMOS IC chip manufacturing. Bipolar and BiCMOS chips use wafers with ⟨111⟩ orientation. Wafers used in IC processing are doped wafers, either n- or p-type, with typical substrate dopant concentrations of 1×10^{15} atoms/cm^3. Before the mid-1970s, pMOS IC chips were made with n-type wafers. After the introduction of ion implantation processes in the mid-1970s, nMOS IC chips were manufactured with p-type wafers. Although both n- and p-type wafers can be used for CMOS IC fabrication, more fabs use p-type wafers than n-type wafers. The reason is mainly historical. Driven by the demands of digital logic IC chips with low power consumption, high noise immunity, and high thermal stability, the CMOS process was developed based on the nMOS process in the late 1970s. Naturally, p-type wafers, which were used to make nMOS ICs, were chosen as the substrate for early CMOS IC fabrication.

The simplest nMOS IC process had five masking steps: activation area, gate, contact, metal, and bonding pad. Early CMOS IC processing required eight masking steps: n-well (for p-type substrate), activation, gate, n-S/D, p-S/D, contact, metal, and bonding pad. Figure 13.1(a) illustrates an nMOS chip cross section, and Fig. 13.1(b) shows a cross section of an early CMOS chip.

Bipolar transistors and BiCMOS chips always require wafers with an epitaxial silicon layer to form a heavily doped buried layer. Certain power devices need substrates with resistance so high that only wafers made with a floating zone (FZ) method can meet the requirement. When the clock frequency of CMOS IC chips is not very high, the chips do not need an epitaxial layer. However, silicon wafers with an epitaxial silicon layer are required for high-speed CMOS chips. Silicon wafers made by the Czochralski (CZ) method always have some oxygen, and oxygen can reduce carrier lifetime and slow down device speed. The epitaxial silicon layer can create an oxygen-free substrate and help to achieve high device speed.

RCA cleaning processes are commonly used to remove contaminants from the silicon wafer surface before epitaxy. Anhydrate HCl dry cleaning helps to remove

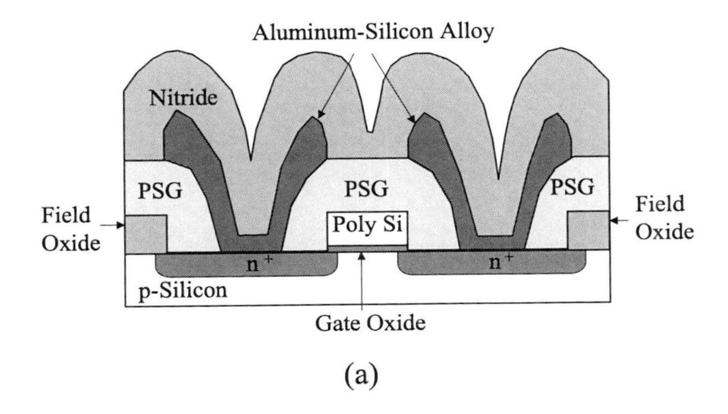

(a)

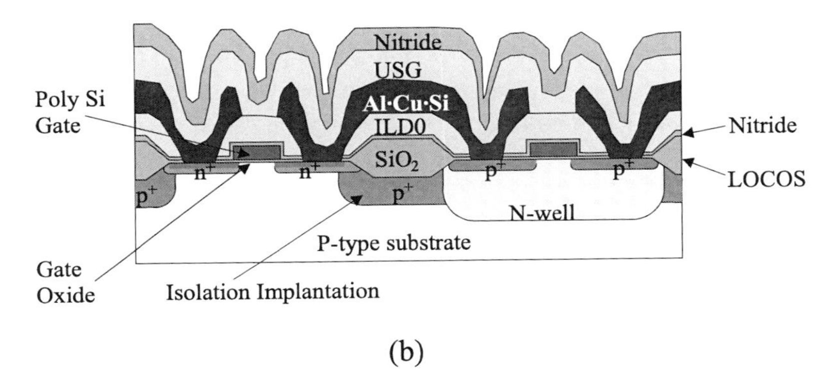

(b)

Figure 13.1 Cross section of (a) nMOS and (b) early CMOS chips.

mobile ions and native oxide. Epitaxial silicon layer growth usually is a high-temperature (>1000 °C) CVD process with silane (SiH_4), DCS (SiH_2Cl_2), or TCS ($SiHCl_3$) as the main processing gas. Hydrogen is commonly used in the epitaxy growth process as a secondary processing gas or carrier and purge gas. Arsine (AsH_3) or phosphine (PH_3) is used as the n-type dopant gas, and diborane (B_2H_6) is used as the p-type dopant gas.

Advanced CMOS IC chips normally use p-type ⟨100⟩ single-crystal silicon wafers with p-type epitaxial layers.

13.3 Isolations

Blanket field oxide, local oxidation of silicon (LOCOS), and STI are the three isolation techniques that can be used in IC fabrication. p-type doped junctions also have been used to electronically isolate neighboring transistors.

13.3.1 Blanket field oxide

Blanket field oxide was used in the early years of the IC industry. It is a simple and straightforward process. Blanket field oxide can be formed by growing an oxide

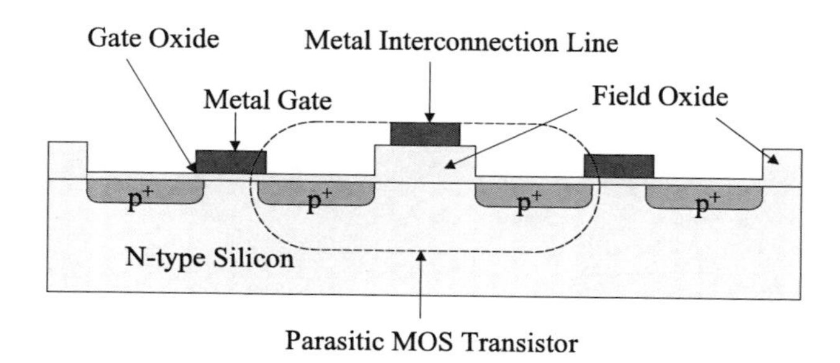

Figure 13.2 pMOS IC with blanket-field oxide isolation.

layer with a desired thickness on a flat silicon surface, then patterning and etching the activation windows on the oxide layer. The thickness requirement of a field oxide is determined by the field threshold voltage, noted as V_{FT}, which needs to be high enough ($V_{FT} \gg V$) to prevent crosstalk between the neighboring transistors. While the circuit power supply voltage can switch on and off MOS transistors on the IC chips ($V > V_T$), it cannot turn on the parasitic MOS transistors and cause chip malfunction. Figure 13.2 illustrates a pMOS IC chip with a blanket-field oxide as isolation. The thickness of the blanket-field oxide normally is about 10,000 to 20,000 Å.

13.3.2 Local oxidation of silicon

Blanket field oxides were used until about 1970. Although the process is simple, it has some disadvantages. One is that the active window has a high oxide step with a sharp edge, which is very difficult to cover in the subsequent metal deposition process. Another disadvantage is that the channel stop doping must occur before oxidation, requiring the field oxide to align with the isolation doping region. This requirement causes difficulties in feature size reduction.

LOCOS has been used in IC production since the1970s. One of the advantages of LOCOS is that the silicon dioxide is grown after channel stop implantation; thus, the field oxide is self-aligned with the isolation doping area. By using channel stop implantation, field threshold voltage V_{FT} can remain unchanged, while the field oxide thickness decreases. Compared to blanket-field oxide, the step height between the activation area and LOCOS field oxide is lower, and the sidewall is sloped. This makes sidewall coverage much easier for the later metal or polysilicon layer deposition. The thickness of LOCOS oxide is 5,000 to 10,000 Å. The LOCOS isolation process is illustrated in Fig. 13.3.

LPCVD silicon nitride is used as an oxidation mask, which only allows thick silicon dioxide (LOCOS) to grow at designated areas. The activation areas where transistors will be built are covered by nitride and do not grow the thick oxide. Pad oxide is needed to relieve the strong tensile stress of the LPCVD nitride. Plasma etch with fluorine chemistry is employed for the nitride patterned etch, and hot phosphoric acid is used to strip the nitride layer after LOCOS formation.

One of the major disadvantages of LOCOS is the "bird's beak." Silicon dioxide grows in all directions, causing encroachment in a lateral direction underneath the

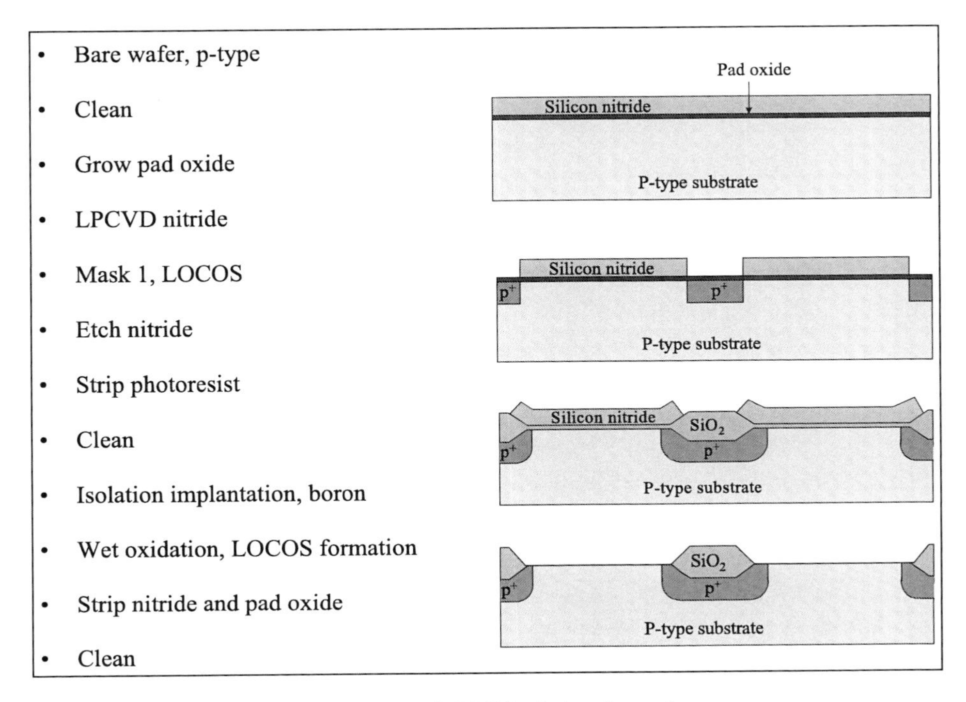

Figure 13.3 LOCOS isolation formation.

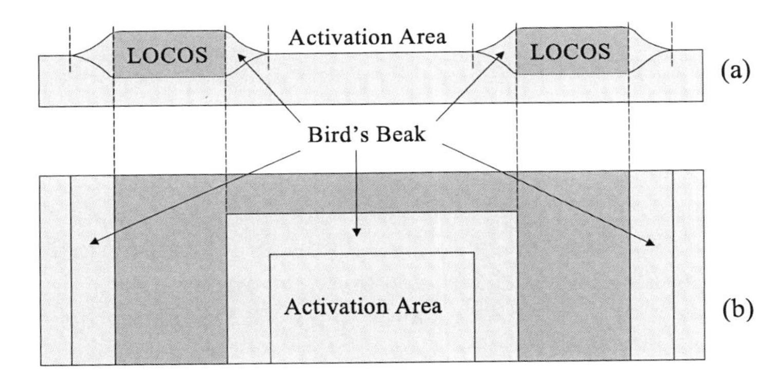

Figure 13.4 Bird's beak of LOCOS: (a) cross section view and (b) top view.

silicon nitride layer, as shown in Fig. 13.3. Bird's beak is caused by the isotropic diffusion of oxygen inside the silicon dioxide layer during the thermal oxidation process. The dimension of LOCOS encroachment is about the same as the oxide thickness on both sides for 5000-Å oxide; the bird's beak is about 0.5 μm on both sides. This requires a considerable amount of silicon surface area and creates difficulties when increasing the transistor packing density, as shown in Fig. 13.4.

To reduce the bird's beak of LOCOS, several improvements have been made. Poly-buffered LOCOS (PBL) is the most popular. Depositing a polysilicon buffer layer before LPCVD silicon nitride can reduce bird's beak from 0.1 to 0.2 μm. PBL

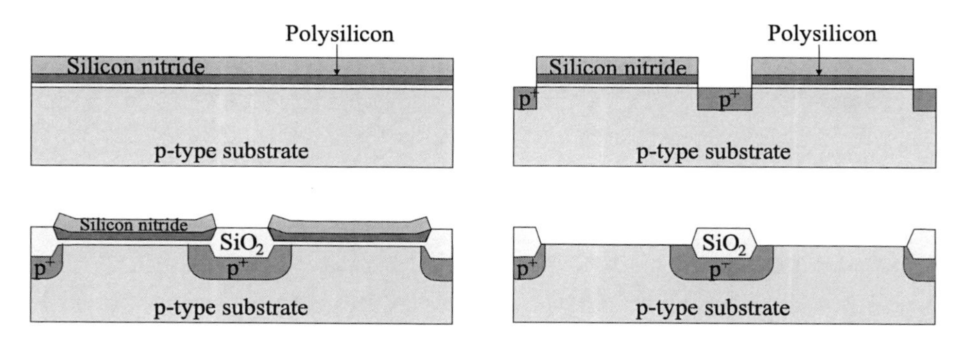

Figure 13.5 Processing steps of PBL formation.

reduces bird's beak due to the consumption of the lateral diffusion oxygen by the polysilicon layer. The process is illustrated in Fig. 13.5.

13.3.3 Shallow trench isolation

LOCOS and improved PBL work very well until the feature size reduces to less than a half micron. Just like a very thick wall to separate neighboring houses would take up too much real estate, the bird's beak of LOCOS takes up too much precious space from the silicon surface that could be used to build transistors. Keep in mind that transistors are surrounded by isolation oxide. In real life, narrow fences are used to separate neighboring houses; in IC fabrication, their equivalent is STI.

To reduce oxide encroachment, silicon etching with a nitride mask and oxidation of the trench was researched. The STI process was then developed with CVD oxide trench fill instead of thermal oxidation. An early STI process is illustrated in Fig. 13.6.

In early STI processes, an LPCVD silicon nitride layer is used as a hard mask for single-crystal silicon etching and the oxide etching stop layer. Before CVD oxide fills the trench etch on the silicon surface, a thin layer of barrier oxide is grown, and boron ions are implanted at the bottom of the trench to form a channel stop junction. Channel stop implantation can help reduce the required isolation oxide thickness, which reduces the depth of the trench. Photoresist etchback planarization is applied to remove CVD oxide from the wafer surface. By choosing a suitable CF_4/O_2 ratio, an approximately 1:1 etch rate ratio of photoresist and oxide can be achieved. Etchback processing stops at the nitride layer; this endpoint occurs automatically by the appearance of a C-N line radiation signal in the glow of the plasma. After stripping the nitride with hot phosphoric acid, oxide left inside the trenches serves as the isolation blocks for electrically isolating the neighboring devices.

Although STI has several advantages over LOCOS, it did not immediately replace LOCOS. The processing sequence of LOCOS has fewer steps and was production proven. LOCOS continued to be used in IC fabrication until the mid-1990s, when the minimum feature size became smaller than 0.35 μm, and the bird's beak of LOCOS was no longer tolerable. The smaller geometry also required

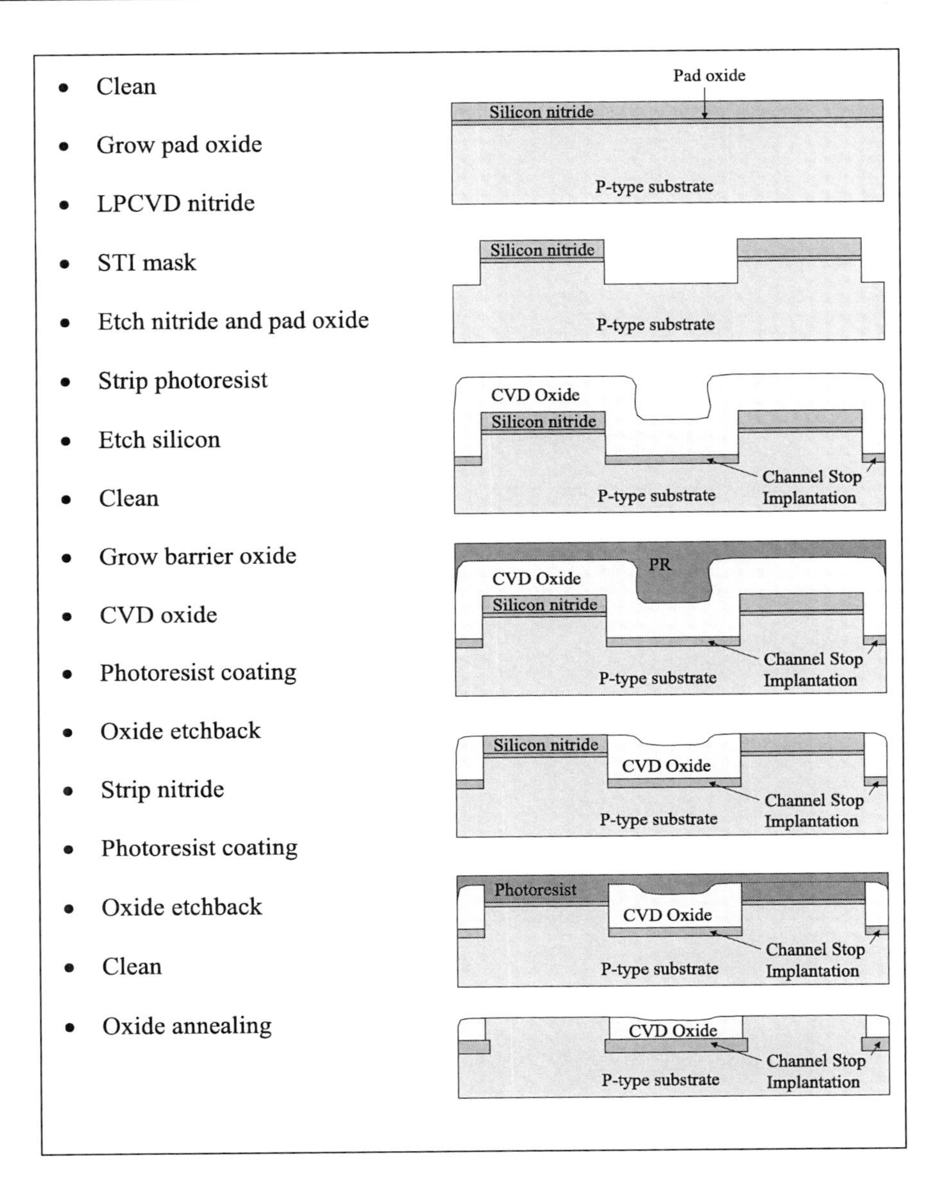

Figure 13.6 An early STI process.

highly planarized surfaces to ensure the photolithography resolution because of the DOF requirement. LOCOS always has a step of 2500 Å or higher between the active area and oxide surface, a step that is too large to precisely pattern 0.25-μm features. STI processes with oxide CMP have been developed and applied in IC fabrication. An advanced STI processing flow is illustrated in Fig. 13.7.

Channel stop ion implantation is no longer necessary when the power supply voltage of an IC device reduces to 1.8 V or lower. The trench fill process can also be achieved with thermal CVD processing with O_3-TEOS chemistry at atmospheric

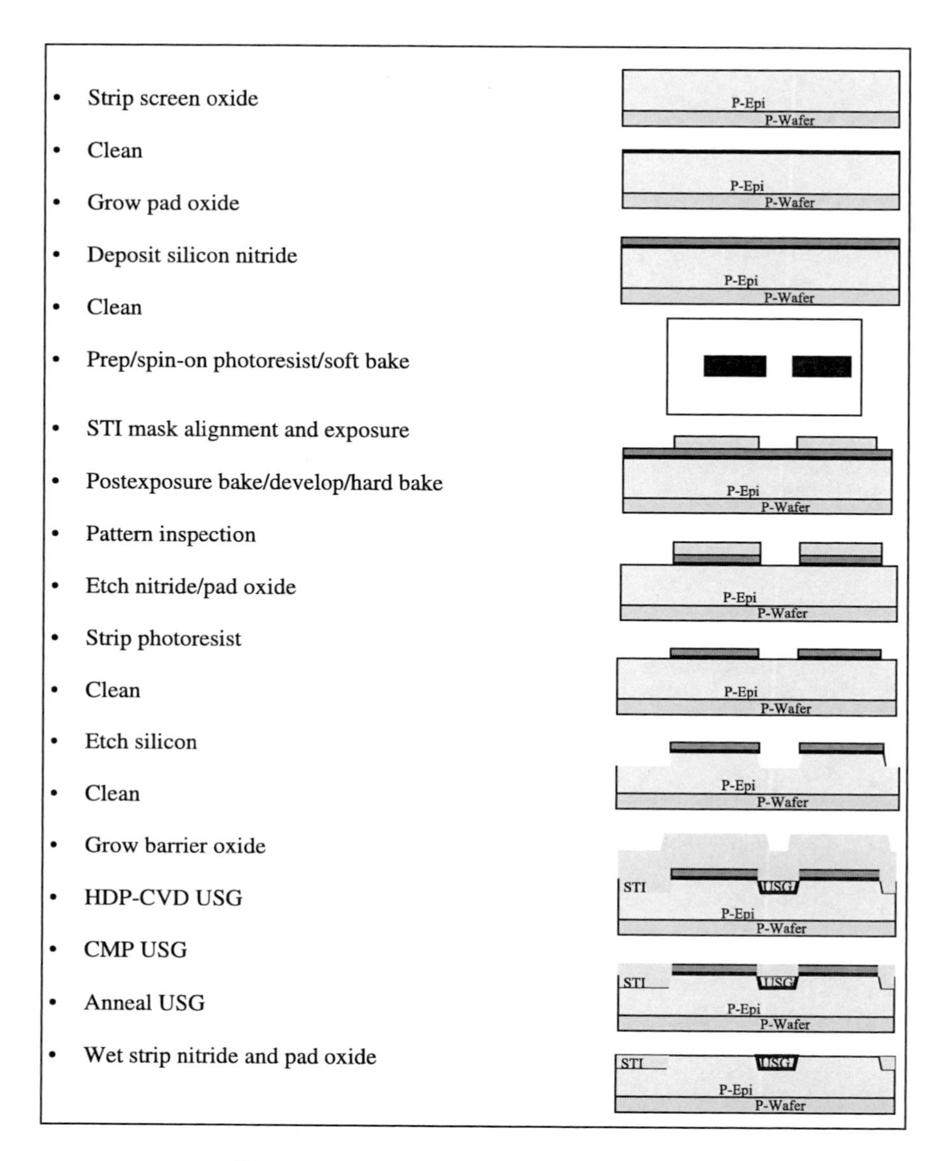

Figure 13.7 Advanced STI processing flow.

or subatmospheric pressure. Oxide deposited with a high-density plasma (HDP) process does not require thermal annealing because the oxide is very dense due to the heavy ion bombardment during deposition. Oxide deposited with O_3-TEOS chemistry has to anneal at more than 1000 °C in the oxygen ambient to densify the film.

STI formation involves many processes, such as oxidation, nitride deposition, nitride/oxide etching, silicon etching, oxide CVD, oxide CMP, oxide annealing, and nitride stripping. The main challenges of STI processes as device feature sizes continue to shrink are single-crystal silicon etching, oxide CVD, and oxide CMP.

Recently, STI oxide has also been used to induce stress to strain silicon in the active area to improve device speed.

13.3.4 Self-aligned shallow trench isolation

For flash memory devices, the most commonly used isolation technique is the self-aligned STI. Figure 13.8 shows a layout of NAND flash memory with self-aligned STI. The process starts with wafer cleaning, followed by gate oxide growth, floating gate polysilicon, and nitride hard mask depositions, as shown in Fig. 13.9(a). After patterning with an AA mask, a sequential process etches nitride, polysilicon, gate oxide, and silicon substrate, as shown in Fig. 13.9(b). After wafer cleaning, CVD oxide fills the gap [Fig. 13.9(c)], and a CMP process removes oxide from the wafer surface, endpointing at the nitride layer [Fig. 13.9(d)]. Stripping of the nitride layer finishes the formation of the self-aligned STI, as shown in in Fig. 13.9(e).

13.4 Well Formation

13.4.1 Single well

Early CMOS IC processing only required a single well, either n- or p-, depending on the type of wafers used. Well formation is accomplished by high-energy, low-current ion implantation and thermal anneal/drive-in. Single-well processing steps are illustrated and listed in Fig. 13.10.

A masking step indicates a photolithography process, which includes photoresist coating, alignment and exposure, photoresist development, and pattern inspection. Processing steps for p-well formation on an n-type wafer are almost the same as the n-well formation on a p-type wafer. Figure 13.11 shows a CMOS with n- and p-wells.

13.4.2 Self-aligned twin wells

A twin-well structure allows IC designers more flexibility in designing CMOS IC circuits. To save a photolithography masking step, a self-aligned twin-well process has been developed and used in the well formation process. The self-aligned twin well is achieved by using LPCVD silicon nitride. Nitride is a very dense layer. It can prevent oxidation on p-wells by blocking oxygen diffusion, and it also prevents n-type ions from penetrating to the p-well during ion implantation. The thick oxide growth on an n-well region can block p-well boron ion implantation. A pad oxide layer is needed to relieve the high tensile stress of the LPCVD nitride, otherwise the high stress can break down the wafers. Nitride can be stripped by hot phosphoric acid with a high selectivity to oxide. The processing steps of a self-aligned twin well are listed and illustrated in Fig. 13.12.

The advantage of a self-aligned twin well is that it eliminates a photolithographic masking step, reducing production cost and improves chip yield. The disadvantage of a self-aligned twin well is that the silicon wafer surface is no longer flat. When silicon dioxide grows on an n-well region, the oxide layer grows into the silicon

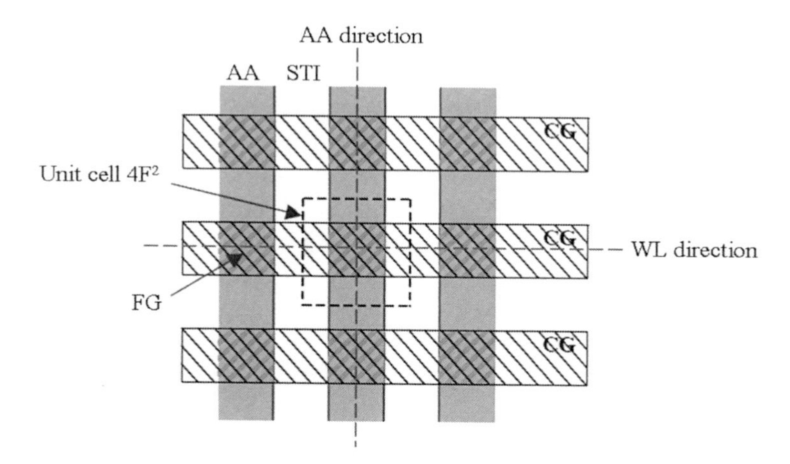

Figure 13.8 Layout NAND flash with self-aligned STI. FG indicates the floating gate, and CG indicates the control gate.

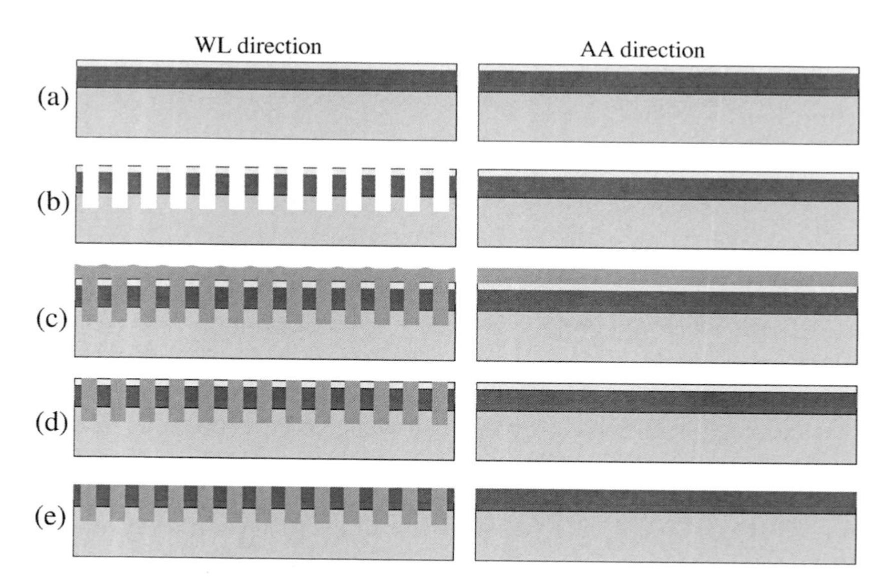

Figure 13.9 Self-aligned STI process of a NAND flash: (a) gate oxidation, polysilicon, and nitride hard mask deposition; (b) pattern etch of nitride hard mask, polysilicon, gate oxide, and silicon substrate; (c) CVD oxide; (d) oxide CMP; and (e) stripping of the nitride hard mask.

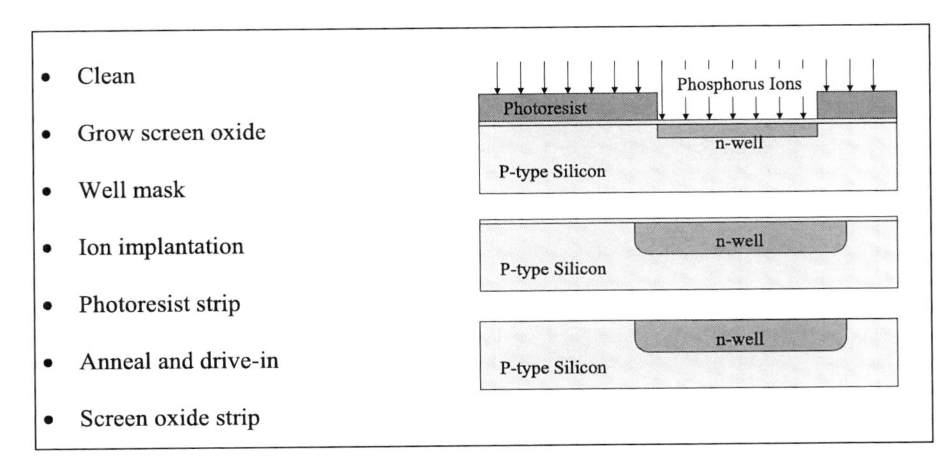

Figure 13.10 Step formation of n-wells.

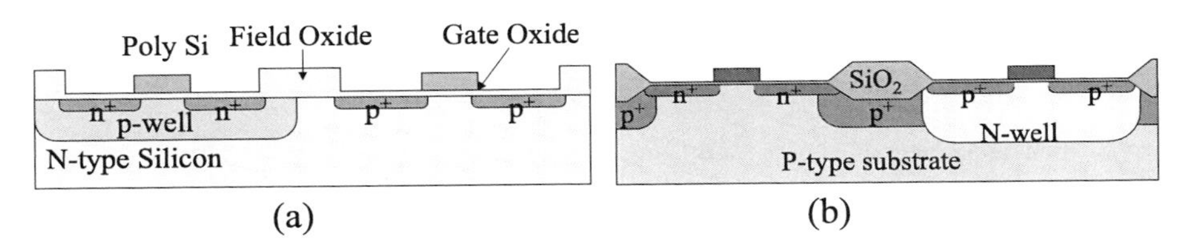

Figure 13.11 CMOS with (a) p-well and (b) n-well.

substrate. Therefore, the n-well is always lower than the p-well in a self-aligned twin well.

Twin-well structures have better control of substrates and allow more freedom for CMOS IC design. In twin-well formation process, the n-well is always implanted before the p-well. Phosphorus has a lower diffusion rate in single silicon than boron has. If a p-well is implanted first, boron in the p-well can diffuse out of control during n-well annealing and dopant drive-in.

13.4.3 Twin wells

The p- and n-wells formed by self-aligned twin-well processing are not on the same level, which can affect photolithography resolution due to the DOF problem. Double-photomask twin wells are frequently used in advanced CMOS IC chip manufacturing. Their processing steps are listed and illustrated in Fig. 13.13. Both well implantation processes use high-energy, low-current implanters. Furnaces are used for well implantation annealing and drive-in processes.

13.5 Transistor Formation

13.5.1 Metal gate process

Before the mid-1970s, an MOS transistor was formed by a process in which the S/D and gates were not self-aligned. First, the sources and drains were formed

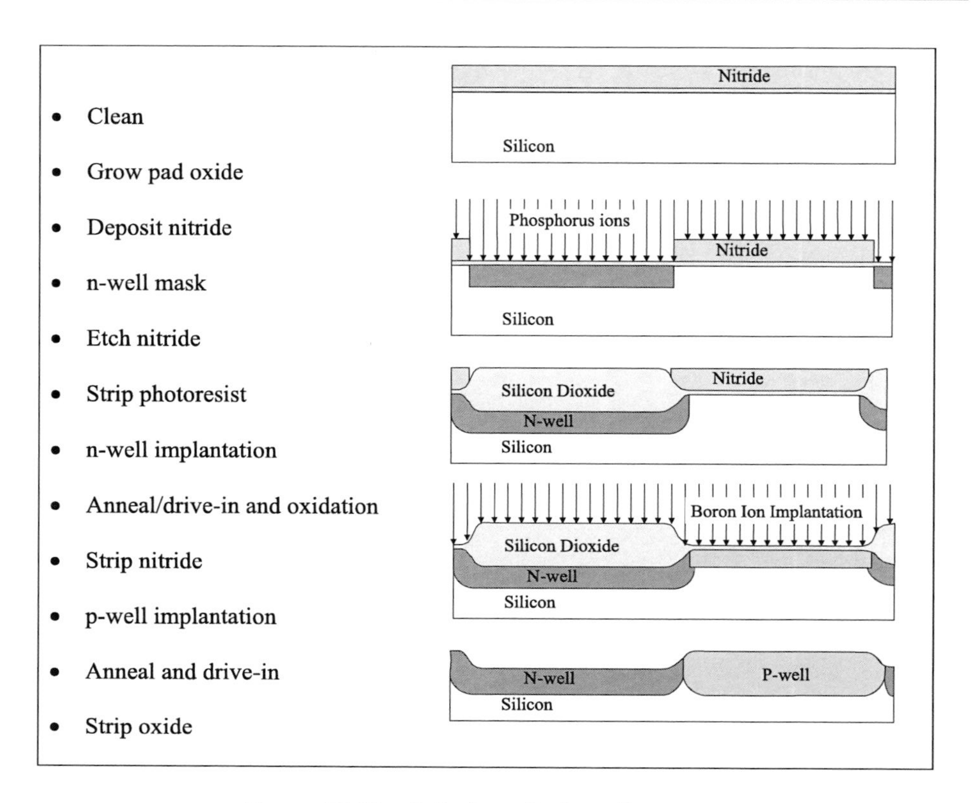

Figure 13.12 Self-aligned twin-well process.

by a diffusion process with thermally grown silicon dioxide as a diffusion mask. Then, oxide in the gate area was etched away, and a thin layer of gate oxide was grown. A third mask defined the contact holes, and a fourth mask formed the metal gates and interconnections. A final mask defined the bonding pad. After etching of the bonding pads and photoresist stripping, the wafers were ready for testing and packaging. Table 3.1 lists the pMOS transistor process without self-alignment, and Figs. 3.30 and 3.31 illustrate the processing steps. A gate usually is designed wider than the distance between the S/D to ensure full coverage of the S/D by the gate. This makes it very difficult to scale down the device feature size. Currently, only educational laboratories manufacture MOSFETs using this process.

13.5.2 Self-aligned source/drain process

With the introduction of ion implantation, processes for manufacturing MOSFETs have changed to self-aligned S/D processes. Polysilicon has replaced aluminum as the gate material because aluminum alloys cannot sustain the high temperature required by the postimplantation thermal annealing. The process starts with an activation mask with an etched window on the field oxide to define the area where the transistors will be built. After wafer cleaning, gate oxide growth, and polysilicon deposition, a gate mask defines the gate and local interconnection.

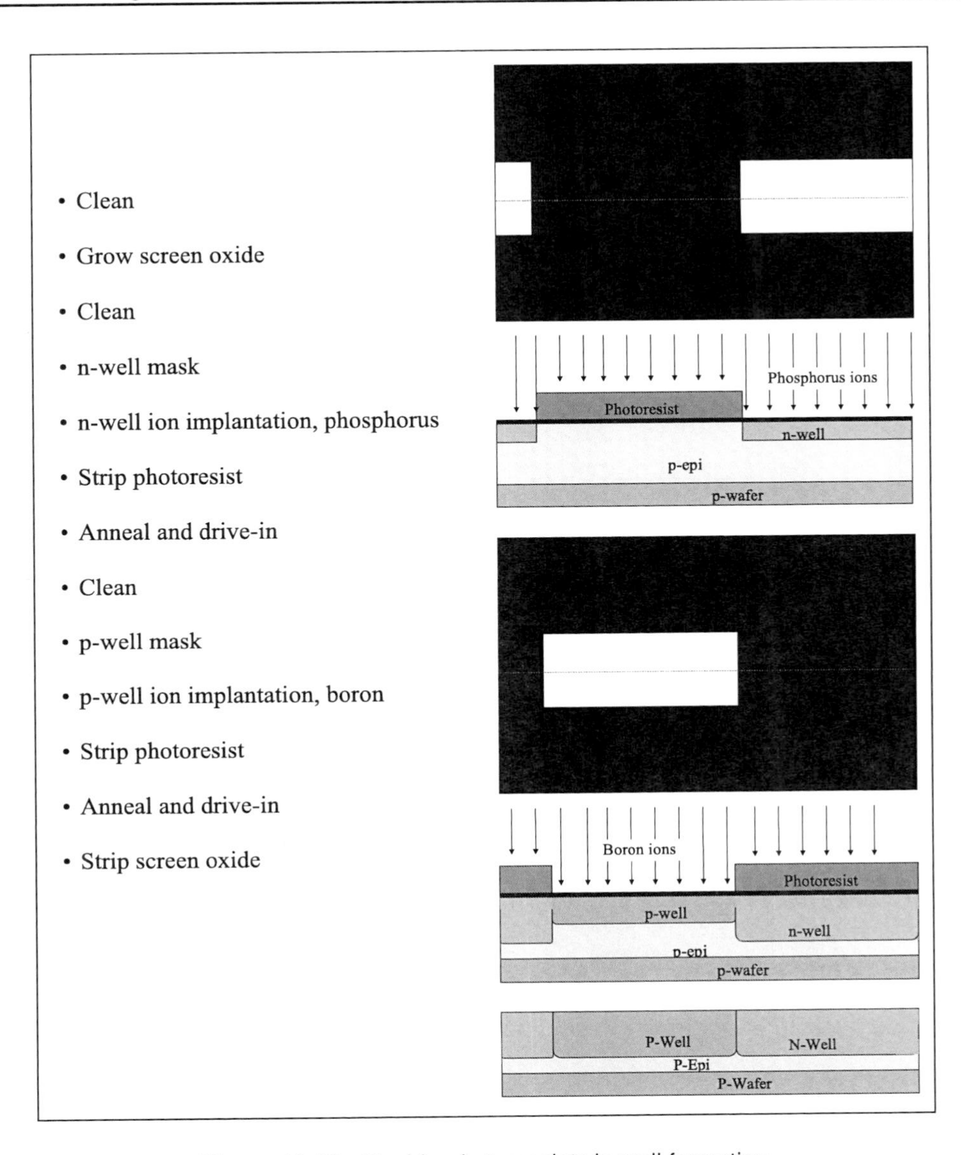

Figure 13.13 Double-photomask twin-well formation.

After S/D ion implantation and thermal annealing, MOSFETs are formed. Figure 13.14 illustrates the self-aligned S/D process for an nMOS transistor.

Self-aligned gate processes are the basic transistor formation processes in IC fabrication. Almost all transistor formation processes for advanced IC chips are developed based on this process.

13.5.3 Lightly doped drain

When a gate width is smaller than 2 μm, the vertical component of the electric field induced by the bias voltage between the S/D can be high enough to accelerate

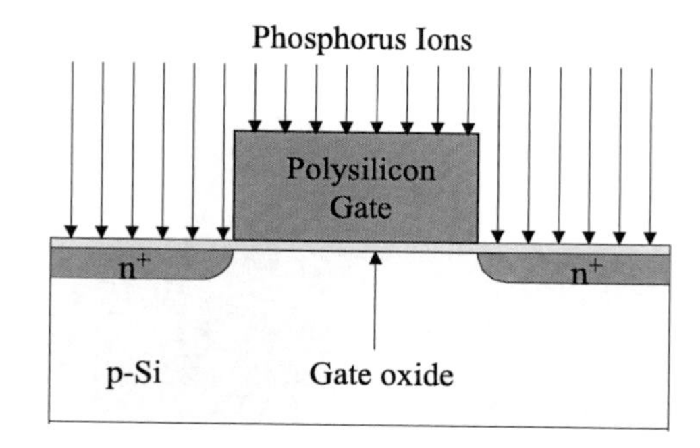

Figure 13.14 Formation of an nMOS with self-aligned S/D.

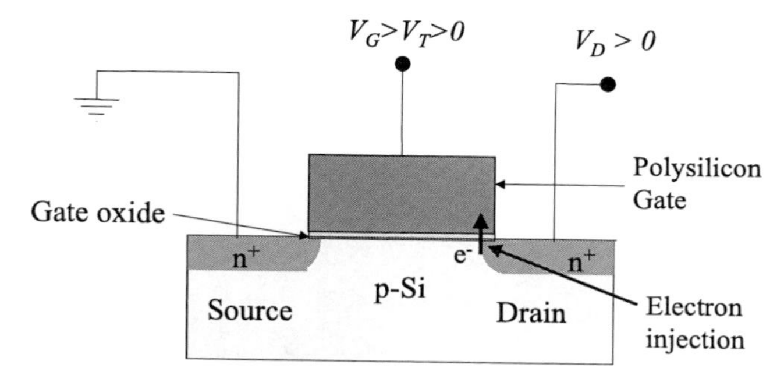

Figure 13.15 Hot electron effect of a MOS transistor.

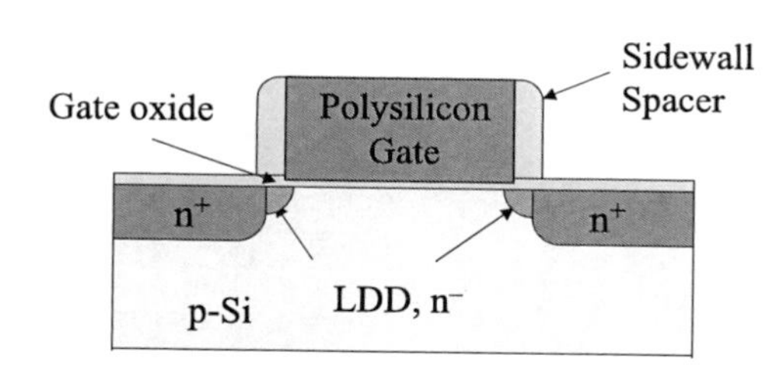

Figure 13.16 LDD of a MOS transistor.

electrons tunneling through the thin gate oxide layer. This is called the hot electron effect, which can affect transistor performance due to gate leakage, and cause reliability problems for IC chips due to the trapping of electrons in the gate oxide. Figure 13.15 illustrates the hot electron effect of a MOS transistor.

The most widely used method to suppress the hot electron effect is called a lightly doped drain (LDD) or source-drain extension (SDE), as shown in Fig. 13.16.

LDD/SDE junctions can be formed by using low-energy, low- to medium-current ion implantation. It is a shallow junction with very low to medium dopant concentration extended just underneath the gate. By depositing conformal dielectric layers and etching them back vertically, sidewall spacers can be formed on both sides of the polysilicon gate. High-current, low-energy ion implantation forms a heavily doped S/D, which is kept apart from the gate by the sidewall

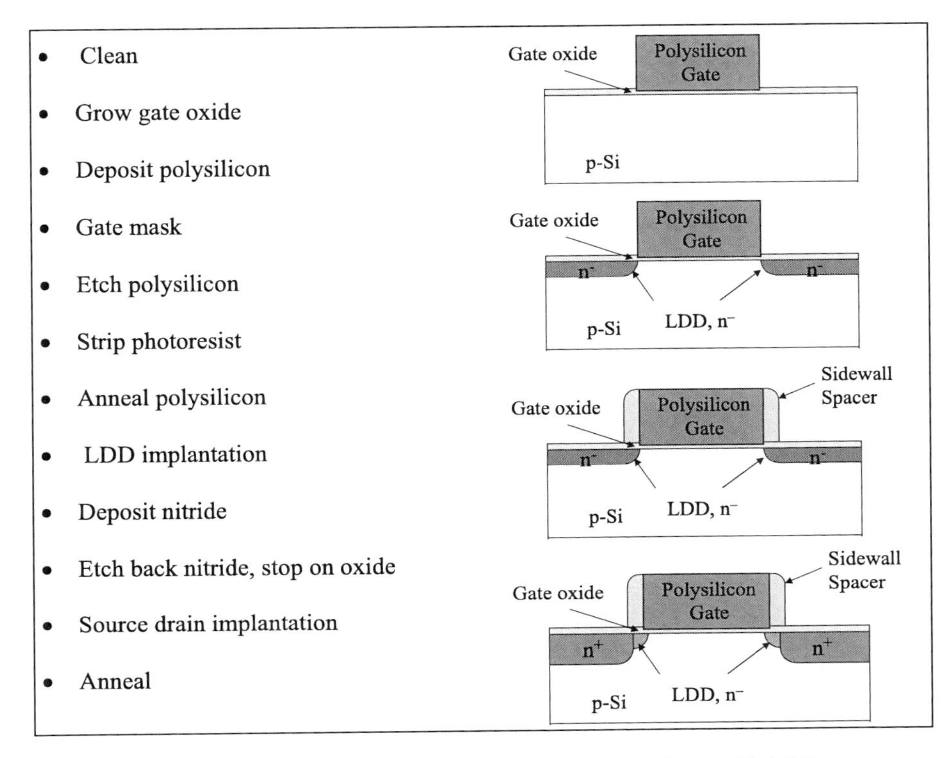

Figure 13.17 Processing steps for a MOS transistor with LDD.

space. This reduces the vertical component of the S/D bias-induced electric field, and reduces the available electrons for tunneling, thus suppressing the hot electron effect. Transistors with LDD can be made using the processing steps listed and illustrated in Fig. 13.17.

When feature sizes reduce to sub-0.18 μm, the dosage of this ion implantation process is no longer considered to be light; therefore, the process is referred to as SDE implantation.

13.5.4 Threshold adjustment

Threshold adjustment implantation controls the threshold voltage of MOSFETs. This ion implantation process ensures that the power supply voltages of electronic systems can turn on and off MOSFETs in IC chips. The threshold adjustment implantation is a low-energy, low-current implantation process and is usually performed before gate oxide growth. Figure 13.18 illustrates this implantation process.

For CMOS IC chips, two threshold adjustment implantation processes are needed: one for p-type and one for n-type. With feature size shrinking, well implantation depth can be achieved with high-energy ion implantation, and the drive-in process is no longer required. Thus, V_T-adjust implantation can be performed with well implantation.

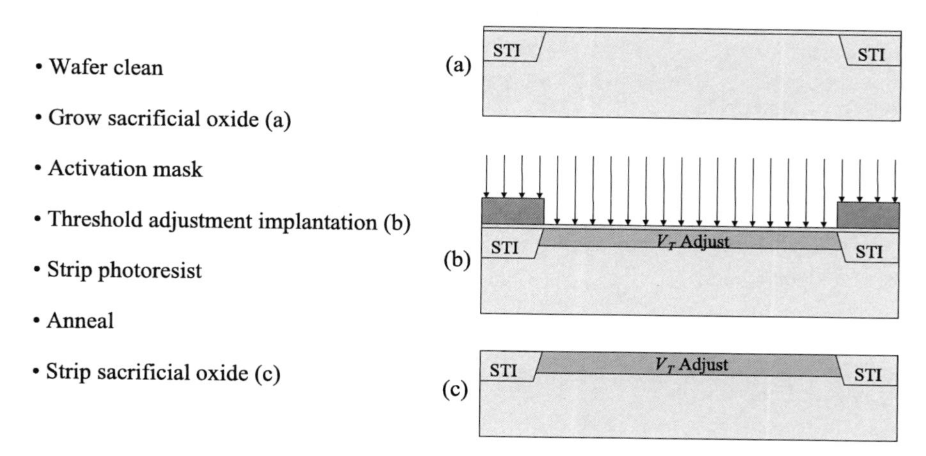

Figure 13.18 Threshold adjustment implantation process.

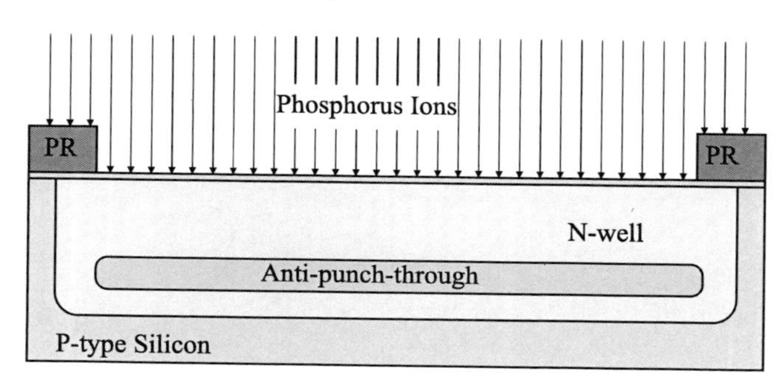

Figure 13.19 Anti-punch-through implantation process.

13.5.5 Anti-punch-through

Punch-through is an effect where depletion regions of the S/D short each other under the influence of both gate substrate bias and S/D bias. Anti-punch-through implantation, which is a medium-energy, low-current implantation process, protects transistors against this effect. Anti-punch-through implantation is normally performed with well implantation. Figure 13.19 illustrates the anti-punch-through implantation process.

Another implantation process commonly used to prevent punch-through effect is halo implantation. It is a low-energy, low-current implantation process with a large incident angle, such as 45 deg. The halo junction formed in this implantation process can help to suppress punch-through effect. Figure 13.20 illustrates the halo implantation process.

13.6 Metal-Oxide-Semiconductor Field-Effect Transistors with High-κ and Metal Gates

As device feature sizes continue to shrink, gate oxide thicknesses will become too thin for MOS transistor to operate reliably, even at 1 V. High-κ dielectrics

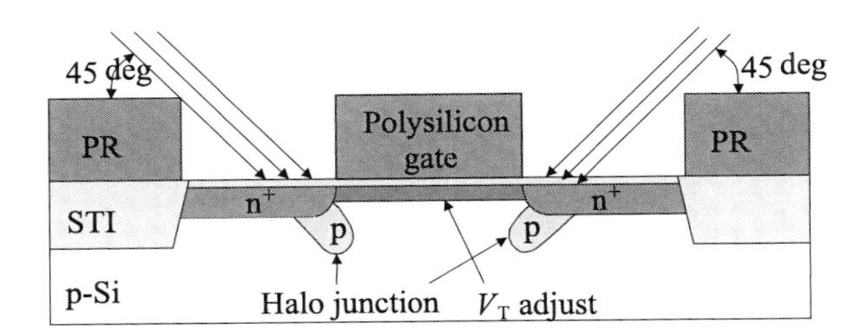

Figure 13.20 Threshold adjustment implantation process.

started to replace standard silicon dioxide or silicon oxynitride as gate dielectric materials when device dimensions reached 45 nm. By using a high-κ dielectric, the thickness of gate dielectrics can be increased to prevent tunneling leakage and dielectric breakdown. It also helps to maintain a large enough gate capacitance while gates sizes shrink. The gate capacitance must be able to hold enough charge to cause carrier inversion underneath the gate dielectric, which forms a minority carrier channel to turn on the MOS transistor.

To improve device speed, metal will likely be used again as a gate electrode for the MOS transistor because metals such as aluminum have significantly lower resistivity than polysilicon and silicide. Also, when a gate is biased to turn on a MOSFET, polysilicon forms an inversion layer at the polysilicon–oxide interface, which acts just like a dielectric layer, effectively increasing the thickness of the gate oxide when the MOSFET is turned on, and slowing down the device speed. Using metal gates to replace polysilicon gates solves the poly-inversion layer issue.

There are several high-κ metal gate (HKMG) integration approaches, such as gate-first, gate-last, and hybrid methods (nMOS gate-first and pMOS gate-last methods).

13.6.1 Gate-first process

Figure 13.21 shows a gate-first process of an nMOS on SOI substrate. This process consists of a thin layer (~4 Å) of silicon dioxide growth, high-κ dielectric deposition, followed by deposition of cap dielectric layers [Fig. 13.21(a)]. The cap layer of nMOS is different from the cap layer of pMOS for controlling the work functions and achieving desired threshold voltages. A thin metal layer, usually titanium nitride (TiN), is then deposited, followed by polysilicon and hard mask deposition [Fig. 13.21(b)]. After patterning and etching the hard mask, the photoresist is stripped, and a gate pattern is etched with the hard mask [Fig. 13.21(c)]. The rest of the process is very similar to normal SiON-polysilicon gate processes, with SDE implantation, sidewall spacer formation, S/D implantations [Fig. 13.21(d)], RTP, silicide formation [Fig. 13.21(e)], ILD0 CVD and CMP, contact etch and cleaning [Fig. 13.21(f)], glue layer and tungsten deposition, and WCMP to form the contact plugs [Fig. 13.21(g)]. For the 32-nm technology node, the high-κ dielectric is hafnium dioxide (HfO_2)-based material, and metal gate usually uses titanium nitride (TiN).

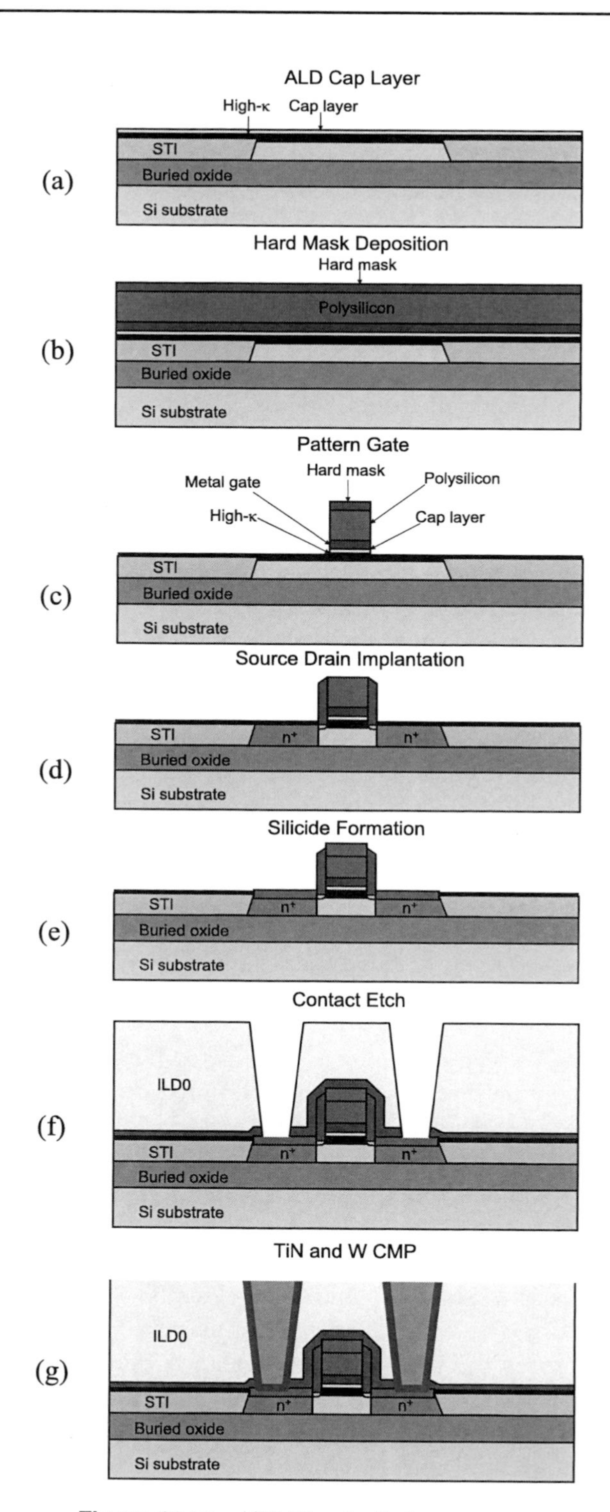

Figure 13.21 HKMG gate-first processes.

Since the gate-first HKMG process is very similar to a regular SiON-polysilicon process, many tools can be shared by both processes. Gate-first HKMG also has fewer processing steps compared to gate-last HKMG, thus total cost can be lower, which is very important for cost-sensitive fabs. In gate-first HKMG processes, high-κ and metal materials must be able to sustain the high-temperature annealing processes; this makes the choice of candidates very limited.

The main challenges for a gate-first process are polysilicon and oxide stripping, high-κ dielectric deposition, and metal gate formation. Polysilicon and oxide stripping must remove all of the oxide on the silicon surface and avoid undercutting caused by overetch. A hafnium-oxide-based dielectric is the most promising of the high-κ dielectric materials. A thin layer (<10 Å) of silicon oxide or silicon oxynitride grown on the silicon surface prior to high-κ dielectric deposition is needed to prevent high interstate charge. High-κ gate dielectrics are formed by ALD. The requirements for high-κ dielectric thin-film deposition are excellent bottom coverage and uniform deposition across the wafer.

13.6.2 Gate-last process

The gate-last process of high-κ metal gates is the first HKMG process used in IC products in the 45-nm technology node. Although it has more processing steps compared to gate-first approaches, it has many advantages, such as a wider range of choice of high-κ and metal gate materials, because HKMGs are formed after S/D and silicide annealing. For the gate-first approach, HKMG materials must be able to sustain the high-temperature annealing processes. Figure 13.22 illustrates the gate-last HKMG processing steps.

The first half of the process is similar to a regular CMOS process, with gate oxidation, polysilicon deposition and gate patterning [Fig. 13.22(a)], spacer formation, and selective epitaxy formation of nMOS S/D [Fig. 13.22(b)] and pMOS S/D [Fig. 13.22(c)]. Advanced CMOS devices use selective epitaxy growth to form the S/D to achieve stronger channel strain and better junction profile control. Silicide formation and ILD deposition [Fig. 13.22(d)] finish the first half of the gate-last HKMG process. Dielectric CMP opens the polysilicon [Fig. 13.22(e)], and wet etch selectively removes the dummy polysilicon gate and gate oxide [Fig. 13.22(f)]. High-κ gate dielectric ALD and metal ALD form the HKMG [Fig. 13.22(g)], and bulk metal deposition and CMP are the last of the gate-last HKMG process [Fig. 13.22(h)].

Different metals are needed for nMOS and pMOS to adjust the work function and threshold voltage. One method deposits TiN for pMOS, then deposits a protective layer of tantalum (Ta). Using a photoresist mask to protect pMOS and remove Ta from nMOS, a layer of TiAl alloy is deposited that reacts with the TiN underneath to form TiAlN after thermal annealing. Because of the Ta barrier layer, there is no chemical reaction between TiN and TiAl on the pMOS side. TiAlN works well for nMOS, while TiN works well for pMOS.

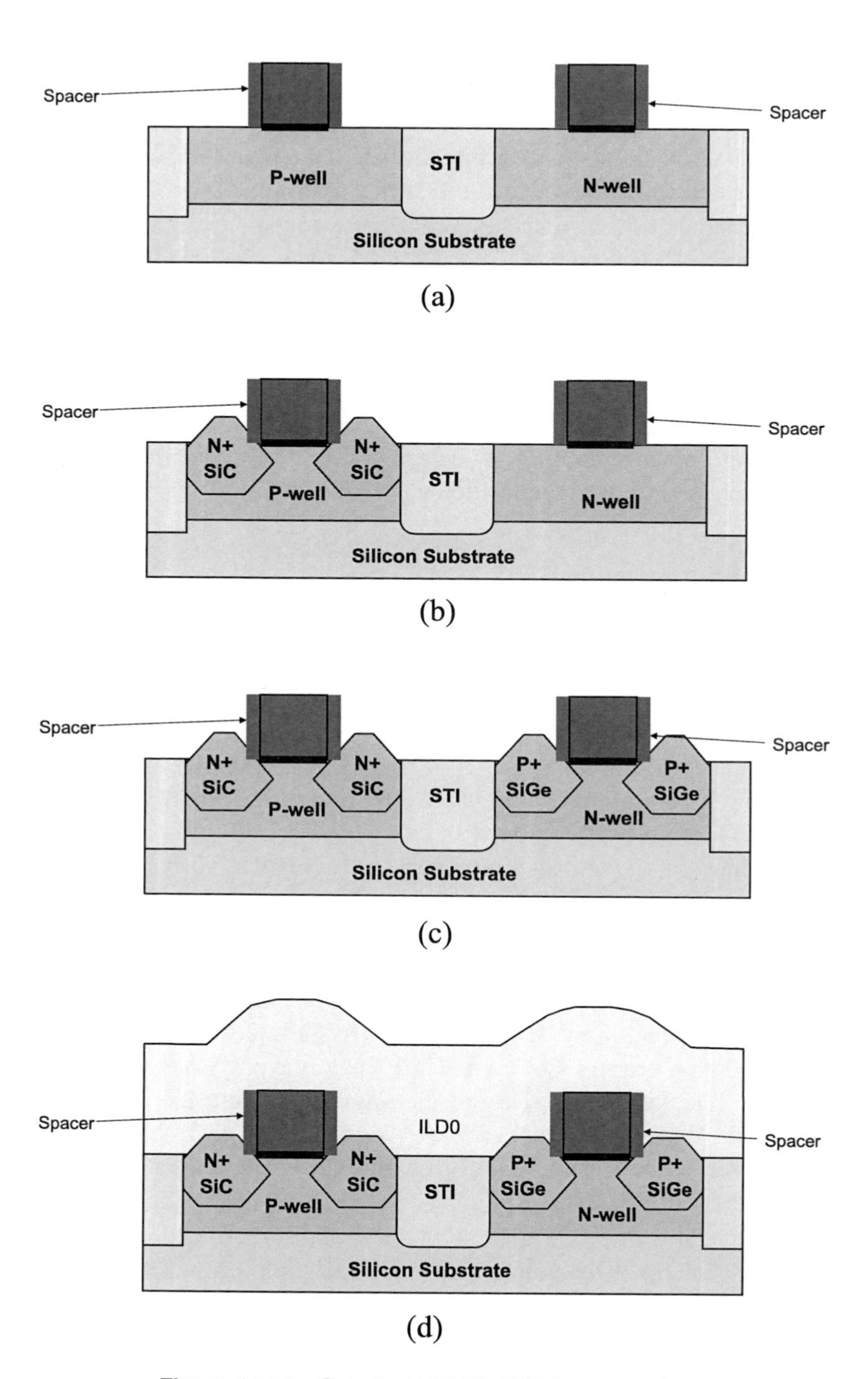

Figure 13.22 Gate-last HKMG CMOS process flow.

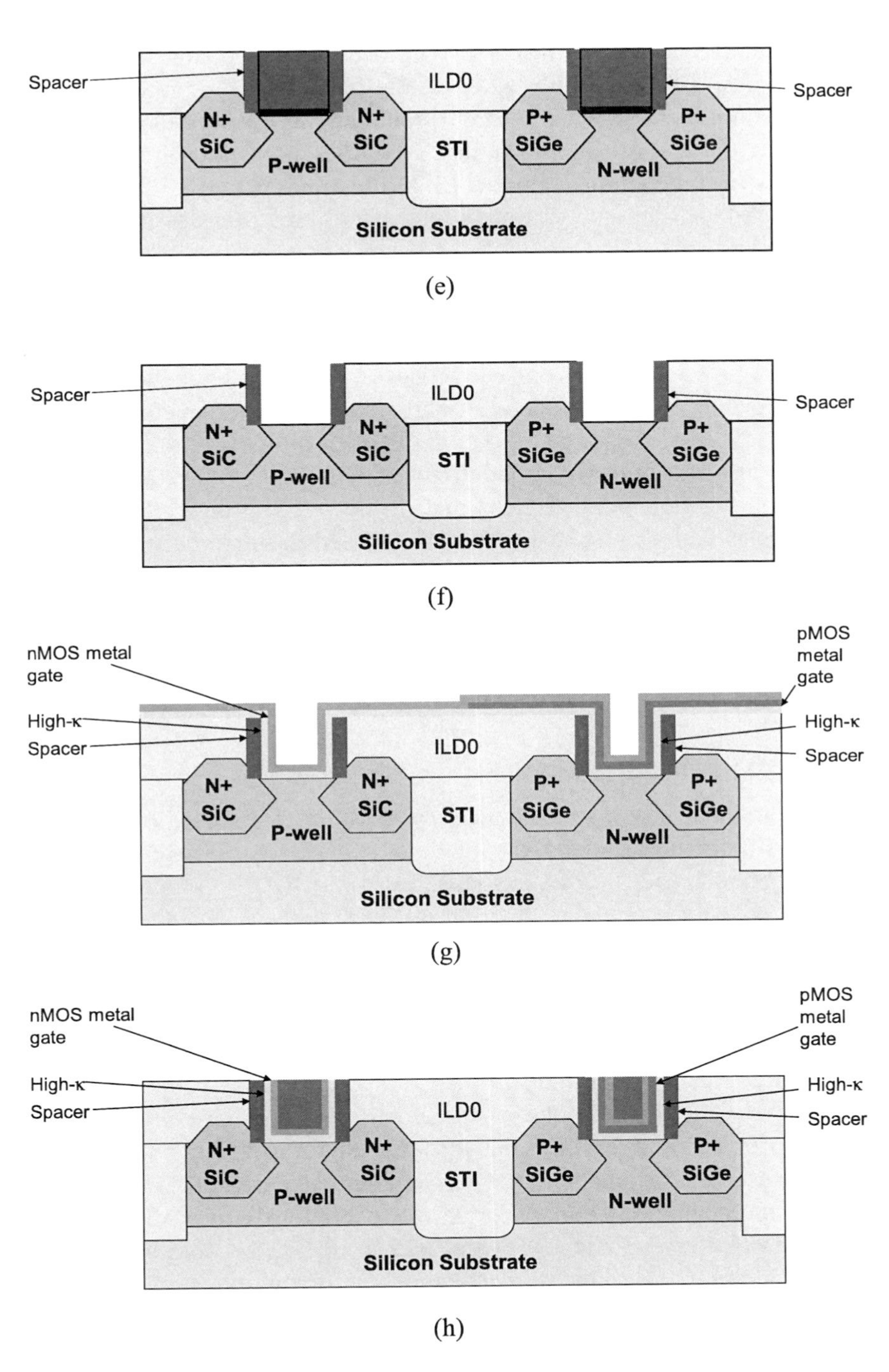

Figure 13.22 *(continued)*

13.6.3 Hybrid high-κ metal gates

One big advantage of the gate-last HKMG process is the enhancement of channel strain by removal of the dummy gate, a step that is more significant for pMOS performance improvement when selective epitaxial SiGe is used to form the pMOS S/D. Therefore, a hybrid approach that combines both gate-first and gate-last processes has been developed. Figure 13.23 illustrates a hybrid HKMG process. The nMOS uses gate-first integration, while pMOS uses gate-last integration.

13.7 Interconnections

After MOS transistors are formed on the wafer surface, the wafer process is approximately half done. The front-end process is finished, while the back-end process has just begun. For an advanced IC chip, multiple metal layers with dielectric material in between are required to route millions or billions of transistors. Silicide is commonly used to improve conductivity of local interconnections and reduce contact resistance. For many devices that do not require more than a gigahertz (GHz) of clock frequency (such as most DRAM and flash memory devices), tungsten and aluminum alloys are still widely used in metallization processes, along with undoped silicate glass (USG) as the dielectric. At the turn of the century, interconnection technology made a transition from traditional tungsten and aluminum alloy interconnections to copper interconnections for high-speed (clock frequency >1 GHz) CMOS logic devices.

13.7.1 Local interconnections

A local interconnection is the connection between neighboring transistors. It is formed by polysilicon or a polysilicon-silicide stack. Tungsten silicide and tungsten/tungsten nitride are extensively used in DRAM local interconnection applications. For flash memory devices, cobalt silicide is widely used for local interconnections. For CMOS logic devices, frequently used silicides for local interconnects are titanium silicide (>180 nm), cobalt silicide (250 to 90 nm), and nickel silicide (65 nm and beyond).

Tungsten silicide used in DRAM chips is deposited by a CVD process, with WF_6 as a tungsten precursor and SiH_4 as a silicon precursor. Titanium silicide is formed by sputtering titanium onto a silicon surface, then thermally annealing the wafer to induce chemical reactions between the titanium and silicon. Cobalt silicon is frequently used in flash memory devices and has been used in CMOS devices for 250- and 90-nm technology nodes. For 65-nm technology nodes, silicide materials had to make another major change, and nickel silicide became widely used. A process flow for tungsten silicide gates and local interconnections is listed and illustrated in Fig. 13.24.

Titanium silicide is used in local interconnection processes in some less-advanced flash memory devices and CMOS IC chips. It is formed by a self-aligned silicide (salicide) process. Compared with tungsten silicide, titanium silicide has lower resistivity.

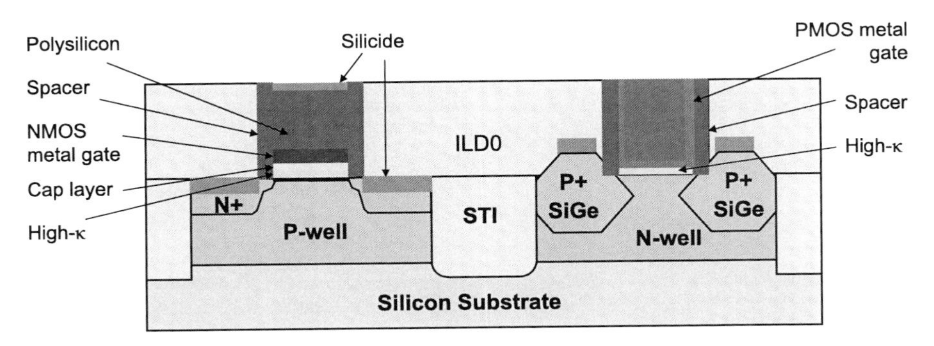

Figure 13.23 Hybrid HKMG CMOS with gate-first integration for nMOS and gate-last integration for pMOS.

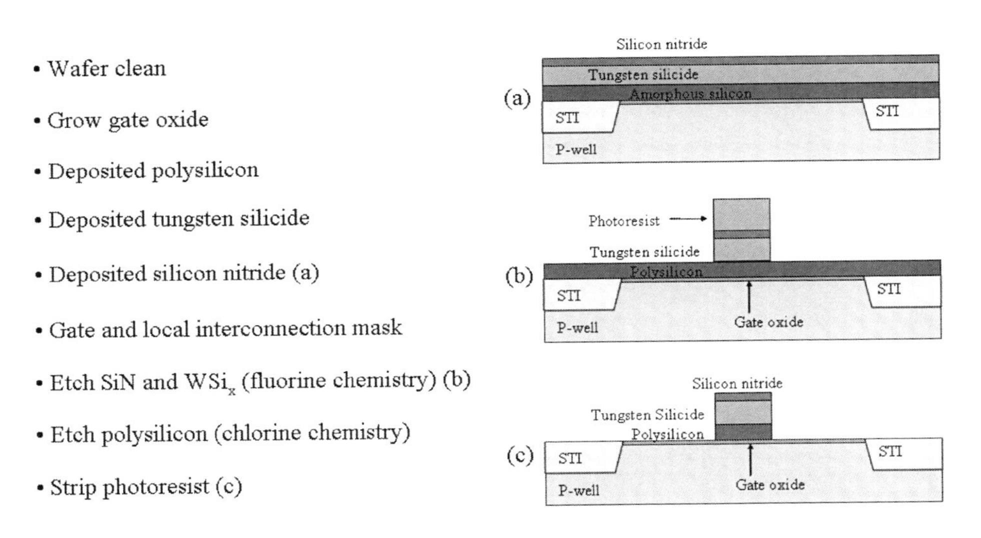

Figure 13.24 Tungsten silicide gate and local interconnection formation.

The grain size of low-resistance, C-54 phase titanium silicide is about 0.2 μm. When gate widths become smaller than 0.2 μm, titanium silicide cannot be used. Therefore, cobalt silicide was introduced into local interconnection applications. Cobalt silicide has low resistivity, and it can also be formed in a self-aligned silicide process. Both $TiSi_2$ and $CoSi_2$ formation require ~750 °C annealing, which is too high for devices with feature sizes 65 nm or smaller; thus, nickel silicide processing has been developed and implemented in CMOS IC manufacturing. The annealing temperature of NiSi is about 450 °C, significantly lower than the 750 °C required by $CoSi_2$. Figure 13.25 lists and illustrates nickel silicide processing steps.

Tungsten has been used to form local interconnections, which can significantly reduce resistance and improve device speed. Tungsten local interconnections use damascene processes similar to the tungsten plug process. First, trenches are etched in the silicate glass layer; then titanium and titanium nitride, and diffusion barrier

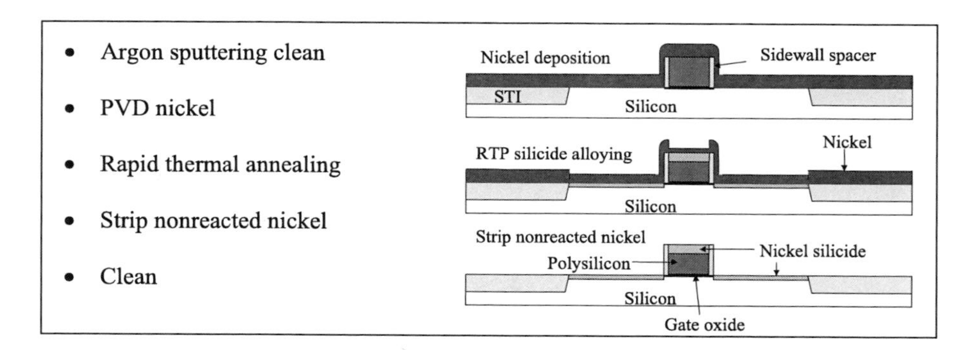

Figure 13.25 Nickel silicide formation.

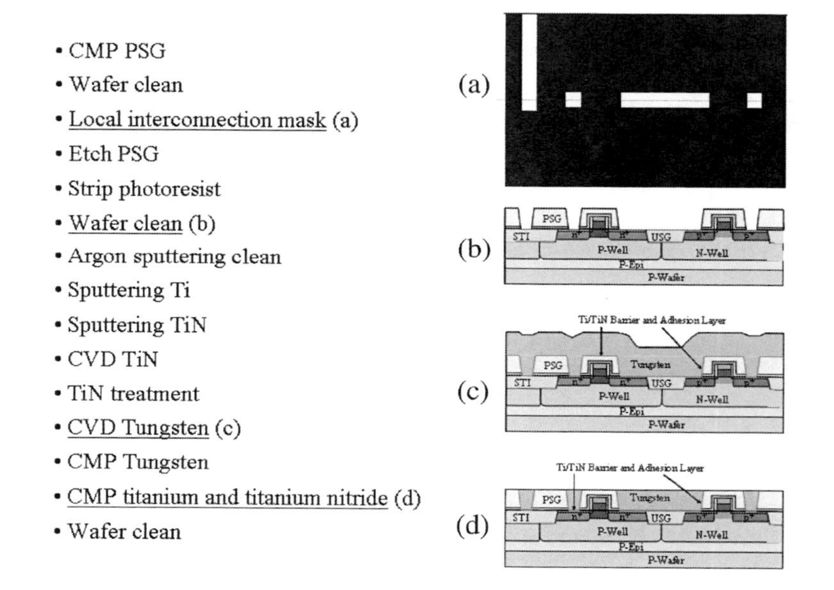

Figure 13.26 Tungsten local interconnection processing steps.

and tungsten adhesion layers, are deposited. A WCVD process fills trenches and contract holes with tungsten, and a CMP process removes the bulk tungsten from the wafer surface, leaving tungsten in the trenches and contact holes. A tungsten local interconnection process is listed and illustrated in Fig. 13.26.

13.7.2 Early interconnections

The steps of early global interconnection processing are oxide CVD, oxide etch, metal PVD, and metal etch. Oxide etching forms contact or via holes, and metal etching forms interconnection lines. Figure 13.27 lists and illustrates the early interconnection process.

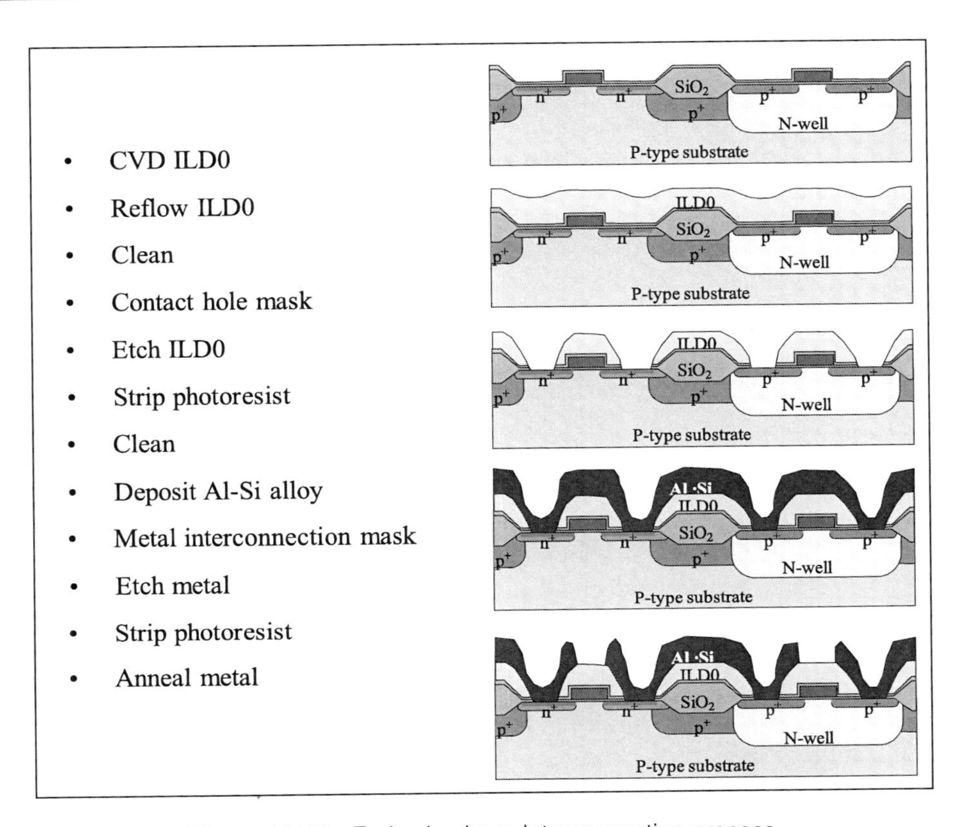

Figure 13.27 Early aluminum interconnection process.

13.7.3 Aluminum alloy multilevel interconnections

When the number of transistors increased, one layer of metal interconnections could not route all of the transistors on the chip; thus, multilayer metal interconnections began to be used. The earlier interconnection processes always left a rough surface, which caused problems for photolithography and metal deposition. As device feature size scaled down, there was no room for widely tapered contact holes, allowing PVD aluminum to cover the bottom of the holes. The tungsten CVD process was introduced to fill the narrow contact and via holes. The basic interconnection processing steps are dielectric CVD, dielectric planarization, dielectric etch, tungsten CVD, bulk tungsten removal, metal stacked PVD, and metal etch. A commonly used dielectric is silicate glass, both doped (such as PSG or BPSG for ILD0) and undoped (USG for IMD or ILD*x*). Dielectric planarization can be achieved with thermal flow (only for ILD0), etchback, and CMP. Bulk tungsten removal also can be achieved by both etchback and CMP. CMP became more popular in dielectric planarization and bulk tungsten removal applications in the late 1990s. The dielectric etch forms the contact and via holes. A frequently used metal stack includes a titanium welding layer, aluminum–copper alloy as the main conductor, and titanium nitride as an ARC. Metal etch defines the metal interconnection lines. Figure 13.28 lists and illustrates interconnection processes commonly used in multilevel IC chips. The figure describes the processes between metal 1 and metal 2; the steps for metal 3, metal 4, and more are almost identical to these processing steps. Of course, the feature size or pattern CD increases with the number of metal layers.

13.7.4 Copper interconnections

Copper has lower resistivity and higher EMR than aluminum alloy. Copper is very hard to dry etch due to the lack of simple gaseous copper chemical compounds. This disadvantage has delayed its application in IC interconnections, since traditional interconnection processes require a metal etching process.

In the 1990s, CMP technology developed and matured quickly. It is still widely used in bulk tungsten removal processes to form tungsten plugs. The copper interconnection process is very similar to a tungsten plug formation process. Instead of via holes, trenches are etched on the dielectric surface. Then copper is deposited into the trenches, and a subsequent copper CMP process removes the bulk copper layer on the wafer surface, leaving the copper lines embedded in the dielectric layers. By using this damascene process, a metal etch process becomes unnecessary. The dual damascene process, which combines via and trench etching before metal deposition, is the most commonly used method for copper metallization. The dual damascene process can eliminate some metal deposition and CMP processes, unlike two single damascene processes.

The fundamental difference between Al–Cu and copper interconnection processes is that the Al–Cu process requires one dielectric etch and one metal etch, while the dual damascene copper process requires two dielectric etches and no metal etch. The challenges in Al–Cu interconnection processes are void-free dielectric CVD, dielectric planarization, via hole etch, and metal etch. The main challenges of dual damascene copper processes are dielectric etch, metal deposition, and metal CMP. Figure 13.29 lists and illustrates the single damascene copper interconnection processing steps of metal 1 for CMOS IC, and Fig. 13.30 illustrates a via-first dual damascene copper low-κ interconnection process.

A thin nitride cap layer (100 to 500 Å, depending on the technology node) can be deposited by a PECVD process with silane (SiH_4), ammonia (NH_3), and nitrogen (N_2). This cap layer is important for preventing copper from diffusing into the oxide layer and eventually diffusing into the silicon substrate, causing unstable performance in the transistor. A cap nitride layer can also prevent the oxidation of copper during oxide deposition. Unlike aluminum oxide, copper oxide is a loose layer that cannot prevent the further oxidation of copper. PECVD silicon-rich nitride or oxynitride (SiON) can be used for this layer, which can also serve as an ARC. ILD0 can be USG, fluorinated silicate glass FSG, or a low-κ dielectric such as SiCOH, and ULK such as porous SiCOH, depending on the technology node. Nitride can also serve as an etch-stop nitride layer for metal trench etching; this layer can provide nitrogen light emission for indication of the etch endpoint.

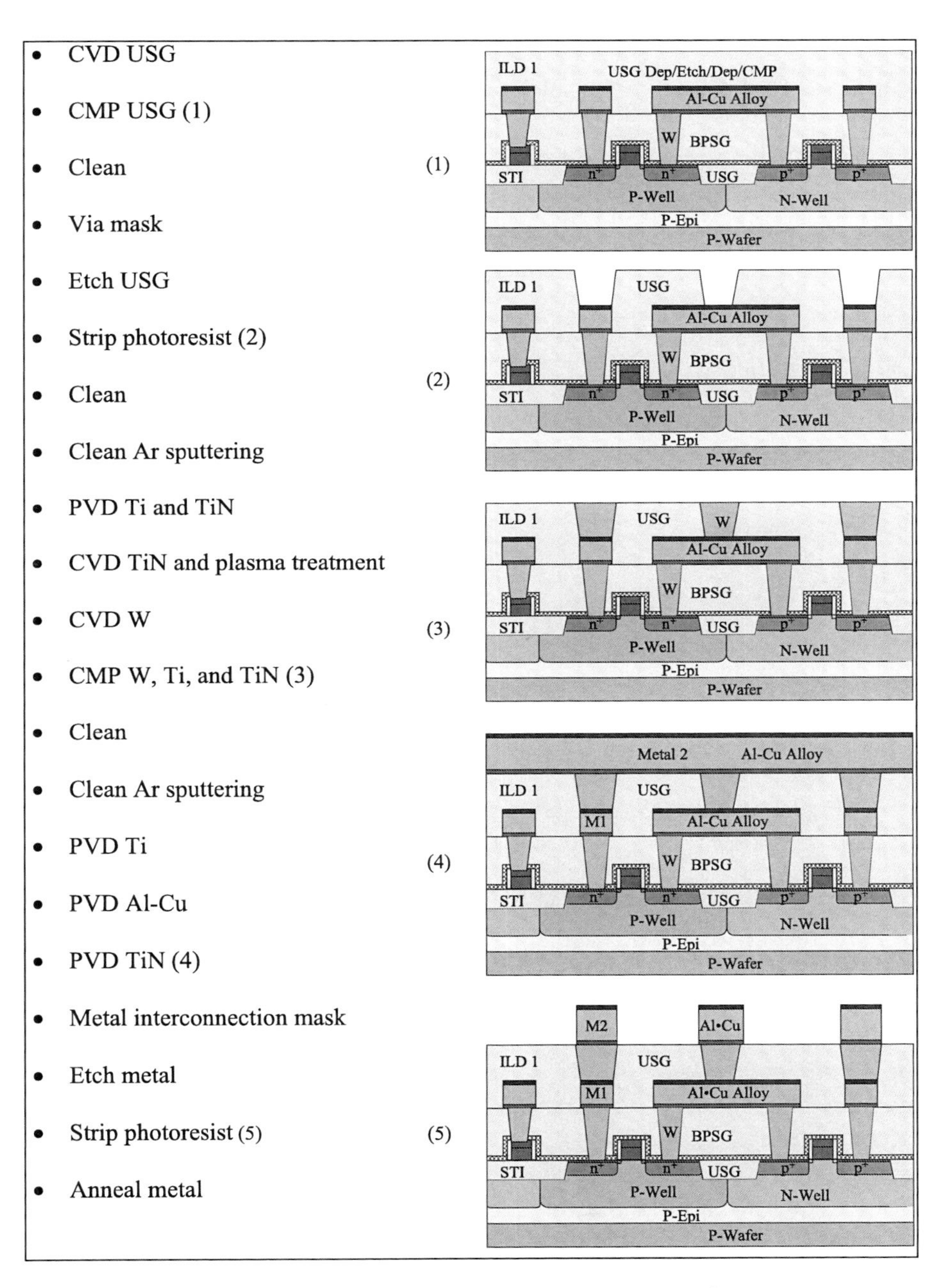

Figure 13.28 Aluminum alloy interconnection processing steps.

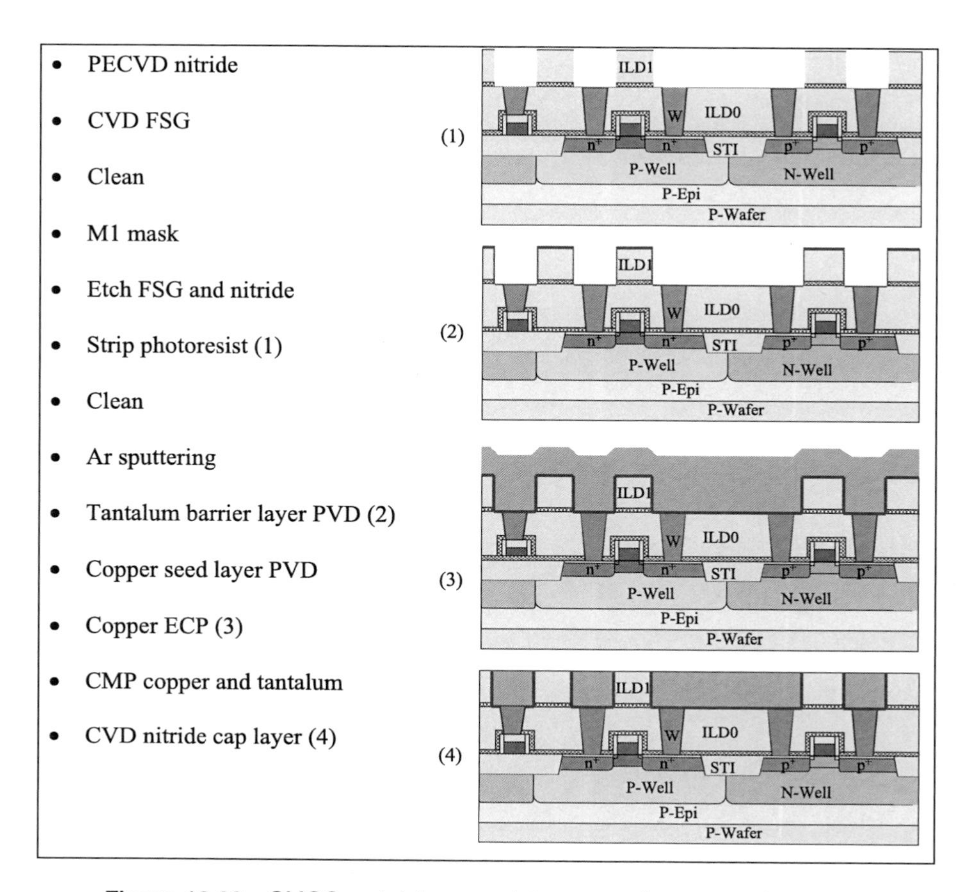

Figure 13.29 CMOS metal-1 copper interconnection processing steps.

13.7.5 Copper and low-κ dielectrics

The combination of copper and low-κ dielectrics can further increase IC chip speed. Several silicon-based low-κ dielectric materials have been developed. They can be etched with fluorine-based chemistry, similar to silicate glass etching. The most commonly used low-κ dielectric in IC manufacturing is SiCOH, or carbon-doped oxide and porous SiCOH, which has an even lower κ value due to pores inside the film. Figure 13.30 lists and illustrates the via-first copper and low-κ interconnection processing steps.

For a via-first dual damascene process, each via needs to stop on the cap layer, which can be broken through using a highly selective etching process after trench formation. It is very important to have the via holes stop on the thin cap layer. Underetching will cause open circuits because breakthrough etching is highly selective of the cap layer, not the low-κ film, while overetching will cause copper corrosion during breakthrough and lead to high contact resistance.

Because the oxygen plasma used to strip the photoresist can oxidize carbon in low-κ film and increase the κ value of the dielectric, a TiN hard mask is used

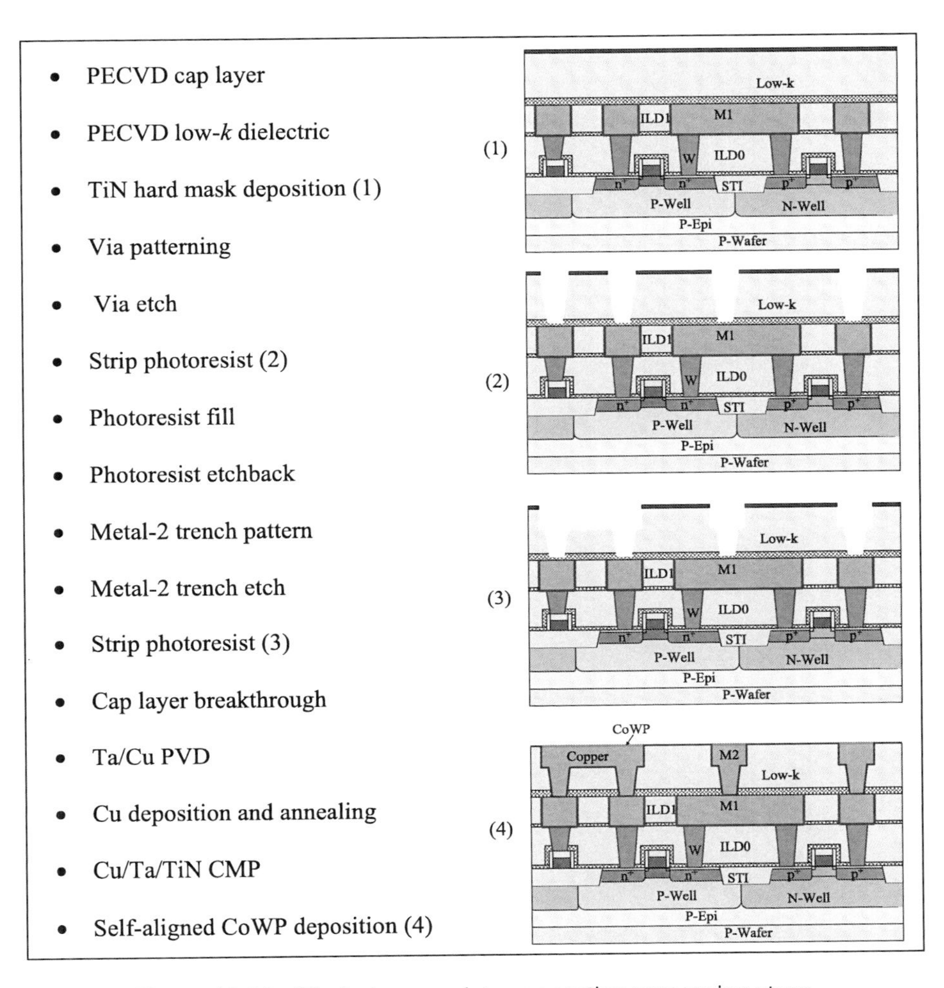

Figure 13.30 Via-first copper interconnection processing steps.

to cover the surface to protect the film. A CoWP cap layer helps prevent copper diffusion and reduces electromigration to improve device reliability. It can be deposited only on the copper surface with an electroless planting process.

Figure 13.31 illustrates a trench-first copper and ULK interconnection process with SEG SiGe pMOS, gate-last HKMG, and trench silicide, which only forms at the bottom of a trench-style contact. Because it is a metal gate, there is no need to form silicide on top of the gate.

13.8 Passivation

After the last metal layer formation, a passivation layer is deposited to protect the IC chip from moisture and other contaminants such as sodium. Silicon nitride is the most commonly used final passivation material in the IC industry. A silicon oxide layer is deposited as a stress buffer before silicon nitride deposition. A silane-based PECVD chamber is used to deposit both oxide and nitride in situ with one

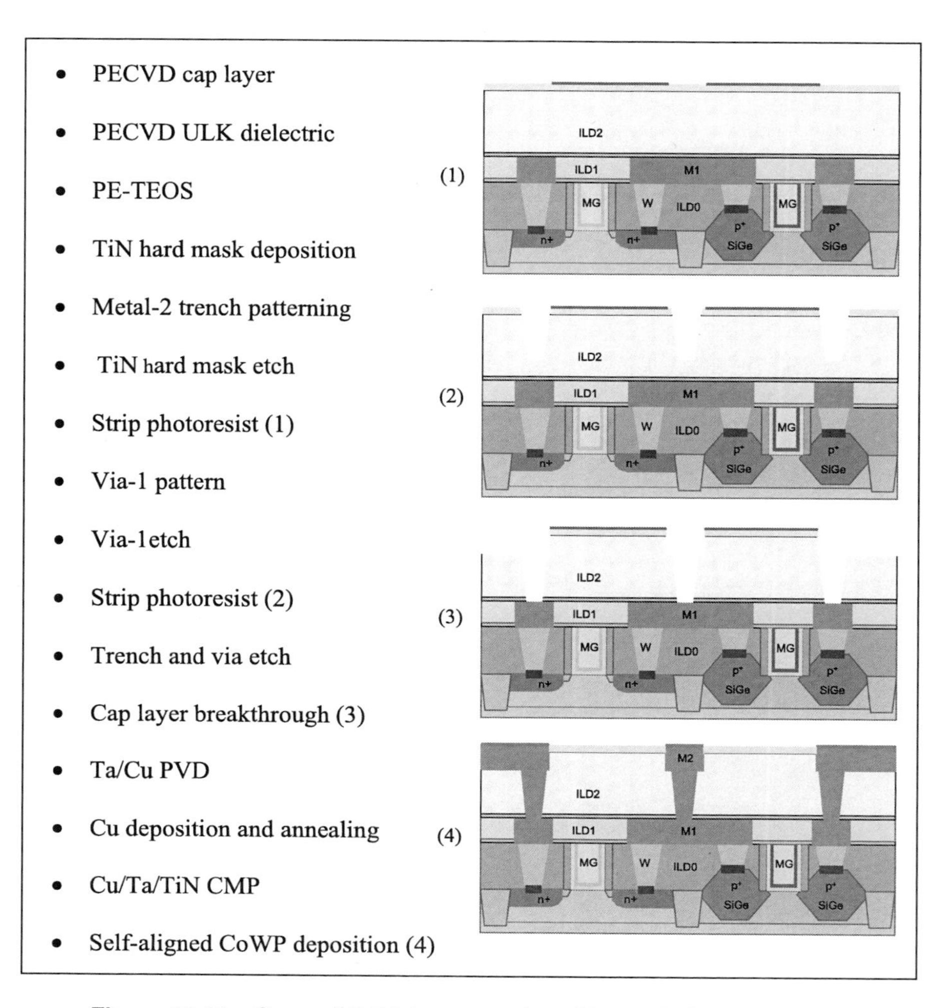

Figure 13.31 Copper/ULK interconnection with trench-first integration.

processing sequence. After nitride deposition, the last photolithography process defines the openings for the bonding pads or connecting bumps. A fluorine-based nitride/oxide etch and photoresist stripping finishes the silicon wafer processing. Figure 13.32 illustrates the passivation processing steps for bonding pad packaging.

13.9 Summary

- Blanket field oxide, LOCOS, and STI are three isolation methods used for IC chips.
- The three-well formation processes of CMOS IC chips are single well, self-aligned twin well, and double-photomask twin well.
- Sidewall spacers are formed by deposition and etchback of a conformal dielectric thin film. They are used to form S/D extensions, selective epitaxy growth, and self-aligned silicide.

- The threshold V_T adjustment implantations control the threshold voltage of MOS transistors.
- Al, polysilicon-silicide, and TiN have been used for gate and local interconnections.
- W, Ti, TiN, and Al–Cu are commonly used in Al–Cu interconnection processes.
- Basic processing steps for copper metallization are dielectric deposition, dielectric etches, metal deposition, and metal CMP.
- Ta and TaN are frequently used barrier layers for Cu metallization.
- SiCOH ($\kappa \sim 2.5$ to 2.8) and porous SiCOH (κ to 2) are commonly used low-κ dielectrics.
- CVD SiON, CVD SiCN, and self-aligned CoWP are used as cap layers to prevent Cu diffusion.

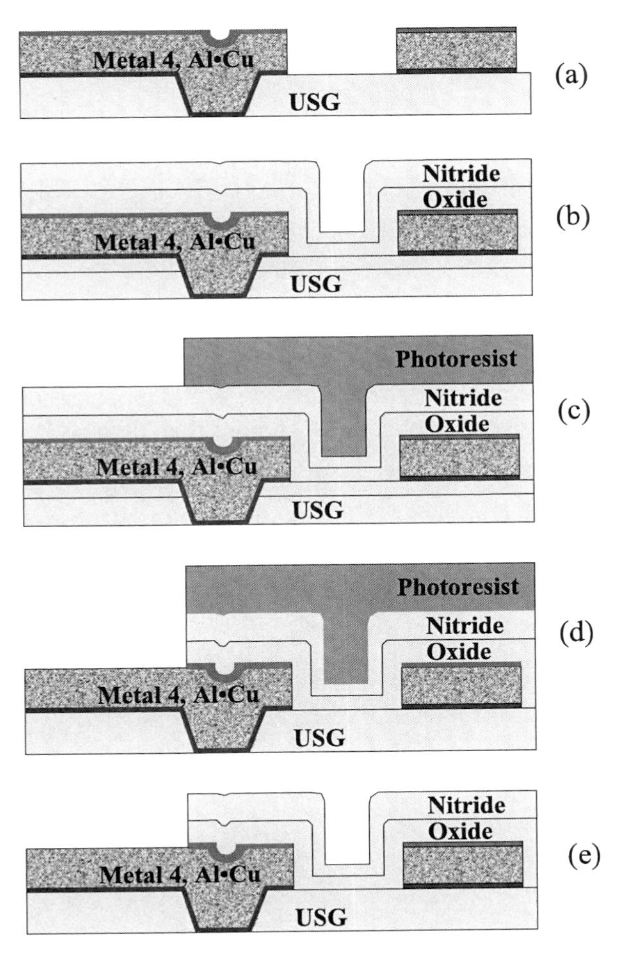

Figure 13.32 Passivation process: (a) metal annealing, (b) PECVD oxide and nitride, (c) bonding pad mask exposure and development, (d) oxide, nitride, and TiN etch, and (e) photoresist strip.

- Silicon nitride is the most commonly used passivation materials in IC processing.

13.10 Bibliography

M. Fukasawa et al., "BEOL process integration with Cu/SiCOH (k = 2.8) low-k interconnects at 65-nm groundrules," *Proc. IEEE 2005 Intl. Interconnect Technol. Conf.*, 9–11 (2005).

A. Grill, "Plasma enhanced chemical vapor deposited SiCOH dielectrics: from low-k to extreme low-k interconnect materials," *J. Appl. Phys.* **93**, 1785–1790 (2003).

L. Shon-Roy, A. Wiesnoski, and R. Zorich, *Advanced Semiconductor Fabrication Handbook*, Integrated Circuit Engineering Corp., Scottsdale, AZ (1998).

S. Wolf, *Silicon Processing for the VLSI Era, Vol. 2, Process Integration*, Lattice Press, Sunset Beach, CA (1990).

13.11 Review Questions

1. What is the main advantage of epitaxy layers in an CMOS IC chip?
2. What are the main differences between self-aligned twin-well and double-photomask twin-well processes?
3. Explain the advantages of STI over LOCOS.
4. What is the difference between LDD (or SDE) and S/D implantation?
5. Explain the self-aligned silicide process. Why did nickel silicide replace cobalt silicide when technology nodes grew smaller than 65 nm?
6. Why is tungsten used for interconnection processes?
7. What are the applications of titanium and titanium nitride?
8. List the basic processing steps for both copper and traditional interconnections. Explain the differences.
9. Explain the application of high-κ dielectrics.
10. Explain the application of low-κ dielectrics. What major impact occurs in the process when SiCOH replaces FSG?

Chapter 14
Integrated Circuit Processing Technologies

Objectives

After finishing this chapter, the reader will be able to:

- list the major processing technology changes from the 1980s to the 2010s
- explain the differences between DRAM and NAND flash memory devices.

14.1 Introduction

CMOS-based chips are the most common IC chips in the electronics industry. The digital revolution that produced personal computers, Internet networks, and telecommunications strongly drives the demands of CMOS IC chips. In this chapter, four complete CMOS processes are examined. First is the CMOS process from the early 1980s, with only one layer of aluminum alloy interconnection. Next is the CMOS with a four-layer aluminum alloy interconnection used in the 1990s. After that is an advanced CMOS process with copper and low-κ interconnections, used in the first decade of the 2000s, and the last is the state-of-the-art CMOS technology with high-κ metal gates (HKMGs), stress engineering, and copper low-κ interconnections.

Memory chips are one of most important parts of IC products. They are also important drivers of IC technological developments. The manufacturing processes of array cells of DRAM and NAND flash chips are quite different from CMOS processes. Thus, two sections of this chapter are dedicated to describing their processes. The periphery devices of both DRAM and NAND flash chips are very similar to normal CMOS processes.

14.2 Complementary Metal-Oxide-Semiconductor Process Flow of the Early 1980s

In the mid-1970s, IC processing technology leapt forward after the introduction of ion implantation, which replaced diffusion for semiconductor doping. Self-aligned S/D formation became a standard MOS transistor-making process.

Polysilicon gates replaced metal gates due to the high temperature requirement of postimplantation annealing. Because electrons move much faster than holes, nMOS is much faster than pMOS at the same size and dopant concentration. nMOS replaced pMOS shortly after the introduction of ion implantation, which, unlike diffusion, can easily dope n-types at high concentrations.

In the 1980s, CMOS IC processing technology developed rapidly, driven by the demands of digital logic electronic devices such as watches, calculators, personal computers, mainframe computers, etc. Applications of liquid crystal displays (LCDs) also accelerated the transition from nMOS ICs to lower-power-consumption CMOS ICs. Minimum feature size shrank from 3 μm to 0.8 μm, while wafer size increased from 100 mm (4 in.) to 150 mm (6 in.).

In the early 1980s, LOCOS was used with isolation ion implantation to electronically isolate neighboring transistors. PSG was used as ILD0, which was reflowed at about 1100 °C. Al–Si alloy thin films deposited with thermal or e-beam evaporators were used on tapered contact holes to form metal interconnections. Horizontal furnaces were used for oxidation, LPCVD, postimplantation annealing and drive-in, and PSG reflow. Plasma etchers were used for some patterns such as gate etching, while other larger patterns still used wet etching. Projection alignment and exposure systems were used in photolithography processes. Positive photoresist started to replace negative photoresist to meet photolithography resolution requirements. Most processing tools were batch systems. Figure 14.1 shows a CMOS process flow with early 1980s technology; the minimum feature size is about 3 μm. The figure includes steps for isolation and well formation, transistor formation, and interconnections, as well as 17 illustrations.

Figure 14.2 illustrates a cross-section of a CMOS device using LOCOS isolation, reflowed PSG as ILD0, tapered contacts, and Al-Si alloy as interconnects.

14.3 Complementary Metal-Oxide-Semiconductor Process Flow with 1990s Technology

In the 1990s, CMOS IC processing technology developed rapidly, driven by the continued demands of digital logic electronics such as personal computers, telecommunication devices, and the Internet. Minimum feature size shrank from 0.8 to 0.18 μm, while wafer size increased from 150 mm (6 in.) to 300 mm (12 in.).

A complete CMOS process flow with mid-1990s technology is provided in this section. Minimum feature size of the CMOS chip is about 0.25 μm. Figure 14.3 illustrates epitaxial-layer growth on a p-type wafer. This kind of wafer preparation is usually performed by wafer manufacturers before the wafers are sold and shipped to IC chip manufacturers.

Figure 14.4 illustrates the rest of the CMOS process flow, including STI, well formation, transistor formation, global interconnections, and passivation and bonding pad formation. The figure lists the processing steps and includes 57 illustrations. This ends the wafer process. The wafer is ready for testing, die sawing, sorting, packaging, and final testing.

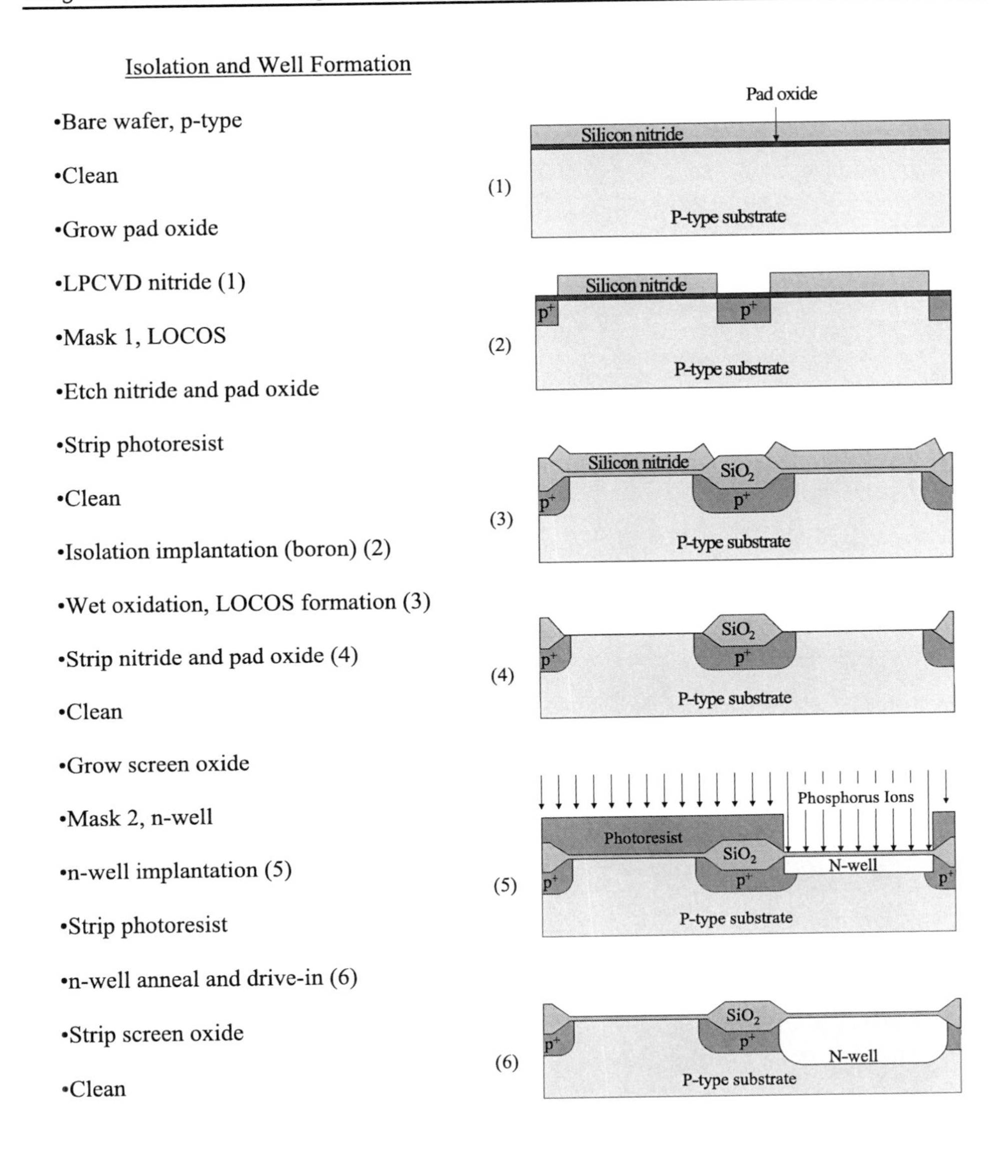

Figure 14.1 CMOS process flow of the early 1980s (*continued on following two pages*).

14.3.1 Comments

Several major developments in CMOS IC chip processing technology occurred during the 1990s. Silicon wafers sliced from single-crystal ingots made by the CZ method always have trace amounts of oxygen and carbon from the crucible materials. To eliminate these impurities and to improve chip performance, epitaxy silicon began to be used for advanced CMOS IC chips, as shown in Fig. 14.3. Shallow trench isolation, shown in Fig. 14.4, replaced LOCOS as an isolation block for preventing crosstalk between neighboring transistors. The sidewall spacer

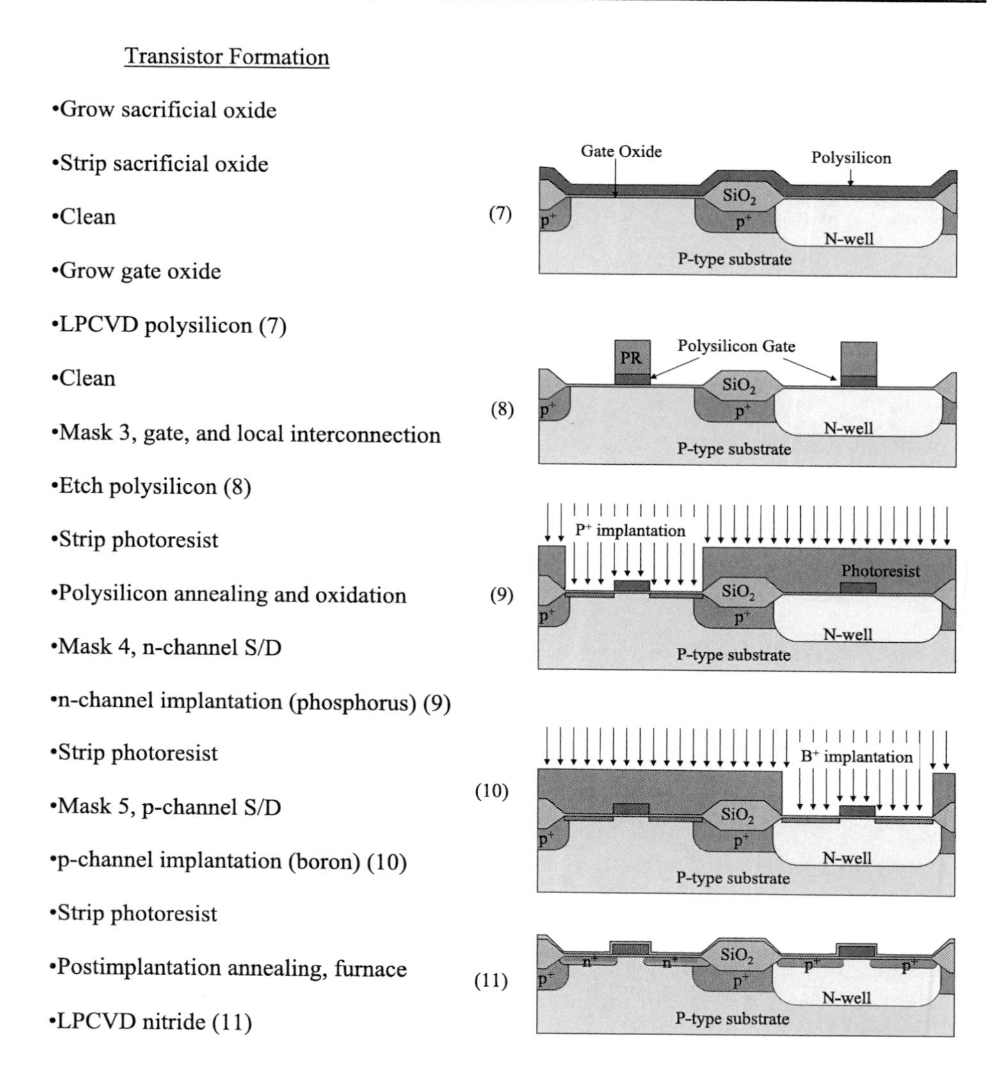

Figure 14.1 *(continued)*

was introduced to form LDD to suppress the hot electron effect for submicron devices and to form self-aligned silicide (salicide) to reduce resistance of gates and local interconnections. Silicide became widely used with polysilicon to form gates and local interconnections. Because silicide has a much lower resistivity than polysilicon, it improved device speed and reduced power consumption. The most commonly used silicides in the 1990s were tungsten and titanium. In that period, IC chip power supply voltage was gradually reduced from 12 to 3.3 V; therefore, threshold voltage V_T adjustment implantation processes were required to make sure that a normal-off nMOS could be turned on, and a normal-on pMOS could be turned off at the operation voltage. The process flow in Fig. 14.4

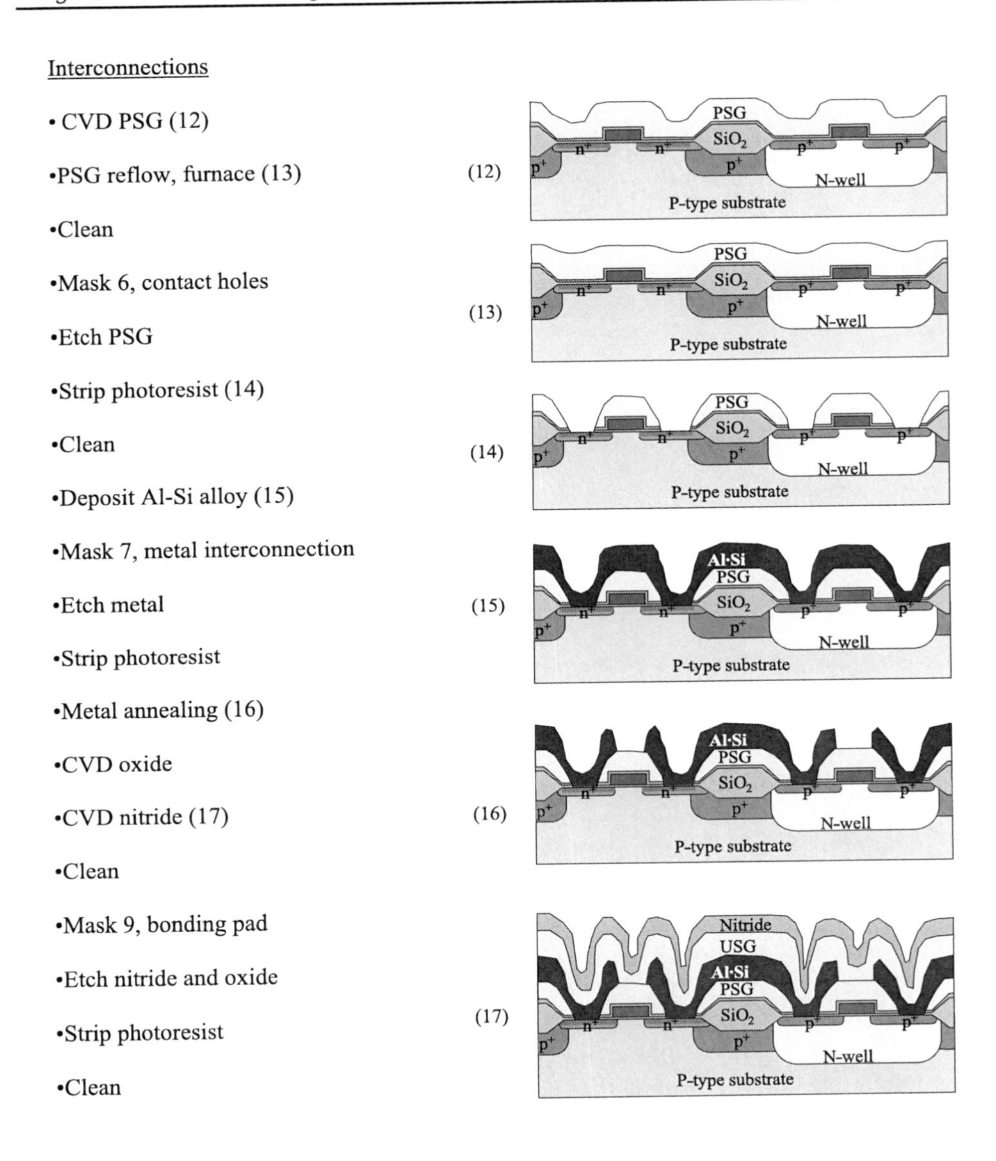

Figure 14.1 *(continued)*

shows a salicide process that forms titanium silicide on top of the polysilicon gate and S/D at the same time. The silicide at the S/D helps to reduce contact resistance.

Before the 1990s, most IC manufacturers made their own processing tools and developed IC processes themselves. In the 1990s, semiconductor equipment companies rapidly developed. They not only provided processing tools, they also provided integrated processes to the IC fabs. Cluster tools that were capable of running different processes in the same mainframe became very popular in the IC industry. Single-wafer processing systems became widely used because they have

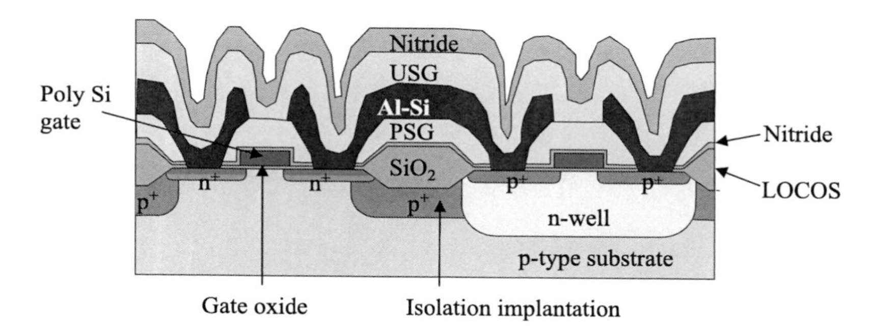

Figure 14.2 Cross-sectional view of a CMOS chip with early 1980s technology.

- Bare wafer, p-type
- Clean
- Deposit p-type epitaxial silicon (shown)

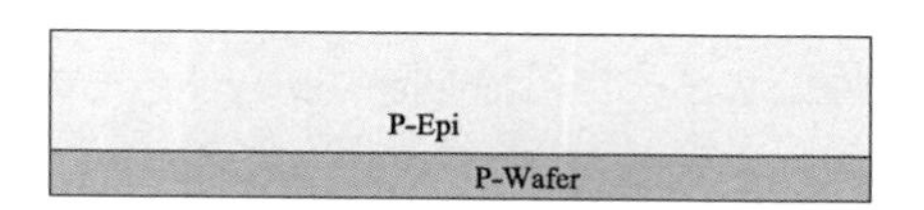

Figure 14.3 Epitaxial-layer growth on a p-type wafer.

better WTW uniformity control; however, batch systems are still currently used in many noncritical processes because of their high throughput.

In the 1990s, photolithography exposure wavelengths shifted from UV to deep ultraviolet (DUV) of 248 nm. Positive photoresists were used in photolithography. Steppers replaced other alignment and exposure systems, and integrated track-stepper systems, which perform photoresist coating, baking, alignment and exposure, and development in one processing sequence, were commonly used. All patterned etches were plasma etch processes, whereas wet etches were still widely applied in blanket film stripping and CVD film quality control. Vertical furnaces became dominant because of their smaller footprints and better contamination control. Rapid thermal processing (RTP) systems became more popular in postimplantation annealing and silicide formation because they are faster and have better processing and thermal budget control. Sputtering replaced evaporation and became the process of choice for metal deposition, and dc magnetron sputtering systems are now common in metal PVD in IC fabs.

Because the number of transistors increased so dramatically, one layer of metal was no longer enough to connect all of the microelectronic devices made on the silicon surface. Thus, multilayer metal interconnections were created to route the circuits. CVD tungsten was used to fill the narrow contact and via holes in the form of plugs to connect the different conducting layers. Titanium and titanium nitride became popular as both barrier and adhesion layers for tungsten. Titanium was also used as a welding layer for aluminum–copper alloys to reduce contact resistance, and titanium nitride was used as an ARC to help the photolithography process pattern metal wires.

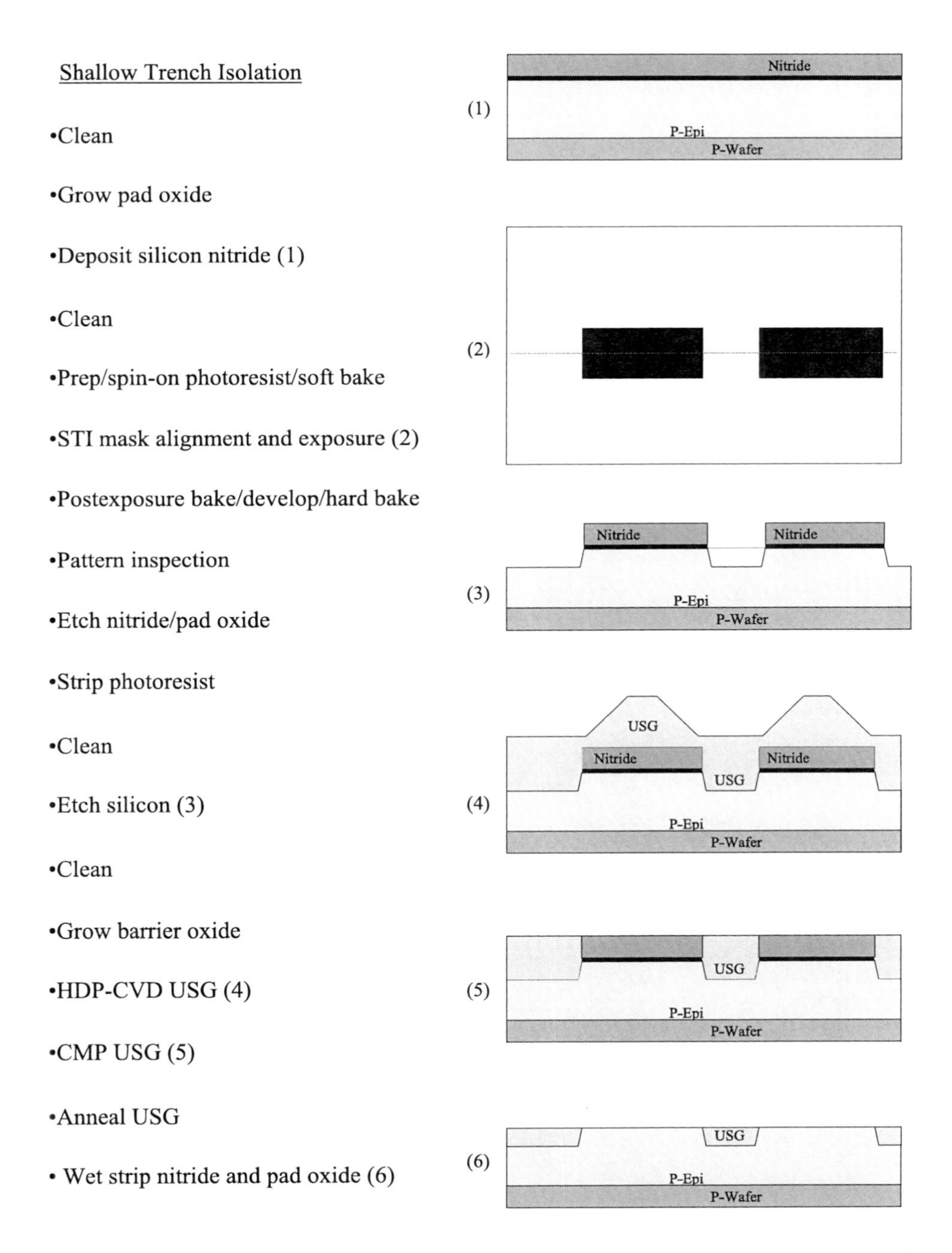

Figure 14.4 CMOS process with multilayer Al alloy interconnects (*continued on following 11 pages*).

Well Formation

•Clean

•Grow screen oxide

•Clean

•Prep/spin-on phtoresist/soft bake

•n-well mask alignment and exposure (7)

•Postexposure bake/develop/hard bake

•Inspection

•n-well ion implantation, phosphorus (8)

•Strip photoresist

•Clean

•Anneal and drive-in

•Clean

•Prep/spin-on photoresist/soft bake

•p-well mask alignment and exposure (9)

•Postexposure bake/develop/hard bake

•Inspection

•p-well ion implantation, boron (10)

•Strip photoresist

•Anneal and drive-in (11)

(7)

(8)

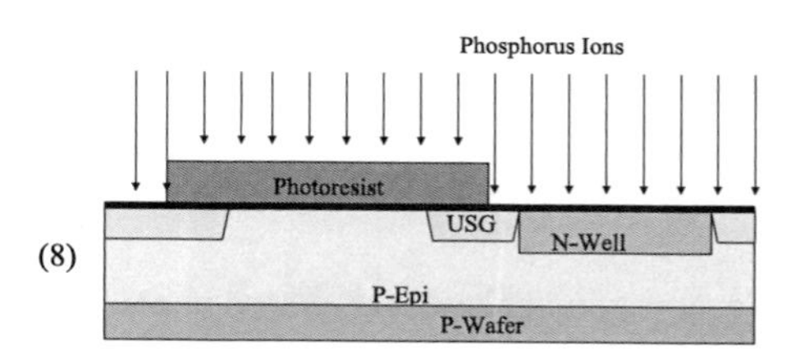

(9)

(10)

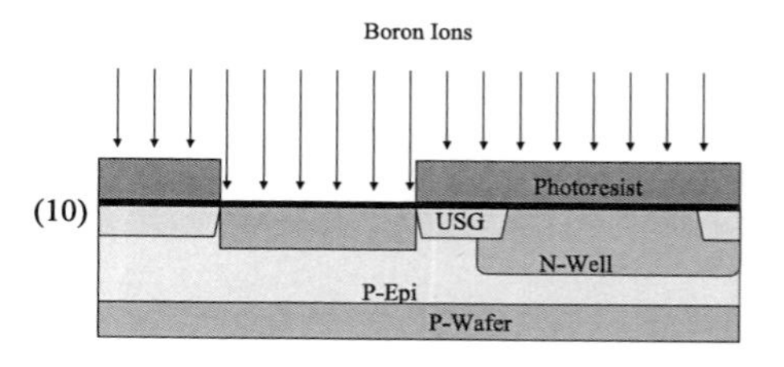

(11)

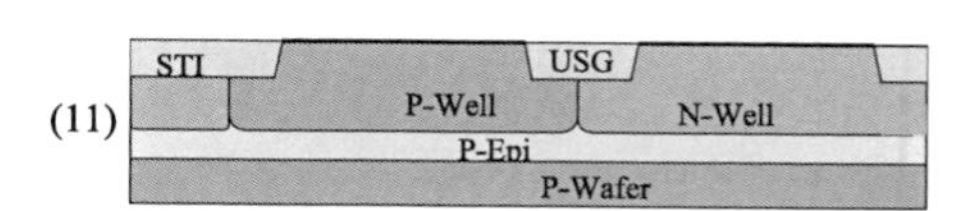

Figure 14.4 *(continued)*

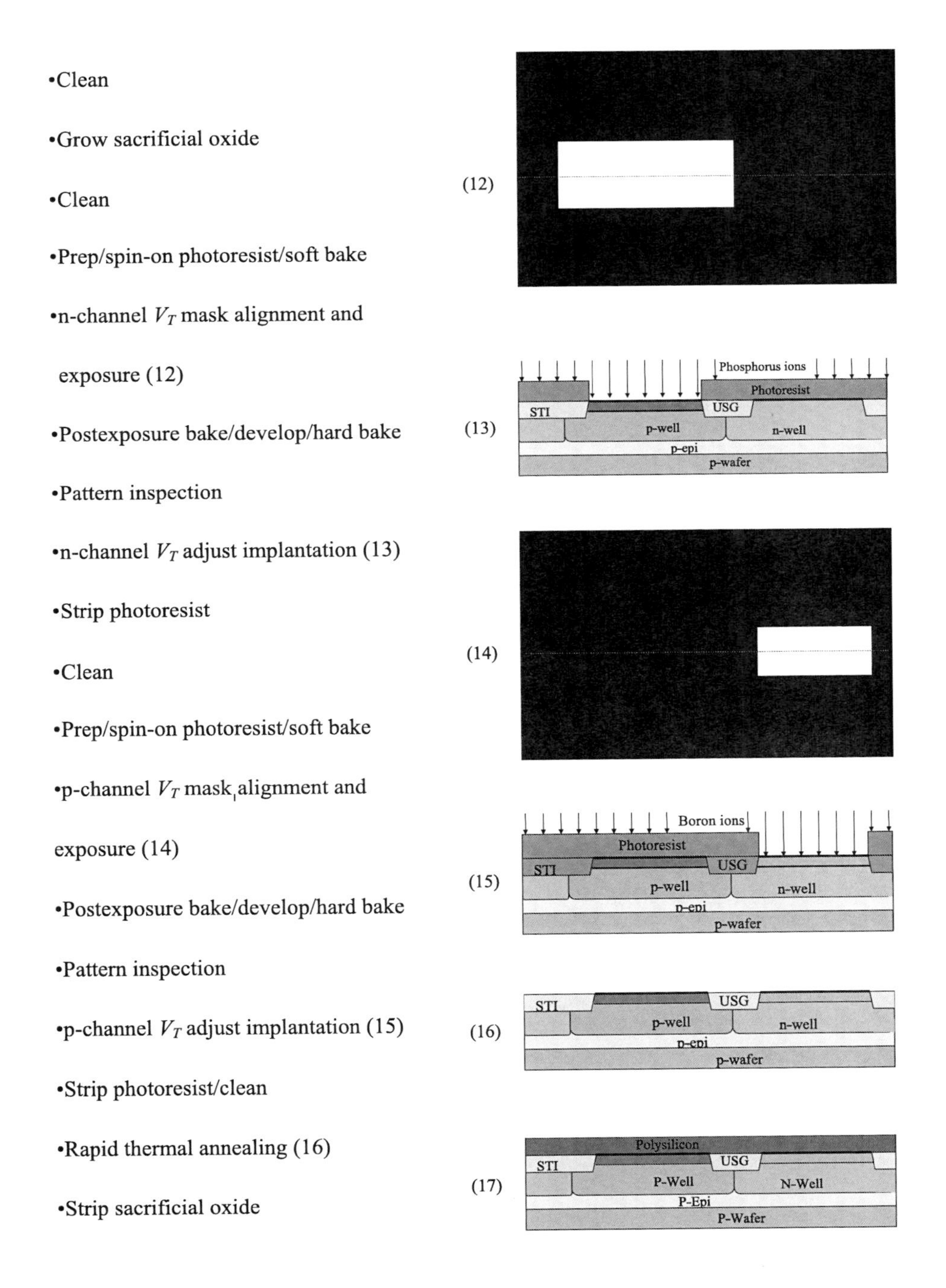

Figure 14.4 *(continued)*

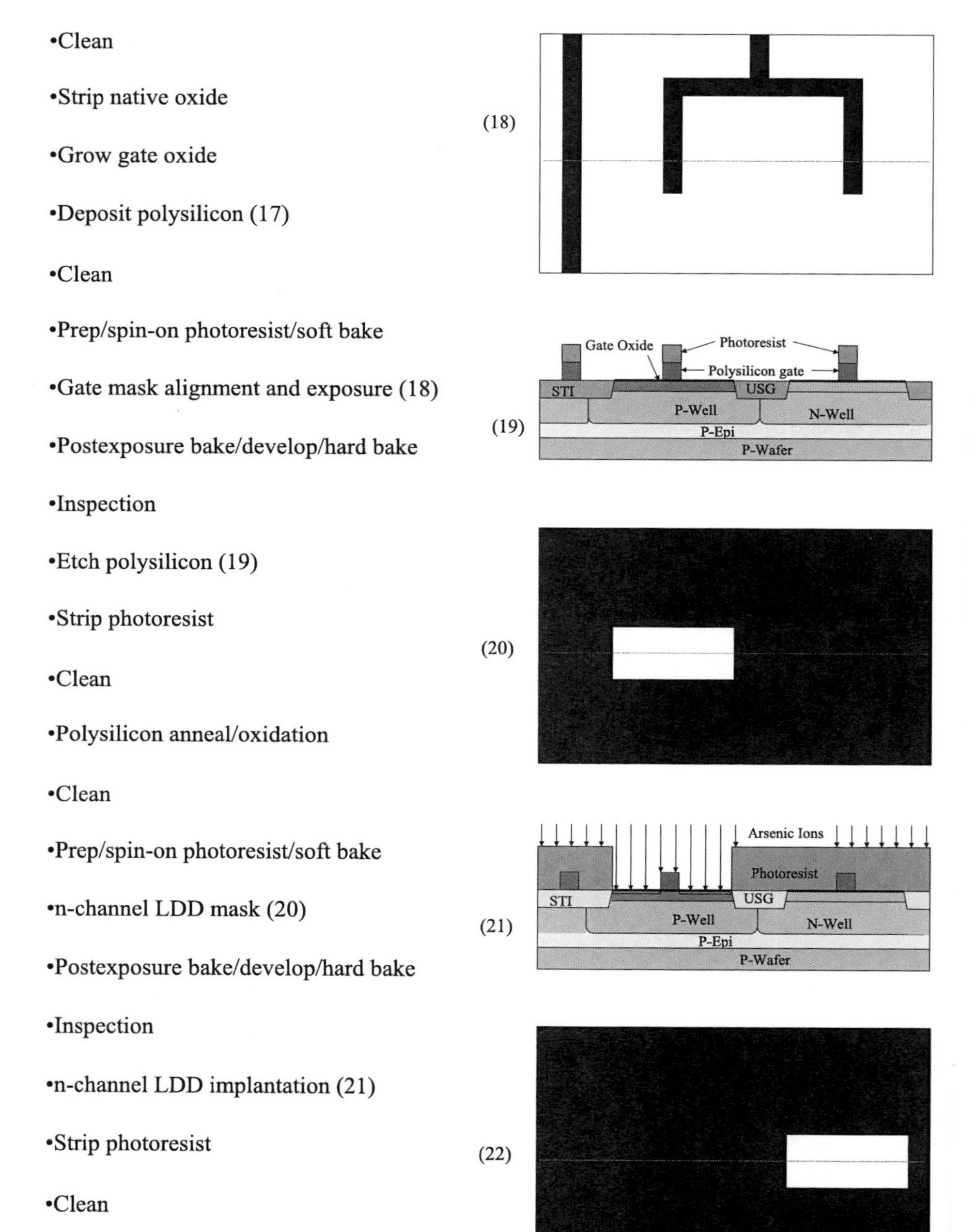

Figure 14.4 *(continued)*

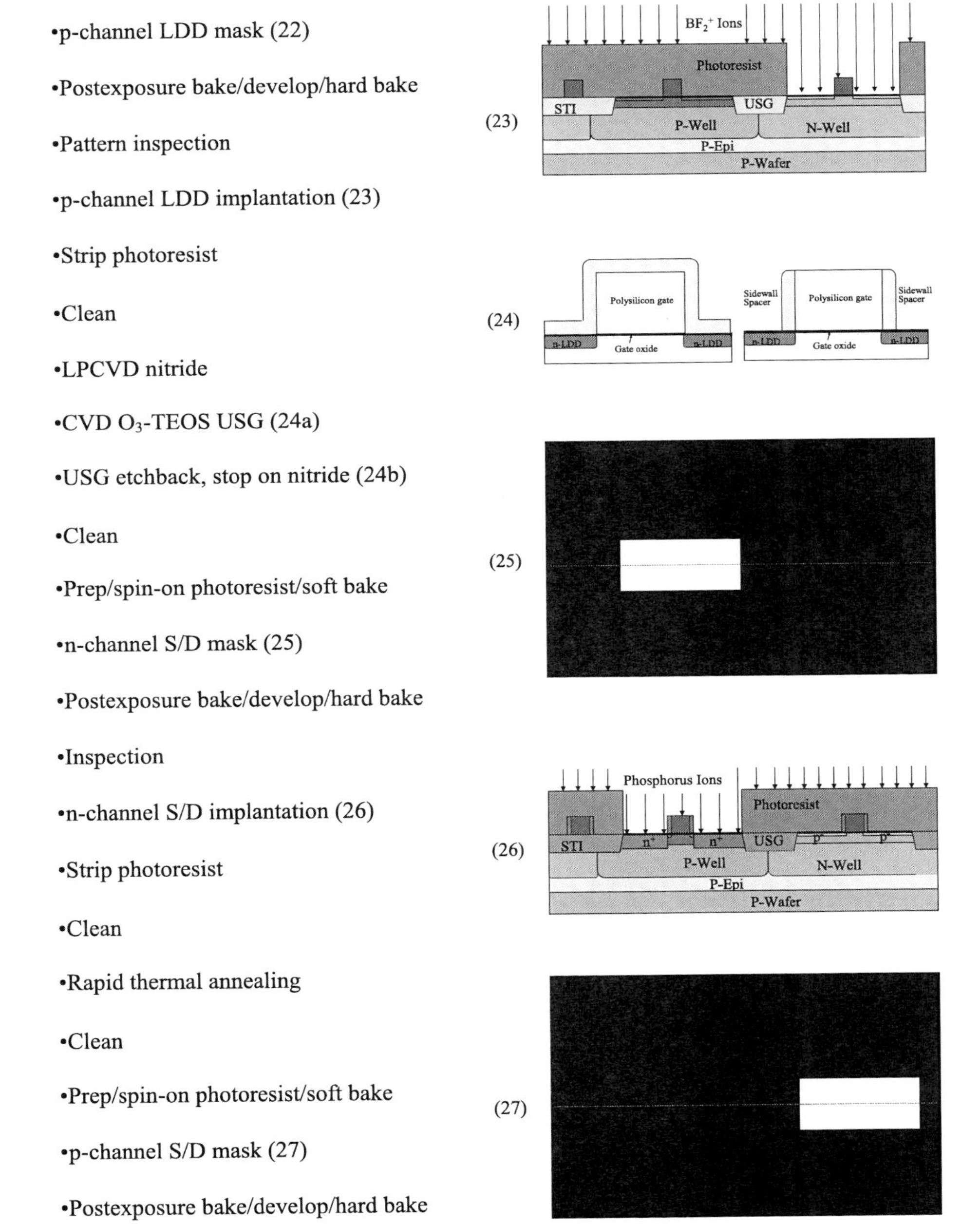

Figure 14.4 *(continued)*

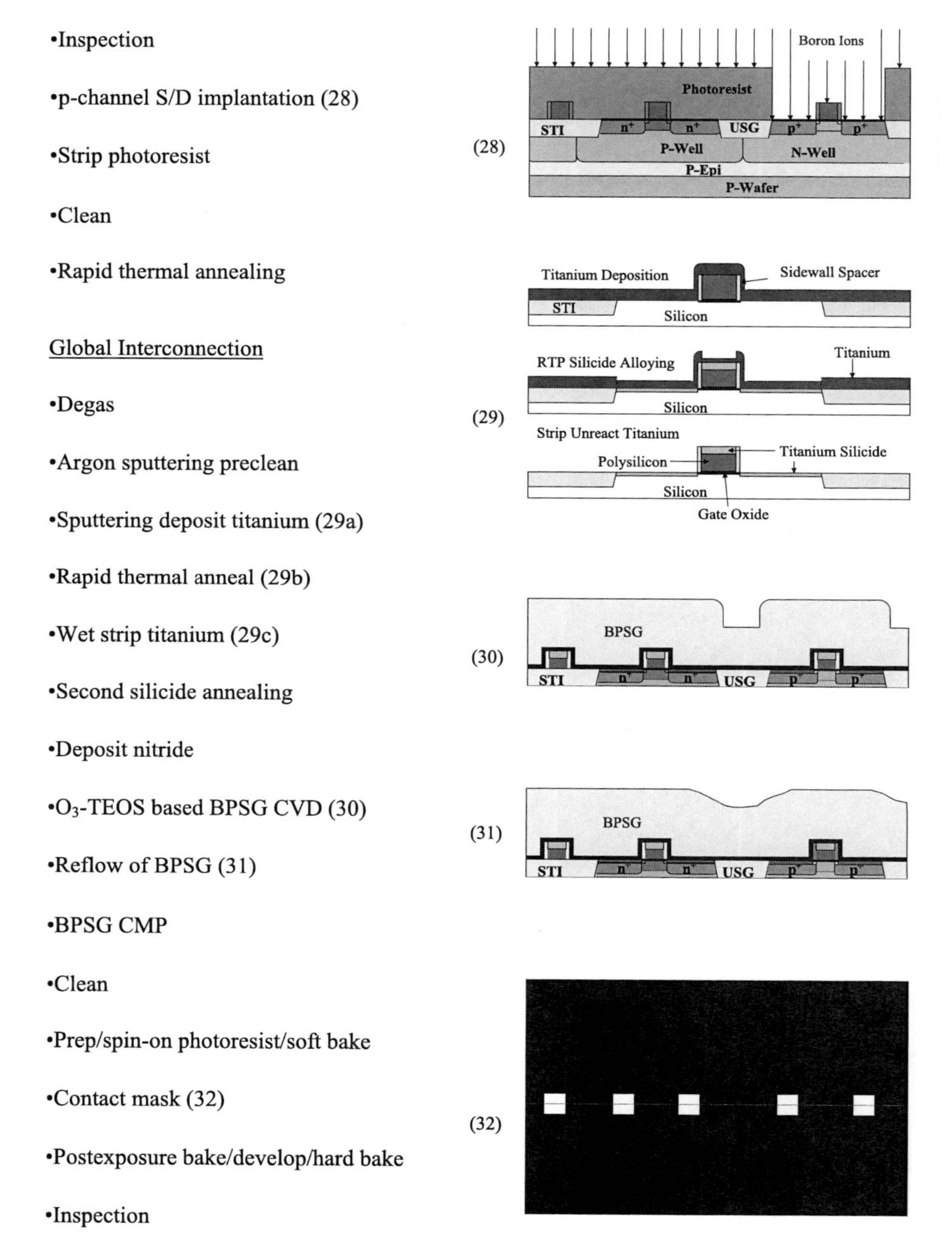

Figure 14.4 *(continued)*

•Etch BPSG, Stop on silicide surface

•Strip photoresist (33)

•Clean

•Degas

•Argon sputtering pre-PDV clean

•Deposit Ti/TiN by both PVD and CVD

•CVD tungsten (34)

•Polish tungsten/TiN/Ti

•Clean

• Ar sputtering pre-PDV clean

•PVD Ti

•PVD of Al-Cu alloy

•PVD of TiN ARC layer (35)

•Clean

•Prep/spin-on photoresist/soft bake

•Metal-1 mask (36)

•Postexposure bake/develop/hard bake

•Inspection

•Etch metal

•Strip photoresist (37)

(33)

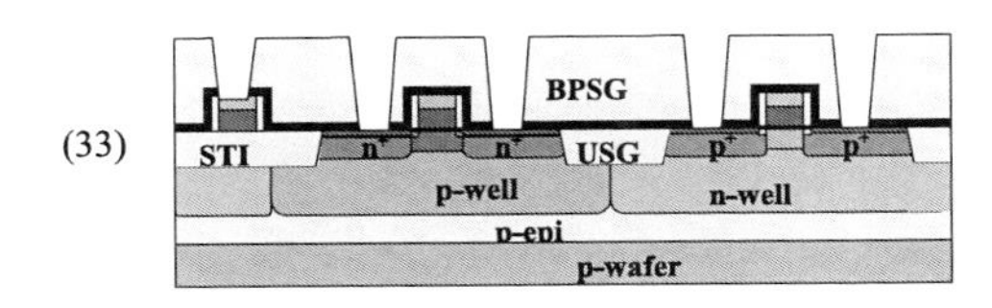

(34)

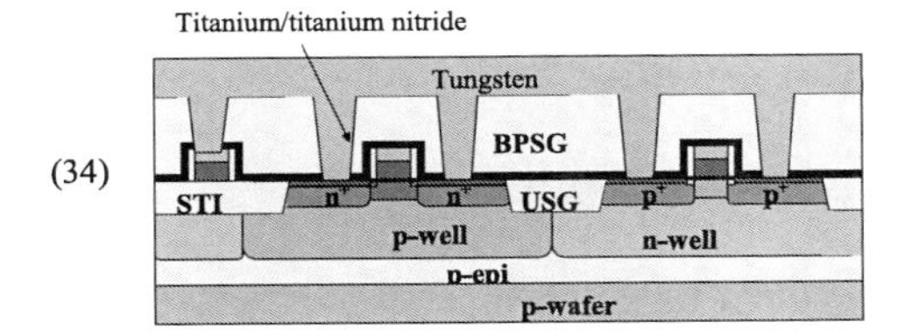

(35)

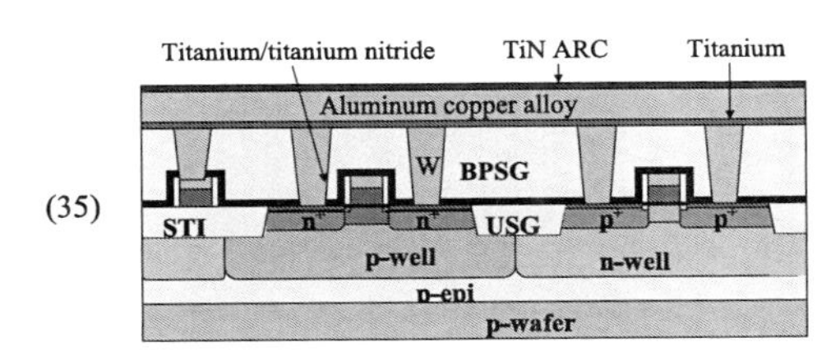

(36)

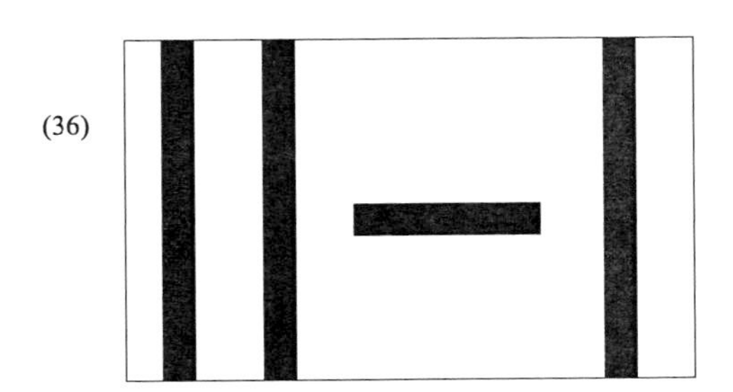

(37)

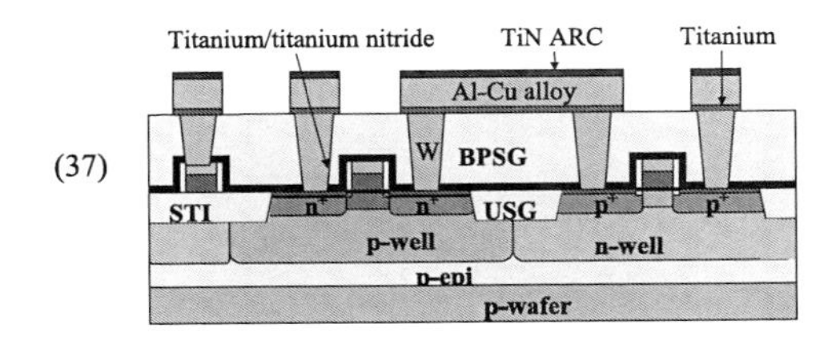

Figure 14.4 *(continued)*

•Metal anneal

•CVD USG

•Sputtering etchback USG

•CVD USG

•CMP USG (38)

(38)

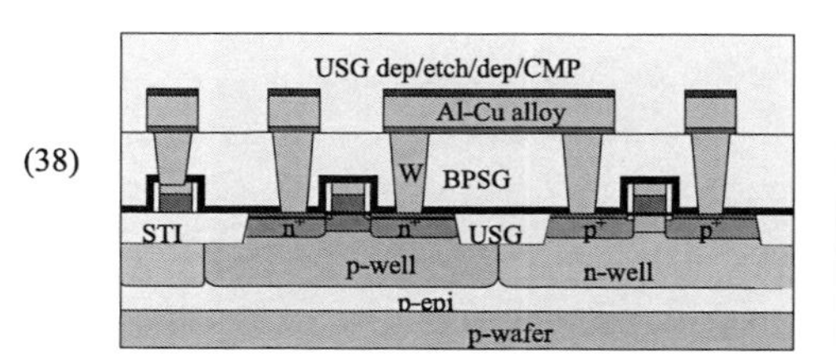

•Clean

•Prep/spin-on photoresist/soft bake

•Via-1 mask (39)

•Postexposure bake/develop/hard bake

•Inspection

(39)

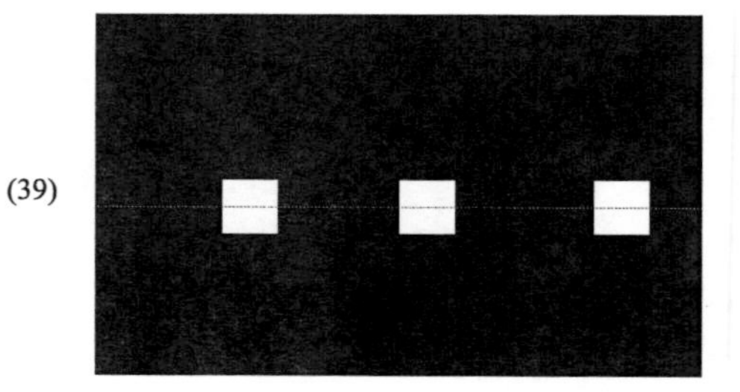

•Etch USG

•Strip photoresist (40)

•Degas

•Ar^+ sputtering preclean

(40)

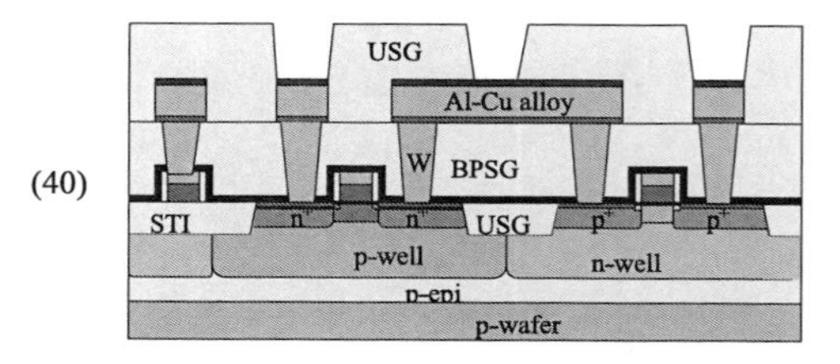

•Deposit Ti/TiN

•Deposit tungsten

•Polish tungsten/TiN/Ti

•Ar^+ sputtering pre-clean

•Deposit Ti

•Deposit Al•Cu

•Deposit TiN (41)

(41)

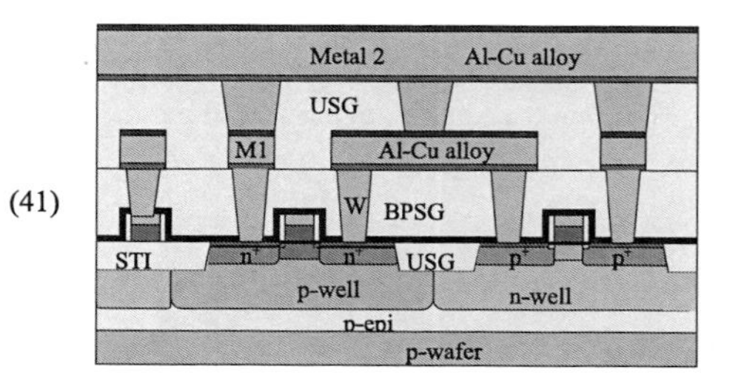

Figure 14.4 *(continued)*

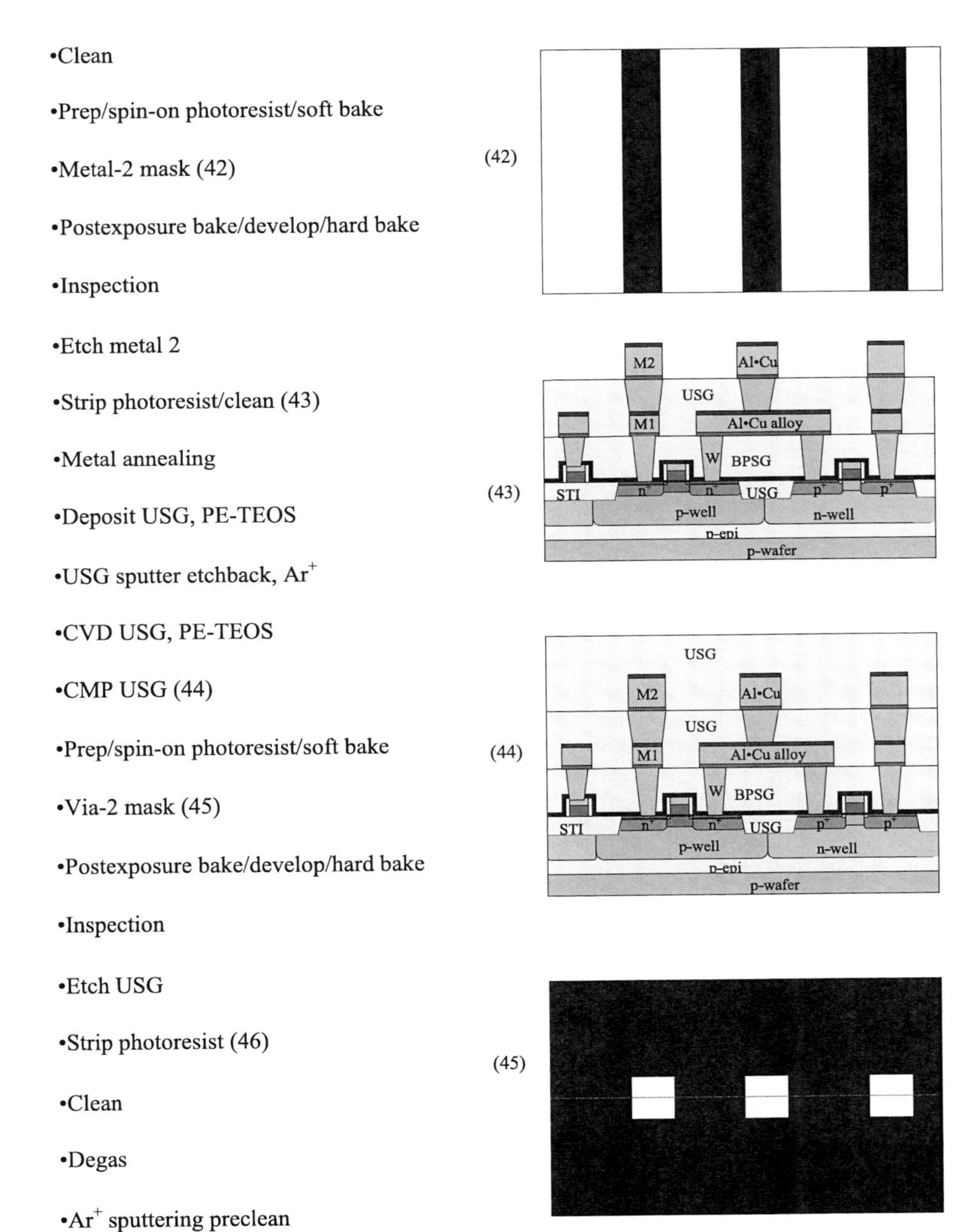

Figure 14.4 *(continued)*

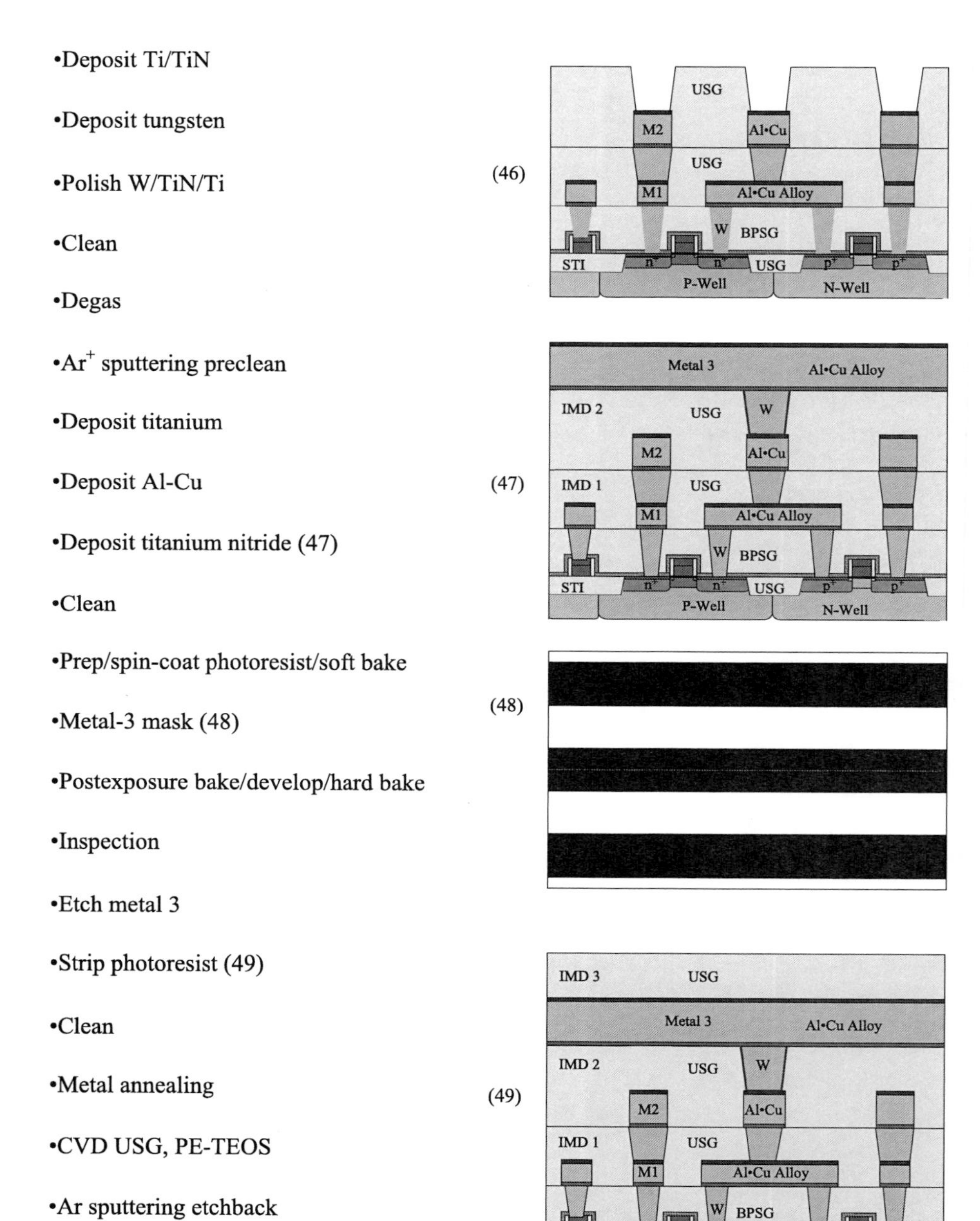

Figure 14.4 *(continued)*

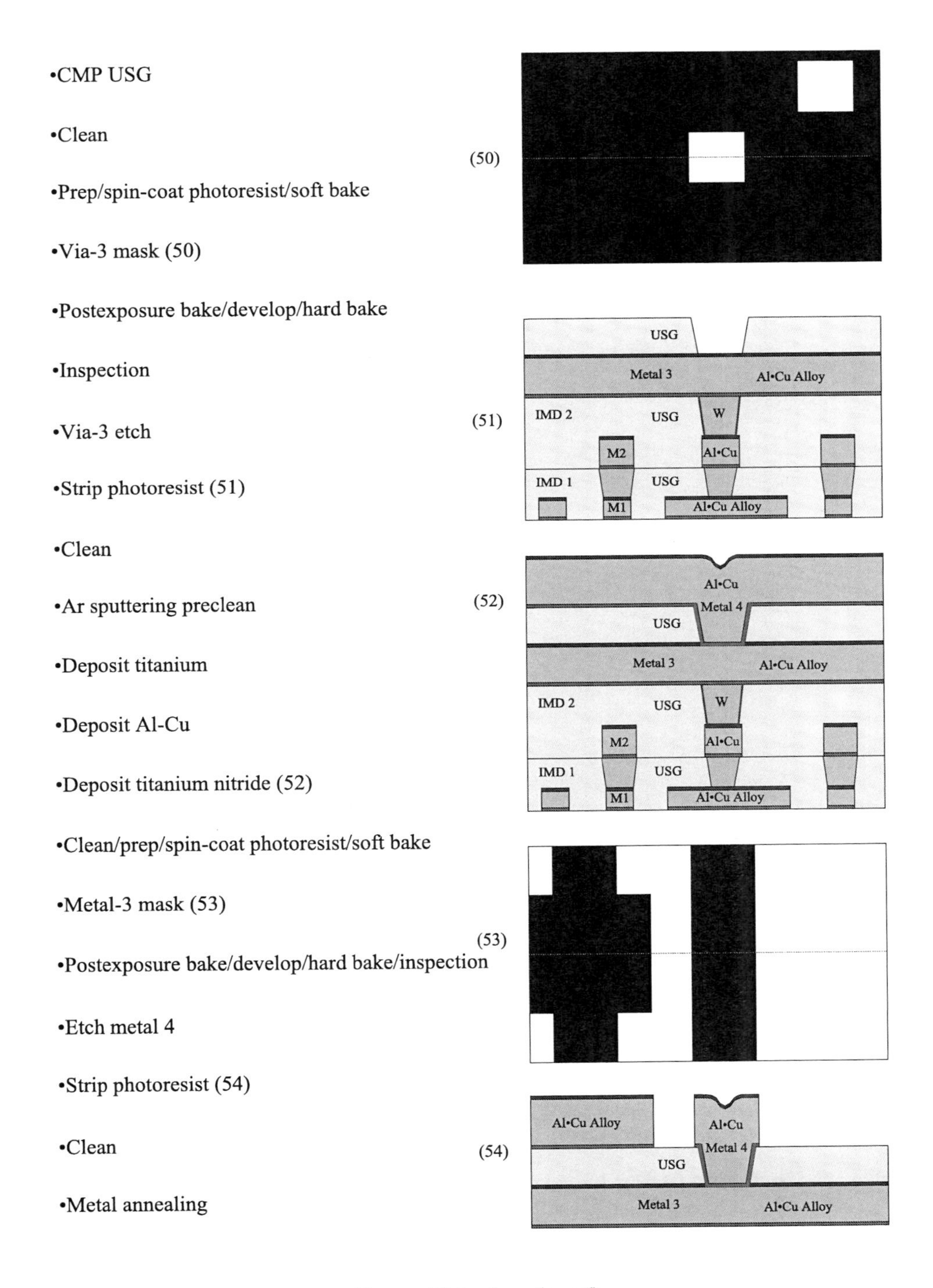

Figure 14.4 *(continued)*

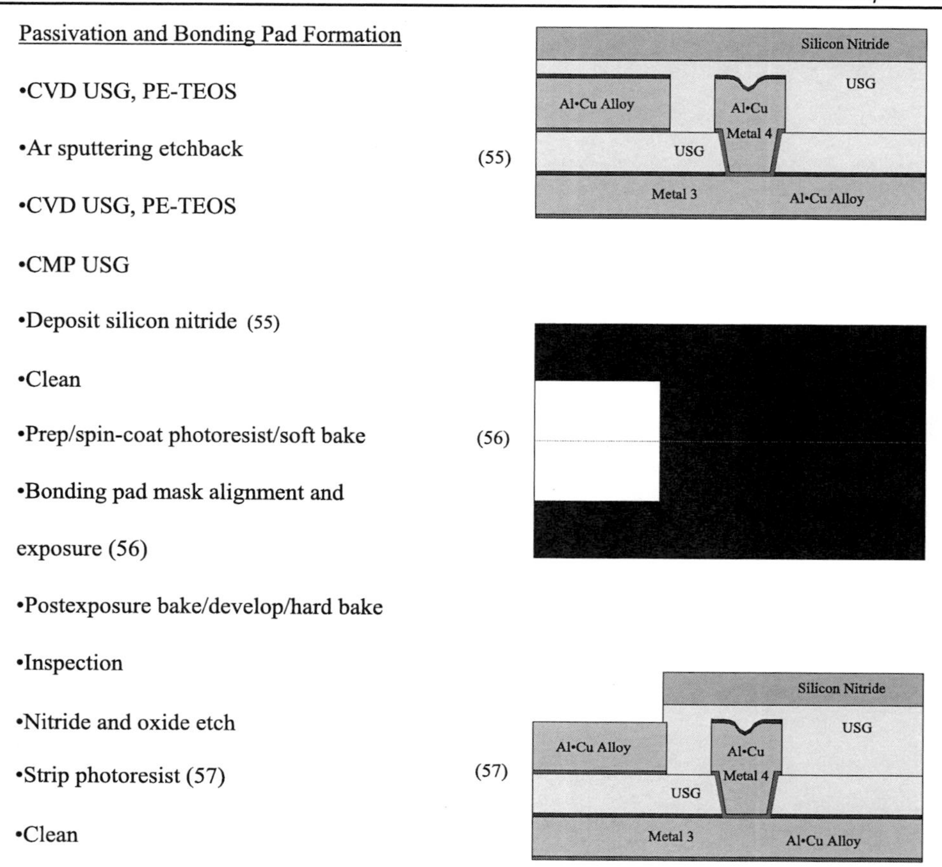

Figure 14.4 *(continued)*

BPSG was used as a premetal dielectric. By adding boron into the silicate glass, glass reflow temperature reduced from approximately 1100 °C for PSG to as low as 800 °C. This helped to save the precious thermal budget, which dwindled as feature size shrank. PE-TEOS and O_3-TEOS processes were widely employed in dielectric CVD processes for STI, sidewall spacers, and interconnection applications. The CMP process was frequently applied to remove bulk CVD tungsten on the wafer surface during tungsten plug formation. CMP was also used to planarize silicate glass surfaces to achieve better photolithography resolution and to make the succeeding metal deposition processes easier. Figure 14.4 shows the manufacturing process flow of a CMOS IC chip with mid-1990s processing technology. Figure 14.5 shows a cross section of a CMOS IC chip made with mid-1990s processing technology.

14.4 Complementary Metal-Oxide-Semiconductor Process Flow with Technology after 2000

Two factors affect CMOS IC speed: gate delay and interconnection delay. Gate delay is the time required to turn the MOSFET on and off, and interconnection

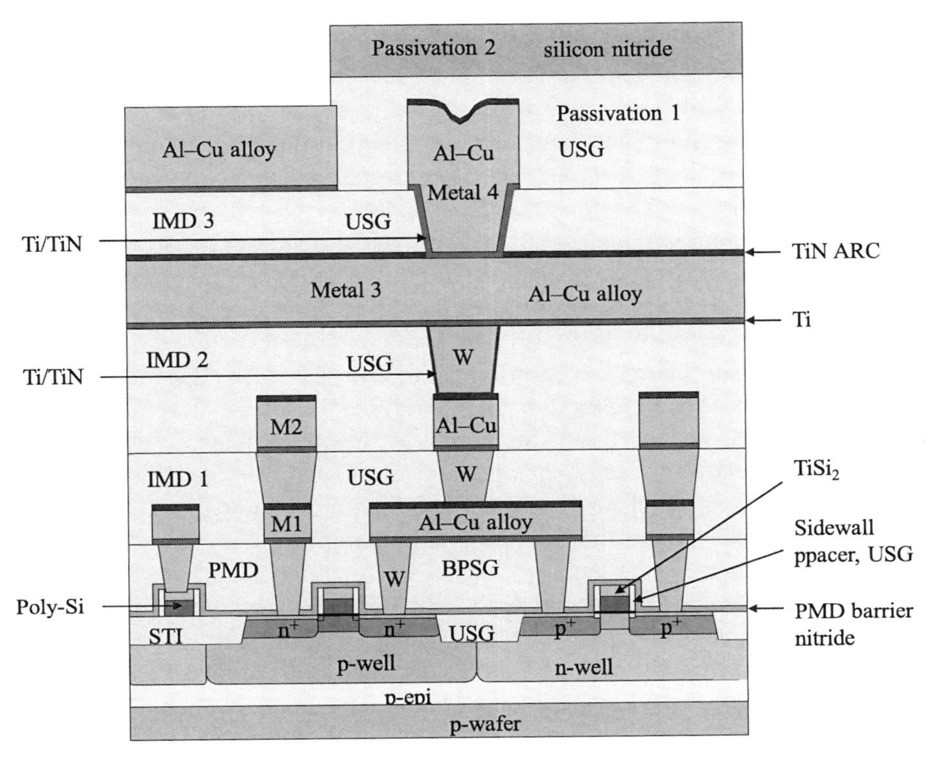

Figure 14.5 Cross section of a CMOS chip with mid-1990s technology.

delay is determined by chip design, processing technology, and the conducting and dielectric materials used for the interconnection.

Gate delay is determined by two factors: the time required to accumulate enough charge to turn on the MOS transistor, and the time required for carriers (electrons for nMOS and holes for pMOS) to pass through the channel underneath the gate between the S/D. The MOS structure of the MOSFET also forms a capacitor with the conducting gate as one electrode, the semiconductor substrate under it as a second electrode, and an insulating layer (gate oxide) between them. MOS capacitance should be large enough to hold the charges (carriers) that form the conducting channel underneath the gate between the S/D when gate voltage exceeds V_T. This allows the MOSFET to turn on. Reducing the gate capacitance can reduce the time required to form the channel and increase the switching speed. However, if the capacitance is too low, the MOSFET becomes unstable because any small noise such as background radiation can turn on or off the transistor and cause a soft error (described in Chapter 8). The distance between the source and drain of a MOSFET is called the channel length, and carriers need to pass through the channel to carry the current. Reducing gate width reduces time for carriers to pass through the channel and increases device speed. However, it also reduces gate capacitance and can cause reliability problems in the device because the MOS capacitance is already designed at the lowest possible level. To further increase

IC chip speed, a substrate with high noise resistance is required to allow smaller feature sizes. Silicon-on-isolator (SOI) is one promising approach, which isolates transistors on the silicon surface from the bulk silicon substrate; therefore, SOI almost completely eliminates the chance of radiation-induced soft errors.

A combination of SOI and STI allows total isolation of the neighboring microelectronic devices and prevents crosstalk between them, a feature that enables chip designers to increase the packing density of transistors on an IC chip. IC chips designed on SOI substrate have been used in high-radiation environments, such as space shuttles, rockets, and scientific research laboratories. Another approach is using a strained silicon channel with a bulk silicon wafer.

Resistance of the interconnection wires and the parasitic capacitance between these wires determines an interconnection delay, or RC delay. To reduce RC delay, metals with lower resistivity and insulators with lower dielectric constant (low-κ) are preferred for forming the interconnection. Copper has lower resistivity than aluminum–copper alloy, so using copper can reduce power consumption and increase IC speed. Traditional aluminum-copper alloy interconnections require one dielectric etch and one metal etch, whereas a copper interconnection usually employs the dual damascene process, which needs two dielectric etches but no metal etch. Dual damascene uses metal CMP instead of metal etch to form the interconnection wires. This is the major difference between copper and aluminum–copper alloy interconnections. The main challenges for copper interconnections are dielectric etch, metal deposition, and metal CMP.

Several low-κ dielectric materials were developed with two competing approaches: CVD and spin-on dielectric (SOD). The advantages of a SiCOH-based CVD low-κ dielectric are familiar technologies, existing processing equipment, and experience. One important advantage of SOD was its extendibility to very low dielectric constants ($\kappa < 2$) with porous silica. However, reliability issues of SOD during chip packaging processes eventually determined CVD SiCOH as the best material for low-κ dielectrics used for mass production of advanced IC chips.

Bonded SOI uses two wafers: one is implanted with high-current hydrogen ions to form a hydrogen-rich layer under the silicon surface, and the other wafer grows a silicon dioxide layer on the surface (Fig. 14.6). Then, the two wafers are pressed together face to face at a high temperature to bond them, with silicon dioxide in between. At high temperatures, the hydrogen atoms in wafer A in Fig. 14.6 react with silicon atoms to form gaseous byproducts ($4H + Si \rightarrow SiH_4$), which induce voids in wafer A. This voided hydrogen-rich layer has a very high wet etch rate; therefore, wafer A can be easily separated from the bonded wafer in a wet etch process. Then, a CMP process is applied to remove defects and roughness on the silicon surface, and this also makes it highly flat and smooth (see Fig. 14.7). The thickness of the silicon layer above the buried silicon dioxide is controlled by hydrogen implantation energy and CMP time. It ranges from several hundred to approximately ten nanometers, determined by device requirements.

Another way to make SOI wafers is to use very-high-current oxygen ion implantation to form the oxygen-rich layer beneath the silicon surface. High-temperature (>1200 °C) annealing is performed, leading to the formation of a

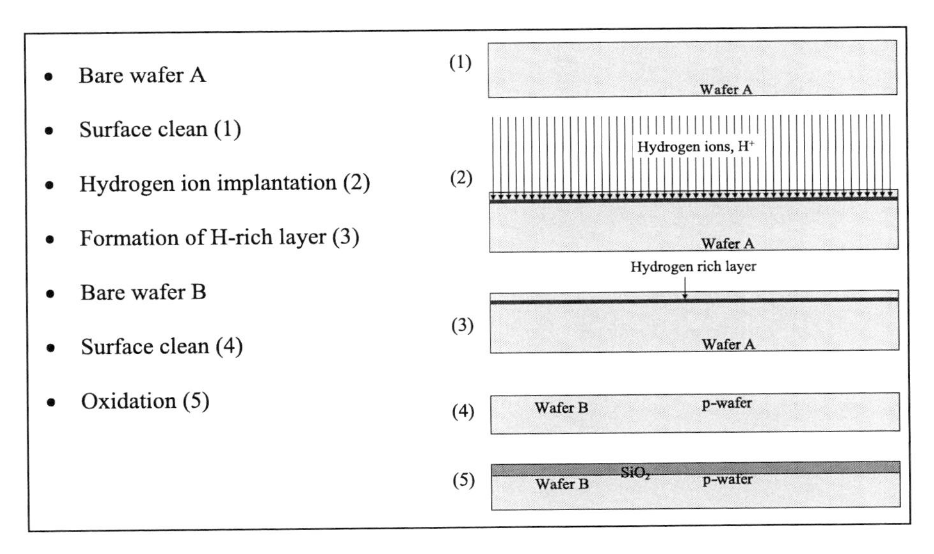

Figure 14.6 Bonded SOI (wafer A): Preparation of wafers.

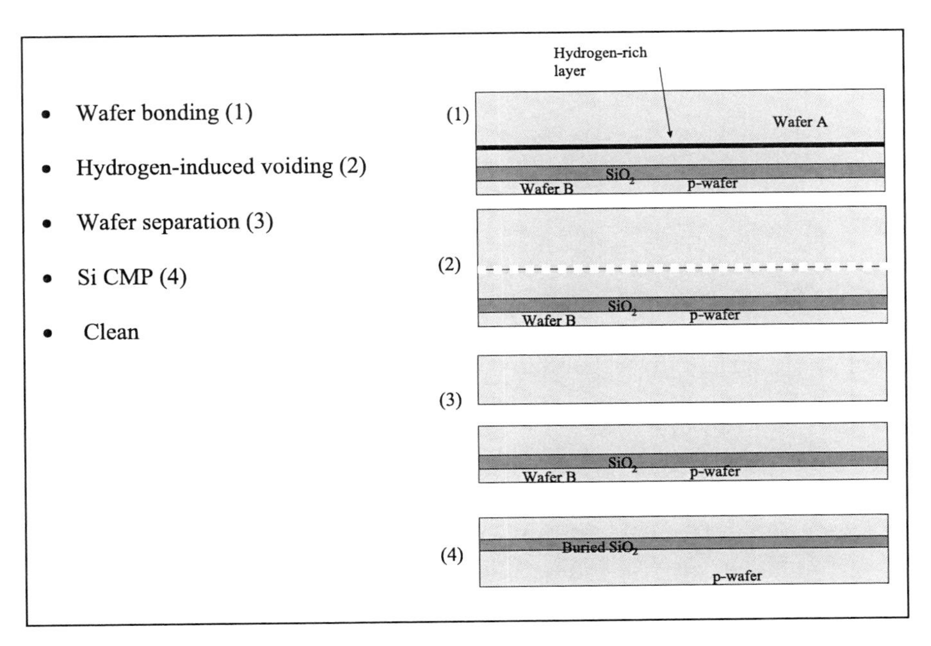

Figure 14.7 Bonded SOI (wafer B): wafer bond, wafer separation, and CMP.

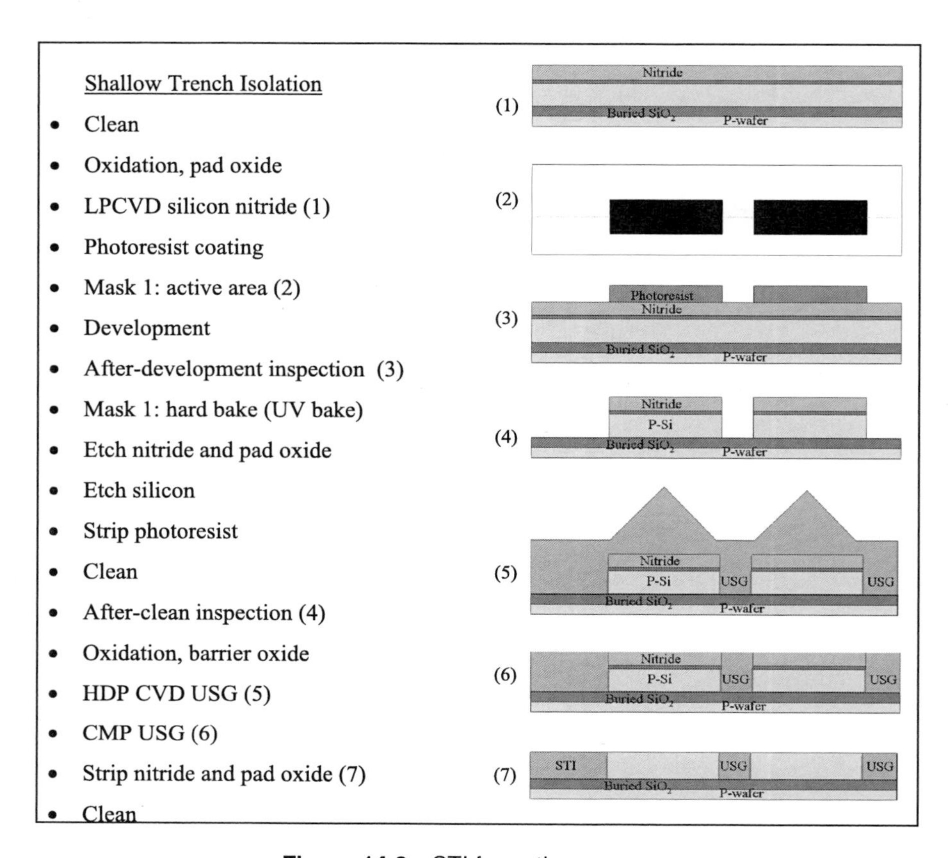

Figure 14.8 STI formation process.

buried oxide layer underneath a thin single-crystal silicon layer. Epitaxial silicon is deposited on the wafer surface, preventing oxygen in the silicon from slowing down the device.

Because activation areas are surrounded by trenches that were etched all the way to the buried silicon dioxide [see Fig. 14.8, illus. (4)], they are completely isolated by silicon dioxide after CVD trench fill and oxide CMP. This total isolation completely eliminates any possibility of crosstalk between neighboring devices and allows transistors to be densely packed. It also solved the radiation-induced soft error issue to allow further shrinking of the device.

Because feature size shrinks, the junction depths of both n- and p-wells are reduced. Therefore, existing high-energy ion implanters can directly implant dopant ions at the required depths, and a drive-in process is no longer needed for well formation. Several implantations at different energy levels are required to achieve well formation.

Without the requirement of well junction drive-in (in which thermal diffusion drives dopant atoms deep into the substrate at high temperatures), engineers can use the same photomask to perform well and V_T adjustment implantations, as shown in Fig. 14.9. Since ion implanters can precisely select the desired ion species with

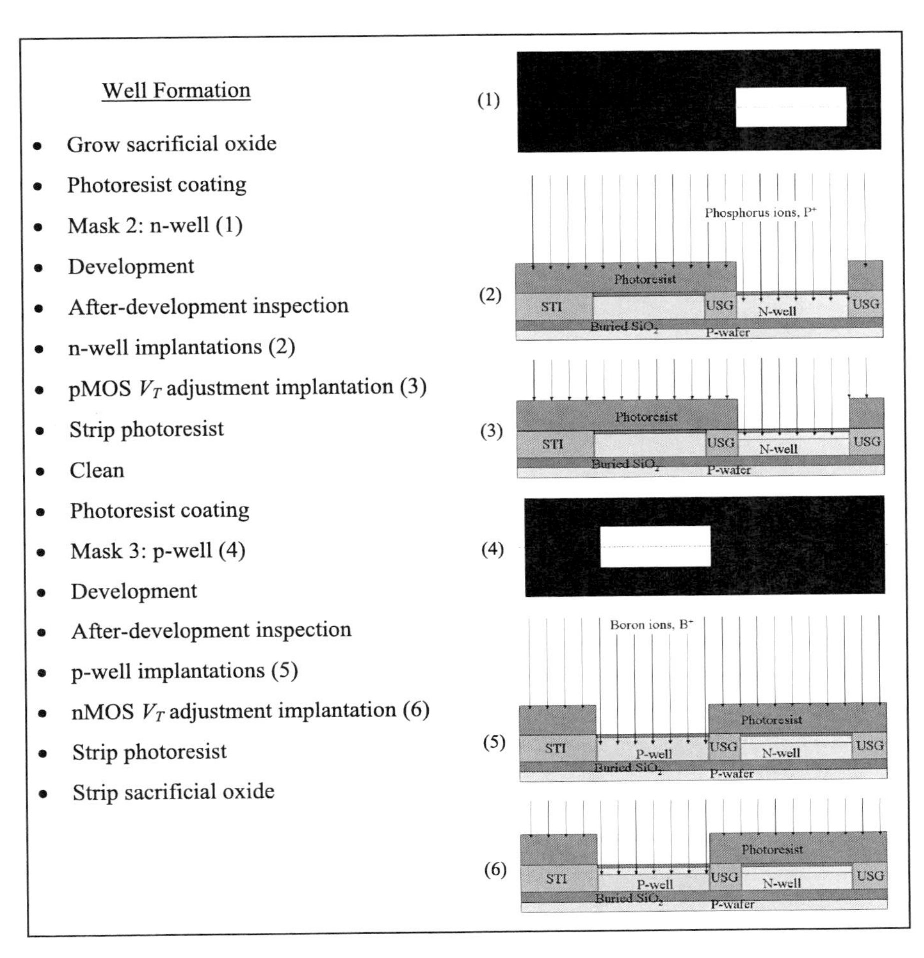

Figure 14.9 Well and threshold adjustment implantations.

their magnetic mass analyzer, all of these implantation processes can be finished with a high-energy, low-current implanter in one processing sequence.

To better control the threshold voltage of the MOSFET, an nMOS polysilicon gate needs to be heavily doped in n-type, and pMOS polysilicon needs to be heavily doped in p-type. By blanket implantation of n-type dopant and selective implantation of a heavier p-type counterdopant, doping nMOS polysilicon with n-type and pMOS polysilicon with p-type can be achieved with one photomask. This is preferred, especially for cost-sensitive DRAM manufacturers, because it helps reduce manufacturing costs and improve device yield. Figure 14.10 shows the polysilicon doping and gate formation processes.

Polysilicon consists of many single-crystal units called grains. Large grain size is preferred because it has less of a grain boundary to deflect electrons, leading to lower resistivity. However, large grain size can cause substantial surface roughness on the sidewalls of the polysilicon lines after etching. For small gates, etching amorphous silicon (α-Si), then annealing it to form polysilicon is preferred. Heavy

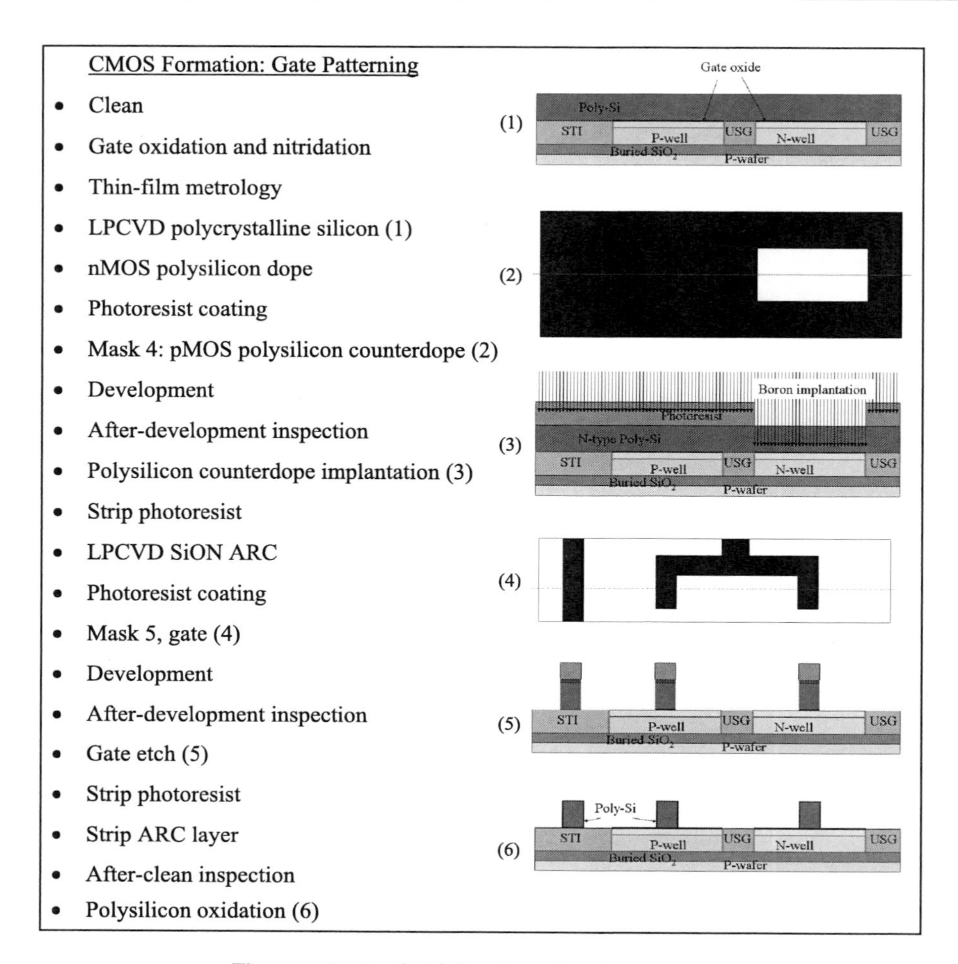

Figure 14.10 CMOS gate patterning process.

polysilicion doping implantation can create amorphous silicon, which allows better etching profile control than polysilicon. The grain size of the polysilicon annealed from α-Si is also more uniform than that of LPCVD polysilicon. After gate etching, plasma-induced gate oxide damage can be repaired by polysilicon oxidation during the annealing process.

Heavy ions are used to form the shallow junctions of the S/D extension shown in Fig. 14.11, usually BF_2^+ for pMOS SDE and As^+ or even heavier Sb^+ for nMOS SDE.

For spacer formation, both nitride and oxide are used. In Fig. 14.12, CVD oxide is used as the etchstop layer, and LPCVD nitride forms the main body of the sidewall spacer.

Figure 14.13 illustrates the halo junction and S/D junction formation of a CMOS. Halo implantation is a large tilt-angle implantation process. It usually requires two or four implantations, depending on whether the MOSFET is placed

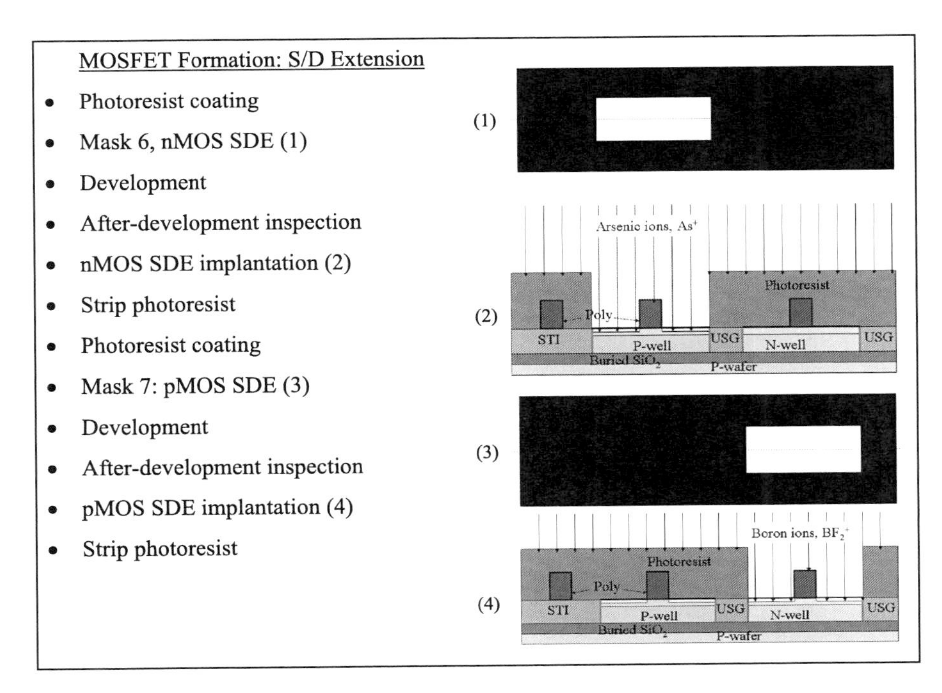

Figure 14.11 Formation of the S/D extension.

in one direction or two. Halo implantation is used to prevent punch-through of the device.

The grain size of titanium silicide must be larger than 0.2 μm to achieve lower resistance. If the gate width is narrower than 0.2 μm, a titanium silicide application would be in trouble. Cobalt silicide has started to replace titanium silicide for gate applications after the 0.18-μm technology node. Because cobalt is very reactive and easily forms cobalt oxide when in contact with air or moisture, titanium nitride is needed to cap the cobalt to prevent contact with the atmosphere. Cobalt and titanium nitride are deposited in different PVD chambers in the same cluster tool in an integrated processing sequence.

As device dimensions shrank deep into nanometer technology nodes, the annealing temperature of $CoSi_2$ (~750 °C) became too high for the thermal budget of a tiny MOSFET. Nickel silicide (NiSi), which can form at temperatures lower than 500 °C, became widely used after devices reached the 65-nm technology node and smaller.

Figure 14.14 illustrates the self-aligned NiSi formation process. At first, an argon sputtering etch is required to remove native oxide from the silicon surface before nickel deposition; otherwise, contact resistance can be too high due to poor silicide formation, and this can cause electronic failure of IC chips.

Because NiSi is not thermally stable, Ni tends to further react with silicon and diffuse across the junction, causing severe device leakage. Platinum (Pt) can be alloyed with the PVD target to form NiPtSi on the wafer surface to achieve better

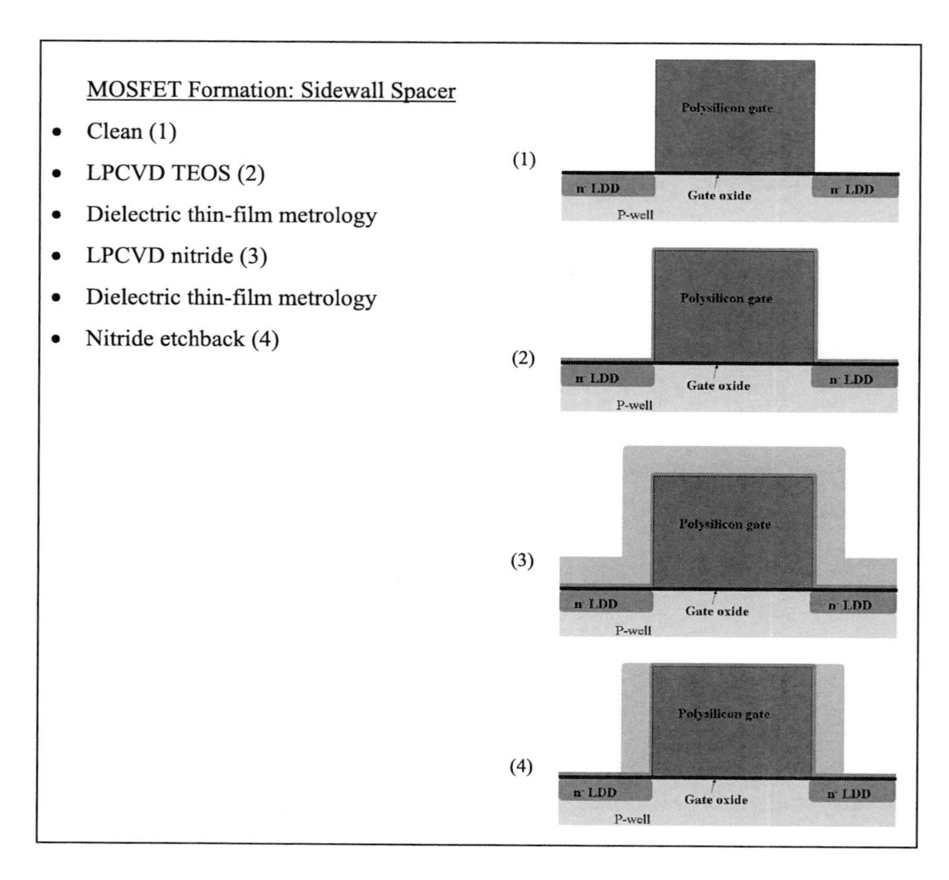

Figure 14.12 Formation of a sidewall spacer.

stability of the silicide. The yield-impacting Ni diffusion or encroachment issue can be captured and monitored with electron beam inspection (EBI) systems.

Figure 14.15 shows the contact module of a sub-90nm CMOS process. After NiSi formation, a nitride layer is necessary to prevent phosphorus from PSG diffusing into the active area. Because of thermal budget limitations, PECVD nitride deposited at a lower temperature (<580 °C) can be used to replace LPCVD nitride, which is normally deposited at 750 °C. For small devices (<0.18 μm), there is no thermal budget for ILD0 thermal reflow; therefore, boron is no longer needed in silicate glass, and PSG replaces BPSG as the material of choice for ILD0. PSG is planarized with CMP instead of reflow. Tungsten is only used for local interconnections and as plugs between the metal and S/D and silicide. Titanium and titanium nitride are still used as barrier and glue layers for tungsten.

For some advanced-technology-node CMOS processes, USG is used for ILD0; the nitride layer is used as a stress liner to strain the channel, enhance carrier mobility, and improve MOSFET performance.

The contact module is very critical because it is the connection between the devices on the wafer surface and the metal interconnect wires above. If the contact

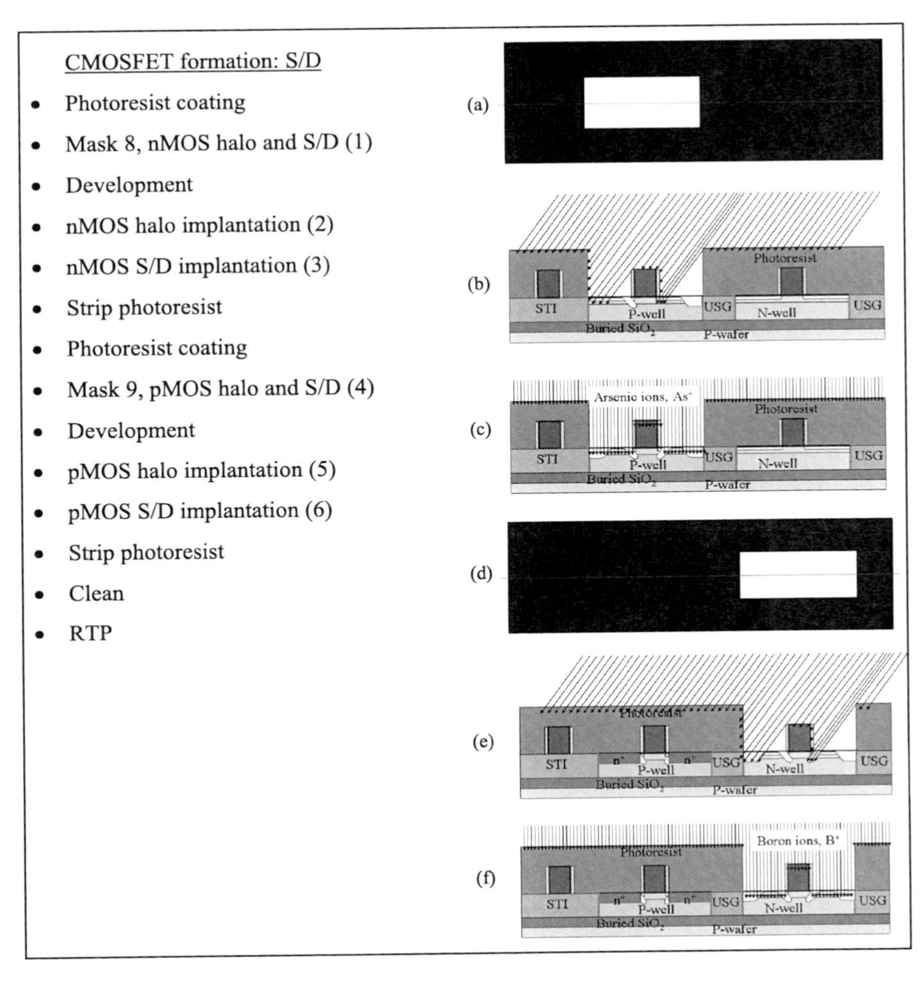

Figure 14.13 S/D implantations.

holes are not etched all the way through, the metal wires will not connect to the devices, and overall yield will suffer.

To reduce resistance, PVD titanium is widely used as a wetting layer for contacts. Titanium nitride serves as a glue layer for tungsten. Without TiN, tungsten film does not stick well to the wafer surface; it will crack and peel off, generating large numbers of particles on the wafer and along the path of the wafer. TiN can be deposited with both PVD and CVD. As device feature size shrinks, the aspect ratio of the contact hole becomes larger, and PVD processes can no longer offer sufficient step coverage; thus, CVD TiN processes have become more popular.

WCMP is one of most important layers for applications of EBI, which allows engineers to both capture device leakage issues and locate contacts that did not open. Application of EBI can help accelerate yield education, reduce the technology development cycle, and shorten yield ramp time.

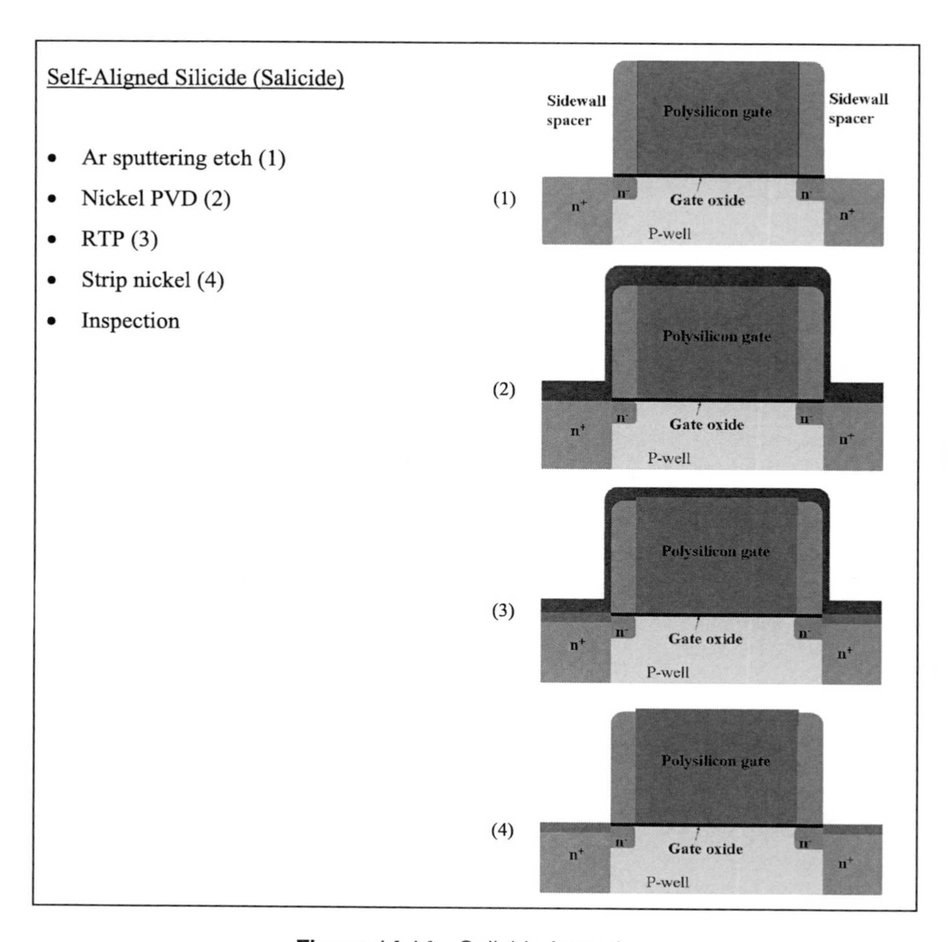

Figure 14.14 Salicide formation.

Figure 14.16 shows the metal-1 copper interconnect with a single damascene process. Silicon carbon nitride (SiCN) is a dense material that can replace silicon nitride as a barrier layer for preventing copper diffusion. It also can serve as an etch-stop layer (ESL) in copper interconnect processes. It has a lower dielectric constant (κ = 4 to 5) than silicon nitride (κ = 7 to 8); therefore, using SiCN reduces the overall dielectric constant of ILD layers. The most commonly used low-κ dielectric material is PECVD SiCOH, which is widely used in interconnect processes. Tantalum barrier and copper seed layers are deposited with a PVD process, usually with ionized metal plasma processing to enhance bottom coverage. Because a low-κ dielectric has less mechanical strength than silicate glass, the downforce of the copper CMP process with a low-κ dielectric must be lower than that of USG or FSG. After metal-1 CMP, an electric test can be performed for the first time on the test structures. Tiny probes contact the pads of the test structures and apply voltage or current to test the electrical properties of the devices. To avoid yield loss from copper oxidation, there is a time limit between copper CMP and cap layer deposition. Defect inspections with optical and electron beam inspection

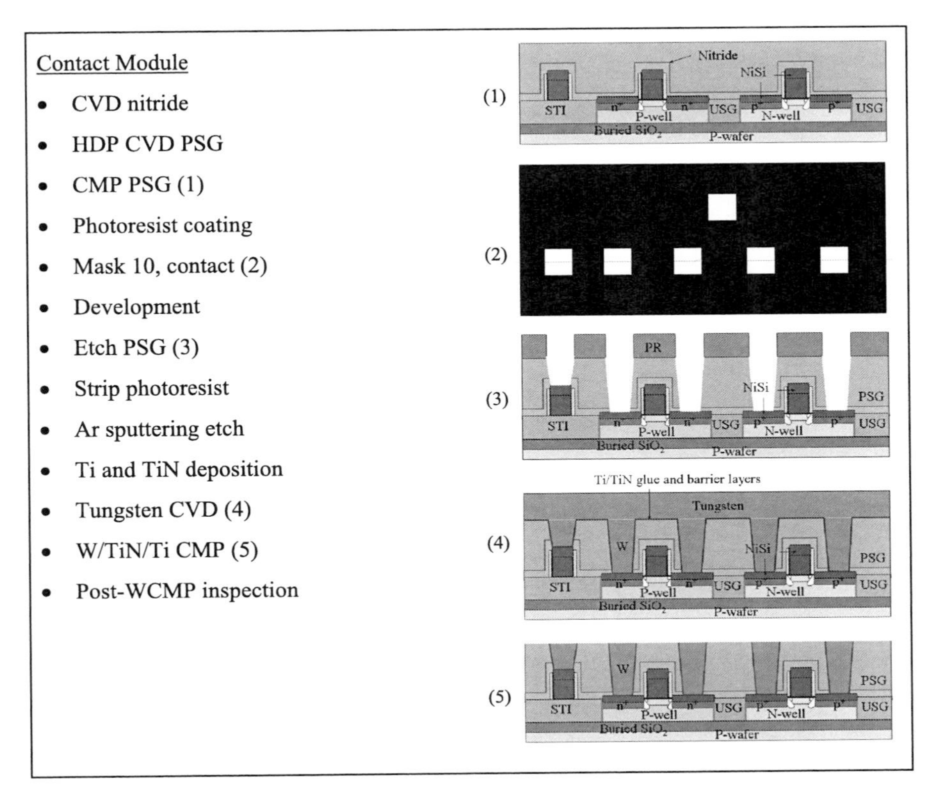

Figure 14.15 Contact module of a CMOS device.

systems are performed after cap layer deposition because of this queued time restriction.

Dual damascene processing is commonly used for copper metallization, which needs two dielectric etch processes. There are at least three different approaches for forming the trench–via structures of dual damascene copper metallization processes. One method etches the trench first, then etches the via, as shown in Fig. 14.17. Another method is the buried hard mask, which etches via holes on the ESL first, then forms both trench and via at the same time using a trench mask, as shown in Fig. 14.18. In Fig. 14.19, the via-first approach is illustrated. Both via-first and trench-first methods are used in copper metallization in IC fabrications.

The ESL helps to define the depth of trench and via holes by providing etch byproducts in the plasma to emit light signals. These signals indicate the endpoint during low-κ dielectric etch processes. PE-TEOS USG and SiCN can be etched with F/O chemistry. For buried-hard-mask dual damascene etching (see Fig. 14.18), high selectivity to SiCN of the low-κ dielectrics is required.

Tantalum (Ta), tantalum nitride (TaN), and the combination of both can be used as a barrier layer for copper to prevent diffusion through the dielectric layer into the silicon substrate, possibly ruining transistors. A copper seed layer is required for

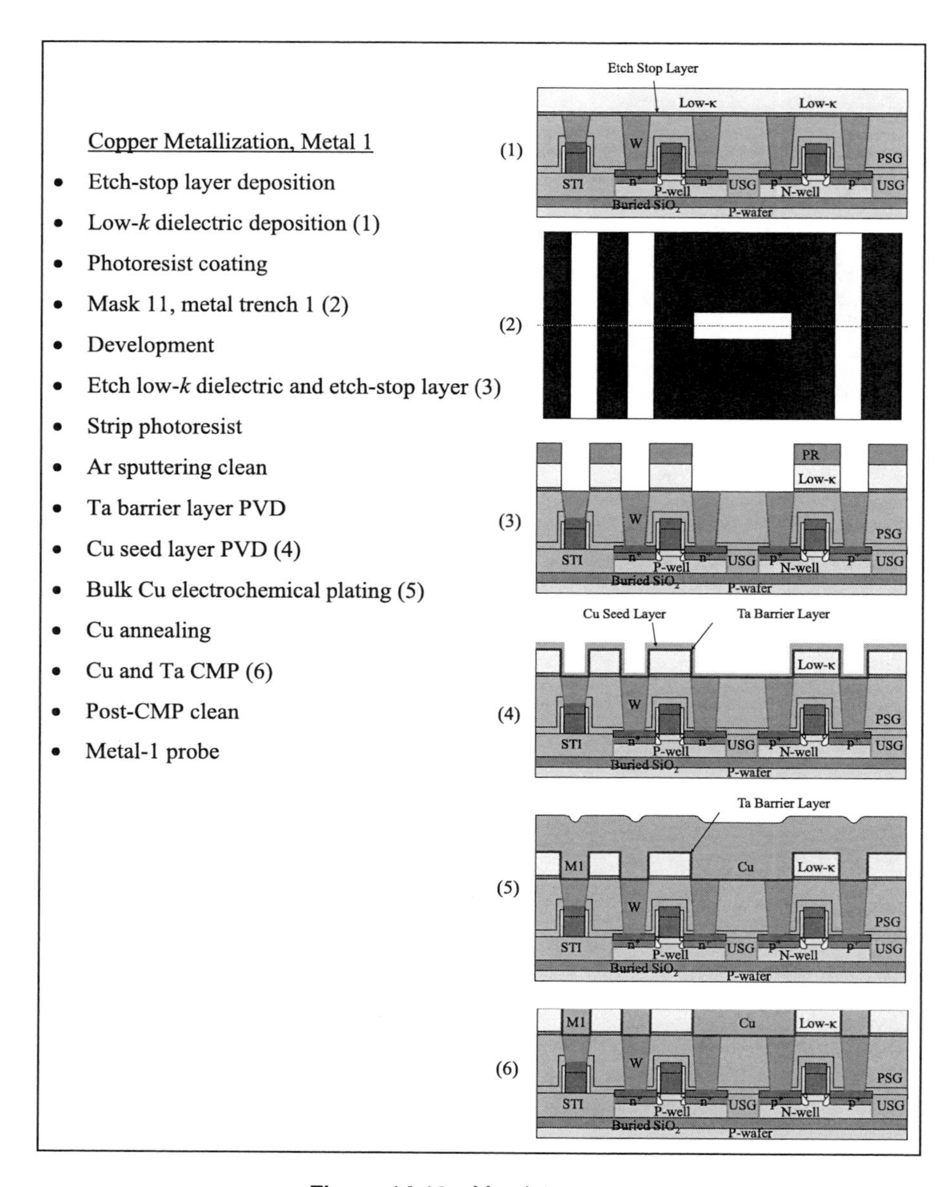

Figure 14.16 Metal-1 process.

bulk copper electrochemical plating (ECP). After the wafer is seeded, it is plated with bulk copper in the ECP process, which fills the narrow trenches and via holes. It is very important to plate the bulk copper soon after copper seed layer deposition because even at room temperature copper can quickly self-anneal. Annealed copper seeds have larger grains and rough surfaces, which can cause voids in the trench and vias during ECP and induce yield loss.

After copper plating, the wafer is annealed in a furnace at about 250 °C to increase grain size and reduce resistivity. Cu and Ta are then removed from the

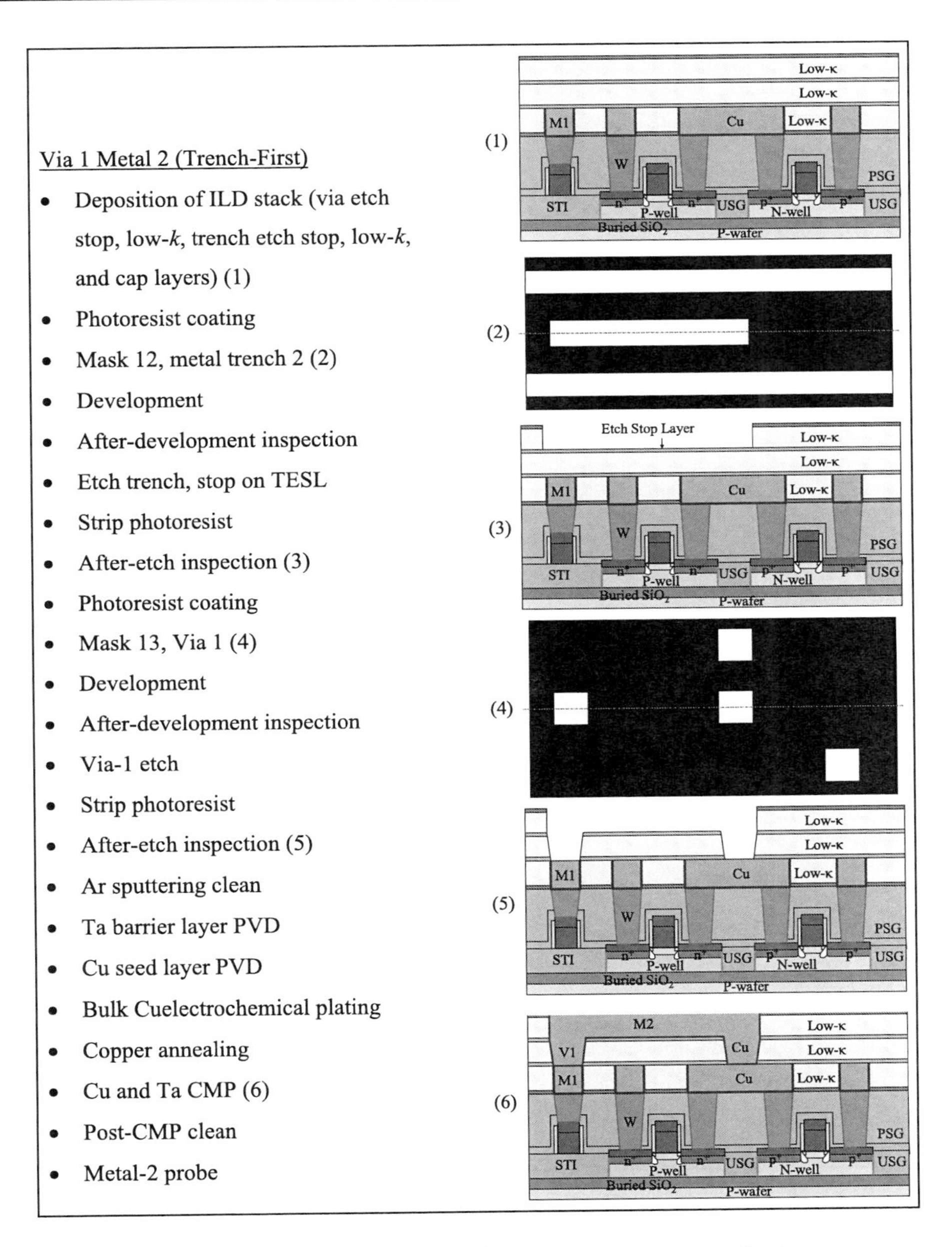

Figure 14.17 Trench-first dual damascene copper metallization.

wafer surface by a CMP process, and only metals in the trenches and via holes are left behind to form the circuit interconnections. Reflectivity of the wafer surface is used to determine the endpoint of the metal CMP because most metals are very shiny with high reflectivity. When metal is polished away from the surface, its reflectivity reduces dramatically, a reduction that indicates the endpoint.

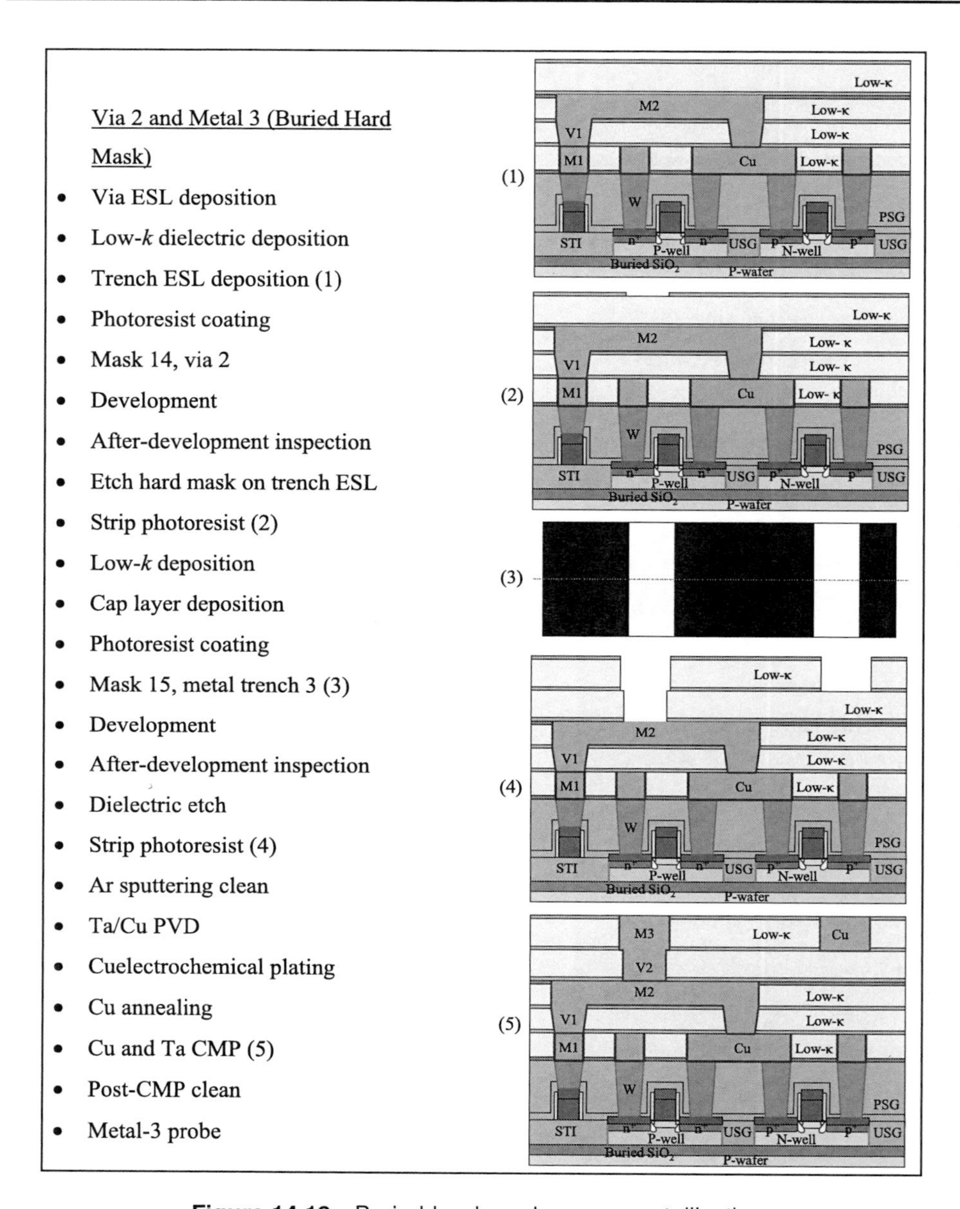

Figure 14.18 Buried-hard-mask copper metallization.

For via-first processes, the ILD of the via ESL, low-κ dielectric, trench ESL, low-κ dielectric, and cap layer are deposited, as shown in Fig. 14.19, illus. (1). The via ESL and trench ESL can be silicon nitride (SiN) or silicon carbon nitride (SiCN). The via mask [Fig. 14.19, illus. (2)] defines the pattern, and via etching stops on the via ESL [Fig. 14.19, illus. (3)]. Before trench patterning, a dummy photoresist layer is coated on the wafer surface to fill via holes and protect via ESL from the trench etching process [Fig. 14.19, illus. (4)]. After trench etching, the photoresist strip process also removes dummy photoresist in the via holes. A wet

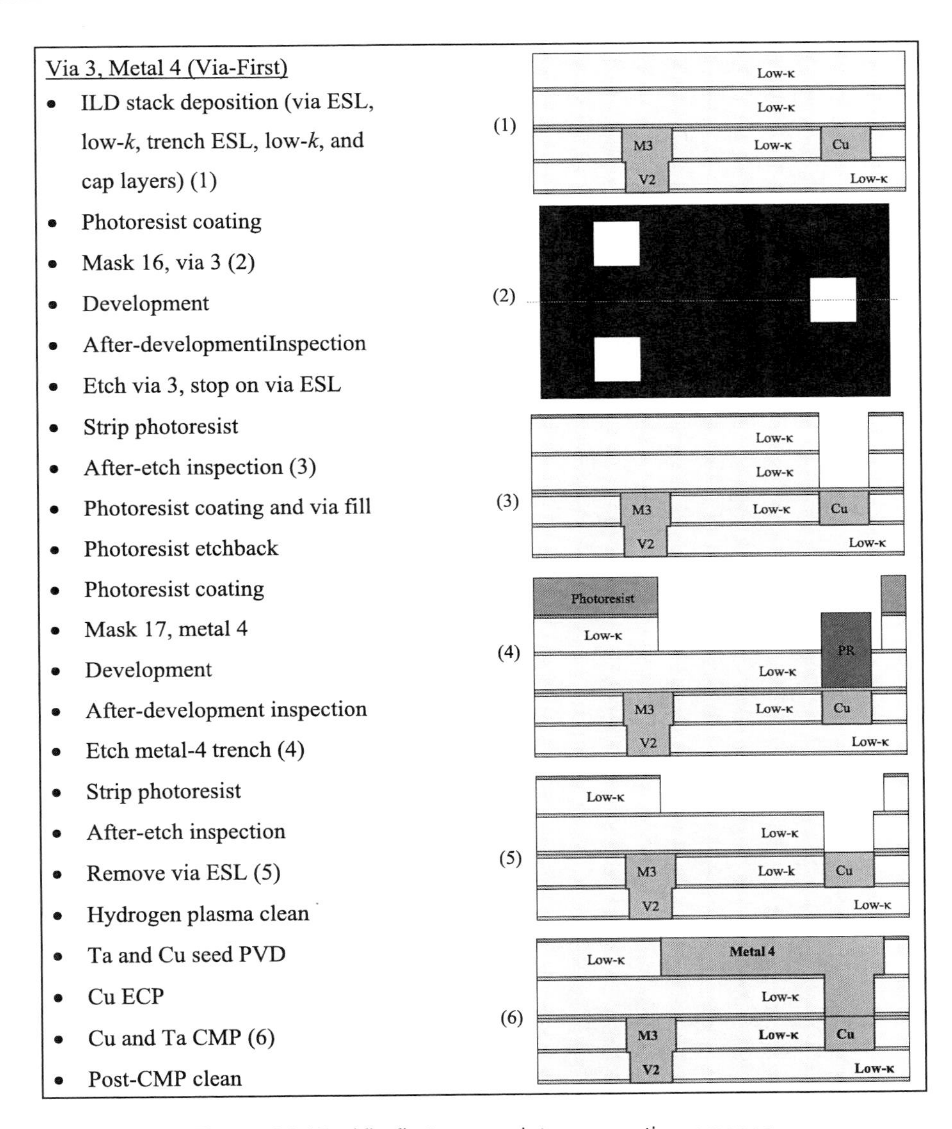

Figure 14.19 Via-first copper interconnection process.

cleaning process removes via ESL at the bottom of via holes [Fig. 14.19, illus. (5)]. After Ta/Cu PVD and Cu ECP, a metal CMP process removes Cu and Ta from the wafer surface and completes the dual damascene copper metallization of this metal layer. Figure 14.20 illustrates via first dual damascene copper metallization process. It is another approach many IC fabs used to form copper interconnection.

Heavy ion bombardment during an argon sputtering etch sometimes can sputter small amounts of copper from the bottom of the hole and deposit it on the sidewall of the via. Copper atoms diffuse very quickly in most dielectrics, such as silicate glass and porous silica. If copper atoms reach the silicon substrate, they can cause unstable performance of the microelectronic devices. Therefore, argon sputtering

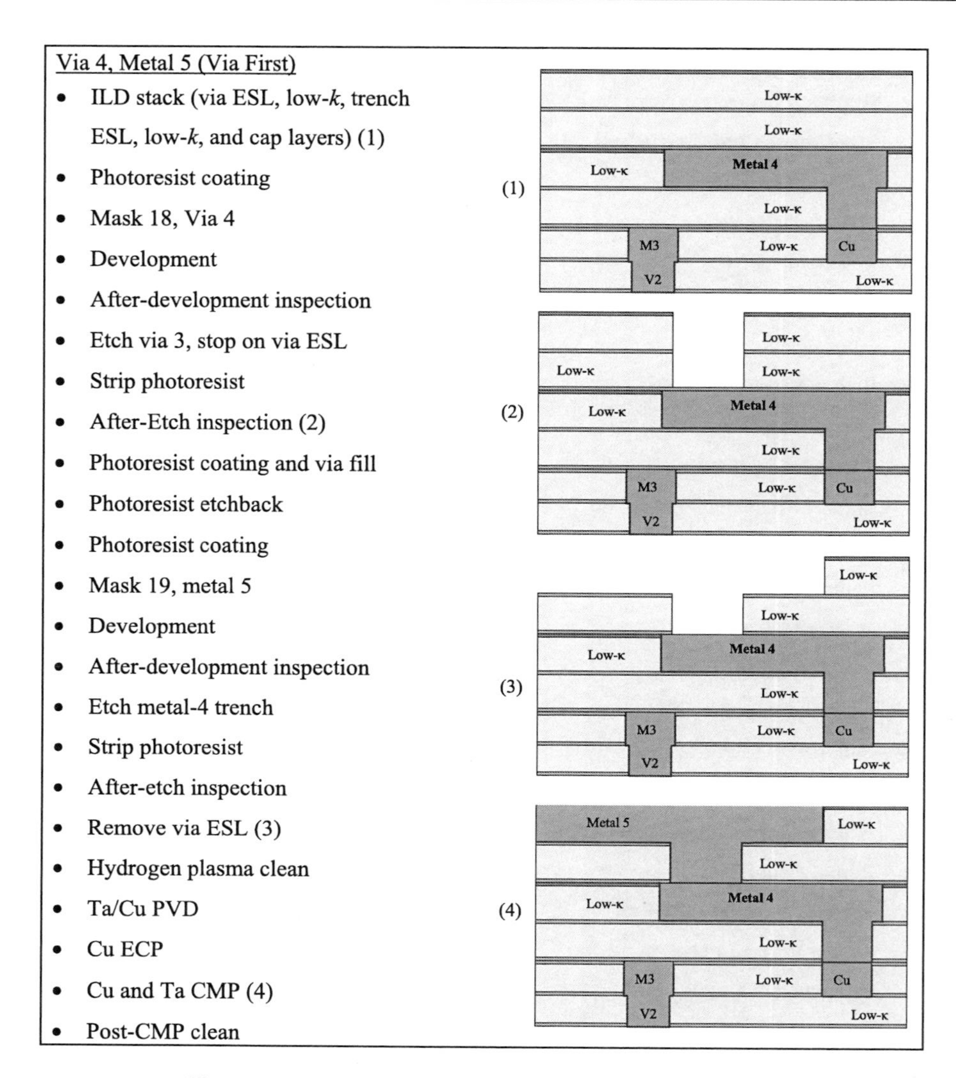

Figure 14.20 Via-first copper interconnection process.

etch prior to metal PVD can cause long-term reliability problems in IC chips. Hydrogen plasma cleaning uses plasma to generate hydrogen free radicals, which react with copper oxide to form copper and water vapor. This effectively removes the native oxide from the copper surface at the bottom of the via without ion-bombardment-induced copper sputtering.

Figure 14.21 shows the final metal layering and passivation processes. Silicon nitride is a very dense material that can prevent moisture and impurities such as sodium from diffusing into the chip and causing device damage. It is commonly used as a passivation dielectric for chips with plastic packaging to protect them from chemical contamination and mechanical damage during die testing, separation, sorting, and packaging. An oxide layer such as PSG is deposited

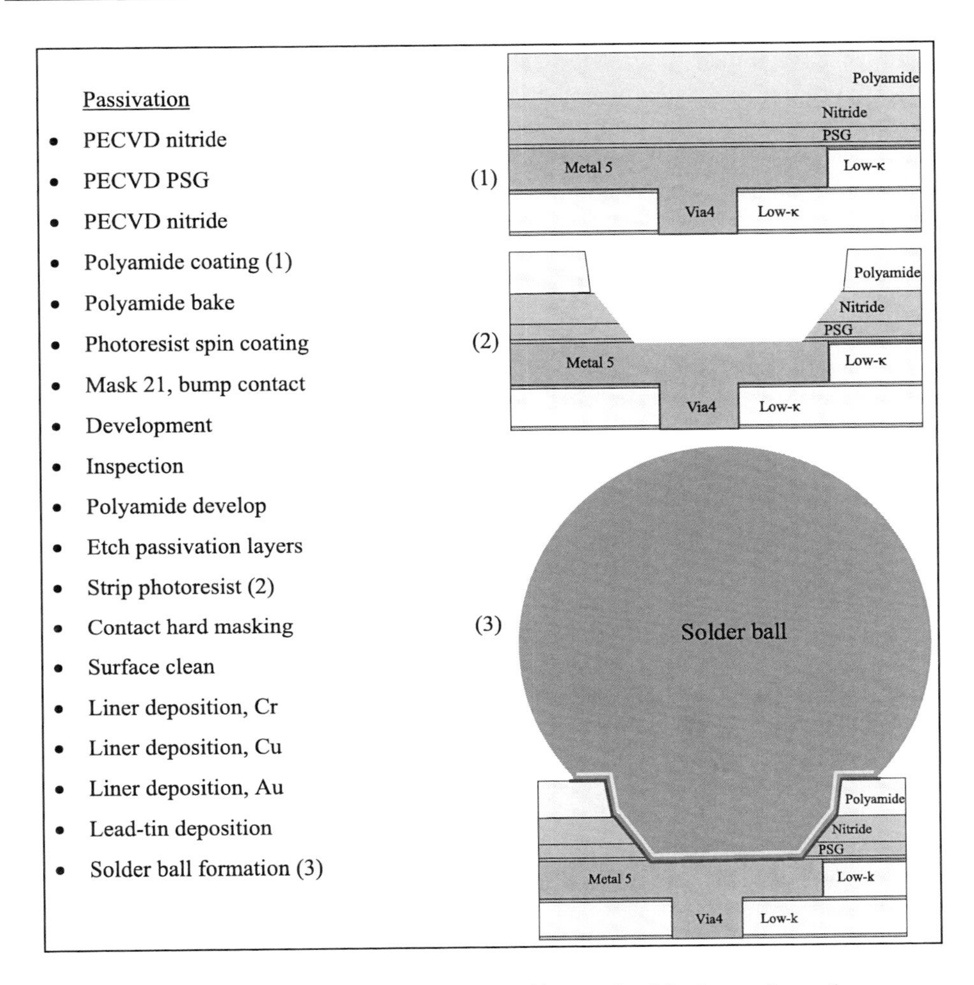

Figure 14.21 Passivation layer deposition and solder bump formation.

before thick passivation nitride deposition to provide a stress buffer. For chips using ceramic packaging (which is a better barrier than plastic packaging but more expensive), a CVD silicon dioxide or silicon oxynitride layer is commonly used as a passivation dielectric. After passivation layer deposition, a layer of polyamide is coated, followed by photoresist coating, baking, and development. Polyamide is etched during photoresist development. A polyamide coating provides protection from mechanical scratches during wafer shipping, and it can also protect microelectronic devices from damage caused by background radiation, such as alpha radiation. After photoresist stripping, the wafer processing is almost finished.

Illustration (3) in Fig. 14.21 shows the bump formation of flip-chip packaging. Bump formation is the last stage of wafer processing, usually performed in a different facility. Because its feature size is very large (about 50 to 100 μm), it does not require a high-grade cleanroom. Chrome, copper, and gold are used as liners to achieve low-contact resistance. Chrome prevents copper, gold, and lead from diffusing into the substrate and causing heavy metal contamination, and gold helps lead–tin alloys stick to the wafer surface. A metal contact mask is mounted on the wafer surface during metal deposition, allowing only metal in the bump openings. Using masked metal deposition helps to avoid photolithography and etch, therefore reducing production cost. Bumps are formed on the wafer surface after the reflow of a lead–tin alloy. The wafer is now ready for die testing, separation, sorting, and packaging.

Copper interconnection technology is relatively new. After intensive research and development, IC chips with copper interconnections came to production in limited quantities for the first time in 1999. Copper interconnections have been applied to IC chips with minimum feature sizes smaller than 0.18 μm. They have been used in logic IC chip fabrication since the 0.13-μm technology node and have also been applied in memory chip manufacturing. Because the dual damascene process simplified the interconnection process by reducing the number of steps, the cost of copper interconnections became lower than that of traditional tungsten/aluminum-copper alloy interconnections. Copper interconnections have become mainstream in advanced IC chip fabrication.

By combining SOI substrate and copper/low-κ interconnections, fabs can manufacture faster, more powerful, and more reliable IC chips with higher noise resistance and lower power consumption. Figure 14.22 shows a cross section of a CMOS IC with SOI substrate and copper/low-κ dielectric interconnections.

14.5 Complementary Metal-Oxide-Semiconductor Process Flow with Technology after 2010

As devices shrank to 45 nm and beyond, such as 40, 32/28 nm, and 22/20 nm, gate oxide thicknesses reached a point where they could no longer be reduced, due to severe gate-to-substrate leakage. High-κ gate dielectrics have been developed to replace the commonly used silicon dioxide (SiO_2) and nitrided oxide (SiON). To further improve device performance, metal gates have been developed to replace polysilicon gates. Strain engineering has been widely used to enhance the mobility of electrons and holes, and improve device speed. Selective epitaxial growth (SEG) of silicon germanium (SiGe) and carbon-doped silicon (SiC) has been implemented in CMOS manufacturing to achieve desirable channel strain. Self-aligned electroless plating of CoWP has been developed to cap copper surfaces after copper CMP to prevent copper diffusion and reduce electromigration, thus improving the reliability of copper interconnections. Metal (TiN) hard masks are used for low-κ dielectric etching.

193-nm immersion lithography and double-pattern processing are used to pattern line-space features, and source-mask optimization (SMO) techniques are employed to pattern contact and via holes. Design teams must work closely with lithography and processing teams to optimize the design (called design for manufacturing, or DFM) to achieve a manufacturable yield.

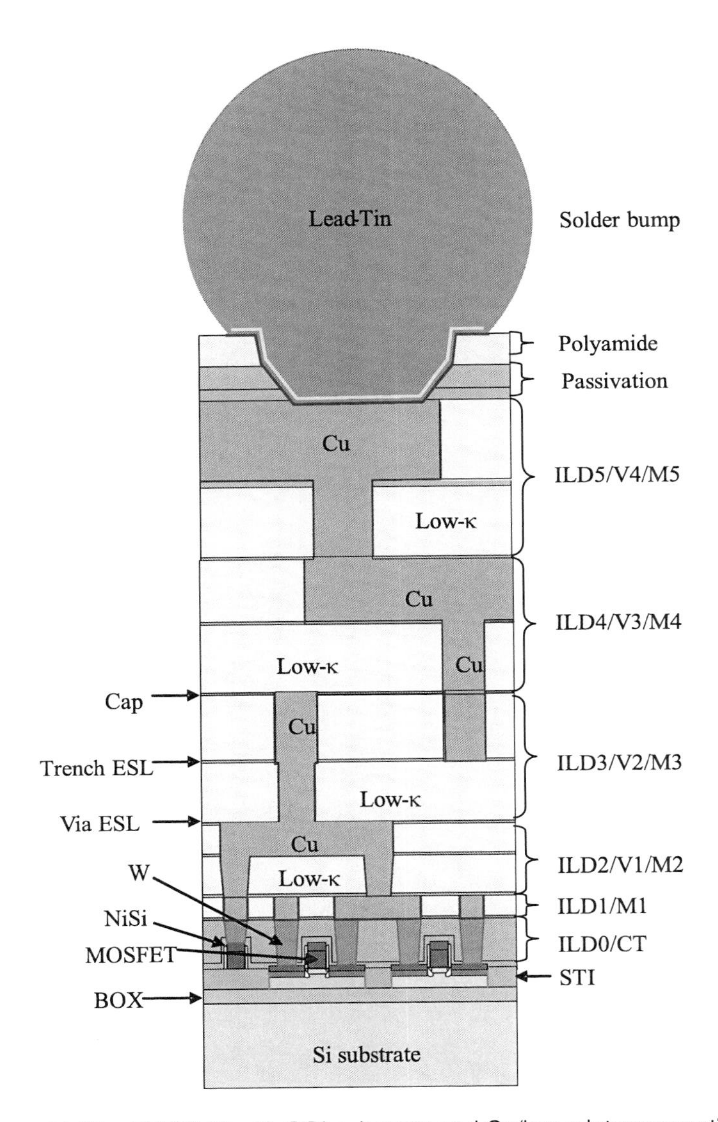

Figure 14.22 CMOS IC with SOI substrate and Cu/low-κ interconnections.

Figure 14.23 illustrates a cross section of a 32/28-nm CMOS with gate-last HKMG, SEG SiGe S/D, stress memory technique (SMT), copper/ultralow-κ interconnections, and lead-free solder bump.

Figure 14.24 illustrates the STI process. It looks almost the same as the STI process illustrated in Fig. 14.8. However, due to the shrinkage of device dimensions, many details are different. For example, the device illustrated in Fig. 14.8 requires stress-free STI fill, while the device in Fig. 14.24 can require extra stress from the STI oxide to help further strain the channel.

Figure 14.25 shows the well and V_T adjustment implantations that form the twin wells required for a CMOS. Due to diminishing device dimensions, the depth of

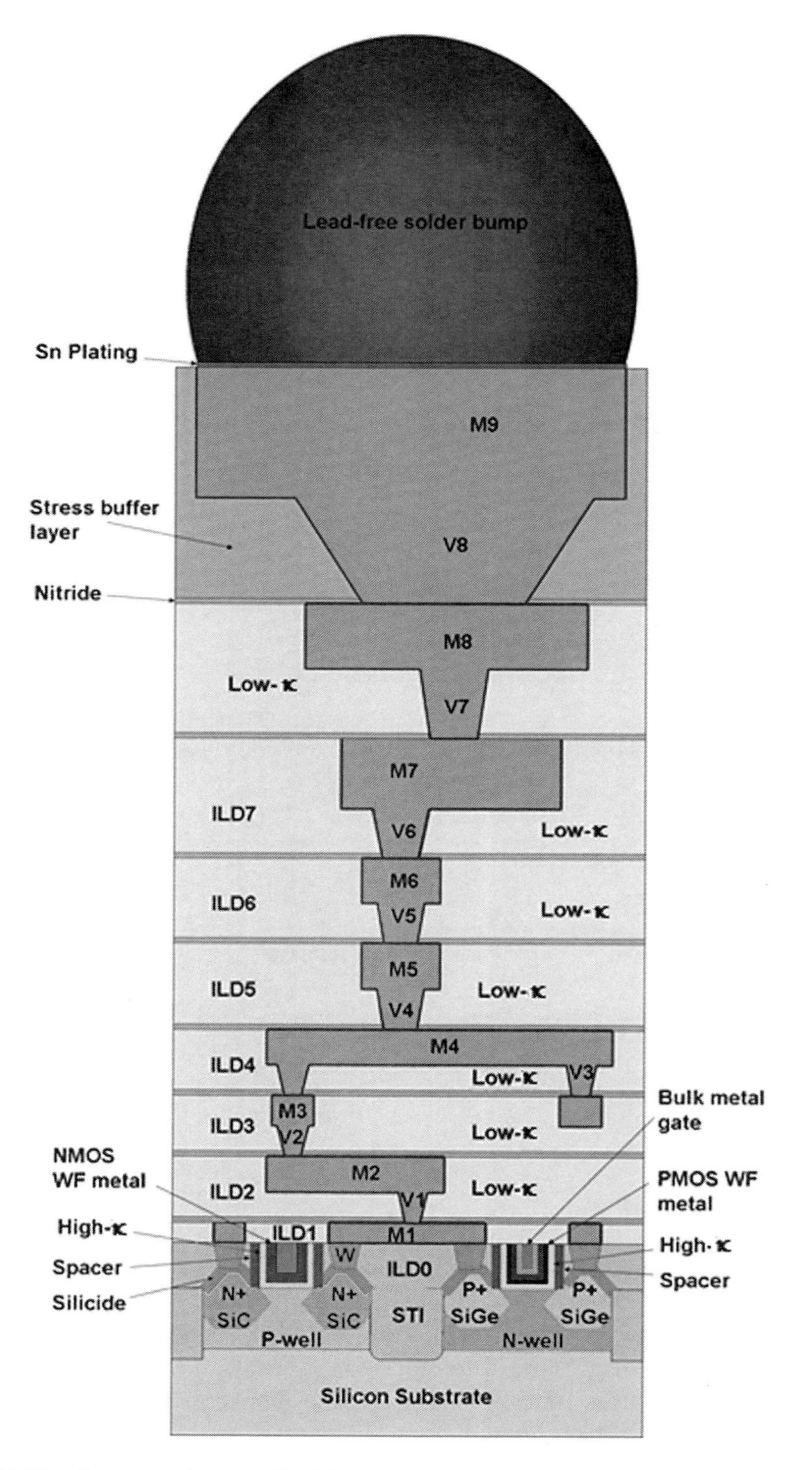

Figure 14.23 Cross section of a CMOS with HKMG, SMT, SEG SiGe S/D, copper/ultralow-κ interconnections, and lead-free solder bump.

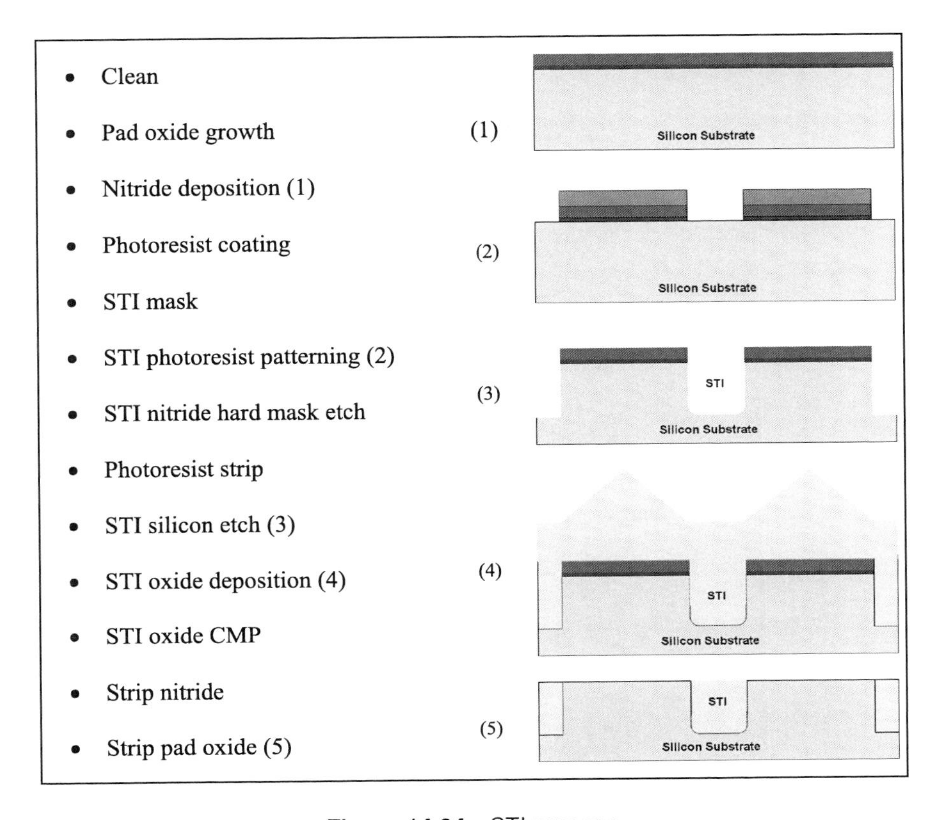

Figure 14.24 STI process.

well junctions becomes shallower compared to the well implantation illustrated in Figure 14.9.

Figure 14.26 illustrates the formation of a gate, nMOS SDE, and sidewall spacer. Because the pMOS S/D will be formed by the SEG of SiGe (which is heavily p-type doped during the SEG process), pMOS does not need SDE or SD implantations. Different from previous processes, this polysilicon gate is just a dummy gate, which will be replaced by high-κ and metal gates later on. It does not require poly-dope implantation, and the gate work function is controlled by different metal materials of the pMOS and nMOS gates.

The major differences of this advanced process are shown in Fig. 14.27. The block layer deposited in Fig. 14.27, illus. (1) is needed so that the SEG can be grown in the designated area. Figure 14.27, illus. (3) shows a KOH etch of silicon, which is highly selective to the ⟨111⟩ crystal orientation of silicon. By using a KOH etch, the silicon etch profile can be precisely controlled, and optimal device performance can be achieved due to the enhanced channel strain. pMOS S/D doping is accomplished with in-situ doping during the SEG process [Fig. 14.27, illus. (4)]. nMOS S/D doping still uses traditional low-energy, high-current ion implantation and thermal annealing [Fig. 14.27, illus. (5) and (6)]. By performing deep amorphous ion implantation and spike annealing on the nMOS

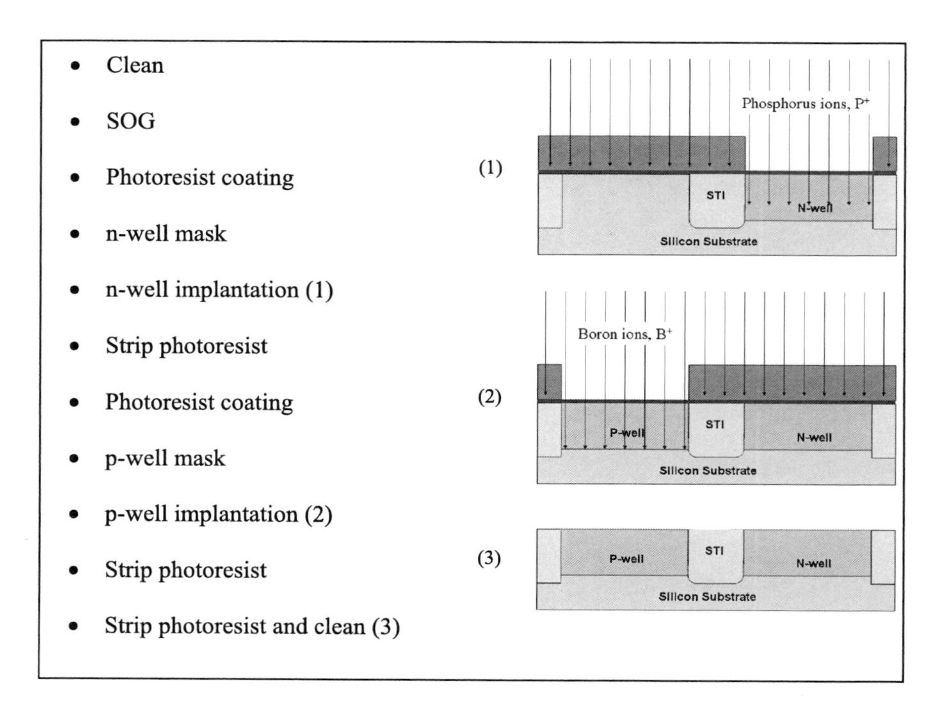

Figure 14.25 Twin-well formation.

S/D, a strong tensile strain on the nMOS channel can be created, increasing electron mobility and improving nMOS speed. Because tensile stress will continue after S/D formation, this technique is a SMT. Due to the limited thermal budget of this tiny device, annealing must be performed with spike or laser methods, or the combination of both.

Figure 14.28 shows a HKMG process with gate-last integration. Figure 14.28, illus. (1) shows ILD0 deposition, which has at least two layers—nitride-based liner/ESL (not shown in the figures) and an oxide layer. Figure 14.28, illus. (2) shows the poly-open CMP, which basically is overpolished ILD0 that exposes the dummy poly gate. Figure 14.28, illus. (3) illustrates the dummy gate removal process, which uses a highly selective etch process to remove polysilicon while having very little effect on ILD0 and sidewall spacers. After HF dip removes the oxide above the channel, a thin layer of silicon oxide is formed, followed by hafnium-based high-κ dielectric deposition [Fig. 14.28, illus. (4)]. HfO_2 is deposited by an atomic layer deposition (ALD) process. Titanium nitride (TiN) and tantalum (Ta) layers are deposited, also with ALD processes. TiN is commonly used for pMOS work-function metal, while nMOS requires a different work-function metal, which is formed in later process. Ta is the barrier layer that protects TiN on the pMOS. Photolithography and etching processes are needed to remove Ta from the nMOS region [Fig. 14.28, illus. (5)]. Titanium–aluminum (TiAl) alloy is deposited, and bulk aluminum (Al) is used to fill the gap. In an annealing process, TiAl reacts with TiN to form TiAlN, which serves as the nMOS work-function

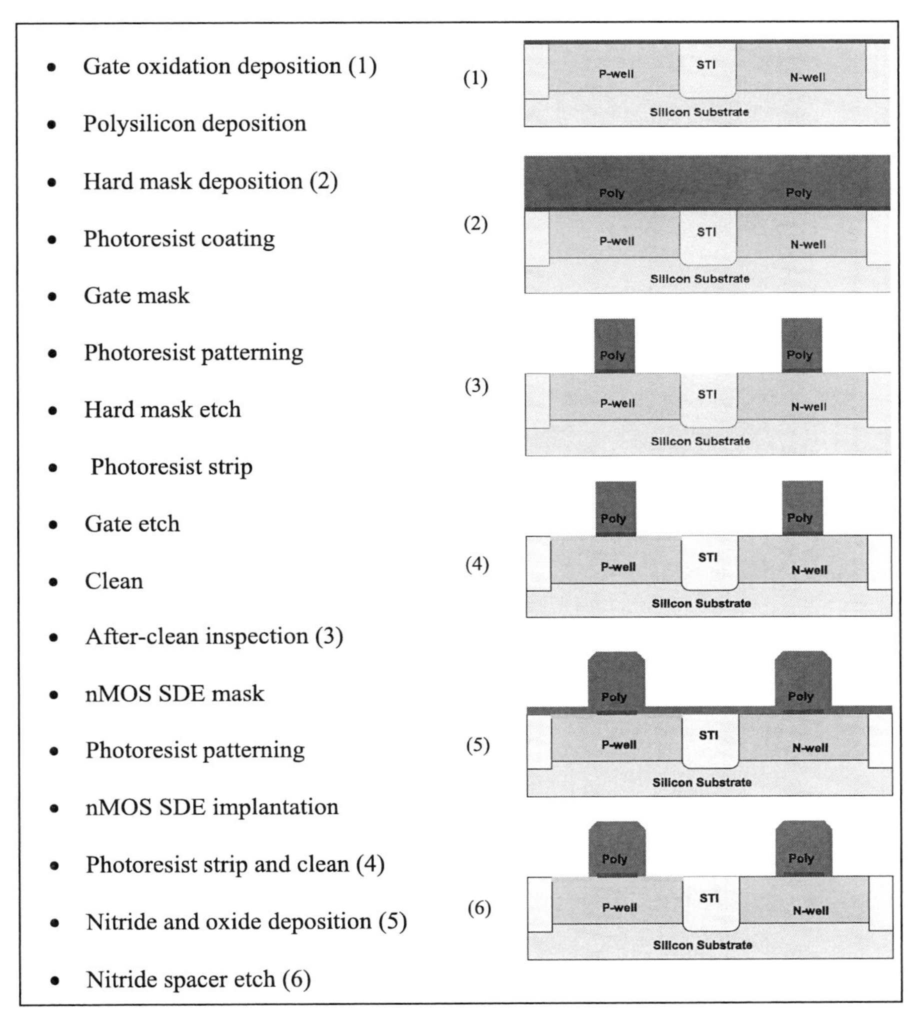

Figure 14.26 Dummy gate patterning, nMOS SDE formation, and sidewall spacer formation.

metal. The metal CMP process shown in Fig. 14.28, illus. (6) finishes the gate-last HKMG process.

While removing the dummy polysilicon gates, both pMOS and nMOS channel strains increase significantly, giving them higher carrier mobility and higher device speed. This is similar to removing the hard stop of a spring, allowing the spring to be further stressed by either compressive or tensile force. This is one advantage of gate-last HKMG integration over the gate-first approach.

Figure 14.29 shows a middle end of the line (MEoL) process. There are several differences in this contact module. One of them is the trench silicide, which is formed after trench contact etching and is not the normal salicide that was formed after S/D formation (before ILD0 deposition). The trench silicide is only formed

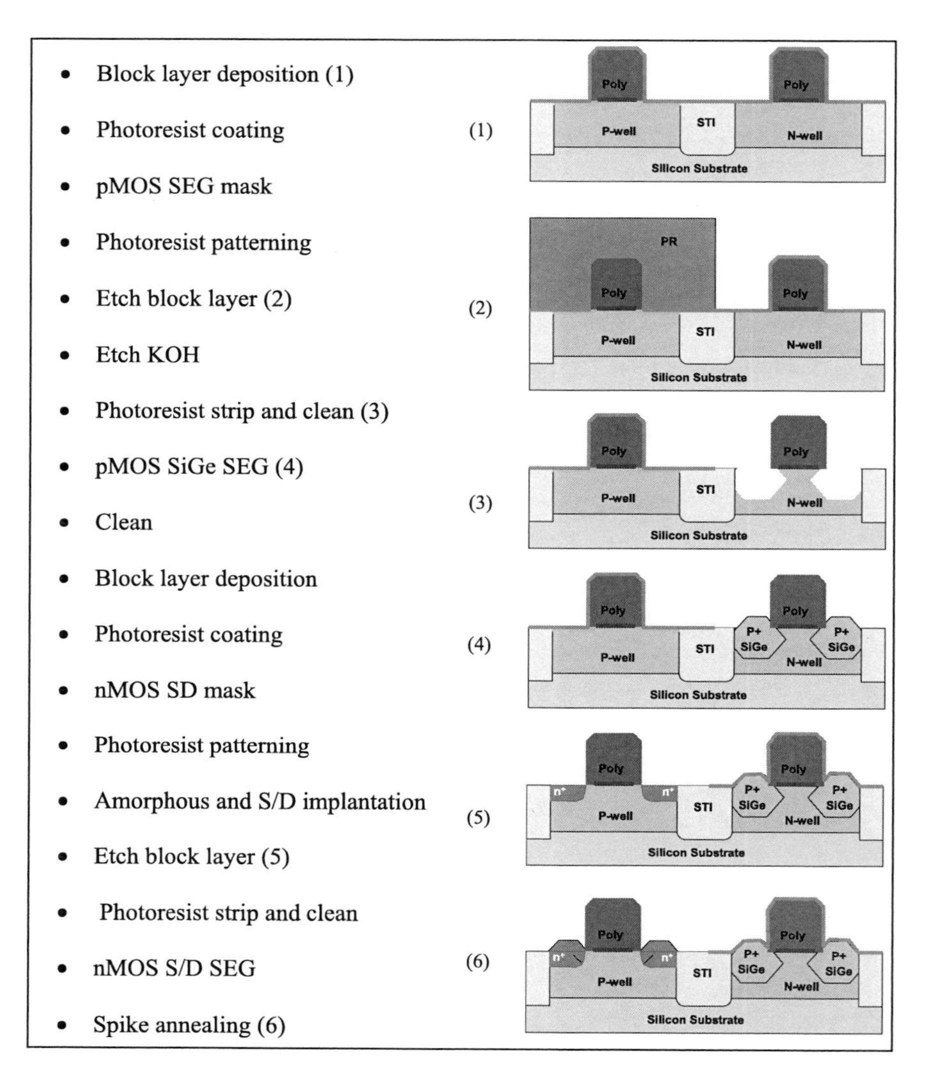

Figure 14.27 SEG processes that form the S/D.

at the bottom of a contact trench [Fig. 14.29, illus. (3)]. Because the gate is formed with metal, it does not need silicide. A small amount of plutonium (Pt) is alloyed with nickel; its silicide, NiPtSi, is more stable than NiSi. Another difference is that the tungsten (used to fill the contact trenches) is polished to the same level as the metal gate [Fig. 14.29, illus. (5)]. Because the contact trenches only connect to the raised S/Ds, there is only one depth of contact trench, simplifying overetch control. From a layout point of view, the trench style contact replaces the round and oval contact holes, simplifying the photoresist patterning process. However, this replacement can cause tungsten spiking due to overetching into the STI oxide during contact etch. Because the length of the tungsten plug is significantly shorter, resistance of the plug is also reduced.

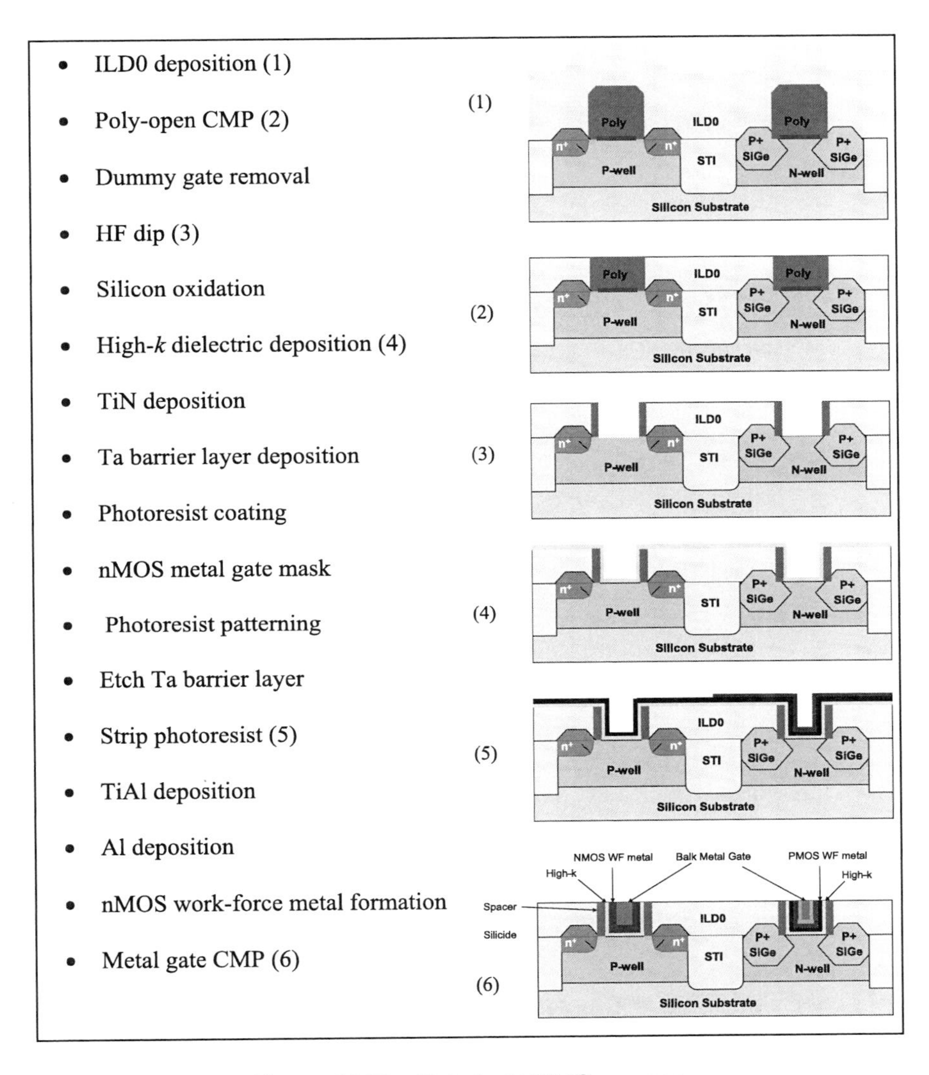

Figure 14.28 Gate-last HKMG process.

The metal-1 (M1) process is illustrated in Fig. 14.30. A hard mask (usually TiN on a TEOS cap) is used to protect the porous low-κ dielectric from damage caused by the photoresist stripping process. The porous low-κ dielectric has a lower κ-value (~2.2 to 2.5) than the κ-value of carbon-doped oxide or carbonated silicate glass (CSG, κ ~ 2.7 to 2.9). A porous low-κ dielectric can be formed by PECVD carbon-doped silicon oxide film with <2-nm pores and up to 40% porosities introduced by adding porogens into the gas flow. CVD precursors can be trimethylsilanes or tetramethylsilanes, and porogen species can be norbornene or α-terpinene.

Ta barrier and copper seed layers are deposited in the PVD process with an ionized metal plasma process. Bulk copper is deposited with the ECP process.

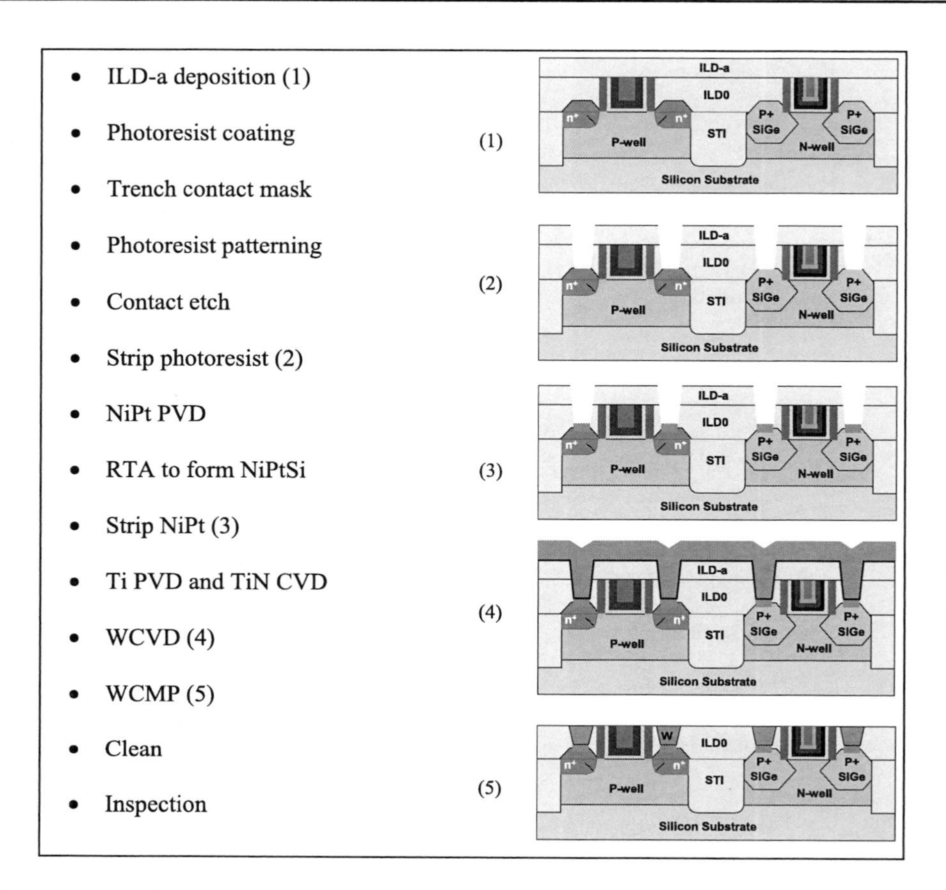

Figure 14.29 MEoL (contact module) processes: (1) sacrificial ILD-a deposition, (2) trench contact etch, (3) trench silicide formation, (4) WCVD, and (5) WCMP.

After copper annealing, the metal CMP process removes the bulk Cu layer, Ta barrier layer, and TiN hard mask. The CMP stops in the TEOS cap layer, which protects the porous low-κ dielectric from contact and damage by CMP slurry.

A dual damascene copper/low-κ dielectric interconnection process with trench-first integration is illustrated in Fig. 14.31.A metal hard mask (usually TiN) and TEOS PECVD oxide, or TEOS, cap layer are used to protect the porous low-κ dielectric from damage from photoresist stripping and metal CMP processes. The self-aligned electroless plating of CoWP can be used to prevent copper diffusion and to enhance EMR, improving reliability of the IC chip.

Figure 14.32 shows the copper/low-κ dielectric interconnections of the metal layers from M3 to M9. They are basically multiple repetitions of the trench-first process described in detail in Fig. 14.31.

Lead (Pb) was widely used for solder bumps. It is well known that lead is a contaminant that can affect the development of the heart, bones, intestines, kidneys, and nervous system, and as such it is particularly toxic to children. Every year several tons of used IC chips from outdated electronics go into landfills

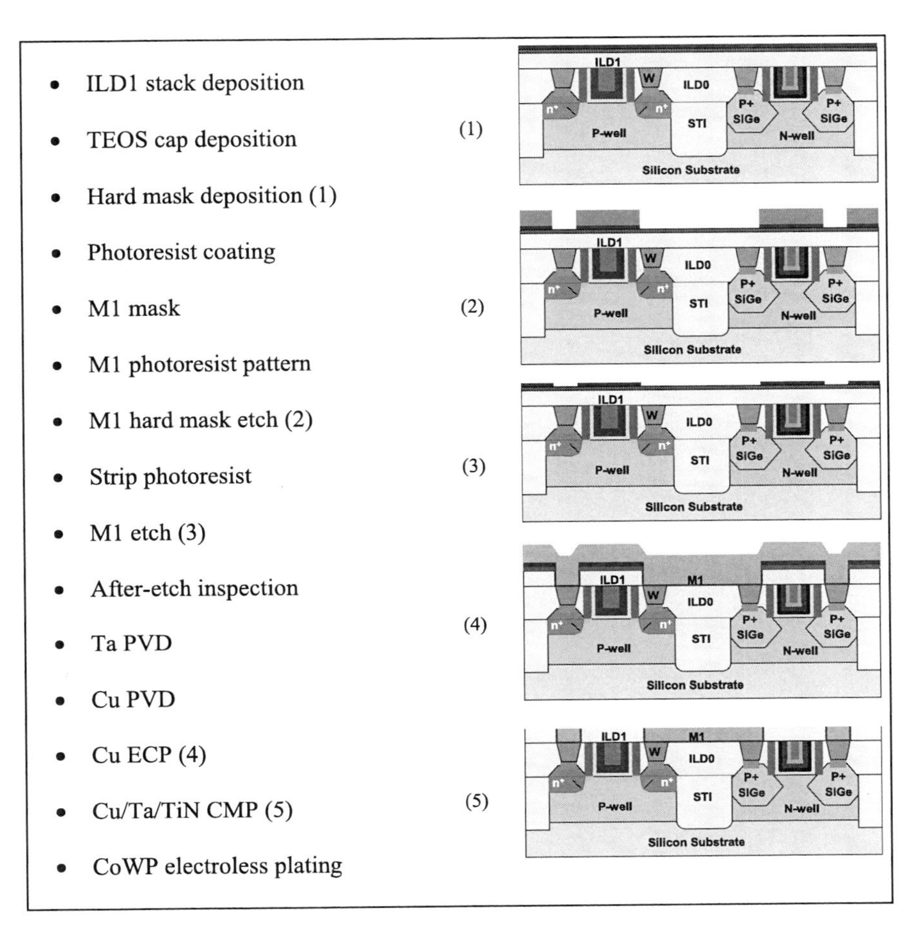

Figure 14.30 M1 process.

and create potential risks for lead contamination in the environment. Therefore, many countries such as Japan, Europe, and China have legislation or regulations restricting or eliminating the use of lead in the semiconductor industry and all electronics industries. Figure 14.33 shows a lead-free solder bump.

14.6 Memory Chip Manufacturing Processes

Memory chips play important roles in driving IC technology and its markets. Two major memory products on the market are DRAM and NAND flash. In a computer, whether it is a desk top PC, laptop, or tablet, generated data are always stored in DRAM first before being written in a nonvolatile memory (NMV) device such as a hard disk drive (HDD) or a solid-state drive (SSD) for permanent storage. The amount of memory bits in a desktop or laptop PC has multiplied many times in the last few years. In 1993, a typical PC, say a desktop 486, had 4-Mb DRAM, which could be upgraded to 8 Mb for \$100. It cost about \$30 to purchase 4-Gb DRAM for a desktop PC in 2009. With increasing demands

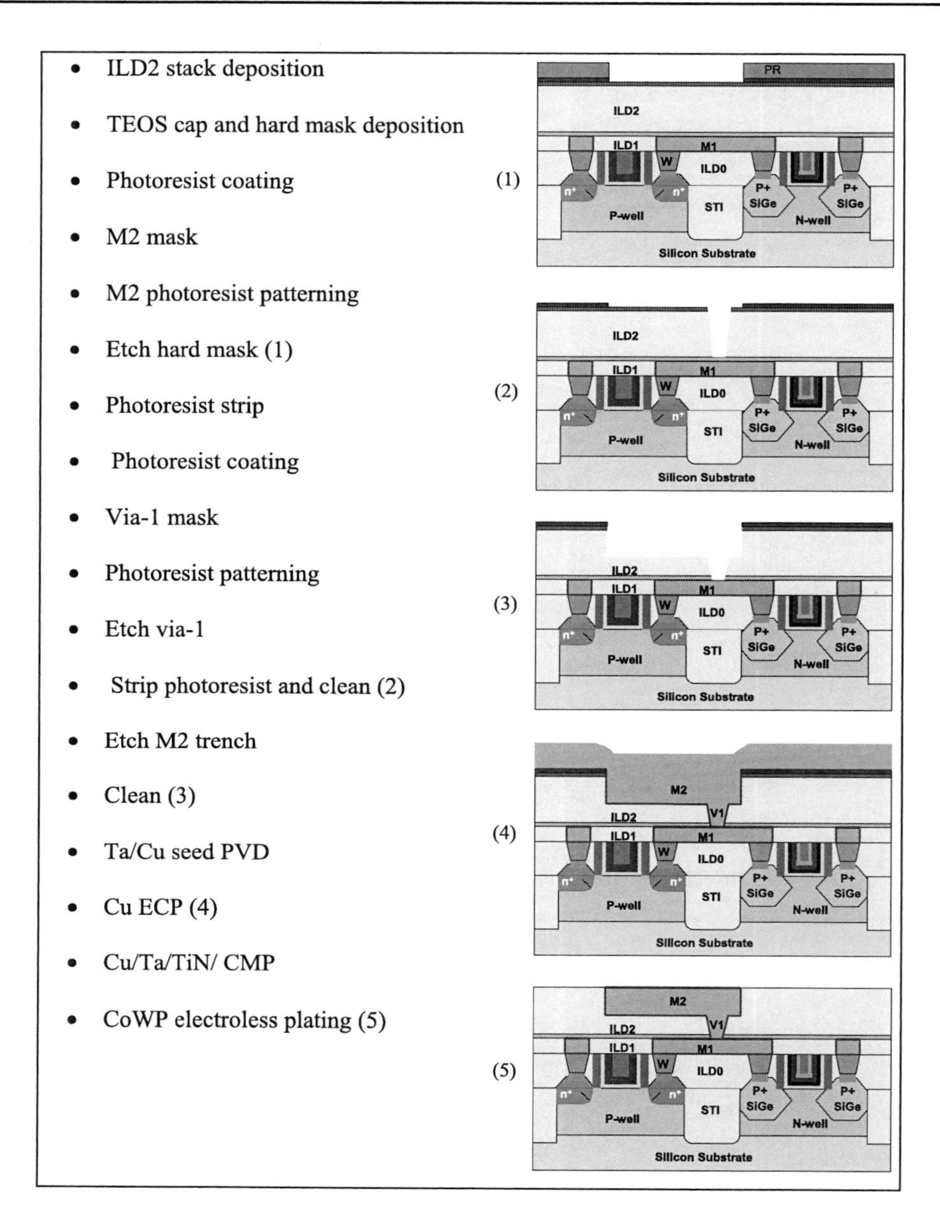

Figure 14.31 Trench-first copper/low-κ dielectric interconnection of via 1 metal 2 (V1M2).

for more graphics, especially 3D graphics, the required amount of DRAM for a computer will further increase and will continue to drive DRAM manufacturing technologies.

Different from DRAM (which requires a constant power supply to keep the stored data), a NAND flash is a nonvolatile memory that can keep stored data without a power supply for years. NAND flash is widely used for data storage in mobile digital electronic products, such as MP3 players, digital cameras, cell

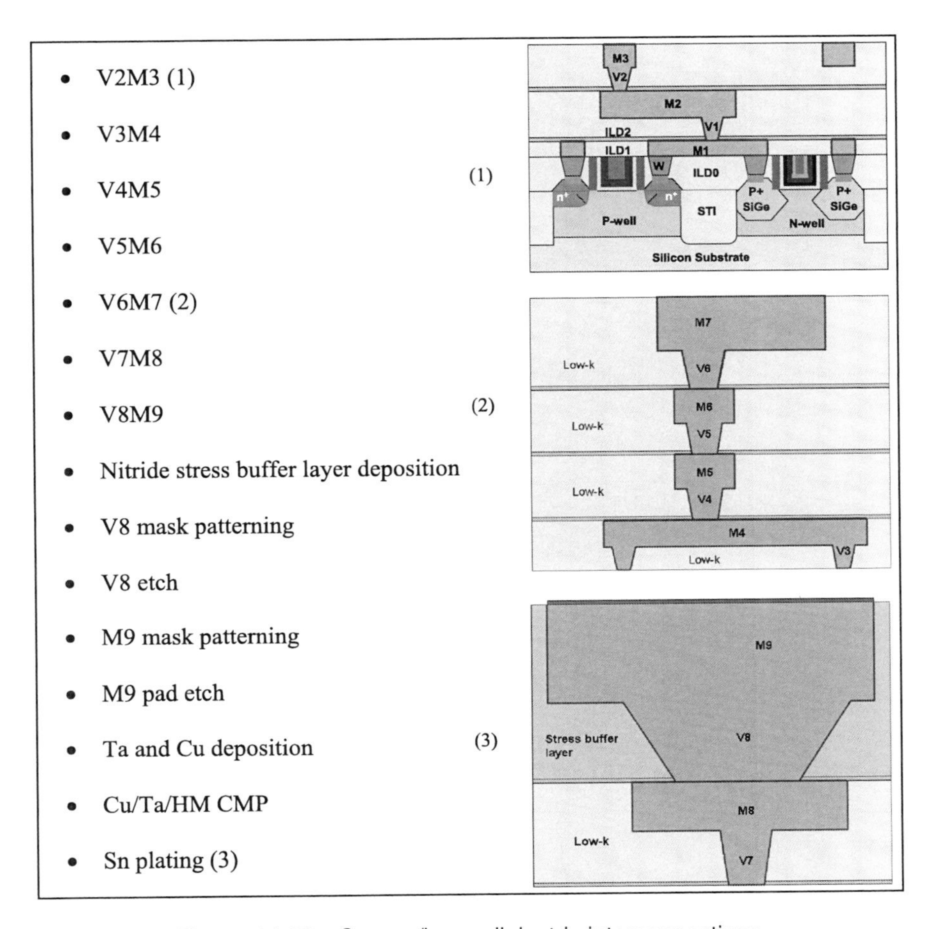

Figure 14.32 Copper/low-κ dielectric interconnections.

phones, and high-end laptop PCs. With more graphic and video applications in mobile electronic devices, the demands for NAND flash memory will further increase. NAND flash is also used in hybrid disk drives to combine the benefits of fast data access of a SSD and the low cost of a HDD.

14.6.1 Dynamic random access memory process flow

DRAM played a very important role in driving IC markets and technologies. A DRAM cell consists of an nMOS and a storage capacitor, as shown in Fig. 14.34.

There are two types of DRAM processes: one is a stacked DRAM, which stacks storage capacitors on top of select transistors (nMOS); the other is deep trench DRAM, which builds storage capacitors in a deep trench on the silicon surface alongside the nMOS. Stacked DRAM is similar to creating storage by building an attic, and deep trench DRAM is similar to creating storage by building a cellar.

Figure 14.35(a) illustrates a stacked DRAM and includes the self-aligned contact (SAC), bitline contact (BLC), wordline (WL), bitline (BL), storage node (SN, which is the storage capacitor), and storage node contact (SNC).

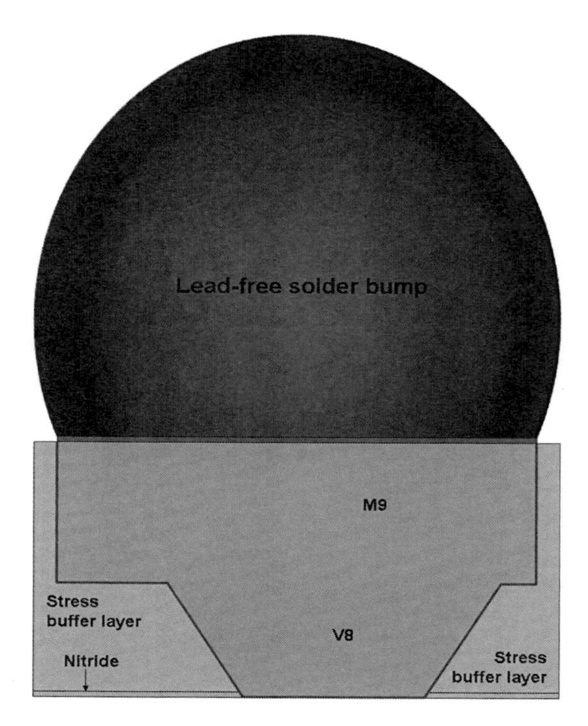

Figure 14.33 Lead-free solder bump formation.

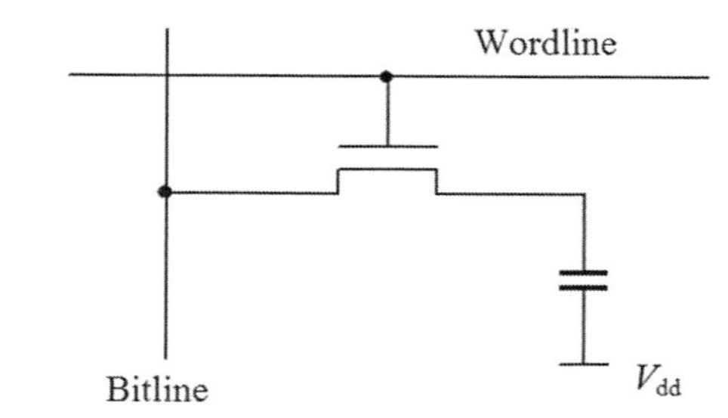

Figure 14.34 Circuit of a DRAM memory cell.

Figure 14.35(b) illustrates a deep trench DRAM. Because the trench capacitor is very deep with an aspect ratio more than 50, the depth of the drawing was reduced to fit the figure onto the page. Deep trench DRAM has fewer requirements for the metal interconnections above the silicon surface, making it more compatible with normal CMOS backend processes, and a better candidate for embedded DRAM for system-on-chip (SOC) applications. However, because it needs precious silicon real estate to build its storage capacitors, its packing density is not as scalable as the stacked DRAM, which does not require any of the silicon surface to build its storage capacitors. General-purpose DRAM chips are commodity products and thus are very sensitive to cost. Because stacked DRAM has a small advantage

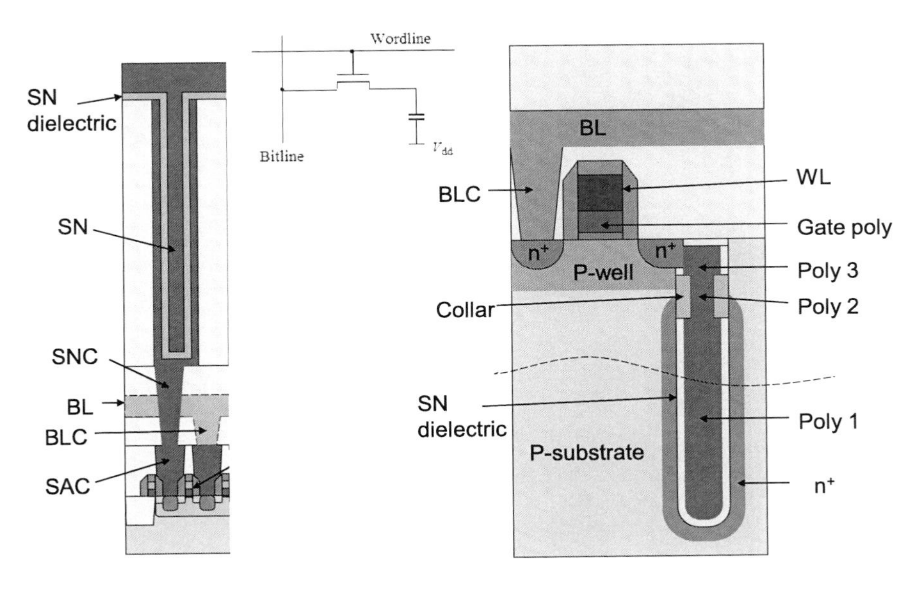

Figure 14.35 (a) Stacked DRAM and (b) deep trench DRAM.

in cost over deep trench DRAM, it dominates the DRAM market. Only stacked DRAM processes are described in this chapter.

14.6.2 Stacked dynamic random access memory process flow

Most DRAM chips in computers and other digital electronic devices are stacked DRAM chips. Figure 14.36 shows a cross section of a stacked DRAM chip. The left side of Fig. 14.36 shows a cross section of an array area with four unit memory cells. A 2-Gb DRAM made with 30-nm technology has two billion such unit cells. The periphery circuits are logic devices to control write, read, and input/out operations of the DRAM chip. Devices in the periphery usually are larger than those in the cell array area. Processes of the periphery devices are very similar to the CMOS processing technologies described earlier. This section focuses on the process flow in the cell area.

Figure 14.37 illustrates STI and well formations of a stacked DRAM cell. Figure 14.37(a) shows the layout of the AA, and the dashed line indicates the location of the cross section. Figure 14.37(b) illustrates the cross section after AA etching, Fig. 14.37(c) shows the cross section after STI formation, and Fig. 14.37(d) shows the cross section after p-well formation. The STI and p-well formation processes are simultaneously performed in the periphery area. It is important to note that there is a p-well masking step for p-well formation. It has finer patterns in the periphery area, and it is blank in the cell area. n-well formation is only performed in the periphery area, not in the cell area, because there is only nMOS in a DRAM cell.

Figure 14.38 illustrates WL, LDD, sidewall spacer, and S/D formations of a stacked DRAM cell. Figure 14.38(a) shows the layout of the WL that overlaps

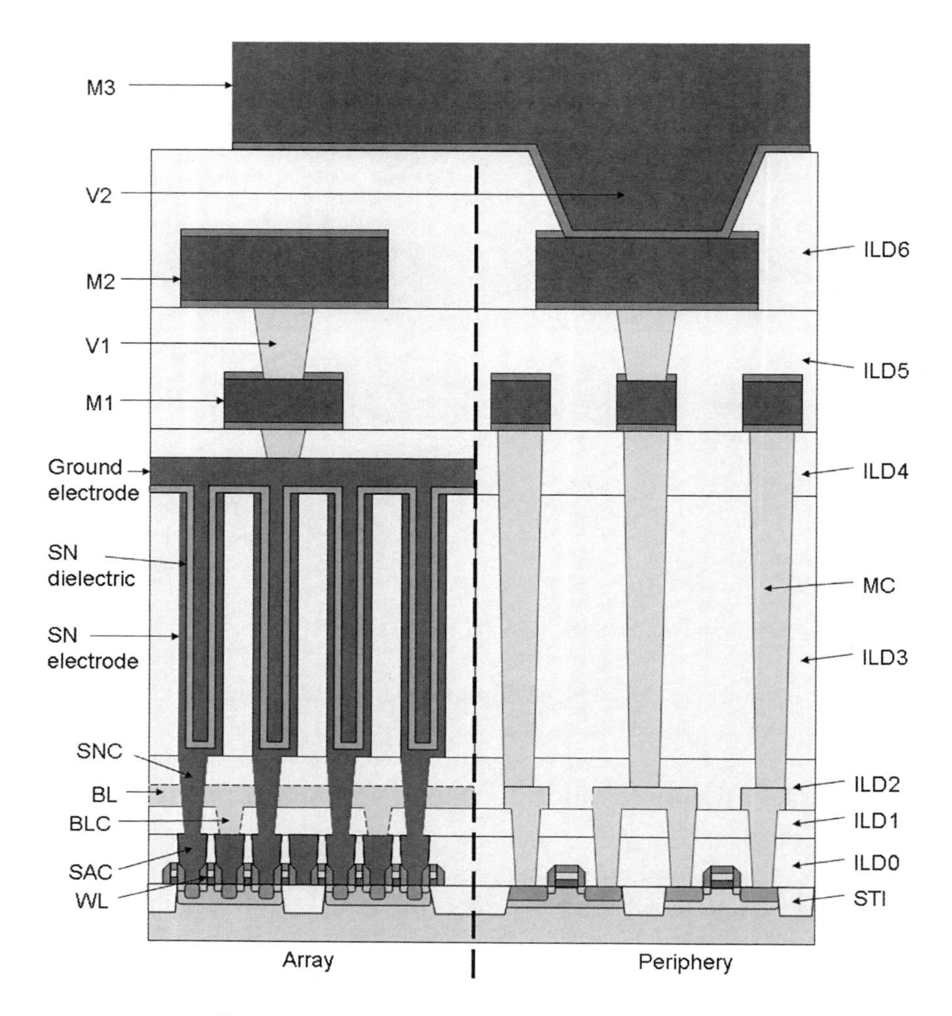

Figure 14.36 Cross section of a stacked DRAM.

with the AA layer, and the dashed line indicates the location of the cross section. Figure 14.38(b) shows the cross-sectional view of the formation of the DRAM cell nMOS gate, which is also the WL. Figure 14.38(c) shows the lightly doped drain (LDD) formation, Fig. 14.38(d) illustrates the sidewall spacer formation, and Fig. 14.38(e) shows S/D formation. There are two masking steps not shown in Fig. 14.38, the pMOS LDD and pMOS S/D formations in the periphery area. These are shown on the right side of Fig. 14.39. Cobalt silicide is also formed in periphery devices to reduce contact resistance.

Figure 14.40(a) shows the first contact, the SAC, of a stacked DRAM. This module is also called the landing pad contact (LPC) after etch and landing pad poly (LPP) after the polysilicon CMP. Both unopened contact holes and shorts to the WL are killer defects, and it is very challenging to etch contact holes through ILD0, which is BPSG in between dense WLs that must reach every S/D of the cell

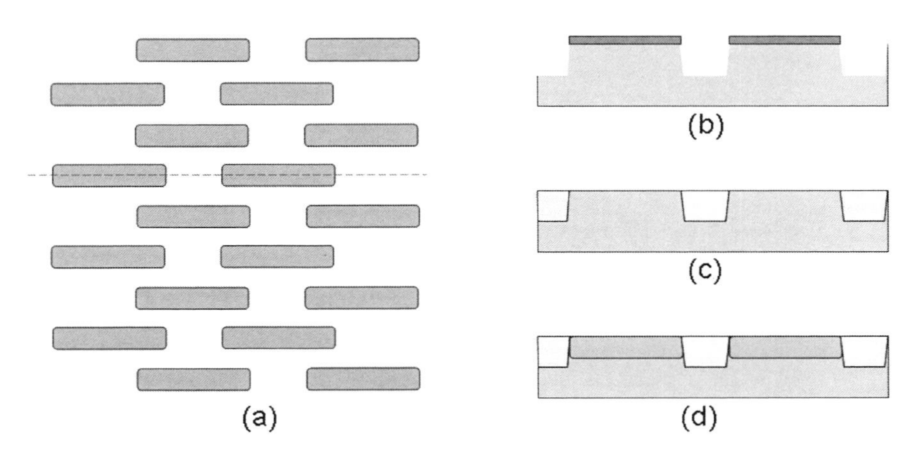

Figure 14.37 (a) Stacked DRAM AA layer layout, (b) cross section along the dashed line after AA etch, (c) after STI formation, and (d) after p-well formation.

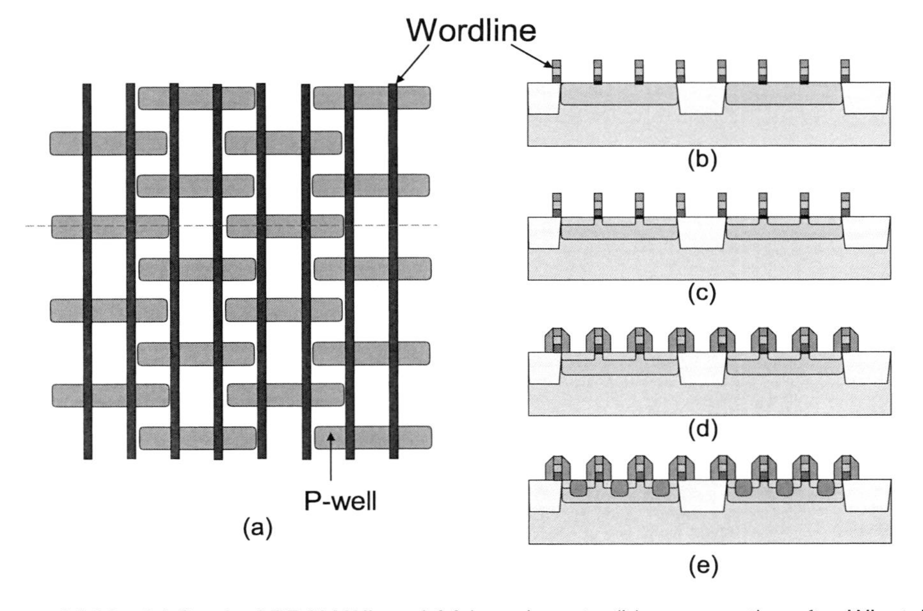

Figure 14.38 (a) Stacked DRAM WL and AA layer layouts, (b) cross section after WL etch, (c) after LDD formation, (d) after spacer formation, and (e) after S/D formation.

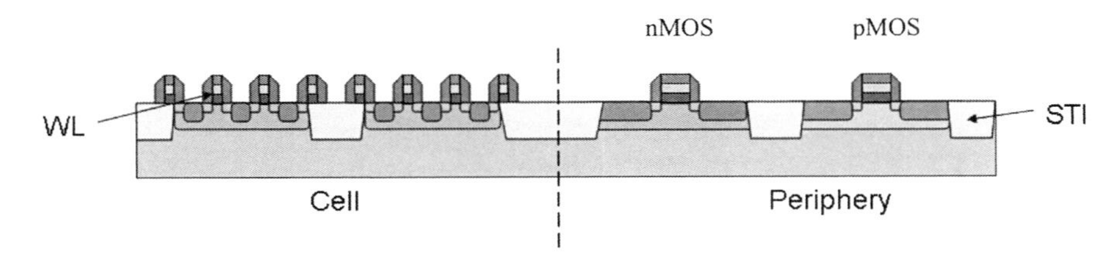

Figure 14.39 Cross section of DRAM cell and periphery devices.

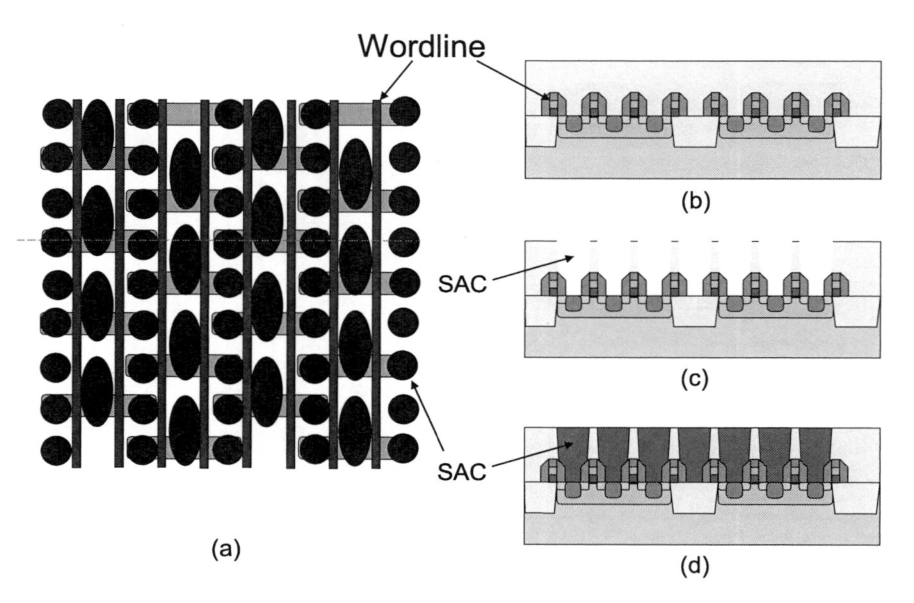

Figure 14.40 (a) Stacked DRAM SAC layout overlapped with WL and AA layers, (b) cross section after ILD0 CMP, (c) after SAC etch, and (d) after SAC poly CMP.

nMOS. Therefore, self-aligned contact processes were developed. By maintaining a nitride hard mask on top of the WLs, and with nitride spacers on both sides, WLs are surrounded with nitride. When the SAC etching process has high enough selectivity between BPSG and nitride, the etching process becomes self-aligned, allowing the contact holes to go through the dense WLs and reach the silicon surface without electrically shorting them.

Before polysilicon is deposited to fill the SAC holes, contact ion implantation (which is usually high-dose n-type ion implantation) is performed to reduce contact resistance. E-beam inspections are applied in this module, either after etch or after polysilicon CMP, to ensure the capture of defects such as unopened contacts and contact plug-to-WL shorts at the earliest stages. The SAC process only happens in the array area.

Figure 14.41 shows the BLC module of a stacked DRAM. Figure 14.41(a) shows that BLCs are landed in the SAC plugs that connect to the middle sections of the AA. Every BLC is shared by two DRAM cells. Figure 14.41(b) shows a cross section after ILD1 deposition and CMP, and Fig. 14.41(c) is the cross-sectional view after BLC etch. ILD1 is usually BPSG for a stacked DRAM.

BLC in the periphery area can be patterned with the BLC in the cell array area. Because the feature size and depth of the BLCs in the array and periphery areas are quite different, most fabs separate these two contact processes.

Figure 14.42(a) shows the BL module of a stacked DRAM. The BLs are connected to the middle sections of the array AA through BLC plugs that are landed in the SAC plugs. Tungsten is the most commonly used metal for BL formation. After Ti/TiN barrier/glue layer deposition, tungsten is deposited in a CVD process to fill BLC holes and form a thin film on the surface of the wafer.

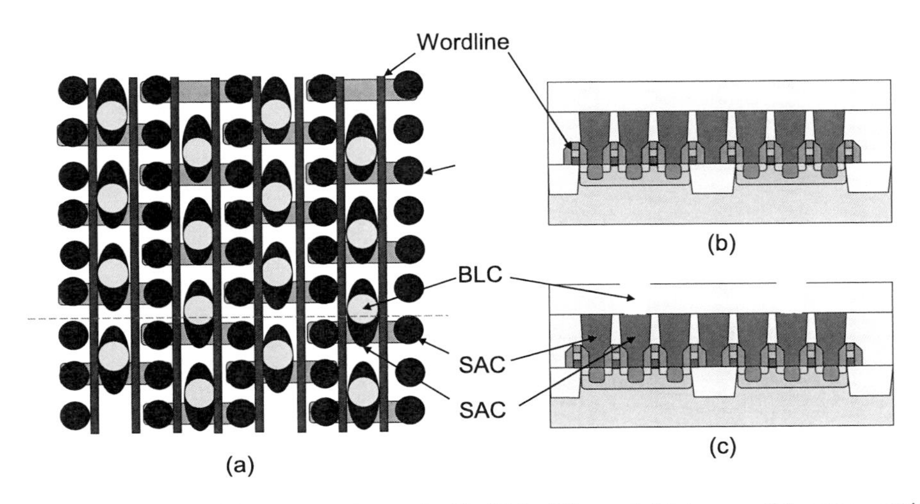

Figure 14.41 (a) BLC layout overlapped with SAC, WL, and AA layers; (b) cross section view after ILD1 deposition; and (c) after BLC etch.

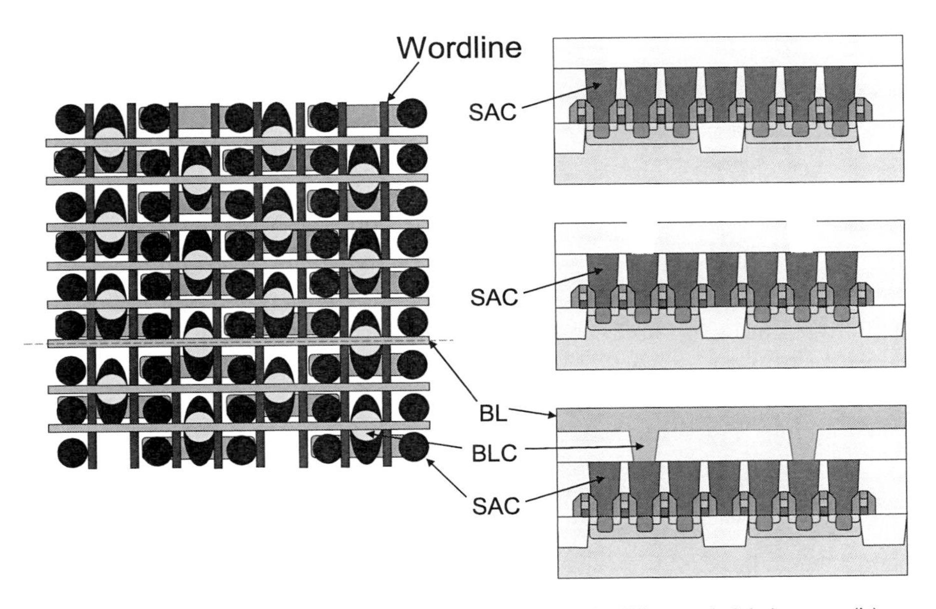

Figure 14.42 (a) BL layout overlapped with BLC, SAC, WL, and AA layers, (b) cross section view after BL metal deposition, and (c) after BL etch.

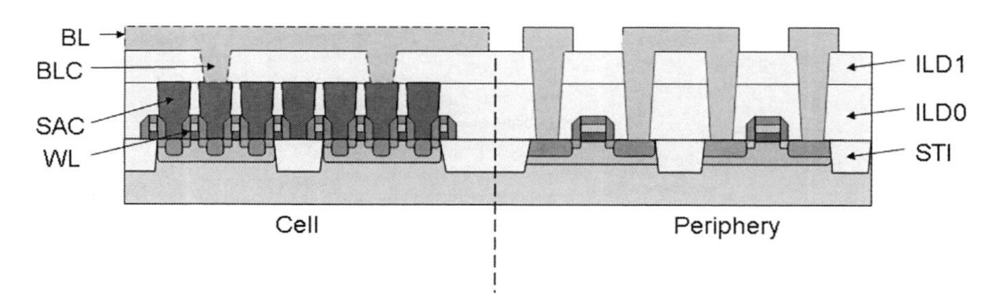

Figure 14.43 Cross section view of a stacked DRAM in both cell and periphery areas after BL formation.

A BL mask defines the metal wires of the BL in both cell and periphery areas, and a metal etching process forms the BL pattern. Figure 14.43 illustrates a cross-section of both cell and periphery areas after BL and BLC formation. To prevent the contact shorting to BL, a spacer is formed on the sidewall of the BL.

Figure 14.44(a) shows the SNC module of a stacked DRAM. The SNC holes connect to the side sections of the array AA through SAC plugs. Conducting plugs of SNC can be formed with polysilicon or tungsten, depending on the technology node. Before conductor deposition, a thin layer of silicon nitride is deposited and etched back to form a liner on the sidewall of the SNC holes to prevent the conducting plugs from shorting the BLs.

Figure 14.45(a) shows the SN layout, and Figure 14.45(b) shows the cross section of SN holes along the dashed line in Figure 14.45(a). The SN holes are connected to the SNC plugs, which are landed on the SAC plugs connected to the two side sections of the array AA.

To form the storage capacitors, two conducting layers are needed to form the two electrodes and a dielectric layer, which is sandwiched between them. Figure 14.46 shows the SN capacitor formation process. After SN hole etch and cleaning, a conductor layer such as polysilicon or titanium nitride (TiN) is deposited [Fig. 14.46(a)]. Because the aspect ratio of the SN hole is very high, this conducting layer needs to have very good sidewall and bottom step coverage. Photoresist fills in the SN holes after electrode layer deposition to protect the conducting film in the holes, and to allow an etchback process that removes the conducting film on the surface [Fig. 14.46(b)]. The SN electrodes are connected to the SNC plugs, which are connected to the two side sections of the AA via the SAC plugs. After removal of the photoresist from the SN holes, a layer of dielectric is deposited on the surface and into the SN holes [Fig. 14.46(c)]. Uniform and conformal sidewall and bottom step coverage is required for this capacitor dielectric film. Figure 14.46(d) shows the conductor deposition that forms the ground electrode of the SN capacitor. This conducting film will be removed in the periphery area in the next photomask step, which finishes the device portion of the DRAM process and starts the BEoL interconnect processes.

With the diminishing technology nodes, SN hole dimensions are becoming smaller. To keep ~30-pF capacitance of the SN capacitor so it can reliably hold enough charge to maintain the memory, the aspect ratio of the SN hole will further

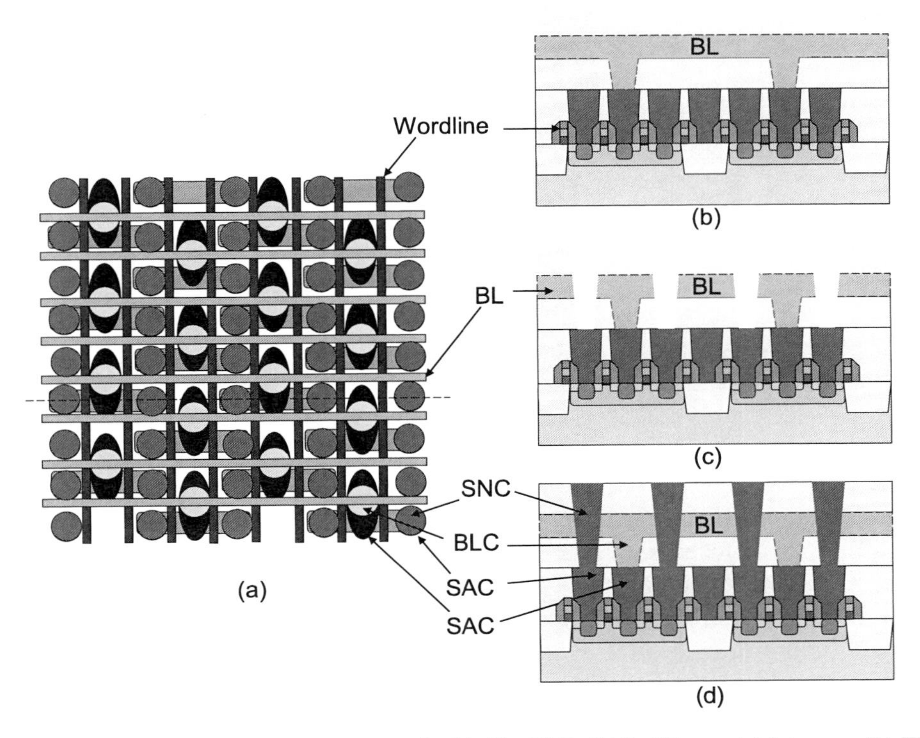

Figure 14.44 (a) SNC layout overlapped with BL, BLC, SAC, WL, and AA layers. (b) The cross section along the dashed line in (a) after ILD2 CMP, (c) after SNC etch, and (d) after SNC polysilicon CMP. The BL with the dashed outline indicates that it is behind the cross section, as shown in (a).

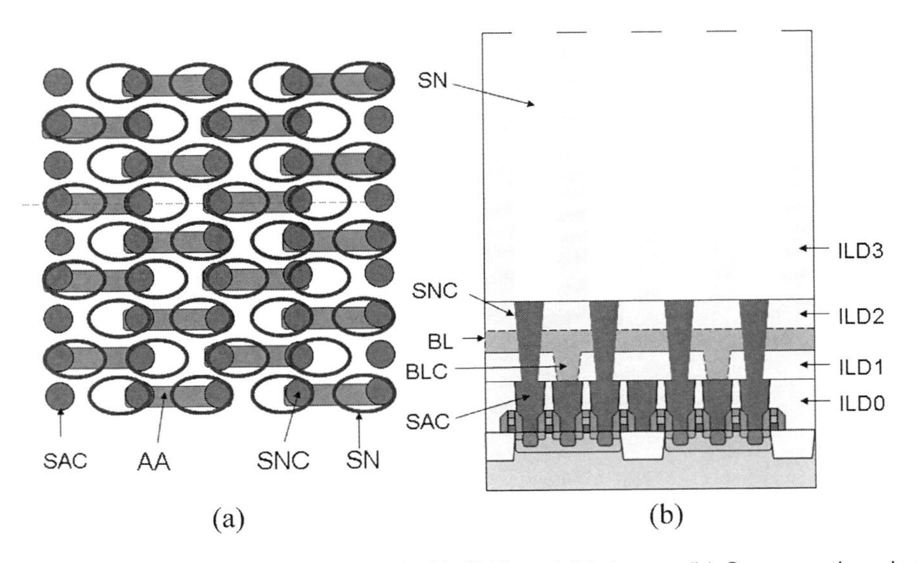

Figure 14.45 (a) SN layout overlapped with SNC and AA layers. (b) Cross section along the dashed line in (a) after SN etch.

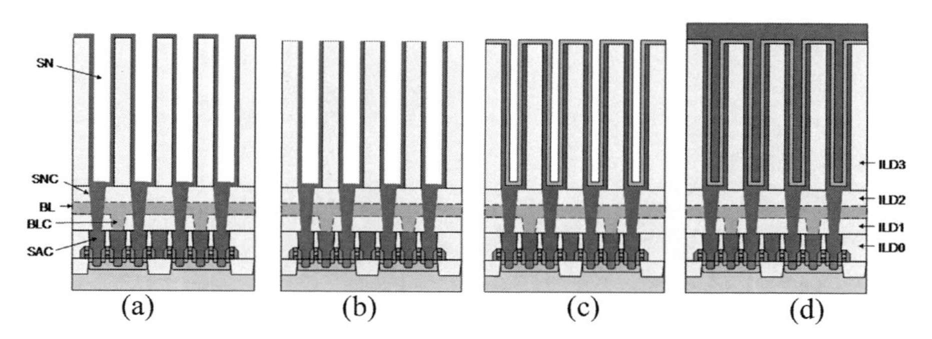

Figure 14.46 SN capacitor formation process: (a) SN electrode conductor deposition, (b) removal of the SN electrode conductor film from the surface, (c) SN dielectric deposition, and (d) ground electrode conductor deposition.

increase if the capacitor structure and capacitor dielectric materials are kept the same. Several methods have been developed to reduce the aspect ratio of SN holes. Using high-κ dielectrics can help to reduce the height of the hole and the aspect ratio. As before, silicon oxide, silicon nitride, and silicon oxide stacking have been used. High-κ materials such as aluminum oxide (Al_2O_3), hafnium dioxide (HfO_2), and zirconium dioxide (ZrO2) have been studied and applied to SN capacitors. Other methods to reduce the height of SN holes have recessed the ILD3 after SN electrode formation; this allows the ground electrode to form at both sides of the SN electrode. Also, polysilicon was widely used early on as an electrode material. More advanced DRAM chips started to use TiN as SN electrodes. Figure 14.47 illustrates a stacked DRAM with a recessed gate (RG) array transistor, TiN SN electrodes, recessed ILD3, high-κ capacitor dielectric, and metal ground electrodes with three layers of metal interconnects. RGs are used to reduce the short channel effect (SCE) of the array nMOS that becomes stronger with smaller feature sizes.

14.6.3 NAND flash process

Flash chips are NVM chips widely used for data storage applications in electronic devices, especially mobile electronics such as digital cameras, MP3 players, cell phones, global positioning systems (GPSs), high-end notebook computers, tablet computers, etc. Compared with HDDs, flash memory has shorter data access time, consumes less power, and is more reliable because it does not have any moving parts.

Almost all flash memory chips on the market are charge trap devices based on the floating-gate structure described in Chapter 3. The basic device structure is illustrated in Fig. 14.48 and is very similar to an nMOS. Depending on the circuit architecture, there are two types of flash devices, NAND and NOR (see Fig. 14.49).

Figures 14.49(a) and 14.49(b) show the NOR flash circuit and its cross section, respectively. Figures 14.49(c) and 14.49(d) show a 64-bit string NAND flash circuit and the correspondence cross sections, respectively. A NOR flash is equivalent to a 1-bit string NAND flash that does not require select gates. NOR flash can achieve

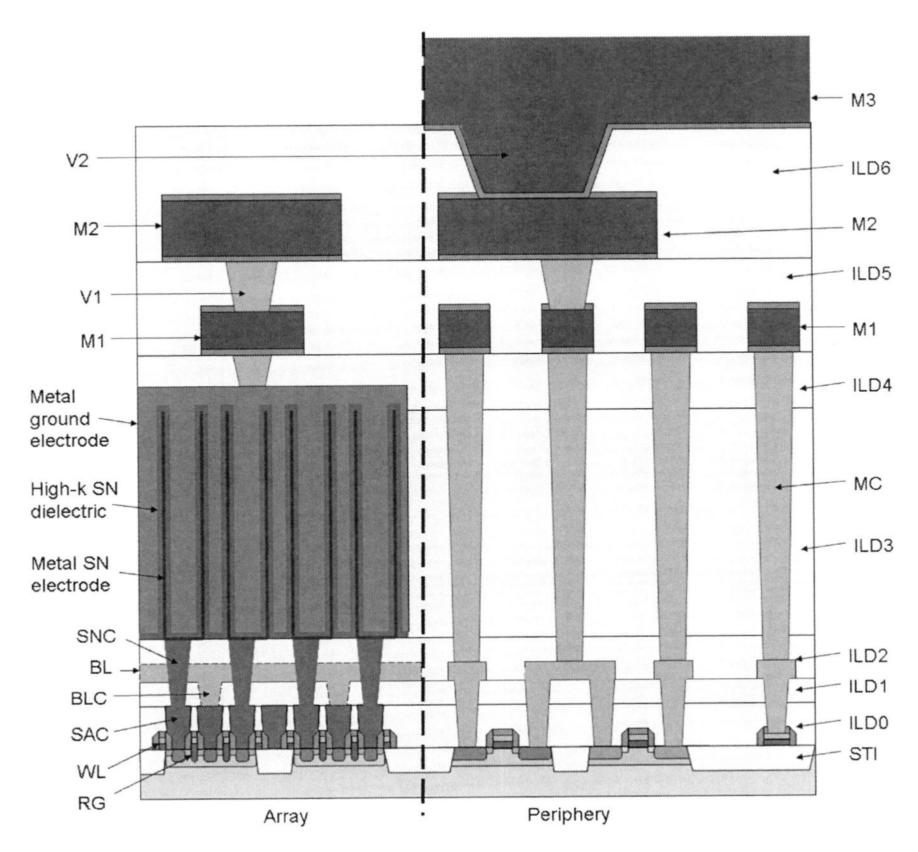

Figure 14.47 Cross-section view of a stacked DRAM with RG array transistor, TiN SN electrodes, recessed ILD3, high-κ capacitor dielectric, and metal ground electrodes.

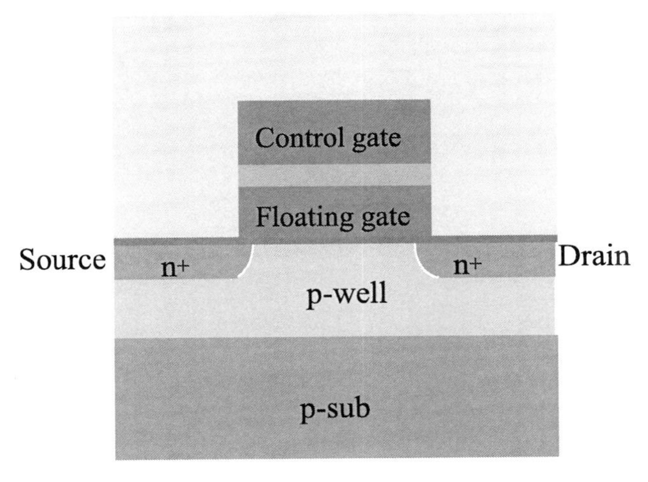

Figure 14.48 Basic structure of a floating-gate NVM device.

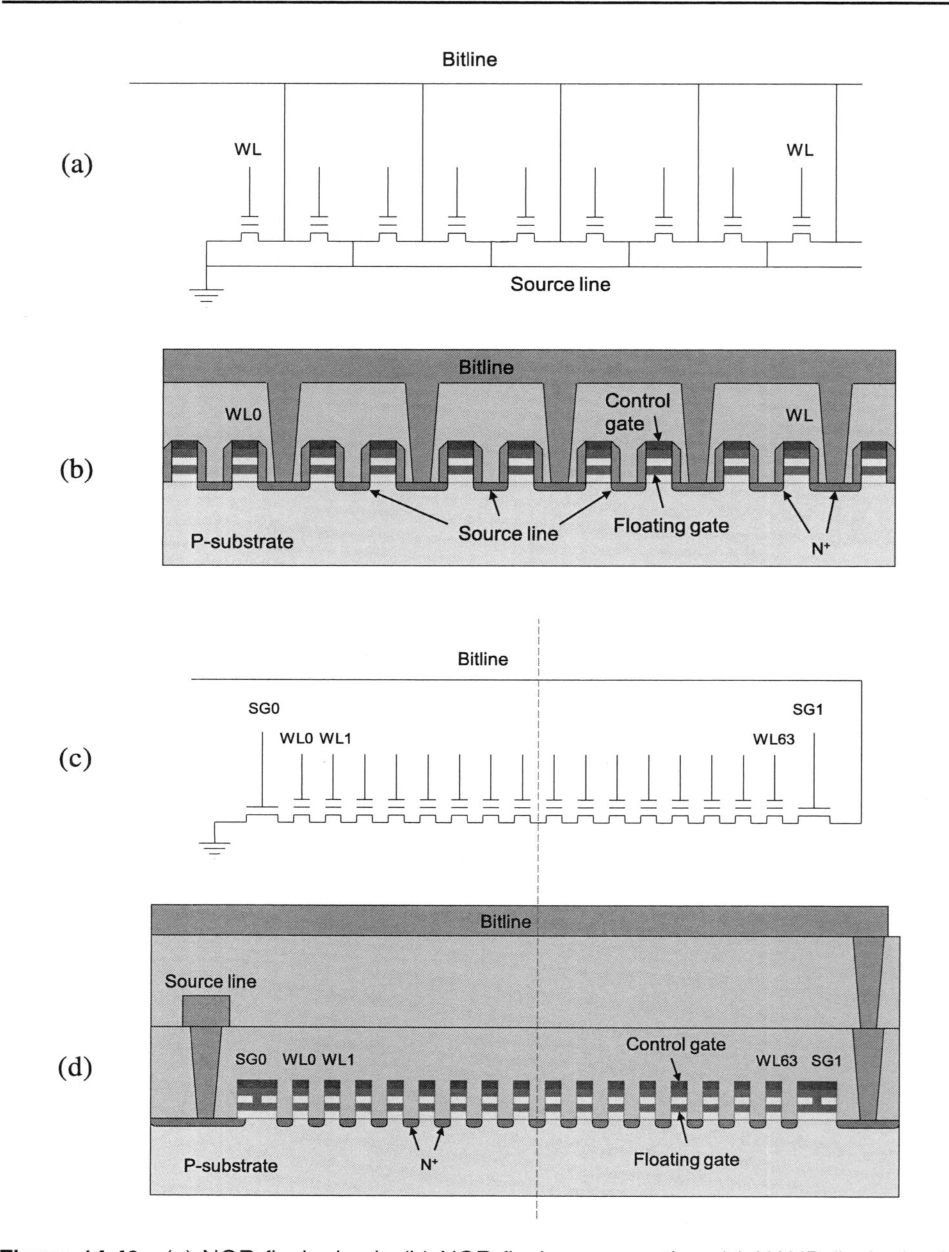

Figure 14.49 (a) NOR flash circuit, (b) NOR flash cross section, (c) NAND flash circuit, and (d) NAND flash cross section.

random access of any memory cell, while NAND flash cannot. Although NOR flash has a shorter reading time than NAND flash, it has much longer write and erase times. It also costs significantly more than NAND flash because of its lower packing density due to the dense BL contacts. Since the majority of flash memory devices are NAND flash chips, only NAND flash processes are discussed in this section.

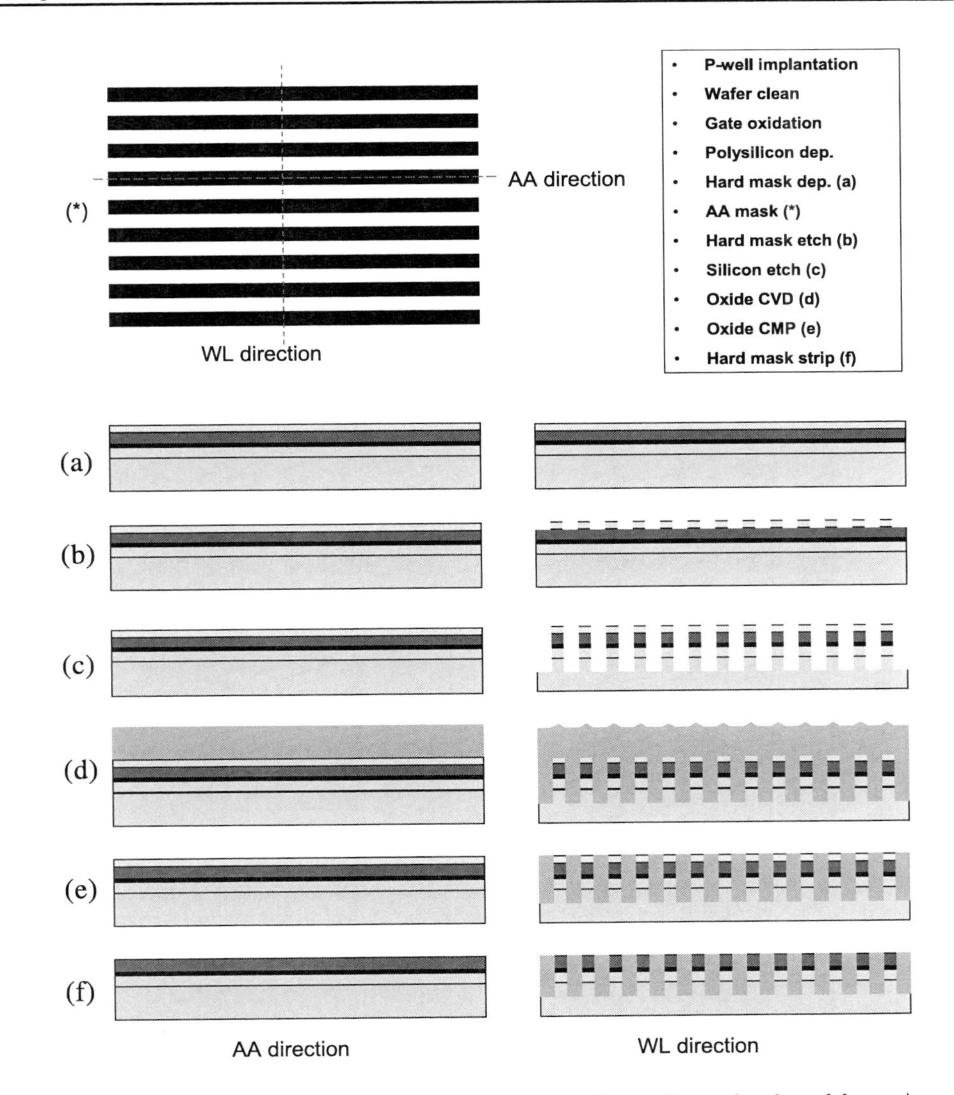

Figure 14.50 AA module of the NAND flash. The top is a schematic of an AA mask, and the left and right sides of illus. (1) through (6) are the cross sections of the SA-STI processing steps (described in the textbox) in AA and WL directions, respectively.

Figure 14.50 illustrates the self-aligned shallow trench isolation (SA-STI) processing module. After p-well implantation, gate oxide is grown and deposits a floating-gate layer with a hard mask on top. An AA mask is then used to pattern a hard mask for etching the floating gate, gate oxide, and silicon substrate, to form active areas. Silicon nitride or silicon oxynitride is the most commonly used hard mask material, and polysilicon is the most commonly used floating-gate material. After the silicon trench etch, high plasma density (HPD) CVD oxide is used to fill the trenches, and a CMP process removes the oxide and stops on the hard mask. A stripping process removes the hard mask and finishes the SA-STI process.

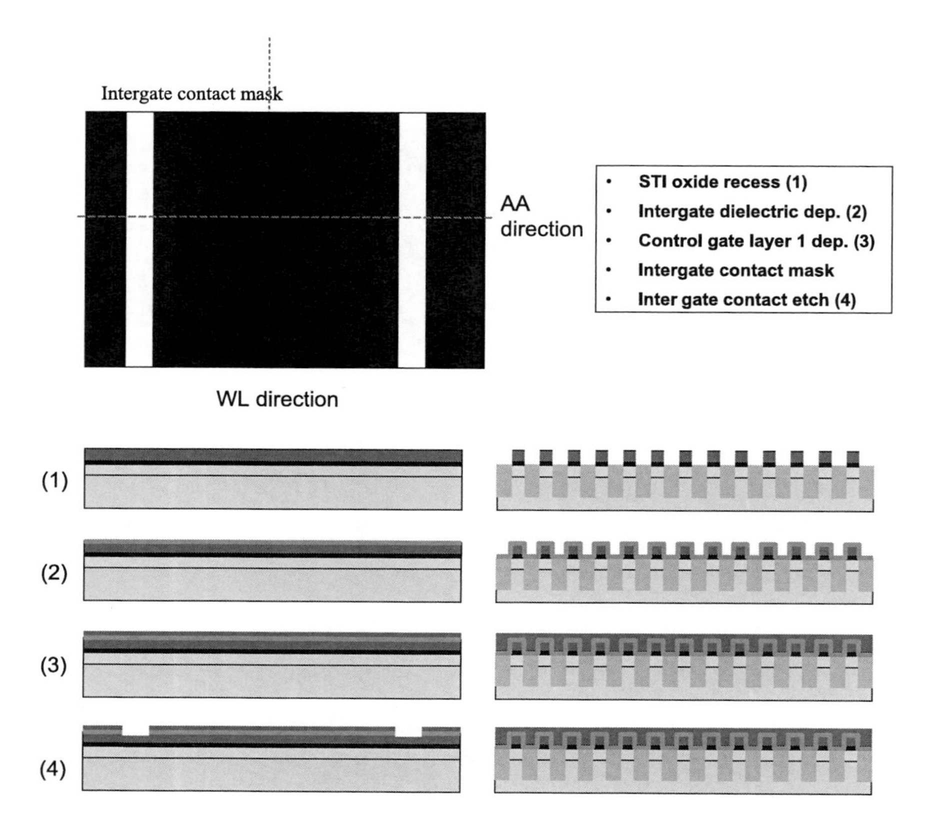

Figure 14.51 Intergate contact module. At the top is the intergate mask in the array area. The left and right sides of illus. (1) through (4) are cross sections of the intergate contact steps of the WL process (described in the textbox) in AA and WL directions, respectively.

Figure 14.51 illustrates the intergate contact process, which is a unique processing module of floating-gate NVM devices. The MOSFETs of selected gates in the array and control circuits in the periphery regions require intergate contact to electrically short the floating-gate layer and control the gate layer. Floating gates are made with polysilicon, the intergate dielectric is stacked oxide-nitride-oxide (ONO), the first layer of the control gate is also polysilicon, and the second control gate layer is usually metal, such as tungsten silicide or tungsten. The recessed STI oxide shown in Fig. 14.51, illus. (1) allows an increase in coupling between the control and floating gates, an increase that is necessary when feature size shrinks.

Figure 14.52 illustrates the WL process module. Metal deposition through intergate contact shorts the floating and control gates of the selected gates in the array area and all MOSFET gates in the periphery area. Line/space patterning of the WL layer has the highest pattern density in IC production. The square box inside the WL mask (shown in the top image of Fig. 14.52) is the area of a unit cell of NAND flash memory, which is $4F^2$, where F is the minimum feature size of the structure, and $4F^2$ is the highest pattern density it can achieve. For NAND flash, the line/space ratios of AA and WL are both 1:1; therefore, F is the CD of

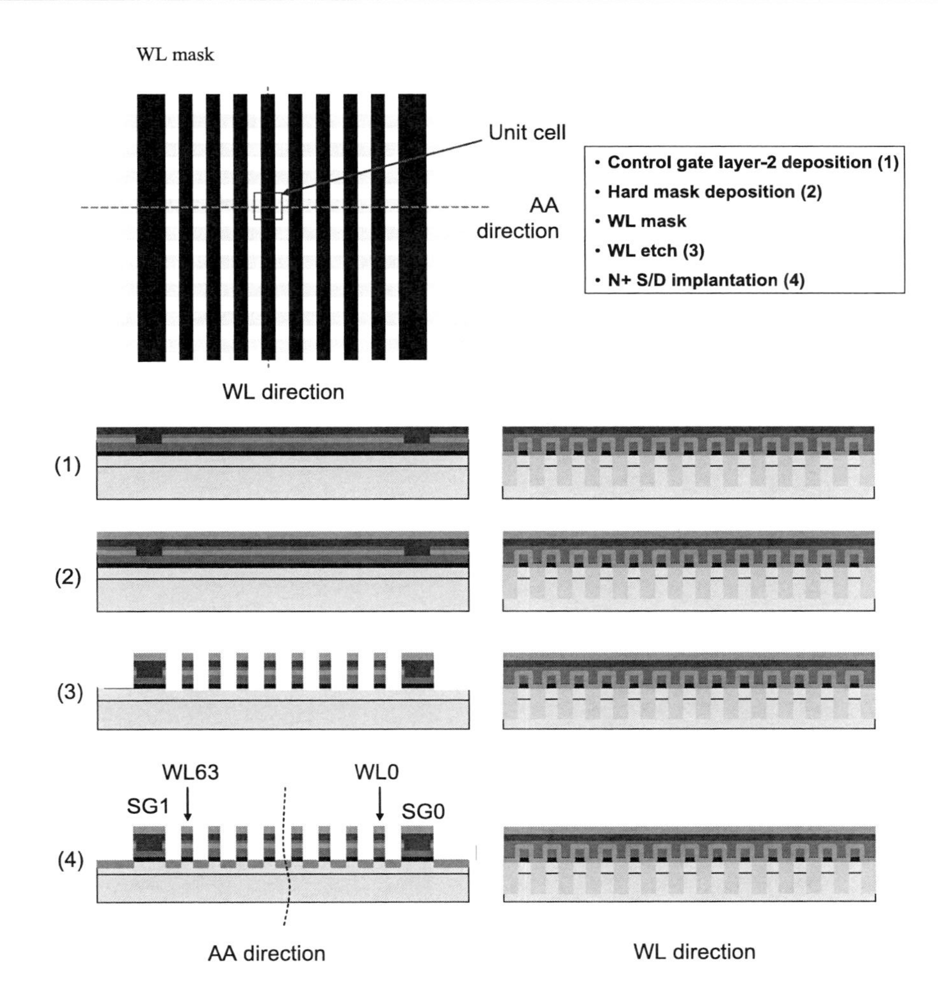

Figure 14.52 WL module. The top schematic shows the WL mask of the array area. The left and right sides of illus. (1) through (4) are the cross sections of the WL processing steps (described in the textbox) in AA and WL directions, respectively.

the AA and WL. $F = 25$ nm for a 25-nm NAND flash device, which has 25-nm CDs for both AA and WL, and the unit area of the memory cell is 2500 nm^2, or 0.0025 μm^2. A self-aligned double-patterning technique is suitable for patterning the WL layer and also can be applied to AA and BL layers when F becomes too small for a single exposure to handle.

Figure 14.53 shows the CB1 process. In the array area, contact holes are very dense in the WL direction, but very sparse in the AA direction because there is no contact in the 32- or 64-bit strings between the two select gates (see the top schematic in Fig. 14.53). Although the CB and contact-to-source line (CS) look the same in Fig. 14.53, their requirements are quite different. For CB holes, both hole-to-hole shorts and holes that do not open are killer defects, while only nonopen

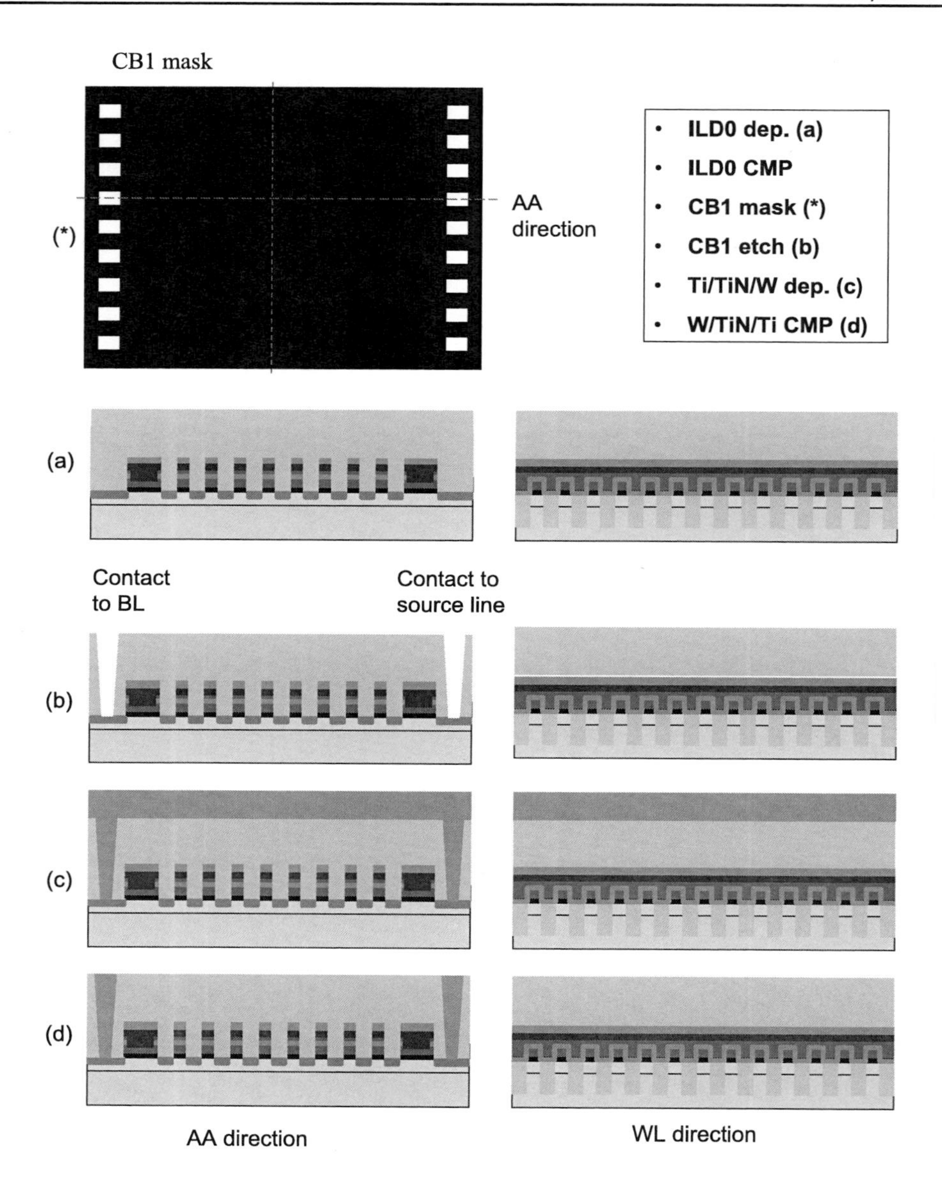

Figure 14.53 The first contact-to-BL (CB1) process. The top schematic shows the CB1 mask of the array area. The left and right sides of illus. (1) through (4) are the cross sections of the CB1 processing steps (described in the textbox) in AA and WL directions, respectively.

holes are killers for CSs. All CSs are connected to source lines that lead to a common ground; therefore, shorts between CSs are not killer defects.

Because the requirements of CB and CS holes are different, some designers use different hole patterns for CB and CS, as shown in Fig. 14.54. The curved dash line in Fig. 14.54 indicates that the distance between the CB and CS holes is far greater than what is shown in the figure.

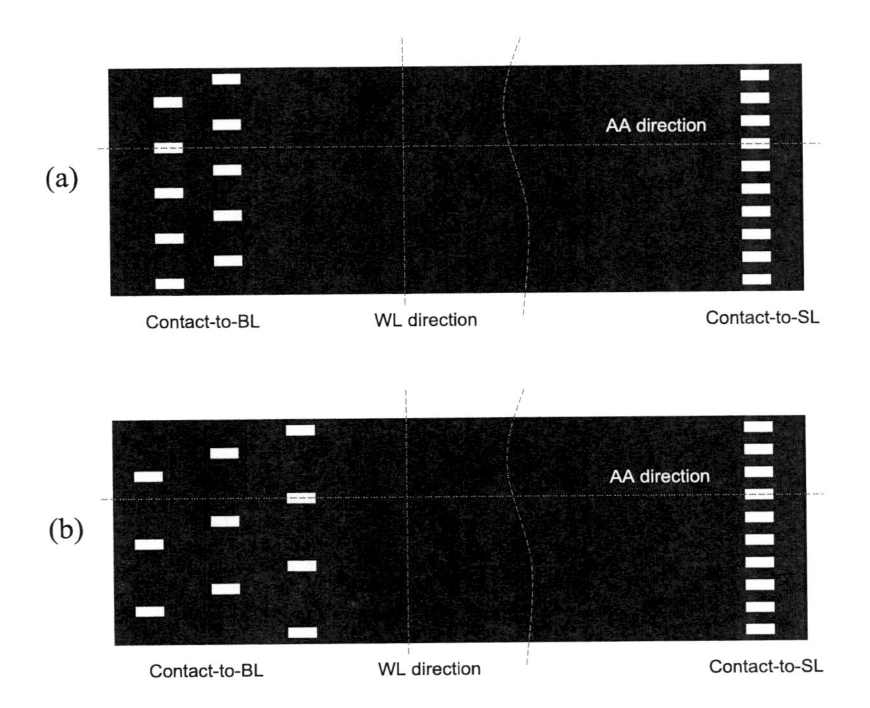

Figure 14.54 CB1 layout of an advanced NAND flash with different CB and CS hole patterns: (a) two-row and (b) three-row split of CB.

For 90-nm technology nodes, the CB hole CD is 90 nm, and the space between the CB holes is also 90 nm. 193-nm optical lithography can resolve the 90-nm hole patterns illustrated in Fig. 14.54. 193-nm high-NA immersion lithography can resolve 5×-nm hole patterning without row splits. With further diminishment of feature sizes, CB hole patterns can be split into two rows, providing more room and allowing immersion optical lithography to resolve them. Figure 14.54(a) shows a two-row split of CB holes widely used in 3×- to 2×-nm nodes. At low 2×-nm (such as 20 nm) and high 1×-nm (18 nm) technology nodes, a two-row split can no longer be resolved by immersion optical lithography, and a three-row split of CB holes must be used, as shown in Fig. 14.54(b). Theoretically, CB holes can be split into four or even more rows to allow optical lithography to pattern low 1×-nm contact holes. However, this would take a large amount of silicon surface from real devices. With a much shorter wavelength (13.5 nm), EUV lithography can resolve CB hole patterns without row splitting, a capability that can significantly increase device density.

The M0 process described in Fig. 14.55 is the module to form the source line and CB pads that the second contact-to-bitline (CB2) will land on. The process is a single inlay process, similar to contact and local interconnect processes for CMOS. Because the source line is very wide, oxide supports are extensively placed into it to avoid the dishing effect of metal CMP. To maintain a constant pattern density

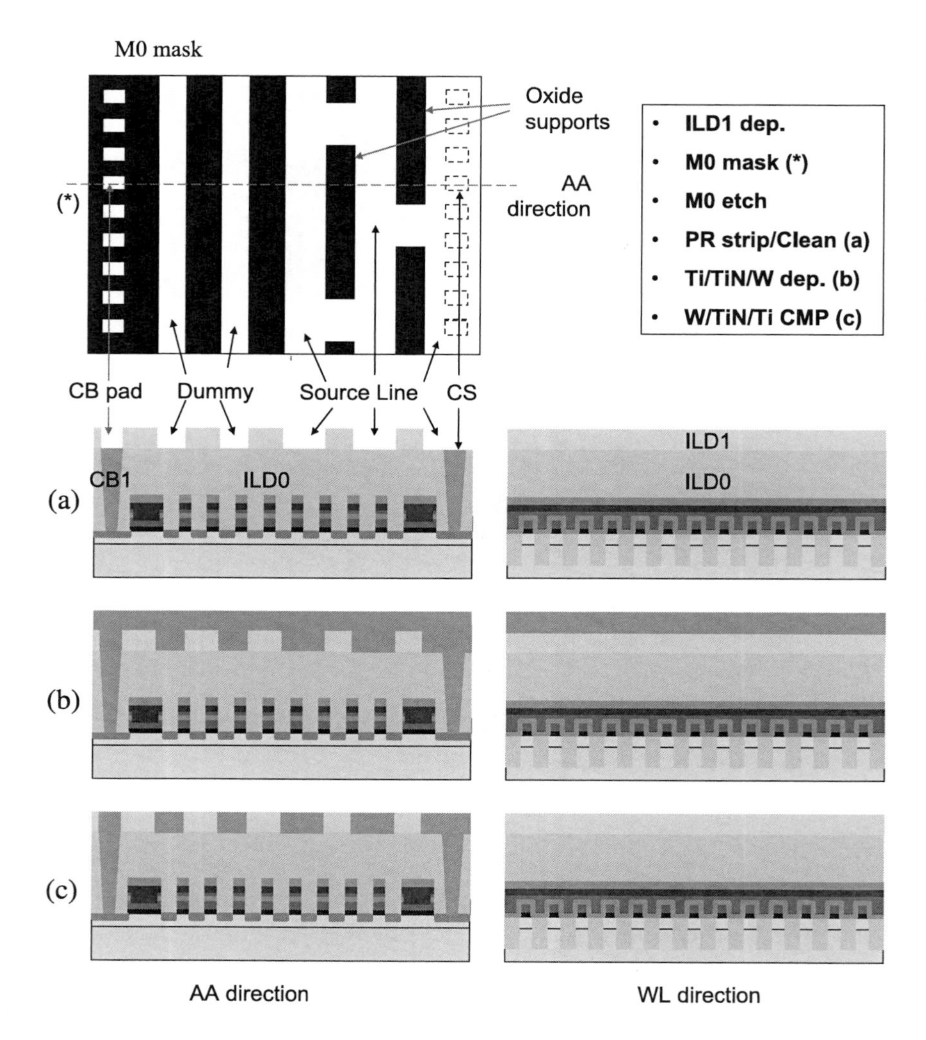

Figure 14.55 Metal 0 (M0) processes. The top schematic shows a M0 mask of the array area. The left and right sides of illus. (1), (2), and (3) are cross sections of the M0 processing steps (described in the textbox) in AA and WL directions, respectively.

to avoid erosion effects during metal CMP, dummy patterns are also designed into the M0 layer.

Figure 14.56 shows CB2 processes. ILD2 consists of an ESL and a bulk dielectric layer. The ESL is either silicon nitride, silicon oxynitride, or silicon-nitrocarbide (SiNC), and the bulk layer is silicon oxide, without or with carbon doping. After the ILD2 deposition process, a CB2 mask is used to pattern the wafer, and dielectric etching forms the CB2 holes that land on the CB pads, which connect to the CB1 plugs.

Figure 14.57 shows the M1 processes. In an array area, M1 is the BL, and in a periphery area, it is the local interconnection. Tungsten is commonly used in M1. Silicon oxide, either without or with carbon doping, is used for the ILD2.

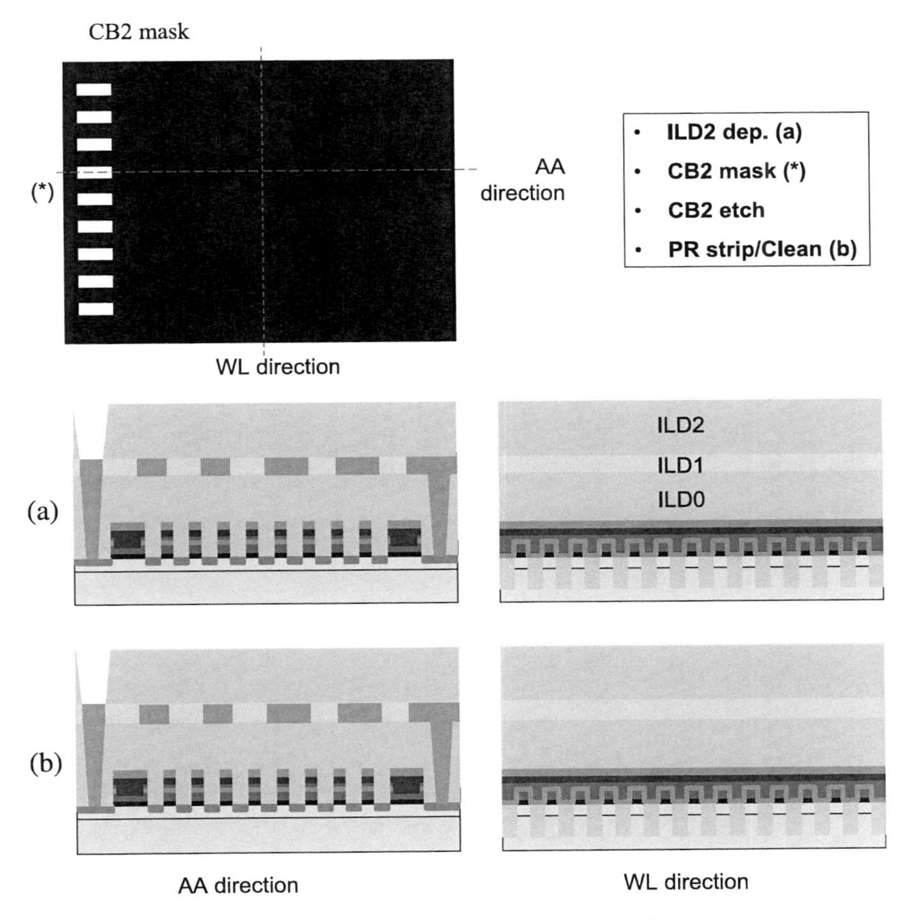

Figure 14.56 CB2 processes. The top schematic shows a CB2 mask of the array area. The left and right sides of illus. (1) and (2) are the cross sections of the CB2 processing steps (described in the textbox) in AA and WL directions, respectively.

Tantalum or tantalum nitride is widely used as a barrier layer for copper, and an ionized metal plasma process is frequently used to deposit the barrier and copper seed layers. ECP is used to deposit the bulk copper.

Figure 14.58 shows a cross section of a 64-bit string NAND flash with both array and periphery areas. There are intergate contacts in both nMOS and pMOS in the periphery area, and these devices have larger feature sizes than the devices in the array area. The CB1 in the array area is the contact in the periphery area. M0 forms the CB pad and source line in the array area and local interconnects in the periphery area. CB2 in the array area is the V1 in the periphery area, and M1 forms the BL in the array area, and the first global metal interconnect in the periphery area.

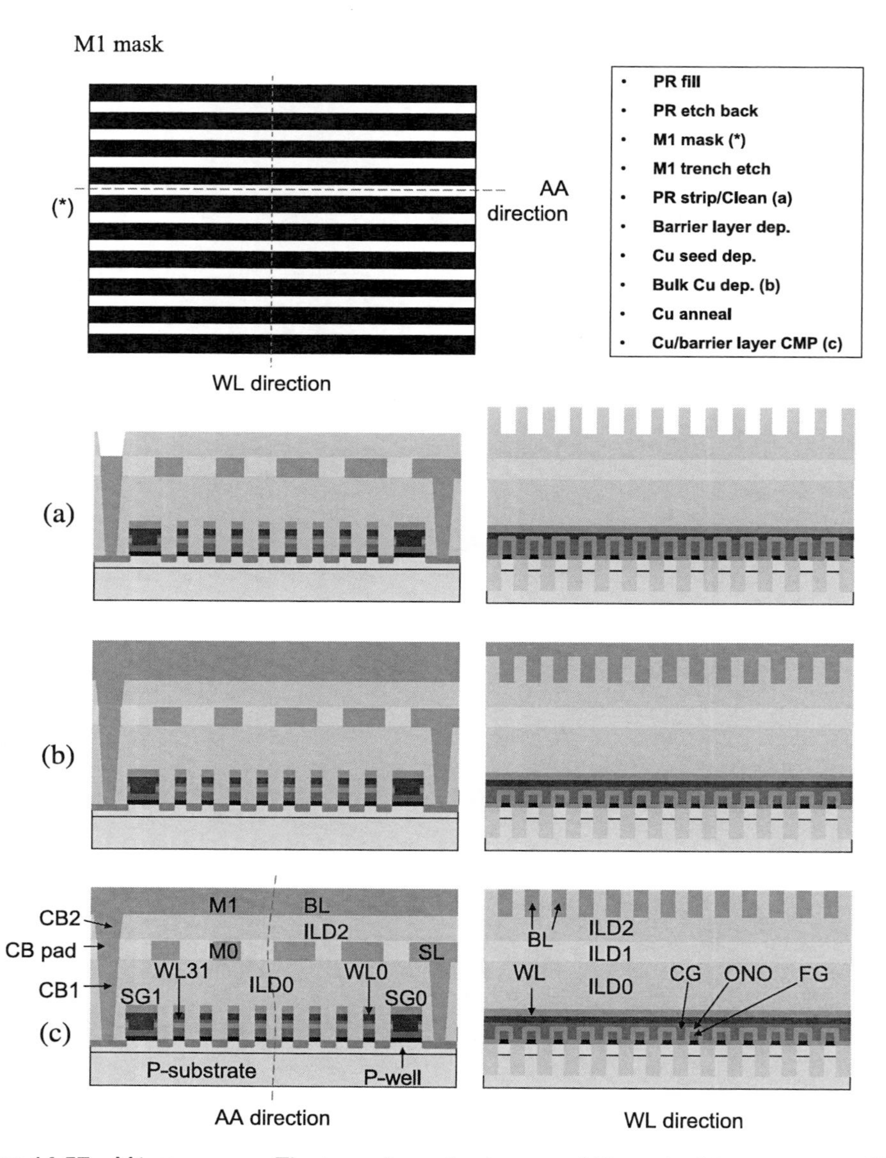

Figure 14.57 M1 processes. The top schematic shows an M1 mask of the array area. The left and right sides of illus. (1), (2), and (3) are the cross sections of the M1 processing steps (described in the textbox) in AA and WL directions, respectively.

14.7 Summary

- CMOS IC chips have dominated the semiconductor industry since the 1980s, when minimum feature size shrank from 3 μm to submicron levels.
- The 1980s saw the introduction of tungsten CVD and dielectric PECVD processes; sidewall spacers, plasma etching, steppers, and projection systems became common.
- In the 1990s, minimum feature size shrank from 0.8 to 0.18 μm.

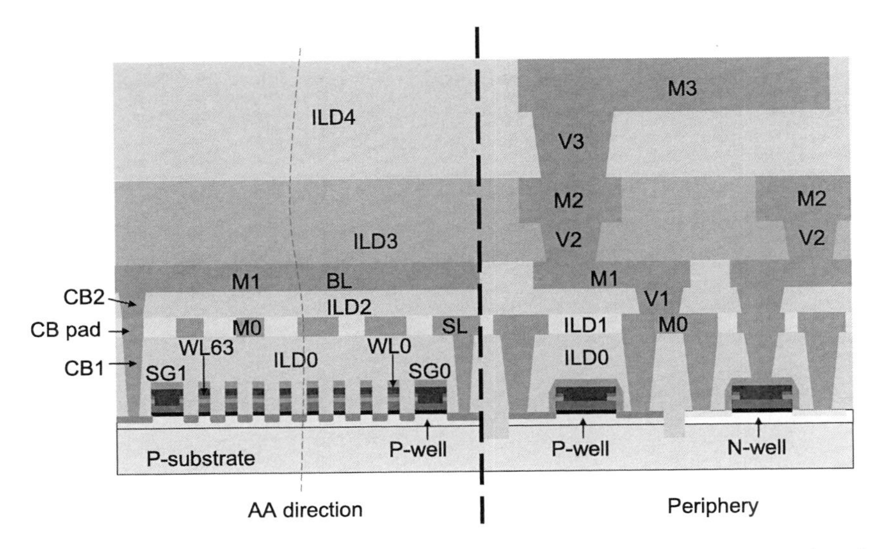

Figure 14.58 Cross section of a 64-bit string NAND flash with both array and periphery areas.

- Technologies such as RTP, silicides, CMP, HDP CVD, O_3-TEOS-based dielectric CVD, and ECP processes were introduced or improved.
- In the first decade of 2000, IC technology nodes reduced to 25 nm in NAND flash memory devices, and to 28 nm in CMOS logic devices.
- Immersion 193-nm lithography, double patterning, nickel silicide, and SEG SiGe have all contributed to process advancements.
- High-κ dielectric and metal gates have been used in high-end microprocessors and other high-performance logic IC chips, and copper with porous low-κ dielectric interconnections have been used in high-volume production.
- DRAM and NAND flash devices are the most important memory devices in IC production. With the development of 3D graphics, stream video, and other applications, further technological developments will be seen in memory chip manufacturing.

14.8 Bibliography

C. Y. Chang and S.M. Sze, *ULSI Technologies*, McGraw-Hill, New York (1996).

K. Kim, "Technology for sub-50-nm DRAM and NAND flash manufacturing," *IEDM Tech. Dig.*, 323–326 (2005).

K. Y. Lim et al., "Novel stress-memorization-technology (SMT) for high electron mobility enhancement of gate last high-k/metal gate devices," *IEDM Tech. Dig.*, 229–232 (2010).

K. Mistry et al., "A 45-nm logic technology with high-k + metal gate transistors, strained silicon, 9 Cu interconnect layers, 193-nm dry patterning, and 100% Pb-free packaging," *IEDM Tech. Dig.*, 247–250 (2007).

S. Natarajan et al., "32-nm logic technology featuring 2nd-generation high-k + metal-gate transistors, enhanced channel strain and 0.171m^2 SRAM cell size in a 291-Mb array," *IEDM Tech. Dig.*, 941–943 (2008).

K. Prall and K. Parat, "25-nm 64-Gb MLC NAND technology and scaling challenges," *IEDM Tech. Dig.*, 102–105 (2010).

L. Shon-Roy, A. Wiesnoski, and R. Zorich, *Advanced Semiconductor Fabrication Handbook*, Integrated Circuit Engineering Corp., Scottsdale, AZ (1998).

H. Xiao, "Method for Forming Memory Cell Transistor," U.S. Patent Application 12553067 - Filed on Sep 2, 2009.

14.9 Review Questions

1. What was the isolation material used in CMOS IC chips in the 1980s? In the late 1990s?
2. What were the metal materials used in CMOS IC chips in the early 1980s and mid-1990s? What were metal materials that most likely will be used in the 2010s?
3. What materials are used for the final passivation of CMOS chips?
4. List the silicon nitride layers used in copper metallization processes. Try to explain their functions.
5. Silicon nitride is a good barrier for copper diffusion. However, no one uses silicon nitride as the main dielectric in copper metallization. Why?
6. What is the main difference between titanium silicide and cobalt silicide?
7. Why has nickel silicide replaced cobalt silicide since the 65-nm technology node?
8. What is the main difference between USG and FSG?
9. Compare the ILD0 processing steps listed in Secs.14.2 and 14.3. What are the differences?
10. Discuss the advantages and disadvantages of hydrogen plasma premetal cleaning.
11. What are the advantages of a dual damascene process compared to an aluminum alloy interconnection process?
12. What is the benefit of high-κ dielectrics?
13. Why does a pMOS created with a gate-last HKMG process have a higher speed than a pMOS created with a gate-first HKMG process?
14. Why do IC manufacturers need to generate lead-free solder bumps?
15. List the two types of DRAM capacitor structures. Which type is more popular in DRAM manufacturing?
16. List at least three contact layers that are in a stacked DRAM cell array area. Which one has the highest hole density, and which one has the lowest?
17. Draw a sketch of a basic flash device. What is the main difference between it and an nMOS?
18. What are the main differences between NOR and NAND flash devices?

Chapter 15
Future Trends and Summary

In the first decade of the 21st century, IC processing technologies saw some fundamental changes from the traditional processing technologies of the 20th century.

Copper replaced aluminum–copper alloys and tungsten to form metal interconnection lines in advanced CMOS logic IC chips because of its lower resistivity and high EMR. By using copper interconnections, device speed can be increased, power consumption can be reduced, and reliability can be improved. Low-κ and ULK dielectrics replaced USG and FSG for the interlayer dielectric applications. PECVD organosilicate glass, or SiCOH, and porous OSG processes were developed and integrated with copper to form interconnections. There are very few choices for reducing RC (resistance multiplied by capacitance) delay. To change metal to reduce R, the only choice is silver (Ag), which has a bulk resistivity of 1.6 μΩ·cm, just a little lower than the 1.7 μΩ·cm of copper. It is unlikely that the gain from switching to silver would justify the cost of the materials and process development, making it unattractive to IC manufacturers. New materials with even lower resistivity on a nanometer scale, such as carbon nanotubes (CNTs), have been studied to be applied to IC interconnections. To reduce C, air gap is one approach to further reduce the κ-value. In the future, interconnections that integrate CNTs and air gaps could receive some attention; however, cost will determine whether these interconnections will be implemented in mass production.

Metals made a remarkable comeback after being replaced by polysilicon as the gate electrode materials of MOSFETs in the mid-1970s. For CMOS logic devices, TiN has been used as a gate electrode for gate-first approaches. Replacement metal gate, or the gate-last process, is becoming mainstream for metal gate formation in HKMG CMOS manufacturing. Because metal has a lower resistivity than polysilicon/silicide stacks, metal gates and local interconnections can help reduce power consumption and increase device speed. TiN and W have been used as gate electrodes for array transistors in bWL DRAM chips in which TiN serves as a work-function metal, and W as a bulk metal for the gate as well as the WL.

To conduct electric current between the source and drain, a conductive channel must be formed underneath the gate by reversion of the minority carrier there. Minority carrier reversion can be achieved by charging a metal–oxide interface with applied gate voltage. Therefore, capacitance of the MOS gate capacitor must be large enough to hold the charges. With feature size reduction, the area of the capacitor electrode decreases quickly, and so does the gate capacitance. To maintain enough capacitance, the distance between the two electrodes (which is the thickness of the gate dioxide layer) must be reduced. For 0.18-μm IC chips, the gate oxide is about 35 Å. It needs to be approximately 25 Å for 0.13-μm minimum-feature-size IC chips, and about 15 Å for 90-nm devices. The problem is, if the gate oxide thickness further decreases, gate-to-substrate leakage will increase significantly, and IC chip reliability and low power performance will be affected. Nitridated silicon oxide with $\kappa \sim 5$ has been used in 90-, 65-, 45-, and even in 32/28-nm devices. High-κ dielectrics such as hafnium dioxide (HfO_2, $\kappa \sim 25$) have been developed and implemented in HKMG CMOS manufacturing, starting from the 45-nm technology node.

Many other high-κ dielectrics such as titanium dioxide (TiO_2, $\kappa \sim 80$), tantalum pentoxide (Ta_2O_5, $\kappa \sim 26$), and zirconium dioxide (ZrO_2, $\kappa \sim 25$) have also been researched and could be implemented in IC chip manufacturing in the future.

Research and development of the deposition and etching of dielectric materials with very high-κ values, such as BST ($Ba_{1/2}Sr_{1/2}TiO_3$, κ up to 600), has continued for a long time. It could be applied in DRAM chip manufacturing as a dielectric material for storage node capacitors if a stable, integrative, and manufacturable deposition process can be developed.

Planar MOSFETs have been used as basic IC building blocks for more than half a century. Because traditional scaling is rapidly approaching its physical limit, and the cost to scale technology nodes in the deep nanometer range has skyrocketed, fabs are actively looking for other approaches to extend technology development. One promising technology is the 3D multiple-gate field effect transistor, or MuGFET. Figure 15.1 shows a MuGFET on a SOI substrate. By folding the active area (AA) of a planar MOSFET into a thin strip like a fin, device dimensions can be dramatically reduced without feature size shrinkage. This is shown in Fig. 15.1(b), which folds the AA with channel width Z from Fig. 15.1(a) into a fin, with $2h+w = Z$. Because the MuGFET shown in Fig. 15.1(b) has gate electrodes on three sides of the fin, it is called a triple-gate FET, or simply a trigate. The trigate device illustrated in Fig. 15.1(b) could perform similarly to the planar MOSFET shown in Fig. 15.1(a). By using a trigate, fabs can use the same lithographic technology to reduce transistor dimensions while maintaining the same performance, or to improve device performance without increasing device density. With trigates, IC manufacturers can further improve device performance by making the fin taller; however, that will pose great challenges for etch processing development, polysilicon deposition, polysilicon dummy gate etch, SEG, dummy gate removal, trench cleaning, HKMG deposition, etc.

Silicon fin etching has a nitride hard mask with a pad oxide underneath. By keeping this hard mask while forming the MuGFET, the gate electrode is only on two sides of the fin, not on the top or bottom. Fabs usually call this double-gate FET a finFET. For finFET devices, channel length only relates to the height of the fin ($2h$) and is no longer related to the fin CD (w), which is the smallest CD of the device. From a processing point of view, a finFET is easier to build than a trigate. Intel's 22-nm IC chips are built with trigate devices.

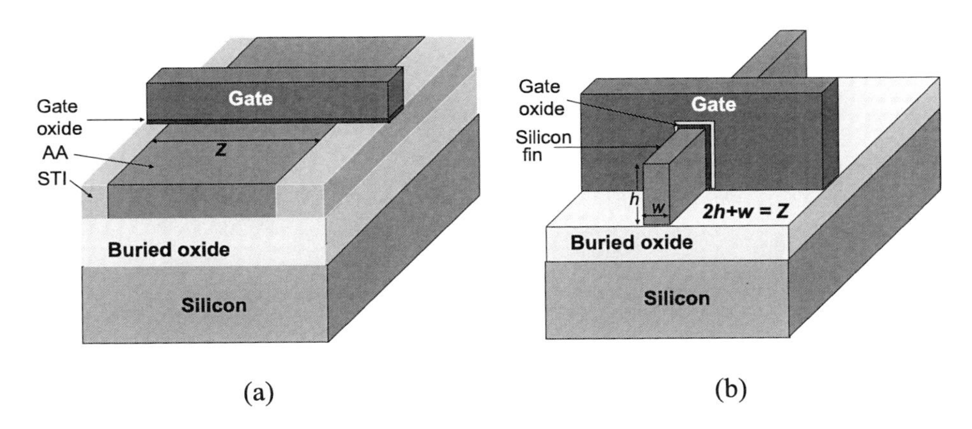

Figure 15.1 (a) A 3D illustration of a planar MOSFET with SOI substrate, and (b) an equivalent trigate FET with SOI substrate.

Using semiconductor materials with higher carrier mobilities than silicon is another way to improve device performance. For example, germanium (Ge) has much higher hole mobility than silicon, 1900 $cm^2 V^{-1} s^{-1}$ of Ge versus 430 $cm^2 V^{-1} s^{-1}$ Si, and III-V compound semiconductors such as gallium arsenide (GaAs) have much higher electron mobility than silicon (8000 $cm^2 V^{-1} s^{-1}$ versus 1630 $cm^2 V^{-1} s^{-1}$). If Ge is used for pMOS channels and GaAs for nMOS channels, both the hole mobility of the p-channel and electron mobility of the n-channel can be significantly improved. Because the driving current of MOSFET is proportional to carrier mobility, CMOS with epitaxial Ge pMOS channels and GaAs nMOS channels can be used in future IC manufacturing to enhance device performance without aggressively scaling the feature size. Other new materials, such as CNTs and graphene, have received a large amount of research attention for creating nanometer-scale electronic devices for silicon eras in the future.

Figure 15.2 shows a timeline of IC technology nodes and the start of new technologies and materials in IC chip manufacturing. With innovations such as double patterning and source-mask optimization (SMO), fabs can extend optical lithography to 14 nm. With quadruple patterning, optical lithography could even stretch to a 10-nm technology node. The most promising next-generation lithography technology, EUV lithography, is still in development and inching toward production readiness. It could compete with quadruple patterning of immersion 193-nm optical lithography for mass production at a 10-nm technology node.

Although the semiconductor industry has been developing for more than 60 years, it is still not a mature industry like the automobile industry. New technologies are frequently introduced, and device structure and materials change for every nanometer technology node. Manufacturing tools worth multimillions of dollars when they are brand new become obsolete in just few years and must be replaced if manufacturers want to keep up with technological developments. The skyrocketing costs of an advanced IC fab keep many IC manufacturers from pursuing the most advanced technology nodes. For example, a 300-mm fab of

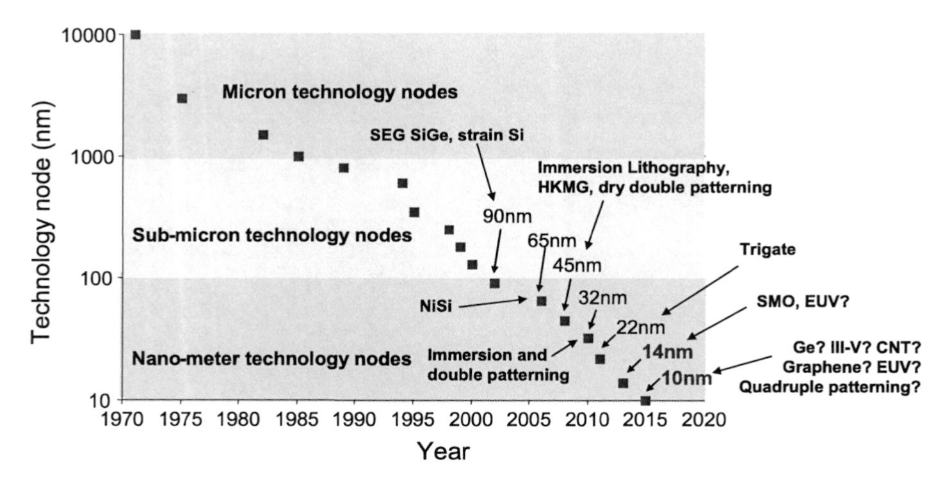

Figure 15.2 Technology nodes and introductions of new technologies and materials.

65-nm nodes with production capability of 50,000 wafers per month would cost approximately three billion dollars. It would cost approximately 4.5 billion dollars to build a 300-mm 32/28-nm fab of that capacity. The cost would be even higher for 20- and 14-nm node fabs.

As processing technology continues to develop, the cost of the semiconductor fab rapidly increases. At the same time, IC chip prices continue to decrease, which drives the manufacturers to turn to larger wafer sizes to reduce cost. The 300-mm wafer has become mainstream for advanced wafer fabs, and the first 450-mm wafer fabs could appear by the end of the second decade of the 21st century. The cost of a 450-mm fab could be more than ten billion dollars! Predictably, only few highly profitable IC manufacturers could afford this fab.

After the 10-nm technology node, silicon will run out of atoms to form a reliable MOSFET. The quantum tunneling effect between source and drain increases dramatically when the channel length is less than 10 nm, unless the drain voltage V_D drops significantly, to maybe 0.3 V. Tunneling-induced S/D leakage, I_{off}, is very sensitive to channel length, which means that IC manufacturers will need to control the gate CD and SDE junction profile very tightly, so tightly that it could become unmanufacturable.

Even before physics stops the scaling of IC devices, there is a good chance that economics could stop the process. In the good old days, shrinking the feature size made the device faster, less expensive, and less power hungry, which gave IC manufacturers enough incentive to invest in research and development for decreasing the device geometry. Now the skyrocketing research and development costs to scale nanometer devices have forced many IC manufacturers to stay away from leading edge technologies, leaving only a few financially sound companies to lead the charge to the bleeding edge.

Since planar scaling is rapidly approaching physical and economical limits, 3D integration is becoming more attractive. For a memory chip, if two dies are stacked together, the amount of storage is doubled with the same chip area, which is equivalent to shrinking the feature size by one generation. If four dies can be stacked, the amount of storage will be quadrupled, which is equivalent to reducing the feature size by half, or two technology nodes. With wire bonding, fabs have already stacked 8 chips in mass production and 16 chips in development. Through-silicon via (TSV) technology recently has received much attention. Numerous resources have been dedicated to TSV technology development, and it will help continue the goal of "more Moore" or "more than Moore" to continue putting more devices on a chip, even though planar scaling reaches its limit.

DRAM had been an important driver of IC technology, and it still plays an important role in the IC industry. After the long competition between the two capacitor technologies—stack and deep trench—DRAM manufacturers are working on bWL technology, which can reduce a very challenging contact layer compared to a normal stacked DRAM, and allows $6F^2$ array transistor density. F is the smallest feature size of the transistor. By making a gate-all-around transistor in the vertical direction and using buried-wordline-buried-bitline (BWBB) technology (shown Fig. 15.3), the density of array transistors can further reduce to $4F^2$.

NAND flash chips have grown rapidly, driven by the demands for storage in mobile digital devices, such as cell phones, digital cameras, digital camcorders, MP3 players, USB drives, and tablet PCs. NAND-flash-based solid-state drives (SSDs) have been used in high-end laptops. They have also been used with traditional magnetic HDDs to form hybrid drives, which combine the advantages of a fast-reading SSD and the low cost of an HDD. NAND flash technology developed rapidly, driven by market demands and furious competition between the last few manufacturers. One of the leading manufacturers, Toshiba/SanDisk, makes 64-Gb (or 8-GB) NAND flash memory with 19-nm technology.

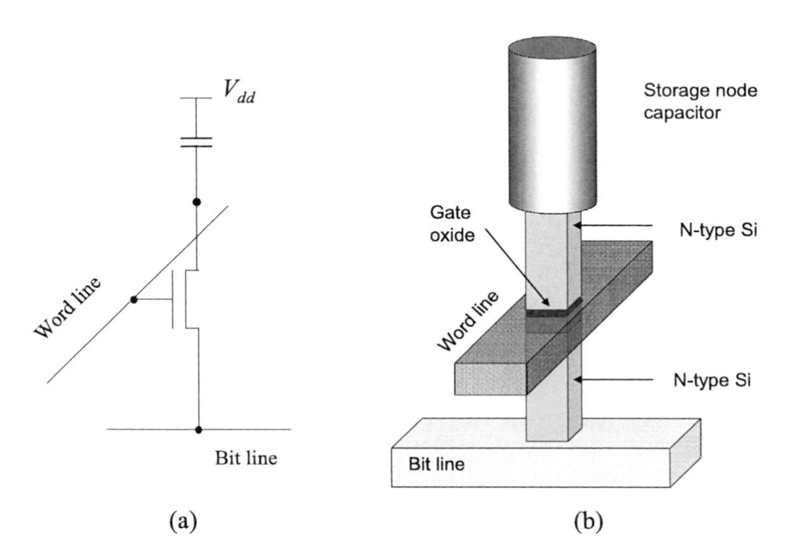

Figure 15.3 (a) Circuit of a DRAM unit cell and (b) 3D illustration of bWL, buried bitline $4F^2$ DRAM.

Traditional NAND flash is formed by strings of 32 or 64 series-connected flash devices on a silicon surface. By making the device string in the vertical direction, packing density dramatically increases. Figure 15.4 illustrates a birds-eye view and top view of a 3D NAND flash. In the illustration, the string only stacks four flash devices plus upper and lower select gates. In mass production, 16 flash devices can be stacked in a string and connected to another string from the bottom to form a 32-bit string; both select gates are located in the upper layer.

DRAM is fast and volatile, while NAND flash is nonvolatile but slow. For a long time, a fast NVM device has been sought that can be used for applications of both DRAM and NAND flash. Many types of novel memory devices have been proposed and studied. One promising candidate, phase-change memory (PCM), is already in production. Another candidate, magnetoresistive random access memory (MRAM), is still in intensive research and development, and could be integrated in system-on-chip IC manufacturing.

3D device approaches have been developed and applied in the production of CMOS logic IC, DRAM, and NAND flash chips. 3D devices will begin to replace planar devices in advanced technology nodes. Also, 3D packaging is already being used in CMOS image sensor chip manufacturing. It will eventually be applied in other IC productions.

Decreasing IC chip prices drives down the price of consumer electronics, such as flat-panel TVs, Blu-ray Disc™ players, MP3 players, smartphones, and PCs, including desktops, laptops, and tablets. The continuously dropping price of consumer electronics and the steady economic growth of developing countries, especially China and India (which together have more than one-third of the world's population), will dramatically increase the purchasing power and demand for consumer electronics.

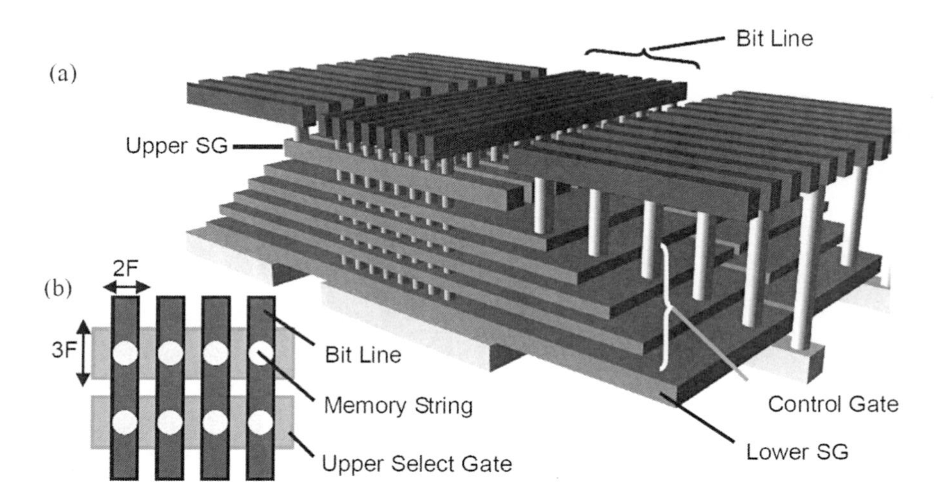

Figure 15.4 (a) Birds-eye view and (b) top view of a 3D flash memory array. (Y. Fukuzumi, et al.)

In addition to the electronics industry, demands for IC chips from other industries such as the automobile and medical/healthcare industries are also increasing rapidly. More chips will be used in motor vehicles to make them safer, more comfortable, have higher fuel efficiency, and easier to drive. Gadgets such as GPSs, smart routing and traffic management, collision warning radar, and voice-activated Internet access to audio on-demand services could become standard features in automobiles in the near future. By miniaturizing test probes and analysis circuits, medical IC chips have been developed for DNA testing, and fast, accurate diagnostics of DNA-related diseases. Medical and healthcare applications of IC chips, such as lab-on-chip (LoC) systems, will grow quickly in the future. Miniature fiber optics and photonics devices could be integrated with silicon IC chips to form interconnections between dies in a same-chip package to dramatically increase the data transmission rate. The development and popularization of 3D video in gaming and entertainment industries provide an additional driving force for more computational power and memory capability. Robotic pets have already been used in senior centers to tirelessly give comfort to lonely seniors, and robotic housemaids could serve ordinary households in the foreseeable future. Worldwide development of the wireless Internet and telecommunications combined with smartphones and tablets will continue to be an important driving force that feeds the rapid development of the IC industry in the near future.

Technologies that branched out from the semiconductor industry also have pushed into other industries, such as microelectromechanical systems (MEMS), light-emitting diodes (LEDs), and solar industries, which have helped drive down the cost of these products to make them more affordable for consumers worldwide.

It is hard to accurately forecast the demands for IC chips, a challenge that partly contributes to the periodic boom-bust nature of the IC industry. However, the trend is very clear in the foreseeable future. Global demands for IC chips will steadily grow, as will global demands for skilled, knowledgeable, innovative, creative, and hardworking technicians, engineers, and scientists in the semiconductor industry.

15.1 Bibliography

Y. Fukuzumi et al., "Optimal integration and characteristics of vertical array devices for ultra-high density, bit-cost scalable flash memory," *IEDM Tech. Dig.*, 449–452 (2007).

S. Lai, "Current status of the phase change memory and its future," *IEDM Tech. Dig.*, 255–258 (2003).

S. Lai, "Non-volatile memory technologies: the quest for ever-lower cost," *IEDM Tech. Dig.*, 11–16 (2008).

Index

Dr. Hong Xiao is a principal engineer of KLA-Tencor Corp. Previously, he was a technical marketing specialist at Hermes-Microvision, Inc. and a technical manager at Hermes Epitek. Other past positions include a working as a consultant of semiconductor process technology, a senior process engineer at Motorola Semiconductor Production Sector, and an associate professor at Austin Community College. After receiving his Ph.D., Dr. Xiao worked at Applied Materials as a senior technical instructor with expertise in dielectric thin-film deposition, semiconductor process integration, and plasma physics.

Dr. Xiao has authored and co-authored over 30 journal and conference papers. He is the author of *Introduction to Semiconductor Manufacturing Technology*, published by Prentice Hall in 2000. He has six US patents and more than ten patents in the application process. He has been a member of SPIE since 2005.